T0388126

Nanotechnology in Paper and Wood Engineering

Nanotechnology in Paper and Wood Engineering

Fundamentals, Challenges and Applications

Edited by

RAJEEV BHAT

Food By-Products Valorization Technologies (ERA- Chair in VALORTECH), Estonian University of Life Sciences, Tartu, Estonia, EU

ASHOK KUMAR

Department of Biotechnology and Bionformatics, Jaypee University of Information Technology, Waknaghat, India

TUAN ANH NGUYEN

Institute for Tropical Technology, Vietnam Academy of Science and Technology, Hanoi, Vietnam

SWATI SHARMA

University Institute of Biotechnology, Chandigarh University, Mohali, India

ELSEVIER

Elsevier
Radarweg 29, PO Box 211, 1000 AE Amsterdam, Netherlands
The Boulevard, Langford Lane, Kidlington, Oxford OX5 1GB, United Kingdom
50 Hampshire Street, 5th Floor, Cambridge, MA 02139, United States

British Library Cataloguing-in-Publication Data
A catalogue record for this book is available from the British Library

Library of Congress Cataloging-in-Publication Data
A catalog record for this book is available from the Library of Congress

ISBN: 978-0-323-85835-9

For Information on all Elsevier publications
visit our website at https://www.elsevier.com/books-and-journals

Publisher: Matthew Deans
Acquisitions Editor: Simon Holt
Editorial Project Manager: Gabriela D. Capille
Production Project Manager: Debasish Ghosh
Cover Designer: Greg Harris

Typeset by MPS Limited, Chennai, India

Contents

List of contributors

S. Abiramasundari
Centre for Research, Department of Biotechnology, Kamaraj College of Engineering and Technology, Madurai, India

Nadia Afsheen
Department of Biochemistry, Riphah International University, Faisalabad, Pakistan

Ishtiaq Ahmed
School of Medical Science, Menzies Health Institute Queensland, Griffith University, Southport, QLD, Australia

Mahboob Alam
Division of Chemistry and Biotechnology, Dongguk University, Gyeongju, Republic of Korea

Farman Ali
Department of Chemistry, Hazara University, Dhodial, Pakistan

Nisar Ali
Key Laboratory for Palygorskite Science and Applied Technology of Jiangsu Province, National & Local Joint Engineering Research Centre for Deep Utilization Technology of Rock-salt Resource, Faculty of Chemical Engineering, Huaiyin Institute of Technology, Huai'an, P.R. China

Tayyub Ali
Department of Chemical Engineering, Ryerson University, Toronto, ON, Canada

Tibor Alpár
Simonyi Károly Faculty of Engineering, University of Sopron, Sopron, Hungary

Nowshad Amin
Institute of Sustainable Energy, Universiti Tenaga Nasional (@The National Energy University), Kajang, Malaysia

Nilofar Asim
Solar Energy Research Institute, Universiti Kebangsaan Malaysia, Bangi, Malaysia

Marzieh Badiei
Independent Researcher, Mashhad, Iran

Ram Naresh Bharagava
Laboratory for Bioremediation and Metagenomics Research (LBMR), Department of Environmental Microbiology (DEM), Babasaheb Bhimrao Ambedkar University (A Central University), Lucknow, India

Rajeev Bhat
ERA-Chair in VALORTECH, Estonian University of Life Sciences, Tartu, Estonia, European Union

Muhammad Bilal
School of Life Science and Food Engineering, Huaiyin Institute of Technology, Huai'an, P.R. China

Ivana Cesarino
Department of Bioprocesses and Biotechnology, School of Agriculture (FCA), São Paulo State University (UNESP), Botucatu, São Paulo, Brazil

Ram Chandra
Department of Environmental Microbiology, School for Environmental Science, Babasaheb Bhimrao Ambedkar University (A Central University), Lucknow, India

Paweł Chmielarz
Department of Physical Chemistry, Faculty of Chemistry, Rzeszow University of Technology, Rzeszow, Poland

Yaser Dahman
Department of Chemical Engineering, Ryerson University, Toronto, ON, Canada

Otavio Augusto Titton Dias
Centre for Biocomposites and Biomaterials Processing, John H. Daniels Faculty of Architecture, Landscape, and Design, University of Toronto, Toronto, ON, Canada

Luiz Fernando Romanholo Ferreira
Graduate Program in Process Engineering, Tiradentes University (UNIT), Aracaju, Brazil; Institute of Technology and Research (ITP), Tiradentes University (UNIT), Aracaju, Brazil

Thomas Geiger
Cellulose and Wood Materials Laboratory, Swiss Federal Laboratories for Materials Science and Technology (Empa), Dübendorf, Switzerland

Nishil Gosalia
Department of Chemical Engineering, Ryerson University, Toronto, ON, Canada

Nikita Goswami
University Institute of Biotechnology, Chandigarh University, Mohali, India

Sarminiyy Lenga Gururuloo
Infectomics Cluster, Advanced Medical and Dental Institute Universiti Sains Malaysia, Kepala Batas, Malaysia

K. M. Faridul Hasan
Simonyi Károly Faculty of Engineering, University of Sopron, Sopron, Hungary

Katrin Greta Hoffmann
Cellulose and Wood Materials Laboratory, Swiss Federal Laboratories for Materials Science and Technology (Empa), Dübendorf, Switzerland

Péter György Horváth
Simonyi Károly Faculty of Engineering, University of Sopron, Sopron, Hungary

Asim Hussain
Department of Biochemistry, Riphah International University, Faisalabad, Pakistan

Nishanth Ignatius
Department of Chemical Engineering, Ryerson University, Toronto, ON, Canada

Hafiz M.N. Iqbal
Tecnologico de Monterrey, School of Engineering and Sciences, Monterrey, Mexico

Zara Jabeen
Department of Biochemistry, Riphah International University, Faisalabad, Pakistan

Mohammad Jawaid
Universiti Putra Malaysia, Serdang, Malaysia

Nurul Huda Abd Kadir
Faculty of Science and Marine Environment, Universiti Malaysia Terengganu, Kuala Nerus, Malaysia

George N. Karuku
Department of Land Resources and Agricultural Technology, University of Nairobi, Nairobi, Kenya

Adnan Khan
Institute of Chemical Sciences, University of Peshawar, Peshawar, Pakistan

Aleksa Krunic
Department of Chemical Engineering, Ryerson University, Toronto, ON, Canada

Adarsh Kumar
Department of Environmental Microbiology, School for Environmental Science, Babasaheb Bhimrao Ambedkar University (A Central University), Lucknow, India

Gulshan Kumar
University School of Basic and Applied Sciences, Guru Gobind Singh Indraprastha University, New Delhi, India

Ritesh Kumar
University School of Basic and Applied Sciences, Guru Gobind Singh Indraprastha University, New Delhi, India

Tushar Kumar
University Institute of Biotechnology, Chandigarh University, Mohali, India

Vineet Kumar
Department of Botany, School of Life Science, Guru Ghasidas Vishwavidyalaya (A Central University), Bilaspur, India

Chin Wei Lai
Nanotechnology & Catalysis Research Centre (NANOCAT), Institute for Advanced Studies (IAS), University of Malaya (UM), Kuala Lumpur, Malaysia

Alcides Lopes Leão
Department of Bioprocesses and Biotechnology, School of Agriculture (FCA), São Paulo State University (UNESP), Botucatu, São Paulo, Brazil

Ming Liu
Cellulose and Wood Materials Laboratory, Swiss Federal Laboratories for Materials Science and Technology (Empa), Dübendorf, Switzerland

Peter Ma
Department of Chemical Engineering, Ryerson University, Toronto, ON, Canada

G. Madhubala
Centre for Research, Department of Biotechnology, Kamaraj College of Engineering and Technology, Madurai, India

Sumeet Malik
Institute of Chemical Sciences, University of Peshawar, Peshawar, Pakistan

Damaris Mbui
Department of Chemistry, University of Nairobi, Nairobi, Kenya

Shweta Mishra
Shobhaben Pratapbhai Patel School of Pharmacy and Technology Management, SVKM's NMIMS (Deemed-to-be-University), Mumbai, India

Masita Mohammad
Solar Energy Research Institute (SERI), National University of Malaysia, Bangi Selangor, Malaysia

Sikandar I. Mulla
Department of Biochemistry, School of Applied Sciences, REVA University, Bangalore, India

Ashok Kumar
Department of Biotechnology and Bioinformatics, Jaypee University of Information Technology, Solan, India

Tuan Anh Nguyen
Institute for Tropical Technology, Vietnam Academy of Science and Technology, Hanoi, Vietnam

Gustav Nyström
Cellulose and Wood Materials Laboratory, Swiss Federal Laboratories for Materials Science and Technology (Empa), Dübendorf, Switzerland

V.C. Padmanaban
Centre for Research, Department of Biotechnology, Kamaraj College of Engineering and Technology, Madurai, India

Anil M. Pethe
Datta Meghe College of Pharmacy, Datta Meghe Institute of Medical Sciences, Deemed-to-be-University, Sawangi (Meghe), Wardha, India

Anthony Poblete
Department of Chemical Engineering, Ryerson University, Toronto, ON, Canada

Sonal Prasad
Department of Bio-Sciences, Institute of Bio-Sciences and Technology, Shri Ramswaroop Memorial University, Barabanki, India

Prerna
Department of Environmental Microbiology, School for Environmental Science, Babasaheb Bhimrao Ambedkar University (A Central University), Lucknow, India

Hamza Rafeeq
Department of Biochemistry, Riphah International University, Faisalabad, Pakistan

Abbas Rahdar
Department of Physics, Faculty of Science, University of Zabol, Zabol, Iran

Tahir Rasheed
Interdisciplinary Research Center for Advanced Materials, King Fahd University of Petroleum and Minerals, Dhahran, Saudi Arabia

Sheel Ratna
Department of Environmental Microbiology, School for Environmental Science, Babasaheb Bhimrao Ambedkar University (A Central University), Lucknow, India

Komal Rizwan
Department of Chemistry, University of Sahiwal, Sahiwal, Pakistan

Kiplangat Rop
Department of Chemistry, University of Nairobi, Nairobi, Kenya

Nelson Pynadathu Rumjit
Nanotechnology & Catalysis Research Centre (NANOCAT), Institute for Advanced Studies (IAS), University of Malaya (UM), Kuala Lumpur, Malaysia

Nurul Asma Samsudin
Institute of Sustainable Energy, Universiti Tenaga Nasional (@The National Energy University), Kajang, Malaysia

Swati Sharma
University Institute of Biotechnology (UIBT), Chandigarh University, Mohali, India

Farooq Sher
Department of Engineering, School of Science and Technology, Nottingham Trent University, Nottingham, United Kingdom

Thayvee Geetha Bharathi Silvaragi
Integrative Medicine Cluster, Advanced Medical and Dental Institute, Universiti Sains Malaysia, Kepala Batas, Malaysia

Ajay Kumar Singh
Department of Environmental Microbiology, School for Environmental Science, Babasaheb Bhimrao Ambedkar University (A Central University), Lucknow, India

Palakjot K. Sodhi
University Institute of Biotechnology, Chandigarh University, Mohali, India

Kamaruzzaman Sopian
Solar Energy Research Institute, Universiti Kebangsaan Malaysia, Bangi, Malaysia

Milena Chanes de Souza
Department of Bioprocesses and Biotechnology, School of Agriculture (FCA), São Paulo State University (UNESP), Botucatu, São Paulo, Brazil

K.R. Talluri Rameshwari
Division of Microbiology, Department of Water & Health, Faculty of Life Sciences, JSS Academy of Higher Education and Research, Mysuru, India.

Mohammad Torkashvand
Fouman Faculty of Engineering, College of Engineering University of Tehran, Tehran, Iran

Vivek Yadav
State Key Laboratory of Crop Stress Biology in Arid Areas, College of Horticulture, Northwest A&F University, Yangling, P. R. China

Izabela Zaborniak
Department of Physical Chemistry, Faculty of Chemistry, Rzeszow University of Technology, Rzeszow, Poland

Preface

The field of nanotechnology is an exciting multidisciplinary arena that synchronizes basic science with technological features to produce novel improved materials that can find wide applications in the day-to-day human life. Some of the latest developments on the application of nanotechnology in various fields (agriculture, food, medicine, electronics, material sciences, wood engineering, etc.) have indicated the technology to be a potential driving force contributing significantly to the regional economy.

This book is designed with an innovative approach aimed to provide an overview of the most recent developments, and a wide range of applications of nanotechnology with a focus on paper and wood engineering industries. The basic idea of editing this book was motivated by the fact of the uniqueness and excellent properties of the nanopapers (such as high thermal stability, mechanical strength, durability, glossiness, lowering of UV transmission, fine printability, and lightweight) that can overcome the limitations of the conventional papers produced from cellulose, lignin, or hemicellulose.

This book *Nanotechnology in Paper and Wood Engineering: Fundamentals, Challenges, and Applications* describes the fundamental concepts and processes of nanopaper synthesis and nanoadditives involved in large-scale production. Some of the key features of the book include providing information on the production of nanopapers and identifying their applications, explaining some of the vital techniques of synthesis and designing concepts of cellulosic/wooden based nanomaterials for potential industrial applications, and assessing certain sustainability challenges of nanotechnology-based manufacturing systems in paper and wood engineering.

The book has been divided into two sections: Section I focuses on the fundamental concepts of the nanopapers and current research developments, while Section II describes various types of materials employed in the production of nanopapers. The topics covered in Section I deal with the fundamentals of nanotechnology in pulp and paper processing, synthesis and applications of nanofibers, nanofiller pigments and cellulose-based nano-crystals, use of nanofibrils in paper industries, as well as pectin-based nanomaterials additives. Further, the application of nanotechnology for water treatment and dye removal in the paper and pulp industries, and wastepaper recycling are also discussed as individual chapters in this section. Additionally, in this section, the application of nanotechnology in wood engineering, nanofillers/nanoadditives in wood furniture making, nanocoatings and varnish for wooden commodities, nanocomposites of wood for oil—water separation and filtration, nanodimensional transparent wood and papers, lignin-based nanomaterials, nanotechnology for waste wood recycling, and photocatalytic nanotreatments of wood are also discussed as individual chapters. Going ahead, Section II mainly covers the application part of nanotechnology focusing on the fabrication of nanopapers and nanowoods, integration of nanosensors/nanodevices into the conventional papers, nanoscale characterization of nanopapers and nanowoods, and the use of nanotechnology in food packaging and drugs. In addition, this section includes pharmaceutical, biocatalysis, photocatalytic, and energy storage applications. The last two chapters in Section II covers the environmental and health impacts of nanowood/nanopapers and future perspectives of this invigorating field.

The individual chapter contribution for this book comes from outstanding leading experts in the field who have penned their thought very meticulously. This book is expected to be an imperative reference material for researchers, engineers, industry personnel, and postgraduate students who are engaged in understanding the concept of nanotechnology and their applications to design more competent manufacturing processes in paper and/or wood industries.

We, the editors are highly grateful to all the contributing authors for their vital contributions and the entire team of Elsevier Sciences (Academic Press) publishers. The encouragement and support provided by our respective universities/institutions are gratefully acknowledged. In particular, Rajeev personally acknowledges Profs. Ülle Jaakma and Toomas Tiirats; Ashok to Profs. S.S. Kanwar, V.K. Thakur, and Su Shuing Lam; and Swati to Profs. Rajan Jose and Susheel Kalia.

Finally, without the support of the family, it would not have been possible to gain the anticipated success in venturing to the publication of this book, and in this regard, all of us (as editors) dedicate this book to our family with much love and affection.

Rajeev Bhat
Ashok Kumar
Tuan Anh Nguyen
Swati Sharma

PART I

Fundamentals

CHAPTER 1

Nanotechnology in paper and wood engineering: an introduction

Ashok Kumar[1], Tuan Anh Nguyen[2], Swati Sharma[3] and Rajeev Bhat[4]
[1]Department of Biotechnology and Bioinformatics, Jaypee University of Information Technology, Solan, India
[2]Institute for Tropical Technology, Vietnam Academy of Science and Technology, Hanoi, Vietnam
[3]University Institute of Biotechnology (UIBT), Chandigarh University, Mohali, India
[4]ERA-Chair in VALORTECH, Estonian University of Life Sciences, Tartu, Estonia, European Union

1.1 Introduction

Nanotechnology plays a vital role in the development of material science and manufacturing of various materials of industrial use. The development of nanoscale manufacturing of polymeric materials of biological origin such as wood has transformed the paper, pulp, furniture, and electronics industry to a large scale. Various innovative products have been developed by using plant-based biopolymers such as cellulose, pectin, and lignin (Sharma et al., 2020). These products from renewable sources synthesized with the aid of nanotechnology are the prerequisite for pharmaceutical, electronics, and bioenergy industry. The furniture and timber industries also demand for the micro and nanoscale derivatives of the cellulose and lignin for use as bioadhesive and biolubricants. The nanogels and nanoadditives of cellulosic origin have also key role in the biocatalysis for the immobilization of the enzymes and carry out the biotransformations (Kumar & Kanwar, 2012b,c; Kumar et al., 2015; Patel et al., 2017; Rahman, Culsum, Kumar, Gao, & Hu, 2016). Several derivatives of wood with peculiar properties have been generated using nanotechnology-based approaches. The cellulosic nanomaterials in various forms, such as fibers, gel, particle, and microcrystalline cellulose, have been prepared from various plant products such as flexbast, cotton, hemp, and kraft pulp (Bhatnagar & Sain, 2005; Qua, Hornsby, Sharma, Lyons, & McCall, 2009; Roohani et al., 2008). The synthesis of nanomaterials from wood derivatives has been facilitated by mechanochemical, chemo-enzymatic, and microbial methods (Henriksson, Henriksson, Berglund, & Lindström, 2007; Iwamoto, Nakagaito, Yano, & Nogi, 2005; Tsuchida & Yoshinaga, 1997; Wang, Cheng, Rials, & Lee, 2006). The plant biopolymers based nanostructured materials have been also used to immobilize the enzymes such as pectinase, cellulase, and lipase for various catalytic applications (Aggarwal, Sharma, Kamyab, & Kumar, 2020; Kondaveeti et al., 2019; Thakur, Kumar, & Chauhan, 2018; Zhang et al., 2020). The cellulose and its composites in nanoscale have been used for the immobilization

Nanotechnology in Paper and Wood Engineering
DOI https://doi.org/10.1016/B978-0-323-85835-9.00015-5

enzyme and to carry out the synthesis of various industrially important products (Kumar & Kanwar, 2012a; Sharma & Kumar, 2021; Thakur, Kumar, & Kanwar, 2012). In the wood and paper industry nanotechnology has generated various research and entrepreneurship opportunities with the development of various smart materials such as fiber sheets, membrane, or particle-based composites (Bhushan, 2017; Schulte, 2005). The nanocoating of wood materials make these waterproof, resistant against the microbial attack, and stable at varied temperature and moisture conditions. Thus earlier used conventional and unsophisticated methods can be replaced with newly developed nanoengineering techniques (Wegner & Jones, 2009). Nanotechnology-based systems become rapidly prevalent in paper and pulp industry, because of the requirement of less energy, small area, less time, and lesser amount of material (Kamel, 2007). In this review chapter, the use and development of nanotechnology-based processes for the wood processing and paper industry have been highlighted.

1.2 Applications of nanotechnology in the paper and pulp industry

A diverse array of ultra-small-scale materials, including metal oxides, ceramics, and polymeric materials, and wide-ranging processing methods, including techniques that use "self-assembly" on a molecular scale, are either in use today or are being groomed for commercial-scale use.

Paper manufacturing and pulp industries holds high prominence, mainly owing to their impact on the regional economy. As per the available statistical data, it was estimated that the global production of paper and cardboard to be ~420 million metric tons in 2018 (https://www.statista.com/topics/1701/paper-industry/). Further, in 2019, the market size for paper and pulp industry was projected to be ~US$348 billion, and by 2027, this is predicted to reach up to US$370 billion (https://www.statista.com/statistics/1073451/global-market-value-pulp-and-paper/).

The paper and pulp industries rely entirely on natural plant-based wood materials to produce paper, pulp, cellulose-based materials, and much more. Recent advancements witnessed in these industries have laid their focus mainly toward production of good-quality papers (e.g., smooth with lower ink absorption, good barrier properties, etc.) that are produced in a sustainable and cost-effective way (Marcus, 2005; Shen, Song, Qian, & Liu, 2009a,b; Zaieda & Bellakhal, 2009).

Of late, nanoadditives have been realized to induce a wide range of distinctive properties and affect the overall quality of the paper. Nevertheless, owing to low particle size, during the production stages of paper, these additives need to be carefully held within the fiber matrix, and this is a much challenging task (Lee et al., 2013). During manufacturing process of the paper, nanoadditives have been confirmed to possess multiple uses, such as enhancement in opacity of the paper, improving gas, and water absorption barrier properties, as a bleaching agent, as an antimicrobial agent, and much more. There are a wealth of literature indicating on the preparation and characterization of nanomaterials, which can be explored during papermaking process (Ghosh, Kundu, Majumder, & Chowdhury, 2020; Vaseghi & Nematollahzadeh, 2020). Some of the selected reports on application of nanoadditives in paper production are discussed in the ensuing text.

Nanocalcium silicate, hydrated calcium silicate, and calcium carbonate have been successfully used to enhance paper opacity (e.g., in newspaper print), reduce high-quality images on printing along with preventing unwanted spread of the printing ink on paper (El-Sherbiny, El-Sheikh, & Barhoum, 2015; Liu, Yin, & Xu, 2013; Shen, Song, Qian, & Liu, 2009a; Johnston & Schloffer, 2009). Superhydrophobic high-opacity papers have been produced using titanium-di-oxide (TiO_2) nanoparticles (Huang, Chen, Lin, Yang, & Gerhardt, 2011; Wu et al., 2021; Wang et al., 2021), while antimicrobial (antifungal) properties have been established postnanocoating the papers with TiO_2 and zinc oxide (ZnO) (Costa, Gonçalves, Zaguete, Mazonb, & Nogueira, 2013). Besides, paper sheets fabricated with titania nanowires have been studied for their optical properties (Chauhan, Chattopadhyay, & Mohanty, 2013). In addition, nanosilica (nano-SiO_2) has been observed to provide the paper with antibacterial effects, besides delivering high-quality printing effects, as well as enhancing the opacity (optical properties) and color performances (Morsy, El-Sheikh, & Barhoum, 2019; Zhang et al., 2013). In many of the instances, nano-SiO_2 has been obtained from agricultural wastes/biomass (Carmona, Oliveira, Silva, Mattoso, & Marconcini, 2013; Chen et al., 2010).

On the other hand, nanopigments (e.g., nanokaoline) as a binding agent has been explored as a coating agent in papermaking process, which provided good water barrier properties with better mechanical quality attributes (Basilio, 2013; Ouyang, Xu, Lo, & Sham, 2011; Saroha, Dutt, & Bhowmick, 2019). In addition, nanopolymers that are readily water soluble have been used in papermaking. Nanopolymers have been successfully synthesized from natural plant based nanoproteins and nanopolysacchrides. Development of nanocellulose-based papers from cheap, reliable, and renewable resources like that of agri-food wastes/byproducts and wood wastes has seen much success too (Bahloul et al., 2021; Deutschle et al., 2014; Hu et al., 2021; Isogai, Saito, & Fukuzumi, 2011; Jin, Tang, Liu, Wang, & Ye, 2021; Sacui et al., 2014; Sharma, Thakur, Bhattacharya, Mandal, & Goswami, 2019). Nanocellulosic materials possessing appreciable thermo-mechanical properties have been produced (e.g., carboxylated nanocellulose) via use of 2,2,6,6-tetramethylpiperidine-1-oxyl mediated oxidation process (Jiang & Hsieh, 2014; Sanchez-Salvador, Campano, Negro, Monte, & Blanco, 2021). Both cellulose and ligno-cellulose hold high potential to be used as nanomaterials owing to their nanofibrillar structure, are lightweight with high strength, and have good adhesion properties (Gardner, Oporto, Mills, & Samir, 2008; Nakagaito & Yano, 2005; Wegner & Jones, 2006).

High-quality paper, which had good gas barrier properties and ability to absorb indoor pollutants, has also been reproduced by using nanozeolite (Ichiura, Kubota, Zonghua, & Tanaka, 2001; Ichiura, Kitaoka, & Tanaka, 2003; Ozcan & Dogan, 2020). On the other note, high-quality food packaging material with good water barrier properties has been produced by use of nanoclay and SiO_2 nanoparticles (Bumbudsanpharoke & Ko, 2019; Gaikwad & Ko, 2015). Nevertheless, owing to broader aspect ratio, low levels of nanoparticles or nanoadditives are adequate to alter the properties of paper, especially when used as a packaging material (Lei, Hoa, & Ton-That, 2006).

From the available reports, it is evident that nanoadditives can find wide applications in paper industries that can improve the overall qualities of the final product. However, there is always a word of caution, wherein selection of nanoparticles as a nanoadditive needs to be carefully considered, especially when they are of low particle size and when consumers are concerned about safety and pollution aspects.

1.3 Applications of nanotechnology in the wood industry

The unique morphology of the nanoscale derivatives such as nanocrystal fibbers, nanowhiskers, and monocrystals make these useful as filler materials and additive in the manufacturing of wooden and electrical goods (Siqueira, Bras, & Dufresne, 2009). Nanofibrillated cellulose has been reported to be synthesized from waste newspapers and, agricultural residues such as waste plants (Josset et al., 2014; Siró & Plackett, 2010). Both chemical as well as enzymatic methods were used in the synthesis of fibrillated cellulose at nanoscale (Henriksson et al., 2007; Henriksson, Berglund, Isaksson, Lindstrom, & Nishino, 2008; Hubbe, Rojas, Lucia, & Sain, 2008; Zimmermann, Bordeanu, & Strub, 2010; Zimmermann, Pöhler, & Schwaller, 2005). Nanotechnology plays an important role in wood industry. Wood is also composed of lignocellulosic fibers and it is as strong as metallic steel. Harvesting these lignocellulosic nanofibrils would benefit in sustainable construction as renewable materials. Wood is known as multifaceted material that finds various applications in the fields such as sports equipment, shipbuilding, stationary, construction, fencing, and furniture (Hunt, 2012; Mazzanti, Togni, & Uzielli, 2012; Švajlenka & Kozlovská, 2020). Wood is widely used in the construction and manufacturing the goods because of its biodegradability, appealing appearance, eco-friendliness, and durability. Also, wood gets easily decayed by the attack of insects and fungi (Ajuong, Pinion, & Bhuiyan, 2018; Bari et al., 2019). To increase wood durability against weather and biological attacks, treatment of wood with the help of nanotechnology-based products had proved to be quite useful (da Silva et al., 2019; Mantanis, Terzi, Kartal, & Papadopoulos, 2014; Papadopoulos, Bikiaris, Mitropoulos, & Kyzas, 2019), as nanoadditives can deeply penetrate the wood and improves its durability (De Filpo, Palermo, Rachiele, & Nicoletta, 2013; Goffredo et al., 2017).

Nano material are used as fire retardants such as mineral materials (nanowollastonite or nanoclay) UF-SiO, that enhances the flame resistance and hardness of treated wood. Also, Nanowollastonite ($CaSiO_3$) has been used to improve fire retardancy in medium-density fiber board (Chigwada, Jash, Jiang, & Wilkie, 2005). Nanowood-composites are low-density wood that can be converted to high value material by impregnating with nanoparticle-blended resins. They are impregnated under vacuum/pressure. Resin/polymers such as phenol formaldehyde, melamine formaldehyde, furfuryl alcohol, polyvinyl alcohol, nanoclay, and nano-SiO_2 have been reported to be used widely in the treatment of wood (Wei, Li, Shen, Zhang, & Wu, 2020). Nanofillers improve the properties of natural fiber reinforced thermoplastic composite and offer improved heat resistance, bio resistance, and flame retardation (Iamareerat, Singh, Sadiq, & Anal, 2018). Nanoclay along with wood derivatives is generally used (e.g., montmorillonite, bentonite) in the manufacturing of automobile parts, outdoor decking, and interior applications like wall cladding, paneling, packaging, etc. Nanomaterials can be used to enhance the wood stability by just coating them with nano-ZnO or nano-TiO_2titanium oxide. It can enhance its functionality: durability, fire resistance, and ultraviolet (UV) absorption which makes the wood live longer (De Filpo et al., 2013). The UV radiations have the capability of photochemically degrading the components of wood like lignin, cellulose, and hemicellulose (Teacă, Roşu, Bodîrlău, & Roşu, 2013). To improve the mechanical and physical properties in wood composites some other nanoparticles such as nano-ZnO (da Silva et al., 2019), nano silver (Taghiyari & Norton, 2014), nano-Al_2O_3 (Candan & Akbulut, 2015), nanowollastonite (Taghiyari, Ghamsari, & Salimifard, 2018), nanocellulose, and nano-SiO_2 (Candan & Akbulut, 2015) have been widely used. Nanomaterials with better chemical, physical, and biological properties can be developed with the help of nanoscale cellulose. Nanocellulose has many peculiar features like high strength and stiffness, high strength to weight ratio, electromagnetic responses, and a large surface area to volume ratio that makes it an interesting material for wood applications (Shatkin, Wegner, Bilek, & Cowie, 2014). In wood industries, the nanocellulose can be applied as a reinforcing agent in pulp, paper, and wood composites or as a coating material in wood. Due to its versatility, nanocellulose may be used as stabilizer (Shatkin et al., 2014). To improve the printing quality of the paper and water resistance nanocellulose has been used as an additive. Furthermore, nano-SiO_2 coated paper has shown better dimensional stability, optical density, and print quality than other papers (Liu, Xu, Lv, & Li, 2011; Wu, Jing, Zhou, & Dai, 2011). Lignin nanoparticles—an ecofriendly and versatile resource—are used as hybrid nanocomposites reinforcement filler (Bian et al., 2018), antibacterial agent, UV absorbents, and antioxidant formulations, in biomedical field, production of hydrogel, flocculent and coagulant, and in drug delivery system (Phan, Nguyen, Nguyen, Kumar, & Nguyen, 2020). It is used in environmental bioremediation and as carbon precursor and nontoxic material in the various industries. Transparent nanowood composites, provides fascinating architecture (optical anisotropy),

Table 1.1 Recent biosensors developed using nanocellulose.

Substrate	Reagents	Application	References
Bacterial nanocellulose	Curcumin	Albumin assay kit	Naghdi et al. (2019)
Cotton cellulose nanocrystals	Peptide	Detection of human neutrophil elastase and inflammatory diseases	Ling et al. (2019)
Polyaniline/crystalline nanocellulose/ionic modified screen-printed electrode	Cholesterol oxidase	As a novel and sensitive electrochemical cholesterol biosensor	Abdi et al. (2019)
Cellulose nanocrystals	Quartz crystal microbalance	Humidity sensor	Yao et al. (2020)
Microfibrillated cellulose/ polyvinyl alcohol	Nitrogen modified carbon quantum dot	Tartrazine sensor	Ng, Lim, and Leo (2020)

has energy efficiency, (low thermal conductivity), high optical transmittance, antiglare (high haze), tough and shatter-proof, and light weight (low density wood and polymer). The other potential applications are energy-efficient buildings, interior decorative lighting, solar panels, optoelectronics, and smart windows (Zhu et al., 2016). Nanozeolite and nano-SiO_2 are some other nanomaterials that can be used for the production of paper (Julkapli & Bagheri, 2016).

With the addition of nanoadditives or nanomaterials performance of paper products can be improved. Scientists are being attracted to nanocellulose for its use in the wood industry because of its availability in abundance. It also helps in improving the internal bonding that leads to a reduction in opacity, air permeability, higher density, and it increases the dry tensile strength (Bajpai, 2016). On the other hand, nanocellulose has been used in various types of biosensors (Table 1.1).

Common coatings which serve both purposes, decorative and protective, are applied on the wood surfaces like paints, varnishes, and lacquers. Coatings show some limitations such as loss of strength, inferior abrasion resistance, limited flexibility, less durability, and disproportionate adhesion between substrate and the coating layer.

1.4 Conclusion

In conclusion, various kinds of nanomaterials, such as nanofibrils, nanoparticles, nanoadditives, and nanogels have been used in papermaking and wood processing. Nanotechnology created the new categories for wood and paper products, namely nanopapers and nanowoods. For emerging application, nanopapers have many advantages over the conventional papers, such as (1) higher thermal stability, (2) better electrical property, (3) higher mechanical strength, (4) higher durability, (5) better glossiness, (6) fine printability, and (7) lightweight. These excellent properties render nanopapers to be a key resource for fabrication of nanosensors, nanodevices (lab-on-a-paper), energy devices, artistic works, and smart materials/structures. In case of nanowoods, nanotechnology provides the better weatherability and leach resistance of wood products (against UV and moisture exposures). Besides the advantages of nanopapers and nanowoods, their toxic risks and environmental impacts should be considered and evaluated carefully.

References

Abdi, M. M., Razalli, R. L., Tahir, P. M., Chaibakhsh, N., Hassani, M., & Mir, M. (2019). Optimized fabrication of newly cholesterol biosensor based on nanocellulose. *International Journal of Biological Macromolecules*, *126*, 1213–1222.

Aggarwal, J., Sharma, S., Kamyab, H., & Kumar, A. (2020). The realm of biopolymers and their usage: An overview. *Journal of Environmental Treatment Techniques*, *8*, 1005–1016.

Ajuong, E., Pinion, L., & Bhuiyan, M. S. (2018). *Degradation of wood. Reference module in materials science and materials engineering*. Elsevier. Available from http://doi.org/10.1016/B978-0-12-803581-8.10537-5.

Bahloul, A., Kassab, Z., El Bouchti, M., Hannache, H., Oumam, M., & El Achaby, M. (2021). Micro- and nano-structures of cellulose from eggplant plant (*Solanum melongena* L) agricultural residue. *Carbohydrate Polymers*, *253*, 117311.

Bajpai, P. (2016). *Pulp and paper industry: Nanotechnology in forest industry*. Elsevier.

Bari, E., Jamali, A., Nazarnezhad, N., Nicholas, D. D., Humar, M., & Najafian, M. (2019). An innovative method for the chemical modification of Carpinus betulus wood: A methodology and approach study. *Holzforschung*, *73*, 839–846.

Basilio, C. (2013). Kaolin-based reagent improves waste paper deinking. *Paper*, *8*(5), 22–23, 360.

Bhatnagar, A., & Sain, M. (2005). Processing of cellulose nanofiber-reinforced composites. *Journal of Reinforced Plastics and Composites*, *24*, 1259–1268.

Bhushan, B. (2017). *Introduction to nanotechnology. Springer handbook of nanotechnology* (pp. 1–19). Springer.

Bian, H., Wei, L., Lin, C., Ma, Q., Dai, H., & Zhu, J. (2018). Lignin-containing cellulose nanofibril-reinforced polyvinyl alcohol hydrogels. *ACS Sustainable Chemistry & Engineering*, *6*, 4821–4828.

Bumbudsanpharoke, N., & Ko, S. (2019). Nanoclays in food and beverage packaging. *Journal of Nanomaterials*, *2019*, 13. Available from https://doi.org/10.1155/2019/8927167, Article ID 8927167.

Candan, Z., & Akbulut, T. (2015). Physical and mechanical properties of nanoreinforced particleboard composites. *Maderas Ciencia y Tecnología*, *17*, 319–334.

Carmona, V. B., Oliveira, R. M., Silva, W. T. L., Mattoso, L. H. C., & Marconcini, J. M. (2013). Nanosilica from rice husk: Extraction and characterization. *Industrial Crops and Products*, *43*, 291–296.

Chauhan, I., Chattopadhyay, S., & Mohanty, P. (2013). Fabrication of titania nanowires incorporated paper sheets and study of their optical properties. *Mater Express*, *3*(4), 343–349.

Chen, H., Wang, F., Zhang, C., Shi, Y., Jin, G., & Yuan, S. (2010). Preparation of nano-silica materials: The concept from wheat straw. *Journal of Non-Crystalline Solids*, *356*(50–51), 2781–2785.

Chigwada, G., Jash, P., Jiang, D. D., & Wilkie, C. A. (2005). Fire retardancy of vinyl ester nanocomposites: Synergy with phosphorus-based fire retardants. *Polymer Degradation and Stability*, *89*, 85–100.

Costa, S. V., Gonçalves, A. S., Zaguete, M. A., Mazonb, T., & Nogueira, A. F. (2013). ZnO nanostructures directly grown on paper and bacterial cellulose substrates without any surface modification layer. *Chemical Commmunications*, *49*, 8096–8098.

da Silva, A. P. S., Ferreira, B. S., Favarim, H. R., Silva, M. F. F., Silva, J. V. F., dos Anjos Azambuja, M., et al. (2019). Physical properties of medium density fiberboard produced with the addition of ZnO nanoparticles. *BioResources*, *14*, 1618–1625.

De Filpo, G., Palermo, A. M., Rachiele, F., & Nicoletta, F. P. (2013). Preventing fungal growth in wood by titanium dioxide nanoparticles. *International Biodeterioration & Biodegradation, 85*, 217–222.

Deutschle, A. L., Romhild, K., Meister, F., Janzon, R., Riegert, C., & Saake, B. (2014). Effects of cationic xylan from annual plants on the mechanical properties of paper. *Carbohydrate Polymers, 102*(1), 627–635.

El-Sherbiny, S., El-Sheikh, S. M., & Barhoum, A. (2015). Preparation and modification of nano calcium carbonate filler from waste marble dust and commercial limestone for papermaking wet end application. *Powder Technology, 279*, 290–300.

Gaikwad, K. K., & Ko, S. (2015). Overview on in polymer-nano clay composite paper coating for packaging application. *Journal of Material Sciences & Engineering, 4*(151), 2169.

Gardner, D. J., Oporto, G. S., Mills, R., & Samir, M. A. S. A. (2008). Adhesion and surface issues in cellulose and nanocellulose. *Journal of Adhesion Science and Technology, 22*(5–6), 545–567.

Ghosh, R., Kundu, S., Majumder, R., & Chowdhury, M. P. (2020). Hydrothermal synthesis and characterization of multifunctional ZnO nanomaterials. *Materials Today: Proceedings, 26*, 77–81.

Goffredo, G. B., Accoroni, S., Totti, C., Romagnoli, T., Valentini, L., & Munafò, P. (2017). Titanium dioxide based nanotreatments to inhibit microalgal fouling on building stone surfaces. *Building and Environment, 112*, 209–222.

Henriksson, M., Berglund, L. A., Isaksson, P., Lindstrom, T., & Nishino, T. (2008). Cellulose nanopaper structures of high toughness. *Biomacromolecules, 9*, 1579–1585.

Henriksson, M., Henriksson, G., Berglund, L., & Lindström, T. (2007). An environmentally friendly method for enzyme-assisted preparation of microfibrillated cellulose (MFC) nanofibers. *European Polymer Journal, 43*, 3434–3441.

Hu, J., Li, R., Zhang, K., Meng, Y., Wang, M., & Liu, Y. (2021). Extract nano cellulose from flax as thermoelectric enhancement material. *Journal of Physics: Conference Series, 1790*(1), 012087.

Huang, L., Chen, K., Lin, C., Yang, R., & Gerhardt, R. A. (2011). Fabrication and characterization of superhydrophobic high opacity paper with titanium dioxide nanoparticles. *Journal of Materials Science, 46*(8), 2600–2605.

Hubbe, M. A., Rojas, O. J., Lucia, L. A., & Sain, M. (2008). Cellulosic nanocomposites: A review. *BioResources, 3*, 929–980.

Hunt, D. (2012). Properties of wood in the conservation of historical wooden artifacts. *Journal of Cultural Heritage, 13*, S10–S15.

Iamareerat, B., Singh, M., Sadiq, M. B., & Anal, A. K. (2018). Reinforced cassava starch based edible film incorporated with essential oil and sodium bentonite nanoclay as food packaging material. *Journal of Food Science and Technology, 55*, 1953–1959.

Ichiura, H., Kitaoka, T., & Tanaka, H. (2003). Removal of indoor pollutants under UV irradiation by a composite TiO_2–zeolite sheet prepared using a papermaking technique. *Chemosphere, 50*(1), 79–83.

Ichiura, H., Kubota, Y., Zonghua, W., & Tanaka, H. (2001). Preparation of zeolite sheets using a papermaking technique. *Journal of Materials Science, 36*(4), 913–917.

Isogai, A., Saito, T., & Fukuzumi, H. (2011). TEMPO-oxidized cellulose nanofibers. *Nanoscale, 3*, 71–85.

Iwamoto, S., Nakagaito, A. N., Yano, H., & Nogi, M. (2005). Optically transparent composites reinforced with plant fiber-based nanofibers. *Applied Physics A, 81*, 1109–1112.

Jiang, F., & Hsieh, Y. L. (2014). Super water absorbing and shape memory nanocellulose aerogels from TEMPO-oxidized cellulose nanofibrils via cyclic freezing-thawing. *Journal of Materials Chemistry A, 2*(2), 350–359.

Jin, K., Tang, Y., Liu, J., Wang, J., & Ye, C. (2021). Nanofibrillated cellulose as coating agent for food packaging paper. *International Journal of Biological Macromolecules, 168*, 331–338.

Johnston, J.H., & Schloffer, A. (2009). The use of nano-structured calcium silicate as the functional component in paper coating for high quality ink-jet printing. In: *63rd Appita annual conference and exhibition* (p. 217). Melbourne, Australia.

Josset, S., Orsolini, P., Siqueira, G., Tejado, A., Tingaut, P., & Zimmermann, T. (2014). Energy consumption of the nanofibrillation of bleached pulp, wheat straw and recycled newspaper through a grinding process. *Nordic Pulp & Paper Research Journal, 29*, 167–175.

Julkapli, N. M., & Bagheri, S. (2016). Developments in nano-additives for paper industry. *Journal of wood science*, *62*, 117–130.
Kamel, S. (2007). Nanotechnology and its applications in lignocellulosic composites, a mini review. *Express Polymer Letters*, *1*, 546–575.
Kondaveeti, S., Pagolu, R., Patel, S. K., Kumar, A., Bisht, A., Das, D., et al. (2019). Bioelectrochemical detoxification of phenolic compounds during enzymatic pre-treatment of rice straw. *Journal of Microbiology and Biotechnology*, *29*(11), 1760–1768.
Kumar, A., & Kanwar, S. (2012a). An innovative approach to immobilize lipase onto natural fiber and its application for the synthesis of 2-octyl ferulate in an organic medium. *Current Biotechnology*, *1*, 241–248.
Kumar, A., & Kanwar, S. S. (2012b). An efficient immobilization of *Streptomyces* sp. STL-D8 lipase onto photo-chemically modified cellulose-based natural fibers and its application in ethyl ferulate synthesis. *Trends in Carbohydrate Research*, *4*(4), 13–23.
Kumar, A., & Kanwar, S. S. (2012c). Catalytic potential of a nitrocellulose membrane-immobilized lipase in aqueous and organic media. *Journal of Applied Polymer Science*, *124*, E37–E44.
Kumar, A., Zhang, S., Wu, G., Wu, C. C., Chen, J., Baskaran, R., et al. (2015). Cellulose binding domain assisted immobilization of lipase (GSlip–CBD) onto cellulosic nanogel: Characterization and application in organic medium. *Colloids and Surfaces B: Biointerfaces*, *136*, 1042–1050.
Lee, J. Y., Kim, Y. H., Lee, S. R., Kim, C. H., Joo Sung, Y., Lim, G. B., ... Park, J. H. (2013). Effect of precipitated calcium carbonate on paper properties and drying energy reduction of duplex-board. *Journal of Korea Technical Association of the Pulp and Paper Industry*, *45*(6), 24–29.
Lei, S. G., Hoa, S. V., & Ton-That, M.-T. (2006). Effect of clay types on the processing properties of polypropylene nanocomposites. *Composites Science and Technology*, *66*, 1274–1279.
Ling, Z., Xu, F., Edwards, J. V., Prevost, N. T., Nam, S., Condon, B. D., et al. (2019). Nanocellulose as a colorimetric biosensor for effective and facile detection of human neutrophil elastase. *Carbohydrate Polymers*, *216*, 360–368.
Liu, Q. X., Xu, W. C., Lv, Y. B., & Li, J. L. (2011). Application of precipitated silica in low basis weight newspaper. *Advanced Materials Research*, 1107–1111.
Liu, Q. X., Yin, Y. N., & Xu, W. C. (2013). Study on application of hydrated calcium silicate in paper from wheat straw pulp. *Advanced Materials Research*, *774*, 1277–1280.
Mantanis, G., Terzi, E., Kartal, S. N., & Papadopoulos, A. (2014). Evaluation of mold, decay and termite resistance of pine wood treated with zinc-and copper-based nanocompounds. *International Biodeterioration & Biodegradation*, *90*, 140–144.
Marcus, W. (2005). How to reconcile environmental and economic performance to improve corporate sustainability: Corporate environmental strategies in the European paper industry. *Journal of Environmental Management*, *76*(2), 105–118.
Mazzanti, P., Togni, M., & Uzielli, L. (2012). Drying shrinkage and mechanical properties of poplar wood (Populus alba L.) across the grain. *Journal of Cultural Heritage*, *13*, S85–S89.
Morsy, F. A., El-Sheikh, S. M., & Barhoum, A. (2019). Nano-silica and $SiO_2/CaCO_3$ nanocomposite prepared from semi-burned rice straw ash as modified papermaking fillers. *Arabian Journal of Chemistry*, *12*(7), 1186–1196.
Naghdi, T., Golmohammadi, H., Vosough, M., Atashi, M., Saeedi, I., & Maghsoudi, M. T. (2019). Lab-on-nanopaper: An optical sensing bioplatform based on curcumin embedded in bacterial nanocellulose as an albumin assay kit. *Analytica Chimica Acta*, *1070*, 104–111.
Nakagaito, A. N., & Yano, H. (2005). Novel high-strength biocomposites based on microfibrillated cellulose having nano-order-unit web-like network structure. *Applied Physics A*, *80*(1), 155–159.
Ng, H. M., Lim, G., & Leo, C. (2020). N-modified carbon quantum dot in 3D-network of microfibrillated cellulose for building photoluminescent thin film as tartrazine sensor. *Journal of Photochemistry and Photobiology A: Chemistry*, *389*, 112286.
Ouyang, D., Xu, W., Lo, T. Y., & Sham, J. F. C. (2011). Increasing mortar strength with the use of activated kaolin by-products from paper industry. *Construction and Building Materials*, *25*(4), 1537–1545.
Ozcan, A., & Dogan, T. (2020). The effect of zeolite on inkjet coated paper surface properties and deinking. *Nordic Pulp & Paper Research Journal*, *35*(3), 432–439.

Papadopoulos, A. N., Bikiaris, D. N., Mitropoulos, A. C., & Kyzas, G. Z. (2019). Nanomaterials and chemical modifications for enhanced key wood properties: A review. *Nanomaterials*, *9*, 607.

Patel, S. K., Singh, R. K., Kumar, A., Jeong, J.-H., Jeong, S. H., Kalia, V. C., et al. (2017). Biological methanol production by immobilized *Methylocella tundrae* using simulated biohythane as a feed. *Bioresource Technology*, *241*, 922–927.

Phan, P. T., Nguyen, B.-S., Nguyen, T.-A., Kumar, A., & Nguyen, V.-H. (2020). Lignocellulose-derived monosugars: A review of biomass pre-treating techniques and post-methods to produce sustainable biohydrogen. *Biomass Conversion and Biorefinery*, 1–15.

Qua, E., Hornsby, P., Sharma, H. S., Lyons, G., & McCall, R. (2009). Preparation and characterization of poly (vinyl alcohol) nanocomposites made from cellulose nanofibers. *Journal of Applied Polymer Science*, *113*, 2238–2247.

Rahman, M. A., Culsum, U., Kumar, A., Gao, H., & Hu, N. (2016). Immobilization of a novel cold active esterase onto $Fe_3O_4 \sim$cellulose nano-composite enhances catalytic properties. *International Journal of Biological Macromolecules*, *87*, 488–497.

Roohani, M., Habibi, Y., Belgacem, N. M., Ebrahim, G., Karimi, A. N., & Dufresne, A. (2008). Cellulose whiskers reinforced polyvinyl alcohol copolymers nanocomposites. *European Polymer Journal*, *44*, 2489–2498.

Sacui, I. A., Nieuwendaal, R. C., Burnett, D. J., Stranick, S. J., Jorfi, M., Weder, C., . . . Gilman, J. W. (2014). Comparison of the properties of cellulose nanocrystals and cellulose nanofibrils isolated from bacteria, tunicate, and wood processed using acid, enzymatic, mechanical, and oxidative methods. *ACS Applied Materials & Interfaces*, *6*(9), 6127–6138.

Sanchez-Salvador, J. L., Campano, C., Negro, C., Monte, M. C., & Blanco, A. (2021). Increasing the possibilities of TEMPO-mediated oxidation in the production of cellulose nanofibers by reducing the reaction time and reusing the reaction medium. *Advanced Sustainable Systems*, *5*(4), 2000277.

Saroha, V., Dutt, D., & Bhowmick, A. (2019). PVOH modified nano-kaolin as barrier coating material for food packaging application. *AIP Conference Proceedings*, *2162*(1), 020001.

Schulte, J. (2005). *Nanotechnology: Global strategies, industry trends and applications*. John Wiley & Sons.

Sharma, A., Thakur, M., Bhattacharya, M., Mandal, T., & Goswami, S. (2019). Commercial application of cellulose nano-composites—A review. *Biotechnology Reports*, *21*, e00316.

Sharma, S., Sharma, A., Mulla, S. I., Pant, D., Sharma, T., & Kumar, A. (2020). Lignin as potent industrial biopolymer: An introduction. In S. Sharma, & A. Kumar (Eds.), *Lignin* (pp. 1–15). Springer.

Sharma, T., & Kumar, A. (2021). Bioprocess development for efficient conversion of CO_2 into calcium carbonate using keratin microparticles immobilized *Corynebacterium flavescens*. *Process Biochemistry*, *100*, 171–177.

Shatkin, J. A., Wegner, T. H., Bilek, E. T., & Cowie, J. (2014). Market projections of cellulose nanomaterial-enabled products-Part 1: Applications. *Tappi Journal*, *13*(5), 9–16.

Shen, J., Song, Z., Qian, X., & Liu, W. (2009a). Modification of precipitated calcium carbonate filler using sodium silicate/zinc chloride based modifiers to improve acid-resistance and use of the modified filler in papermaking. *BioResources*, *4*(4), 1498–1519.

Shen, J., Song, Zb, Qian, X., & Liu, W. (2009b). A preliminary investigation into the use of acid-tolerant precipitated calcium carbonate fillers in papermaking of deinked pulp derived from recycled newspaper. *BioResources*, *4*(3), 1178–1189.

Siqueira, G., Bras, J., & Dufresne, A. (2009). Cellulose whiskers vs microfibrils: Influence of the nature of the nanoparticle and its surface functionalization on the thermal and mechanical properties of nanocomposites. *Biomacromolecules*, *10*, 425–432.

Siró, I., & Plackett, D. (2010). Microfibrillated cellulose and new nanocomposite materials: A review. *Cellulose*, *17*, 459–494.

Švajlenka, J., & Kozlovská, M. (2020). Evaluation of the efficiency and sustainability of timber-based construction. *Journal of Cleaner Production*, *259*, 120835.

Taghiyari, H. R., Ghamsari, F. A., & Salimifard, E. (2018). Effects of adding nano-wollastonite, date palm prunings and two types of resins on the physical and mechanical properties of medium-density fibreboard (MDF) made from wood fibres. *Bois & Forets Des Tropiques*, *335*, 49–57.

Taghiyari, H. R., & Norton, J. (2014). Effect of silver nanoparticles on hardness in medium-density fiberboard (MDF). *iForest-Biogeosciences and Forestry*, *8*, 677.

Teacă, C. A., Roşu, D., Bodîrlău, R., & Roşu, L. (2013). Structural changes in wood under artificial UV light irradiation determined by FTIR spectroscopy and color measurements—A brief review. *BioResources*, *8*, 1478—1507.

Thakur, A., Kumar, A., & Kanwar, S. (2012). Production of n-propyl cinnamate (musty vine amber flavor) by lipase catalysis in a non-aqueous medium. *Current Biotechnology*, *1*, 234—240.

Thakur, S. S., Kumar, A., & Chauhan, G. S. (2018). Cellulase immobilization onto zirconia-gelatin-based mesoporous hybrid matrix for efficient cellulose hydrolysis. *Trends in Carbohydrate Research*, 10.

Tsuchida, T., & Yoshinaga, F. (1997). Production of bacterial cellulose by agitation culture systems. *Pure and Applied Chemistry*, *69*, 2453—2458.

Vaseghi, Z., & Nematollahzadeh, A. (2020). *Nanomaterials: Types, synthesis, and characterization. Green synthesis of nanomaterials for bioenergy applications* (pp. 23—82). Wiley.

Wang S., Cheng Q., Rials T.G., Lee S.-H. Cellulose microfibril/nanofibril and its nanocompsites. In: *Proceedings of the 8th Pacific rim bio-based composites symposium*, Kuala Lampur, Malaysia. 2006.

Wang, X., Shu, Hong, Hailan, Lian, Xianxu, Zhan, Mingjuan, Cheng, Zhenhua, Huang, Maurizio, Manzo, Liping, Cai, Ashok Kumar, Nadda, Quyet, Van Le, & Changlei, Xia (2021). Photocatalytic degradation of surface-coated tourmaline-titanium dioxide for self-cleaning of formaldehyde emitted from furniture. *Journal of Hazardous Materials*, *420*, 126565. Available from http://doi.org/10.1016/j.jhazmat.2021.126565.

Wegner, T. H., & Jones, P. (2006). Advancing cellulose-based nanotechnology. *Cellulose*, *13*, 115—118. Available from https://doi.org/10.1007/s10570-006-9056-1.

Wegner, T. H., & Jones, E. P. (2009). *A fundamental review of the relationships between nanotechnology and lignocellulosic biomass. The nanoscience and technology of renewable biomaterials* (pp. 1—41). Hoboken, NJ: Wiley.

Wei, S., Li, X., Shen, Y., Zhang, L., & Wu, X. (2020). Study on microscopic mechanism of nano-silicon dioxide for improving mechanical properties of polypropylene. *Molecular Simulation*, *46*, 468—475.

Wu, W., Jing, Y., Zhou, X., & Dai, H. (2011). Preparation and properties of cellulose fiber/silica core—shell magnetic nanocomposites. In: *Proceedings of 16th international symposium on wood, fiber and pulping chemistry*, ISWFPC (pp. 1277—1282), Tiajin, China.

Wu, Y., Liang, Y., Mei, C. T., Cai, L., Nadda, A. K., Le, Q. V., Peng, Y., Lam, S. S., Sonne, C., & Xia, C. (2021). Advanced nanocellulose-based gas barrier materials: Present status and prospects. *Chemosphere*, *286*, 131891. Available from http://doi.org/10.1016/j.chemosphere.2021.131891.

Yao, Y., Huang, X.-H., Zhang, B.-Y., Zhang, Z., Hou, D., & Zhou, Z.-K. (2020). Facile fabrication of high sensitivity cellulose nanocrystals based QCM humidity sensors with asymmetric electrode structure. *Sensors and Actuators B: Chemical*, *302*, 127192.

Zaieda, M., & Bellakhal, N. (2009). Electrocoagulation treatment of black liquor from paper industry. *Journal of Hazardous Materials*, *163*(2), 995—1000.

Zhang, S., Bilal, M., Zdarta, J., Cui, J., Kumar, A., Franco, M., et al. (2020). Biopolymers and nanostructured materials to develop pectinases-based immobilized nano-biocatalytic systems for biotechnological applications. *Food Research International*, 109979.

Zhang, X., Zhao, Z., Ran, G., Liu, Y., Liu, S., Zhou, B., & Wang, Z. (2013). Synthesis of lignin-modified silica nanoparticles from black liquor of rice straw pulping. *Powder Technology*, *246*, 664—668.

Zhu, M., Li, T., Davis, C. S., Yao, Y., Dai, J., Wang, Y., et al. (2016). Transparent and haze wood composites for highly efficient broadband light management in solar cells. *Nano Energy*, *26*, 332—339.

Zimmermann, T., Bordeanu, N., & Strub, E. (2010). Properties of nanofibrillated cellulose from different raw materials and its reinforcement potential. *Carbohydrate Polymers*, *79*, 1086—1093.

Zimmermann, T., Pöhler, E., & Schwaller, P. (2005). Mechanical and morphological properties of cellulose fibril reinforced nanocomposites. *Advanced Engineering Materials*, 7, 1156—1161.

CHAPTER 2

Nanofibers for the paper industry

Paweł Chmielarz and Izabela Zaborniak
Department of Physical Chemistry, Faculty of Chemistry, Rzeszow University of Technology, Rzeszow, Poland

2.1 Paper industry: challenges

The pulp and paper industry constitutes global manufacturing with a wide range of application in almost all industry branches around the world. As paper is a biodegradable material for potential replacing of plastics bags and packages, the papermaking process needs rapid development to create production processes maintaining environmental aspects, cost-effectiveness and high quality of the received products depending on the applications. Therefore the paper industry is a challenge and is continuously developing in response to different issues: (1) constantly increasing requirements for high-quality materials with improved mechanical, physical, and printing characteristics; (2) the need to maintain a high quality of recycled fibers despite the increasing recycle rate; and (3) needs for implementation of cost-effective production processes (Delgado-Aguilar, González, et al., 2015).

Among the challenges of the paper industry, paper requirements should also be considered in the point of the different issues developed in the few recent years such as the digitization of information, needs for sustainable development, and the increasing demand for packaging for different industries, for example, there is a significant increase in demand packaging for food (c. 51%), beverages (c. 18%) and less in cosmetics, healthcare, and packaging for industry (Azeredo, Rosa, & Mattoso, 2017; Boufi et al., 2016; Ehman, Lourenço, et al., 2020). Towards these aspects, the implementation of new strategies in paper manufacturing is strongly studied in recent years. New products are continuously developed and the products available on the market are modified to receive materials for more integrated industries. Paramount importance is focused on the production of recycled paper (Delgado-Aguilar, Tarrés, Pèlach, Mutjé, & Fullana-i-Palmer, 2015). It is a result of the greater ecological awareness of both people and companies and, as mentioned above, significantly higher demand for packaging papers. Among the advantages of paper from recycles, for example, environmental, economic, and social benefits, there are also some disadvantages. In this context, there are two main challenges including the deterioration of material properties due to the hornification phenomena and the problem associated with the accumulation of dissolved and colloidal substances (Fernandes Diniz, Gil, & Castro, 2004). Different strategies have been adopted to avoid this inconvenience. Among biorefining

Nanotechnology in Paper and Wood Engineering
DOI: https://doi.org/10.1016/B978-0-323-85835-9.00002-7

processes with the use of enzymes, dry strengthening agents, and application of soft mechanical refining of the recycled fibers, the addition of nanofibers is considered to improve the properties of recycled paper. For example, the use of cellulosic nanofibers (CNF), microfibrillated cellulose (MFC), or lignocellulosic nanofibers (LCNF) for paper manufacturing is one of the approaches for reaching those objectives. The mentioned above nanofibers are received from cellulosic pulp with the various content of lignin and hemicelluloses. More precisely, CNF is obtained from bleached pulps, namely, lignin-free raw materials, while LCNF—is obtained from a substrate with high percentage lignin content (unbleached pulp) (Azeredo et al., 2017; Boufi et al., 2016; Ehman, Felissia, et al., 2020). These nanofibers applied in paper production, improve the quality and performance of paper, reduce the paper grammage, and are beneficial in food packaging—enhance barrier properties (Vallejos et al., 2016).

2.2 Nanofibers: characteristics

Nanotechnology is a discipline for the development of materials at nano levels. It is currently the most rapidly developing scientific discipline because it allows for creating novel materials with advanced applications. This technology significantly influenced various engineering disciplines, for example, electronics, polymer engineering, and material science, including paper manufacturing. Part of this discipline is the preparation of nanofibers. Nanofibers are fibers with very small diameters, c. 50−500 nm. They are characterized by a considerable length and a small cross-section; the diameter is about 100 times smaller than their length. Nanofibers are found in both natural materials (e.g., collagen fibers in tissues) and man-made materials. They can act as matrix strengthening materials in composites or constitute the main component of the material (fabric, nonwoven fabric). Nanofibers continuously attract much attention due to their unique properties. Compared to conventional fibrous structures, nanofibers exhibit remarkable characteristics, including lightweight with small diameters, have controllable pore structures, and due to the large surface area to weight ratio, they have a large specific surface which gives high strength properties (Kenry & Lim, 2017; Subbiah, Bhat, Tock, Parameswaran, & Ramkumar, 2005; Vallejos et al., 2016). Considering the paper industry, the nanofibers are added to the pulp or created due to the fibrillation of fibers during pulp refining as the papermaking process proceeds. Fibrillation of the fibers by mechanical refining is the most often applied approach to enhance paper strength due to a variety of changes that occur as a result of refining, for example, external and internal fibrillation, fines formation, fiber shortening or cutting. The processes increase bonding in the nanofibers' structure, therefore improve tensile strength and stiffness (Afra, Yousefi, Hadilam, & Nishino, 2013; Gharehkhani et al., 2015). However, multiple refining cycles can damage the fibers, thus reducing their utilization. To prevent this phenomenon, the nanofibers are used in paper furnishes and are coated on the paper surface, or the refining with enzymes is used (Gharehkhani et al., 2015).

2.3 Cellulose nanofibers

CNFs are created by the smallest structural unit that represents cellulose fibers, going through aggregates reaching higher dimensions, for example, cellulose fiber, microfibrils, and finally receiving CNFs with diameter between 3 and 5 nm (Fig. 2.1). As the CNFs are characterized by such a tiny diameter, the specific area of the nanomaterials is huge. Therefore CNF is able to bond adjacent fibers significantly better than cellulose fibers that are characterized by higher dimensions. They have also a high aspect ratio and are able to reach lengths of about 70 μm, thus possess high intrinsic mechanical properties (Delgado-Aguilar, Tarrés, et al., 2015).

Nanomaterial attracts many attention to scientists in both completely novel or well-established technological and scientific domains, due to the high potential as a sustainable reinforcement in material sciences (Shaghaleh, Xu, & Wang, 2018). Bearing in mind the unique properties of CNF, a significant number of studies considered this nanomaterial for paper reinforcement in the form of paper additive with great strengthening potential (Ahola, Österberg, & Laine, 2008; Alcalá, González, Boufi, Vilaseca, & Mutjé, 2013;

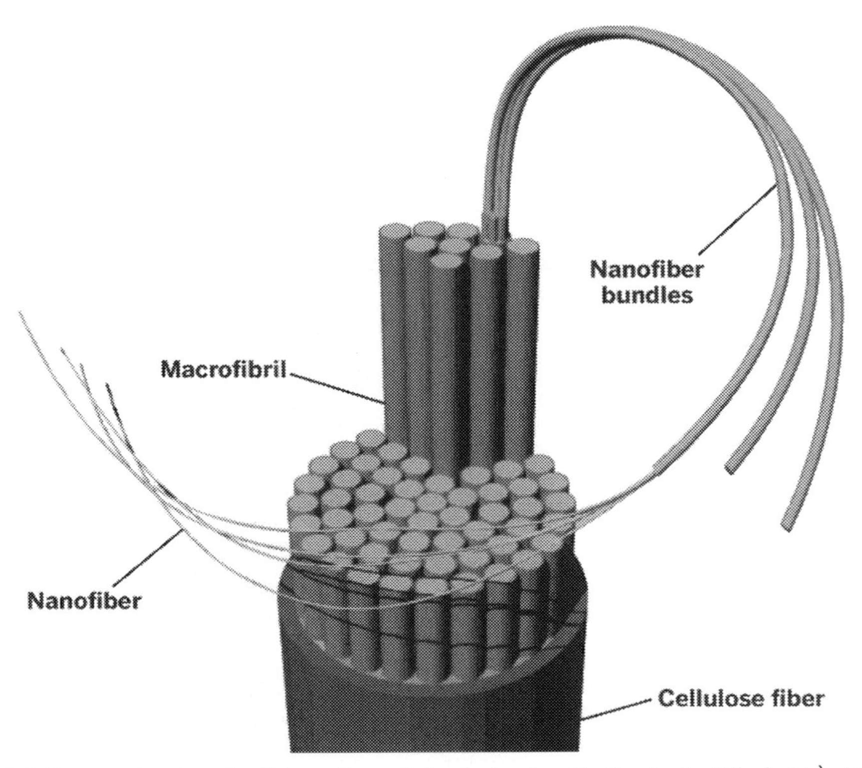

Figure 2.1 Structural units of cellulose. *From Delgado-Aguilar, M., Tarrés, Q., Pèlach, M.À., Mutjé, P., & Fullana-i-Palmer, P. (2015). Are cellulose nanofibers a solution for a more circular economy of paper products?* Environmental Science & Technology, 49(20), 12206–12213. *https://doi.org/10.1021/acs.est.5b02676.*

Delgado-Aguilar, González, et al., 2015; Merayo, Balea, de la Fuente, Blanco, & Negro, 2017; Tarrés et al., 2016). Delving into this topic of CNF using as a new type of paper strengthening component, there are different reasons that motivate this application in the context of paper manufacturing: (1) specific area of the CNF is expanded by the nanoscale lateral dimension of nanomaterial; (2) micrometer dimension of the length; (3) extended cellulose chains that create semicrystalline structure; (4) combination of unique intrinsic mechanical strength with good flexibility properties; and (5) ability of hydrogen bonding along the cellulosic fibers, thus the high potential for the interaction and formation a strong entangled network (Azeredo et al., 2017; Boufi et al., 2016). Considering the possibility of recycling paper, adding CNFs to the paper pulp results in paper products that can be recycled more often than regular paper—up to seven times of recycling (Delgado-Aguilar, Tarrés, et al., 2015). As the addition of the CNF to the wood pulp is rapidly developing, several reviews on these topics have been already published, including the review of the highlights in recent progress in the field of the usefulness of CNF as an additive for papermaking including wood pulps, agricultural wastes, and recycled paper, CNF reinforcing effect mechanism, and their application as a coating material (Boufi et al., 2016; Brodin, Gregersen, & Syverud, 2014; Osong, Norgren, & Engstrand, 2016).

2.3.1 Types of CNF in paper manufacturing

The production of CNF was firstly developed by Herrick, Casebier, Hamilton, & Sandberg, (1983) and Turbak, Snyder, and Sandberg (1983), producing nanomaterials from wood pulp fibers that were passed several times through a high-pressure homogenizer. It is one of the mechanical methods that have been applied to obtain fibrillation among the approaches as follows: microfluidization (Zimmermann, Pöhler, & Geiger, 2004), microgrinding (Iwamoto, Nakagaito, & Yano, 2007), refining (Nakagaito & Yano, 2004), or cryocrushing (Taniguchi & Okamura, 1998). Mechanical pretreatments reduce the risk of blockage during fibrillation. For many years these approaches have been widely used for CNF receiving, however recently there has been an exponential growth in the methods of obtaining these nanomaterials in a more economical and sustainable way. The commercial interest in mechanical pretreatment methods to produce CNF was poor for many years due to the high energy consumption involved during the mechanical disintegration of the fibers into nanofibers, even 12,000–65,000 kWh ton^{-1} has been reported (Naderi, Lindström, & Sundström, 2014; Spence, Venditti, Rojas, Habibi, & Pawlak, 2011). Therefore, their scale-up production was significantly limited and failed to meet the expectation. This high energy demand is generated by the high pressure (200–800 bar) that is necessary to receive a high fibrillation degree. As the energy consumption during the fibrillation process is strongly dependent on the used fibers pretreatment method, there is rapid development in the elaborating novel approaches. The solution for the high energy demand problem during the production of CNF is the implementation of some kind of chemical pretreatment. This strategy is currently the most popular and efficient pretreatment method. The mechanism of chemical pretreatment relies on the formation of ionic or ionizable groups within the fibers facilitating the break-up of the fibers network. It allowed for a drastic reduction of energy consumption during CNF fabrication down to 500–1500 kWh ton^{-1} (Delgado-Aguilar, González, et al., 2015; Klemm et al., 2011). Chemical pretreatment can be performed using: (1) 2,2,6,6-tetramethylpiperidine-1-oxyl radical (TEMPO)-mediated oxidation process—the introduction of a significant amount of carboxylate and aldehyde groups into native cellulose by regioselective conversion of the primary hydroxyl groups to carboxylate ones (Isogai & Bergström, 2018; Isogai, Saito, & Fukuzumi, 2011; Saito, Kimura, Nishiyama, & Isogai, 2007) (Fig. 2.2); (2) carboxymethylation—the presence of carboxymethyl groups on

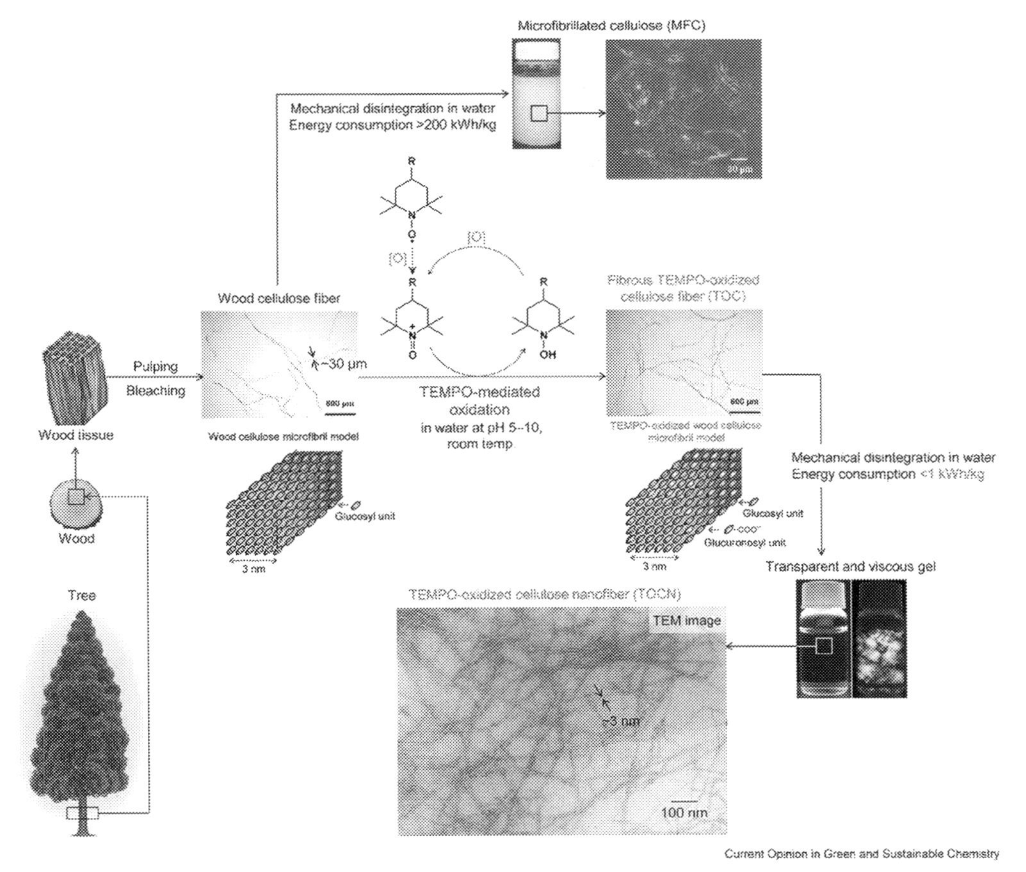

Figure 2.2 Synthetic route for the preparation of 2,2,6,6-tetramethylpiperidine-1-oxyl radical (TEMPO)-oxidized cellulose nanofibers by TEMPO-mediated oxidation followed by gentle mechanical disintegration of oxidized cellulose in water. *From Isogai, A., & Bergström, L. (2018). Preparation of cellulose nanofibers using green and sustainable chemistry.* Current Opinion in Green and Sustainable Chemistry, 12, *15–21. https://doi.org/10.1016/j.cogsc.2018.04.008.*

cellulose chains fibers facilitate the production of the fibers with nanosize dimensions (Naderi et al., 2014; Su et al., 2019); (3) sulfonation with sodium bisulfate (Buzała, Przybysz, Rosicka-Kaczmarek, & Kalinowska, 2015; Sirviö, Ukkola, & Liimatainen, 2019); and (4) periodate oxidation—selective stereospecific reaction relays on the cleavage of C_2-C_3 bond, the glucopyranose unit, thus two aldehyde groups per glucose unit are formed (Errokh, Magnin, Putaux, & Boufi, 2018), and the received dialdehyde could be easily transformed into carboxylic (Liimatainen, Visanko, Sirviö, Hormi, & Niinimaki, 2012), sulfonic (Liimatainen, Visanko, Sirviö, Hormi, & Niinimäki, 2013), imine, amino (Sirviö, Visanko, Laitinen, Ämmälä, & Liimatainen, 2016), hydroxyl, or quaternized ammonium groups (Visanko et al., 2015). The content of the ionic groups in cellulose is crucial to effectively release the CNFs, breaking down the cell wall.

Chemical pretreatment methods, especially the most widely used TEMPO-mediated oxidation process possess a few drawbacks. Among them, the use of the highly negative charge of the CNF prepared with the use of TEMPO often results in filler flocculation problems, and thus this phenomenon diminishes its retention in the paper matrix (Martin-Sampedro et al., 2012). The alternative to chemical pretreatment strategies is an enzymatic approach that enhances both the accessibility and reactivity of the fiber wall. This solution was firstly used by Pääkkö et al. (2007) and Henriksson, Henriksson, Berglund, and Lindström (2007) to produce CNF by enzymatic hydrolysis. The enzyme cellulase covers two subgroups: cellohydrolases, which attack the crystalline region of cellulose, and endoglucanase which attack the amorphous (disordered region) of cellulose reducing the length of cellulose chains. Cellulases are complex enzymes formed by a catalytic domain that modified the cellulose by hydrolyzing the β-1,4-glycosidic bonds, while noncatalytic domains contain fibronectin-like type 3 domains and cellulose-binding domains. These noncatalytic domains are responsible for highly specific binding to the substrate (Cadena et al., 2010). Enzymatic pretreatment was widely studied by researchers as an approach to produce both MFC and CNF (González et al., 2013; Martin-Sampedro et al., 2012). Using this approach, the CNF in the range of 20–100 nm and lengths up to 100 μm have been received, and the influence of the enzymatic CNF additive to improve reinforcing properties in the paper manufacturing process has been investigated. It has been found that the additive increases the strength of pulp handsheets up to 40%–60%. However, considering the industrial scale, some studies have already been addressed to study the complex interactions between CNF, cellulosic pulps, mineral fillers, and additives that are used for paper production (Tenhunen et al., 2018).

Considering the above-mentioned approaches for CNF production, the combination of mechanical, chemical, and enzymatic treatments is currently mostly used. The most appropriate solution is a combination of chemical oxidation and mechanical fragmentation, which gives more efficient results and allows to receive nanofibers with a diameter of several nanometers. However, the most ecological and prospective approach constitutes enzymatic hydrolysis. It does not require high financial outlays and does not cause the production of substances that are difficult to dispose of (Qing et al., 2013; Syverud, Chinga-Carrasco, Toledo, & Toledo, 2011).

The addition of CNF into paper structure enhances its strength properties, reduces its porosity, and also increases density. A significant improvement in both physical and mechanical properties of prepared paper material after the addition of CNF is a result of the increase in specific surface area as a consequence of CNF combination with papermaking suspensions. As the specific surface area is boosted, the fiber—fiber bonds are formed, therefore density, tensile strength, rigidity, and strength are significantly increased. The addition of CNF also modifies light scattering properties. The improvement of the above-mentioned properties is influenced by several factors, for example, the content of CNF, available specific surface area of CNF, degree of fibrillation of the nanofibers, use and addition strategy of retention agents, and pulp's refined degree.

2.3.2 The mechanism of CNF strengthening properties

The strength properties of the paper with the CNF additives are mostly dependent on the number of fiber—fiber bonds that were formed during the consolidation and drying of the fiber network processes. In this context, a huge impact has the effect of Laplace pressure resulting from the curvature of the meniscus of the liquid bridge in the fiber boundary. This phenomenon causes the fibers to come close together during the drying process, the hydrogen bonds are formed, and thus the cohesion between neighboring fibers is significantly increased. There are known various factors that influenced the paper strength characteristics, including (1) the length and strength of the fiber; (2) the strength of specific bond and bonded area; (3) formation of the sheet; and (4) distribution-residual stresses. The strength of the paper can be enhanced by few approaches, that is, refining—known as a mechanical beating of pulp, is a common procedure to improve the mechanical properties of the final product as a result of increased density and to better bonding between fibers (Kumar, Pathak, & Bhardwaj, 2020; Motamedian, Halilovic, & Kulachenko, 2019) (Fig. 2.3), an application of wet or dry strengthening additives, for example, polyacrylamides, polysaccharides, starch, and fibers (Vega et al., 2013; Wang, He, Wang, & Song, 2015), and chemical modification of the fibers (John & Anandjiwala, 2008). All the proposed solutions are based on chemical additive application or physical treatment to facilitate the capacity of the fibers, thus the strengthening effect can be considered by two potential mechanisms. The first one covers the behavior of CNF as an adhesion promoter that favors the fiber—fiber bonding simultaneously enhancing the bonded area, while the other concept includes the action of the CNF as a network that is embedded among other fibers with higher dimension filling in voids and pores among each fiber, contributing to boosting the load-bearing capacity of the paper. Due to this phenomenon the final strength properties are affected by both CNF used as an additive and other fibers.

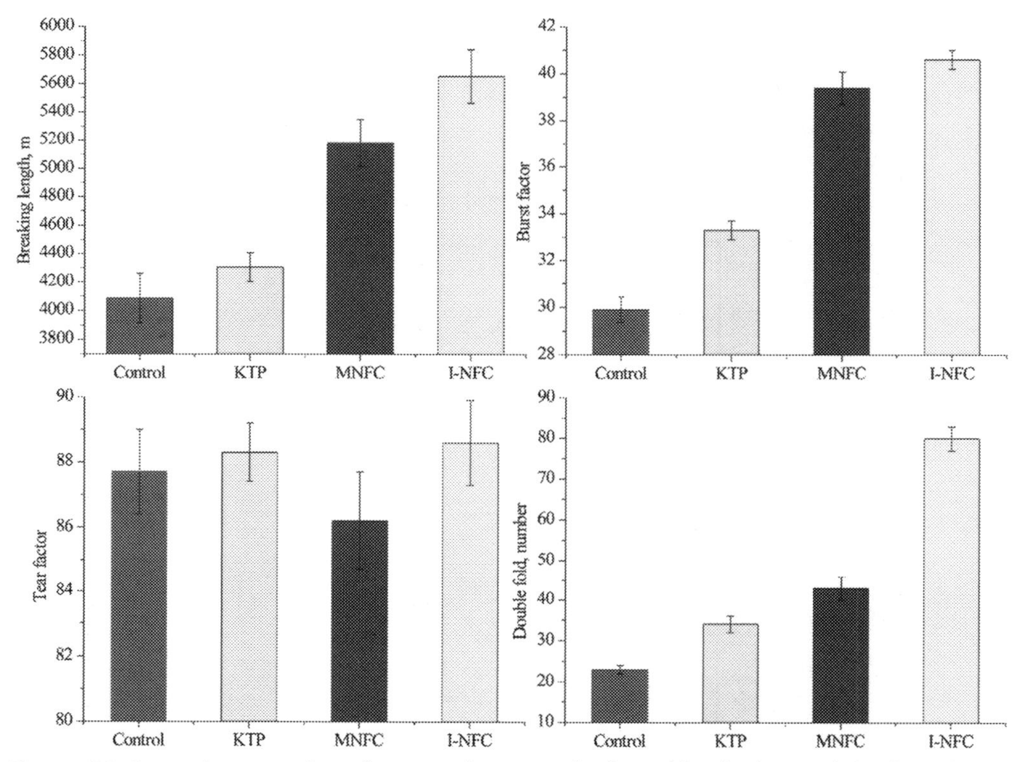

Figure 2.3 Strength properties of paper sheets made from bleached mixed hardwood pulp (BMHWP) after the addition of the potassium hydroxide treated pulp (KTP), micronanofibrillated cellulose (MNFC), and imported NFC (I-NFC). *From Kumar, V., Pathak, P., & Bhardwaj, N. K. (2020). Facile chemo-refining approach for production of micro-nanofibrillated cellulose from bleached mixed hardwood pulp to improve paper quality.* Carbohydrate Polymers, 238, *116186. https://doi.org/10.1016/j.carbpol.2020.116186.*

The effect of different types of CNF on the strength properties of paper was investigated by Taipale, Österberg, Nykänen, Ruokolainen, and Laine (2010). The carboxymethylated CNF significantly increased tensile strength and also gave low drainage resistance, caused by the negatively charged CNF that were deposited on the fiber surface promoted by cationic starch as a retention aid. While, treatment of the CNF with a cationic polyelectrolyte—poly(amide amine) epichlorohydrin—improved both wet and dry strengths of paper, as a result of formation of a nonuniform and more rigid layer (Ahola et al., 2008). It proved that CNF without retention aid is mostly a filler in the pores of the fibers, and form the bridges between fibers, increasing the bond area of the sheets, thus the strength, and also enhancing the drainage resistance of the sheets. An addition of retention aid gives a softer surface.

2.3.3 CNF as an additive in paper industry

Many researchers studied the use of CNF as a strength additive. González et al. (2013) improved the paper mechanical strength using an additive amount of CNF while providing an economic process (Cadena et al., 2010). They reduced energy demand during pulp refining by the use of enzymatic pretreatment combined (biobeating) with CNF addition, enhancing the mechanical and physical properties of a papermaking pulp. The biobeating process combined with CNF addition improved strength properties without modifying drainability, and the received length at break values were similar to those observed in commercial printing/writing paper, making this approach a substituent for a mechanical beating. Biobeating process presents essential advantages, for example, this process can improve beatability of the pulp, simultaneously increasing drainability; using cellulases to treat a dried pulp enhances the relative bonded area of the fibrous paper network, and beneficial in influencing some paper properties; this type of paper modification improves the bonding ability between fibers, increasing its strength; above others it creates cost-effective process reducing beating intensity, for example, by 33% (Lecourt, Meyer, Sigoillot, & Petit-Conil, 2010), keeping the breaking length values similar to beaten pulps. CNFs can be also considered as a material that improves binding properties of fillers, increasing the filler content of paper (Torvinen, Kouko, Passoja, Keränen, & Hellen, 2014). It was observed that CNF improved the strength properties of sheets elaborated in a semipilot trial using different fiber grades and various fillers, that is, precipitated calcium carbonate, cationic polyacrylamide, and starch. The binding of the filler with CNF allowed for an increase of the filler content of papers from 30% to 40%, simultaneously reducing variable costs in supercalendered-paper production.

CNF was also investigated in combination with clay in layered thermo-mechanical pulp fractions (TMP) paper product (Mörseburg & Chinga-Carrasco, 2009). A composition of TMP, CNF, and clay was studied in the context of mechanical and optical properties. Used separately, nanofibrillated cellulose enhances the strength properties of paper and fillers improve the optical properties. The results indicate a significant improvement in both z-strength and tensile index after substitution of bleached softwood kraft pulp fibers with CNF additive. The filler addition may cause a reduction of the product strength, but this study showed that an appropriate addition of CNF counteracted this phenomenon. Considering the estimation of an overall quality index, considering five variables presented in the study, it was stated that placing the fillers and TMP accepts fraction in the surface, and CNF location in the center of the sheets in combination with the refined TMP rejects fraction, give the best sheet construction.

The use of modified CNF in papermaking opens new opportunities by improving the properties or imparting quite new characteristics to paper products. Such a procedure can enhance hydrophobicity, compatibility, wettability, and ability to interact with a matrix.

As cellulose and hemicellulose are biopolymers with ubiquitous hydroxyl groups in fibers' structure, the paper is characterized by a great tendency to absorb water from both environment and the nearby products, reducing its properties and usefulness. Therefore the main concept of the use of modified CNF relays on the introduction to the paper CNF with improved hydrophobic properties.

Missoum, Martoïa, Belgacem, and Bras (2013) modified CNF with alkyl ketene dimer (AKD) nanoemulsion, and studied the influence of the modified CNF on mechanical and barrier reinforcements of paper, but also an internal sizing of paper. The used AKD is widely known to endow hydrophobic characteristics to the paper material, that is, the so-called sizing effect. The CNF for the experiments was received as a result of bleached wood pulp enzymatic pretreatment, and then it was mechanically disintegrated. The prepared CNF—modified and unmodified—was investigated in various content in pulp slurry (5%–50% of CNF) without using any retention agent to avoid the loss of CNF fraction during the filtration process. The results showed that above the appropriate concertation value of CNF, the type of CNF used did not affect the retention properties, because the nanofibers did not retain in the pulp slurry. The chapter presents the ability to use AKD-modified CNF in the role of an internal sizing agent, that allows for the production of the paper with hydrophobic properties, and the application of CNF or modified CNF (50% content) provided more packed paper due to the CNF behavior as a binder (Fig. 2.4). In summary, the received paper exhibited multilevel characteristics, namely hydrophobicity, improved mechanical strength, and air permeability.

Paper strength and barrier properties were also enhanced by the application of acetylated CNF as an additive (Mashkour, Afra, Resalati, & Mashkour, 2015), which means the hydroxyl groups (-OH) of cellulose were substituted with acetyl (CH_3CO) moieties. Therefore such a type of modification increases the hydrophobic characteristics of the paper. Acetylation is a cost-effective and environmentally friendly process that can be conducted in two mechanisms. The first one is fibrous (heterogeneous) acetylation while the cellulose acetate is insoluble and the process preserves the fiber morphological structure, and the other one is solution (homogeneous) acetylation while the product is dissolved during the reaction. The study presents the preparation by heterogeneous acetylation process avoiding the use of catalyst into the pulp, preserving the core of the cellulose. As the full substitution of cellulose hydroxyl groups can cause a decrease in paper mechanical strength, the partial acetylation of CNF was applied. Unmodified CNF added to the pulp increased density, tensile strength, and air resistance without preventing water absorption, while acetylated CNF used as a substituent to CNF, avoid the water absorption c. 23% compared to unmodified CNF-based paper sheets.

Hollertz, Durán, Larsson, and Wågberg (2017) used three different types of CNF as paper strength additive, that is, carboxymethylated CNF, periodate-oxidized

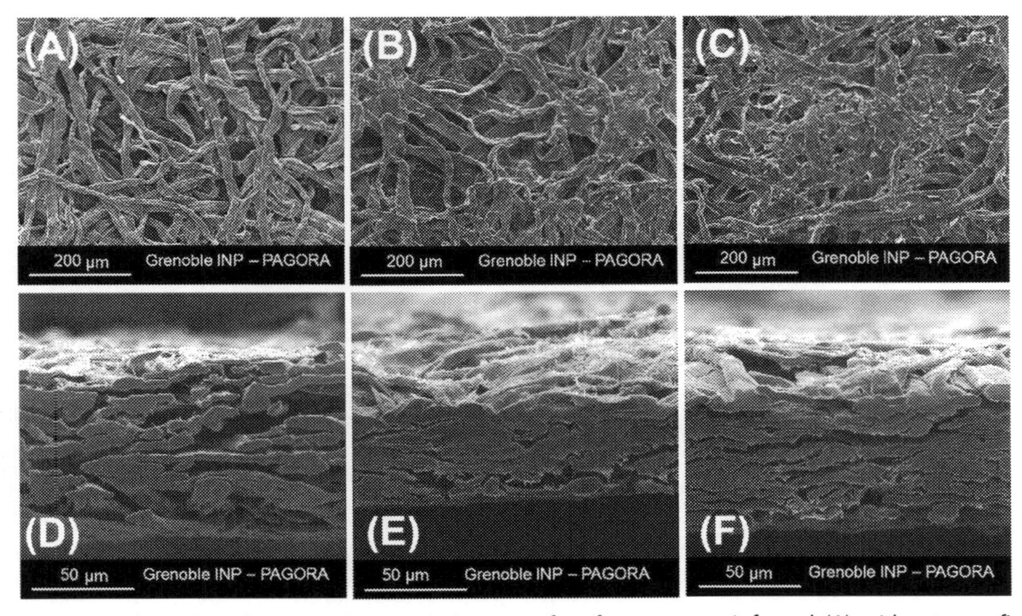

Figure 2.4 Scanning electron microscopic images of surface paper reinforced (A) without nanofibrillated cellulose (NFC, reference), (B) with NFC, (C) with modified NFC and cross-section of (D) without NFC, (E) with NFC, and (F) with modified NFC. *From Missoum, K., Martoïa, F., Belgacem, M. N., & Bras, J. (2013). Effect of chemically modified nanofibrillated cellulose addition on the properties of fiber-based materials.* Industrial Crops and Products, 48, *98–105. https://doi.org/10.1016/j.indcrop.2013.04.013.*

carboxymethylated CNF, and dopamine-grafted carboxymethylated CNF in the context of the influence of the CNF type on mechanical properties, the fibril retention, sheet density, and sheet morphology of the final paper sheets. The use of native CNF usually increases the tensile strength, Young's modulus, and strain-at-break properties of the paper sheets. While the modified CNF is added, the underlying mechanisms behind the strengthening effect can be considered. Considering the above-mentioned CNF types, the best strength properties were showed for periodate-oxidized CNF in combination with retention aid as polyvinyl amine (PVAm) or poly(dimethyldiallylammonium chloride), receiving almost 89% increase in tensile strength at a 15 wt.% addition, comparing the results to reference with only retention aid. With the use of PVAm wet-strong sheets with a high value of the wet tensile index (30 $kNm\,kg^{-1}$) were also received, as a result of the formation of cross-links between the functional groups located in the CNF—hydroxyl groups, modified CNF—aldehyde groups introduced by the periodate oxidation, and the primary amines present in PVAm. When the dopamine-grafted carboxymethylated CNF were used, the received nanomaterial exhibited a strong tendency to form films and a significant improvement of the mechanical properties of the sheets was observed. The study presents the approach of the chemical modification of CNF to produce sheets with much stronger mechanical properties compared to paper with unmodified CNF, using a lower amount of the additive, without an increase of density.

2.3.4 CNF as coating material in papermaking

The CNF is also considered as films and coating material for a wide range of potential applications as transparent and biodegradable packaging films exhibiting high barrier properties. There are few approaches to prepare CNF coating on the paper surface: bar coating, roll coating, size press coating, spray coating, and foam coating. Depending on the applied technique, a varying amount of coating material is applied to the paper. Lavoine, Desloges, and Bras (2014) studied the influence of the applied coating techniques (bar coating and size press coating) on the number of CNF layers formed on the paper surface, receiving 14 g m^{-2} versus 3 g m^{-2} of CNF, respectively. Moreover, the effects of the bar coating method on compressive strength, bending stiffness, and barrier properties (oxygen and air permeability) of the most often used fiber-based packaging material (cardboard) were also studied. The used CNF coating approach impacts notably the cardboard characteristics in two opposite ways. First, it decreases the compressive strength of the modified cardboard because it negatively influences the structural cohesion. On the other hand, applying of CNF layer on cardboard enhances bending stiffness due to an increase in the paper-based material thickness. The addition of CNF layer counterbalanced the negative impact of the used coating technique, as the presence of CNF improves bending stiffness and compressive strength by 30% in the machine direction. However, the oxygen and air permeability properties were not significantly affected after the addition of the CNF coating on the paperboard surface, also the water absorption of modified material was inconsiderable.

Spray coating method is advantageous to prepare a thin coating layer that can be uniformly distributed. However, among the benefits, this approach has drawbacks, for example, the coating formulation should be characterized by low viscosity, thus the low solid content is necessary that limits the amount of CNF as a coating material that can be applied using this method (Vega et al., 2013; Wang et al., 2017). Foam coating approach makes it possible to spread a thicker coating layer over the paper surface, thus more even distribution of CNF at low coating is achieved. Kinnunen-Raudaskoski, Hjelt, Kenttä, and Forsström (2014) used a mixture of 2.9% CNF with a combination of an anionic surfactant creating a foam, followed by vigorous stirring in a foam generator together with compressed air, resulting in stable foams (80%–95% air). The proposed method provided coat weights c. 1 g m^{-2} for a single layer coating and 2.6 g m^{-2} for a double-layer coating. The method did not enable full coverage of the paper surface, however, the surface properties were enhanced, for example, higher surface smoothness, reduced air permeability, and lower contact angle.

The CNF coatings are usually used to improve the barrier properties of the paper. Syverud and Stenius (2008) examined strength properties and air permeability of the paper from unrefined softwood pulp coated with various content of CNF (0%–8% of total basis weight) prepared by shearing disintegration. The strength of the prepared paper materials was studied by measuring tensile index values, indicating a slight increase in this parameter value after the addition of 8% CNF coat (35–40 N mg^{-1}). While the air permeability changed dramatically after the application of CNF as a coating (from 6.5 to 360 Pa^{-1} s^{-1} for the paper samples with 0% and 8% of CNF coat, respectively). The significant reduction in air permeability is a consequence of an increase in surface porosity caused by nanofibers. Similar results were published by Lavoine, Desloges, Khelifi, and Bras (2014). Experiments indicated a substantial increase of air permeability by the addition of an MFC coat onto the paper surface. CNF coat was also used in combination with shellac to create a layer of film on the paper and paperboard using a bar coater or a spray coating method (Hult, Iotti, & Lenes, 2010). The prepared paper-based materials with CNF coat were characterized by a significant decrease in air permeability, oxygen transmission rate, and water-vapor transmission. Therefore it constitutes a highly interesting material for barrier packaging.

The CNF was applied as a coating material to replace synthetic binders in board coating with pilot-scale coating and printing trials (Pajari, Rautkoski, & Moilanen, 2012). The main aim of the study was to investigate runnability and the properties of the colors of the coating by periodically adding appropriate portions of CNF instead of latex binder. The use of CNF as a partial substituent of latex binder resulted in enhanced viscosity of the coating colors particularly at low shear rates, while the increase in viscosity was inconsiderably affected at higher shear rates. Additionally, the use of CNF improved air permeance and resulted in lower gloss. Among the advantages, two drawbacks were noticed, namely, the process was wastefull because the use of CNF at low consistencies, the time of dryness was extended, and the cost of CNF compared to latex binders was higher.

Nanocellulose was applied for the production of nanofiltration membranes by the approach mimicking the papermaking process, creating a fiber mat (Mautner et al., 2014). More precisely, the formation of nanopapers was provided in the presence of trivalent ions, therefore, inducing flocculation of nanofibrils. The proposed solution applied an aqueous suspension of nanocellulose, while eliminating vast amounts of organic solvents widely utilized for the production of conventional organic solvent nanofiltration (OSN) polymer membranes. The efficiency of filtration properties of the prepared membrane was strongly affected by the hydrophilicity of the solvents used—water, tetrahydrofuran, and n-hexane were studied. The permeance of the nanopaper was affected by the dimensions of the nanofibrils and the grammage of the nanopapers. The diameter of the nanofibrils also influenced molecular weight cut-off, so it is possible to adjust the membrane performance by selecting nanofibrils of varying dimensions, thus producing materials with a wide range and advanced potential applications. The paper presents a simple way of preparing a stable and efficient OSN membrane using a naturally derived substrate.

2.4 Lignocellulosic nanofibers

Due to the excellent strength properties and thus great potential for use in the paper industry, the topic of CNF becomes an attractive and widely studied issue among the scientific and technological community. However, the CNF production process poses many challenges and difficulties mainly caused by the economical demand of the CNF production, which significantly limits its rapid industrialization. Considering the processes of CNF formation, most of the research was focused on chemical pretreatments, for example, TEMPO-oxidized nanocelluloses, to receive CNF, generating high costs of energy and chemicals, and also it is unfavorable from environmental aspects. In this context, a new type of nanocellulose materials—lignin-containing cellulose nanofibers (LCNFs)—are increasingly developed as a substituent of CNF materials (Ahola et al., 2008; Alcalá et al., 2013; Delgado-Aguilar, González, et al., 2015; Merayo et al., 2017; Tarrés et al., 2016). The row material for CNF production contains hemicellulose and lignin, which are removed from the substrate by the delignification processes. Skipping this technological process, thus significantly removing the costs of the procedures, the LNCF is received. The lignin content in the papermaking material gives a characteristic appearance, namely, brown color. It is a limitation to use LCNF as a reinforcing agent in white papers because the whiteness and brightness of the paper product are significantly reduced. Nevertheless, these nanomaterials found an application in brown packaging papers, which is a growing sector. The recent studies of LCNF are focused on the application of nanomaterials as brown paper strength enhancement.

Tarrés, Pellicer, et al. (2017) proposed the production of LCNF from pine sawdust and also nonwood source—triticale (Tarrés et al., 2020; Tarrés, Ehman, et al., 2017). Pine sawdust from sawmills constitutes an enormous byproduct that is generated during the wood transformation. The LCNF was produced from pine sawdust applying only mechanical treatments (Tarrés, Pellicer, et al., 2017). The received LCNF material was mixed with recycled cardboard furnish, and the paper sheets were prepared and examined in the context of mechanical properties, due to a potential application for the production of paper bags. The proposed approaches make properly treated and bleached pine sawdust promising and cost-effective materials with comparable reinforcing properties as other nanomaterials; moreover, an addition of LCNF to paper increased the mechanical properties of products while insignificantly affected pulp drainability. The presented results make an opportunity to use LCNF in the production of brown-line papers, using an economic and environmentally friendly solution due to the use of natural resources, saving production and transport costs. Triticale straws were also applied as a substrate to produce LCNF. The substrate was digested and gradually delignified by treatment with sodium hydroxide followed by bleaching in presence of sodium hypochlorite, preserving as many hemicelluloses as possible (Tarrés, Ehman, et al., 2017). The received LCNF was utilized as a paper strength additive and the results showed that LCNF is characterized by similar paper reinforcing potential as CNF received by costly chemical pretreatment (TEMPO-mediated oxidation). Tarrés et al. (2020) recently also proposed an efficient system for the retention of LCNF produced from triticale in the production of paper with industrial water. The paper presented a significant influence of the retention capacity on received paper properties, namely a greater increase in density and thus decrease in porosity are caused by the water with a high density of charge. The drainage capacity was reduced while the water conductivity and charge density were increased. When the cation starch was added, the retention of nanomaterials was decreased as a result of the anionic character of the trash. The application of the LCNF using a retention system with anionic trash proves a great potential for use on an industrial scale.

The presented results indicate the LCNF as a cost-effective, eco-friendly substituent of the CNF for the application as an additive in the paper industry, however considering the color aspect, it is limited for brown paper.

2.5 Conclusions and future prospective

The implementation of nanofibers in paper manufacturing is increasingly studied in recent years. This is related to the growing need to develop novel products or modify existing products to meet the expectations of various rapidly evolving issues, such as the digitization of information, requirements for sustainable development in this sector and the growing demand for packaging for different industries. Therefore, it is of paramount importance to enhance the strength of the paper-based product and increase its barrier properties. The use of nanofibers, especially CNFs in papermaking, helps to cope with these requirements. The unique properties of CNF, such as high strength and stiffness, give it an important role in paper manufacturing. Undoubtedly an advantageous feature is also the high specific surface area that is crucial in improving the bond between the fibers and nanofibrils. Therefore the CNF was repeatedly studied as a strength additive in paper sheets and paperboard and for barrier coating applications. Despite all the beneficial characteristics of CNF in papermaking processes, a few issues are continuously considered. The major point in this context is focused on the costs of CNF manufacturing. Over many years pretreatment methods have been drastically developed, going from mechanical and chemical pretreatments to cost-effective and more eco-friendly enzymatic approaches. The recent development of nanofibers for the paper production is also concentrated in LCNFs, which remove the necessity of the delignification processes stages, thus the technology is cost-effective and energy saving. As the main disadvantage of this approach is the brown color of products, it is mainly used in the production of brown and recycled paper. To summarizing, the paper industry is constantly developing to receive high-quality materials with improved mechanical, physical, and printing characteristics. It also strives to maintain the high quality of recycled fibres despite the increasing recycling rate and to elaborate cost-effective production processes.

References

Afra, E., Yousefi, H., Hadilam, M. M., & Nishino, T. (2013). Comparative effect of mechanical beating and nanofibrillation of cellulose on paper properties made from bagasse and softwood pulps. *Carbohydrate Polymers*, *97*(2), 725–730. Available from https://doi.org/10.1016/j.carbpol.2013.05.032.

Ahola, S., Österberg, M., & Laine, J. (2008). Cellulose nanofibrils—Adsorption with poly(amideamine) epichlorohydrin studied by QCM-D and application as a paper strength additive. *Cellulose*, *15*(2), 303–314. Available from https://doi.org/10.1007/s10570-007-9167-3.

Alcalá, M., González, I., Boufi, S., Vilaseca, F., & Mutjé, P. (2013). All-cellulose composites from unbleached hardwood kraft pulp reinforced with nanofibrillated cellulose. *Cellulose*, *20*(6), 2909–2921. Available from https://doi.org/10.1007/s10570-013-0085-2.

Azeredo, H. M. C., Rosa, M. F., & Mattoso, L. H. C. (2017). Nanocellulose in bio-based food packaging applications. *Industrial Crops and Products*, *97*, 664–671. Available from https://doi.org/10.1016/j.indcrop.2016.03.013.

Boufi, S., González, I., Delgado-Aguilar, M., Tarrès, Q., Pèlach, M. À., & Mutjé, P. (2016). Nanofibrillated cellulose as an additive in papermaking process: A review. *Carbohydrate Polymers*, *154*, 151–166. Available from https://doi.org/10.1016/j.carbpol.2016.07.117.

Brodin, F. W., Gregersen, Ø. W., & Syverud, K. (2014). Cellulose nanofibrils: Challenges and possibilities as a paper additive or coating material – A review. *Nordic Pulp & Paper Research Journal*, *29*(1), 156–166. Available from https://doi.org/10.3183/npprj-2014-29-01-p156-166.

Buzała, K., Przybysz, P., Rosicka-Kaczmarek, J., & Kalinowska, H. (2015). Comparison of digestibility of wood pulps produced by the sulfate and TMP methods and woodchips of various botanical origins and sizes. *Cellulose*, *22*(4), 2737–2747. Available from https://doi.org/10.1007/s10570-015-0644-9.

Cadena, E. M., Chriac, A. I., Pastor, F. I. J., Diaz, P., Vidal, T., & Torres, A. L. (2010). Use of cellulases and recombinant cellulose binding domains for refining TCF kraft pulp. *Biotechnology Progress*, *26*(4), 960–967. Available from https://doi.org/10.1002/btpr.411.

Delgado-Aguilar, M., González, I., Pèlach, M. A., De La Fuente, E., Negro, C., & Mutjé, P. (2015). Improvement of deinked old newspaper/old magazine pulp suspensions by means of nanofibrillated cellulose addition. *Cellulose*, *22*(1), 789–802. Available from https://doi.org/10.1007/s10570-014-0473-2.

Delgado-Aguilar, M., Tarrés, Q., Pèlach, M. À., Mutjé, P., & Fullana-i-Palmer, P. (2015). Are cellulose nanofibers a solution for a more circular economy of paper products? *Environmental Science & Technology*, *49*(20), 12206–12213. Available from https://doi.org/10.1021/acs.est.5b02676.

Ehman, N. V., Felissia, F. E., Tarrés, Q., Vallejos, M. E., Delgado-Aguilar, M., Mutjé, P., & Area, M. C. (2020). Effect of nanofiber addition on the physical–mechanical properties of chemimechanical pulp handsheets for packaging. *Cellulose*, *27*(18), 10811–10823. Available from https://doi.org/10.1007/s10570-020-03207-5.

Ehman, N. V., Lourenço, A. F., McDonagh, B. H., Vallejos, M. E., Felissia, F. E., Ferreira, P. J. T., ... Area, M. C. (2020). Influence of initial chemical composition and characteristics of pulps on the production and properties of lignocellulosic nanofibers. *International Journal of Biological Macromolecules*, *143*, 453–461. Available from https://doi.org/10.1016/j.ijbiomac.2019.10.165.

Errokh, A., Magnin, A., Putaux, J.-L., & Boufi, S. (2018). Morphology of the nanocellulose produced by periodate oxidation and reductive treatment of cellulose fibers. *Cellulose*, *25*(7), 3899–3911. Available from https://doi.org/10.1007/s10570-018-1871-7.

Fernandes Diniz, J. M. B., Gil, M. H., & Castro, J. A. A. M. (2004). Hornification—Its origin and interpretation in wood pulps. *Wood Science and Technology*, *37*(6), 489–494. Available from https://doi.org/10.1007/s00226-003-0216-2.

Gharehkhani, S., Sadeghinezhad, E., Kazi, S. N., Yarmand, H., Badarudin, A., Safaei, M. R., & Zubir, M. N. M. (2015). Basic effects of pulp refining on fiber properties—A review. *Carbohydrate Polymers*, *115*, 785–803. Available from https://doi.org/10.1016/j.carbpol.2014.08.047.

González, I., Vilaseca, F., Alcalá, M., Pèlach, M. A., Boufi, S., & Mutjé, P. (2013). Effect of the combination of biobeating and NFC on the physico-mechanical properties of paper. *Cellulose*, *20*(3), 1425–1435. Available from https://doi.org/10.1007/s10570-013-9927-1.

Henriksson, M., Henriksson, G., Berglund, L. A., & Lindström, T. (2007). An environmentally friendly method for enzyme-assisted preparation of microfibrillated cellulose (MFC) nanofibers. *European Polymer Journal*, *43*(8), 3434–3441. Available from https://doi.org/10.1016/j.eurpolymj.2007.05.038.

Herrick, F. W., Casebier, R. L., Hamilton, J. K., & Sandberg, K. R. (1983). Microfibrillated cellulose: Morphology and accessibility. *Journal of Applied Polymer Science: Applied Polymer Symposium*, *37*(9), 797–813.

Hollertz, R., Durán, V. L., Larsson, P. A., & Wågberg, L. (2017). Chemically modified cellulose micro- and nanofibrils as paper-strength additives. *Cellulose*, *24*(9), 3883–3899. Available from https://doi.org/10.1007/s10570-017-1387-6.

Hult, E.-L., Iotti, M., & Lenes, M. (2010). Efficient approach to high barrier packaging using microfibrillar cellulose and shellac. *Cellulose*, *17*(3), 575–586. Available from https://doi.org/10.1007/s10570-010-9408-8.

Isogai, A., & Bergström, L. (2018). Preparation of cellulose nanofibers using green and sustainable chemistry. *Current Opinion in Green and Sustainable Chemistry*, *12*, 15–21. Available from https://doi.org/10.1016/j.cogsc.2018.04.008.

Isogai, A., Saito, T., & Fukuzumi, H. (2011). TEMPO-oxidized cellulose nanofibers. *Nanoscale*, *3*(1), 71–85. Available from https://doi.org/10.1039/C0NR00583E.

Iwamoto, S., Nakagaito, A. N., & Yano, H. (2007). Nano-fibrillation of pulp fibers for the processing of transparent nanocomposites. *Applied Physics A*, *89*(2), 461–466. Available from https://doi.org/10.1007/s00339-007-4175-6.

John, M. J., & Anandjiwala, R. D. (2008). Recent developments in chemical modification and characterization of natural fiber-reinforced composites. *Polymer Composites*, *29*(2), 187–207. Available from https://doi.org/10.1002/pc.20461.

Kenry., & Lim, C. T. (2017). Nanofiber technology: Current status and emerging developments. *Topical Volume on Advanced Polymeric Materials*, *70*, 1–17. Available from https://doi.org/10.1016/j.progpolymsci.2017.03.002.

Kinnunen-Raudaskoski, K., Hjelt, T., Kenttä, E., & Forsström, U. (2014). Thin coatings for paper by foam coating. *Tappi Journal*, *13*(7), 9–19. Available from https://imisrise.tappi.org/TAPPI/Products/14/JUL/14JUL09.aspx.

Klemm, D., Kramer, F., Moritz, S., Lindström, T., Ankerfors, M., Gray, D., & Dorris, A. (2011). Nanocelluloses: A new family of nature-based materials. *Angewandte Chemie International Edition*, *50*(24), 5438–5466. Available from https://doi.org/10.1002/anie.201001273.

Kumar, V., Pathak, P., & Bhardwaj, N. K. (2020). Facile chemo-refining approach for production of micro-nanofibrillated cellulose from bleached mixed hardwood pulp to improve paper quality. *Carbohydrate Polymers*, *238*, 116186. Available from https://doi.org/10.1016/j.carbpol.2020.116186.

Lavoine, N., Desloges, I., & Bras, J. (2014). Microfibrillated cellulose coatings as new release systems for active packaging. *Carbohydrate Polymers*, *103*, 528–537. Available from https://doi.org/10.1016/j.carbpol.2013.12.035.

Lavoine, N., Desloges, I., Khelifi, B., & Bras, J. (2014). Impact of different coating processes of microfibrillated cellulose on the mechanical and barrier properties of paper. *Journal of Materials Science*, *49*(7), 2879–2893. Available from https://doi.org/10.1007/s10853-013-7995-0.

Lecourt, M., Meyer, V., Sigoillot, J.-C., & Petit-Conil, M. (2010). Energy reduction of refining by cellulases. *Holzforschung*, *64*(4), 441–446. Available from https://doi.org/10.1515/hf.2010.066.

Liimatainen, H., Visanko, M., Sirviö, J. A., Hormi, O. E. O., & Niinimaki, J. (2012). Enhancement of the nanofibrillation of wood cellulose through sequential periodate–chlorite oxidation. *Biomacromolecules*, *13*(5), 1592–1597. Available from https://doi.org/10.1021/bm300319m.

Liimatainen, H., Visanko, M., Sirviö, J., Hormi, O., & Niinimäki, J. (2013). Sulfonated cellulose nanofibrils obtained from wood pulp through regioselective oxidative bisulfite pre-treatment. *Cellulose*, *20*(2), 741–749. Available from https://doi.org/10.1007/s10570-013-9865-y.

Martin-Sampedro, R., Filpponen, I., Hoeger, I. C., Zhu, J. Y., Laine, J., & Rojas, O. J. (2012). Rapid and complete enzyme hydrolysis of lignocellulosic nanofibrils. *ACS Macro Letters*, *1*(11), 1321–1325. Available from https://doi.org/10.1021/mz300484b.

Mashkour, M., Afra, E., Resalati, H., & Mashkour, M. (2015). Moderate surface acetylation of nanofibrillated cellulose for the improvement of paper strength and barrier properties. *RSC Advances*, *5*(74), 60179–60187. Available from https://doi.org/10.1039/C5RA08161K.

Mautner, A., Lee, K.-Y., Lahtinen, P., Hakalahti, M., Tammelin, T., Li, K., & Bismarck, A. (2014). Nanopapers for organic solvent nanofiltration. *Chemical Communications*, *50*(43), 5778–5781. Available from https://doi.org/10.1039/C4CC00467A.

Merayo, N., Balea, A., de la Fuente, E., Blanco, Á., & Negro, C. (2017). Synergies between cellulose nanofibers and retention additives to improve recycled paper properties and the drainage process. *Cellulose*, *24*(7), 2987–3000. Available from https://doi.org/10.1007/s10570-017-1302-1.

Missoum, K., Martoïa, F., Belgacem, M. N., & Bras, J. (2013). Effect of chemically modified nanofibrillated cellulose addition on the properties of fiber-based materials. *Industrial Crops and Products*, *48*, 98–105. Available from https://doi.org/10.1016/j.indcrop.2013.04.013.

Mörseburg, K., & Chinga-Carrasco, G. (2009). Assessing the combined benefits of clay and nanofibrillated cellulose in layered TMP-based sheets. *Cellulose*, *16*(5), 795–806. Available from https://doi.org/10.1007/s10570-009-9290-4.

Motamedian, H. R., Halilovic, A. E., & Kulachenko, A. (2019). Mechanisms of strength and stiffness improvement of paper after PFI refining with a focus on the effect of fines. *Cellulose*, *26*(6), 4099–4124. Available from https://doi.org/10.1007/s10570-019-02349-5.

Naderi, A., Lindström, T., & Sundström, J. (2014). Carboxymethylated nanofibrillated cellulose: Rheological studies. *Cellulose*, *21*(3), 1561–1571. Available from https://doi.org/10.1007/s10570-014-0192-8.

Nakagaito, A. N., & Yano, H. (2004). The effect of morphological changes from pulp fiber towards nano-scale fibrillated cellulose on the mechanical properties of high-strength plant fiber based composites. *Applied Physics A*, *78*(4), 547–552. Available from https://doi.org/10.1007/s00339-003-2453-5.

Osong, S. H., Norgren, S., & Engstrand, P. (2016). Processing of wood-based microfibrillated cellulose and nanofibrillated cellulose, and applications relating to papermaking: A review. *Cellulose*, *23*(1), 93–123. Available from https://doi.org/10.1007/s10570-015-0798-5.

Pääkkö, M., Ankerfors, M., Kosonen, H., Nykänen, A., Ahola, S., Österberg, M., ... Lindström, T. (2007). Enzymatic hydrolysis combined with mechanical shearing and high-pressure homogenization for nanoscale cellulose fibrils and strong gels. *Biomacromolecules*, *8*(6), 1934–1941. Available from https://doi.org/10.1021/bm061215p.

Pajari, H., Rautkoski, H., & Moilanen, P. (2012). Replacement of synthetic binders with nanofibrillated cellulose in board coating: Pilot scale studies. *TAPPI International conference on nanotechnology for renewable materials 2012* (pp. 409–425). TAPPI Press. Available from https://cris.vtt.fi/en/publications/replacement-of-synthetic-binders-with-nanofibrillated-cellulose-i.

Qing, Y., Sabo, R., Zhu, J. Y., Agarwal, U., Cai, Z., & Wu, Y. (2013). A comparative study of cellulose nanofibrils disintegrated via multiple processing approaches. *Carbohydrate Polymers*, *97*(1), 226–234. Available from https://doi.org/10.1016/j.carbpol.2013.04.086.

Saito, T., Kimura, S., Nishiyama, Y., & Isogai, A. (2007). Cellulose nanofibers prepared by TEMPO-mediated oxidation of native cellulose. *Biomacromolecules*, *8*(8), 2485–2491. Available from https://doi.org/10.1021/bm0703970.

Shaghaleh, H., Xu, X., & Wang, S. (2018). Current progress in production of biopolymeric materials based on cellulose, cellulose nanofibers, and cellulose derivatives. *RSC Advances*, *8*(2), 825–842. Available from https://doi.org/10.1039/C7RA11157F.

Sirviö, J. A., Ukkola, J., & Liimatainen, H. (2019). Direct sulfation of cellulose fibers using a reactive deep eutectic solvent to produce highly charged cellulose nanofibers. *Cellulose*, *26*(4), 2303–2316. Available from https://doi.org/10.1007/s10570-019-02257-8.

Sirviö, J. A., Visanko, M., Laitinen, O., Ämmälä, A., & Liimatainen, H. (2016). Amino-modified cellulose nanocrystals with adjustable hydrophobicity from combined regioselective oxidation and reductive amination. *Carbohydrate Polymers*, *136*, 581–587. Available from https://doi.org/10.1016/j.carbpol.2015.09.089.

Spence, K. L., Venditti, R. A., Rojas, O. J., Habibi, Y., & Pawlak, J. J. (2011). A comparative study of energy consumption and physical properties of microfibrillated cellulose produced by different processing methods. *Cellulose*, *18*(4), 1097–1111. Available from https://doi.org/10.1007/s10570-011-9533-z.

Su, L., Ou, Y., Feng, X., Lin, M., Li, J., Liu, D., & Qi, H. (2019). Integrated production of cellulose nanofibers and sodium carboxymethylcellulose through controllable eco-carboxymethylation under mild conditions. *ACS Sustainable Chemistry & Engineering*, 7(4), 3792–3800. Available from https://doi.org/10.1021/acssuschemeng.8b04492.

Subbiah, T., Bhat, G. S., Tock, R. W., Parameswaran, S., & Ramkumar, S. S. (2005). Electrospinning of nanofibers. *Journal of Applied Polymer Science*, *96*(2), 557–569. Available from https://doi.org/10.1002/app.21481.

Syverud, K., Chinga-Carrasco, G., Toledo, J., & Toledo, P. G. (2011). A comparative study of Eucalyptus and *Pinus radiata* pulp fibres as raw materials for production of cellulose nanofibrils. *Carbohydrate Polymers*, *84*(3), 1033–1038. Available from https://doi.org/10.1016/j.carbpol.2010.12.066.

Syverud, K., & Stenius, P. (2008). Strength and barrier properties of MFC films. *Cellulose*, *16*(1), 75. Available from https://doi.org/10.1007/s10570-008-9244-2.

Taipale, T., Österberg, M., Nykänen, A., Ruokolainen, J., & Laine, J. (2010). Effect of microfibrillated cellulose and fines on the drainage of kraft pulp suspension and paper strength. *Cellulose*, *17*(5), 1005–1020. Available from https://doi.org/10.1007/s10570-010-9431-9.

Taniguchi, T., & Okamura, K. (1998). New films produced from microfibrillated natural fibres. *Polymer International*, *47*(3), 291–294, https://doi.org/10.1002/(SICI)1097-0126(199811)47:3<291::AID-PI11>3.0.CO;2-1.

Tarrés, Q., Area, M. C., Vallejos, M. E., Ehman, N. V., Delgado-Aguilar, M., & Mutjé, P. (2020). Lignocellulosic nanofibers for the reinforcement of brown line paper in industrial water systems. *Cellulose*, *27*(18), 10799–10809. Available from https://doi.org/10.1007/s10570-020-03133-6.

Tarrés, Q., Delgado-Aguilar, M., Pèlach, M. A., González, I., Boufi, S., & Mutjé, P. (2016). Remarkable increase of paper strength by combining enzymatic cellulose nanofibers in bulk and TEMPO-oxidized nanofibers as coating. *Cellulose*, *23*(6), 3939–3950. Available from https://doi.org/10.1007/s10570-016-1073-0.

Tarrés, Q., Ehman, N. V., Vallejos, M. E., Area, M. C., Delgado-Aguilar, M., & Mutjé, P. (2017). Lignocellulosic nanofibers from triticale straw: The influence of hemicelluloses and lignin in their production and properties. *Carbohydrate Polymers*, *163*, 20–27. Available from https://doi.org/10.1016/j.carbpol.2017.01.017.

Tarrés, Q., Pellicer, N., Balea, A., Merayo, N., Negro, C., Blanco, A., ... Mutjé, P. (2017). Lignocellulosic micro/nanofibers from wood sawdust applied to recycled fibers for the production of paper bags. *International Journal of Biological Macromolecules*, *105*, 664–670. Available from https://doi.org/10.1016/j.ijbiomac.2017.07.092.

Tenhunen, T.-M., Pöhler, T., Kokko, A., Orelma, H., Gane, P., Schenker, M., & Tammelin, T. (2018). Enhancing the stability of aqueous dispersions and foams comprising cellulose nanofibrils (CNF) with $CaCO_3$ particles. *Nanomaterials (Basel, Switzerland)*, *8*(9), 651. Available from https://doi.org/10.3390/nano8090651.

Torvinen, K., Kouko, J., Passoja, S., Keränen, J. T., & Hellen, E. (2014). Cellulose micro and nanofibrils as a binding material for high filler content papers. *Proceedings of the Tappi Papercon 2014 conference* (Vol. 1, pp. 348–361). TAPPI Press. Available from https://cris.vtt.fi/en/publications/cellulose-micro-and-nanofibrils-as-a-binding-material-for-high-fi.

Turbak, A. F., Snyder, F. W., & Sandberg, K. R. (1983). Microfibrillated cellulose, a new cellulose product: Properties, uses, and commercial potential. *Journal of Applied Polymer Science: Applied Polymer Symposium; (United States)*, *37*. Available from https://www.osti.gov/biblio/5062478-microfibrillated-cellulose-new-cellulose-product-properties-uses-commercial-potential.

Vallejos, M. E., Felissia, F. E., Area, M. C., Ehman, N. V., Tarrés, Q., & Mutjé, P. (2016). Nanofibrillated cellulose (CNF) from eucalyptus sawdust as a dry strength agent of unrefined eucalyptus handsheets. *Carbohydrate Polymers*, *139*, 99–105. Available from https://doi.org/10.1016/j.carbpol.2015.12.004.

Vega, B., Wondraczek, H., Zarth, C. S. P., Heikkilä, E., Fardim, P., & Heinze, T. (2013). Charge-directed fiber surface modification by molecular assemblies of functional polysaccharides. *Langmuir: The ACS Journal of Surfaces and Colloids*, *29*(44), 13388–13395. Available from https://doi.org/10.1021/la402116j.

Visanko, M., Liimatainen, H., Sirviö, J. A., Mikkonen, K. S., Tenkanen, M., Sliz, R., ... Niinimäki, J. (2015). Butylamino-functionalized cellulose nanocrystal films: Barrier properties and mechanical strength. *RSC Advances*, *5*(20), 15140–15146. Available from https://doi.org/10.1039/C4RA15445B.

Wang, M., He, W., Wang, S., & Song, X. (2015). Carboxymethylated glucomannan as paper strengthening agent. *Carbohydrate Polymers*, *125*, 334–339. Available from https://doi.org/10.1016/j.carbpol.2015.02.060.

Wang, Z., Ma, H., Chu, B., & Hsiao, B. S. (2017). Fabrication of cellulose nanofiber-based ultrafiltration membranes by spray coating approach. *Journal of Applied Polymer Science*, *134*(11). Available from https://doi.org/10.1002/app.44583.

Zimmermann, T., Pöhler, E., & Geiger, T. (2004). Cellulose fibrils for polymer reinforcement. *Advanced Engineering Materials*, *6*(9), 754–761. Available from https://doi.org/10.1002/adem.200400097.

CHAPTER 3

Role of laccase in the pulp and paper industry

Asim Hussain[1], Muhammad Bilal[2], Hamza Rafeeq[1], Zara Jabeen[1], Nadia Afsheen[1], Farooq Sher[3], Vineet Kumar[4], Ram Naresh Bharagava[5], Luiz Fernando Romanholo Ferreira[6,7] and Hafiz M.N. Iqbal[8]

[1]Department of Biochemistry, Riphah International University, Faisalabad, Pakistan
[2]School of Life Science and Food Engineering, Huaiyin Institute of Technology, Huai'an, P.R. China
[3]Department of Engineering, School of Science and Technology, Nottingham Trent University, Nottingham, United Kingdom
[4]Department of Botany, School of Life Science, Guru Ghasidas Vishwavidyalaya (A Central University), Bilaspur, India
[5]Laboratory for Bioremediation and Metagenomics Research (LBMR), Department of Environmental Microbiology (DEM), Babasaheb Bhimrao Ambedkar University (A Central University), Lucknow, India
[6]Graduate Program in Process Engineering, Tiradentes University (UNIT), Aracaju, Brazil
[7]Institute of Technology and Research (ITP), Tiradentes University (UNIT), Aracaju, Brazil
[8]Tecnologico de Monterrey, School of Engineering and Sciences, Monterrey, Mexico

3.1 Introduction

The paper and pulp processing involves the degradation of cellulose fibers and the elimination of lignin. Lignin covalent attached to carbohydrate moieties is one of the leading components that give the pulp a dark brown color, and during this process, a large amount of lignin is extracted. This yellow/brown kraft pulp needs to be bleached before processing the paper (Sailwal, Banerjee, Bhaskar, & Ghosh, 2020). A multistage bleaching method is used to remove lignin residues from Cl, ClO_2, and NaOH. The extent of raw materials processed, that is wood resources, is much high and therefore the use of Cl-dependent compounds is also extensive. This leads to an extremely persistent and bioaccumulative development of organochlorine, causing hormonal disturbances and reproductive defects, and thus raising concerns about the unacceptable levels of highly toxic compounds found in the atmosphere. As a result, totally chlorine free (TCF) pulp processing implicates non-chlorinated chemical reagents, such as O_2, H_2O_2, or O_3. Though these compounds are not as successful as Cl reagents in bleaching pulps, they often appear to aggravate problems with pitch originating from the lipophilic part of the raw material that endures pulping and bleaching (Elakkiya & Niju, 2020). A step-by-step overview of the pulping procedure is illustrated in Fig. 3.1.

Microbial enzymes can be used to improve the characteristics of pulp and waste disposal. The use of traditional methods for bleaching and pitch removal is slow, expensive, and environmentally unsustainable. This requires the use of hemicellular

Nanotechnology in Paper and Wood Engineering
DOI: https://doi.org/10.1016/B978-0-323-85835-9.00006-4

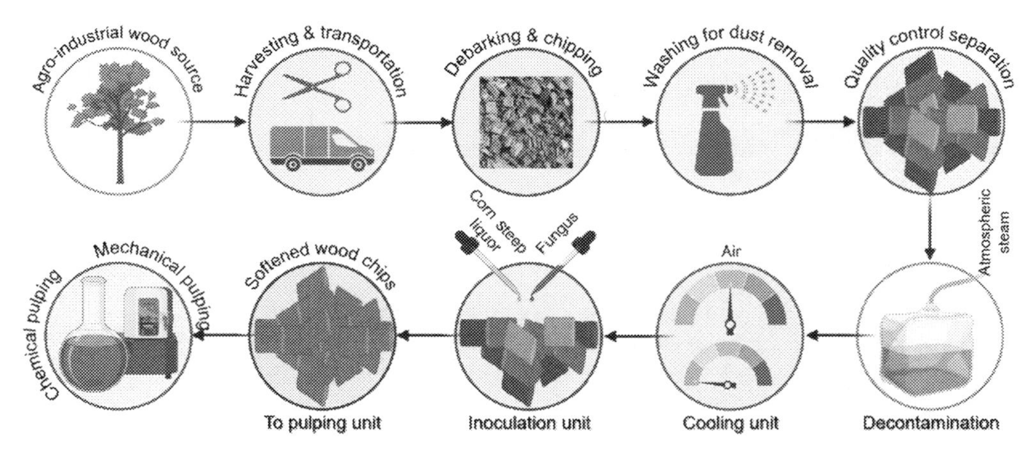

Figure 3.1 A step-by-step overview of the pulping procedure.

enzymes, such as xylanases and ligninolytic enzymes, in particular laccases. The efficacy of xylanase has been thoroughly revealed. Xylan is closely linked to lignin and cellulose, and its interruption facilitates the elimination of lignin during bleaching and reduces the use of decolorizing substances. Laccases are oxidative enzymes with a wide range of effects on the paper and pulp industries (Bajaj & Mahajan, 2019).

Papermaking requires a number of steps in the production of pulp, bleach, and paper. Woody as well as nonwoody fibers are used as pulp to produce paper. Wood pulp is made of spruce, pine, fir and hemlock wood, and hardwoods, including eucalyptus, aspen, and birch. The lower but significant amount of pulp consists of nonwood plant fibers, including agricultural waste (i.e., wheat straw and bagasse), and other plant fibers, such as bamboo, and fiber crops (such as flax). Overall, nonwood plant fibers are cheaper for collection and processing than wood fibers and account for only 10% of the total volume of paper fibers used. Cloth fiber, such as linen, cotton, abaca, kenaf, jute, or sisal, is used to produce high-quality pulp for speciality paper products such as tea bags, filters, tobacco, bibles, condensers, etc. (Fedyaeva & Vostrikov, 2019).

Laccase enzyme was primarily used in the biobleaching of eucalyptus kraft pulp. Eucalyptus is the world's largest solitary market source of pulp. Low production costs in specific areas, largely due to increased forestry and high pulp yields, are of great importance to eucalyptus timber. Kraft pulping, which comprises "cooking" sodium solution wood in a heated digestive vessel, is the extensively used process for pulping as it makes the most solid sheet of paper. The process of lignozym uses a mediator system, such as N-hydroxybenzotriazole or 2,2-azinobis(3-ethylbenzothiazoline-6-sulfonic acid) (ABTS), can delineate the pulp and therefore improve the luminosity of the pulp, making the paper "white." The lipophilic extractives deposited on the machine, thus reducing the productivity of the mill operations, can be removed from the pulp (Saleem, Khurshid, & Ahmed, 2018). Various possible roles of laccase-mediator system (LMS) are shown in Fig. 3.2.

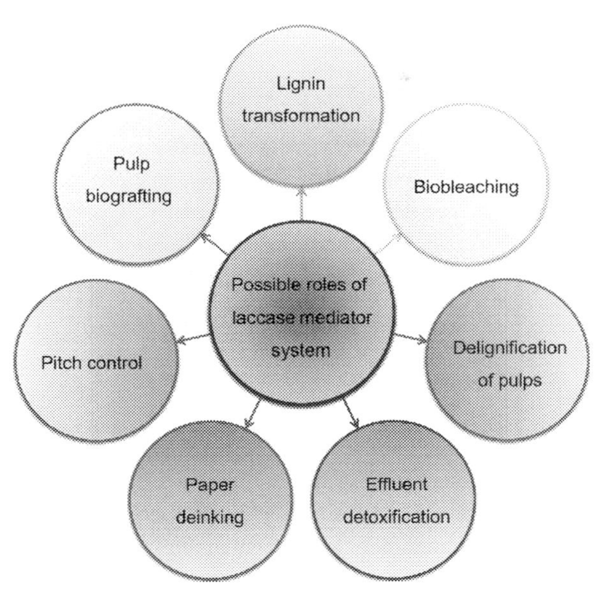

Figure 3.2 Various possible roles of laccase-mediator system.

Laccases may activate fiber-bound lignin during composite production resulting in the creation, without the use of toxic synthetic adhesives, of fiber boards with some strong mechanical properties. A variety of phenolic acid derivatives are generated during laccase-assisted grafting on kraft pulp fibers. This capacity could be used for fiber surfaces to fasten chemical compounds that could lead to new fiber materials. Laccases have also been tested for the deinking of old paper along with cellulases and hemicellulases and play a key role in the decontamination of waste pulp and paper mill effluent. Laccase is one of the most promising enzymes for potential applications in the pulp and paper industry (Sharma, Chaudhary, Kaur, & Arya, 2020).

3.2 Laccases, redox potential, and delignification

Laccases are multicopper oxides found in fungi, bacteria, plants, and insects. A wide variety of organic and inorganic compounds can be oxidized under moderate and severe environmental conditions. As industrial biocatalysts, they are desirable due to their safe reaction mechanism with H_2O as a byproduct (Guimarães et al., 2017). Laccases consist of three different forms of Cu groups separated by ultraviolet/visible and electron paramagnetic resonance spectroscopy. In the active sites, Cu type 1 catalyzes and transfers electrons, molecular oxygen can be activated by type 2, and oxygen can be consumed by type 3. Enzymes without the Cu atom liable for the blue colors are referred to as "yellow" or "white" laccases (Morsi, Bilal, Iqbal, & Ashraf, 2020).

The enzymes containing Cu as a cofactor are divided into three groups (low, medium, and high) based on their redox potential (E0). E0 below 430 mV indicates low potential laccases, while E0 above 710 mV indicates high potential laccases. The medium potential laccases lie in between them. Laccases can only oxidize those compounds whose ionization potential lies below the E0 of type 1's catalytic site (Mateljak et al., 2019). Aromatic compounds are part of the substrates of laccase. A lot of these are phenolic lignin particles. Lignin polymerization and depolymerization are the primary functions of natural laccases. The cleaving of nonphenolic lignin moieties based on the E0 of the type 1 core is known to be a lacquer for successful delignification. Nonphenolic lignin residues, such as veratryl alcohol, 1,2-dimethoxybenzene, etc., have a high E0 ($>$ 1400 mV), and oxidation of these compounds by laccase has been reported, but the mechanism is not clear yet. With these redox mediators, laccase can oxidize nonphenol lignin compounds to a large degree and reduces the amount of pulp kappa (Angural et al., 2020).

The typical E0 range is normally between 500 and 800 mV for laccase activity against common hydrogen electrodes, which can only attack phenolic moieties containing less than 10% of the total natural lignin polymer content. A large number of nonphenolic lignin units possess E0 above 1300 mV. The redox ability of the laccases recorded was less than that of nonphenolic compounds so that these enzymes could oxidize these rigid substances. This limits the variety of substrates that laccases can oxidize and thus restricts pulp delignification (Bagewadi, Mulla, & Ninnekar, 2017). The efficacy of mediator E0 is based on the reactive nature of the radicals formed during the first stage of oxidation and the catalytic efficacy of lignin oxidation. Considerable biotechnological potential for biobleaching pulp is provided by the possibility of extending its oxidation spectrum through redox mediators. Use of redox mediators is particularly important as these radicals will continue to oxidize other molecules, comprising such molecules that are not directly used as substrates in the enzyme, until they have been oxidized to stable radicals by laccases (Hilgers, Kabel, & Vincken, 2020; Hilgers, Van Dam, Zuilhof, Vincken, & Kabel, 2020; Hilgers, van Erven, et al., 2020).

3.3 Laccases-assisted biobleaching/delignification of pulps

Pulp bleaching use huge quantities of Cl-containing chemicals until the end of the 20th century, but ClO_2 is now an elemental Cl-free decolorizing agent for high-quality white paper processing in most pulp and paper industries worldwide (Jahan, Uddin, & Kashem, 2017). The high organic content in combination with ClO_2, particularly in the wood-based pulp, results from the treatment of organochlorine compounds, which are then released into water systems as bleaching effluents. Genetic and reproductive losses of these organochlorine compounds (as measured as adsorptive organic halogens) in aquatic and terrestrial organisms, including humans, have been identified. While more

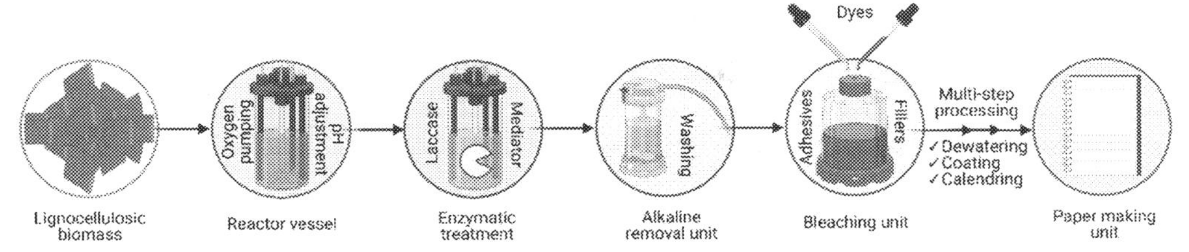

Figure 3.3 Laccase-mediated processing of biomass for paper making.

environmentally safe solutions are available for pulp bleaching in ClO_2 paper mills, such as prolonged-cooking, oxygen, H_2O_2, and O_3-based delignification processes, these approaches require process modifications (Yang, Fang, Huang, Zhao, & Liu, 2018).

Cl-containing reagents and other blood toxic compounds used in the paper industry have been reduced using a safer and more cost-effective biocatalyst approach. Enzymes have a simple solution to achieve a higher luminosity ceiling. In the last few years, the use of biocatalysts in pulp whitening has been greatly appreciated. Lignin phenol components may be specifically targeted using white−red laccases. These enzymes are unique to lignin; cellulose is not damaged or lost, and chemical savings on a wide scale can be achieved in contrast to xylanase (Boruah, Sarmah, Das, & Goswami, 2019). Various white−red fungi and bacteria have developed laccases that have demonstrated their potential to delignify kraft or pulp and lift their shine and hazardous chemicals. While different fungi produce laccases with different E0s between 0.43 and 0.78 V, a laccase that can directly oxidize nonphenolic lignin subunits is yet to be found. However, these phenolic nonunits, when used in conjunction with radical mediator species, including 1-hydroxybenzotriazole (HBT) and ABTS, will be indirectly oxidized by laccases (Vermani, Chauhan, Yadav, Roy, & Singh, 2020). Laccase-mediated processing of biomass for papermaking is shown in Fig. 3.3.

3.4 Laccase mediators

3.4.1 Natural mediators

These mediators are well known for their commercial accessibility, but natural phenols are derived from pulp liquors, effluent streams, and plant materials. Such compounds have been found to function in the same or better way as synthetic mediators. In general, it does not cause the enzyme to be inactivated and is a promising option for environmentally safe pulp delignification. Such natural phenolic compounds include acetovanillone, vanillin, acetosyringone (AS), p-coumaric acid (PC), coniferaldehyde (CLD), syringaldehyde (SA), sinapic acid, FA, and sinapyl aldehyde. Phenolic mediators in ortho-positioned methoxy groups have been found to be the least likely to experience grafting reactions and to experience the most likely polymerization.

Table 3.1 Use of natural mediators in laccase-mediator system.

Compounds	Laccase source	Function	References
Acetosyringone (1.5)	*Bacillus* sp. FNT	Pine kraft	Gupta et al. (2020)
Syringaldehyde	*Bacillus* sp. FNT	Soft and hardwood pulp	Gupta et al. (2020)
Acetosyringone	*Bacillus* sp. SS4	Decolorization of the indigo carmine	Singh, Kaur, et al. (2019), Singh, Singh, et al. (2019)
Vanillin	*Bacillus* sp. SS4	Decolorization of the indigo carmine	SinghKaur, et al. (2019), Singh, Singh, et al. (2019)
Orcinol	*Bacillus* sp. SS4	Decolorization of the indigo carmine	Singh, Kaur, et al. (2019), Singh, Singh, et al. (2019)
Veratraldehyde	*Bacillus* sp. SS4	Decolorization of the indigo carmine	Singh, Kaur, et al. (2019), Singh, Singh, et al. (2019)
Feruloyl esterase	*Bacillus megaterium*	Pulp bleaching	Ozer et al. (2020)

In particular, the power of the mediator's highest occupied molecular orbital (EHOMO) has been shown to be linearly dependent on these shifts. The lowest EHOMO (PC) phenolic mediator resulted in the highest increase in kappa numbers and the lowest decrease in shine. Instead, there is a slight increase in kappa and a major loss of shine due to syringyl derivatives (i.e., SA) with high EHOMO values (Gupta, Kapoor, & Shukla, 2020; Ozer, Sal, Belduz, Kirci, & Canakci, 2020; Singh, Kaur, Khatri, & Arya, 2019; Singh, Singh, Kaur, Arya, & Sharma, 2019). Some examples of natural mediators are summarized in Table 3.1.

3.4.2 Artificial mediators

Effective delignification of pulp is only promising in the presence of redox mediators and laccases. The optimal mediator for pulp bleaching should be a strong laccase substrate. A smaller mediator must be potent (in the oxidized form) enough to generate stable radicals, unlikely to degrade enzyme, and its subsequent recycling without its degeneration must be possible (Dillies et al., 2020). ABTS was the first active artificial or synthetic compound to act as a mediator for oxidizing nonphenolic lignin compounds (Table 3.2). As laccases begin to oxidize ABTS, cationic radicals ($ABTS^{\bullet+}$) are formed, which later become dicationic ($ABTS^{2+}$) after further oxidation. The laccase–ABTS mixture can be obtained by oxidizing the compounds and organic colors via the electron transfer pathway (Hilgers, Van Dam, et al., 2020).

NN-H mediators, such as violuric acid, N-hydroxyphthalimide (HPI), HBT, or N-hydroxyacetanilide are the most important laccase mediators for the oxidation of recalcitrant

Table 3.2 Use of ABTS in the laccase-mediator system.

Laccase source	Function	References
Trametes versicolor	Depolymerization	(Dillies et al., 2020)
T. versicolor	Depolymerization of the three major industrial lignins (organosolv lignin, kraft lignin, and sodium lignosulfonate)	Dillies et al. (2020)
T. versicolor	Cleavage of a β-O-4′ linkage	Hilgers, Kabel, et al. (2020), Hilgers, Van Dam, et al. (2020), Hilgers, van Erven, et al. (2020)
T. versicolor	Phenolic lignin dimer (GBG)	Hilgers, Vincken, Gruppen, and Kabel (2018)
Commercial Laccase	Dyes decolorization	Liu, Geng, Yan, and Huang (2017)
T. versicolor	Bisphenol A degradation	Zeng, Qin, and Xia (2017), Zeng, Zhao, and Xia (2017)
Bacillus sp. MSK-01	Decolorization	Sondhi, Kaur, Kaur, and Kaur (2018)
Kocuria sp. PBS-1	Pulp bleaching	Boruah et al. (2019)
NS51002 (L1) and NS51003 (L2)	Demethylation of kraft lignin	Wang, Zhao, and Li (2018)
T. versicolor	Lignin degradation	Pan et al. (2018)
T. versicolor	Degradation of Congo red dye	Saha and Mukhopadhyay (2020)
T. versicolor	Lignin conversion to aromatic monomers	Stevens et al. (2019)
Trametes hirsuta Bm-2	Indigo carmin decolorization	Ancona-Escalante et al. (2018)
Bacillus sp. SS4	Decolorization of indigo carmine	Singh, Kaur, et al. (2019), Singh, Singh, et al. (2019)

aromatic compounds, such as nonphenolic lignin materials (Table 3.3). The laccase oxidation of this kind of mediators produces an extremely reactive nitroxyl radical (NN-O•), which helps remove an electron followed by the release of a proton in an enzymatic reaction. The hydrogen atomic transfer mechanism of nitroxyl radicals oxidizes the target substrate (Kontro et al., 2020; Munk, Andersen, & Meyer, 2018; Wang et al., 2018).

3.5 Lignin degradation by laccase-mediator system

Biodegradation of lignin by LMS is an enzyme-catalyzed oxidative process which generates reactive radicals. These radicals degrade the phenolic and nonphenolic structure of lignin in the presence of mediators. The initial attack on phenolic lignin (<20% of total lignin) was indicated, and the nonphenolic benzyl structures were destroyed. On degradation, phenolic

Table 3.3 Utilization of HBT, HPI, and NHA mediators in laccase-mediator system.

Compounds	Laccase sources	Function	References
HBT	*Trametes versicolor*	Degradation of the herbicide isoproturon	Zeng, Qin, et al. (2017), Zeng, Zhao, et al. (2017)
HBT	NS51002 (L1) and NS51003 (L2)	Demethylation of kraft lignin	Wang et al. (2018)
HBT	*Bacillus* sp. SS4	Decolorization of the indigo carmine	Singh, Kaur, et al. (2019), Singh, Singh, et al. (2019)
HBT	*Aquifex aeolicus*	Eucalyptus kraft	Gupta et al. (2020)
HBT	*T. versicolor*	Cleavage of a β-O-4′ linkage	Hilgers, Kabel, et al. (2020), Hilgers, Van Dam, et al. (2020), Hilgers, van Erven, et al. (2020)
HPI	*Obba rivulosa*	Oxidation of biorefinery lignin	Kontro et al. (2020)
NHA	*O. rivulosa*	Oxidation of biorefinery lignin	Kontro et al. (2020)
HPI	*T. versicolor* and *Myceliophthora thermophila*	Radical formation in lignin	Munk et al. (2018)
HTB	*T. versicolor* and *M. thermophila*	Radical formation in lignin	Munk et al. (2018)

Notes: HBT, 1-hydroxybenzotriazole; *HPI*, N-hydroxyphthalimide; *NHA*, N-hydroxyacetanilide.

contents of lignin generated low molecular weight phenolic residues with side chains (aldehydes, ketones, or carboxylic acids). These residues penetrate to the bulk lignin polymer and oxidize the nonphenolic content of lignin (Longe et al., 2018). LMS oxidation can be performed in three distinct oxidation processes: (1) electron transfer, (2) hydrogen radical transfer, and (3) ionic process. Fig. 3.4 shows the first two oxidation pathways. These pathways are demonstrated by the patterns of nonphenolic lignin degradation and the effects of intramolecular kinetic isotopes. The use of LMS was found to be autonomous of the enzyme characteristics in the oxidation of nonphenolic LMS. Nonphenolic LMS oxidation by laccase—ABTS occurred via the electron transport route. Laccase—HBT and HPI reactions support the radical mechanism of action. The radical HBT has been proposed to form an intermediate coupling with lignin. The intermediate product then degrades the release of a reduced form of benzotriazole (BT) or creates a stable complex linking some HBTs with lignin through covalent bonding (Hilgers, Kabel, et al., 2020; Hilgers, Van Dam, et al., 2020; Hilgers, van Erven, et al., 2020).

Treatment of laccase—HBT with softwood lignin resulted in a higher p-hydroxyphenyl structure content and a lowering of G-units and aromatic lignin structures. Laccase—HBT oxidation interrupts LMS bonds, such as hydrobenzoin dimers, to form carboxylic acid groups in treated pulp lignin. In Fig. 3.5, nonphenol β-O-4-linked

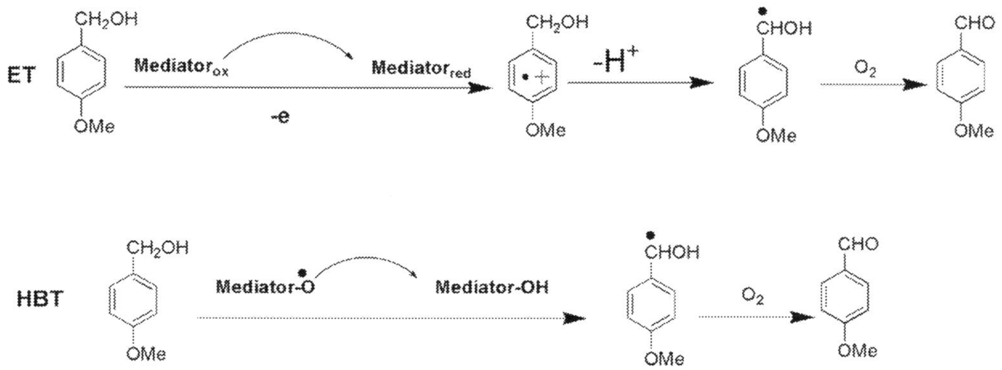

Figure 3.4 Mechanisms of oxidation of 4-methoxybenzile alcohol by laccase-mediator system. *ET*, electron transfer; *HBT*, 1-hydroxybenzotriazole (Christopher, Yao, and Ji, 2014). *Reprinted from Christopher, L.P., Yao, B., & Ji, Y. (2014). Lignin biodegradation with laccase-mediator systems. Frontiers in Energy Research, 2, 12 with permission under the terms of the Creative Commons Attribution License (CC BY).*

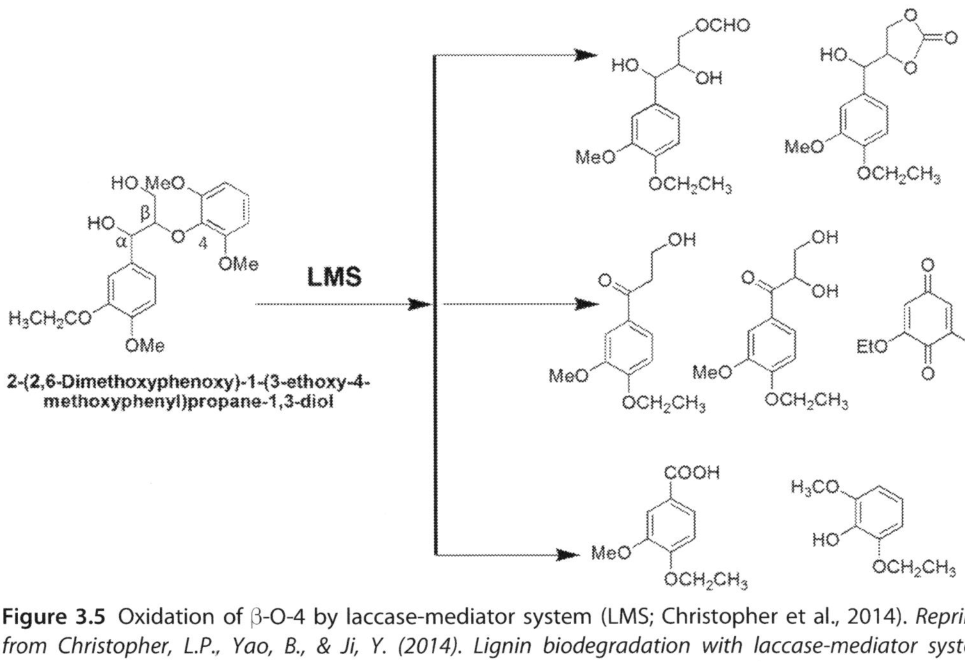

Figure 3.5 Oxidation of β-O-4 by laccase-mediator system (LMS; Christopher et al., 2014). *Reprinted from Christopher, L.P., Yao, B., & Ji, Y. (2014). Lignin biodegradation with laccase-mediator systems. Frontiers in Energy Research, 2, 12 with permission under the terms of the Creative Commons Attribution License (CC BY).*

laccase–HBT oxidation of LMSs is seen in aromatic ring openings (phenol electron oxidation), oxidation and β-ether oxidation, as well as in the development of aromatic carbonyl compounds and carboxylic acids of Cα–Cβ (Hilgers, Kabel, et al., 2020; Hilgers, Van Dam, et al., 2020; Hilgers, van Erven, et al., 2020).

The electron shifting process generates the β-aryl radical cation or benzylic-radical intermediate (Cα) radicals during oxidation reactions. The β-ether breakdown of the β-O-4 lignin residue results from a benzylic-radical reaction of the intermediate CIII-peroxy. The nominal molecular weight was reduced by 5 and 4 times, respectively, when *Trametes villosa* laccase—HBT was used to oxidate phenolic and nonphenolic contents. Cα—Cβ bond clipping occurred in both LMSs, resulting in a degradation of approximately 10% of its substructures (Hilgers, Kabel, et al., 2020; Hilgers, Van Dam, et al., 2020; Hilgers, van Erven, et al., 2020). Lignin degradation products usually include 2,6-dimethoxy-4-((E)-prop-1-enyl)benzaldehyde, 4-ethyl-2,6-dimethoxybenzaldehyde, and 2,6-dimethoxy-4-methylbenzaldehyde. The molecular weight and phenolic quality of laccases and LMS are believed to impact the reaction routes. Parallel polymerization and depolymerization reactions during LMS therapy were proposed, as phenolic groups (particularly lignin low molecular weight [LMW]) could be used as lignin polymerization sites that would prevent lignin lysis (Longe et al., 2018).

3.6 Biobleaching by laccase-mediator system

LMS was initially proposed for the safe and ecofriendly bleaching of pulp and paper. The mechanism oxidizes and breaks down phenolic and nonphenolic lignin units that are bleached during chemical bleaching, resulting in reduced kappa and increased pulp brightness. This method was first developed as biobleaching and xylanase-assisted kraft pulp bleaching (Kumar, 2020). Biobleaching is caused by partial xylan hydrolysis, which is repositioned to the surface of the fiber in the kraft pulp process and improves access to residual lignin from bleaching chemicals. Although xylanase indirectly helps to remove lignin from the pulp, LMS targets lignin that specifically contributes to the generation of LMW fragments of oxidized lignin that are likely to degrade further and become soluble in alkaline (E) during chemical bleaching (Colonia, Woiciechowski, Malanski, Letti, & Soccol, 2019).

Degradation and removal of LMS-catalyzed lignin from bleached pulp allow environmentally friendly chemicals that contain less hazardous chlorine, such as Cl, ClO_2, and NaOCl. For example, in a comparative analysis, a kraft pulp was identified with HBT as a mediator with several fungal laccases. *Pleurotus ostreatus* laccase with 13% reduction in lignin, *Trametes versicolor* laccase 22%, and *Ganoderma colossum* were the most productive laccase with 40%. A *Fusarium proliferatium* laccase—ABTS system with large G-units and methylated side-chain phenolic compounds of residual lignin may oxidize and extract 19% of industrial kraft pine lignin (induline AT) mainly in the form of ketones (10%) and acids (2%) (Chaurasia & Bhardwaj, 2019).

The unbleached eucalyptus kraft pulp was decomposed using two LMS, *Pycnoporus cinnabarinus* laccase—HBT and *Myceliophthora thermophila* laccase—methyl syringes. Pyrolysis—gas chromatography—mass spectrometry and nuclear magnetic resonance spectroscopy experiments have shown that both systems have oxidized Cα in lignin, while Cα−Cβ has been oxidized with an aromatic ring of xylan−lignin aggregation fractions consisting of uncondensed S and G units. The use of *Streptomyces cyanaus* and *Pseudomonas stutzeri* bacterial laccases with ABTS and HBT as mediators was also investigated for the biobleaching of kraft eucalyptus pulp (Loaiza, Alfaro, López, García, & García, 2019). High reactivity, alkaline nature, and thermo-stability make LMS an effective approach for lignin degradation. The degree to which pulp delignification with LMS is performed, based on the nature of the pulp and mediator, the pulp and laccase source, and the enzyme and mediator used. Pine kraft, for example, has been delignified between 19% and 40%, depending on the preparation of laccase (from several fungal strains) and the choice of mediators (HBT and ABTS) (Sharma et al., 2020). Laccase has been shown to be effective with HBT on fiber-derived pulp (hardwood, softwood, and bagasse) produced using various pulping techniques (sulfite pulping force), resulting in up to 60% delineation. The first test of the production plant for oxygen-delignification kraft pulp with laccase−HBT (Lignozym-process) attained 55% delignification. However, LMS is still used to a large extent. The mediator's costs and environmental concerns regarding the potential toxicity of mediators are the main obstacles (Wang et al., 2018).

3.7 Effect of laccase and xylanase on biobleaching

The effects of xylanase and laccase are different; formerly, the bleaching influence is increased by forming the enzyme pulp more sensitive to bleaching substances. The lignin was removed directly during laccases acting specifically on the lignin polymer. In concurrent therapy, the combination of xylanase and LMS increases the pulp shine and the ductile potential of the fibers. Xylanase and laccase were obtained from *Bacillus pumilus* and *Ganoderma* sp. rckk-02, respectively. Eucalyptus kraft pulp on treatment with xylanase resulted in 1.8 units increase in brightness. Unit luminosity has been increased by 1.8 units. On the other hand, when the pulp was treated with laccase (60 Ug^{-1} pulp at 50°C, pH 8.0, HBT of 0.1%, and reaction time 4.0 h), the brightness was increased by 2.5 units (Sharma et al., 2020).

ClO_2 was decreased individually by 15% as well as by 25% by xylanase and laccase, and ClO_2 decreased by 35% during concomitant treatment. Laccase and xylanase were coproduced by *Bacillus* sp. and *Bacillus halodurans* during solid-state fermentation. The same pH and temperatures of 9.0 and 70°C were also found for both enzymes and added to the biobleaching of *eucalyptus* kraft pulp. Due to laccase and xylanase treatments, the luminosity after alkali extraction increased by 13%, the whiteness increased by 106%, and the kappa decreased by 15% (Nagpal, Bhardwaj, & Mahajan, 2020). Delignification of the orange tree pruning soda anthraquinone pulp with *T. villosa* laccase in the presence of HBT was observed in biobleached pulp due to the delignification and enhance ophthalmic properties. Preprocessing xylanase has not been added to improve some of the pulp properties before laccase and alkaline extraction, possibly due to low HexA material (Fillat et al., 2017).

3.8 Laccase utilization for pulp biografting

Due to its ability to supply pulp fibers with antimicrobial, antioxidant properties and improve the hydrophobicity of pulp and paper materials, the biografting technique has gained considerable popularity. This accidental adsorption of the active material can often has a detrimental influence on the strength of the applied coating if the grafted compounds are absorbed into the substance and are not washable despite a thorough washing process (Ni, Liu, Fu, Gao, & Qin, 2021). It is important to understand the distinction between true covalent binding and adsorption, but it is still confusing. Fig. 3.6 illustrates a scheme of laccase-assisted phenol grafting on lignocellulosic materials/fibers. Phenolic (initiation) laccase generated radical reactions with oxidizable moiety (propagation) or other radical reactions (termination). The hand-paper of *Trametes Pubescens* and phenolic compounds were generated by holding the hand-paper in the presence of laccases at pH 5.0, at 100 rpm, and at 50°C for 18 h. With 4.0 mM, the hand-paper produced in the presence of caffeic acid or P-hydroxybenzoic acid by the antibacterial surface treated with laccase (LASP)

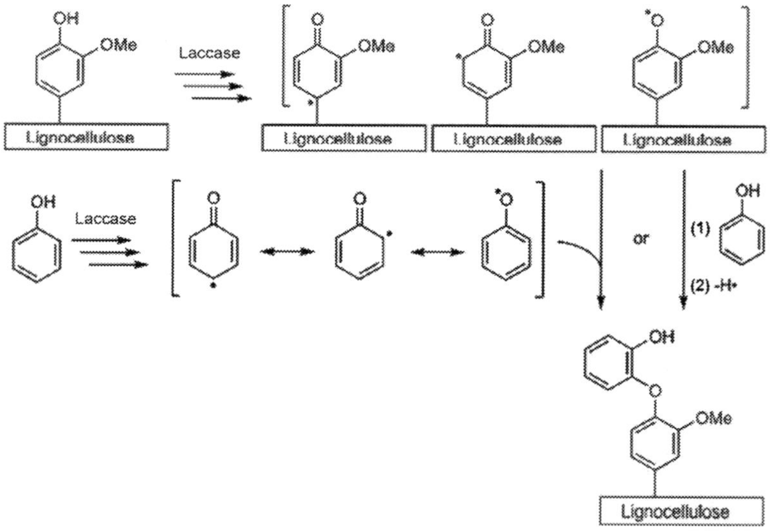

Figure 3.6 Scheme of laccase-mediated phenol grafting on lignocellulosic materials/fibers (Slagman, Zuilhof, and Franssen, 2018). *Reprinted from Slagman, S., Zuilhof, H., & Franssen, M.C. (2018). Laccase-mediated grafting on biopolymers and synthetic polymers: A critical review. Chembiochem, 19(4), 288–311 with permission under the terms of the Creative Commons Attribution-NonCommercial-NoDerivs License.*

showed good bactericidal activity against the *Staphylococcus aureus*. However, a sophisticated amount of natural phenols was needed to destroy *Escherichia coli*. Both gram-positive and gram-negative bacteria were satisfactory with LASP dopamine (Mohit, Tabarzad, & Faramarzi, 2020).

In the presence of phenols, unbleached flax fibers were acquired with laccases of *P. cinnabarinus* (40 UG^{-1} pulp). AS and the p-coumaric acid (PCA) grafted fibers demonstrated the effective antibacterial action against *Klebsiella pneumonia*, while the AS treatment showed the most resistance to *Pseudomonas aeruginosa* by killing 97% bacteria. *T. villosa* laccase (40 Ug^{-1} pulp) and FA were used to treat the sisal pulp at 3.5% (wt./wt.), pH 4.0, 50°C, and 30 rpm for 4 h. The kappa of the sisal pulp improved due to the laccase–FA coupling contact. The biografted mixture of poly(3-hydroxybutyrate) (P(3HB)) and ethyl cellulose (EC) has been shown to increase the thermal and mechanical properties significantly. The hydrophobic properties of the grafted composites are greater than those of the pristine composites. The P(3HB) ratio: EC (50:50) was combined at 25°C with *Saccharomyces cerevisiae* recombinant laccases for 30 min (Singh & Arya, 2019).

3.9 Pitch control by laccases

Lipophilic wood extractives are known as resins that chemically refer to alkanes, fatty alcohols, fatty acids, sterols, terpenoids, sterol conjugations, triglycerides, or waxes. These resins account for 10% of the total wood weight but create significant complications in the processing of paper, such as pore production and the breaking of paper due to sticky deposits on drying rolls. Various wood pulp types have a high pitch content, particularly in pines, including sulfite pulp and mechanical pulp (Moreno, Ibarra, Eugenio, & Tomás-Pejó, 2020). During the 1990s, two biotechnological methods for field checks were discovered and tested on an industrial scale in various parts of the world. The fungal process was used at the biopulping stage to regulate the pitch of the albino strain *Ophiostome piliferum*. The second approach is to use lipase to regulate the pitch. Ideally, triglycerides should be eliminated in all scenarios. Positive experiments in conjunction with appropriate redox mediators on the use of laccases have been published. LMS is selective, but its function in removing lipophilic compounds has also been identified (Singh & Arya, 2019).

P. cinnabarinus laccase, HBT and tween 20 were treated with different lipid-forming compounds and different reaction periods. In the laccase–HBT system, the response time of 30 min was sufficient to reduce all unsaturated lipids by 40%–100%. After 5 min of laccase treatment, abiatic acid and trilinolein were completely oxidized. Cholesteryl linoleate required a maximum reduction (to 95%) for only 5 min, while cholesterol and linoleic acid required 30 min. On the other hand, sitosterol was reduced by up to 75% after 1 h, and a complete sterol transition was achieved in 2 h. It has been suggested from this analysis that laccases may also bind the phenolic substrate to the unsaturated lipid substrate. The nature of lipid oxidation products with laccase alone and LMS provided little insight into the laccase—HBT pathways of lipid oxidation. Free sterols and sterols esters were extracted from the *Eucalyptus globulus* kraft pulp by *M. thermophila* laccases. Although the mixture of laccases and SA decreased the concentration of sterols, sterol glycosides, and sterol ester (Kumar, Yadav, & Tiruneh, 2020).

3.10 Deinking of waste papers by LMS

Wastepaper reprocessing is a reliable measure to minimize paper degradation and reduce emissions. Other benefits of desiccating recycled paper waste include reduced energy involvements in paper presses, consistent and simple pulp supplies for profitable white paper processing, and low pulp prices compared to the pulp. There are a lot of other advantages. One of the major recycled resources for paper production is the old newsprint (ONP). Deinking is an indispensable phase in reprocessing fiber, but the strength of paper reduces the capacity of recycled fiber for each subsequent deinking with chemical substances (Singh, Arya, Gupta, & Sharma, 2017).

Enzymes accelerated and remodeled to conventional chemically deinking approaches to increase the brightness of secondary pulp (Fig. 3.7). Any paper type, such as newspapers, can be deinked quickly by conventional methods. The United States

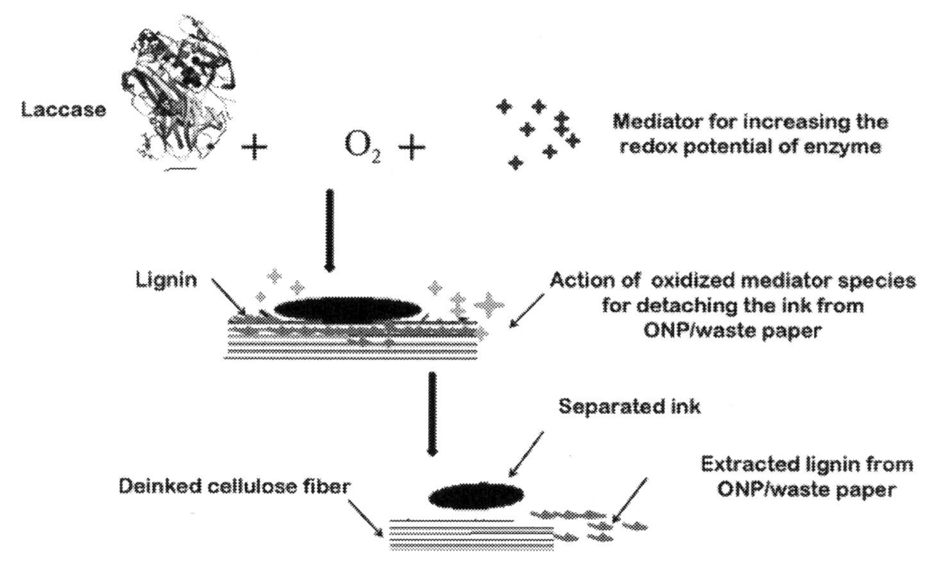

Figure 3.7 Deinking mechanism of wastepaper by laccase-mediator system. *ONP*, old newsprint (Singh and Arya, 2019). *Reprinted from Singh, G., & Arya, S.K. (2019). Utility of laccase in pulp and paper industry: A progressive step towards the green technology. International Journal of Biological Macromolecules, 134, 1070—1084, with permission from Elsevier.*

and other countries are also expected to follow the color printing of the offset lithograph. Cross-linking inks employed in this progression using traditional methods are very problematic to clear. Paper from offices is a good source of fine fiber and can be utilized for good-quality sheet formation by deinking. Mixed office waste is a source of high-quality fiber that can be converted into fine sheets and many other items if deinking improves. Enzyme-assisted deinking is an environmentally sustainable solution to the traditional chemical deinking method (Akbarpour, Ghasemian, Resalati, & Saraeian, 2018).

Deinked xylanase pulp decreased by 47% in effective residual ink production concentration (ERIC) and 62% in laccase, while the ERIC decrease was greater for the combined xylanase/laccase deinked pulp. On the other hand, increases in brightness (21.6%), break time, burst factor, tear factor, viscosity, and cellulose crystallinity were observed along with decreased kappa and chemical use. *B. halodurans* were used together with ABTS for 3 h, with deinking of ONP in *Bacillus* sp. and xylanase. Both enzymes have a synergistic effect that increases their brilliance (11.8%), freedom (17.8%), rupture period (34.8%), explosion (277%), and tearing (2.4%). In contrast to the cellulases that induce the shape of specks on the surface of the final paper products, laccases are considered favorable enzymes for deinking ONPs and other waste materials (Virk, Puri, Gupta, Capalash, & Sharma, 2013).

3.11 Laccase-mediated treatment of pulp and paper industry effluents

The paper industry uses an enormous extent of intensely dyed black liquor to release toxic chlorolignine, chlorophenol, and chloroalyphatic toxic lignin. The paper mill effluents in soil and water sources are highly alkaline and can rapidly alter their pH. In addition to its ability to minimize the need for biological oxygen demand (BOD) and chemical oxygen demand (COD), laccase itself (or laccase generating microbes) has the maximum ability to color black liquor (Patel, Arora, Pruthi, & Pruthi, 2017). Treatment with *Perenniporia* KU-Alk4 of the pulp mill effluent for three days, resulted in a decrease in the pH of the treated effluent from 9.6 to 6.8. In comparison, the reduction in color and COD decreased to 83% and 81%, respectively. Laccase was detectable during the biotreatment operation. The production of fungal and bacterial isolates with high characteristics such as pH, halophilic and organic solvent tolerance can be advantageous for the proper bioremediation of paper mill effluent. Studies and advancements in this field are necessary to establish rapid biodegradation processes that are likely to be economically feasible (Singh & Arya, 2019).

Phlebia brevispora BAFC 633 has been identified with bioremediation potential for discoloration of lignin-rich kraft black liquor. Cultivation of BAFC 633 *P. brevispora* on diluted black liquor kraft (1:30 vol./vol. in water) increased by 1% of glucose wt./vol. at pH 4.7 over 10 days by discoloration tests. The supernatant was observed between 200 and 400 nm wavelengths after the filtration of the fungus' growth compared to media without the use of the fungus. The spectral scan results showed that absorption of the processed sample was significantly lower than that of the control sample. Bioremediation in paper mill effluents was observed within 4 h using laccase for 50 mL effluent at 65°C, 150 rpm, with *B. tequilensis* SN4 thermo-alkali stable lacquer. Color reduction of 83%, BOD of 82%, COD of 77%, phenol content of 62%, lignin content of 74%, and gross dissolved solids of 28% were observed (Fonseca et al., 2018).

3.12 Lignin transformation by laccases

Large amounts of lignin are produced as a byproduct of the paper and pulp industries and converted into paper and paper products by lignocellulose biomass cellulose. Due to its recalcitrant heterogeneous composition, lignin is still the least part of the biomass used. Various natural and engineered laccases and peroxidase substances, such as vanillin, SA, FA, and cinnamic acid, have been investigated to transform lignin into market-oriented monomeric aromatic compounds. Lignin derivatives in dairy, pharmaceutical, and cosmetics are in high demand. Despite extensive study in this field, no solitary enzyme was identified to transform the real lignin medium into monomeric phenolic subunits (Agarwal, Rana, & Park, 2018; Diaz, Bodi-Paul, Gordobil, & Labidi, 2020; Shi, Zhu, Dai, Zhang, & Jia, 2020).

In order to attain the depolymerization of the lignin molecule, the compounds of the phenyl propane units must be disrupted. The main and most important type of linkage is β-O-4 aryl glycerol ether in the native form of lignin comprising 50% of the bonds in softwood and hardwood lignin for lignin macromolecule. Additional main links include 5-5, β-5, α-O-4, β-β, β-1, 4-O-5, and dibenzodioxocin. Organocat (biphasic lignocellulosic fractionation of biogenic oxalic acid as a catalyst) resulted in lignin depolymerization of the enzyme. The biocracking route reported was the first combination of laccase—violuric acid to oxidize the hydroxyl group directly into the structure of lignin β-O-4. The selective cleavage of β-O-4 oxidized bonds with β-etherase and glutathione lyase was subsequently performed with β-O-4 ether. This synergistic biocatalytic method provided an oily fraction of a low molecular mass of aromatic, coniferylaldehyde, and other guaiacyl and syringe elements (Li & Wilkins, 2020). *Caldalkalibacillus thermarum* strain TA2.A1 thermoalkaliphilic recombinant laccase (CtLac) was used in the depolymerization of rice straw, maize stub, and reed lignin. The most important additives in rice straw (homosyringaldehyde, p-hydroxybenzaldehyde, *p*-coumarium acid, vanillin, and SA) were derived from depolymerization-dependent enzymes. The sum of p-hydroxybenzoic and vanillic acids, either indicating biotransformation induced by CtLac or repolymerization of released monomers, is reduced to 69% and 63%, respectively. Further releases of p-hydroxybenzaldehyde and vanillin were increased by 2.3 and 5.6 times, respectively, compared to CtLac-treated organic straw lignin (removes lignin from rice straw after treatment with 60% ethanol, 0.25 M NaOH at 75°C for 3 h with a solid—liquid ratio of 1.25 (g/mL) under stirring) (Singh & Arya, 2019).

An innovation "microbial sink" has been presented to diminish the polymerization of aromatic compounds with LMW after depolymerization by laccases. It was a laboratory study, but its hypothesis was very informative and interesting. Maize lignin stove was depolymerized by laccase from *Pleurotus eryngii* for a maximum of 3 days. However, polymerization was detected over 4 days due to the high molecular weight of lignin. *Pseudomonas putida* KT2440 was grown on the dialysis membrane with lignin and laccase, which were applied externally to the membrane. Depolymerization was improved significantly in the presence of bacteria. These studies have shown that the existence of the bacterium stimulates the depolymerization of lignin, which could partially prevent repolymerization owing to the bacterial catabolism of LMW lignin (Singh & Arya, 2019).

3.13 Recovery of lignin byproducts

Petrochemical goods are very expensive. Their limited finances and uncertain political conditions in most oil-producing countries still call into question their continued supply. The graft of lignin to synthetic polymers is capable of producing a new class of plastics. The pulp and paper industries, as well as other biomass-based industries, produce industrial lignin byproducts. They are harmless, reliable, cheap, and produced at larger scale. Their highly reactive groups might be enzymatically adapted to produce environmentally friendly novel products (Wang et al., 2019). A LMW of vanillic acid, 4,4'-methyleneedi-phenyl diisocyanate, and acrylamide has been reported in the manifestation of *T. versicolor* laccases copolymerization, kraft lignin, and organosolv lignin. Chemo-enzymatic grafting of acrylic compounds, including polyacrylamide and polyacrylic acid-based on laccases has been shown to be a workable way to manufacture newly engineered plastics, thickeners, fillers, and biodegradable adsorbents (Shen et al., 2019).

Many experiments have shown that acrylamide can be effectively copolymerized with softwood organosolve lignin combined with organic peroxide, such as dioxane peroxide. Investigations on the reaction between acrylic and lignin monomers have shown that the lignin and acrylate copolymers described can be produced with chemo-enzyme copolymers. The device controls the molecular weights of the product to such an extent that chemical catalysts are not possible. The grafting of the wood surface with kraft lignin-based on laccase gave the wood antimicrobial properties, which was further enhanced by the incorporation of *Coniophora puteana* and *T. versicolor* (Krall, Serum, Sibi, & Webster, 2018).

3.14 Laccase for biofuels synthesis

It is an economical and feasible idea to turn the pulp and paper industries into potentially integrated biorefinery from an industrial perspective. The most successful case is the Borregaard Company, which operates as one of the most advanced biorefineries in

Norway. Borregaard began producing cellulose pulp in 1889 and began synthesizing hemicellulose ethanol from spruce (softwood) in 1938, mainly mannose present in the spent sulfite liquor, which is the product of the pulping process (Agrawal, Chaturvedi, & Verma, 2018). Various studies suggest that laccases can efficiently convert agricultural waste into food to produce ethanol as a biocatalyst. *Streptomyces ipomoea* E laccase performed more effectively than *T. villosa* fungal laccases in assessing the delignification and detoxification of steam blown wheat straw. The use of bacterial laccases resulted in more sugar than fungal laccases. Inhibitor phenols decreased and increased the efficacy of *S. cerevisiae* during the treatment of ethanol due to laccase treatment for wheat straw. Around 24% of lignin was isolated from paulownia wood by LMS. The major increase in hydrolytic enzymes saccharification was due to the enzyme's delignification (38% glucose and 34% xylose). Added value of aromatic compounds such as vanillin, vanillic acid, SA, and syringic acid has also been detected (Sharma et al., 2020).

In *P. cinnabarinus*, the laccase-treated wheat stroke slurry was produced from the strain *S. cerevisiae* F12 even without an external nitrogen source, and an improved ethanol output of 22 gL^{-1} was recorded. The increase in fermentation was caused by a decrease in the phenolic inhibitor associated with the treatment of laccases. Hydrolyzation and fermentability of steam explosive wheat straw have been enhanced by a new bacterial laccase (MetZyme). The amount of glucose and xylose enzyme hydrolysis increased by 21% and 30%, respectively, in conjunction with the alkaline extraction process and laccases. Inhibitor phenolic compounds were also reduced by up to 21% with laccase therapy. *Kluyveromyces marxianus* CECT 10875 thermotolerant yeast, therefore, performed well without inhibiting its growth. This biocatalytic method focused on laccases was intended to play a key role in implementing cost-effective technology in future biorefineries (Garedew et al., 2020).

3.15 Oxygen role in biobleaching of pulp

Lignin components found in the pulp need to be oxidized by laccases with O_2 to reduce the O_2 molecule, provide a water molecule and an oxidized substrate. ABTs of less than 100–300 KPa, 2 h, reacted with laccase from *T. versicolor* (after the alkaline extraction stage) to a delignification of up to 32%. O_2 pressure was a major factor in the biobleaching of the *T. villosa* eucalyptus kraft pulp. O_2 pressure of 600 KPa increased visibility from 3% to 4% throughout the process. Also, after laccases have lost all activity and only slightly during delignification reactions, a lot of dioxygen is involved inside reactions. The enhanced pulp properties at high O_2 pressure may have been due to the delignification effect of O_2 itself. Increased O_2 levels (4.0–7.0 ppm), resulting in more effective oxidation of flax pulp lignin and reduced Kappa Number, were achieved during mediator therapy in laccases. The presence of an increased volume of O_2 in the medium increases the ability of LMS to alter embedded lignin in the pulp rather than improve the ability of the enzyme to extract the polymer (Nathan, Rani, Rathinasamy, & Dhiraviam, 2017).

3.16 Challenges to implement laccase at industrial level

There is no technology available to successfully recycle laccases after being used in the pulp or paper industry. The enzyme has been highly costly to paper mills without reuse. On the contrary, the pulp and paper industries wanted environmentally stable laccases with highly specialized enzyme action (e.g., beneficial for pulp delignification) with a wide pH range, temperature, and resistance to industrial salts and chemical substances. The dose of laccase for pulp delignification is generally between 10 and 20 Ug^{-1} pulp by most workers. This means that 10 to 20 million laccase units would be needed for the paper industry to delignify 1.0 tons of pulp. The use of a high-specific enzyme will reduce the need for this enormous amount of laccase (Huang, Zhang, Gan, Yang, & Zhang, 2020).

Fungal lacquers, trees, bacteria at their T1Cu site have different E0s. Plant and bacterial laccases E0s are approximately 400 mV, whereas fungal laccases E0 is approximately 800 mV. Catalytic behavior of laccase ranges from substrate to T1Cu site to E0 site. Laccases can oxidize only molecules, the ionization potential of which is not greater than the E0 of the T1Cu core (energy needed to remove an electron from the atom). Nonphenolic lignin components, which are based on the E0 of the T1Cu core, are thought to be capable of cleaving laccases in an appropriate delineation of the pulp. E0s are stronger in nonphenolic structures such as veratryl, 1,2-dimetoxybenzene. However, it has been shown that mediator laccases can oxidize some E0-higher substances than laccases themselves, but the process by which this occurs is not clear. When redox mediators are present, laccase may oxidize nonphenolic lignin model compounds and significantly reduce the number of pulp kappa. The E0 of fungal and bacterial laccases investigated is less than that of nonphenolic compounds, and lignin cannot be oxidized without these enzymes (Singh & Arya, 2019).

To date, ABTS is currently using the best LMS in all applications in the pulp and paper industries at the laboratory level, but their commercial use does not appear to be feasible in the short term as, although produced in bulk, it is expected that the price of pulp bleaching and discoloration at scale-up conditions will be too high for cost-effective industrial use. Some recommendations will minimize mediator costs, such as the use of low mediator dosages by improving mixtures with lignocellulose products, improving pH regulation, and increasing pulp stability, or adding additives, etc. (Froidevaux, 2020).

Lignozym process (LMS) with HBT is the second most common and appears to be cheaper in the pulp and paper industry. The removal of 30%–60% of the pulp with LMS is feasible, and the pulp strength and viscosity losses are small. Many experimental studies in different laboratories have shown the feasibility and benefits of this laccase-mediator device using HBT. The main downside of the LMS is the inactivation of the laccase by oxidized mediator species. HBT has even been shown to inactivate laccase (50% in 4 h), which has a much smaller amount of elevated toxicity. The oxidation of aromatic amino acid residues on the protein surface has inactivated HBT radical. It is also important to find innovative and more operative mediators for efficient lignin degradation (Gupta et al., 2020).

Biological and medicinal chemistry studies have been regularly studied in quantitative structural activity relationship (QSAR), but LMS analysis is much less. This is one reason why QSAR LMS experiments, mainly associated with the variability of usable investigational data, have limits. Numerous reaction conditions, such as the substrate—mediator molar ratio, the source of the laccases, the substrate's nature, and the reaction time, have been used in various works. Besides, there have been various activity rate measures, as well as a small number of mediators. This heterogeneity of the available data is the key downside to which the mediator is suited for comparative purposes if our preference occurs (Bilal, Rasheed, Nabeel, Iqbal, & Zhao, 2019).

3.17 Recombinant laccases in biobleaching of pulps

The production of pulp and paper laccase for continuous supply is an enormous task. Much research is done on the production/induction of fungi and bacterial laccases, but the desired findings remain unrelated to the industrial requirements for pulp and paper. Laccases are often produced as a secondary metabolite in fungi; therefore laccases isolation is triggered only after the release of carbon or nitrogen from the media. In such cases, its industrial productivity was limited by the long laccase production period. Thus high-demand microbes from hyperlaccases can be synthesized to produce the enzyme in a short fermentation cycle with high yields (Singh, Kaur, et al., 2019; Singh, Singh, et al., 2019). There are strong research and development (R&D) interests in the pulp and paper industry to advance the use of laccases. This has been revealed by a significant number of published research articles and patents on this subject, which have been growing every year since 1995. Biotech companies such as Novo Nordisk and Novozymes focus heavily on lacquers' production for the pulp and paper industry. Annual reports have shown that R&D expenditure and sales of enzymes for the pulp and paper industry have been steadily increasing (Lin, Wu, Zhang, Liu, & Nie, 2018).

Some new biotechnological techniques, such as mutagenesis, recombinant DNA, polymerase chain reaction, and encoding, have been shown to increase the catalyst for improved pulp delignification. The *Aspergillus oryzae* laccase producing strain was developed by transforming *A. oryzae* host strain B711 with two plasmids namely pRaMB17.WT and pToC90. The laccases gene is aligned with the DNA sequences of the regulator, promoter, and terminator. Preparation for laccase is advertised as a "Flavourstar" trademark. *P. cinnabarinus* laccase was examined in two hosts (*A. oryzae* and *Aspergillus niger*) and compared to the indigenous enzymes by their ability to blacken. Wild *P.* cinnabarinus laccase and HBT *A. niger*, as a redox mediator, had a delignification of up to 75%, while *A. oryzae's* recombinant laccase could not delineate the pulp. Three laccases were diverse, and variations were introduced during protein processing of each fungal strain (Preethi et al., 2020).

Ravalason et al. (2009) synthesized a chimeric laccase to enhance pulp laccase treatment by fusing *P. cinnabarinus* laccase lac1 into the *A. niger* cellobiohydrolase B carbohydrate binding module (CBM). The chemical protein was analyzed for its softwood kraft pulp biobleaching ability concerning the native compartment. CBM has dramatically enhanced the properties of laccase delignification by providing the chimeric protein with the ability to bind to a cellulosic substrate. A recombinant laccase from *Thermus thermophilus* was added for the biobleaching of wheat straw pulp. Around 3.0 U laccase g^{-1} dry pulp at 90°C, pH 4.5, 8% consistency for 1.5 h, was optimized for biobleaching. Pulp luminosity increased by 3.3% ISO and the number of pulps kappas decreased by 5.6 U. In the presence of 5 mM ABTS, pulp biobleaching further improved the luminosity of the pulp by 1.5% ISO.

The direct development of the basidiomycete gene PM1, expressed in *S. cerevisiae*, was carried out and concluded with a 34,000-fold increase in the laccase operation experiment. Guided synthesis of laccase from *P. cinnabarinus* has been documented, and overall enzyme activity has improved by 8,000 times. Advances in the pre–pro factor have increased secretions by 40 times, and numerous transformations in mature laccase have increased by 13.7 times in kcat. Homodimeric laccase has recently been produced in *S. cerevisiae* from *Cerrena* unicolor BBP6. Three advanced fusion variant clones with improved thermostability and wider pH profiles were 31–37 times higher in total laccase activity. The α-factor preleader developed increased laccases by up to 2.4. Adequate laccase production to meet the needs of the paper industries is only possible if the industrial scale of enzyme production is combined with conventional and new technologies. Fig. 3.8 reveals the synergistic effect of numerous (traditional and contemporary) methods of development when integrated, with a variety of laccase synthesis and alteration opportunities (Bertrand, Martínez-Morales, & Trejo-Hernández, 2017; Preethi et al., 2020).

3.18 Conclusion and perspectives

Extensive literature reports have evidenced that fungal and bacterial laccases present promising applications in the pulp and paper industry for delignification, pitch removal, pulp graft, and ONP deinking. Nevertheless, further research is required to address constraints such as low manufacturing and mediator costs that imped the industrialization of laccases. A key requirement is developing laccases with improved properties, such as increased redox capabilities, the stabilization of small molecules and ion inhibitors, and efficient catalysis in harsh environments via genetic and protein engineering. Natural mediators that are novel, cost-effective, and nontoxic to

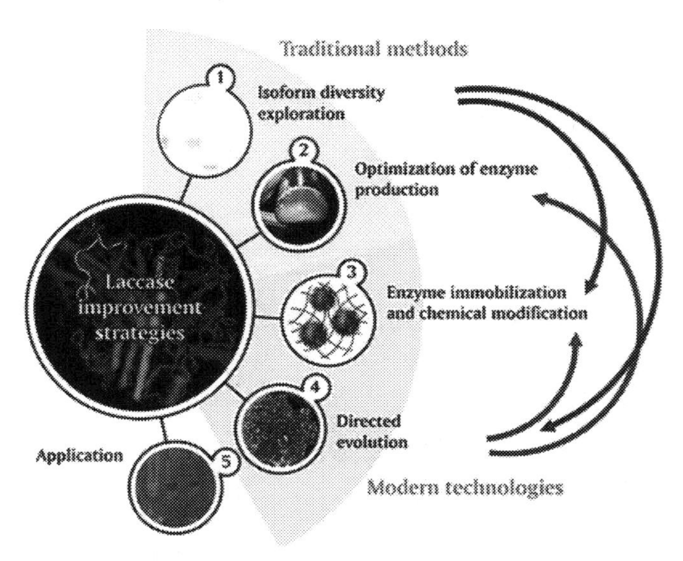

Figure 3.8 Synergistic effect of numerous (traditional and contemporary) methods of development (Bertrand et al., 2017). *Reprinted from Bertrand, B., Martínez-Morales, F., & Trejo-Hernández, M.R. (2017). Upgrading laccase production and biochemical properties: Strategies and challenges. Biotechnology Progress, 33(4), 1015–1034 with permission from John Wiley and Sons.*

the environment should be sought out and used effectively in the pulping process. Progress in research and development would reduce the use of toxic materials in pulp production significantly.

Acknowledgment

Consejo Nacional de Ciencia y Tecnología (MX) is thankfully acknowledged for partially supporting this work under Sistema Nacional de Investigadores (SNI) program awarded to Hafiz M.N. Iqbal (CVU: 735340).

Conflict of interests

The author(s) declare no conflicting interests.

References

Agarwal, A., Rana, M., & Park, J. H. (2018). Advancement in technologies for the depolymerization of lignin. *Fuel Processing Technology*, *181*, 115–132.

Agrawal, K., Chaturvedi, V., & Verma, P. (2018). Fungal laccase discovered but yet undiscovered. *Bioresources and Bioprocessing*, *5*(1), 1–12.

Akbarpour, I., Ghasemian, A., Resalati, H., & Saraeian, A. (2018). Biodeinking of mixed ONP and OMG waste papers with cellulase. *Cellulose*, *25*(2), 1265–1280.

Ancona-Escalante, W., Tapia-Tussell, R., Pool-Yam, L., Can-Cauich, A., Lizama-Uc, G., & Solís-Pereira, S. (2018). Laccase-mediator system produced by *Trametes hirsuta* Bm-2 on lignocellulosic substrate improves dye decolorization. *3 Biotech*, *8*(7), 1–8.

Angural, S., Rana, M., Sharma, A., Warmoota, R., Puri, N., & Gupta, N. (2020). Combinatorial bioble-aching of mixedwood pulp with lignolytic and hemicellulolytic enzymes for paper making. *Indian Journal of Microbiology*, *60*(3), 383–387.

Bagewadi, Z. K., Mulla, S. I., & Ninnekar, H. Z. (2017). Optimization of laccase production and its application in delignification of biomass. *International Journal of Recycling of Organic Waste in Agriculture*, *6*(4), 351–365.

Bajaj, P., & Mahajan, R. (2019). Cellulase and xylanase synergism in industrial biotechnology. *Applied Microbiology and Biotechnology*, *103*(21), 8711–8724.

Bertrand, B., Martínez-Morales, F., & Trejo-Hernández, M. R. (2017). Upgrading laccase production and biochemical properties: Strategies and challenges. *Biotechnology Progress*, *33*(4), 1015–1034.

Bilal, M., Rasheed, T., Nabeel, F., Iqbal, H. M., & Zhao, Y. (2019). Hazardous contaminants in the environment and their laccase-assisted degradation—A review. *Journal of Environmental Management*, *234*, 253–264.

Boruah, P., Sarmah, P., Das, P. K., & Goswami, T. (2019). Exploring the lignolytic potential of a new laccase producing strain *Kocuria* sp. PBS-1 and its application in bamboo pulp bleaching. *International Biodeterioration & Biodegradation*, *143*, 104726.

Chaurasia, S. K., & Bhardwaj, N. K. (2019). Biobleaching-An ecofriendly and environmental benign pulp bleaching technique: A review. *Journal of Carbohydrate Chemistry*, *38*(2), 87–108.

Christopher, L. P., Yao, B., & Ji, Y. (2014). Lignin biodegradation with laccase-mediator systems. *Frontiers in Energy Research*, *2*, 12.

Colonia, B. S. O., Woiciechowski, A. L., Malanski, R., Letti, L. A. J., & Soccol, C. R. (2019). Pulp improvement of oil palm empty fruit bunches associated to solid-state biopulping and biobleaching with xylanase and lignin peroxidase cocktail produced by *Aspergillus* sp. LPB-5. *Bioresource Technology*, *285*, 121361.

Diaz, R. H., Bodi-Paul, L., Gordobil, O., & Labidi, J. (2020). Fast methods for the identification of suitable chemo-enzymatic treatments of Kraft lignin to obtain aromatic compounds. *Biofuels, Bioproducts and Biorefining*, *14*(3), 521–532.

Dillies, J., Vivien, C., Chevalier, M., Rulence, A., Châtaigné, G., Flahaut, C., ... Froidevaux, R. (2020). Enzymatic depolymerization of industrial lignins by laccase-mediator systems in 1,4-dioxane/water. *Biotechnology and Applied Biochemistry*, *67*(5), 774–782.

Elakkiya, E., & Niju, S. (2020). Application of microbial fuel cells for treatment of paper and pulp industry wastewater: Opportunities and challenges. *Environmental Biotechnology*, *2*, 125–149.

Fedyaeva, O. N., & Vostrikov, A. A. (2019). Processing of pulp and paper industry wastes by supercritical water gasification. *Russian Journal of Physical Chemistry B*, *13*(7), 1071–1078.

Fillat, U., Martin-Sampedro, R., Gonzalez, Z., Ferrer, A. N. A., Ibarra, D., & Eugenio, M. E. (2017). Biobleaching of orange tree pruning cellulose pulp with xylanase and laccase mediator systems. *Cellulose Chemistry and Technology*, *51*(1–2), 55–65.

Fonseca, M. I., Molina, M. A., Winnik, D. L., Busi, M. V., Fariña, J. I., Villalba, L. L., & Zapata, P. D. (2018). Isolation of a laccase-coding gene from the lignin-degrading fungus *Phlebia brevispora* BAFC 633 and heterologous expression in *Pichia pastoris*. *Journal of Applied Microbiology*, *124*(6), 1454–1468.

Garedew, M., Lin, F., Song, B., DeWinter, T. M., Jackson, J. E., Saffron, C. M., ... Anastas, P. T. (2020). Greener routes to biomass waste valorization: Lignin transformation through electrocatalysis for renewable chemicals and fuels production. *ChemSusChem*, *13*(17), 4214–4237.

Guimarães, L. R. C., Woiciechowski, A. L., Karp, S. G., Coral, J. D., Zandoná Filho, A., & Soccol, C. R. (2017). *Laccases. Current developments in biotechnology and bioengineering* (pp. 199–216). Elsevier.

Gupta, G. K., Kapoor, R. K., & Shukla, P. (2020). *Advanced techniques for enzymatic and chemical bleaching for pulp and paper industries. Microbial enzymes and biotechniques* (pp. 43–56). Singapore: Springer.

Hilgers, R., Kabel, M. A., & Vincken, J. P. (2020). Reactivity of p-coumaroyl groups in lignin upon laccase and laccase/HBT treatments. *ACS Sustainable Chemistry & Engineering*, *8*(23), 8723–8731.

Hilgers, R., Van Dam, A., Zuilhof, H., Vincken, J. P., & Kabel, M. A. (2020). Controlling the competition: Boosting laccase/HBT-catalyzed cleavage of a β-O-4′ linked lignin model. *ACS Catalysis*, *10*(15), 8650–8659.

Hilgers, R., van Erven, G., Boerkamp, V., Sulaeva, I., Potthast, A., Kabel, M. A., & Vincken, J. P. (2020). Understanding laccase/HBT-catalyzed grass delignification at the molecular level. *Green Chemistry*, *22*(5), 1735–1746.

Hilgers, R., Vincken, J. P., Gruppen, H., & Kabel, M. A. (2018). Laccase/mediator systems: Their reactivity toward phenolic lignin structures. *ACS Sustainable Chemistry & Engineering*, *6*(2), 2037–2046.

Huang, W., Zhang, W., Gan, Y., Yang, J., & Zhang, S. (2020). Laccase immobilization with metal-organic frameworks: Current status, remaining challenges and future perspectives. *Critical Reviews in Environmental Science and Technology*, 1–42.

Jahan, M. S., Uddin, M. M., & Kashem, M. A. (2017). Modification of chlorine dioxide bleaching of *Gmelina arborea* (gamar) pulp. *Bangladesh Journal of Scientific and Industrial Research*, *52*(4), 247–252.

Kontro, J., Maltari, R., Mikkilä, J., Kähkönen, M., Mäkelä, M. R., Hilden, K., ... Sipilä, J. (2020). Applicability of recombinant laccases from the white-rot fungus *Obba rivulosa* for mediator-promoted oxidation of biorefinery lignin at low pH. *Frontiers in Bioengineering and Biotechnology*, *8*, 604497.

Krall, E. M., Serum, E. M., Sibi, M. P., & Webster, D. C. (2018). Catalyst-free lignin valorization by acetoacetylation. Structural elucidation by comparison with model compounds. *Green Chemistry*, *20*(13), 2959–2966.

Kumar, A. (2020). Biobleaching: An eco-friendly approach to reduce chemical consumption and pollutants generation. *Physical Sciences Reviews*, *1*. (ahead-of-print).

Kumar, A., Yadav, M., & Tiruneh, W. (2020). Debarking, pitch removal and retting: Role of microbes and their enzymes. *Physical Sciences Reviews*, *5*(10).

Li, M., & Wilkins, M. (2020). Lignin bioconversion into valuable products: Fractionation, depolymerization, aromatic compound conversion, and bioproduct formation. *Systems Microbiology and Biomanufacturing*, *1*, 1–20.

Lin, X., Wu, Z., Zhang, C., Liu, S., & Nie, S. (2018). Enzymatic pulping of lignocellulosic biomass. *Industrial Crops and Products*, *120*, 16–24.

Liu, Y., Geng, Y., Yan, M., & Huang, J. (2017). Stable ABTS immobilized in the MIL-100 (Fe) metal-organic framework as an efficient mediator for laccase-catalyzed decolorization. *Molecules*, *22*(6), 920.

Loaiza, J. M., Alfaro, A., López, F., García, M. T., & García, J. C. (2019). Optimization of laccase/mediator system (LMS) stage applied in fractionation of *Eucalyptus globulus*. *Polymers*, *11*(4), 731.

Longe, L. F., Couvreur, J., Leriche Grandchamp, M., Garnier, G., Allais, F., & Saito, K. (2018). Importance of mediators for lignin degradation by fungal laccase. *ACS Sustainable Chemistry & Engineering*, *6*(8), 10097–10107.

Mateljak, I., Monza, E., Lucas, M. F., Guallar, V., Aleksejeva, O., Ludwig, R., ... Alcalde, M. (2019). Increasing redox potential, redox mediator activity, and stability in a fungal laccase by computer-guided mutagenesis and directed evolution. *ACS Catalysis*, *9*(5), 4561–4572.

Mohit, E., Tabarzad, M., & Faramarzi, M. A. (2020). Biomedical and pharmaceutical-related applications of laccases. *Current Protein and Peptide Science*, *21*(1), 78–98.

Moreno, A. D., Ibarra, D., Eugenio, M. E., & Tomás-Pejó, E. (2020). Laccases as versatile enzymes: From industrial uses to novel applications. *Journal of Chemical Technology & Biotechnology*, *95*(3), 481–494.

Morsi, R., Bilal, M., Iqbal, H. M., & Ashraf, S. S. (2020). Laccases and peroxidases: The smart, greener and futuristic biocatalytic tools to mitigate recalcitrant emerging pollutants. *Science of The Total Environment*, *714*, 136572.

Munk, L., Andersen, M. L., & Meyer, A. S. (2018). Influence of mediators on laccase catalyzed radical formation in lignin. *Enzyme and Microbial Technology*, *116*, 48–56.

Nagpal, R., Bhardwaj, N. K., & Mahajan, R. (2020). Synergistic approach using ultrafiltered xylano-pectinolytic enzymes for reducing bleaching chemical dose in manufacturing rice straw paper. *Environmental Science and Pollution Research*, *27*(35), 44637–44646.

Nathan, V. K., Rani, M. E., Rathinasamy, G., & Dhiraviam, K. N. (2017). Low molecular weight xylanase from *Trichoderma viride* VKF3 for bio-bleaching of newspaper pulp. *BioResources*, *12*(3), 5264–5278.

Ni, S., Liu, N., Fu, Y., Gao, H., & Qin, M. (2021). Laccase mediated phenol/chitosan treatment to improve the hydrophobicity of Kraft pulp. *Cellulose*, 1–13.

Ozer, A., Sal, F. A., Belduz, A. O., Kirci, H., & Canakci, S. (2020). Use of feruloyl esterase as laccase-mediator system in paper bleaching. *Applied Biochemistry and Biotechnology, 190*(2), 721–731.

Pan, Y., Ma, H., Huang, L., Huang, J., Liu, Y., Huang, Z., … Yang, J. (2018). Graphene enhanced transformation of lignin in laccase-ABTS system by accelerating electron transfer. *Enzyme and Microbial Technology, 119*, 17–23.

Patel, A., Arora, N., Pruthi, V., & Pruthi, P. A. (2017). Biological treatment of pulp and paper industry effluent by oleaginous yeast integrated with production of biodiesel as sustainable transportation fuel. *Journal of Cleaner Production, 142*, 2858–2864.

Preethi, P. S., Gomathi, A., Srinivasan, R., Kumar, J. P., Murugesan, K., & Muthukailannan, G. K. (2020). *Laccase: Recombinant expression, engineering and its promising applications. Microbial enzymes: Roles and applications in industries* (pp. 63–85). Singapore: Springer.

Ravalason, H., Herpoël-Gimbert, I., Record, E., Bertaud, F., Grisel, S., de Weert, S., … Sigoillot, J. C. (2009). Fusion of a family 1 carbohydrate binding module of *Aspergillus niger* to the *Pycnoporus cinnabarinus* laccase for efficient softwood kraft pulp biobleaching. *Journal of Biotechnology, 142*(3–4), 220–226.

Saha, R., & Mukhopadhyay, M. (2020). Elucidation of the decolorization of Congo Red by *Trametes versicolor* laccase in presence of ABTS through cyclic voltammetry. *Enzyme and Microbial Technology, 135*, 109507.

Sailwal, M., Banerjee, A., Bhaskar, T., & Ghosh, D. (2020). *Integrated biorefinery concept for Indian paper and pulp industry. Waste Biorefinery* (pp. 631–658). Elsevier.

Saleem, R., Khurshid, M., & Ahmed, S. (2018). Laccases, manganese peroxidases and xylanases used for the bio-bleaching of paper pulp: An environmental friendly approach. *Protein and Peptide Letters, 25*(2), 180–186.

Sharma, D., Chaudhary, R., Kaur, J., & Arya, S. K. (2020). Greener approach for pulp and paper industry by xylanase and laccase. *Biocatalysis and Agricultural Biotechnology, 25*, 101604.

Shen, X. J., Chen, T., Wang, H. M., Mei, Q., Yue, F., Sun, S., … Sun, R. C. (2019). Structural and morphological transformations of lignin macromolecules during bio-based deep eutectic solvent (DES) pretreatment. *ACS Sustainable Chemistry & Engineering, 8*(5), 2130–2137.

Shi, Y., Zhu, K., Dai, Y., Zhang, C., & Jia, H. (2020). Evolution and stabilization of environmental persistent free radicals during the decomposition of lignin by laccase. *Chemosphere, 248*, 125931.

Singh, G., & Arya, S. K. (2019). Utility of laccase in pulp and paper industry: A progressive step towards the green technology. *International Journal of Biological Macromolecules, 134*, 1070–1084.

Singh, G., Arya, S. K., Gupta, V., & Sharma, P. (2017). Enzyme technology for lignocellulosic biomass conversion and recycling to valuable paper and other products: Challenges ahead. *Journal of Molecular Biology and Techniques, 2*, 105.

Singh, G., Kaur, S., Khatri, M., & Arya, S. K. (2019). Biobleaching for pulp and paper industry in India: Emerging enzyme technology. *Biocatalysis and Agricultural Biotechnology, 17*, 558–565.

Singh, G., Singh, S., Kaur, K., Arya, S. K., & Sharma, P. (2019). Thermo and halo tolerant laccase from *Bacillus* sp. SS4: Evaluation for its industrial usefulness. *The Journal of General and Applied Microbiology, 65*(1), 26–33.

Slagman, S., Zuilhof, H., & Franssen, M. C. (2018). Laccase-mediated grafting on biopolymers and synthetic polymers: A critical review. *Chembiochem, 19*(4), 288–311.

Sondhi, S., Kaur, R., Kaur, S., & Kaur, P. S. (2018). Immobilization of laccase-ABTS system for the development of a continuous flow packed bed bioreactor for decolorization of textile effluent. *International Journal of Biological Macromolecules, 117*, 1093–1100.

Stevens, J. C., Das, L., Mobley, J. K., Asare, S. O., Lynn, B. C., Rodgers, D. W., & Shi, J. (2019). Understanding laccase–ionic liquid interactions toward biocatalytic lignin conversion in aqueous ionic liquids. *ACS Sustainable Chemistry & Engineering*, 7(19), 15928–15938.

Vermani, M., Chauhan, V., Yadav, S. S., Roy, T., & Singh, S. (2020). *Role of glycoside hydrolases in pulp and paper industries. Industrial applications of glycoside hydrolases* (pp. 191–215). Singapore: Springer.

Virk, A. P., Puri, M., Gupta, V., Capalash, N., & Sharma, P. (2013). Combined enzymatic and physical deinking methodology for efficient eco-friendly recycling of old newsprint. *PLoS One, 8*(8), e72346.

Wang, H. M., Ma, C. Y., Li, H. Y., Chen, T. Y., Wen, J. L., Cao, X. F., ... Sun, R. C. (2019). Structural variations of lignin macromolecules from early growth stages of poplar cell walls. *ACS Sustainable Chemistry & Engineering*, *8*(4), 1813–1822.

Wang, M., Zhao, Y., & Li, J. (2018). Demethylation and other modifications of industrial softwood kraft lignin by laccase-mediators. *Holzforschung*, *72*(5), 357–365.

Yang, L., Fang, L., Huang, L., Zhao, Y., & Liu, G. (2018). Evaluating the effectiveness of using ClO^2 bleaching as substitution of traditional Cl2 on PCDD/F reduction in a non-wood pulp and paper mill using reeds as raw materials. *Green Energy & Environment*, *3*(3), 302–308.

Zeng, S., Qin, X., & Xia, L. (2017). Degradation of the herbicide isoproturon by laccase-mediator systems. *Biochemical Engineering Journal*, *119*, 92–100.

Zeng, S., Zhao, J., & Xia, L. (2017). Simultaneous production of laccase and degradation of bisphenol A with *Trametes versicolor* cultivated on agricultural wastes. *Bioprocess and Biosystems Engineering*, *40*(8), 1237–1245.

CHAPTER 4

Nanotechnology for waste wood recycling

K. M. Faridul Hasan, Péter György Horváth and Tibor Alpár
Simonyi Károly Faculty of Engineering, University of Sopron, Sopron, Hungary

4.1 Introduction

Wood wastes (W@Ws) are generated from renewable resources which could be recycled easily for the production of particle boards/composites or energy productions (Kim & Song, 2014). W@W is a major part of waste streams entailing whole trees, pruned branches, sawdust from companies, composites, used wood products, and so on (K.F. Hasan, Horváth, & Alpár, 2021; K.M.F. Hasan, Horváth, & Alpár, 2021; Tibor, Péter, & Hasan, 2021). But, the primary components of W@Ws include trees, branches, trims, debris, laminated timber, and debris from constructions and demolitions. Wood-based wastes are collected from various sources such as wood processing in mills, agricultural wastes, different fiber plants, and so on (Onishchenko et al., 2013; Hasan et al., 2020b). However, W@Ws are also mainly coming from renovation, packaging, construction, and demolitions as well. Besides, huge amounts of W@Ws are generated from the large urban areas which may create a huge burden to the environment. The recycling of W@W could minimize the burden from the environment besides reducing the consumptions of virgin raw materials (from plants to energy and water usage during the manufacturing processes) in contrast to the fresh raw materials (Sommerhuber et al., 2017). Conversely, the main advantage of recycling W@W is to meet the constant challenges for increasing oil prices through reducing the exhaustion of fossil foil and reduction in natural disasters consequence of the increasing trends of global warming and significant climate changes (Burnley, Phillips, & Coleman, 2012; Kim & Song, 2014; Sormunen & Kärki, 2019).

Woods are broadly classified into two categories such as hard wood and soft wood. Woods derived from the plants like oak, sal, poplar, maple, and mango are termed hardwood, whereas pine, fir, spruce, cedar, etc. are renowned as softwood materials (Khan et al., 2020; Richter & Rein, 2020). These species of woods are different from each other in terms of physical and chemical properties, so the composites made from different wood materials vary significantly. The mechanical properties and chemical compositions of different wood species are provided in Tables 4.1 and 4.2. The usage of plant-based natural fibers is also increasing from the last few decades significantly as reinforcement material for producing biocomposites (Hasan, Horváth, & Alpár, 2020c;

Nanotechnology in Paper and Wood Engineering
DOI: https://doi.org/10.1016/B978-0-323-85835-9.00014-3

Table 4.1 Tensile properties and densities of different hard and softwoods (Ahmad, Bon, & Abd Wahab, 2010; Aydin, Yardimci, & Ramyar, 2007; Bendtsen & Ethington, 1975; Clemons, 2008; Jain & Gupta, 2018; John & Anandjiwala, 2008; Khan, Srivastava, & Gupta, 2020; Kretschmann, 2010; Lukmandaru, 2015; Saxena & Gupta, 2019; Shekar & Ramachandra, 2018).

Plants	Tensile strength (MPa)	Young's modulus (MPa)	Density (g cm^{-3})
Birch	—	37.90	1.2
Elm	186.2	9.24	0.68
Sal	39.64	1.54	0.72
Isora	500—600	—	1.2—1.3
Oak	77.90	10.41	0.63
Keruing	41.2	11.2	0.74
Kedongdong	38.7	13.4	0.7
Bintangor	38.8	11.9	0.74
Maple	108.2	9.58	0.54
Teak	95	1.4	0.55
Poplar	154.4	5.86	0.33
Hornbeam	—	15.26	0.37
Bamboo	23.65	140—230	0.52
Beech	86.2	9.51	0.56
Mango	140—230	11—17	0.6—1.1
Spruce	84.8	18—40	1.4
Pine	12—27	0.1—0.5	0.35
Cedar	78.6	7.2	0.35
Fir	77.9	13.81	0.33

Table 4.2 Chemical compositions of different wood species (Agnantopoulou et al., 2012; Danish & Ahmad, 2018; Jain & Gupta, 2018; John & Anandjiwala, 2008; John & Thomas, 2008; Khan & Ali, 1992; Khan et al., 2020; Kim et al., 2011; Okoroigwe, Enibe, & Onyegegbu, 2016).

Wood material	Cellulose (%)	Lignin (%)	Hemicellulose (%)	Others (%)
Beech	44—46	21—23	30—35	—
Sal	40—50	15—30	12—25	—
Isora	74	23	—	1.09
Oak	42—47	25—27	20—27	3—4
Maple	47	21	30	2
Teak	49.28	30.33	29.67	—
Poplar	44—46	18—21	33—36	—
Bamboo	26—43	21—31	30	—
Mango	40.19	35.96	11.47	—
Spruce	44—46	27—29	25—27	—
Pine	44—46	27—29	25—28	4—5
Cedar	37	34	19	3
Eucalyptus	57.3	25.9	16.8	—
Wattle	62	25.4	25.4	—
Almond tree	33.7	25	20.1	—
Acacia mangium	46.5	27.64	16.03	9.83

Hasan, Horváth, Bak, Mucsi, & Alpár, 2021; Hasan, Horváth, Kóczán, & Alpár, 2021). Plant-based composites are also termed green composites (Hasan, Péter, K., & L.A., 2021; Mahmud, et al., 2021) as they could be reproduced from recycled plastics or W@W. However, there are some production limitations for the incompatibility between the hydrophobic polymer and hydroscopic plant materials. The poor interactions could result in the poor mechanical performances of final biocomposites if the W@Ws are not uniformly dispersed throughout the matrix (Masoudifar, Nosrati, & Mohebbi Gargari, 2018). Nanotechnologies have brought a new route for the improved mechanical properties of plant-based polymeric composites. Several studies have reported that nanoparticles could significantly enhance the mechanical properties of the composites (Ahmed et al., 2018; Amiri, Rahmaninia, & Khosravani, 2019; Kale et al., 2016; Masoudifar et al., 2018; Robles et al., 2016; Sabazoodkhiz, Rahmaninia, & Ramezani, 2017). The unique surface properties of nanoparticles accelerate the fundamental processes of reaction phases in the matrix system. The positive influences are generated in matrix systems for the development of larger interfaces in nanomaterial dispersed polymer composites. The molecular bonding is enhanced in the polymer crosslink of the matrix system for the larger surface area of nanofillers which assists in increased composites strength and stiffness. Various metallic nanoparticles like titanium dioxide (TiO_2), silicon dioxide (SiO_2), zinc oxide (ZnO), silver (Ag), carbon nanotube (CNT), reduced graphene oxide (rGO), and so on are used as nanofillers to improve the wood-based composites which also could have the similar significance for the effective usage of W@W as well (Ghorbani, Biparva, & Hosseinzadeh, 2018; Lin et al., 2011).

Recycled fiber materials are another important source of nanobiocomposites productions. Although the reinforcement of natural fiber-based composites has started its journey in 1900s (Saba et al., 2016), the potentiality of recycled W@Ws for the reinforcement in composites along with nanofiller is still in the preliminary stage. In Malaysia, 6.93 million tons of empty fruit bunch from oil palm is generated during industrial processing (Saba et al., 2015, 2016). There are some reports found for utilizing different fiber-based wastes such as jute (Ahuja, Kaushik, & Singh, 2018; Baheti, Militky, & Marsalkova, 2013), flax (Jia et al., 2019), cotton (Dobircau et al., 2009; Thambiraj & Shankaran, 2017), rice husk (Liou & Wang, 2020), coir (Rout, Tripathy, & Ray, 2018), sugarcane bagasse (Ferreira et al., 2018), pineapple leaf (dos Santos et al., 2013), for the reinforcement with nanofillers to form nanobiocomposites. The utilization of fiber-based wastes through recycling could accelerate the production of fancy biocomposites and reduce the burden of disposals on environment. Besides, the performance of reused W@W could be enhanced significantly by using the advanced concept of applying nanoparticles (Saba et al., 2019; Visakh et al., 2012).

Woods as well as the W@Ws are composed of different biopolymers, as shown in Table 4.2, such as cellulosic carbohydrate polymers, lignin-based phenolic polymers, and hemicellulose along with some minor extractives (oil, wax, protein, fat, and so on) (Deveci et al., 2018). The fibrous property of wood has made it a prominent natural fiber with diversified potentialities. The recycled W@W-based composites incorporated with nanofiller could provide prominent mechanical performances and functionalities to the biocomposites by ensuring sustainable productions. Composites made from W@Ws could be utilized to a larger extent for different outdoor usages such as tiles and pavements by ensuring adequate water stability. This chapter reports about the development of different potential applications of recycled W@Ws, feasible incorporation of nanoparticles, sustainable production processes, and so on.

4.2 Wood waste materials

The W@Ws are the most attractive and sustainable raw materials which could be utilized for particle composites or energy production through recycling. The Food and Agricultural Organizations of United Nations reported that just only 28% of trees are used for saw timber whereas the remaining 72% becomes wastes which could be utilized for particle boards, or productions or some other potential usage (Ramirez et al., 2017). One of the studies has revealed that particle boards produced from W@Ws generate −428 kg of CO_2-eq in contrast to the virgin wood-based particle boards (Kim & Song, 2014). The same study also has claimed that the average life span of W@W-based composites is 14 years (Kim & Song, 2014). The produced particle boards could be utilized for attractive furniture in office rooms, facades, building constructions, kitchen cabinets, bath cabinets, door components, household, etc. A production process of particle board utilizing W@W is shown in Fig. 4.1, where hot-press-oriented technology implementation is highlighted to produce fiber/particle board. W@Ws are highly competitive materials which exists water content by less than 25% and density ranging from 0.5 to 0.9 g/cm^3, thus consequences for less possibility of bacterial degradations (Kim & Song, 2014). The timber industries produce pallets and packaging products. Wood-based pallets and packaging materials demand consumption of 20% of sawn woods throughout Europe (Moreno & Saron, 2017).

Wood-based products are renewable and easy to recycle, thus generates less emission of greenhouse gases. On the other hand, wood particles/flours are biodegradable and available throughout the world. However, the continuous usage of virgin woods is creating a shortage of fresh woods in so many regions of the universe. So, awareness is rising to develop new technologies for the effective utilizations W@Ws to the manufactures and researchers for reducing the pressures on consuming virgin woods (Nourbakhsh & Ashori, 2010). There are numerous studies reported on virgin wood material-based composites (Ferdosian et al., 2017; Partanen & Carus, 2016); however, the incorporation of nanoparticles into the biocomposites (Castro et al., 2018; Kaymakci, Birinci, & Ayrilmis, 2019) is not a very much old concept, although there

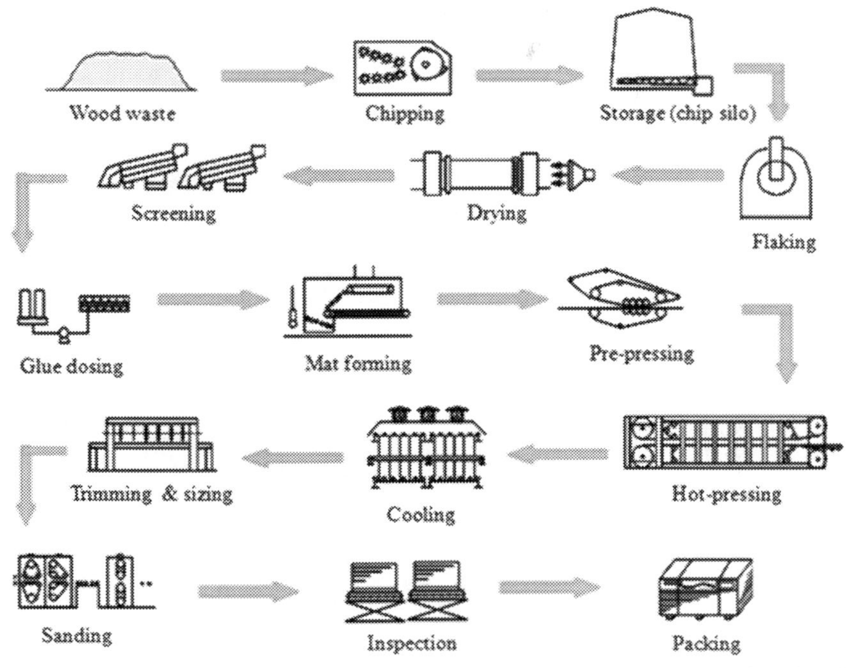

Figure 4.1 Composite panel manufacturing process (Kim & Song, 2014). *Reproduced with permission from Elsevier (Kim, M.H. and H.B. Song, Analysis of the global warming potential for wood waste recycling systems. Journal of Cleaner Production, 2014. 69:199–207), courtesy also goes to Dongwha holdings (http://www.dongwha.co.kr). Copyright Elsevier, 2014.*

is significant interest going on for manufacturing recycled plant material–based composites in terms of environmental sustainability and economical aspects.

Sometimes, the industries are using lignocellulosic fiber materials discharged after production as disposal/wastes for manufacturing recycled fiber composite panels. A research was conducted for utilizing the byproduct of a plywood manufacturing company (containing polypropylene-based composites) for enhancing the mechanical properties by using organonanoclay (Kajaks et al., 2015). Besides, a large amount of fiber-based wastes are generated throughout the world from various sectors (construction and building, furniture, packaging, and agriculture). Although different fiber waste–based biocomposites reinforced with cement were developed (Hospodarova et al., 2018; Savastano, Warden, & Coutts, 2000; Wisky Silva et al., 2016), the incorporation of nanomineral with the cement is still limited which could enhance the mechanical and physical performances of wastage fiber composites (Hosseinpourpia et al., 2012). On the other hand, the usage of waste natural fibers on a nanoscale range to form nanobiocomposites is also getting popular. da Silva, Kano, and Rosa (2019) have reported the development of green technology to produce lignocellulosic nanofibers from the sawdust of eucalyptus plant to reinforce with poly (butylene adipate-co-terephthalate) nanocomposites.

4.3 Nanotechnology

Nanotechnology is going to be the important element in materialistic chemistry for developing a brand new universe through bringing remarkable features in diversified sectors. The era of nanotechnology is considered to have begun from 1959 through a speech of Richard Feynmam in a meeting of the American Physical Society (Pacheco-Torgal & Jalali, 2011). Since then, enormous researches are going on to functionalize natural and synthetic fibers by using greenly synthesized nanoparticles for versatile applications (composites, constructions, packaging, biomedical, coloration, textiles, automotive, and so on) (Hasan et al., 2019, 2020a,b; Jiang et al., 2020; Khan et al., 2020; Mahmud et al., 2019; Siddiqui et al., 2020; Zhang et al., 2020; Zhou & Fu, 2020; Hasan, Horváth, Miklos, & Alpár, 2021). The term nanotechnology is defined by several researchers; however, the most accurate definition was given by Drexler in 1981, such as production within the precise range of 0, 1, and 100 nm dimensions (Drexler, 1981). The investment for nanotechnology-based researches is also getting increased day by day.

There are several popular methods to synthesize metal nanoparticles which could be classified as physical, chemical, and biological routes (Mirzaei & Darroudi, 2017; Pielichowski & Majka, 2018; Shamaila et al., 2016). In the case of chemical synthesis, metal precursors are dissolved in an aqueous solution with a suitable reducing agent. The electrochemical reduction, hydrothermal synthesis, reduction of chemicals in aqueous solutions, and sonochemistry synthesis are some of the popularly used techniques. In some cases, the size of the nanoparticles is also controlled by using stabilizing agent during their reactions. The agglomeration of the nanoparticles is also minimized by using the stabilizer. Physical methods are popular for producing gold (Au) and Ag nanoparticles. Different organic solvents, supercritical fluids, water, and ionic liquids are mostly used, which were mentioned by the researchers (Amendola, Polizzi, & Meneghetti, 2006; Anjana et al., 2019; Cao & Matsuda, 2016; Tsuji et al., 2017; Verma, Ebenso, & Quraishi, 2019). On the other hand, biological agents (plant extract, fungi, yeasts, and bacteria) are also getting attention for the biosynthesis of metallic nanoparticles (Hasan, Horváth, Horváth, & Alpár, 2021; Hasan et al., 2021). Metallic nanomaterials like Ag and Au are also produced by using the biological approach. However, some intermolecular actions are important for the production of nanomaterials by using this route such as Van der Waals force, hydrogen bond, metal coordination, dipolar interactions, electrostatic interactions, and so on (Pielichowski & Majka, 2018).

Nanoscience and technology are going to play a significant role in the development of W@W-based composites manufacturing. As the usage of compatible resin for composites production could have the chance of having formaldehyde content, so the usage of nanotechnology could help for minimizing the emission of formaldehyde content from particle boards or W@W-based composites (Candan & Akbulut, 2013). The usage of nano-ZnO, nano-SiO_2, and nano-calcium carbonate ($CaCO_3$) could significantly decrease formaldehyde emissions (Candan & Akbulut, 2013). Besides, the building and construction sector is experiencing tremendous potentiality for nanotechnology-implemented products. Carbon nanotube and nanominerals also play a significant role in enhancing the durability and strength of the cement-based nanobiocomposites. Nanomaterials also assist in increasing thermal insulation properties of composite panels (Pacheco-Torgal & Jalali, 2011). Besides, nanotechnology has supreme potentiality in terms of sustainability through reducing waste, resources, energy, etc. from the manufacturing units, as shown in Fig. 4.2.

Nanotechnology is showing a potentiality to be the driver of world's economy with the technical development of products in multidisciplinary areas through fabricating and designing the products feature by using nanofillers. The outstanding properties of nanomaterial are also getting huge attention from researchers and industrialists as well. The development of nano-based technology is also getting significant investment which was nearly $6 billion in 2010 throughout the universe (Candan & Akbulut, 2013). The market volume of world nanotechnology was within US$45–50 billion in 2018, which is expected to exceed by 13% from 2019 to 2025 (Inshakova & Inshakova, 2020).

The dispersion of nanofillers in the W@W could play a significant influence on the production of the successful composite with an expected outcome. However, the interfacial activity between the W@Ws and nanofiller is a critical factor for attaining projected outcomes from the composites. Some commonly used nanoparticles for the production of wood-based composites are montmorillonite, ZnO, SiO_2, CNT, alumina, $CaCO_3$, rGO, and so on. These nanoparticles provide outstanding features [thermal stability, antibacterial property, ultraviolet (UV)-resistance property, corrosion resistance, dielectric behavior, better tribological property, mechanical performance, and adhesion to substrates] to the

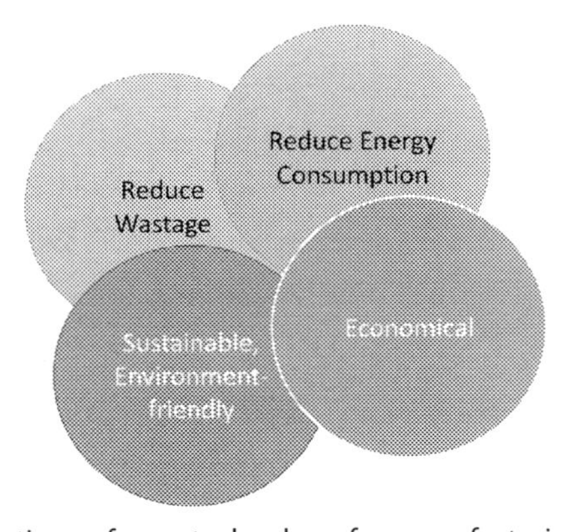

Figure 4.2 Potential contributions of nanotechnology for manufacturing units and the world.

composite panels (Deka & Maji, 2012; Devi & Maji, 2013; Dong et al., 2017; Hazarika & Maji, 2014; Hosseini, 2020; Kumar et al., 2013; Wang et al., 2011).

4.3.1 Nanographene

The influences of nanographene content were studied by Beigloo et al. (2017) in terms of mechanical, thermal, and physical performances where wood flours (30%), nanographene (0, 0.5, 1.5, and 2.5 wt.%), and 70% high density polyethylene (HDPE) were used to produce composite panels. They have claimed that the flexural properties and impact strengths of the composites were increased upto a certain percentage of nanographene loading (0.5%), whereas the performance decreases after 2.5% of the loading. The scanning electron microscopic (SEM) images exhibited the uniform and homogeneous distribution of the nanographene in the composites as shown in Fig. 4.3.

However, another study (Sheshmani, Ashori, & Fashapoyeh, 2013) has revealed that the increased loading of nanographene (3%–5%) creates agglomeration and blocks the transferred stresses. The same study also revealed that the water uptake percentage

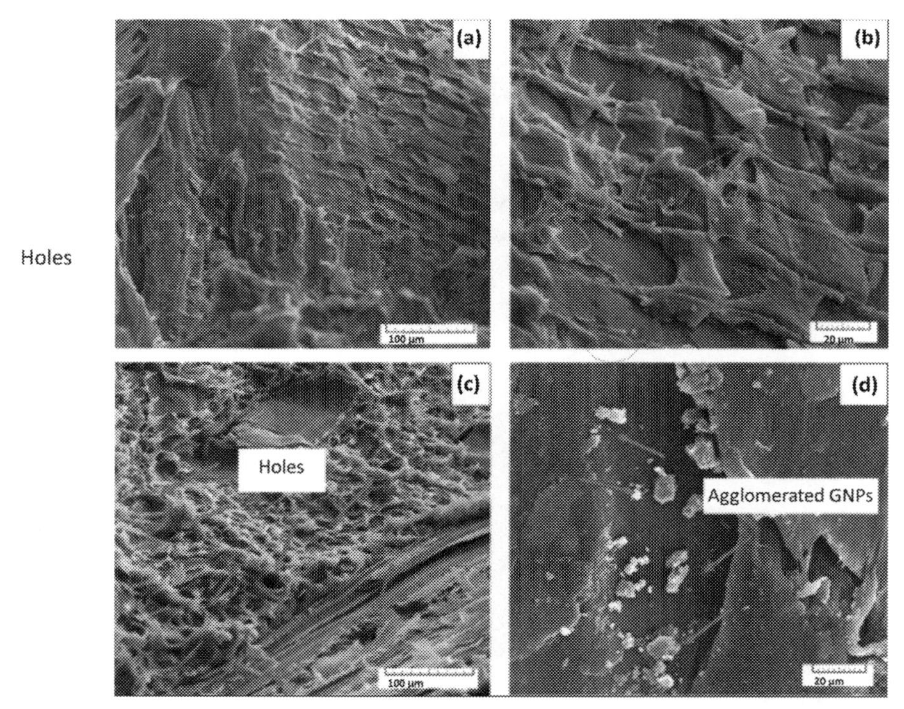

Figure 4.3 Scanning electron microscopic images of control and graphene nanoparticle-loaded nanobiocomposites: (a) control (without graphene loading), (b) 2 wt.% graphene loading, (c) 3 wt.% graphene loading, and (d) 5 wt.% graphene loading (Sheshmani et al., 2013). *Reproduced with permission from Elsevier (Sheshmani, S., A. Ashori, and M.A. Fashapoyeh, Wood plastic composite using graphene nanoplatelets. International Journal of Biological Macromolecules, 2013. 58:1–6). Copyright Elsevier, 2013.*

was decreased by 35% and thickness swelling by 30% with the increased loading of graphene nanoparticles (Sheshmani et al., 2013). The SEM photographs also represent the loading of nano-treated composites as illustrated in Fig. 4.3A, where (A) is the control sample SEM image, whereas Fig. 4.3B is 1% graphene incorporated sample. There are few traces observed when the filler particles were pulling-out. The fibers are strongly bonded with the matrix polymer, hence these are not separated. However, the holes shown in Fig. 4.3C indicate that the fibers were not strongly bonded, and the composite is comparatively weak maybe due to usage 5% graphene nanoparticle. On the other hand, excessive use of nanofillers (8%) creates agglomeration, as shown in Fig. 4.3D, which is a common characteristic of nanoparticles that are used for nanobiocomposites productions. Thus the voids were generated between the polymer matrix and fibers.

4.3.2 Nanotitanium dioxide

TiO_2 exhibits outstanding features in enhancing the performances of polymer composites. Masoudifar et al. (2018) have reported that the pretreatment of cellulosic fibers through using mercerization, acrylization, permanganate, peroxide, silane treatment, and so on, could improve the interface between the wood-based materials and TiO_2 nanopolymer matrix. This study also has revealed that the incorporation of TiO_2 in the wood saw dust-based composites could enhance tensile, impact, and flexural strengths but upto a certain level of loading (3 phc). If the nano-TiO_2 loading is further increased above certain level, the performance characteristics start to decline. The metal TiO_2 is highly efficient nanofiller in providing resistivity against corrosion, higher stiffness, chemical neutrality, thermal conductivity, and afterall for the low-cost features. Awang, Mohd, and Sarifuddin (2019) also claimed that the incorporation of TiO_2 in polypropylene-reinforced rice husk composites could significantly improve the thermal conductivity and mechanical characteristics.

Deka and Maji (2011) have manufactured a wood polymer nanocomposite where they have used TiO_2 and nanoclay as nanofillers. They have loaded various concentrations of nanoclay and TiO_2 ranging from 1 to 5 phr as illustrated in Fig. 4.4, where black dots indicated the TiO_2 and dark lines represented the silicate layers from nanoclay. The agglomeration of nanoparticles was observed when the higher TiO_2 (5 phr) was loaded in the composites. However, an even dispersion was seen at lower loading ranging from 1 to 3 phr.

4.3.3 Nanosilicon dioxide

Nourbakhsh, Baghlani, and Ashori (2011) have studied different weight levels (0%—4%) of nano-SiO_2 loading on 55%—58wt.% lignocellulosic fibers (grounded with beech bark and rice husk) where the polypropylene-grafted malic anhydride was used as coupling

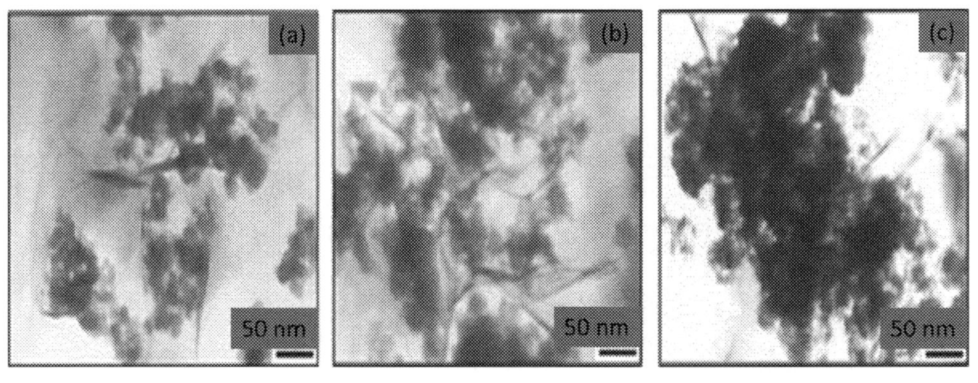

Figure 4.4 Transmission electron microscopic images of (A) 1 phr titanium dioxide (TiO_2)-loaded nanocomposite; (B) 3 phr TiO_2-loaded nanocomposite; and (C) 5 phr TiO_2-loaded nanocomposite (Deka & Maji, 2011). *Reproduced with permission from Elsevier (Deka, B.K. and T.K. Maji, Effect of TiO_2 and nanoclay on the properties of wood polymer nanocomposite. Composites Part A: Applied Science and Manufacturing, 2011. 42(12):2117–2125). Copyright Elsevier, 2011.*

agent. The same study has found that the elongation at break and flexural properties increase with the increase of nanofiller loading although the tensile and impact strengths decrease (Nourbakhsh et al., 2011). But the composites provide satisfactory mechanical performances, whereas the composites reinforced with rice husks provide higher strengths compared to beech bark. Another significant finding was that the water absorption and thickness swelling also increase with the increase of nanofiller loading (Nourbakhsh et al., 2011).

Salari et al. (2013) have reported that nano-SiO_2 has a significant impact on the oriented strand board made from pawlonia wood and urea formaldehyde (UF) resin. The loading of a small amount of nano-SiO_2 (3 phc) into UF resin enhances the performance of composite panels. This study also revealed that the formaldehyde emission is significantly reduced by using nano-SiO_2 incorporation along with an increase in water resistance and mechanical properties. The X-ray diffraction (XRD) test for different ratios of nanosilica loading (1%, 3%, and 5%) analysis showed that the performance of the composites increases with the addition of nanosilica, as the nanoparticles are uniformly dispersed in the matrix system (Salari et al., 2013). The sharp peaks in the XRD patterns reflect the presence of amorphous structure, whereas the narrow peaks reflect crystallization structure (Fig. 4.5).

4.3.4 Nano ZnO_2

Ye et al. (2016) have developed a nanocomposite from wood fiber, polypropylene, and ZnO nanoparticles with significant improvement in thermal stability, water resistance, and mechanical properties. They also reported that the incorporation of nanofiller in composite has increased 90% tensile strength and flexural properties (Ye et al., 2016).

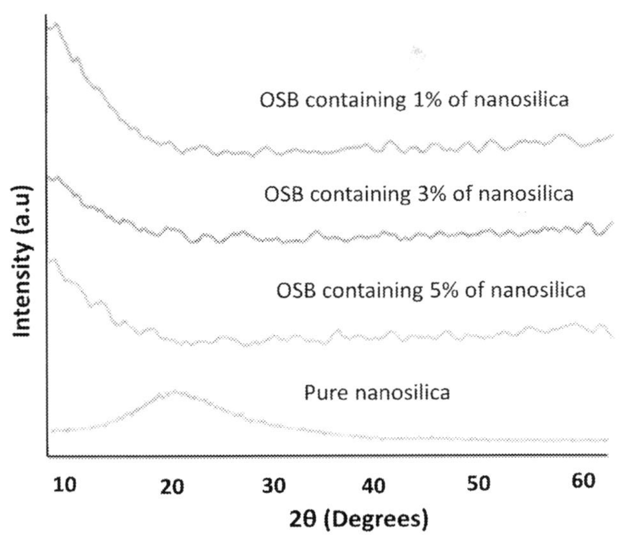

Figure 4.5 X-ray diffraction pattern of oriented strand board (OSB) for different proportions of nanosilica (Salari et al., 2013). *Reproduced with permission from Elsevier (Salari, A., et al., Improving some of applied properties of oriented strand board (OSB) made from underutilized low quality paulownia (Paulownia fortunie) wood employing nano-SiO$_2$. Industrial Crops and Products, 2013. 42:1–9). Copyright Elsevier, 2013.*

The thermal stability of the composites was also enhanced when the nanoparticle was loaded as shown in Fig. 4.6. The nanocomposites exhibit two types of decomposition, initially at 207°C for wood materials and second at 342°C for polypropylene, as shown in Fig. 4.6. It is also seen from the thermogravimetry curve that nano assembled composites provide better thermal stability compared to the pure composites (without nano assemblage). However, the loading of ZnO nanoparticles (upto a certain level—3%) provide better resistance against thermal degradation of polymer composites and after that it starts to decline again (5% ZnO-loaded composites). On the other hand, all the nano-reinforced composites are showing better performance than the control one.

Dang et al. (2017) have developed nanocomposite from wood fiber, polyethylene, and ZnO nanoparticles with a significant improvement for flexural strength (58 MPa), modulus of elasticity (9656 MPa), internal bonding (0.88 MPa), and dimensional stability with 9% thickness swelling. They have also reported that more incorporation of nanofiller in composite gradually increases storage modulus (Dang et al., 2017). The UV radiation energy is one of the significant agents responsible for the damage of composite materials surface in the outdoor environment due to mechanical and chemical changes. The color of the composites becomes faded and mechanical properties decrease in weathering conditions. To overcome such problems, Rasouli et al. (2016) have reported an investigation through producing composites from HDPE, wood

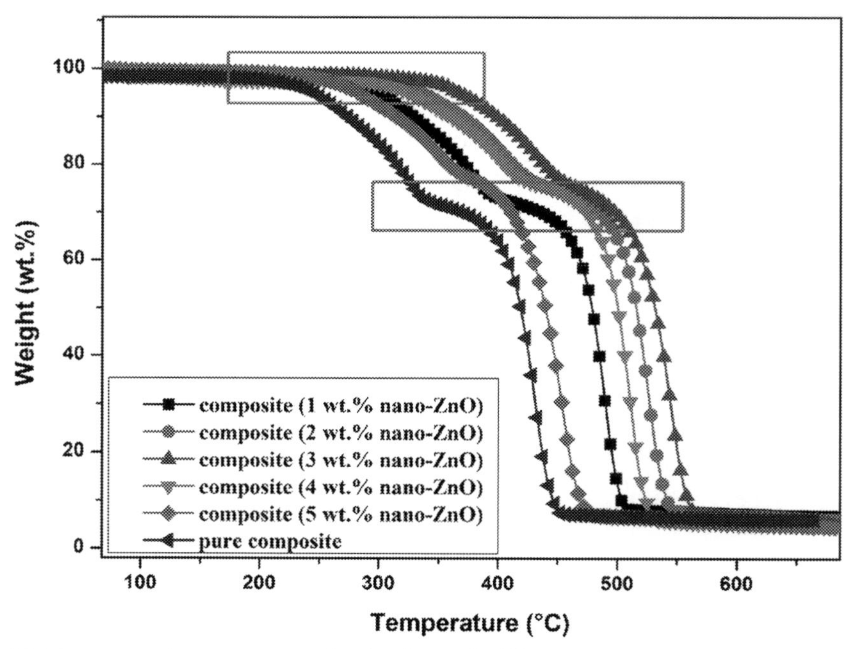

Figure 4.6 Thermogravimetry curves for loading of different percentages of nano-zinc oxide (nano-ZnO) (Ye et al., 2016). *Reproduced with permission from Elsevier (Ye, X., et al., The interface designing and reinforced features of wood fiber/polypropylene composites: Wood fiber adopting nano-zinc-oxide-coating via ion assembly. Composites Science and Technology, 2016. 124:1–9). Copyright Elsevier, 2016.*

flour (beech), and nano-ZnO where they found less tensile strength loss and less cracking with a decreased degradation in weathering conditions (Rasouli et al., 2016). Besides, ZnO also function as a very good UV observer.

4.3.5 Carbon nanotube

Carbon nanotubes have superior mechanical and electronic property; hence, provides 10–100 times more strength compared to steel. Besides, CNT does not exhibit fragile properties and is highly resistant to stress (Ashori, Sheshmani, & Farhani, 2013; Chen, Ozisik, & Schadler, 2010). So, carbon nanomaterial-based composites are getting interesting for researchers and manufacturers in terms of attaining higher tensile strength, thermal stability, higher modulus, and reduction in water absorption. In this regard, multiwalled CNTs of various weight percentages (0, 1, 2, and 3) were used to compound with polypropylene and pine flours through a twin-screw extruder for producing biocomposites (Kaymakci et al., 2019). The wood-to-polypropylene mass ratio was 50/50 (wt./wt.) and samples were prepared by an injection molding machine. They have found that the increased incorporation of multiwalled CNT also increases

Table 4.3 Flexural properties of control and carbon nanotube reinforced composites (Jin & Matuana, 2010).

Type of cap layer	Flexural strength (MPa)	Flexural modulus (MPa)
Control (none)	44.24 ± 2.31	4518 ± 224
Rigid poly(vinyl chloride) (unfilled)	54.75 ± 3.03	3842 ± 236
5% carbon nanotube filled poly(vinyl chloride) (rigid)	65.04 ± 0.89	4915 ± 237

the surface roughness of the composites; however, the wettability of the samples decreases (Kaymakci et al., 2019).

Jin and Matuana (2010) have investigated the effects of CNT on wood plastic composites and found that incorporation of 5% CNT reinforces poly(vinyl chloride) on the core and cap layers of composites could enhance the flexural properties significantly in contrast to the control composites (without nano-loading), as illustrated in Table 4.3. Both the flexural strength (65.04 ± 0.89 MPa) and modulus (4915 ± 237 MPa) for CNT-loaded composites are greater than the rest two composites.

Sadare, Daramola, and Afolabi (2020) have investigated the effects of incorporating CNT with soy-based adhesives to produce nanocomposites. They claimed that the shear strength of the nanocomposite increased to 6.91 MPa (incorporated with 0.3% CNT) from 3.48 MPa (without adding any CNT). The same study also found that the thermal stability of the composites enhances with the addition of CNTs.

4.4 W@W-based nanocomposites

A huge quantity of wood-based materials is used industrially for diversified applications (constructions, furniture, packaging, interior decoration, exterior decorations, automotive, and so on) throughout the world which produce a significant amount of wood-based wastes that appeals for effective and sustainable recycling (Brostow et al., 2016). Besides, there is a significant demand for alternative wood-based compatible substitutes to meet the constant demands of reduction in prices. In this regard, recycled W@W could be the most prominent selection. As virgin wood and wood fibers have limited availability with increased cost, recycled W@Ws are getting popular among the manufacturers and consumers for sustainable applications. On the other hand, W@W and waste papers contribute a major portion of solid wastes for urban disposals (Brostow et al., 2016).

The processing technologies of wood polymer composites play a significant role in the ultimate composite panels. A probable flow process of nanocomposite is shown in

Fig. 4.7 through using hot-press method. The composites production from wood-based materials is categorized into two main ways: (1) compounding and (2) forming. A screw extruder is used to blend the polymers with wood materials for uniform mixing through high temperatures to soften the polymers. This is a challenging approach as it requires a longer time to process the blending if the fiber dimension is high. However, the hydroscopic and hydrophilic nature of the wood materials also makes it difficult to be dispersed uniformly. But such problems could be overcome by controlling the processing temperatures within 180°C.

Wood materials as well as W@W contain 40%–50% cellulose content, whereas other constituents are hemicellulose, pectin, lignin, and little bit impurities (Chen et al., 2020; Dufresne, 2013; Isogai, 2013). So, a hydrogen crosslinking is formed in the polymer composites. The stiffness, strength, toughness, and thermal conductivity of nanobiocomposites become relatively higher than the traditional biocomposites (Hasan et al., 2020c; Khan, Madhu, & Sailaja, 2017; Masoudifar et al., 2018). The superior performance of nanocomposites has made them feasible to be used in automotive, packaging, paper, and biomedical applications. Zikeli et al. (2019) have produced lignin nanoparticles from W@Ws (Iroko sawdust and Norway spruce) for protecting wood and some other materials from weathering effects. Another report has shown that ZnO nanoparticles (up to 5%) reinforced with byproducts of wood sawmill significantly increase the mechanical strength (impact and flexural) of the composites (Abdel-Rahman, Awad, & Fathy, 2020).

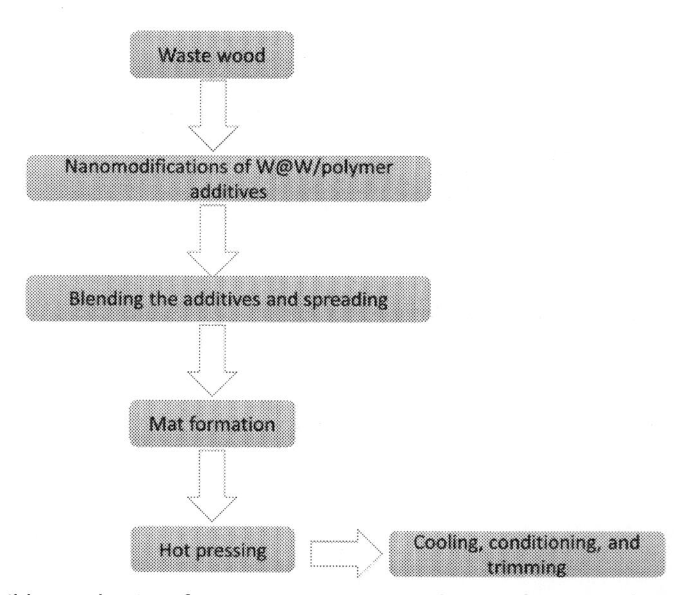

Figure 4.7 Possible mechanism for nanocomposites production from wood wastes (W@W) and nanofillers.

Wood-based forestry wastes are also another important renewable source of biodegradable materials that could be utilized for higher strength functionalized (flame-retardant) products. Wood auto hydrolysates are hygroscopic in nature, so provide less strength, which could be improved by producing nanocomposites through applying suitable nanomaterials. In this regard, a flame-retardant nanocomposite was reported from wood autohydrolysates (containing cellulose and lignin) through incorporating montmorillonite and rGO (Chen et al., 2018). Although hemicellulose-oriented films are hydrophilic in nature, their reported nanocomposite was hydrophobic maybe for using the rGO.

4.5 Summary

The recycling of W@W covers significant interests from different points of view: engineering, environment, economy, and social perspective to ensure sustainable productions. Recycled W@W could reduce the volume of consumption for virgin wood materials and minimize the costs of transportation, logging, and disposals (landfilling or incineration). Besides, W@W could be utilized for sustainable design of hybrid and nanoparticles-based products, providing excellent performances in various sectors such as packaging, construction, biomedical, electronics, and environmental monitoring. Nanotechnology-mediated features in these sectors could offer significant advantages through reducing toxicity, enhancing mechanical performances, improving functional properties, and costs through ensuring reliability and efficiency.

To achieve the best performance from the reinforced nanofillers, proper dispersion of the nanoparticles is essential in the matrix. The better dispersion also facilitates enhancing the interfacial bonding between the W@W and polymers in the composites. As the nanofillers contain a high surface volume, the addition of a small amount of nanofillers to the composites improves the mechanical properties significantly. On the other hand, greenly synthesized nanomaterial-based technologies could facilitate the expansion of innovative and sustainable solutions through generating significant positive feedback in life cycle assessment for complex engineered nanoparticle-based products. So, it is expected that the recycled W@W-based technology is going to be very much popular and useful in the coming years.

References

Abdel-Rahman, H. A., Awad, E. H., & Fathy, R. M. (2020). Effect of modified nano zinc oxide on physico-chemical and antimicrobial properties of gamma-irradiated sawdust/epoxy composites. *Journal of Composite Materials*, *54*(3), 331–343.

Agnantopoulou, E., et al. (2012). Development of biodegradable composites based on wood waste flour and thermoplastic starch. *Journal of Applied Polymer Science*, *126*(S1), E273–E281.

Ahmad, Z., Bon, Y., & Abd Wahab, E. (2010). Tensile strength properties of tropical hardwoods in structural size testing. *International Journal of Basic and Applied Sciences*, *10*(03), 1–6.

Ahmed, L., et al. (2018). Application of polymer nanocomposites in the flame retardancy study. *Journal of Loss Prevention in the Process Industries*, *55*, 381–391.

Ahuja, D., Kaushik, A., & Singh, M. (2018). Simultaneous extraction of lignin and cellulose nanofibrils from waste jute bags using one pot pre-treatment. *International Journal of Biological Macromolecules*, *107*, 1294–1301.

Amendola, V., Polizzi, S., & Meneghetti, M. (2006). Laser ablation synthesis of gold nanoparticles in organic solvents. *The Journal of Physical Chemistry B*, *110*(14), 7232–7237.

Amiri, E., Rahmaninia, M., & Khosravani, A. (2019). Effect of chitosan molecular weight on the performance of chitosan-silica nanoparticle system in recycled pulp. *BioResources*, *14*(4), 7687–7701.

Anjana, P., et al. (2019). Antibacterial and electrochemical activities of silver, gold, and palladium nanoparticles dispersed amorphous carbon composites. *Applied Surface Science*, *479*, 96–104.

Ashori, A., Sheshmani, S., & Farhani, F. (2013). Preparation and characterization of bagasse/HDPE composites using multi-walled carbon nanotubes. *Carbohydrate Polymers*, *92*(1), 865–871.

Awang, M., Mohd, W. R. W., & Sarifuddin, N. (2019). Study the effects of an addition of titanium dioxide (TiO_2) on the mechanical and thermal properties of polypropylene-rice husk green composites. *Materials Research Express*, *6*(7), 075311.

Aydin, S., Yardimci, M. Y., & Ramyar, K. (2007). Mechanical properties of four timber species commonly used in Turkey. *Turkish Journal of Engineering and Environmental Sciences*, *31*(1), 19–27.

Baheti, V., Militky, J., & Marsalkova, M. (2013). Mechanical properties of poly lactic acid composite films reinforced with wet milled jute nanofibers. *Polymer Composites*, *34*(12), 2133–2141.

Beigloo, J. G., et al. (2017). Effect of nanographene on physical, mechanical, and thermal properties and morphology of nanocomposite made of recycled high density polyethylene and wood flour. *BioResources*, *12*(1), 1382–1394.

Bendtsen, B. A., & Ethington, R. L. (1975). *Mechanical properties of 23 species of eastern hardwoods* (Vol. 230). Madison, WI: U.S. Department of Agriculture, Forest Service, Forest Products Laboratory. Available from https://www.fpl.fs.fed.us/documnts/fplrn/fplrn230.pdf.

Brostow, W., et al. (2016). Recycled HDPE reinforced with sol–gel silica modified wood sawdust. *European Polymer Journal*, *76*, 28–39.

Burnley, S., Phillips, R., & Coleman, T. (2012). Carbon and life cycle implications of thermal recovery from the organic fractions of municipal waste. *The International Journal of Life Cycle Assessment*, *17*(8), 1015–1027.

Candan, Z., & Akbulut, T. (2013). Developing environmentally friendly wood composite panels by nanotechnology. *BioResources*, *8*(3), 3590–3598.

Cao, C., & Matsuda, T. (2016). Biocatalysis in organic solvents, supercritical fluids and ionic liquids. *Organic synthesis using biocatalysis* (pp. 67–97). Elsevier.

Castro, D. O., et al. (2018). The use of a pilot-scale continuous paper process for fire retardant cellulose-kaolinite nanocomposites. *Composites Science and Technology*, *162*, 215–224.

Chen, C., et al. (2020). Highly strong and flexible composite hydrogel reinforced by aligned wood cellulose skeleton via alkali treatment for muscle-like sensors. *Chemical Engineering Journal*, 125876.

Chen, G.-G., et al. (2018). Fabrication of strong nanocomposite films with renewable forestry waste/montmorillonite/reduction of graphene oxide for fire retardant. *Chemical Engineering Journal*, *337*, 436–445.

Chen, L., Ozisik, R., & Schadler, L. S. (2010). The influence of carbon nanotube aspect ratio on the foam morphology of MWNT/PMMA nanocomposite foams. *Polymer*, *51*(11), 2368–2375.

Clemons, C. (2008). Raw materials for wood–polymer composites. *Wood–polymer composites* (pp. 1–22). Elsevier.

da Silva, C. G., Kano, F. S., & Rosa, D. S. (2019). Lignocellulosic nanofiber from eucalyptus waste by a green process and their influence in bionanocomposites. *Waste and Biomass Valorization*, 1–14.

Dang, B., et al. (2017). Fabrication of a nano-ZnO/polyethylene/wood-fiber composite with enhanced microwave absorption and photocatalytic activity via a facile hot-press method. *Materials*, *10*(11), 1267.

Danish, M., & Ahmad, T. (2018). A review on utilization of wood biomass as a sustainable precursor for activated carbon production and application. *Renewable and Sustainable Energy Reviews*, *87*, 1–21.

Deka, B. K., & Maji, T. K. (2011). Effect of TiO_2 and nanoclay on the properties of wood polymer nanocomposite. *Composites Part A: Applied Science and Manufacturing*, *42*(12), 2117–2125.

Deka, B. K., & Maji, T. K. (2012). Effect of silica nanopowder on the properties of wood flour/polymer composite. *Polymer Engineering & Science*, *52*(7), 1516–1523.

Deveci, I., et al. (2018). Effect of SiO_2 and AL_2O_3 nanoparticles treatment on thermal behavior of oriental beech wood. *Wood Research*, *63*(4), 573–582.

Devi, R. R., & Maji, T. K. (2013). Interfacial effect of surface modified TiO_2 and SiO_2 nanoparticles reinforcement in the properties of wood polymer clay nanocomposites. *Journal of the Taiwan Institute of Chemical Engineers*, *44*(3), 505–514.

Dobircau, L., et al. (2009). Wheat flour thermoplastic matrix reinforced by waste cotton fibre: Agro-green-composites. *Composites Part A: Applied Science and Manufacturing*, *40*(4), 329–334.

Dong, Y., et al. (2017). In-situ chemosynthesis of ZnO nanoparticles to endow wood with antibacterial and UV-resistance properties. *Journal of Materials Science & Technology*, *33*(3), 266–270.

dos Santos, R. M., et al. (2013). Cellulose nanocrystals from pineapple leaf, a new approach for the reuse of this agro-waste. *Industrial Crops and Products*, *50*, 707–714.

Drexler, K. E. (1981). Molecular engineering: An approach to the development of general capabilities for molecular manipulation. *Proceedings of the National Academy of Sciences*, *78*(9), 5275–5278.

Dufresne, A. (2013). Nanocellulose: A new ageless bionanomaterial. *Materials Today*, *16*(6), 220–227.

Ferdosian, F., et al. (2017). Bio-based adhesives and evaluation for wood composites application. *Polymers*, *9*(2), 70.

Ferreira, F., et al. (2018). Isolation and surface modification of cellulose nanocrystals from sugarcane bagasse waste: From a micro-to a nano-scale view. *Applied Surface Science*, *436*, 1113–1122.

Ghorbani, M., Biparva, P., & Hosseinzadeh, S. (2018). Effect of colloidal silica nanoparticles extracted from agricultural waste on physical, mechanical and antifungal properties of wood polymer composite. *European Journal of Wood and Wood Products*, *76*(2), 749–757.

Hasan, K. F., Brahmia, F. Z., Bak, M., Horváth, P. G., Markó, G., Dénes, L., & Alpár, T. (2020). Effects of cement on lignocellulosic fibres. *9TH HARDWOOD PROCEEDINGS* (pp. 99–103). Sopron, Hungary: University of Sopron Press.

Hasan, K. F., Horváth, P. G., Alpár, T., et al.T. Alpár. (2021). Development of lignocellulosic fiber reinforced cement composite panels using semi-dry technology. *Cellulose*, *28*, 3631–3645.

Hasan, K., et al. (2019). A novel coloration of polyester fabric through green silver nanoparticles (G-AgNPs@ PET). *Nanomaterials*, *9*(4), 569.

Hasan, K., Horváth, P. G., & Alpár, T. (2020c). Potential natural fiber polymeric nanobiocomposites: A review. *Polymers*, *12*(5), 1072.

Hasan, K. F., et al. (2020a). Coloration of aramid fabric via in-situ biosynthesis of silver nanoparticles with enhanced antibacterial effect. *Inorganic Chemistry Communications*, 108115.

Hasan, K. F., et al. (2020b). Wool functionalization through AgNPs: Coloration, antibacterial, and wastewater treatment. *Surface Innovations*, *9*(1), 1–10.

Hasan, K. F., Horváth, P. G., Bak, M., Mucsi, Z. M., & Alpár, T. (2021). Rice straw and energy reeds fiber reinforced phenol formaldehyde resin hybrid polymeric composite panels. *Cellulose*, *28*, 7859–7875.

Hasan, K. F., Horváth, P. G., Horváth, A., & Alpár, T. (2021). Coloration of woven glass fabric using biosynthesized silver nanoparticles from Fraxinus excelsior tree flower. *Inorganic Chemistry Communications*, *126*, 1–7. Available from https://doi.org/10.1016/j.inoche.2021.108477.

Hasan, K. F., Horváth, P. G., Kóczán, Z., & Alpár, T. (2021). Thermo-mechanical properties of pretreated coir fiber and fibrous chips reinforced multilayered composites. *Scientific Reports*, *11*(1), 1–13.

Hasan, K. F., Horváth, P. G., Kóczán, Z., Bak, M., Horváth, A., & Alpár, T. (2021). Coloration of flax woven fabric using *Taxus baccata* heartwood extract mediated nanosilver. *Colouration Technology*, *00*, 1–11. Available from https://doi.org/10.1111/cote.12578.

Hasan, K. F., Horváth, P. G., Miklos, B., & Alpár, T. (2021). Hemp/glass woven fabric reinforced laminated nanocomposites via in-situ synthesized silver nanoparticles from *Tilia cordata* leaf extract. *Composite Interfaces*, In press.

Hasan, K. M. F., Horváth, P. G., & Alpár, T. (2021). Lignocellulosic Fiber Cement Compatibility: A State of the Art Review. *Journal of Natural Fibers*. Available from https://doi.org/10.1080/15440478.2021.1875380.

Hasan, K. M. F., Péter, G. H., K., Zsolt, & L.A., Tibor (2021). Design and fabrication technology in biocomposite manufacturing. *value-added biocomposites: technology, innovation and opportunity* (1st, pp. 158–188). Boca Raton, USA: CRC Press.

Hazarika, A., & Maji, T. K. (2014). Properties of softwood polymer composites impregnated with nanoparticles and melamine formaldehyde furfuryl alcohol copolymer. *Polymer Engineering & Science*, *54*(5), 1019–1029.

Hospodarova, V., et al. (2018). Investigation of waste paper cellulosic fibers utilization into cement based building materials. *Buildings*, *8*(3), 43.

Hosseini, S. B. (2020). Natural fiber polymer nanocomposites. *Fiber-reinforced nanocomposites: Fundamentals and applications* (pp. 279–299). Elsevier.

Hosseinpourpia, R., et al. (2012). Production of waste bio-fiber cement-based composites reinforced with nano-SiO2 particles as a substitute for asbestos cement composites. *Construction and Building Materials*, *31*, 105–111.

Inshakova, E., & Inshakova, A. (2020). Nanomaterials and nanotechnology: prospects for technological re-equipment in the power engineering industry. *IOP Conference Series: Materials Science and Engineering*, *709*, 033020.

Isogai, A. (2013). Wood nanocelluloses: fundamentals and applications as new bio-based nanomaterials. *Journal of wood science*, *59*(6), 449–459.

Jain, N. K., & Gupta, M. (2018). Hybrid teak/sal wood flour reinforced composites: Mechanical, thermal and water absorption properties. *Materials Research Express*, *5*(12), 125306.

Jia, X., et al. (2019). A highly sensitive gas sensor employing biomorphic SnO_2 with multi-level tubes/pores structure: Bio-templated from waste of flax. *RSC Advances*, *9*(35), 19993–20001.

Jiang, J., et al. (2020). Improved mechanical properties and hydrophobicity on wood flour reinforced composites: Incorporation of silica/montmorillonite nanoparticles in polymers. *Polymer Composites*, *41* (3), 1090–1099.

Jin, S., & Matuana, L. M. (2010). Wood/plastic composites co-extruded with multi-walled carbon nanotube-filled rigid poly (vinyl chloride) cap layer. *Polymer International*, *59*(5), 648–657.

John, M. J., & Anandjiwala, R. D. (2008). Recent developments in chemical modification and characterization of natural fiber-reinforced composites. *Polymer Composites*, *29*(2), 187–207.

John, M. J., & Thomas, S. (2008). Biofibres and biocomposites. *Carbohydrate Polymers*, *71*(3), 343–364.

Kajaks, J., et al. (2015). Some exploitation properties of wood plastic hybrid composites based on polypropylene and plywood production waste. *Open Engineering*, *1*. (open-issue).

Kale, B. M., et al. (2016). Coating of cellulose-TiO_2 nanoparticles on cotton fabric for durable photocatalytic self-cleaning and stiffness. *Carbohydrate Polymers*, *150*, 107–113.

Kaymakci, A., Birinci, E., & Ayrilmis, N. (2019). Surface characteristics of wood polypropylene nanocomposites reinforced with multi-walled carbon nanotubes. *Composites Part B: Engineering*, *157*, 43–46.

Khan, M. A., & Ali, K. I. (1992). Studies of physico–mechanical properties of wood and wood plastic composite (WPC). *Journal of Applied Polymer Science*, *45*(1), 167–172.

Khan, M. A., Madhu, G., & Sailaja, R. (2017). Reinforcement of polymethyl methacrylate with silane-treated zinc oxide nanoparticles: fire retardancy, electrical and mechanical properties. *Iranian Polymer Journal*, *26*(10), 765–773.

Khan, M. Z. R., Srivastava, S. K., & Gupta, M. (2020). A state-of-the-art review on particulate wood polymer composites: Processing, properties and applications. *Polymer Testing*, 106721.

Kim, M. H., & Song, H. B. (2014). Analysis of the global warming potential for wood waste recycling systems. *Journal of Cleaner Production*, *69*, 199–207.

Kim, T.-W., et al. (2011). Effect of silane coupling on the fundamental properties of wood flour reinforced polypropylene composites. *Journal of Composite Materials*, *45*(15), 1595–1605.

Kretschmann, D. (2010). Mechanical properties of wood. *Wood handbook: Wood as an engineering material* (Vol. 190, pp. 5.1–5.46). Madison, WI: U. S. Department of Agriculture, Forest Service, Forest Products Laboratory. Available from https://www.fpl.fs.fed.us/documnts/fplgtr/fpl_gtr190.pdf.

Kumar, A., et al. (2013). Influence of aluminum oxide nanoparticles on the physical and mechanical properties of wood composites. *BioResources*, *8*(4), 6231–6241.

Lin, X., et al. (2011). Platinum nanoparticles using wood nanomaterials: Eco-friendly synthesis, shape control and catalytic activity for p-nitrophenol reduction. *Green Chemistry*, *13*(2), 283–287.

Liou, T.-H., & Wang, P.-Y. (2020). Utilization of rice husk wastes in synthesis of graphene oxide-based carbonaceous nanocomposites. *Waste Management*, *108*, 51–61.

Lukmandaru, G. (2015). Chemical characteristics of teak wood attacked by *Neotermes tectonae*. *BioResources*, *10*(2), 2094–2102.

Mahmud, S., et al. (2019). In situ synthesis of green AgNPs on ramie fabric with functional and catalytic properties. *Emerging Materials Research*, *8*(4), 1–11.

Mahmud, S., Hasan, K. F., Jahid, M. A., Mohiuddin, K., Zhang, R., & Zhu, J. (2021). Comprehensive review on plant fiber-reinforced polymeric biocomposites. *Journal of Materials Science*, *56*, 7231–7264.

Masoudifar, M., Nosrati, B., & Mohebbi Gargari, R. (2018). Effect of surface treatment and titanium dioxide nanoparticles on the mechanical and morphological properties of wood flour/polypropylene nanocomposites. *International Wood Products Journal*, *9*(4), 176–185.

Mirzaei, H., & Darroudi, M. (2017). Zinc oxide nanoparticles: Biological synthesis and biomedical applications. *Ceramics International*, *43*(1), 907–914.

Moreno, D. D. P., & Saron, C. (2017). Low-density polyethylene waste/recycled wood composites. *Composite Structures*, *176*, 1152–1157.

Nourbakhsh, A., & Ashori, A. (2010). Wood plastic composites from agro-waste materials: Analysis of mechanical properties. *Bioresource Technology*, *101*(7), 2525–2528.

Nourbakhsh, A., Baghlani, F. F., & Ashori, A. (2011). Nano-SiO_2 filled rice husk/polypropylene composites: Physico-mechanical properties. *Industrial Crops and Products*, *33*(1), 183–187.

Okoroigwe, E. C., Enibe, S., & Onyegegbu, S. (2016). Determination of oxidation characteristics and decomposition kinetics of some Nigerian biomass. *Journal of Energy in Southern Africa*, *27*(3), 39–49.

Onishchenko, D., et al. (2013). Promising nanocomposite materials based on renewable plant resources. *Metallurgist*, *56*(9–10), 679–683.

Pacheco-Torgal, F., & Jalali, S. (2011). Nanotechnology: Advantages and drawbacks in the field of construction and building materials. *Construction and Building Materials*, *25*(2), 582–590.

Partanen, A., & Carus, M. (2016). Wood and natural fiber composites current trend in consumer goods and automotive parts. *Reinforced Plastics*, *60*(3), 170–173.

Pielichowski, K., & Majka, T. M. (2018). *Polymer composites with functionalized nanoparticles: Synthesis, properties, and applications*. Woodhead Publishing.

Ramirez, A., et al. (2017). Production and characterization of activated carbon from wood wastes. *Journal of Physics: Conference Series*, *935*(1), 012012.

Rasouli, D., et al. (2016). Effect of nano zinc oxide as UV stabilizer on the weathering performance of wood-polyethylene composite. *Polymer Degradation and Stability*, *133*, 85–91.

Richter, F., & Rein, G. (2020). Reduced chemical kinetics for microscale pyrolysis of softwood and hardwood. *Bioresource Technology*, *301*, 122619.

Robles, E., et al. (2016). Lignocellulosic-based multilayer self-bonded composites with modified cellulose nanoparticles. *Composites Part B: Engineering*, *106*, 300–307.

Rout, S. K., Tripathy, B. C., & Ray, P. K. (2018). Significance of nano-silver coating on the thermal behavior of parent and modified agro-waste coir fibers. *Journal of Thermal Analysis and Calorimetry*, *131*(2), 1423–1436.

Saba, N., et al. (2015). Preparation and characterization of fire retardant nano-filler from oil palm empty fruit bunch fibers. *BioResources*, *10*(3), 4530–4543.

Saba, N., et al. (2016). Dynamic mechanical properties of oil palm nano filler/kenaf/epoxy hybrid nanocomposites. *Construction and Building Materials*, *124*, 133–138.

Saba, N., et al. (2019). Oil palm waste based hybrid nanocomposites: Fire performance and structural analysis. *Journal of Building Engineering*, *25*, 100829.

Sabazoodkhiz, R., Rahmaninia, M., & Ramezani, O. (2017). Interaction of chitosan biopolymer with silica nanoparticles as a novel retention/drainage and reinforcement aid in recycled cellulosic fibers. *Cellulose*, *24*(8), 3433–3444.

Sadare, O. O., Daramola, M. O., & Afolabi, A. S. (2020). Synthesis and performance evaluation of nano-composite soy protein isolate/carbon nanotube (SPI/CNTs) adhesive for wood applications. *International Journal of Adhesion and Adhesives*, 102605.

Salari, A., et al. (2013). Improving some of applied properties of oriented strand board (OSB) made from underutilized low quality paulownia (*Paulownia fortunie*) wood employing nano-SiO_2. *Industrial Crops and Products*, *42*, 1–9.

Savastano, H., Jr, Warden, P., & Coutts, R. (2000). Brazilian waste fibres as reinforcement for cement-based composites. *Cement and Concrete Composites*, *22*(5), 379–384.

Saxena, M., & Gupta, M. (2019). Mechanical, thermal, and water absorption properties of hybrid wood composites. *Proceedings of the Institution of Mechanical Engineers, Part L: Journal of Materials: Design and Applications*, *233*(9), 1914–1922.

Shamaila, S., et al. (2016). Advancements in nanoparticle fabrication by hazard free eco-friendly green routes. *Applied Materials Today*, *5*, 150–199.

Shekar, H. S., & Ramachandra, M. (2018). Green composites: a review. *Materials Today: Proceedings*, *5*(1), 2518–2526.

Sheshmani, S., Ashori, A., & Fashapoyeh, M. A. (2013). Wood plastic composite using graphene nano-platelets. *International Journal of Biological Macromolecules*, *58*, 1–6.

Siddiqui, M., et al. (2020). Synthesis of novel magnetic carbon nano-composite from waste biomass: A comparative study of industrially adoptable hydro/solvothermal co-precipitation route. *Journal of Environmental Chemical Engineering*, *8*(2), 103519.

Sommerhuber, P. F., et al. (2017). Life cycle assessment of wood-plastic composites: Analysing alternative materials and identifying an environmental sound end-of-life option. *Resources, Conservation and Recycling*, *117*, 235–248.

Sormunen, P., & Kärki, T. (2019). Recycled construction and demolition waste as a possible source of materials for composite manufacturing. *Journal of Building Engineering*, *24*, 100742.

Thambiraj, S., & Shankaran, D. R. (2017). Preparation and physicochemical characterization of cellulose nanocrystals from industrial waste cotton. *Applied Surface Science*, *412*, 405–416.

Tibor, L. A., Péter, G. H., & Hasan, K. M. F. (2021). Introduction to biomass and biocomposites. *Value-added biocomposites: technology, innovation and opportunity* (1st, pp. 1–33). Boca Raton, USA: CRC Press.

Tsuji, T., et al. (2017). Morphological changes from spherical silver nanoparticles to cubes after laser irradiation in acetone–water solutions via spontaneous atom transportation process. *Colloids and Surfaces A: Physicochemical and Engineering Aspects*, *529*, 33–37.

Verma, C., Ebenso, E. E., & Quraishi, M. (2019). Transition metal nanoparticles in ionic liquids: Synthesis and stabilization. *Journal of Molecular Liquids*, *276*, 826–849.

Visakh, P., et al. (2012). Crosslinked natural rubber nanocomposites reinforced with cellulose whiskers isolated from bamboo waste: Processing and mechanical/thermal properties. *Composites Part A: Applied Science and Manufacturing*, *43*(4), 735–741.

Wang, Z., et al. (2011). Bonding strength and water resistance of starch-based wood adhesive improved by silica nanoparticles. *Carbohydrate Polymers*, *86*(1), 72–76.

Wisky Silva, D., et al. (2016). Cementitious composites reinforced with kraft pulping waste. *Key Engineering Materials* (668, pp. 390–398).

Ye, X., et al. (2016). The interface designing and reinforced features of wood fiber/polypropylene composites: Wood fiber adopting nano-zinc-oxide-coating via ion assembly. *Composites Science and Technology*, *124*, 1–9.

Zhang, L., et al. (2020). Transparent wood composites fabricated by impregnation of epoxy resin and w-doped VO_2 nanoparticles for application in energy-saving windows. *ACS Applied Materials & Interfaces*.

Zhou, L., & Fu, Y. (2020). Flame-retardant wood composites based on immobilizing with chitosan/sodium phytate/nano-TiO_2-ZnO coatings via layer-by-layer self-assembly. *Coatings*, *10*(3), 296.

Zikeli, F., et al. (2019). Preparation of lignin nanoparticles from wood waste for wood surface treatment. *Nanomaterials*, *9*(2), 281.

CHAPTER 5

Synthesis and characterization of biodegradable cellulose-based polymer hydrogel

Kiplangat Rop[1], George N. Karuku[2] and Damaris Mbui[1]
[1]Department of Chemistry, University of Nairobi, Nairobi, Kenya
[2]Department of Land Resources and Agricultural Technology, University of Nairobi, Nairobi, Kenya

5.1 Introduction

Polymer hydrogels (PHGs) are hydrophilic materials with a three-dimensional network structure. They have capacity to undergo substantial swelling/shrinking in the presence/absence of water and can maintain the swollen state even under pressure (Ahmed, 2015; Baki & Abedi-Koupai, 2018; Hashem, Sharaf, Abd El-Hady, & Hebeish, 2013; Laftah & Hashim, 2014; Laftah, Hashim, & Ibrahim, 2011; Milani, França, Balieiro, & Faez, 2017). The ability to absorb and retain water is attributed to the hydrophilic groups such as -OH, -CONH-, -$CONH_2$, and -SO_3H attached to polymeric chains, while resistance to dissolution is attributed to chemical and physical cross-links between the chains (Ahmed, 2015; Laftah & Hashim, 2014; Laftah et al., 2011). PHG with high absorption capacity, in the order of 10 to 1000 times its own dry weight in water free of ions, is called a superabsorbent polymer (Ahmed, 2015; Ekebafe, Ogbeifun, & Okieimen, 2011; Laftah & Hashim, 2014; Milani et al., 2017; Yu, Liu, Kong, Zhang, & Liu, 2011). These characteristics enable extensive applications of PHG in the food industry as thickening agents, biomedicine as artificial organs, agriculture as soil conditioners, pharmaceuticals in controlled drug delivery, and hygiene as sanitary towels and baby diapers, among others (Baki & Abedi-Koupai, 2018; Milani et al., 2017; Tally & Atassi, 2015; Xiao-Ning, Wen-Bo, & Ai-Qin, 2011; Yu et al., 2011).

Based on the source, PHGs are often classified as either synthetic or natural hydrogels. Synthetic PHGs such as polyacrylate and polyacrylamide (PAM) have excellent water absorbency, long shelf life, and high gel strength (Ahmed, 2015; Qinyuan, Yang, & Xinjun, 2017; Yu et al., 2011). However, due to their nonbiodegradability, high cost of production, and toxicity in the environment, their use in agriculture and other consumer products is limited (Laftah et al., 2011; Yu et al., 2011). For instance, linear and crosslinked PAM is used as a soil conditioner to improve moisture retention

Nanotechnology in Paper and Wood Engineering
DOI: https://doi.org/10.1016/B978-0-323-85835-9.00004-0

and to stabilize soil aggregates in order to minimize crust formation and soil erosion, though its degradation rate in soil is $<10\%$ per year (Hussien et al., 2012). Ecofriendly and sustainable bio-based products are therefore being investigated due to increased environmental consciousness globally. Natural-based PHGs have been developed by amalgamating synthetic and natural substrates where natural material sources include polysaccharides and polypeptides (Ahmed, 2015; Qinyuan et al., 2017; Zohuriaan-Mehr & Kabiri, 2008). Polysaccharides such as pectin, starch, cellulose, chitosan, gum Arabic, among others, have attracted much attention. Cellulose is rated as the best potential candidate due to its exceptional properties such as biodegradability, good mechanical strength, and linear structure of its macromolecule which enable synthesis of PHGs with reinforced networks (Jianzhong, Xiaolu, & Yan, 2015; Qiu & Hu, 2013). Other application properties of cellulose include biocompatibility, hydrophilicity, relative thermo-stabilization, and alterable optical appearance (Jianzhong et al., 2015; Qiu & Hu, 2013; Xiao-Ning et al., 2011).

Nevertheless, it is difficult to make cellulose into other useful materials due to its poor solubility in common solvents such as water, poor dimensional stability, and lack of thermoplasticity (Ambjörnsson, 2013; Roy, Semsarilar, Guthrie, & Perrier, 2009). Therefore the physicochemical properties of cellulose are often modified to achieve the desired properties such as water absorbency, mechanical strength, and hydrophobicity. The major types of approaches in cellulose fiber modification include physical treatment, physicochemical treatment, and chemical modification. Cellulose is an active biopolymer due to the presence of three -OH groups in each of its D-anhydroglucopyranose units, and chemical modifications can therefore be performed on them. The primary -OH at C-6 and two secondary ones at C-2 and C-3 can participate in classical reactions such as esterification, etherification, and oxidation (Ambjörnsson, 2013; Qiu & Hu, 2013; Roy et al., 2009; Sannino, Demitri, & Madaghiele, 2009). Cellulose derivatives have been obtained by reacting some (or all) -OH groups of cellulose repeat unit with an organic species such as methyl and ethyl units to generate water-soluble derivatives. The most commonly produced cellulose derivatives are cellulose ethers such as ethyl, carboxymethyl, hydroxyethyl, and hydroxypropyl cellulose, and cellulose esters such as cellulose acetate and cellulose acetate propionate (Ambjörnsson, 2013; Carlmark, Larsson, & Malmström, 2012; Qiu & Hu, 2013; Sannino et al., 2009). Synthesis of graft copolymers is another way in which physicochemical properties of cellulose can be modified to obtain a composite material with improved mechanical and biodegradable characteristics (Roy et al., 2009; Sannino et al., 2009). Vinyl monomers are often grafted onto cellulose or its derivatives by free radical polymerization due to the technique's low sensitivity to impurities and moderate reaction conditions, as opposed to other techniques such as ring opening polymerization and "living" or controlled radical polymerization (Carlmark et al., 2012; Jianzhong et al., 2015; Rop, 2019; Roy et al., 2009). The free radicals are generated using initiating techniques such as high energy irradiation (γ-ray or electron beams) and redox initiation with Mn or Ce (iv) ion, among others (Ahmed, 2015; Laftah et al., 2011).

The chemical modification of cellulose can be carried out under both homogeneous and heterogeneous reaction mixtures, though reactions are usually performed under heterogeneous mixtures due to poor solubility of native cellulose in common solvents (Carlmark et al., 2012; Qiu & Hu, 2013; Roy et al., 2009). The modifications under heterogeneous mixtures are performed after allowing native cellulose such as cotton and wood fibers to swell in a suitable solvent. Cellulose under this condition can be in the form of cellulose nanocrystals, films or membranes, fibers and particle suspensions; hence, reactions only occur at the surface layer and the gross structure of cellulose can be largely maintained (Ambjörnsson, 2013; Qiu & Hu, 2013). Cellulose modification under homogeneous reaction mixtures can be achieved by dissolving it in nonderivatizing solvents such as N,N-dimethyl acetamide/LiCl, derivatizing solvents such as dinitrogen tetroxide/dimethylformamide (N_2O_4/DMF), or by dissolving its derivatives in appropriate solvents, using substituents as starting, protecting, or leaving groups for consecutive reactions (Ambjörnsson, 2013; Qiu & Hu, 2013) The original supramolecular structure is destroyed under this condition, though the reaction of cellulose chains is higher than that under heterogeneous condition (Gürdağ & Sarmad, 2013; Qiu & Hu, 2013). However, solvents such as N_2O_4/DMF may be toxic and N,N-dimethyl acetamide/LiCl is difficult to remove; hence, modification in this condition is often started with cellulose derivatives (Ambjörnsson, 2013; Carlmark et al., 2012; Qiu & Hu, 2013).

The cellulose derivatives commonly used to synthesize cellulose-based hydrogels by radical polymerization technique include carboxymethyl cellulose (CMC) (Sadeghi, Safari, Shahsavari, Sadeghi, & Soleimani, 2013) and sodium CMC (Demitri, Scalera, Madaghiele, Sannino, & Maffezzoli, 2013; Sadeghi, Soleimani, & Yarahmadp, 2011; Sannino, Mensitieri, & Nicolais, 2004). Polymerization reactions with these derivatives as substrates are performed under homogeneous reaction mixtures because of their solubility in various solvents. The -OH groups in cellulose chains enable grafting of vinyl monomers enhancing hydrophilicity, a feature useful in application prospects such as agriculture where biodegradability and water retention and release are essential. Despite exhibiting admirable properties, the high cost of producing bio-based hydrogels using cellulose derivatives limits their application. Nonetheless, cellulose being the most abundant and renewable biopolymer, can be derived from a wide variety of biomass such as wood, cotton, sisal, flax (Malmström & Carlmark, 2012) as well as water hyacinth (WH or *Eichhornia crassipes*), among others.

The WH is an invasive nonnative species in Kenya that has wreaked havoc in fresh water bodies such as Lakes Victoria and Naivasha, Nairobi dam, and other water masses and wetlands. Due to the hyacinth's high growth rate, it poses a great challenge to manage its invasion and proliferation. The use of herbicides to eliminate this weed is effective, but they are hazardous to the environment. Mechanical and manual harvesting is costly, though the cost of removal is considered to reduce through commercial utilization of WH. It has been locally utilized as compost manure, animal feeds, papermaking and crafts (ropes, baskets, chairs, fiber boards) (Jafari, 2010; Titik et al., 2015), wastewater treatment (Zhu, Zayed, Qian, De Souza, & Terry, 1999), production of biogas (Kivaisi & Mtila, 1997), ethanol (Aswathy et al., 2010), and charcoal briquettes (Rezania et al., 2016). The utilization of WH cellulose in the production of PHGs has not been explored. Activation treatments such as swelling (in acids, bases, etc.) and mechanical pulping, of lignocellulosic material, open the surface cannulae, internal pores, and cavities, disrupt fibrillar aggregations and crystalline order, and break inter- and intra-H bonds, enhancing accessibility and reactivity of -OH groups in a heterogeneous chemical reaction (Gupta, Uniyal, & Naithani, 2013; Pönni, 2014; Roy et al., 2009).

Water absorbency and retention properties of PGHs have attracted much interest in agriculture as slow-release fertilizer (SRF) carriers and soil conditioners (Milani et al., 2017). About 40%–70% N, 80%–90% P, and 50%–70% K of applied conventional fertilizers are lost to the environment through soil erosion and surface runoff, leaching and volatilization causing serious economic losses and environmental pollution (Baki & Abedi-Koupai, 2018; Giroto, Guimarães, Foschini, & Ribeiro, 2017). These losses can be managed by encapsulating the nutrients within a biodegradable PHG as delivery vehicle to synchronize nutrient release with plant uptake subsequently reducing negative environmental impact. PHGs can also improve soil moisture retention capacity thus reducing cost of irrigating crops, particularly in arid and semiarid lands (ASALs). High cost of producing PHGs however limits their application in various fields including agriculture, and efforts have been made to overcome this challenge by developing polymer nanocomposites using inorganic fillers (clay minerals) such as kaolinite (Hussien et al., 2012), montmorillonite (Bortolin, Aouada, Mattoso, & Ribeiro, 2013; Marandi, Mahdavinia, & Ghafary, 2011), bentonite (Shirsath, Hage, Zhou, Sonawane, & Ashokkumar, 2011), attapulgite (palygorskite) (Deng et al., 2012), and clinoptilolite (Rashidzadeh, Olad, Salari, & Reyhanitabar, 2014). These clay minerals are attractive due to their abundance in nature and other exceptional properties such as high specific surface area, high cation exchange capacity, physical-chemical stability, and high intercalation.

The formulation-enhanced efficiency fertilizers mainly involve charging the copolymer with soluble fertilizers such as urea, $(NH_4)_2HPO_4$ and KH_2PO_4. However, high solubility urea and inorganic salts limit slow-release property when chemical interaction does not exist between the nutrients and the delivery material. Available publications on the SRFs formulation focus mainly on N. Few studies consider elements such as P which is prone to erosion loss and fixation into plant unavailable forms by certain soil types such as the Nitisols (WRB, 2015). Recent studies on SRF formulation using nanohydroxyapatite (nano-HA) demonstrate the potential to enhance P use efficiency. (Liu & Lal, 2014) observed better mobility of CMC-stabilized nano-HA in soil compared to nano-HA without CMC, suggesting a more efficient delivery of P to the root system. This nanofertilizer formulation improved growth and yield of soybeans, above those of $CaHPO_4$ treatment. Montalvo, McLaughlin, & Degryse (2015) evaluated efficiency of nano-HA (20 nm) in strong P-sorbing Andisols and Oxisols, relative to bulk HA (600 nm). They observed greater mobility of nano-HA in Andisol than Oxisol and P could not be recovered in the leachates in both soil types for bulk HA due to large particle size. The P uptake and the percentage of P in wheat (*Triticum aestivum*) derived from the fertilizers in both soil types followed the order; Triple super-phosphate (TSP) > nano-HA > bulk HA. Nano-HA can be incorporated into PHG to form a polymer nanocomposite for slow release of phosphate. The active -OH groups on the surface of nano-HA can react with the monomer forming part of the network besides being physically trapped within the copolymer.

In this work, a biodegradable cellulose-based nanocomposite PHG has been developed by heterogeneously grafting acrylic monomers onto cellulose fibers in the presence of nano-HA. The use of cellulose derived from WH could reduce the cost of production, impart biodegradability, and enhance the swelling and mechanical strength while at the same time helping to root out the weed through commercial use. The formulated product was characterized for potential agricultural application as a SRF and soil conditioner, particularly in ASALs and in problematic soils such as Vertisols where water retention and release is essential for plant growth. The release of P from nanocomposite PHG into the soil is considered to depend on microbial degradation of the copolymer and solubilization of nano-HA, and subsequent diffusion of P.

5.2 Materials and methods

5.2.1 Materials

Cellulose was extracted from WH, acrylic acid, and N,N-methylene-bis-acrylamide were obtained from ACROS Organics, Germany. Ammonium persulphate, acetone, toluene, ammonium hydroxide, hydrochloric acid, sulfuric acid, sodium hypochlorite, sodium chloride, calcium chloride, and aluminum chloride were obtained from Loba Chemie, Mumbai, India.

5.2.2 Sample preparation

Fresh WH plants were collected from Nairobi dam, Kenya and washed with clean water to remove dirt and any other material attached to the surface. They were separated into leaves, stems, and roots, and characterized separately. The plant parts were chopped into small pieces and air-dried in the shade. The particle size of dried material was further reduced (to particle size $\leq 450\ \mu m$) through milling so as to increase its surface area for reaction with chemicals. Samples were stored in air-tight polythene bags ready for use in the experiments.

5.2.3 Characterization of water hyacinth

The characterization method was adopted from Abdel-Halim (2014), Kaco et al. (2014), and Titik et al. (2015). The moisture and ash content of the air-dried WH samples were determined gravimetrically. Hemicellulose was isolated by treating the samples with KOH (10% wt./vol.) and the resulting mixture was precipitated with ethanol. For lignin content determination, the air-dried WH sample was first treated with a toluene/ethanol mixture (2:1 vol./vol.) to extract any compound in the WH other than cellulose, hemicellulose, and lignin. The solvent-extracted samples were treated with 72% vol./vol. H_2SO_4 at low temperature to hydrolyze the cellulose and hemicellulose. The resultant extract was solubilized through heating, and insoluble lignin retained. The cellulose content was determined by subtracting the sum of moisture, ash, hemicellulose, and lignin content from 100%.

5.2.4 Isolation of cellulose from water hyacinth

The cellulose fibers were isolated from WH using the method adopted from Titik et al. (2015). The photographs of WH during the isolation of cellulose are presented in Fig. 5.1. Air-dried WH samples were refluxed in a toluene/ethanol solvent mixture

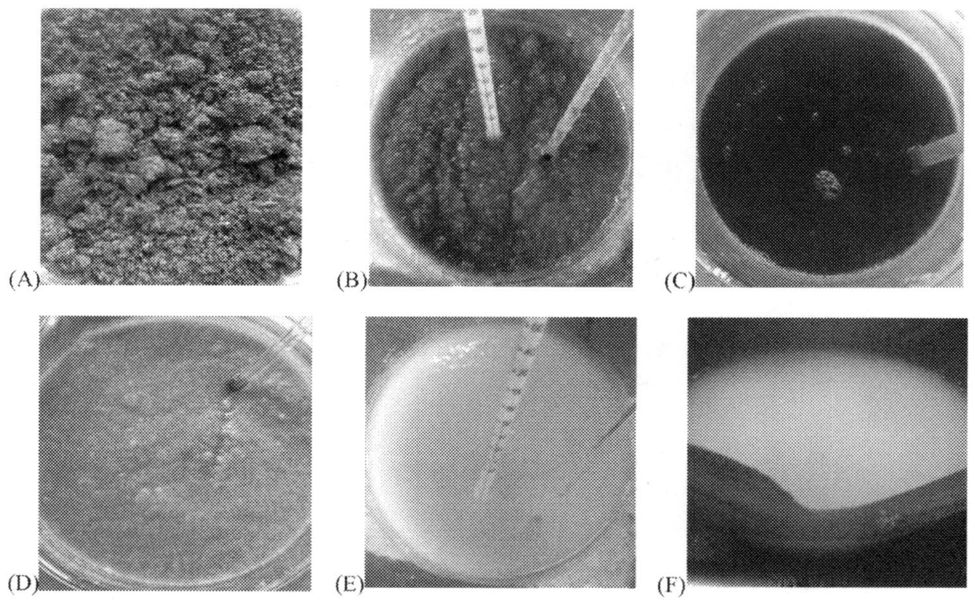

Figure 5.1 Photographs of WH during the isolation of cellulose; (A) solvent–extracted WH, (B) first bleaching with NaOCl, (C) alkaline hydrolysis using NaOH, (D) second bleaching, (E) acid hydrolysis using 1 M HCl, and (F) washing. *WH*, Water hyacinth.

(2:1 vol./vol.) for 3 h and then allowed to cool to room temperature, after which they were filtered and air-dried (Fig. 5.1A). The solvent-extracted samples were bleached to remove lignin using 3% NaOCl in a water bath at 80°C for 2 h (Fig. 5.1B). Hemicellulose was removed by alkaline hydrolysis using 2% NaOH at 60°C for 2 h (Fig. 5.1C). Samples were then bleached for the second time to remove the remaining lignin by heating in 2% NaOCl with stirring at 75°C for 3 h (Fig. 5.1D). The last stage was acid hydrolysis, using 5% HCl as the catalyst at 65°C for 6 h (Fig. 5.1E). The isolated cellulose was washed with water (Fig. 5.1F) and the pH adjusted to 13 by adding 0.1 M NaOH to avoid microbial degradation over time. This alkaline cellulose was neutralized with 0.1 M HCl, washed with distilled water, and then used in subsequent experiments.

5.2.5 Synthesis of water hyacinth cellulose-g-poly(ammonium acrylate-co-acrylic acid) polymer hydrogel

5.2.5.1 Partial neutralization of acrylic acid

The degree of neutralization is defined as molar percentage of carboxylic acid groups that are neutralized by a base, usually NaOH, KOH, or NH_3 (Liu & Lal, 2014). In this study, the carboxyl groups in acrylic acid were partially neutralized using NH_3. For a weak acid, the relationship between the pH and the dissociation equilibrium constant is given by the well-known Henderson−Hasselbach (Eq. 5.1).

$$pH = p \quad K_a + log \quad \left[\frac{\alpha}{1-\alpha}\right] \tag{5.1}$$

where α is the degree of neutralization. From the pK_a value of acrylic acid (4.25 at 25°C) and the desired degree of neutralization, the pH was calculated.

From the value of acrylic acid (4.25 at 25°C) and the desired degree of neutralization, the pH was calculated. The acrylic acid was neutralized using ammonia to a pH value of 4.62 which is equivalent to degree of neutralization of 70% of the carboxyl groups. This is the solution that would be used in the subsequent experiment. The neutralization reaction between NH_3 and acrylic acid is exothermic, thus dropwise addition of NH_3 to acrylic acid was done in a beaker placed in ice-cold water to avoid polymerization of the monomer.

5.2.5.2 Heterogeneous grafting of partially neutralized acrylic acid monomer onto cellulose fibers

The method was adopted from Mohammad and Fatemeh (2012). Around 30 mL of swollen cellulose fibers containing 0.8 g dry weight of cellulose (2.67% wt./vol.) were transferred using a syringe into a three-necked flask. The flask was fitted with a reflux condenser and nitrogen line, and then placed in a thermostated water bath equipped

Figure 5.2 (A) Swollen WH cellulose-g-poly(acrylamide-co-acrylic acid) PHG; (B) soxhlet extraction of homopolymer using acetone. *PHG*, polymer hydrogel; *WH*, water hyacinth.

with a magnetic stirrer. Nitrogen gas was bubbled through the mixture for 10 min, as the temperature was gradually raised to 70°C. Around 0.1 g of ammonium persulphate was added into the mixture and stirred for 30 min so as to generate free radicals. Around 2.7 mL of partially neutralized acrylic acid and 0.25 g N,N-methylene-bis-acrylamide were blended in a beaker, stirred to dissolve and then introduced into the reaction flask. The total volume of the reaction mixture was adjusted to 40 mL by adding distilled water. The mixture was stirred for 1 min after which it was left to stand for 2 h. The PHG formed was allowed to cool to room temperature, removed from the flask, and then cut into small pieces. These pieces were soaked in distilled water and a 1:1 NH_3 solution added dropwise to adjust the pH to 8. They were then washed with water and oven-dried at 60o̲C to a constant weight.

5.2.5.3 Extraction of homopolymer

A 0.5 g dry PHG sample (from Section 5.2.5.2), particle size ≤ 2 mm, was soaked overnight in 100 mL of distilled water in a beaker (Fig. 5.2A). It was then soxhlet extracted with acetone for 2 h to remove the homopolymer and other impurities (Fig. 5.2B). The copolymer obtained was oven-dried at 60°C to a constant weight, pulverized using pestle and mortar, and then stored in air-tight polyethylene bottles for subsequent experiments.

5.2.6 Structural and morphological characterization

5.2.6.1 Fourier transform infrared spectroscopy

A Fourier transform infrared (FTIR) spectrophotometer, Shimadzu IRAffinity-1S, was used to characterize the crude WH, isolated cellulose, and the cellulose-grafted copolymer. The sample holder was cleaned and the background scanned without the sample. The sample was pulverized using pestle and mortar, placed on the sample holder and pressed against the diamond crystal and then scanned between 4000 cm^{-1} and 500 cm^{-1}.

5.2.6.2 Transmission electron microscopy and energy dispersive X-ray analysis

The morphology of copolymers was assessed using high resolution transmission electron microscope, Tecnai G2 F20 X-TWIN MAT instrumentation operating at voltage of 200 kV and fitted with energy dispersive X-ray (EDX) spectrometer.

5.2.6.3 X-ray diffraction analysis

X-ray diffractograms were recorded using D2 PHASER (Bruker) with Cu Kα radiation (1.5418 Å) and a generator working at a voltage of 30 kV and a filament current of 10 mA. The samples were scanned from $2\theta = 8°$ to $40°$ at the rate of $0.090° \text{ min}^{-1}$.

5.2.7 Evaluating the swelling of polymer hydrogel

5.2.7.1 Swelling of polymer hydrogel in water

A 0.1 g powdered PHG sample (250–350 μm) was weighed into a porous bag, immersed in distilled water, and then allowed to attain the swelling equilibrium (SEQ) overnight at room temperature. The surface water was removed by gently dabbing with filter paper and the mass of the swollen PHG determined. The swelling ratio (water absorbency) at equilibrium was determined (Eq. 5.2) (Laftah & Hashim, 2014; Xiao-Ning et al., 2011).

$$SEQ = \left(\frac{M_{eq} - M_{\circ}}{M_{\circ}}\right) \tag{5.2}$$

where M° (g) is the weight of dry PHG, M_{eq} (g) is the weight of swollen PHG and SEQ (g/g) is the swelling ratio at equilibrium.

5.2.7.2 Swelling of polymer hydrogel in salt solution

A 0.2 g powdered PHG sample was weighed into a porous bag and then immersed in different salt concentrations of NaCl, $CaCl_2$, and $AlCl_3$ solutions. It was left overnight at room temperature in order to achieve SEQ. It was then weighed and water absorbency calculated (Eq. 5.2).

5.2.7.3 Influence of the pH on swelling of polymer hydrogel

Several 0.2 g powdered PHG samples were weighed into porous bags and immersed in buffered solutions at pH values of 3, 4, 5, 6, 7, 8, 9, 10, and 11. They were allowed to attain SEQ and then weighed. Potassium dihydrogen phosphate, potassium hydrogen phthalate, and sodium tetraborate were used to prepare the buffer solutions, while HCl and NaOH were used to adjust pH to the desired values.

5.2.7.4 Influence of polymer hydrogel on water holding capacity in soil

Thombare et al. (2018) method was adopted. A soil sample was air-dried in the shade and passed through a 2 mm sieve. Subsamples of air-dried soil weighing 100 g each were transferred into plastic beakers with perforations and filter paper lining at the bottom. The soil subsamples were amended with 0.5, 1.0, and 1.5 g dry PHG (250−350 μm) and mixed to homogenize, while the untreated soil served as the control. The amendments were placed in a water tub and allowed to absorb water through capillarity for 12 h. They were then removed, allowed to drain excess gravitational water, and kept at the same temperature and humidity. The weights of the moistened soil were recorded at intervals of 2 days until there was no noticeable change. Subtracting the weight of empty beaker and filter paper, the water holding capacity (WHC) of soil + treatment was calculated as the following:

$$WHC = \left(\frac{W_{wet} - W_{dry}}{W_{dry}}\right) \times 100 \quad (5.3)$$

where W_{wet} is the weight of wet soil at a particular time interval and W_{dry} is the weight of air-dried soil.

5.2.8 Biodegradation test

5.2.8.1 Biodegradation of polymer hydrogel in soil

Laftah and Hashim (2014) method was adopted, where PHGs samples were buried in the soil for incubation studies to simulate natural degradation conditions. Dry PHG samples (1 g each) ≤2 mm were placed in porous nylon bags that allowed the entry of microorganisms and invertebrates and then buried in normal garden soil moistened to field capacity (30% wt./wt.). Degradation was monitored through unearthing of the samples at 2-week intervals for a period of 14 weeks. The compost materials attached onto the surface of unearthed samples were removed, hand washed with distilled water, and then oven-dried to a constant weight at 60°C. The percentage dry matter remaining at each sampling period was determined:

$$Dry\ matter\ remaining\ (\%) = \left(\frac{W_t}{W_o}\right) \times 100 \quad (5.4)$$

where W_o is the initial weight of the sample and W_t is the weight of the sample at a particular incubation period, *t*.

The decomposition rate constant (*k*) was determined using Olson single exponential model (Eq. 5.5) adopted by Thombare et al. (2018).

$$W_t = W_o\, e^{(-kt)} \quad (5.5)$$

where W_o is the initial weight of the sample, W_t is the weight of the sample at incubation time *t* and *k* is the decomposition rate constant.

The half-life ($t_{1/2}$) was calculated using Eq. (5.6).

$$t_{(1/2)} = \frac{0.693}{k} \tag{5.6}$$

5.2.8.2 Microbial culture and degradation test of the copolymer by soil microbial isolate

The methodology was adopted from Nawaz, Franklin, and Cerniglia (1993) and Nawaz et al (1994). Acrylamide degrading microbes were isolated from the soil with a history of exposure to alachlor (Roundup), in a phosphate-buffered medium (PBM) supplemented with 10 mM acrylamide. Around 5 soil samples were collected from randomly selected points in a coffee farm at the College of Agriculture and Veterinary Science, University of Nairobi, Kenya. Around 5 g of each sample was placed in a 250 mL conical flask containing 50 mL PBM, covered with cotton wool and aluminum foil, and incubated for 10 days at 30°C. Around 2 mL aliquot of microbial suspension was drawn from each of the flasks, transferred to a fresh 50 mL PBM (pH 7.5), and incubated for another 10 days. After five similar transfers, a 10 mL aliquot of each sample was centrifuged at 15,000 $g \times$ 10 min @ 4°C and the supernatant measured colorimetrically for liberated ammonia using indophenol blue method described by Hall (1993). Samples which tested positive for ammonia revealed acrylamide degradation (Nawaz et al., 1993; Nawaz et al. 1994). The pellet (microbial cells) was suspended in 10 mL PBM and after one wash; the suspension was sonicated for 5 min at intervals of 15 s. At the same time, free NH_4^+ ions were leached out of the copolymers using distilled water and oven-dried at 60°C to constant weight. Around 2 mL of microbial suspension was inoculated into a 100 mL PBM supplemented with 0.2–0.6 g of cellulose-grafted copolymer and copolymer without cellulose in separate flasks, then incubated at 30°C. Around 10 mM acrylamide monomer and PBM without the substrate served as the reference and the blank, respectively. A 5 mL aliquot of the suspension was drawn at intervals of 20 h for 12 days, centrifuged, and the supernatant tested for liberated ammonia.

5.2.9 Preparation of nanocomposite polymer hydrogel

5.2.9.1 Synthesis of nanohydroxyapatite

The methodology used in the synthesis was adopted from Kottegoda, Munaweera, Madusanka, and Karunaratne (2011) with some modification by introducing a surfactant. Around 7.716 g $Ca(OH)_2$ was weighed into a beaker and 0.22 mM TX-100 (nonionic surfactant) solution added to make a total volume of 100 mL and the mixture stirred for 30 min using a motorized stirrer. Around 100 mL of 0.6 M H_3PO_4 was added into the suspension of $Ca(OH)_2$ drop-wise (15 mL min^{-1}) from the buret while stirring vigorously at 1000 rpm. After the reaction, the dispersion was stirred for 10 min and then allowed to age for 2 h. It was then oven-dried at 105°C to constant weight and then pulverized into fine powder using pestle and mortar. The surfactant was removed by washing the powder with methanol.

5.2.9.2 Preparation of cellulose-g-poly(ammonium acrylate-co-acrylic acid)/nano-HA composite hydrogel

Around 30 mL of swollen cellulose (2.67% wt./vol.) and 0.25 g nano-HA powder was transferred into a three-necked flask fitted with a reflux condenser and nitrogen. The procedure in Section 5.2.5.2 was then followed using acrylic acid as the monomer, methylene-bis-acrylamide as the cross-linker, and ammonium persulfate as the initiator (Carlmark et al., 2012; Jianzhong et al., 2015; Rop, Karuku, Mbui, Njomo, & Michira, 2019; Rop, Mbui, et al., 2019; Roy et al., 2009). The same procedure was repeated for 0.5, 0.75, 1.0, 1.25, 1.5, 2.5, and 3.0 g of nano-HA. The homopolymer was also extracted following an earlier procedure (Section 5.2.5.3) and the sample oven-dried at 60°C to a constant weight, pulverized using pestle and mortar, and tested for water absorption.

5.2.10 Statistical data analysis

The various data categories were subjected to ANOVA using IBM SPSS Statistics Version 20. Tukey honest significant difference post hoc test was used to compare and assess the significance of the mean values at a probability level of $P \leq .05$.

5.3 Results and discussion

5.3.1 Composition of water hyacinth

The composition of air-dried WH is given in Table 5.1. Cellulose content ranged from 26.1% to 33.3%, with the highest amount having been obtained from the stem, though not significantly different from the amount obtained from the leaves. The moisture content ranged from 9.2% to 9.3%, ash from 13.0% to 24.1%, hemicellulose from 16.1% to 22.0%, and lignin from 20.6% to 23.1%. No significant difference in

Table 5.1 Composition of air-dried water hyacinth (%).

Plant part	Moisture content	Ash	Hemicellulose	Lignin	Cellulose
Leaves	9.3 (0.21)a	13.0 (0.10)a	22.0 (0.45)c	23.1 (0.25)b	31.7 (0.25)bc
Stem	9.3 (0.15)a	20.2 (0.35)c	16.1 (0.40)a	20.8 (0.60)a	33.3 (0.42)c
Roots	9.3 (0.10)a	24.1 (0.45)d	19.7 (0.80)b	20.6 (0.80)a	26.1 (0.49)a
Whole plant	9.2 (0.15)a	18.4 (0.69)b	20.8 (0.40)bc	21.3 (0.26)a	30.5 (1.30)b

Note: Values in the parentheses are standard deviations ($n = 3$), different letters in the same column are significantly different (Tukey test; $P \leq .05$ level)

moisture content was observed amongst the plant parts. Significantly higher ash content ($P \leq .05$) was obtained in the roots, relative to the leaves and stem. Lignin and hemicellulose content was significantly higher ($P \leq .05$) in the leaves, compared to the stem and roots. The results are within the range of values obtained by other workers. For example, Girisuta, Danon, Manurung, Janssen, and Heeres (2008) reported 7.4% moisture, 18.2% ash, 47.7% holocellulose (cellulose + hemicellulose), and 26.7% lignin in leaves. Similarly, Reales-Alfaro, Trujillo-Daza, Arzuaga-Lindado, Castaño-Peláez, and Polo-Córdoba (2013) characterized a 1:1 ratio of leaves to stem and reported 9.3% moisture, 31.7% cellulose, 27.3% hemicellulose, and 3.9% lignin. Saputra, Hapsari, and Pitaloka (2015) obtained values ranging from 8.3% to 9.2% for moisture, 15.2%–19.8% ash, 62.2–64.2% holocellulose, and 7.3–10.0% lignin in the stem. Muchanyerey, Kugara, and Zaranyika (2016) reported 17.4%–18.4% ash and 17%–23% lignin in the roots. The slight variation in composition may be attributed to the different methodologies used and the geographical location (source) of WH.

The isolated cellulose fibers were dispersed in water (Fig. 5.1E and F) due to exposure of hydrophilic -OH groups; however, the dispersion (swelling) was irreversibly lost upon dehydration either by air or oven drying. This irreversible swelling was probably related to the collapse of spaces where hemicellulose and lignin were embedded leading to the restoration of strong inter- and intramolecular H-bonds (Pönni, 2014; Roy et al., 2009). The swelling of cellulose is a vital precondition for rendering the system accessible to the reagents. Bleaching with NaOCl could have decreased the crystallinity of cellulose, thus expanding the amorphous region where most grafting occurs (Swantomo, Rochmadi, Basuki, & Sudiyo, 2013). Alkaline treatment could have also loosened the intermolecular interactions to allow competing interactions with the swelling agent, that is, H_2O, where these interactions are restricted to the amorphous regions and pores (Pönni, 2014; Roy et al., 2009). Consequently, the accessibility and reactivity of cellulose could be improved in heterogeneous chemical reactions (Pönni, 2014).

5.3.2 Mechanism of graft polymerization and extraction of homopolymer

The reaction schemes shown in Fig. 5.3 show a possible mechanistic pathway for the synthesis of WH cellulose-g-poly(ammonium acrylate-co-acrylic acid) PHG. The reaction is thought to initiate through thermal decomposition of ammonium persulphate to generate a sulfate anion radical which then abstracts -H from an alcoholic -OH group in cellulose to form the corresponding macroradical (Fig. 5.3A). Ammonium acrylate and acrylic acid (monomers) then become macroradical receptors. The macroradical initiated monomers in turn donate free radicals to the neighboring molecules (chain propagation) (Fig. 5.3B) and subsequently, the grafting of the monomers onto cellulose chains leads to the formation of the graft copolymer. The chain terminates through combination of two growing polymer chains (Fig. 5.3C) (Sadeghi et al., 2011).

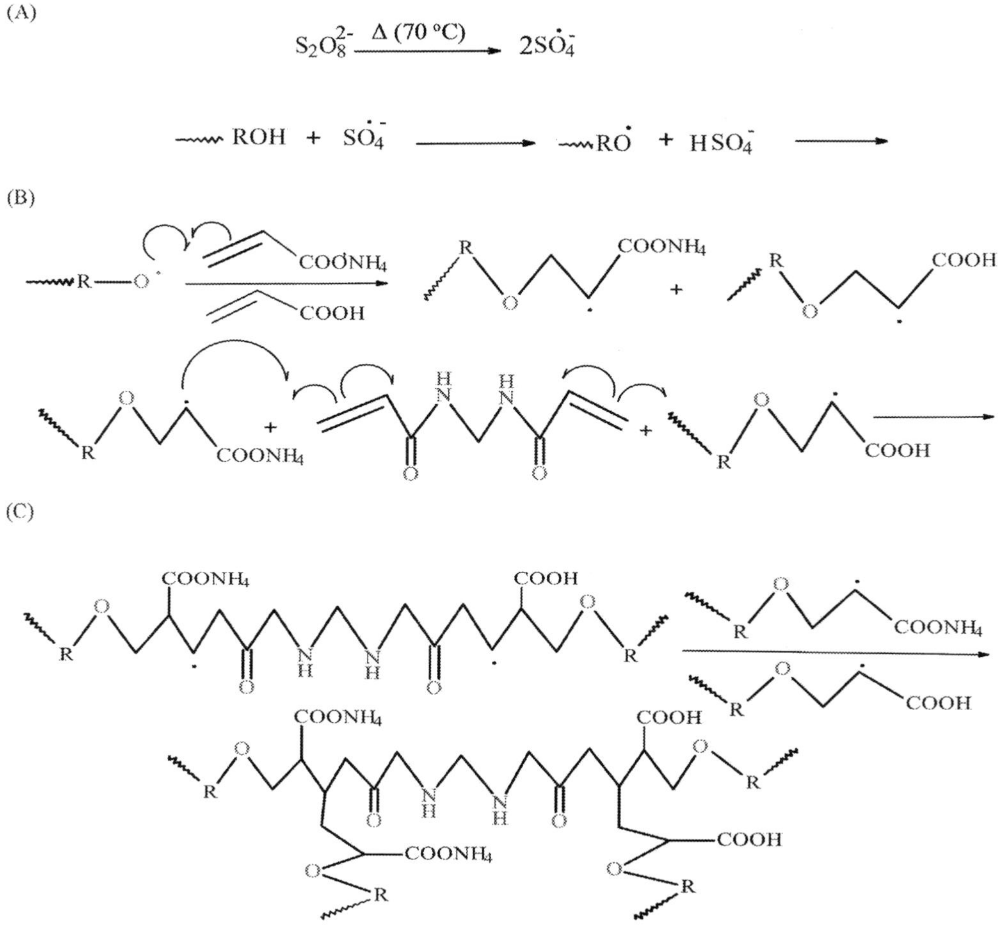

(A)

$$S_2O_8^{2-} \xrightarrow{\Delta\,(70\,^\circ C)} 2SO_4^{\bullet-}$$

$$\sim ROH + SO_4^{\bullet-} \longrightarrow \sim RO^{\bullet} + HSO_4^{-} \longrightarrow$$

(B)

(C)

(R denotes cellulose chain)

Figure 5.3 Proposed mechanistic pathway for the synthesis of water hyacinth cellulose-g-poly (ammonium acrylate-co-acrylic acid) polymer hydrogel: (A) thermal initiation, (B) chain propagation, and (C) termination steps.

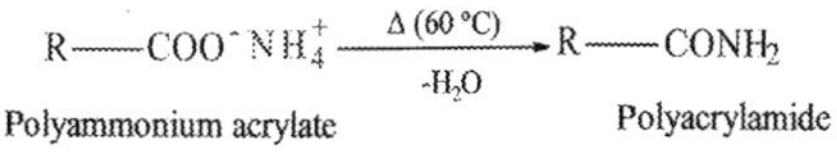

$$\underset{\text{Polyammonium acrylate}}{R{-}COO^{-}NH_4^{+}} \xrightarrow[-H_2O]{\Delta\,(60\,^\circ C)} \underset{\text{Polyacrylamide}}{R{-}CONH_2}$$

Figure 5.4 Dehydration of polyammonium acrylate into polyacrylamide.

The copolymer product was then transformed from cellulose-g-poly(ammonium acrylate-co-acrylic acid) to cellulose-g-poly(acrylamide-co-acrylic acid) (or cellulose-g-PAM-co-AA) upon heating (oven drying), according to Fig. 5.4.

The homopolymer was extracted from the graft copolymer using several solvents including: acetone, water, methanol, ethanol, and an ethanol/acetone mixture. It was found that the use of water alone took long (about 3 days) and alcohols impacted negatively on the hydrophilicity of the PHG material. Dehydration with ethanol or an ethanol/acetone mixture at normal temperatures generated heat, an indication of the formation of an exothermic chemical bond. It was likely that the unneutralized acrylic acid or ammonium acrylate groups in the copolymer network caused the protonation of the -OH group in the alcohol, leading to the formation of an ester according to Fig. 5.5. Acetone was adopted as the solvent of choice as it did not influence the swelling property of the PHG and also shortened the time taken to extract the homopolymer.

Fig. 5.6 shows photographs of oven-dried cellulose-g-PAM-co-AA and acetone-extracted cellulose-g-PAM-co-AA PHGs. The homopolymer was extracted effectively when dry PHG samples were swollen in water to enable the penetration of acetone. The acetone-extracted sample transformed from a tough, black solid material (Fig. 5.6A) to a white sponge-like solid product (Fig. 5.6B), which could be pulverized easily using pestle and mortar into fine powder. The white color observed in Fig. 5.6B may be due to microporous polymeric network structure. This is consistent with observation made by Kabiri and Zohuriaan-Mehr (2004), who induced porosity into PHG composite (kaolin as inorganic component) using $NaHCO_3$, acetone, and a

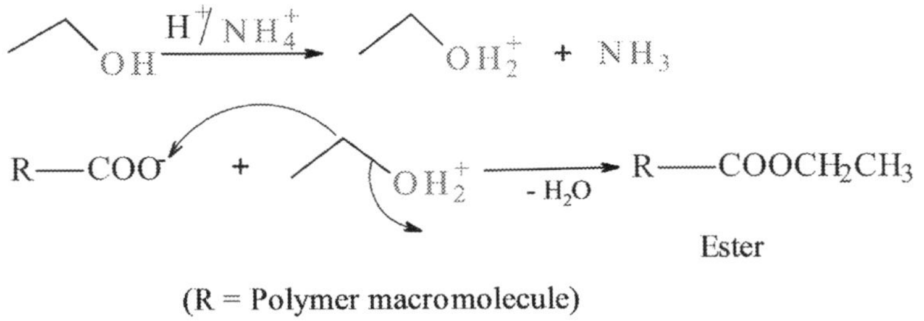

$$CH_3CH_2OH \xrightarrow{H^+/NH_4^+} CH_3CH_2OH_2^+ + NH_3$$

$$R{-}COO^- + CH_3CH_2OH_2^+ \xrightarrow[-H_2O]{} R{-}COOCH_2CH_3$$

Ester

(R = Polymer macromolecule)

Figure 5.5 Protonation of alcohol by unneutralized acrylic acid (or ammonium acrylate) leading to the formation of an ester.

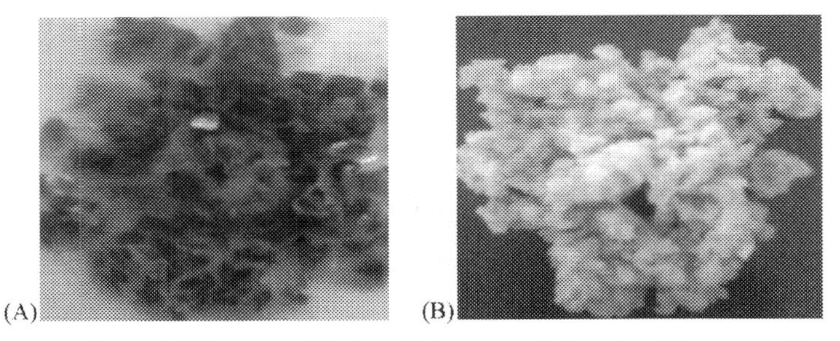

Figure 5.6 Photographs of (A) oven-dried cellulose-g-poly(acrylamide-co-acrylic acid) and (B) acetone-extracted cellulose-g-poly(acrylamide-co-acrylic acid).

combination of both porogens. They obtained a foamy PHG composite with high swelling capacity and swelling rate due to porosity of PHG network. Zhang, Wang, and Wang (2007) obtained higher water absorbency in methanol- and ethanol-dehydrated chitosan-g-poly(acrylic acid)/attapulgite composites compared to acetone-dehydrated sample. They also observed a lower water absorbency of oven-dried PHG composite relative to those of methanol-, ethanol-, and acetone-dehydrated samples. The organic porogens were considered to protect the porous structure easing penetration of water molecules, while oven drying was linked to interaction between polymer chains that collapsed the porous structure, increasing the cross-link density. These workers used NaOH to neutralize acrylic acid, whereas NH_3 was used in this study, implying that protonation of alcohol was most likely due to NH_4^+ ions (Fig. 5.5).

5.3.3 Structural and morphological characteristics of water hyacinth, isolated cellulose, and cellulose-grafted copolymer

5.3.3.1 Fourier transform infrared spectroscopy

The FTIR spectra of WH, isolated cellulose and cellulose-g-PAM-co-AA are displayed in Fig. 5.7. The spectral bands for both crude WH and purified cellulose were similar, except for a strong band at 1608.6 cm^{-1} assigned to the C = C stretch for the aromatic ring, which diminished in isolated cellulose. The delignification possibly removed lignin, a complex polymer derived from p-coumaryl, coniferyl, and sinapyl alcohols coupled by aryl—ether bonds and ether cross-links. The intensity of the spectral bands at 1320—1210 cm^{-1} assigned to C—O stretch for alcohol and ether groups also decreased probably due to removal of lignin and hemicellulose. This indicates that lignin and hemicellulose were effectively removed.

Grafting of the monomer onto cellulose was confirmed by comparing the FTIR spectrum of cellulose with that of cellulose-grafted PHG. The intense spectral band at 1018.4 cm^{-1} (Fig. 5.7A and B) assigned to the C—O stretch for primary or secondary alcohols was drastically weakened and shifted to 1029.9 cm^{-1} in cellulose-g-PAM-co-AA (Fig. 5.7C), implying that most of the -OH groups were involved in grafting. The spectrum of grafted cellulose also showed a broad band between 3700 and 2500 cm^{-1} assigned to N—H stretching and a strong band at 1541 cm^{-1} assigned to N—H bending for primary amides. The moderately strong band at 1697—1647 cm^{-1} is also assigned to C = O axial deformation for the amide group. These spectral bands, characteristic of -$CONH_2$ group, further confirms dehydration of polyammonium acrylate to PAM upon oven drying at 60°C (Fig. 5.4). The broad band extending from 3700 to 2500 cm^{-1} is also assigned to alcoholic and carboxylic -OH. The alcoholic -OH may be attributed to crystalline regions of cellulose that is unlikely to take part in grafting unless disrupted to allow penetration of monomers. The penetration of monomers is restricted to amorphous regions where most grafting occurs, though the reactivity of -OH under heterogeneous conditions may be influenced by steric effects

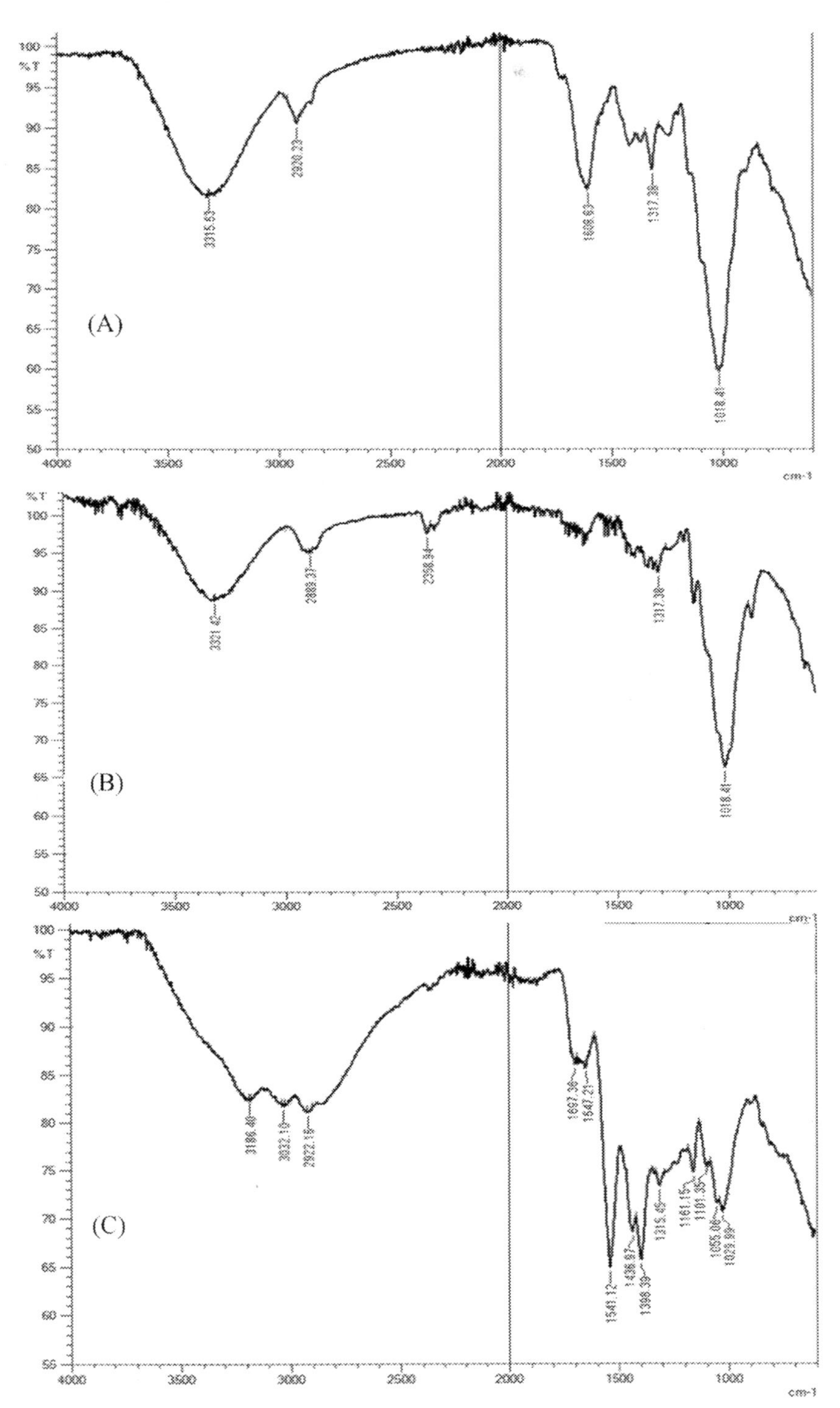

Figure 5.7 FTIR spectra of (A) crude water hyacinth, (B) isolated cellulose, and (C) cellulose-g-poly (acrylamide-co-acrylic acid). *FTIR*, Fourier transform infrared

from the chemical reagent and supramolecular structure of cellulose (Swantomo et al., 2013). The strong spectral band at 1398 cm^{-1} corresponding to carboxylic -OH bending vibration indicated the presence of unneutralized portion of acrylic acid functional groups. Spectral bands at 1315–1100 cm^{-1} assigned to C–O stretch due to ether (also alcoholic and carboxylic groups) and C–O–C bridging due to ether group were more intense relative to that of cellulose (Fig. 5.7B) indicating successful grafting of the monomer onto cellulose.

5.3.3.2 X-ray diffraction analysis

The X-ray diffraction patterns of WH and isolated dry cellulose is presented in Fig. 5.8. The crude WH (Fig. 5.8A) displayed diffraction peaks at $2\theta = 21.6°$ corresponding to (020) crystallographic plane reflections characteristic of cellulose I, and the bands at $2\theta = 26.5°$, 39.5°, and 44° could be related to cellulose Iβ structure. The isolated cellulose (Fig. 5.8B) indicated diffraction peaks at $2\theta = 16°$, 22°, and 34.5° assigned to (110), (200), and (004) crystallographic plane reflections for cellulose I, respectively. The findings are in agreement with those of Ago, Endo, and Hirotsu (2004), Ambjörnsson (2013), Bai, Wang, Zhou, and Zhang (2012), Cheng et al. (2011), Gupta et al. (2013), and Nam, French, Condon, and Concha (2016). The intense peaks observed in isolated (mercerized) dry cellulose (Fig. 5.8B) may be attributed to restructured H-bonds (coalescence) that increased crystallinity of cellulose (Ambjörnsson, 2013; Gupta et al., 2013; Pönni, 2014). The transformation of cellulose I to II has been found to occur at NaOH concentration $> 10\%$ (wt./vol.) (Gupta et al., 2013), and hence, transformation in this study was unlikely because of lower concentration of NaOH (2% wt./vol.) used to isolate cellulose (Section 5.2.4).

The diffractograms of copolymer without cellulose and cellulose-grafted copolymer are displayed in Fig. 5.9. The copolymer without cellulose (Fig. 5.9A) displayed an amorphous pattern. Cellulose-grafted copolymer (Fig. 5.9B) showed a diffraction band of lower intensity at $2\theta = 22°$ (200) for cellulose I crystalline allomorph, relative to the intensity of the same peak in the isolated cellulose (Fig. 5.8B), confirming grafting of monomers onto cellulose fibers. The diffraction band in cellulose-grafted copolymer at $2\theta = 26.5°$ (Fig. 5.9B) is suggestive of the existence of some crystalline cellulose allomorph that was not fully destroyed during alkali treatment, limiting the accessibility of -OH groups during the polymerization reaction.

5.3.3.3 Transmission electron microscopy and energy dispersive X-ray spectroscopy

High resolution transmission electron microscopy (TEM) images of oven-dried cellulose-g-PAM-co-AA and acetone-extracted cellulose-g-PAM-co-AA PHGs are shown in Fig. 5.10. TEM micrographs of the oven-dried PHG showed a smooth dense material (Fig. 5.10A and B), whereas those of acetone-extracted PHG displayed voids and

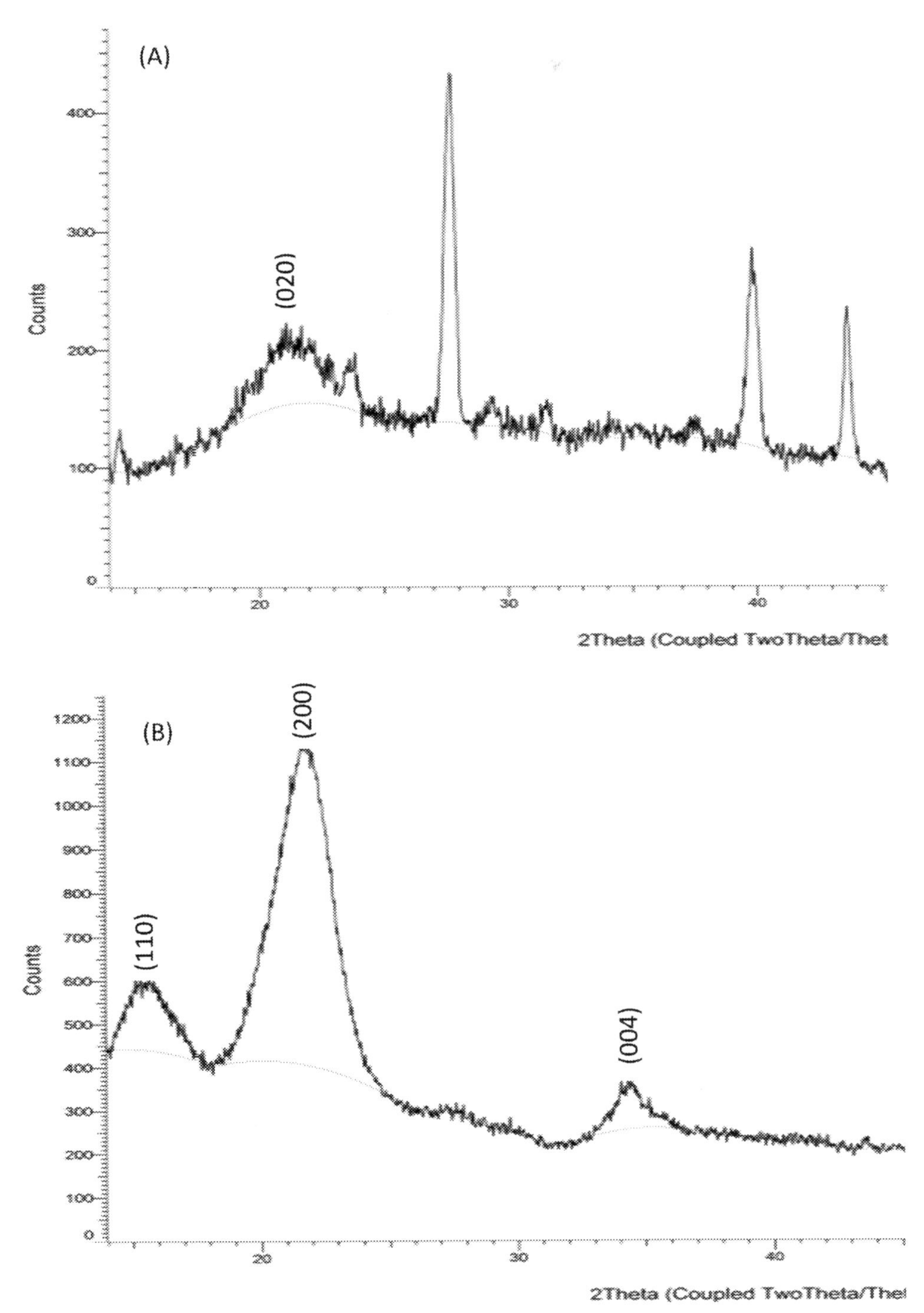

Figure 5.8 Diffractograms of (A) crude water hyacinth and (B) isolated dry cellulose.

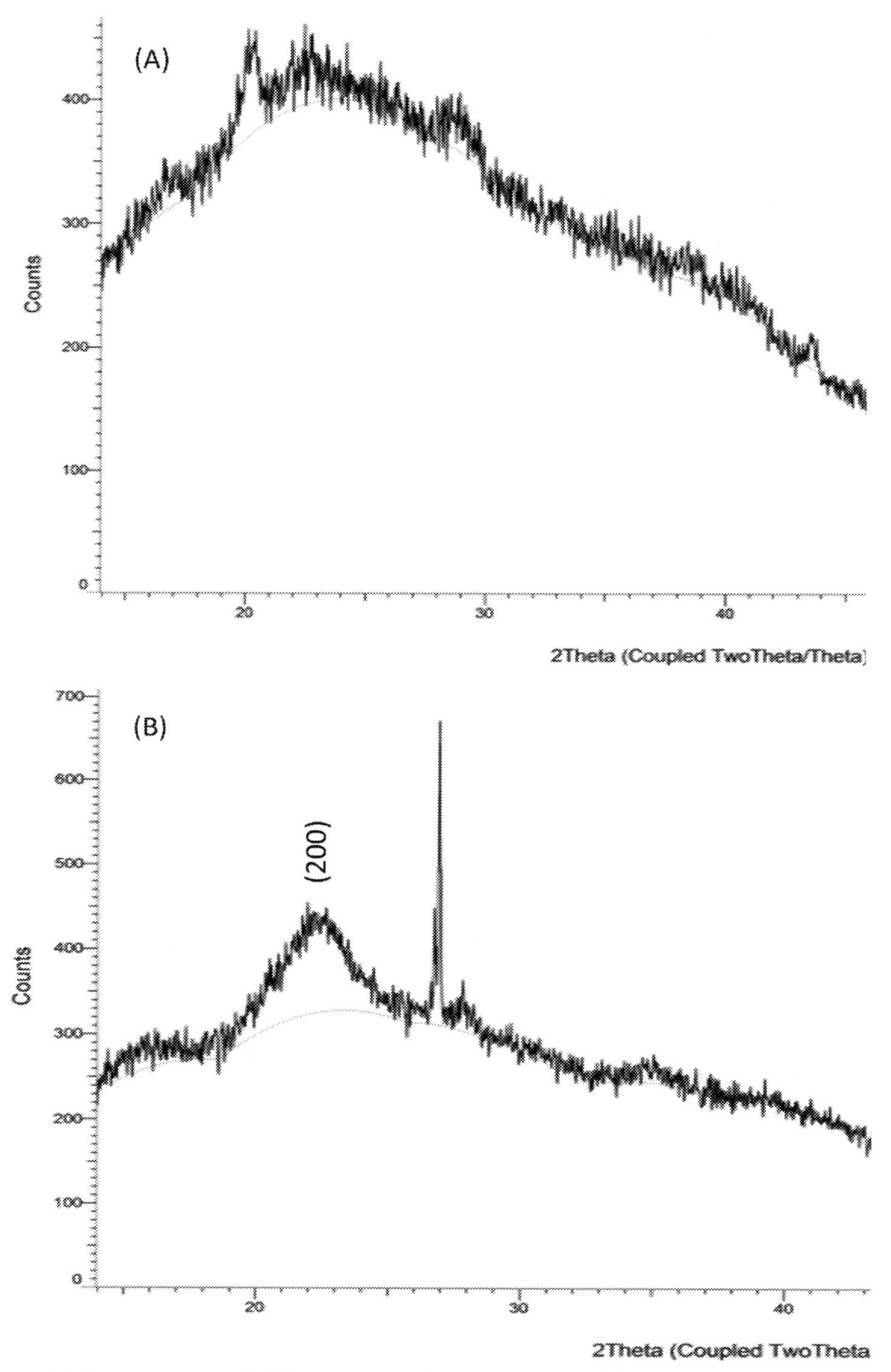

Figure 5.9 Diffractograms of (A) polymer hydrogel without cellulose and (B) cellulose-grafted polymer hydrogel.

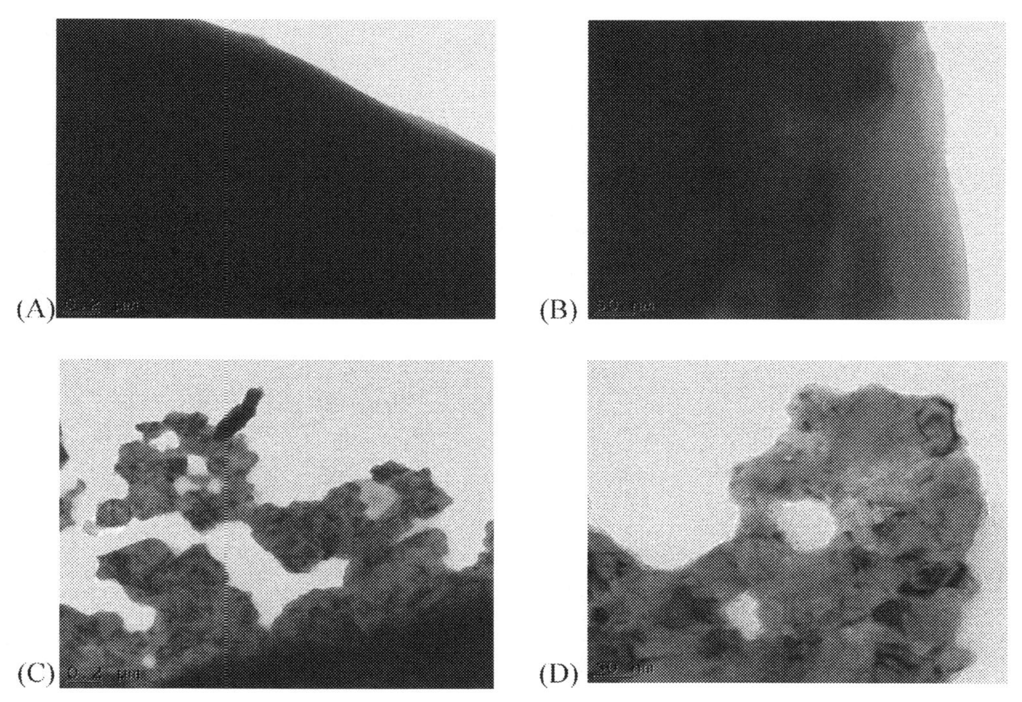

Figure 5.10 TEM micrographs of (A and B) oven-dried cellulose-g-poly(acrylamide-co-acrylic acid), (C and D) acetone-dehydrated cellulose-g-poly(acrylamide-co-acrylic acid), at 0.2 μm and 50 nm scales. *TEM*, Transmission electron microscopy

folding which confirmed the presence of micropores (Fig. 5.10C and D). Acetone-extracted PHG samples were used in water absorption tests due to higher swelling ratio and faster swelling rate compared to oven-dried PHG. This suggested that, heating the samples (at 60°C) enhances cross-link density where unneutralized -COOH and unreacted -OH groups of cellulose could possibly interact through H-bonding and also react forming covalent (ester) bonds. Similar observations were made by Sannino et al. (2004), who used a phase inversion desiccation technique in acetone to dehydrate a PHG based on sodium CMC and hydroxyethyl cellulose cross-linked with divinyl sulfone. This technique induced a microporous structure into the PHG which increased the water absorption and the swelling kinetics due to capillary effects. Simoni et al. (2017) obtained smooth and compact micrographs in air-dried PHG samples that were attributed to water loss through evaporation leading to equilibrium polymeric chain conformation.

The EDX spectra of the cellulose-grafted copolymer is shown in Fig. 5.11. The spectrum of cellulose-grafted copolymer revealed an intense carbon band at 0.2 keV due to cellulose and acrylic chains being the main constituents of the copolymer. Other elements detected were oxygen (0.5 keV) due to cellulose, sulfur (2.3 keV) due

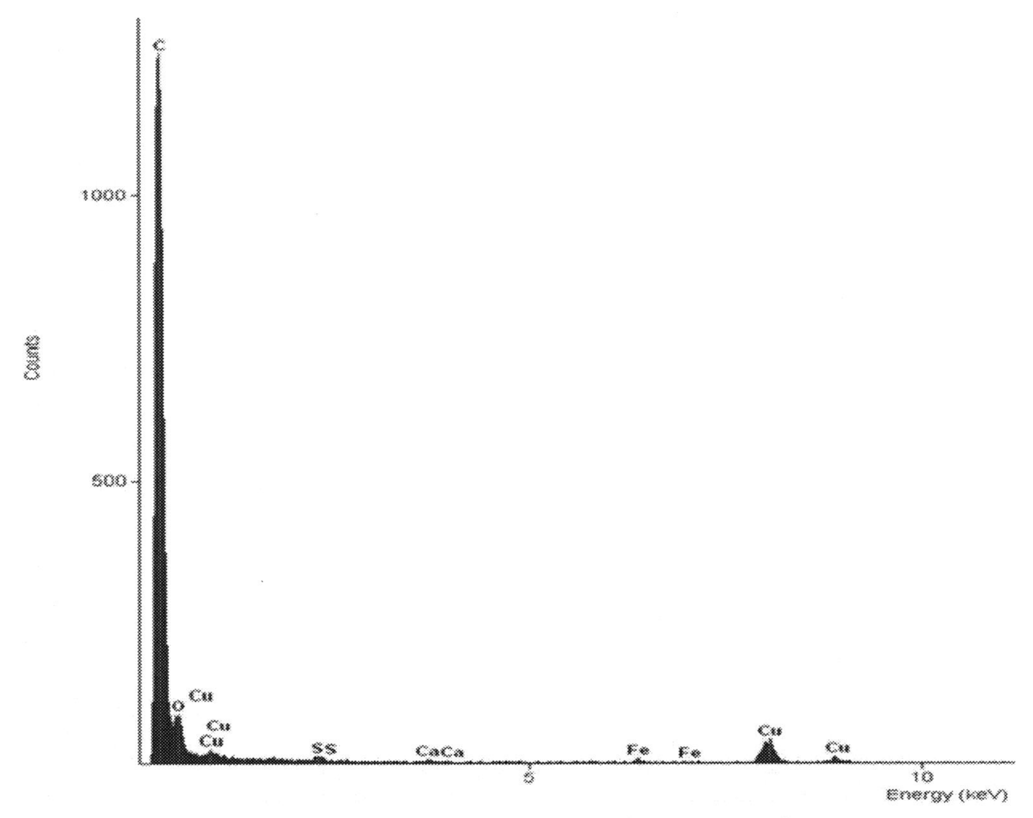

Figure 5.11 EDX of cellulose-g-poly(acrylamide-co-acrylic acid). *EDX*, Energy dispersive X-ray spectra

to ammonium persulphate (radical initiator), Cu (0.9, 8, and 9 keV) due to the grid used during viewing, while Ca (3.7 keV) and Fe (6.4 and 7 keV) could be part of the nutrients absorbed by the WH during growth. The band characteristic of N was not observed as expected and this could be due to absorption by the detector window. The X-ray transmission by Beryllium window is almost 100% for energies >2 keV, but it drops to about zero at 0.5 keV (Liao, 2018); hence, X-ray line such as that of N (0.4 keV) is absorbed. The presence of carbon in the sample is also implicated for low detection of N Kα (X-ray line) because of high mass absorption coefficient value of carbon, 25500 $cm^2\ g^{-1}$ (Liao, 2018).

5.3.4 Evaluation of the factors influencing the swelling of cellulose-grafted polymer hydrogel

5.3.4.1 Influence of salt solutions on water absorbency

The effect of salt concentration and ionic charge on the SEQ is shown in Fig. 5.12. The SEQ decreased in salt solutions, depending on the nature and the concentration

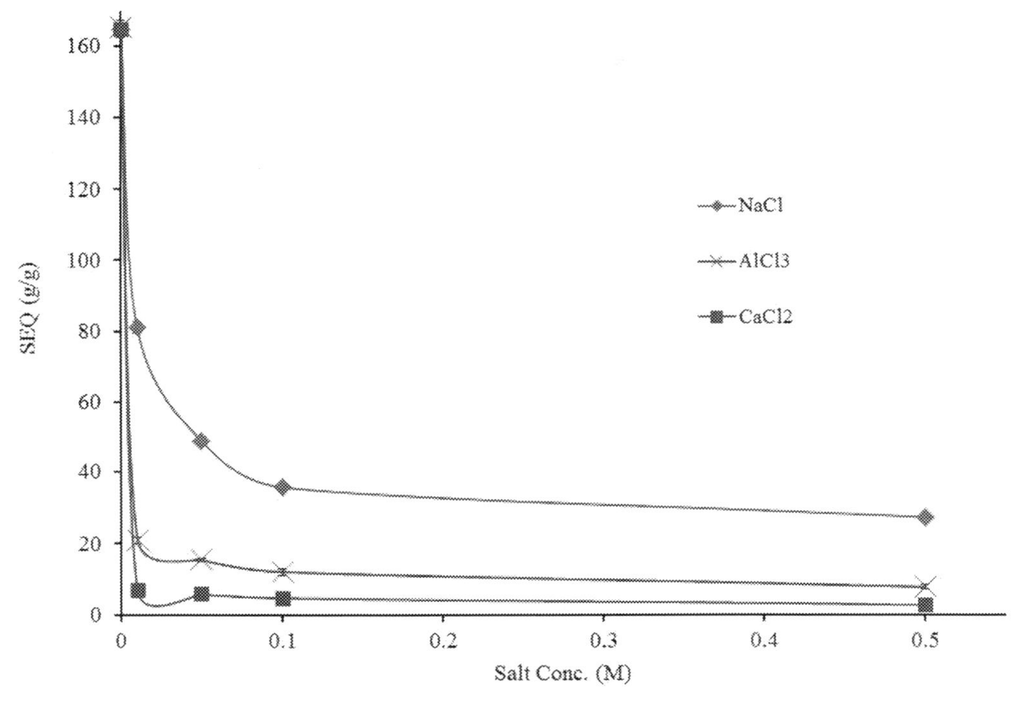

Figure 5.12 Influence of salt concentration and ionic charge of the salt on the swelling of polymer hydrogel. Results are reported as mean ± standard deviation ($n = 3$). *SEQ*, swelling equilibrium.

of the metal cation in the order: $Na^{+} > Al^{3+} > Ca^{2+}$. This was attributed to lowering of the osmotic pressure, the driving force behind the swelling of the PHG (Gupta et al., 2013; Pönni, 2014; Roy et al., 2009). Furthermore, multivalent cations may have neutralized the charge at the surface of the PHG by complexing with carboxamide and carboxylate groups leading to a highly cross-linked network with a small internal free volume as reported by Xiao-Ning et al. (2011) and Livney et al. (2003).

The SEQ was higher in the presence of trivalent Al^{3+} ions compared to divalent Ca^{2+} ions. The observation may be related to charge density, explained by increased electron pair attraction of strongly coordinated water molecules (Livney et al., 2003). Small size and high charge density Al^{3+} ions bind water molecules strongly, coordinating them with the O atoms toward the cation and the H atoms protruding. Due to attraction of electron cloud by the cation, the H atoms of coordinated water molecules become more positive than bulk water molecules; hence, they are more susceptible to formation of H-bonds with electron pairs of polymer dipolar groups such as carboxylic and amide oxygen. On the other hand, large and low charge density Ca^{2+} ions interfere with the water structure without creating an alternative radial structure

(Livney et al., 2003). N. V. Gupta and Shivankumar (2012) observed a decrease in SEQ of poly(methacrylic acid-co-acrylamide) PHG from 155 to 30 g/g in aqueous solution containing Na^+ at concentrations ranging from 0.0001 to 1 M, and also decreased SEQ with increased cationic charge in the order: $Na^+ > Ca^{2+} > Al^{3+}$, slightly different from the copolymer in this study. The decrease was attributed to degree of cross-linking which increased with cationic charge and the lowering of the pH of the swelling medium in the case of $AlCl_3$ salt. Xiao-Ning et al. (2011) also observed decreased SEQ values of guar gum-g-poly(sodium acrylate-co-styrene)/attapulgite PHG in the same order: $Na^+ > Ca^{2+} > Al^{3+}$, a phenomenon attributed to the "charge screening effect" and formation of chemical cross-links through complexation of carboxylate groups by multivalent cations.

5.3.4.2 Influence of pH on water absorbency

The effect of pH on the SEQ of the hydrogel is shown in Fig. 5.13. The SEQ values observed in buffer solutions ranged from 20 to 80 g/g, lower than values of 165 g/g and 120 g/g obtained in distilled and tap water, respectively. This indicates that the PHG was sensitive to charged species in the swelling medium. The SEQ increased with increase in pH of the solution up till pH 10 then declined in extreme alkaline

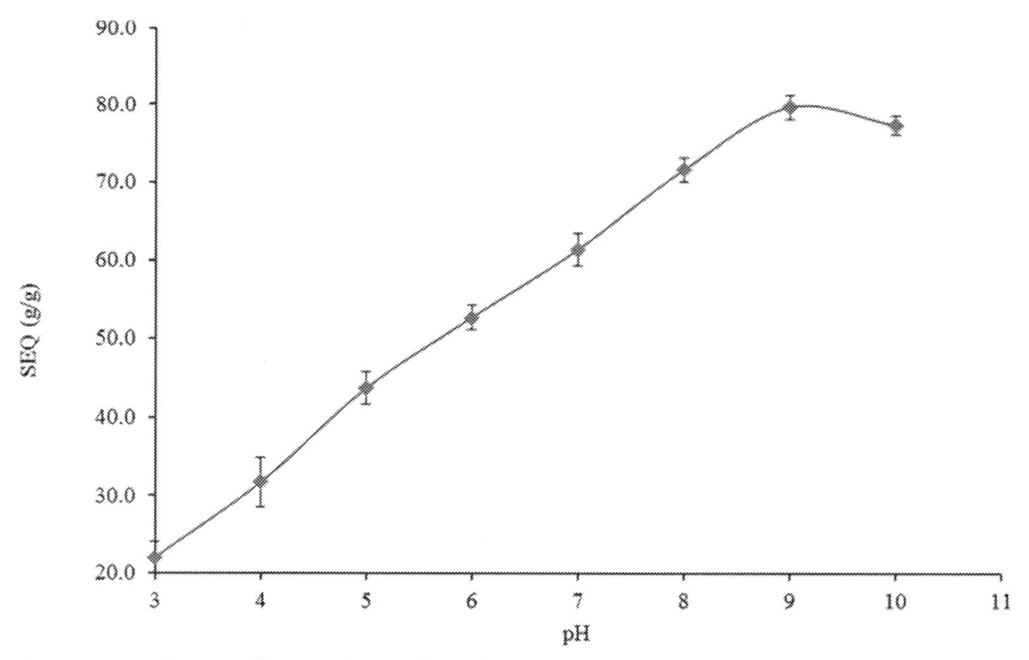

Figure 5.13 Influence of pH on the swelling of PHG. Results are reported as mean ± standard deviation ($n = 3$). *PHG*, Polymer hydrogel; *SEQ*, swelling equilibrium.

range. The observation was attributed to the existence of ionizable (carboxyl) groups which present electrostatic repulsion to each other, expanding free volume to accommodate more water molecules (Mohammad & Fatemeh, 2012). However, excess NaOH used to adjust the pH (Section 5.2.7.4) may cause a "shielding effect" of Na^+ ions on the carboxylate by lowering the repulsion between them, hence decreasing the free volume. The low SEQ at low pH values relates to high proportion of undissociated -COOH groups characterized by low degree of hydrophilicity and high ability to form H-bonds, resulting in a rigid network. Similar observations were made by Demitri et al. (2013) who evaluated the swelling behavior of sodium CMC/hydroxyethyl cellulose PHG at pH values ranging between 2 and 10. The workers observed an increase in SEQ from 37 to 95 g/g and attributed it to dissociation of -COOH groups which depended on the pH of the swelling medium. Mohammad and Fatemeh (2012) obtained a maximum SEQ value of 95 g/g at pH 8 in PHG based on starch-polyacrylate. N. V. Gupta and Shivankumar (2012) obtained higher SEQ value of 160 g/g at pH 7.4. Mahfoudhi and Boufi (2016) observed low SEQ in the pH range of 2–4, which increased from pH 5 to the maximum at pH 7, then declined thereafter.

5.3.4.3 Water holding capacity of polymer hydrogel amended soil

The WHC of sandy-loam soil amended with different amounts of cellulose-grafted copolymer is shown in Fig. 5.14. The water content of the amendments was recorded at an average temperature of 24°C and humidity of 40%. The initial values of the WHC ranged from 35% in the unamended soil to 68% at 1.5% (wt./wt.) copolymer content. A gradual decrease in WHC was observed with time at different rates to near dryness as at 19th day in all the amendments. The copolymer revealed the capacity to absorb and retain water in soil and factors such soil pH and the presence of cations may have influenced its water absorbency. This is an important attribute for efficient use of water in agriculture especially areas with problematic soils such as Vertisols that swell and shrink depending on water content. They can also be used when establishing plants in urban areas with heavy clay soils as well as sandy soils that drain water easily. The water retention in soil is attributed to macromolecular hindrance and hydrophilicity of the copolymer (Demitri et al., 2013), which slows the release of absorbed water. In a similar study, Shahid, Qidwai, Anwar, Ullah, and Rashid (2012) obtained WHC values ranging from 35% to 65% in sandy-loam soil amended with 0.1%–0.4% poly (acrylic acid-co-acrylamide)/$AlZnFe_2O_4$/K humate nanocomposite. The water retention at relatively lower copolymer content was attributed to K humate which enhanced hydrophilicity. Thombare et al. (2018) also obtained WHC values ranging from 39.8% to 51.57% in sandy-loam soil amended with relatively lower copolymer content of 0.1%–0.3% guar gum-based PHG. The WHC values were proportional to the amount of PHG added to the soil.

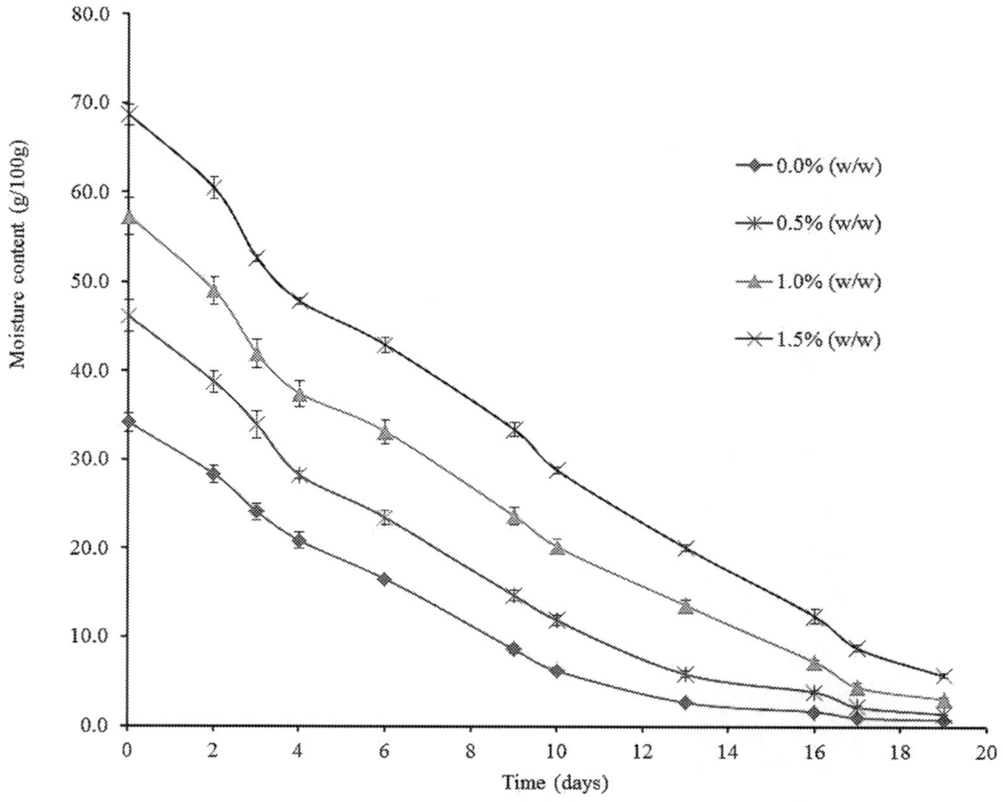

Figure 5.14 Water holding capacity of 100 g PHG-amended sandy-loam soil. Results are reported as mean ± standard deviation ($n = 3$). *PHG*, Polymer hydrogel

5.3.5 Biodegradation of cellulose-grafted copolymer

5.3.5.1 Biodegradation of cellulose-grafted copolymer in soil

The degradation curves of cellulose-g-PAM-co-AA and PHG without cellulose are shown in Fig. 5.15. A slow mass loss was observed in cellulose-grafted PHG in the first 4 weeks followed by a more rapid loss from the 4th to 12th week, after which the curve tends to plateau. No considerable change in mass was observed in the case of PHG without cellulose. After 14 weeks, the mass loss was approximately 25% for cellulose-g-PAM-co-AA and 5% for PHG without cellulose. In a similar study, Laftah and Hashim (2014) obtained a mass loss of about 9.6% in cotton-g-poly (sodium acrylate-co-acrylic acid) PHG and 0.1% in PHG without cotton after 14 weeks incubation in soil. The lower mass loss relative to copolymers in this study was attributed to highly crystalline cotton cellulose and stability of the copolymer. It follows that enhanced degradation of cellulose-g-PAM-co-AA is attributed to low

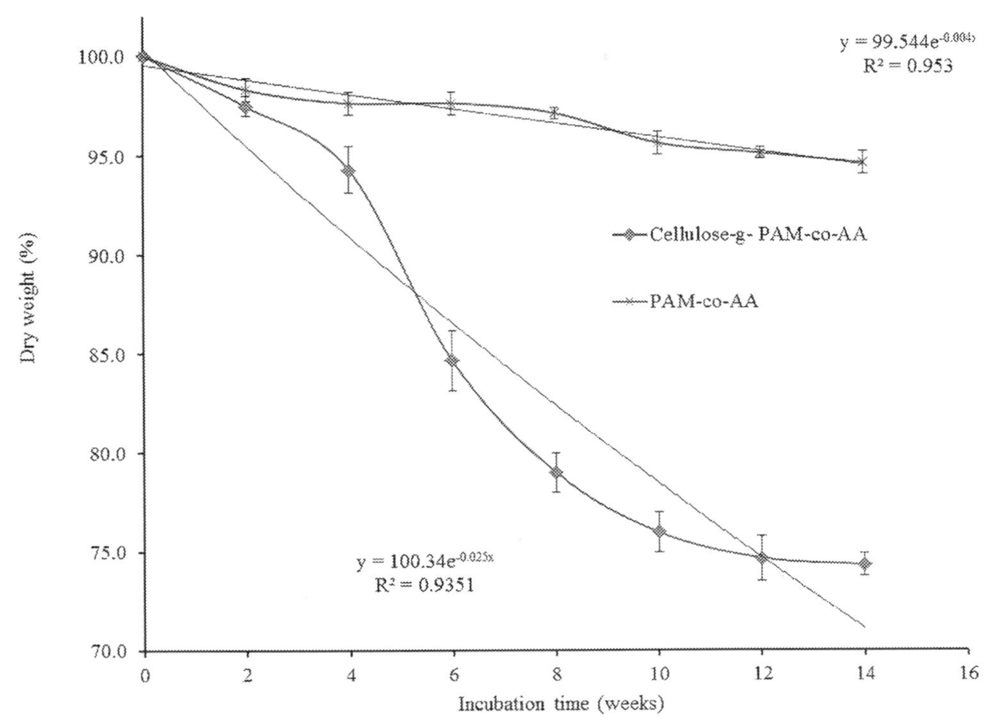

Figure 5.15 Degradation curve of cellulose-grafted PHG and PHG without cellulose. Results are reported as mean ± standard deviation ($n = 3$). *PHG*, Polymer hydrogel.

degree of crystallinity of cellulose due to alkali treatment, in addition to amide-N that acted as source of nourishment to the microbes. The half-life ($t_{1/2}$) of cellulose-g-PAM-co-AA calculated from the rate constant of 0.025 week^{-1} (Fig. 5.15) was found to be 27 weeks, significantly shorter than that of PAM-co-AA of 173 weeks. Thombare et al. (2018) obtained a relatively higher degradation rate constant of 0.009 day^{-1} and ($t_{1/2}$) of 77 days in guar gum-g-polyacrylic acid PHG cross-linked with dimethacrylic acid.

The photographs of swollen PHGs before and after burying in moist soil are displayed in Fig. 5.16. The cellulose-grafted PHG changed from whitish to dark-brown color (Fig. 5.16A and B) after 14 weeks. The brown color of cellulose-grafted PHG (Fig. 5.16B) was darker than that of PHG without cellulose (Fig. 5.16C), an observation which may be revealing microbial degradation of copolymer chains. Oven-dried samples of unearthed cellulose-grafted PHG transformed from the initial tough solid (Fig. 5.16A) to a friable material, whereas PHG without cellulose remained tough. High molecular weight polymer chains are degraded by microbes into monomeric and oligomeric units to enable assimilation (Ahmed, 2015; Ekebafe et al., 2011; Laftah & Hashim, 2014; Milani et al., 2017; Yu et al., 2011). Native soil bacteria (*Pseudomonas*,

Figure 5.16. Photographs of (A) swollen cellulose-g-poly(acrylamide-co-acrylic acid) before burying in moist soil; (B) swollen cellulose-g-poly(acrylamide-co-acrylic acid) after 14 weeks in moist soil; and (C) swollen PHG without cellulose after 14 weeks in moist soil. *PHG*, Polymer hydrogel.

$$R\text{—}CONH_2 \xrightarrow[H_2O]{Amidase} R\text{—}COO^- NH_4^+ \rightleftharpoons R\text{—}COOH + NH_3$$

(R denotes macromolecule)

Figure 5.17 Enzymatic hydrolysis of acrylamide to carboxylic acid and ammonia.

Xanthomonas, *Rhodococcus*, *Klebseilla*) have been found to degrade and utilize PAM as sole source of C and N under both aerobic and anaerobic conditions (Nawaz et al., 1993; Nawaz et al., 1994; Shanker, Ramakrishna, & Seth, 1990; Yu, Fu, Xie, & Chen, 2015). They secrete PAM-specific intracellular and extracellular amidase (amidohydrolase) enzyme to hydrolyze the amide group into acrylic acid and ammonia, according to Fig. 5.17 (Kay-Shoemake, Watwood, Lentz, & Sojka, 1998; Kay-Shoemake, Watwood, Sojka, & Lentz, 1998). Acrylic acid is eventually degraded to carbon dioxide and water as end products that are environmentally friendly. The addition of C sources into PAM has been reported to promote bacterial growth. Yu et al. (2015) observed improved degradation efficiency of PAM in culture medium supplemented with glucose, and in this case, the grafted cellulose could have enhanced the degradation of the copolymer.

5.3.5.2 Microbial culture and degradation of cellulose-grafted copolymer by soil microbial isolates

Of the five soil samples incubated in PBM supplemented with acrylamide, three tested positive for NH_4^+. Acrylamide degrading microbes have been reported in soil contaminated with amides and their derivatives such as herbicides (Nawaz, Franklin, Campbell, Heinze, & Cerniglia, 1991). These microbes transform herbicides such as propanil (acrylamilide) to amides and eventually NH_4^+, to support their growth (Nawaz et al., 1991). They have been isolated, identified and the amidase enzyme purified and characterized (Nawaz et al., 1993, 1994; Shanker et al., 1990; F. Yu et al., 2015). The present study focused on utilizing the soil microbial isolates to degrade acrylamide copolymers, therefore the microbes involved could be more than one species.

Table 5.2 shows the amount of NH_4^+ liberated with time in PBM supplemented with acrylamide and different amounts of the copolymer as the sole source of C and N which are essential in agricultural production. The amount of NH_4^+ accumulated in the copolymer supplemented media increased with time to maximum values at 100 h above which some decline was observed. The decrease after the maxima may be attributed to NH_4^+ utilization by the microbes as N source for their physiological needs. The amount of NH_4^+ accumulated in the media for both copolymers increased significantly ($P \leq .05$) with the copolymer content (P1 to P3 and CP1 to CP3) from 40 to 100 h. The values recorded in 0.05% (wt./vol.) copolymer content (P3 and CP3), were in most instances, not significant from those observed in 0.06% (wt./vol.) (P4 and CP4). Significantly higher ($P \leq .05$) accumulation of NH_4^+ was observed in the cellulose-grafted copolymer CP1 and CP2 from 40 to 100 h, relative to similar content of the copolymer without cellulose P1 and P2. The observation was attributed to easily accessible cellulose-C which enhanced microbial growth and subsequently, increased amount of enzyme that caused considerable hydrolysis of the amide-N.

The degradation of the copolymers was not compared statistically with that of acrylamide (monomer) due to the cross-linked network which must be degraded into monomeric units to enable microbial assimilation. Acrylamide-supplemented medium showed low content of NH_4^+ ranging from 1.1 to 1.9 mg kg^{-1} in the first 60 h, reflecting the lag phase, which then increased considerably from 20 mg kg^{-1} at 80 h to a maximum value of 112.2 mg kg^{-1} at 216 h after which it declined. The decline could be attributed to exhaustion of the substrate by the microbes as observed elsewhere (Karuku & Mochoge, 2016, 2018; Karuku, 2019). In related studies on acrylamide (monomer) degradation by *Pseudomonas* sp., Shanker et al. (1990) observed highest accumulation of NH_4^+ after 6 days, at 30°C. Nawaz et al. (1993) on *Pseudomonas* sp., observed maximum NH_4^+ content after 24 h, at 28°C which decreased to 1.0 mM at 96 h, whereas *Xanthomonas maltophilia* showed highest accumulation of NH_4^+ and acrylic acid at 48 h under the similar conditions. In both cases, the highest acrylic acid content coincided with the disappearance of acrylamide and decreased with cellular growth. Nawaz et al. (1994) in their degradation study of 62 mM acrylamide by *Rhodococcus* sp., observed maximum accumulation of NH_4^+ and acrylic acid after 24 h and a peak cellular growth from 72 to 96 h at 28°C. The variation in the time taken to achieve maximum NH_4^+ accumulation between the present and previous studies on acrylamide may be related to microbial (bacterial) species and diversity involved in the degradation and the initial population of microbes in the culture media.

Table 5.2 Concentration of $NH4^+$ (mg/kg) liberated with time (hours) at different copolymer contents.

Sample code	6 h	20 h	40 h	60 h	80 h	100 h	168 h	216 h	288 h
AM	0.5 (0.1)	1.1(0.1)	1.5 (0.3)	1.9 (0.2)	20.9 (2.0)	54.8 (1.9)	95.8 (3.3)	112.2 (4.2)	93.9 (3.0)
P1	6.7 (1.6)a	11.1 (0.2)a	19.4 (1.4)a	17.4 (1.0)a	20.9 (1.9)a	34.6 (1.6)a	21.7 (2.4)a	21.1 (1.2)a	22.7(0.6)a
P2	10.6 (2.3)ab	14.7 (1.5)ab	31.6 (2.4)b	26.4 (1.8)b	37.8 (0.4)c	50.4 (2.2)c	33.8 (2.9)a	33.8 (1.7)b	34.3(3.1)b
P3	11.7 (2.5)b	22.2 (2.8)cd	41.6 (1.7)c	56.0 (0.9)d	62.8 (1.4)e	66.9 (0.5)d	63.3 (1.6)bc	67.2 (7.4)d	64.1(1.0)de
P4	12.3 (1.6)b	28.3 (1.7)d	56.9 (1.5)cd	61.9 (2.6)e	64.3 (0.8)e	80.3 (1.9)e	80.3 (3.2)d	64.3 (0.6)d	52.3(2.1)c
CP1	12.8 (1.7)b	13.3 (0.5)ab	30.6 (2.4)b	24.1 (1.5)b	34.3 (1.5)b	42.0 (3.7)b	27.7 (3.8)a	28.3 (1.0)ab	32.3(3.2)b
CP2	13.9 (1.2)b	19.2 (1.0)bc	41.8 (1.6)c	38.9 (1.8)c	47.0 (1.0)d	60.7 (1.8)d	53.3 (2.0)b	49.8 (2.6)c	56.3(3.2)cd
CP3	13.5 (1.9)b	21.1 (2.4)c	56.1 (2.1)d	51.8 (2.0)d	64.6 (0.5)e	81.1 (2.5)e	57.4 (10.3)bc	58.2 (0.8)cd	61.7(3.8)de
CP4	14.1 (1.9)b	20.7 (3.3)c	52.5 (2.0)d	53.9 (0.8)d	63.3 (1.6)e	81.7 (2.5)e	68.0 (3.7)cd	66.6 (4.6)d	65.6(3.4)e

Notes: The values in parentheses are standard deviations ($n = 3$), different letters in the same column are significantly different (Tukey test; $P \leq .05$ level). *AM*, Acrylamide monomer, 10 mM [0.07% (wt./vol.)]; *P*, poly(acrylamide-co-acrylic acid), P1 = 0.02, P2 = 0.04, P3 = 0.05, and P4 = 0.06% (wt./vol.); *CP*, cellulose-g-poly(acrylamide-co-acrylic acid), CP1 = 0.02, CP2 = 0.04, CP3 = 0.05, and CP4 = 0.06% (wt./vol.).

5.3.6 Water absorbency of cellulose-g-poly(acrylamide-co-acrylic acid)/nano-HA composite hydrogel

The effect of nano-HA content on water absorbency (SEQ) of cellulose-grafted copolymer is presented in Fig. 5.18. The polymer without nano-HA synthesized by grafting 6.75% (wt./vol.) partially neutralized acrylic acid onto 2% (wt./vol.) cellulose, swelled in distilled water up to 163 g/g (Fig. 5.18). The SEQ initially decreased on introducing nano-HA from 163 to 104 (g/g) followed by a marked increase to an optimal value of 120 (g/g) at 2.5% wt./vol., after which it declined. The decrease was attributed to the rigidity of nano-HA which occupied the voids within the copolymer network, decreasing the free volume for accommodating water molecules. The subsequent increase in SEQ could be attributed to -OH groups at the surface nano-HA which enhanced the hydrophilicity of the copolymer. Nano-HA content higher than the optimum decreased the SEQ, an observation ascribed to (1) generation of extra cross-link points where the -OH at the surface of nano-HA could react with the monomer, reducing flexibility of the copolymer chains; (2) decreased hydrophilic portion of the composite, that is, poly(acrylamide-co-acrylic acid) with increased

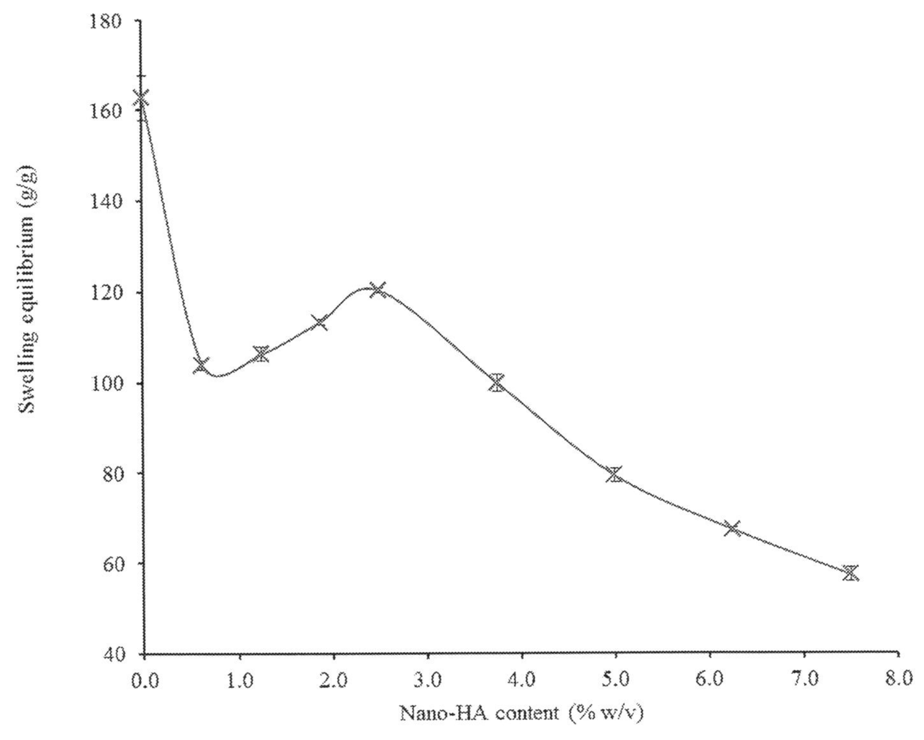

Figure 5.18 Effect of nano-HA content on water absorbency of cellulose-grafted PHG. The values are reported as mean ± standard deviation ($n = 3$). *PHG*, Polymer hydrogel; *nano-HA*, nanohydroxyapatite.

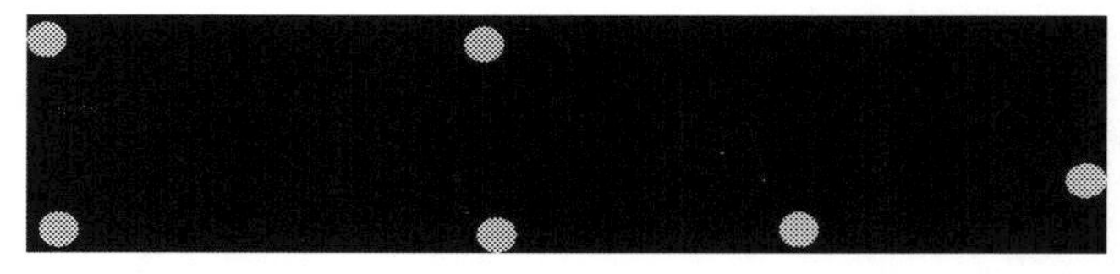

Figure 5.19 Grafting of acrylic acid (monomer) onto nano-HA by chain transfer mechanism. *nano-HA*, Nanohydroxyapatite.

nano-HA content; and (3) hampered growth of polymer chains by chain transfer mechanism according to Fig. 5.19.

Gao, Huang, Lei, and Zheng (2010) in a similar study, grafted methacrylic acid onto HA particle surfaces and observed successful grafting of polymethacrylic acid onto HA particle surfaces. Rashidzadeh et al. (2014) evaluating sodium alginate-g-poly (acrylic acid-co-acrylamide)/clinoptilolite observed increased SEQ from 34 (g/g) to a maximum value of 41 (g/g) after which it declined. The workers attributed initial increase in SEQ to repulsive forces between negative surface charge of clinoptilolite and -COO^- groups of the hydrogel, while the decrease was linked to additional cross-linking and reduced hydrophilicity of the composite with increased clinoptilolite content. Hosseinzadeh and Sadeghi (2012) assessing CMC-g-poly(sodium acrylate)/ kaolin composite observed a decrease in SEQ from 450 to 45 (g/g) and this was attributed to additional cross-linking and hindered growth of polymer chains through chain transfer mechanism. In the current study, nano-HA content influences the SEQ of polymer composite as reported in related studies. Nevertheless, the optimized polymer nanocomposite showed a relatively higher SEQ value of 120 g/g (at 2.5% wt./vol. nano-HA content) and hence exhibited superabsorbent property. This is an important feature for agricultural application as moisture retention aid that can reduce the cost of frequently irrigating crops in ASALs due to prolonged moisture retention in soil. Nano-HA could improve the mechanical strength of copolymer due to possibility of the formation of chemical bonds through its -OH groups, as well as acting as a source of slow-release phosphorous, one of the most limiting and sought after crop nutrients.

5.3.7 Structural and morphological characteristics of cellulose-grafted nanocomposite polymer hydrogel

5.3.7.1 Fourier transform infrared spectroscopy

FTIR spectrum of nano-HA and that of cellulose-g-poly(acrylamide-co-acrylic acid)/ nano-HA composite are shown in Fig. 5.20. Nano-HA displayed bands at 1419 and 875 cm^{-1} (Fig. 5.20A) assigned to CO_3^{2-} ions which is ascribed to the interaction between nano-HA and CO_2 (Iyyappan & Wilson, 2013). The broad and weak band extending from 3600 to 3000 cm^{-1}, and the weak band at 1635 cm^{-1} corresponds

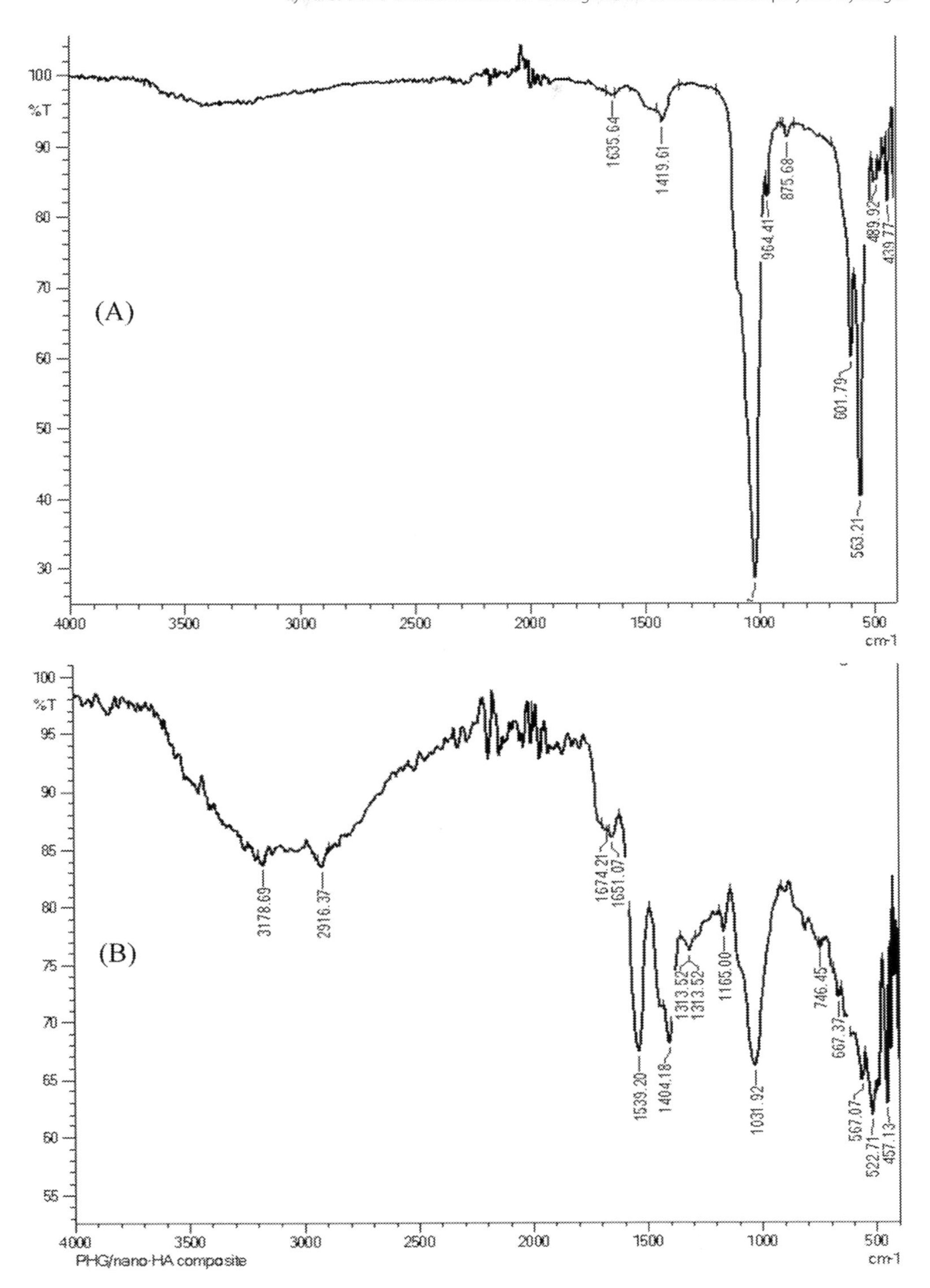

Figure 5.20 FTIR spectra of (A) nano-HA and (B) cellulose graft poly(acrylamide-co-acrylic acid)/nano-HA composite. *FTIR*, Fourier transform infrared; *nano-HA*, nanohydroxyapatite.

to -OH stretching vibrations in the nano-HA lattice and H—O—H of lattice water (Cisneros-Pineda et al., 2014; Costescu et al., 2010; Gayathri, Muthukumarasamy, Velauthapillai, Santhosh, & asokan, 2018). The bands at 1022 and 964 cm^{-1} are assigned to PO_4^{3-} stretching vibrations while those at 601 and 563 cm^{-1} correspond to bending vibrations. Nanocomposite showed strong bands between 1300—1165 assigned to C—O—C and P—O—C bridging (Fig. 5.20B) indicating a possibility of cellulosic and nano-HA -OH groups reacting with the monomer (acrylic acid/ammonium acrylate). The strong band at 1018 cm^{-1} observed in cellulose (Fig. 5.7B) and the very strong band at 1022 cm^{-1} observed in nano-HA (Fig. 5.20A) shifted to 1031 cm^{-1}, further confirming the reaction of alcoholic -OH groups. Spectral band at 1022 cm^{-1} corresponding to PO_4^{3-} for nano-HA (Fig. 5.20A) also decreased in intensity when nano-HA was incorporated into the copolymer (Fig. 5.20B). A strong band at 522 cm^{-1} was observed in polymer nanocomposite due to P = O stretching vibration for the PO_4^{3-} group. The rest of the spectral bands for nano-composite were not different from those observed in the spectrum of cellulose-g-PAM-co-AA PHG (Fig. 5.7C).

5.3.7.2 Transmission electron microscopy

The TEM images of nano-HA and polymer nanocomposite are shown in Fig. 5.21. The images of nano-HA synthesized in presence of TX-100 (nonionic surfactant) displayed more of rod-shaped nanoparticle agglomerates than those synthesized without it (Fig. 5.21A and B). Iyyappan and Wilson (2013) had a similar observation which they attributed it to hydrophobic ring complex formed through ion-dipole interaction between Ca^{2+} and polyoxyethylene group of TX-100 (Fig. 5.22). The formation of this complex is thought to reduce the transfer rate of Ca^{2+} to the growing crystals thereby reducing nanoparticle size under controlled conditions. The formation of nanorods is also attributed to adsorption of TX-100 onto certain planes of formed crystals

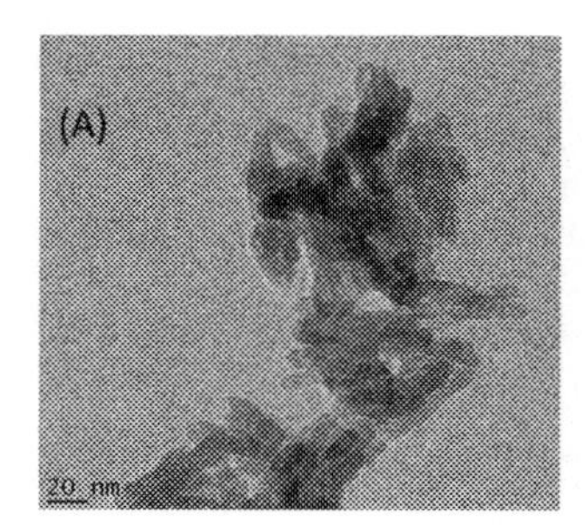

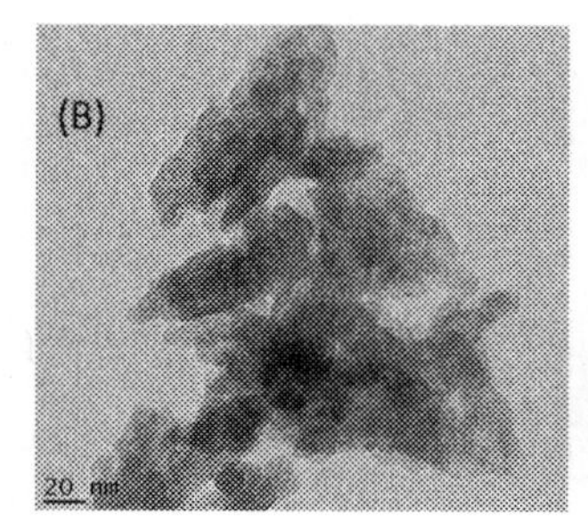

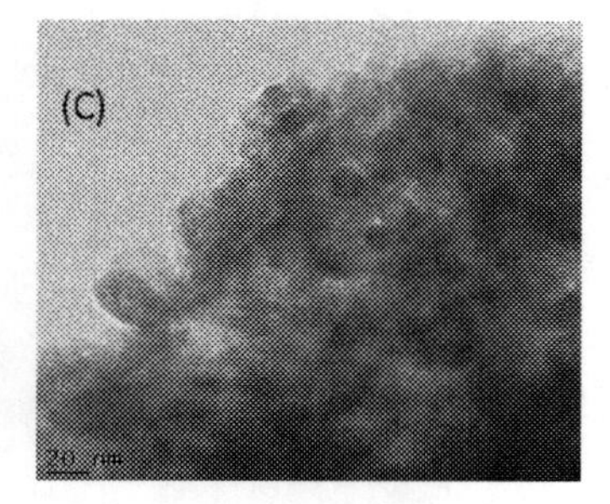

Figure 5.21 TEM images at 20 nm scale for (A) nano-HA synthesized in presence of TX-100; (B) nano-HA synthesized without TX-100; and (C) cellulose-g-poly(acrylamide-co-acrylic acid)/nano-HA composite. *nano-HA*, Nanohydroxyapatite; *TEM*, transmission electron microscopy.

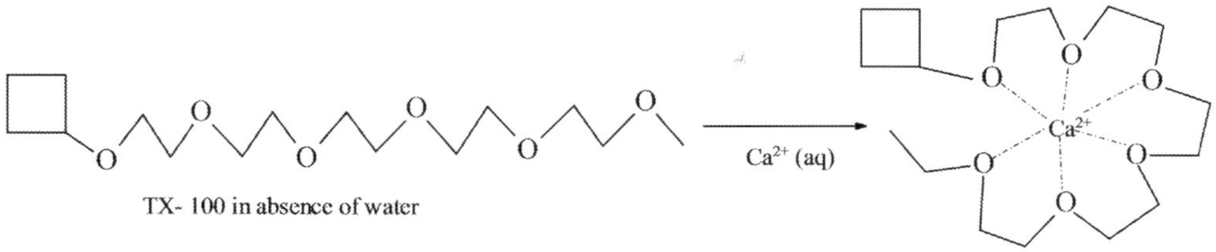

Figure 5.22 The ion-dipole interaction between Ca2 + and polyoxyethylene group of TX-100 leading to formation of hydrophobic complex (Iyyappan & Wilson, 2013).

leading to growth of nanoparticles in a preferential direction. The agglomeration of nanoparticles (Fig. 5.21A and B) was attributed to high specific surface energy that led to aggregation (Nabakumar et al., 2009; Ragu & Sakthivel, 2014). TEM images of nanocomposite (Fig. 5.21C) showed dispersion of nano-HA within the copolymer. Similarly, Ragu and Sakthivel (2014) observed diminished agglomeration of nano-HA in polymethyl methacrylate/nano-HA composite at about 20–50 nm; an observation attributed to reduction of surface energy of nano-HA by the polymer through its pendent PO_4^{3-} group.

5.3.7.3 Energy dispersive X-ray spectroscopy

The EDX spectra of nano-HA and polymer nanocomposite are shown in Fig. 5.23. Nano-HA spectrum (Fig. 5.23A) displayed the main constituents of sample as calcium (3.7 keV), phosphorous (2.0 keV), oxygen (0.5 keV), and carbon (0.2 keV). Ca, P, and O are expected in nano-HA lattice, while C may be attributed to the existence of CO_3^{2-} ions. Cisneros-Pineda et al. (2014) also obtained a peak due to carbon and attributed it to sensitivity of apatite lattice to the substitution environment of the CO_3^{2-} ions. This substitution can either be type A where CO_3^{2-} ions replace -OH ions, or type B where the CO_3^{2-} ions replace PO_4^{3-} ions (Cisneros-Pineda et al., 2014). Similar bands were observed in polymer nanocomposite (Fig. 5.23B), but the peaks were more intense than those observed in nano-HA spectrum due to oxygen and carbon (Fig. 5.23A); an observation attributable to cellulose-grafted copolymer fraction. Sulfur was also detected in polymer nanocomposite due to ammonium persulphate (radical initiator). The rest of peaks due to Cu (0.9, 8.0, and 9.0 keV) may have originated from the grid, Te (3.8 keV) and Si (1.8 keV) are not related to the constituents of the sample and could have probably originated from metal conductive layers and/or impurities.

5.3.7.4 X-ray diffraction analysis

The X-ray diffraction patterns of nano-HA and that of cellulose-grafted nanocomposite are presented in Fig. 5.24. Nano-HA displayed mainly diffraction bands at $2\theta = 26°$, 32°, 33°, 34.3°, 39.5°, 46.5°, and 49.5° (Fig. 5.24A) assigned to (002),

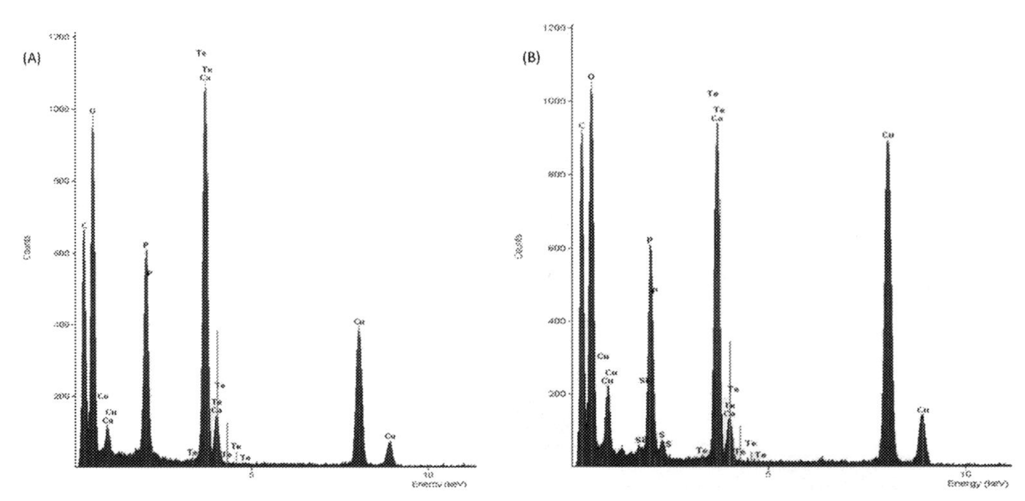

Figure 5.23 EDX spectra of (A) nano-HA and (B) cellulose-g-poly(acrylamide-co-acrylic acid)/nano-HA composite. *EDX, Energy dispersive X-ray;nano-HA*,nanohydroxyapatite

(211), (112), (300), (130), (222), and (213) crystallographic plane reflections, respectively (Brundavanam, Poinern, & Fawcett, 2013; Costescu et al., 2010; Gayathri et al., 2018; Mechay, Feki, Schoenstein, & Jouini, 2012; Ragu & Sakthivel, 2014). The cellulose-grafted polymer/nano-HA composite (Fig. 5.24B) indicated a weak diffraction band at around $2\theta = 22^{\circ}$ (200) attributed to cellulose I allomorph, with the rest of the bands being typical of nano-HA crystalline structure. This observation suggests that nano-HA could be physically entrapped within the copolymer matrix besides forming part of the polymeric network.

Carlmark et al. (2012), Jianzhong et al. (2015), Rop, Karuku, et al. (2019), Rop, Mbui, et al., (2019), and Roy et al. (2009), evaluating agricultural implication of a formulated nanocomposite SRF: cellulose-g-poly(acylamide)/nano-HA/soluble fertilizer composite, observed pH of soil amendments ranging from 5.15–5.33 in the second week of incubation which increased to values ranging from 5.71 to 5.97 as at 16th week. This was largely attributed to the release of Ca^{2+} ions which have the effect of neutralizing the acids. Polymer nanocomposite released solubilized P from nano-HA which is thought to mainly depend on microbial action. Soil bacterial species (e.g., *Pseudomonas*, *Anthrobactor*) and fungi (e.g., *Aspergillus*, *Penicillium*) are implicated for solubilization of insoluble phosphates such as apatite (Alori, Glick, & Babalola, 2017; Khan, Zaidi, & Ahmad, 2014). They secrete low molecular weight organic acids such as citric, gluconic, lactic, oxalic, and acetic acids to lower the pH and chelate mineral ions. The acidification of soil microbial environment as well as rhizospheric environment by plant root exudates under P limitation is considered to solubilize nano-HA according to Fig. 5.25 (Rop, Karuku, et al., 2019; Rop, Mbui, et al., 2019). According to the law of

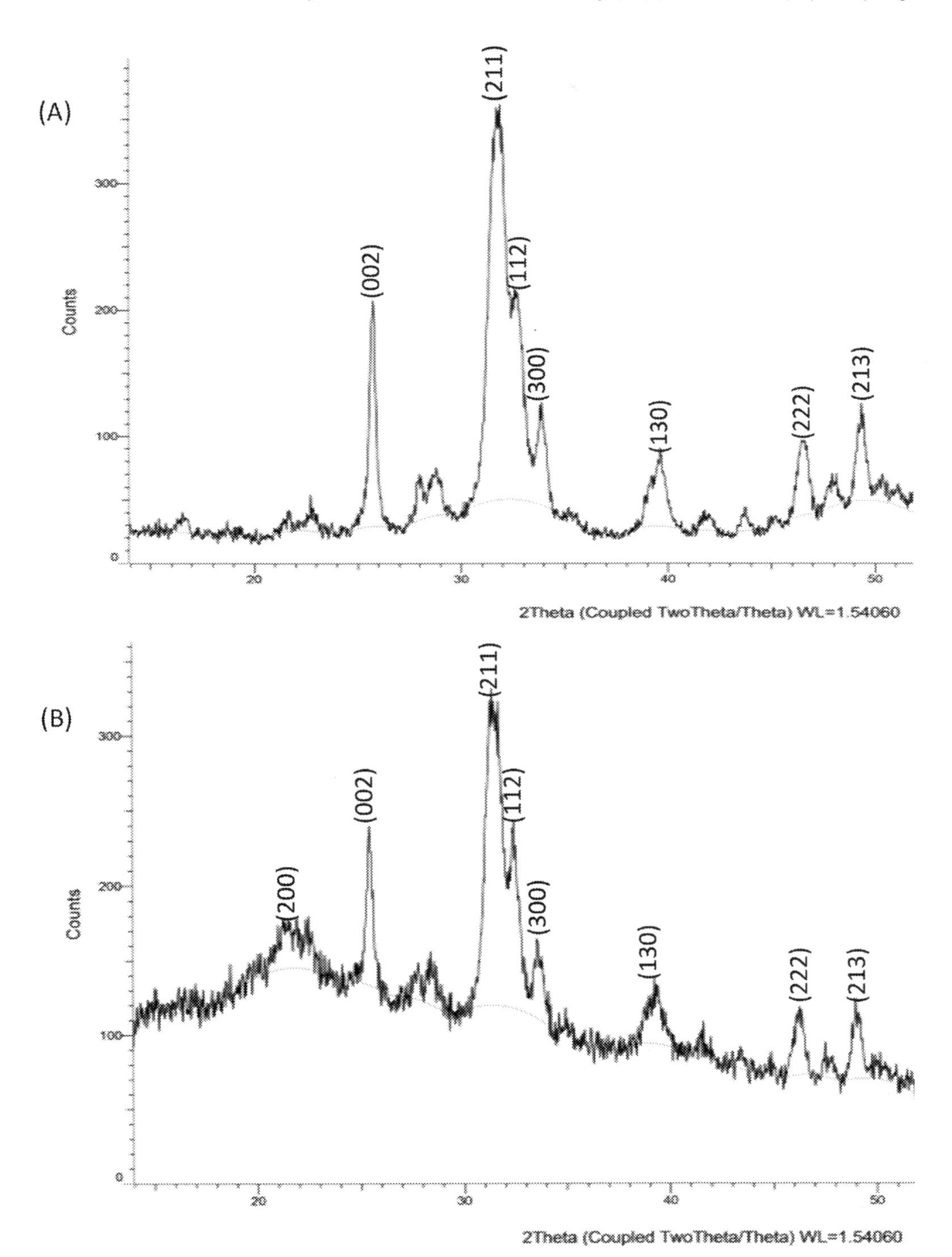

Figure 5.24 Diffractograms of (A) nano-HA and (B) cellulose-g-poly(acrylamide-co-acrylic acid)/nano-HA composite. nano-HA, Nanohydroxyapatite.

$$Ca_{10}(PO_4)_6(OH)_2 + 12H^+ \rightleftharpoons 10Ca^{2+} + 6H_2PO_4^- + 2OH^-$$

Figure 5.25 Neutralization reaction expressing the dissolution of nano-HA. nano-HA, Nanohydroxyapatite.

mass action, equilibrium shift to the right (solubilization) is favored by increased concentration of H^+ and continuous mining of Ca^{2+} and $H_2PO_4^-$ ions by plants.

5.4 Conclusion

Acrylic monomers were grafted onto swollen cellulose isolated from WH by radical polymerization. FTIR spectra revealed purification of cellulose from WH and grafting of the monomer onto cellulose. High resolution TEM images displayed a microporous structure in acetone-dehydrated PHG. The cellulose-grafted copolymer exhibited superabsorbent properties and its swelling was influenced by the pH and by the presence, concentration, and nature of ions present. It also revealed significant hydrolysis of amide-N in the microbial culture, displayed biodegradability, and the potential to absorb and retain water in the soil. These are important features for potential agricultural applications such as formulation of SRFs and soil conditioning. Biodegradable cellulose-based nanocomposite phosphate fertilizer was formulated and characterized. Nano-HA content enhanced hydrophilicity of polymer nanocomposite to an optimal value after which it declined. The synthesis of nano-HA in presence of a surfactant led to the formation of rod-shaped nanocrystals. FTIR analysis indicated existence of chemical interactions between nano-HA and the copolymer structure. The main constituents of the polymer nanocomposite are C, O, Ca, and P. N could not be detected with EDX spectroscopic technique. The X-ray analysis indicated that, dry cellulose was in the form of cellulose I crystalline allomorph. Diffractogram of cellulose-grafted copolymer displayed more of amorphous pattern compared to that of dry WH cellulose. Besides being part of the copolymer network, nano-HA crystals were entrapped within the network structure increasing the mechanical strength and hence making it a feasible source of slow-release P upon degradation and solubilization. This could subsequently enhance P use efficiency and safe handling of nanoparticles, reduce production cost, and environmental contamination as well as condition the soil due to increased moisture-holding capacity.

Acknowledgments

The authors are indebted to DAAD (Germany) for scholarship award and National Research Fund (NRF, Kenya) for facilitating the study. The authors are grateful to Mr. Evans Kimega and Ms. Rose Mutungi of the Department of Chemistry, University of Nairobi, for technical assistance.

References

Abdel-Halim, E. S. (2014). Chemical modification of cellulose extracted from sugarcane bagasse: Preparation of hydroxyethyl cellulose. *Arabian Journal of Chemistry*, 7(3), 362–371. Available from https://doi.org/10.1016/j.arabjc.2013.05.006.

Ago, M., Endo, T., & Hirotsu, T. (2004). Crystalline transformation of native cellulose from cellulose I to cellulose II polymorph by a ball-milling method with a specific amount of water. *Cellulose*, *11*(2), 163–167. Available from https://doi.org/10.1023/B:CELL.0000025423.32330.fa.

Ahmed, E. M. (2015). Hydrogel: Preparation, characterization, and applications: A review. *Journal of Advanced Research*, *6*(2), 105–121. Available from https://doi.org/10.1016/j.jare.2013.07.006.

Alori, E. T., Glick, B. R., & Babalola, O. O. (2017). Microbial phosphorus solubilization and its potential for use in sustainable agriculture. *Frontiers in Microbiology*, *8*, 971. Available from https://doi.org/10.3389/fmicb.2017.00971.

Ambjörnsson, H.A. (2013). *Mercerization and enzymatic pretreatment of cellulose in dissolving pulps* (p. 22) (Dissertation). Karlstad: Karlstad University. <https://www.diva-portal.org/smash/get/diva2:616374/FULLTEXT01.pdf>. Acessed, 23rd Mar. 2018.

Aswathy, U. S., Sukumaran, R. K., Devi, G. L., Rajasree, K. P., Singhania, R. R., & Pandey, A. (2010). Bio-ethanol from water hyacinth biomass: An evaluation of enzymatic saccharification strategy. *Bioresource Technology*, *101*(3), 925–930. Available from https://doi.org/10.1016/j.biortech.2009.08.019.

Bai, H., Wang, X., Zhou, Y., & Zhang, L. (2012). Preparation and characterization of poly(vinylidene fluoride) composite membranes blended with nano-crystalline cellulose. *Progress in Natural Science: Materials International*, *22*(3), 250–257. Available from https://doi.org/10.1016/j.pnsc.2012.04.011.

Baki, M., & Abedi-Koupai, J. (2018). Preparation and characterization of a superabsorbent slow-release fertilizer with sodium alginate and biochar. *Journal of Applied Polymer Science*, *135*, 45966. Available from https://doi.org/10.1002/app.45966.

Bortolin, A., Aouada, F. A., Mattoso, L. H. C., & Ribeiro, C. (2013). Nanocomposite PAAm/methyl cellulose/montmorillonite hydrogel: Evidence of synergistic effects for the slow release of fertilizers. *Journal of Agricultural and Food Chemistry*, *61*(31), 7431–7439. Available from https://doi.org/10.1021/jf401273n.

Brundavanam, R. K., Poinern, G. E. J., & Fawcett, D. (2013). Modelling the crystal structure of a 30 nm sized particle based hydroxyapatite powder synthesized under the influence of ultrasound irradiation from X-ray powder diffraction data. *American Journal of Materials Science*, *3*(4), 84–90. Available from https://doi.org/10.5923/j.materials.20130304.04.

Carlmark, A., Larsson, E., & Malmström, E. (2012). Grafting of cellulose by ring-opening polymerisation—A review. *European Polymer Journal*, *48*(10), 1646–1659. Available from https://doi.org/10.1016/j.eurpolymj.2012.06.013.

Cheng, G., Varanasi, P., Li, C., Liu, H., Melnichenko, Y. B., Simmons, B. A., ... Singh, S. (2011). Transition of cellulose crystalline structure and surface morphology of biomass as a function of ionic liquid pretreatment and its relation to enzymatic hydrolysis. *Biomacromolecules*, *12*(4), 933–941. Available from https://doi.org/10.1021/bm101240z.

Cisneros-Pineda, O. G., Herrera Kao, W., Loría-Bastarrachea, M. I., Veranes-Pantoja, Y., Cauich-Rodríguez, J. V., & Cervantes-Uc, J. M. (2014). Towards optimization of the silanization process of hydroxyapatite for its use in bone cement formulations. *Materials Science and Engineering C*, *40*, 157–163. Available from https://doi.org/10.1016/j.msec.2014.03.064.

Costescu, A., Pasuk, I., Ungureanu, F., Dinischiotu, A., Costache, M., Huneau, F., ... Predoi, D. (2010). Physico-chemical properties of nano-sized hexagonal hydroxyapatite powder synthesized by sol-gel. *Digest Journal of Nanomaterials and Biostructures*, *5*(4), 989–1000, http://www.chalcogen.infim.ro/989_Costescu.pdf.

Demitri, C., Scalera, F., Madaghiele, M., Sannino, A., & Maffezzoli, A. (2013). Potential of cellulose-based superabsorbent hydrogels as water reservoir in agriculture. *International Journal of Polymer Science*, 2013. Available from https://doi.org/10.1155/2013/435073.

Deng, Y., Wang, L., Hu, X., Liu, B., Wei, Z., Yang, S., & Sun, C. (2012). Highly efficient removal of tannic acid from aqueous solution by chitosan-coated attapulgite. *Chemical Engineering Journal*, *181*, 300–306. Available from https://doi.org/10.1016/j.cej.2011.11.082, 182.

Ekebafe, L. O., Ogbeifun, D. E., & Okieimen, F. E. (2011). Polymer applications in agriculture. *Biokemistri*, *23*(2), 81−89.

Gao, J., Huang, B., Lei, J., & Zheng, Z. (2010). Photografting of methacrylic acid onto hydroxyapatite particles surfaces. *Journal of Applied Polymer Science*, *115*(4), 2156−2161. Available from https://doi.org/10.1002/app.31307.

Gayathri, B., Muthukumarasamy, N., Velauthapillai, D., Santhosh, S. B., & asokan, V. (2018). Magnesium incorporated hydroxyapatite nanoparticles: Preparation, characterization, antibacterial and larvicidal activity. *Arabian Journal of Chemistry*, *11*(5), 645−654. Available from https://doi.org/10.1016/j.arabjc.2016.05.010.

Girisuta, B., Danon, B., Manurung, R., Janssen, L. P. B. M., & Heeres, H. J. (2008). Experimental and kinetic modelling studies on the acid-catalysed hydrolysis of the water hyacinth plant to levulinic acid. *Bioresource Technology*, *99*(17), 8367−8375. Available from https://doi.org/10.1016/j.biortech.2008.02.045.

Giroto, A. S., Guimarães, G. G. F., Foschini, M., & Ribeiro, C. (2017). Role of slow-release nanocomposite fertilizers on nitrogen and phosphate availability in soil. *Scientific Reports*, 7(1), 46032. Available from https://doi.org/10.1038/srep46032.

Gupta, N. V., & Shivankumar, H. G. (2012). Investigation of swelling behaviour and mechanical properties of a pH sensitive super-porous hydrogel composite. *Iranian journal of pharmaceutical research (IJPR)*, *11*(2), 481−493.

Gupta, P. K., Uniyal, V., & Naithani, S. (2013). Polymorphic transformation of cellulose I to cellulose II by alkali pretreatment and urea as an additive. *Carbohydrate Polymers*, *94*(2), 843−849. Available from https://doi.org/10.1016/j.carbpol.2013.02.012.

Gürdağ, G., & Sarmad, S. (2013). *Cellulose graft copolymers: Synthesis, properties, and applications. In Polysaccharide based graft copolymers* (pp. 15−57). Berlin Heidelberg: Springer-Verlag, https://doi.org/10.1007/978-3-642−36566-9-2.

Hall, A. (1993). Application of the indophenol blue method to the determination of ammonium in silicate rocks and minerals. *Applied Geochemistry*, *8*(1), 101−105. Available from https://doi.org/10.1016/0883-2927(93)90059-P.

Hashem, M., Sharaf, S., Abd El-Hady, M. M., & Hebeish, A. (2013). Synthesis and characterization of novel carboxymethylcellulose hydrogels and carboxymethylcellulolse-hydrogel-ZnO-nanocomposites. *Carbohydrate Polymers*, *95*(1), 421−427. Available from https://doi.org/10.1016/j.carbpol.2013.03.013.

Hosseinzadeh, H., & Sadeghi, M. (2012). Preparation and swelling behaviour of carboxymethylcellulose-g-poly(sodium acrylate)/kaolin super absorbent hydrogel composites. *Asian Journal of Chemistry*, *24*(1), 85−88, http://www.asianjournalofchemistry.co.in/Journal/ViewArticle.aspx?ArticleID = 24-1-19.

Hussien, R. A., Donia, A. M., Atia, A. A., El-Sedfy, O. F., El-Hamid, A. R. A., & Rashad, R. T. (2012). Studying some hydro-physical properties of two soils amended with kaolinite-modified cross-linked poly-acrylamides. *Catena*, *92*, 172−178. Available from https://doi.org/10.1016/j.catena.2011.12.010.

Iyyappan, E., & Wilson, P. (2013). Synthesis of nanoscale hydroxyapatite particles using triton X-100 as an organic modifier. *Ceramics International*, *39*(1), 771−777. Available from https://doi.org/10.1016/j.ceramint.2012.06.090.

Jafari, N. (2010). Ecological and socio-economic utilization of water hyacinth (*Eichhornia crassipes* Mart Solms). *Journal of Applied Sciences and Environmental Management*, *14*(2). Available from https://doi.org/10.4314/jasem.v14i2.57834.

Jianzhong, M., Xiaolu, L., & Yan, B. (2015). Advances in cellulose-based superabsorbent hydrogels. *RSC Advances*, *5*(73), 59745−59757. Available from https://doi.org/10.1039/C5RA08522E.

Kabiri, K., & Zohuriaan-Mehr, M. J. (2004). Porous superabsorbent hydrogel composites: Synthesis, morphology and swelling rate. *Macromolecular Materials and Engineering*, *289*(7), 653−661. Available from https://doi.org/10.1002/mame.200400010.

Kaco, H., Zakaria, S., Razali, N. F., Chia, C. H., Zhang, L., & Jani, S. M. (2014). Properties of cellulose hydrogel from kenaf core prepared via pre-cooled dissolving method. *Sains Malaysiana*, *43*(8), 1221−1229, http://www.ukm.my/jsm/pdf_files/SM-PDF-43−8−2014/14%20Hatika%20Kaco.pdf.

Karuku, G. N. (2019). Effect of lime, N and P salts on nitrogen mineralization, nitrification process and priming effect in three soil types, andosols, luvisols and ferralsols. *Journal of Agriculture and Sustainability, 12*(1), 74–106.

Karuku, G. N., & Mochoge, B. O. (2016). Nitrogen forms in three Kenyan soils nitisols, luvisols and ferralsols. *International Journal for Innovation Education and Research, 4*, 17–30.

Karuku, G. N., & Mochoge, B. O. (2018). Nitrogen mineralization potential (No) in three Kenyan soils, Nitisols, Ferralsols and Luvisols. *Journal of Agricultural Science, 10*(4), 69. Available from https://doi.org/10.5539/jas.v10n4p69.

Kay-Shoemake, J. L., Watwood, M. E., Lentz, R. D., & Sojka, R. E. (1998). Polyacrylamide as an organic nitrogen source for soil microorganisms with potential effects on inorganic soil nitrogen in agricultural soil. *Soil Biology and Biochemistry, 30*(8–9), 1045–1052. Available from https://doi.org/10.1016/S0038-0717(97)00250-2.

Kay-Shoemake, J. L., Watwood, M. E., Sojka, R. E., & Lentz, R. D. (1998). Polyacrylamide as a substrate for microbial amidase in culture and soil. *Soil Biology and Biochemistry, 30*(13), 1647–1654. Available from https://doi.org/10.1016/S0038-0717(97)00251-4.

Khan, M. S., Zaidi, A., & Ahmad, E. (2014). *Mechanism of phosphate solubilization and physiological functions of phosphate-solubilizing microorganisms. Phosphate solubilizing microorganisms: Principles and application of microphos technology* (pp. 31–62). Springer International Publishing. Available from https://doi.org/10.1007/978-3-319-08216-5_2.

Kivaisi, A. K., & Mtila, M. (1997). Production of biogas from water hyacinth (*Eichhornia crassipes*) (mart) (solms) in a two-stage bioreactor. *World Journal of Microbiology and Biotechnology, 14*(1), 125–131. Available from https://doi.org/10.1023/A:1008845005155.

Kottegoda, N., Munaweera, I., Madusanka, N., & Karunaratne, V. (2011). A green slow-release fertilizer composition based on urea-modified hydroxyapatite nanoparticles encapsulated wood. *Current Science, 101*(1), 73–78, http://www.ias.ac.in/currsci/10jul2011/73.pdf.

Laftah, W. A., & Hashim, S. (2014). Synthesis, optimization, characterization, and potential agricultural application of polymer hydrogel composites based on cotton microfiber. *Chemical Papers, 68*(6), 798–808. Available from https://doi.org/10.2478/s11696-013-0507-5.

Laftah, W. A., Hashim, S., & Ibrahim, A. N. (2011). Polymer hydrogels: A review. *Polymer—Plastics Technology and Engineering, 50*(14), 1475–1486. Available from https://doi.org/10.1080/03602559.2011.593082.

Liao, Y. (2018). *Practical electron microscopy and database*. GlobalSino. Available from http://www.globalsino.com/EM/. Accessed, 20th Nov. 2018.

Liu, R., & Lal, R. (2014). Synthetic apatite nanoparticles as a phosphorus fertilizer for soybean (*Glycine max*). *Scientific Reports, 4*(1), 5686. Available from https://doi.org/10.1038/srep05686.

Livney, Y. D., Portnaya, I., Faupin, B., Ramon, O., Cohen, Y., Cogan, U., & Mizrahi, S. (2003). Interactions between inorganic salts and polyacrylamide in aqueous solutions and gels. *Journal of Polymer Science, Part B: Polymer Physics, 41*(5), 508–519. Available from https://doi.org/10.1002/polb.10406.

Mahfoudhi, N., & Boufi, S. (2016). Poly (acrylic acid-co-acrylamide)/cellulose nanofibrils nanocomposite hydrogels: effects of CNFs content on the hydrogel properties. *Cellulose, 23*(6), 3691–3701. Available from https://doi.org/10.1007/s10570-016-1074-z.

Malmström, E., & Carlmark, A. (2012). Controlled grafting of cellulose fibres—An outlook beyond paper and cardboard. *Polymer Chemistry, 3*(7), 1702–1713. Available from https://doi.org/10.1039/c1py00445j.

Marandi, G. B., Mahdavinia, G. R., & Ghafary, S. (2011). Collagen-g-poly(sodium acrylate-co-acrylamide)/sodium montmorillonite superabsorbent nanocomposites: Synthesis and swelling behavior. *Journal of Polymer Research, 18*(6), 1487–1499. Available from https://doi.org/10.1007/s10965-010-9554-6.

Mechay, A., Feki, H. E. L., Schoenstein, F., & Jouini, N. (2012). Nanocrystalline hydroxyapatite ceramics prepared by hydrolysis in polyol medium. *Chemical Physics Letters, 541*, 75–80. Available from https://doi.org/10.1016/j.cplett.2012.05.047.

Milani, P., França, D., Balieiro, A. G., & Faez, R. (2017). Polymers and its applications in agriculture. *Polimeros, 27*(3), 256–266. Available from https://doi.org/10.1590/0104-1428.09316.

Mohammad, S., & Fatemeh, S. (2012). Synthesis of pH-sensitive hydrogel based on starch-polyacrylate superabsorbent. *Journal of Biomaterials and Nanobiotechnology*, 310−314. Available from https://doi.org/10.4236/jbnb.2012.322038.

Montalvo, D., McLaughlin, M. J., & Degryse, F. (2015). Efficacy of hydroxyapatite nanoparticles as phosphorus fertilizer in andisols and oxisols. *Soil Science Society of America Journal*, *79*(2), 551−558. Available from https://doi.org/10.2136/sssaj2014.09.0373.

Muchanyerey, N. M., Kugara, J., & Zaranyika, M. F. (2016). Surface composition and surface properties of water hyacinth (*Eichhornia crassipes*) root biomass: Effect of mineral acid and organic solvent treatment. *African Journal of Biotechnology*, *15*, 897−909. Available from https://doi.org/10.5897/AJB2015.15068.

Nabakumar, P., Debasish, M., Indranil, B., Kumar, M. T., Parag, B., & Panchanan, P. (2009). Chemical synthesis, characterization, and biocompatibility study of hydroxyapatite/chitosan phosphate nanocomposite for bone tissue engineering applications. *International Journal of Biomaterials*, *2009* (1687−8787), 512417. Available from https://doi.org/10.1155/2009/512417.

Nam, S., French, A. D., Condon, B. D., & Concha, M. (2016). Segal crystallinity index revisited by the simulation of X-ray diffraction patterns of cotton cellulose Iβ and cellulose II. *Carbohydrate Polymers*, *135*, 1−9. Available from https://doi.org/10.1016/j.carbpol.2015.08.035.

Nawaz, M. S., Franklin, W., & Cerniglia, C. E. (1993). Degradation of acrylamide by immobilized cells of a *Pseudomonas* sp. and *Xanthomonas maltophilia*. *Canadian Journal of Microbiology*, *39*(2), 207−212. Available from https://doi.org/10.1139/m93-029.

Nawaz, M. S., Franklin, W., Campbell, W. L., Heinze, T. M., & Cerniglia, C. E. (1991). Metabolism of acrylonitrile by *Klebsiella pneumoniae*. *Archives of Microbiology*, *156*(3), 231−238. Available from https://doi.org/10.1007/BF00249120.

Nawaz, M. S., Khan, A. A., Seng, J. E., Leakey, J. E., Siitonen, P. H., & Cerniglia, C. E. (1994). Purification and characterization of an amidase from an acrylamide- degrading *Rhodococcus* sp. *Applied and Environmental Microbiology*, *60*(9), 3343−3348. Available from https://doi.org/10.1128/aem.60.9.3343-3348.1994.

Pönni, R. (2014). *Changes in accessibility of cellulose for kraft pulps measured by deuterium exchange* (Doctoral Dissertation). Alto University publication series 68/2014. Available from http://urn.fi/URN:ISBN:978-952-60-5686-9. Accessed, 16th Apr. 2017.

Qinyuan, C., Yang, J., & Xinjun, Y. (2017). Hydrogels for biomedical applications: Their characteristics and the mechanisms behind them. *Gels*, *3*(1), 6. Available from https://doi.org/10.3390/gels3010006.

Qiu, X., & Hu, S. (2013). \Smart\ materials based on cellulose: A review of the preparations, properties, and applications. *Materials*, *6*(3), 738−781. Available from https://doi.org/10.3390/ma6030738.

Ragu, A., & Sakthivel, P. (2014). Synthesis and characterization of nanohydroxyapatite with polymethyl methacrylate nanocomposites for bone tissue regeneration. *International Journal of Science and Research (IJSR)*, *3*(11), 2282−2285.

Rashidzadeh, A., Olad, A., Salari, D., & Reyhanitabar, A. (2014). On the preparation and swelling properties of hydrogel nanocomposite based on sodium alginate-g-poly(acrylic acid-co-acrylamide)/Clinoptilolite and its application as slow release fertilizer. *Journal of Polymer Research*, *21*(2). Available from https://doi.org/10.1007/s10965-013-0344-9.

Reales-Alfaro, J. G., Trujillo-Daza, L. T., Arzuaga-Lindado, G., Castaño-Peláez, H. I., & Polo-Córdoba, A. D. (2013). Acid hydrolysis of water hyacinth to obtain fermentable sugars. *CTyF—Ciencia, Tecnologia y Futuro*, *5*(2), 101−112. Available from https://doi.org/10.29047/01225383.60.

Rezania, S., Md Din, M. F., Kamaruddin, S. F., Taib, S. M., Singh, L., Yong, E. L., & Dahalan, F. A. (2016). Evaluation of water hyacinth (*Eichhornia crassipes*) as a potential raw material source for briquette production. *Energy*, *111*, 768−773. Available from https://doi.org/10.1016/j.energy.2016.06.026.

Rop, B.K. (2019). *Development of slow release nano-composite fertilizer using biodegradable superabsorbent polymer* (PhD thesis). University of Nairobi. <http://erepository.uonbi.ac.ke/handle/11295/109205>. Accesed, 15th Jun. 2020.

Rop, K., Karuku, G. N., Mbui, D., Njomo, N., & Michira, I. (2019). Evaluating the effects of formulated nano-NPK slow release fertilizer composite on the performance and yield of maize, kale and capsicum. *Annals of Agricultural Sciences*, *64*(1), 9−19. Available from https://doi.org/10.1016/j.aoas.2019.05.010.

Rop, K., Mbui, D., Njomo, N., Karuku, G. N., Michira, I., & Ajayi, F. R. (2019). Biodegradable water hyacinth cellulose-graft-poly(ammonium acrylate-co-acrylic acid) polymer hydrogel for potential agricultural application. *Heliyon*, *5*(3), e01416. Available from https://doi.org/10.1016/j.heliyon.2019.e01416.

Roy, D., Semsarilar, M., Guthrie, J. T., & Perrier, S. (2009). Cellulose modification by polymer grafting: A review. *Chemical Society Reviews*, *38*(7), 2046–2064. Available from https://doi.org/10.1039/b808639g.

Sadeghi, M., Safari, S., Shahsavari, H., Sadeghi, H., & Soleimani, F. (2013). Effective parameters onto graft copolymer based on carboxymethyl with acrylic monomer. *Asian Journal of Chemistry*, *25*(9), 5029–5032. Available from https://doi.org/10.14233/ajchem.2013.14311b.

Sadeghi, M., Soleimani, F., & Yarahmadp, M. (2011). Chemical modification of carboxymethyl cellulose via graft copolymerization and determination of the grafting parameters. *Oriental Journal of Chemistry*, *27*(3), 967–972.

Sannino, A., Demitri, C., & Madaghiele, M. (2009). Biodegradable cellulose-based hydrogels: Design and applications. *Materials*, *2*(2), 353–373. Available from https://doi.org/10.3390/ma2020353.

Sannino, A., Mensitieri, G., & Nicolais, L. (2004). Water and synthetic urine sorption capacity of cellulose-based hydrogels under a compressive stress field. *Journal of Applied Polymer Science*, *91*(6), 3791–3796. Available from https://doi.org/10.1002/app.13540.

Saputra, A. H., Hapsari, M., & Pitaloka, A. B. (2015). Synthesis and characterization of CMC from water hyacinth cellulose using isobutyl-isopropyl alcohol mixture as reaction medium. *Contemporary Engineering Sciences*, *8*(33–36), 1571–1582. Available from https://doi.org/10.12988/ces.2015.511300.

Shahid, S. A., Qidwai, A. A., Anwar, F., Ullah, I., & Rashid, U. (2012). Effects of a novel poly (AA-co-AAm)/AlZnFe 2O 4/ potassium humate superabsorbent hydrogel nanocomposite on water retention of sandy loam soil and wheat seedling growth. *Molecules (Basel, Switzerland)*, *17*(11), 12587–12602. Available from https://doi.org/10.3390/molecules171112587.

Shanker, R., Ramakrishna, C., & Seth, P. K. (1990). Microbial degradation of acrylamide monomer. *Archives of Microbiology*, *154*(2), 192–198. Available from https://doi.org/10.1007/BF00423332.

Shirsath, S. R., Hage, A. P., Zhou, M., Sonawane, S. H., & Ashokkumar, M. (2011). Ultrasound assisted preparation of nanoclay bentonite-FeCo nanocomposite hybrid hydrogel: A potential responsive sorbent for removal of organic pollutant from water. *Desalination*, *281*(1), 429–437. Available from https://doi.org/10.1016/j.desal.2011.08.031.

Simoni, R. C., Lemes, G. F., Fialho, S., Gonçalves, O. H., Gozzo, A. M., Chiaradia, V., ... Leimann, F. V. (2017). Effect of drying method on mechanical, thermal and water absorption properties of enzymatically crosslinked gelatin hydrogels. *Anais da Academia Brasileira de Ciencias*, *89*(1), 745–755. Available from https://doi.org/10.1590/0001-3765201720160241.

Swantomo, D., Rochmadi, R., Basuki, K. T., & Sudiyo, R. (2013). Synthesis and characterization of graft copolymer rice straw cellulose-acrylamide hydrogels using gamma irradiation. *Atom Indonesia*, *39* (2), 57–64. Available from https://doi.org/10.17146/aij.2013.232.

Tally, M., & Atassi, Y. (2015). Optimized synthesis and swelling properties of a pH-sensitive semi-IPN superabsorbent polymer based on sodium alginate-g-poly(acrylic acid-co-acrylamide) and polyvinylpyrrolidone and obtained via microwave irradiation. *Journal of Polymer Research*, *22*(9). Available from https://doi.org/10.1007/s10965-015-0822-3.

Thombare, N., Mishra, S., Siddiqui, M. Z., Jha, U., Singh, D., & Mahajan, G. R. (2018). Design and development of guar gum based novel, superabsorbent and moisture retaining hydrogels for agricultural applications. *Carbohydrate Polymers*, *185*, 169–178. Available from https://doi.org/10.1016/j.carbpol.2018.01.018.

Titik, I., Nur, R., Richa, R., Metty, M., Slamet, P., & Heru, S. (2015). Cellulose isolation from tropical water hyacinth for membrane preparation. *Procedia Environmental Sciences*, *23*, 274–281. Available from https://doi.org/10.1016/j.proenv.2015.01.041.

WRB. (2015). World reference base for soil resources 2014: International soil classification for naming soils and creating legends for maps. In: *World soil resource reports* (Vol. 106). Rome: FAO. <http://www.fao.org/3/i3794en/I3794en.pdf>. Accessed, 18th Feb. 2018.

Xiao-Ning, S., Wen-Bo, W., & Ai-Qin, W. (2011). Effect of surfactant on porosity and swelling behaviors of guar gum-g-poly(sodium acrylate-co-styrene)/attapulgite superabsorbent hydrogels. *Colloids and Surfaces B: Biointerfaces*, 279–286. Available from https://doi.org/10.1016/j.colsurfb.2011.07.002.

Yu, F., Fu, R., Xie, Y., & Chen, W. (2015). Isolation and characterization of polyacrylamide-degrading bacteria from dewatered sludge. *International Journal of Environmental Research and Public Health, 12*(4), 4214–4230. Available from https://doi.org/10.3390/ijerph120404214.

Yu, Y., Liu, L., Kong, Y., Zhang, E., & Liu, Y. (2011). Synthesis and properties of N-maleyl chitosan-cross-linked poly(acrylic acid-co-acrylamide) superabsorbents. *Journal of Polymers and the Environment, 19*(4), 926–934. Available from https://doi.org/10.1007/s10924-011-0340-2.

Zhang, J., Wang, Q., & Wang, A. (2007). Synthesis and characterization of chitosan-g-poly(acrylic acid)/attapulgite superabsorbent composites. *Carbohydrate Polymers, 68*(2), 367–374. Available from https://doi.org/10.1016/j.carbpol.2006.11.018.

Zhu, Y. L., Zayed, A. M., Qian, J. H., De Souza, M., & Terry, N. (1999). Phytoaccumulation of trace elements by wetland plants: II. Water hyacinth. *Journal of Environmental Quality, 28*(1), 339–344. Available from https://doi.org/10.2134/jeq1999.00472425002800010042x.

Zohuriaan-Mehr, M. J., & Kabiri, K. (2008). Superabsorbent polymer materials: A review. *Iranian Polymer Journal (English Edition), 17*(6), 451–477.

CHAPTER 6

Fabrication of nanowoods and nanopapers

Nikita Goswami, Tushar Kumar and Palakjot K. Sodhi
University Institute of Biotechnology, Chandigarh University, Mohali, India

6.1 Introduction

Currently, the world is facing a huge crisis over the fossil-fuel-based resources. Moreover, the issues have risen to an immense level because of ecological hazards such as greenhouse gas emissions, increased consumption of exhaustible resources, and low degradability of synthetics. Hence, these issues prompt us to substitute the conventional and traditional technologies with more sustainable and environment-friendly replacements.

One such potential solution to these problems is the emerging field of nanotechnology. It possesses the ability to alter and significantly change the materials, currently being used in the 21st century. The traditional and unsophisticated technologies that are currently being used will be substituted by the upcoming and highly effective nanotechnological advancements in the upcoming years (Wegner & Jones, 2009).

Nanotechnology poses more advantages over the conventional technologies in the fact that it requires less material input (substrate), utilizes less area and energy. Moreover, it shows highly effective length scale for assembling, manufacturing, and convergence of living with the nonliving (Kamel, 2007).

The discovery and disclosure of novel materials, procedures, and marvels at the nanoscale and the advancements in experimental and hypothetical methods for research, give new areas of improvements through nanosystems and nanostructured materials (Bhushan, 2017). Diverse manufacturing strategies including the procedures that utilize "self-assembly" on an atomic level are either in current use or being prepped for commercial or business-scale utilization.

The remarkable capacity of nanostructured materials to show exceptional physiochemical properties as compared to the available bulk substances has opened new doors for the creation of novel materials (for new applications and uses) that were not in any way obvious before (Grimsdale & Müllen, 2005).

Forests are a natural resource of immense importance for humankind as they deliver a variety of essential commodities, wood and paper are among these essentials.

Nanotechnology in Paper and Wood Engineering
DOI: https://doi.org/10.1016/B978-0-323-85835-9.00010-6

Uses of wood at the fundamental level are extremely broad. Moreover, wood can likewise be formed into utility materials to profit the arising and key innovative fields. This bulk wood when observed at a nanoscale, is basically a nanostructured matter, comprising nanofibrils and nanocrystals.

The vision of nanotechnology in these woodland products is to economically address the issues of present and future and sustainability by applying nanotech science to enhance the qualities of wood-based lignocellulosic materials (Mohieldin et al., 2011). Some potential applications of nanotechnology in paper-making include: manufacturing new materials, wet end science, nanofiltration in conclusion of water dissemination, coating, sensors microscopy, and nanoscale constructing agents (Puurunen & Vasara, 2007).

6.2 Cellulose and nanocellulose

The prominent constituent of this nanostructure matter is the natural cellulose. It is a sustainable natural polymer that is found in all plants and most algal species. The cellulose of tunicitin-type structures is a major component of shells of some marine animals and is additionally produced by a few microbes (Klemm, Heublein, Fink, & Bohn, 2005). The principal source of cellulose extraction is plants. The content of cellulose however varies from species to species. The advancements in the biotechnological field have led to the spectacular discoveries, and reports regarding minute structural components of cells, constituted a new field (nanotechnology) for never-ending human demands and ecological benefits.

Cellulose is the fibrous component of the plant cellular structure, discovered and named around the late 1830s (discovered in 1837 and terminology in 1839) (Hon, 1994). Cellulose comprises small fibers of approximately 6–7 nm diameter that are presently referred to as nanocellulose. The crystal-like structure of nanocellulose, that is, cellulose nanocrystals were reported with the advancements in the electron microscopy, by Rånby and Ribi (1950).

Nanocellulose finds its merits over the native cellulose in the facts that it is slightly less dense, has higher surface area, exhibits biocompatibility, and biodegradable in nature. It is easily dispersed in aqueous solutions and can improve the polymer barrier property (Favier, Cavaille, Canova, & Shrivastava, 1977). These properties render different kinds of nanocellulose like cellulose nanofibers, cellulose nanocrystals, and bacterial nanocellulose. However, the production and extraction of nanocellulose requires complicated and intricate procedures. Pretreatment of cell is essential to extract the fibers and crystals. The conventional pretreatments include acid (mostly sulfuric acid) or alkali treatment and enzymatic (cello-biohydrolase, cellulose, etc.) and mechanical disruptions (homogenization, ultra-sonication, etc.) (Sharma, Thakur, Bhattacharya, Mandal, & Goswami, 2019).

Nanocellulose is being utilized for the production of new and modified products. One of its most appropriate applications is the synthesis of nanocomposites or nanocellulosic materials. In the recent years, these products have been seeking attention due to their toughness and environmental-friendly properties. The two most promising products of nanocellulose are nanowood and nanopaper.

Nanowood is a highly aligned yet light-weighed cellulosic structure which is mechanically much stronger (than the conventional wood), thermally insulated, and ecofriendly. But its synthesis requires top-down approach. The natural wood is being overused for its handful of applications in several industries, hence nanowood provides a perfect substitute for it. The nanowood overcomes the problems associated with natural wood such as deterioration due to fungal and insect attacks, vulnerability to fire, and dimensional variability to water absorption (Shema, Balarabe, & Alfa, 2018). On the other hand, the nanopaper structure comprises of intertwined cellulosic fibers held together by hydrogen bonds. The strength of the paper increases when the cellulosic fibers are defibrillated, thereby increasing the surface area (Carrasco, Mutje, & Pelach, 1996). These can be viewed as transparent rigid films or sheets. Nanopapers, hence, are excellent for their use on electronic products (Fang et al., 2013).

Thus the mechanical and physical properties of nanopapers and nanowood need to be improvised and hence, more knowledge needs to be acquired in this field in a way that this technology satisfies the human needs and does not pose a threat to the environment.

6.3 Isolation and fabrication of nanocellulose fibrils

The plant cell wall comprises primary and secondary cell walls which are constructed by the cellulose nanofibrils. These nanocellulose particles possess small diameters (usually < 100 nm). The cellulose bundles found in the cell wall consist of millions of micro and nanofibrils exhibiting amorphous and crystalline structures. These densely packed structures are roughly liberated through the various energy-consuming mechanical disruption procedures (Habibi, Lucia, & Rojas, 2010). However, extraction requires pretreatment of these fibrils via chemical exposure (Saito & Isogai, 2004), radiations (Takacs et al., 2000), or enzymatic homogenization (Hayashi, Kondo, & Ishihara, 2005), thereby reducing the expense of the entire process.

Being a remarkable biopolymer, nanocellulose finds its applications in much wider fields, that is, from scientific explorations to the industrial investments in the textile industry, biomedicine areas, nanocellulose-based papermaking industries, printing devices, and other electronic items (Zhu et al., 2016). Nanocellulose hence emerges as a breakthrough in the current era which needs to be worked upon for making a better substitute and thus commercialize it for use in day-to-day life.

Depending upon the types of procedures used for the isolation and extraction of nanocellulose, they can be categorized as nanocellulosic fibers, whiskers, and nanofibrils (Klemm et al., 2009). Other types of isolated nanocellulose include:

1. Cellulose nanocrystals (CNC)
2. Nanofibrillated cellulose (NFC)
3. Bacterial cellulose (BC)
4. Electro spun cellulose nanofibers (ECNF)

Out of these, CNC and NFC are quite commonly prepared. The former is extracted by disruption of cellulose into nanoscale particles, while, the BC and ECNF result from the low-molecular weight polysaccharides which are secreted by bacteria or disseminated cellulose acted upon by electrospinning (Nechyporchuk, Belgacem, & Bras, 2016).

NFCs, also referred to as microfibrillated cellulose or even cellulose nanofibrils (CNF) (sometimes also cellulose filaments or CF), were mechanically broken down and often homogenized (first by Herrick, Casebier, Hamilton, & Sandberg, 1983; Turbak, Snyder, & Sandberg, 1983). The specific surface region and the quantity of the hydrogen bonds following from the surface hydroxyl groups got expanded during fibrillation or delamination. Inferable from this, NFCs were slanted to shape gels indicating expanded consistency. The NFCs nowadays are monetarily created and utilized in wealth applications including composites, coatings, personal care products, and so on (Abitbol et al., 2016).

CNCs, on the other hand, additionally named as nanowhiskers, are round, bar, or needle-shaped glasslike structures. These were first isolated using acid corrosive acid hydrolysis by Ranby in 1949. It was accounted for, that the length and surface charge of CNCs relies upon the acid hydrolysis period (i.e., exposure time) by Dong, Revol, & Gray (1998). Another intriguing perspective was examined by Marchessault, Morehead, & Walter (1959), demonstrating birefringent fluid crystalline stages.

Cellulose is found not only in the plant cell walls, but is also produced by the bacteria naturally. This kind of cellulose and nanocellulose fibrils recovered is referred to as the BC. Sometimes it is also termed as microbial cellulose or biocellulose. Initially, BC was mostly extracted from species like *Acetobacter, Rhizobium, Agrobacterium*, etc. followed by culture media preparation (El-Saied, Basta, & Gobran, 2004).

Besides microbial cellulose, a different category of nanocellulose was extensively produced through electrospinning. The cellulose material that was dissolved in a particular solvent was exposed to high voltage such that the particles acquired charge and hence repelled each other at a critical voltage, overcoming the surface tension of the structure. At a point when it was pushed through the air, the solvent vanished rendering the fibers which were then gathered on an electrically grounded plate (Ramakrishna et al., 2006).

Further to grant more efficient mechanical properties, natural effects and ecofriendly attribute to nanocellulose composites; surface alterations and modifications are being achieved in the composite preparation processes.

6.4 Products of nanocellulose: nanowood and nanopaper

6.4.1 Nanowood

A particular alignment of cellulose nanofiber in a compact structure is called **nanowood**. It is basically nanocellulose-derived product. It is light weight in nature, mechanically strong, ecofriendly, thermally insulated, and has low carbon footprint.

Nanowood is structurally different from natural wood as natural wood contains components like heteropolysaccharide, hemicellulose, paracrystalline cellulose fibril aggregate, lignin, etc., where lignin and hemicellulose are intertwined with each other, but in nanowood, lignin and hemicelluloses are absent and there is only an aligned structure of cellulose nanofibers in the form of aggregated fibrils (Sjostrom, 2013).

Nanofibers have great packaging properties and can be reinforced into polymers. Converging these properties of nanofibers into a new product, that is, the nanowood deduced form wood pulp, a prominent product which overcomes the lacking abilities of wood, can replace the use of natural wood. Literature describes various method of wood modification, wood enhancement with TiO_2, silica, and different wood preservation methods (Kashef & Sabouni, 2010), but instead of preservation, discovering a new material like nanowood, comes out to be the best replacement for natural wood.

Li et al. (2018) were the first who described the structure of nanowood as porous, and hierarchical structure alignment of fibril aggregates which were highly maintained and aligned, and also discussed their preparation. They observed the microporous structure with aligned channels inside the nanowood which were composed of cellulose nanofibers through scanning electron microscope (Li et al., 2018).

6.4.1.1 Fabrication of nanowood

Discovery of this new material called nanowood, accomplished in year 2018 by the team of scientists. They followed a basic strategy of Wicklein et al. (2015) and were the first to innovate an anisotropic nanocomposite by the technique of freeze drying (Li et al., 2018). Freeze drying was the most widely used approach for the drying of nanocellulose particles, and this process involved two major steps: first freezing where solidification of nanocellulose suspension was done and second, the drying in which sublimation of water molecules was achieved (Wang et al., 2019).

The nanowood was primarily prepared from natural American basswood and other types of wood were also used (Li et al., 2018). The wood piece was taken and treated with the mixture of Na_2SO_3 and NaOH till the boiling temperature was reached at.

After boiling, the resultant was treated with H_2O_2 which removed most of the hemicelluloses and lignin from the wood. Throughout the process keen attention was given to maintain the nanoporous structure and hierarchical alignment of wood. The H_2O_2 treatment was further followed by freeze drying process and the resultant product that was obtained was the nanowood (Li et al., 2018). A top-down mechanical disintegration approach was followed for the preparation of nanowood to produce nanowood foams (NWFs). For the preparation as a primary element, the waste wood biomass was used. A simplistic approach of top-down mechanical disintegration of wet grinding and high-pressure homogenization was further followed to get a nanowood colloidal mixture. Then this colloidal mixture was allowed to go through a bottom up approach to generate ion-associated NWFs (Li et al., 2020).

6.4.1.2 Properties of nanowood

Nanowood is a light weight structure that possesses lower mass density. This is due to the removal of intermixed lignin and hemicellulose contributing to approximately 70% mass loss in the wood structure. Being biodegradable in nature, it is devoid of any synthetic material and only constitutes the cellulose nanofibers, which can be easily biodegraded. Some of the other characteristics possessed by the nanowood are mentioned below:

- Anisotropic nature of nanowood

Li et al. (2018) basically explained major four points that would prove the anisotropic nature of nanowood. First, the porous structure of nanowood. The extent of porosity was comparably large from the natural wood and the large porosity resulted in low-thermal conduction. Second, the nanowood lacked the intermixing of lignin and hemicelluloses, which thus reduced the linkage and caused low-thermal conduction. Moreover, the high aspect ratio of aligned nanofibril aggregates resulted in an anisotropic heat flow, and the small void channels of nanosize blocked the pathway of heat and air, making the nanowood thermally insulated (Li et al., 2018).

Hou et al. (2019) also tested the thermal conductivity of hydrophobic nanowood membrane with the help of infrared radiation camera, which resulted in lower backside temperature when compared to the natural wood. When comparing the thermal property of nanowood with other commercially available insulating material like foam, Styrofoam, anisotropic wood, etc., nanowood turned out to exhibit least thermal conduction.

- Mechanical and optical properties of nanowood

Li et al. (2018) compared the mechanical strength of nanowood with the existing insulating materials and it once again, turned out to be much stronger than the rest. The compressive strength of nanowood in both axial and transverse section was also examined with the help of electromechanical tensile and compression testing machine called Tinius Olsen H25KT. The maximum stress approached to be 13 MPa, which represented the highest value of mechanical strength. The reason behind this high mechanical strength was found out to be the maintained orientation of the fibrils and aligned orientation of crystalline cellulose molecules.

It was also observed during the ultraviolet—visible test for testing the optical nature of nanowood, tested on a small sized block, that the nanowood exhibits almost 95%% of reflection and emissivity of around 5%, thereby indicating the effective reflection for thermal energy.

Nanowood-derived product like NWFs also exhibited great compressive properties when combined with certain salts like NWF-Na_2SiO_3. The compressive stress of NWF-Na_2SiO_3 reached 152 kPa, due to the alkaline sodium silicate, acting as a good modifying agent, and thereby cross-linking the fibers making it mechanically much stronger (Li et al., 2020).

6.4.1.3 Applications of nanowood

Nanowood is worth of revolutionizing the building and manufacturing industries as it overcomes the loopholes of the existing building and insulation materials. Construction industry alone contributes 40% in global warming, which results in ozone layer depletion, hence urging the need of an eco-friendly material which can replace the conventionally used wood and other building materials, making nanowood a prominent solution (Shema et al., 2018).

Nanowood proves to be eloquent in sustainable energy saving, temperature maintenance and regulation in buildings, which is a needy requirement of humans for living experience. Existing wood-based materials in the building cause a great loss of energy, they are not significant for thermal management, and also pose some disadvantages. Thus it is necessary to explore new energy-saving materials, and nanowood is an appropriate solution as it comes with very low-thermal conductivity and high heat and light reflection (Zhu, Xiao, Chen, & Fu, 2020). The aerogel particles also help in empowering the nanowood-based materials to achieve high-thermal activity and make materials more flexible when integrated in nanowood composites (Candan & Akbulut, 2013).

Nanowood is used in energy saving devices. Lithium-ion batteries are one of the important energy storage devices and are being used in various commodities like cars, phones, camera, etc. Traditional graphite-based batteries are being used but they do not possess sufficient energy density to meet the demands of energy saving. Thus nowadays nonmaterials like nanowood are replacing the older versions of batteries, which are more flexible in saving energy (Kim, Kim, & Pol, 2019).

Nanowood membrane is efficient in membrane distillation. Hou et al. (2019) developed a hydrophobic membrane for distillation and desalination of water. They fabricated the membrane distillation directly with the wood nonmaterial. This membrane was highly porous in nature and its pore size distribution facilitated water vapor transportation. The membrane also exhibited excellent thermal properties and good water flux (Hou et al., 2019).

Scientist prepare breathable nanowood biofilms for green on-skin electronics in which they prepare two different types of nanowood biofilms, capable of in-plane guiding sweat and heat and also, simultaneously, allowing cross-plane breathability, were developed (Zhou et al., 2019).

Nanowood-derived product, that is, nanowood ion-associated foam was developed which showed high stability, flame retardancy, thermal barrier, etc. They also prepared NWF called NWF-V acted as a three-dimensional (3-D) evaporator which exhibited a high evaporation rate and high conversion efficiency of sea water desalination (Li et al., 2020).

6.4.2 Nanopaper

6.4.2.1 Evolution of paper to nanopaper: an insight

The production of paper began way back around two or three thousand years ago. As man began to feel the need to communicate and keep a track record of the then discoveries and inventions and even when doing business etc., paper rose to become an essential commodity (Gunaratne, 2001). Nowadays, being used in almost every sector, a better alternative to the conventional paper, that is, nanopaper is achieving great deal of appraisal.

Paper is a two-dimensional planar structure which is traditionally derived from the pulp of the woods, an aqueous suspension of the wood fibers is prepared, which is further processed through mechanical and chemical treatments (Banik et al., 2011). It involves extracting paper pulp, bleaching, and finally production of the paper itself. Probably the initial achievement for the mechanical processing of the cellulosic mash or pulp was via the improvement in the Kraft Cycle in 1879 (Hu, Zhang, & Lee, 2018), beginning from which, the paper business has reliably looked into the enhancement of the paper quality. Kraft cycle is a process that dissolves the middle lamella present in the cells by removing the lignin and separating the individual nanocellulosic fibers. As a result, the cells placed in a digester are pressurized through an alkaline liquor, especially sodium hydroxide etc., hence rendering high-quality pulp for further fibers extraction (Smook, 2016).

Paper industry is being associated with research and development, whether it may be the search of new crude materials of high modern profile or adjustment in the extraction process. In the last few decades, the incorporation of nanotechnology in the papermaking industries has led to the lower cost production, and improved paper quality (Barhoum, Samyn, Öhlund, & Dufresne, 2017). Hence, nanotechnology enhances the quality and viability of nanopaper as it involves more proficient utilization of assets by which high-quality papers can be shaped using less amount of raw material. It utilizes auxiliary substances, for example, reused fiber strands with low-grade properties, which can be changed over to nanofibers and additional materials and results in the improvement of new materials, for example, production of paper with unique compatible properties.

Paper in itself gives a magnificent means to create nanomaterials fiber composites for their use in superior printing, food packaging, flexible electronics, medical diagnosis, etc.

Nanopaper, a cutting-edge variant of the traditional paper consists of a network of tiny linked nanocellulosic fibers. The nanopaper pore size ranges from somewhere between 5 and 50 nm with high optical transparency and higher stress tolerance. Possibly, the CNFs are used as paper-coatings due to their high mechanical strength, fire resistance, and water or air impermeability (Aulin & Ström, 2013). Due to their nanoscale dimensions, the CNFs are utilized for the processing of nanocomposite films with tough build networks. Nanocellulose acts as a reinforcing polymer in the paper-production process and this was observed and shown by Ahola, Salmi, Johansson, Laine, and Österberg (2008), and nanofibrils are usually catalyzed with a cationic polyelectrolyte like poly (amide amine) epichlorohydrin, to upgrade their strength.

6.4.2.2 Fabrication of nanopaper

Nanocellulose is a perfect strength enhancing additive in the paper making because of its unique properties. As an additive, nanocellulose acts in a similar manner as that of pulp refining and ensures that the porosity of the paper is reduced (González et al., 2012). These nanocellulose films used in the paper industry decrease the overall paper weight without affecting the mechanical and chemical properties of the paper and hence, enhance the thermal, optical, and antimicrobial properties of the paper (Barhoum et al., 2017). Nanocellulose is an effective replacement for cellulose in the papermaking due to the nanosized structure which forms a network of fibers held together by hydrogen bonds when nanocellulose is immersed in the extracted pulp (Jiang & Hsieh, 2014). Using the nanofibers along with the pulp can also decrease the scattering of light by a significant value, thereby decreasing the opacity and increasing the transparency.

- Fabrication of nanopaper by spraying fluorinated silica/multi-walled carbon nanotubes (SiO_2/MWCNTs) composite

  For the fabrication of the nanopaper, the cellulose nanofibers were synthesized utilizing the 2, 2, 6, 6-tetramethyl-1-piperidinyloxy (TEMPO) for their oxidation. This manufacturing process was initially reported by Tsuguyuki Saito, Kimura, Nishiyama, and Isogai (2007). In this process, the cellulosic fibers were dispersed and dissolved in distilled water which contained TEMPO and sodium bromide already in it. Then, NaClO solution was added slowly and the mixture was stirred through magnetic stirrer (at approximately 500 rpm). The pH of the solution was adjusted using either hydrochloric acid or sodium hydroxide of varying concentrations. The solution color turned to white due to the addition of ethanol. Now, this TEMPO-oxidized cellulose was freed from any excess chemicals through washing using distilled water and stored at a low temperature (usually 4°C) prior to further homogenization via high-pressure microfluidization. Then this nanofiber solution could circulate in a water-vacuum pump (0.1 Mpa vacuum degrees) that resulted in the formation of the nanopaper. This obtained nanopaper was then covered by two sterile glass plates and allowed to dry at room temperature for 2 days. This technique resulted in a nanopaper of thickness around 95–100 nm (Saito et al., 2007). To enhance the transparency and superhydrophobic properties of the nanopaper, fluorinated silica/multiwalled carbon nanotubes (SiO2/MWCNTs) hybrid was used. This hybrid was dispersed in ethanol and sonicated for around 28–30 min. This suspension was then poured into a spray bottle and sprayed over the nanopaper from a distance (coating thickness around 4 um). This amended nanopaper was then freed from excessive moisture by placing in a hot air oven (~50°C for 3–4 min) (Saito et al., 2007). These fluorinated nanoparticles, sized around 80 nm, were immobilized over the nanopaper surface, and this SiO_2/MWCNTs hybrid resulted in a highly organized micronanostructure with a hierarchical texture (Yao, Bae, Jung, & Cho, 2017). This modified nanopaper possessed highly efficient hydrophobicity and water-repelling properties (Saleh & Baig, 2019). The paper possessed remarkable self-cleansing properties too, as it reduced the contact of paper surface with that of sand or dust (Fürstner et al., 2005) and water droplets (Miwa, Nakajima, Fujishima, Hashimoto, & Watanabe, 2000).

- Fabrication of nanopaper from CNF extracted by using supercritical carbon dioxide treatment and acid hydrolysis

  There were several techniques of extraction and isolation of cellulosic nanofibers which had been previously discussed. However, treatment of biomass with supercritical carbon dioxide (SC-CO_2) to render CNFs was considered much more effective (Zheng, Lin, & Tsao, 1998). In this SC-CO_2 process, CO_2 could diffuse deep into the lignocelluloses, when a high pressure was applied to it producing nanofibrous structures (Benito-Román, Rodríguez-Perrino, Sanz, Melgosa, & Beltrán, 2018). For the fabrication of nanopaper, this extracted CNF suspension was used and vacuum filtered. Here, the individual separated CNF fibers combined resulting in a strengthened cluster of fibers held together by hydrogen bonds (Gopakumar et al., 2018). For this purpose, a filtration assembly comprising a cellulosic ester membrane was utilized and finally the paper was peeled off from the filter and allowed to dry in a hot press (~10–15 min). This modified nanopaper possessed a high tensile strength, as that of 75.7 Mpa, as the intermolecular hydrogen bonding increased the overall mechanical strength (Sehaqui, Zhou, Ikkala, & Berglund, 2011).

- Fabrication of nanopaper made up of esterified CNF obtained from corncob cellulose

  The cellulose extracted from corncobs (residue of corn saccharification) was esterified in the presence of hexanoyl chloride by ball milling (Huang et al., 2012). This esterified CNF (E-CNF) was then used to prepare the nanopaper with improvised mechano-chemical and optical properties. For this, the E-CNF suspension was dispersed in N, N-Dimethylformamide and diluted and then sonicated forming a clear suspension. This mixture was then passed through a nylon-6 membrane filter and the content that was retained by the filter paper was dried overnight (50°C–60°C). The quantity of the suspension used determined the viscosity of the nanopaper (Huang et al., 2012).
- Fabrication of porous cellulose nanopaper

In this technique, the NFC hydrogel was utilized and the water content present in it was substituted with ethyl alcohol (usually 95%–96%) when allowed for bath for first 24 h and later in pure ethanol bath for another day. This hydrogel turned alcogel was then allowed to dry in a critical point drying chamber. After this, liquid CO_2 was injected, replacing the ethanol solvent and depressurization released this remnant CO_2, yielding highly porous NFC nanopaper (Henriksson, Berglund, Isaksson, Lindstrom, & Nishino, 2008). The nanopaper prepared by this technique was observed to possess 18%–20% porosity (Sehaqui, Liu, Zhou, & Berglund, 2010).

6.4.2.3 Properties of nanopaper

- Morphology and Hydrophobicity

  According to the reports a liquid droplet reacts onto a solid surface depends upon the structure of the solid material and its surface energy (Lafuma & Quéré, 2003). The SiO_2/MWCNTs hybrid gives micronano hierarchical structural appearance to the nanopaper because of varying particle size (Yao et al., 2017). When the concentration of MWCNTs is increased, the structure reaches high complexity. The fluorinated SiO_2 particles maintain the low surface energy and when combined with MWCNTs exhibit the 3-D structure. The water contact angle determines the water repelling and hydrophobic properties of the nanopaper. For almost every modified nanopaper, the water contact angle is observed to be around 150°C–160°C depicting excellent hydrophobicity (Saleh & Baig, 2019).
- Self-cleansing ability

  Nanopapers show remarkable behavior in case of their self-cleansing property. It is depicted by experimentation, that when small amounts of dust particles (inclined at 20degree angle) (Fürstner et al., 2005) and even water droplets are scattered over the modified nanopaper, the water droplets tend to maintain their original shape and structure, while the dust particles are carried away. This demonstrates that the paper possesses less sliding angle and the structural complexity of the surface further resists the particles to contact with the modified nanopaper (Miwa et al., 2000).

- Thermal properties of nanopaper

  Thermal stability tends to be an essential and robust feature of modified nanopapers, as it increases their shelf life and lowers the cost of production (Faradilla, Lee, Sivakumar, Stenzel, & Arcot, 2019). The nano fibers of the nanopaper are observed to withstand high temperatures. However, if the temperature increases a certain limit (such as $> 200°C$), the fibers begin to degrade (initial degradation temperature, Ti). The fluorinated SiO_2/MWCNTs hybridized nanopaper has a Ti of 208.6°C, much larger as compared to the bare nanopaper. Moreover, the Ti of the modified nanopaper can be controlled by the concentrations of MWCNTs in the paper (Saleh, 2015), hence determining excellent thermal stability.
- Transparency and optical properties

  The transparency of the nanopaper can prove its necessity in their potential applications of the optical devices. As per the Mie hypothesis, when the surface harshness is roughly equivalent to or more than the frequency of light, there is a corresponding connection between the dispersing cross area and the square of the molecule width (Nakajima, Fujishima, Hashimoto, & Watanabe, 1999). Thus the surface roughness of gadgets with great optical properties ought to be a lot more modest than the frequency of obvious light. In this respect, the size of the roughness of the altered nanopaper surface is more moderate than the frequency of light. In this way, the nanopaper covered with fluorinated SiO_2/MWCNTs coats indicates high transparency.
- Electrical conductivity

The electrical conductivity of the modified nanopaper is likewise significant component to widen its potential application zones. With the mass part of MWCNT being expanded, the sheet resistance of the altered nanopaper diminishes. This pattern could be clarified by percolation hypothesis (Semmler, Bley, Klupp Taylor, Stingl, & Vogel, 2019). The electron transport of nanopaper is reliant on the thickness of the conductive MWCNT. Accordingly, the increment in the mass part of MWCNTs would improve the movement of electrons, lessening the sheet resistance, and hence, improving the electrical conductivity.

6.4.2.4 Applications of nanopaper

- Nanopaper in Strain sensors

  Strain sensors are being utilized in various infrastructures and also hold importance for vehicular checking. The existing business items are basically founded on massive advances which are modest but can only identify low strains because of the exceptionally restricted stretch-ability of metal and semiconductors (Barlian, Park, Mallon, Rastegar, & Pruitt, 2009). High-strain sensors are in basic requests for remarkable applications past the current business sectors, for example, wearable health sensors, automated tangible, sensory skin, electronic gloves, etc., which all require high-strain detecting abilities that cannot be accomplished by ordinary innovation. Nanoscale materials are being demonstrated for promising structure blocks for imaginative strain sensors. Graphene nanosheets with exceptional electrical and mechanical properties (Politano et al., 2012); have been broadly considered for strain detecting applications (Kim et al., 2009). The 3-D macro-porous nanopapers made of folded graphene and nanocellulose are inserted in stretchable elastomer matrix to manufacture the strain sensors.

- Nanopaper-based transistor arrays

  Environmentally friendly and cheap cellulose nanopaper is a promising upcomer as a novel substrate for flexible electron gadget applications. Here, a slim and transparent nanopaper-based highly mobile organic thin film transistor cluster is utilized. In the past few years, the flexible electronic gadgets resulting from paper have pulled in significant consideration as a potential next-generation innovation to the presently utilized plastic-based hardware. A wide scope of utilities, for example, brilliant pixels (Andersson et al., 2002), memory gadgets (Martins et al., 2008), printed circuits (Siegel et al., 2010), electrowetting (Kim & Steckl, 2010), batteries (Hu et al., 2013), photovoltaic gadgets (Barr et al., 2011), contact sensors (Mazzeo et al., 2012), etc. have been based on substrates made of traditional paper which is now being replaced by nanopaper, and the techniques utilized in the nanofibrillation and the creation of paper substrates bestow high optical properties to the nanopaper, rendering desired products.

- Nanopapers in the recognition of chiral analytes

  Paper-based sensors open new road to create uncomplicated, fast, cheap, and single-use analytical gadgets for a wide application fields including clinical diagnosis, food investigation, and ecological observation. The discovery of an enantiomer of chiral species can be a valuable pointer for the investigation of food quality (Kaniewska & Trojanowicz, 2011). Likewise, most of the present-day drugs are made from individual enantiomers of chiral species, which may display higher desirable impacts with lower toxic levels than their antipodes (Zor et al., 2017). In the most recent decade, nanomaterial-joined paper-based sensors have pulled in wonderful examination and business consideration. The transparency of the nanopaper additionally makes it a truly reasonable substrate for fuse, or immobilization of optical detecting specialists to be used in optical sensors (Pourreza, Golmohammadi, Naghdi, & Yousefi, 2015). To this point, nanopaper is acquired by environment-friendly approach utilizing BC made of nanofibers, and silver nanoparticles are inserted inside nanopaper by an in-situ strategy.

- Nanopaper-based microcuvette for iodide detection in sea water

  Expendable nanopaper-based sensor is fit for distinguishing iodide (an anion that not just plays a role in organic and inorganic oxidation and maintains the marine climate; Khodari, Bilitewski, & Basry, 2015, but also, additionally has fundamental job in neurological action and thyroid organ work) (Liu et al., 2013). In this manner, the assurance of iodide is basically significant in both the life and ecological sciences. Utilization of BC nanopaper, especially in optical detecting stages is getting progressively famous; and in further aspects for iodide detection, the carbon quantum dots are inserted in paper-based sensor for specific iodide detection. Nitrogen-doped carbon quantum specks are also sometimes inserted into nanopaper to acquire photoluminescence properties. Morales-Narváez et al. (2015) have further assessed the photoluminescent nanoparticles in the nanopaper as detecting platform.

- Nanopaper-based self-powered human interactive systems

The self-controlled and human intelligent electronics have huge applications in anti-theft and antifake for human society. In this study, a nanopaper-based and self-powering system is a good adaptable framework. The framework depends on electrostatic induction instrument. Oneself controlled, self-powering and transparent nanopaper gadgets can be utilized against burglary. Up until now, developers have exhibited adaptable nanopaper semiconductor, organic light-emitting diode, sensory panels, antennas, etc. (Huang et al., 2013). However, all the previously mentioned hardware gadgets require either outside force source or installed batteries. Hence, the self-controlled nanopaper-based hardware frameworks emerge as brilliant and critical innovations; important for both shoppers, military and protection gadgets (Fan, Tian, & Wang, 2012).

6.5 Conclusion

The application of nanoscience to the current day technologies elevates the profitability and potentiality of the products and improvises them. Being biodegradable, biocompatible, and well-assembled, these cellulose-derived materials pose less or no threat to the environment, thereby maintaining the integrity and sustainability of the rendered products. These bioproducts and biopolymers prove to be in favor of the environment issues. In this respect, nanowood turns out to be an eloquent replacement for the conventional wood due to its highly appreciable characteristic properties, which in turn in a way resolves the problem of deforestation (which further chains down to global warming), which we humans are unable to cope with due to the increasing demands in almost every sector of society. Nanopaper, on the other hand, is an emerging nanotechnology which is being used in electronic gadgets and every other biomedical device. However, it needs to be worked upon and modified so that it can substitute the conventional paper, which is yet to be achieved. Thus nanotechnology on the whole, provides promising possibilities in the production of new products that are ecofriendly, along with revolutionizing the already available products such that they satisfy the human needs and also deal efficiently with environmental aspects.

References

Abitbol, T., Rivkin, A., Cao, Y., Nevo, Y., Abraham, E., Ben-Shalom, T., ... Shoseyov, O. (2016). Nanocellulose, a tiny fiber with huge applications. *Current Opinion in Biotechnology, 39*, 76–88.

Ahola, S., Salmi, J., Johansson, L., Laine, J., & Österberg, M. (2008). Model films from native cellulose nanofibrils. Preparation, swelling, and surface interactions. *Biomacromolecules, 9*(4), 1273–1282.

Andersson, P., Nilsson, D., Svensson, P. O., Chen, M., Malmström, A., Remonen, T., ... Berggren, M. (2002). Active matrix displays based on all-organic electrochemical smart pixels printed on paper. *Advanced Materials, 14*(20), 1460–1464.

Aulin, C., & Ström, G. (2013). Multilayered alkyd resin/nanocellulose coatings for use in renewable packaging solutions with a high level of moisture resistance. *Industrial & Engineering Chemistry Research, 52*(7), 2582–2589.

Banik, G., Brückle, I., Daniels, V., Fischer, S., Keller, S. W., Kosek, J. M., ... Whitmore, P. M. (2011). *Paper and water: A guide for conservators* (pp. 419–436). Oxford: Butterworth-Heinemann.

Barhoum, A., Samyn, P., Öhlund, T., & Dufresne, A. (2017). Review of recent research on flexible multifunctional nanopapers. *Nanoscale, 9*(40), 15181–15205.

Barlian, A. A., Park, W. T., Mallon, J. R., Rastegar, A. J., & Pruitt, B. L. (2009). Semiconductor piezoresistance for microsystems. *Proceedings of the IEEE, 97*(3), 513–552.

Barr, M. C., Rowehl, J. A., Lunt, R. R., Xu, J., Wang, A., Boyce, C. M., ... Gleason, K. K. (2011). Direct monolithic integration of organic photovoltaic circuits on unmodified paper. *Advanced Materials, 23*(31), 3500–3505.

Benito-Román, O., Rodríguez-Perrino, M., Sanz, M. T., Melgosa, R., & Beltrán, S. (2018). Supercritical carbon dioxide extraction of quinoa oil: Study of the influence of process parameters on the extraction yield and oil quality. *The Journal of Supercritical Fluids, 139*, 62–71.

Bhushan, B. (2017). Introduction to nanotechnology. In B. Bhushan (Ed.), *Springer handbook of nanotechnology*. Springer, Berlin, Heidelberg: Springer Handbooks. Available from 10.1007/978-3-662-54357-3_1.

Candan, Z., & Akbulut, T. (2013). Developing environmentally friendly wood composite panels by nanotechnology. *BioResources, 8*(3), 3590–3598.

Carrasco, F., Mutje, P., & Pelach, M. A. (1996). Refining of bleached cellulosic pulps: Characterization by application of the colloidal titration technique. *Wood Science and Technology, 30*(4), 227–236.

Dong, X. M., Revol, J. F., & Gray, D. G. (1998). Effect of microcrystallite preparation conditions on the formation of colloid crystals of cellulose. *Cellulose, 5*(1), 19–32.

El-Saied, H., Basta, A. H., & Gobran, R. H. (2004). Research progress in friendly environmental technology for the production of cellulose products (bacterial cellulose and its application). *Polymer-Plastics Technology and Engineering, 43*(3), 797–820.

Fan, F. R., Tian, Z. Q., & Wang, Z. L. (2012). Flexible triboelectric generator. *Nano Energy, 1*(2), 328–334.

Fang, Z., Zhu, H., Preston, C., Han, X., Li, Y., Lee, S., ... Hu, L. (2013). Highly transparent and writable wood all-cellulose hybrid nanostructured paper. *Journal of Materials Chemistry C, 1*(39), 6191–6197.

Faradilla, R. F., Lee, G., Sivakumar, P., Stenzel, M., & Arcot. (2019). Effect of polyethylene glycol (PEG) molecular weight and nanofillers on the properties of banana pseudostem nanocellulose films. *Carbohydrate Polymers, 205*, 330–339.

Favier, V., Cavaille, J. Y., Canova, G. R., & Shrivastava, S. C. (1977). Mechanical percolation in cellulose whisker nanocomposites. *Polymer Engineering and Science, 37*(10), 1732–1739.

Fürstner, R., Barthlott, W., Neinhuis, C., & Walzel, P. (2005). Wetting and self-cleaning properties of artificial superhydrophobic surfaces. *Langmuir: The ACS Journal of Surfaces and Colloids, 21*(3), 956–961.

González, I., Boufi, S., Pèlach, M. A., Alcalà, M., Vilaseca, F., & Mutjé, P. (2012). Nanofibrillated cellulose as paper additive in eucalyptus pulps. *BioResources*, 7(4), 5167–5180.

Gopakumar, D. A., Pai, A. R., Pottathara, Y. B., Pasquini, D., Carlos de Morais, L., Luke, M., ... Thomas, S. (2018). Cellulose nanofiber-based polyaniline flexible papers as sustainable microwave absorbers in the X-band. *ACS Applied Materials & Interfaces, 10*(23), 20032–20043.

Grimsdale, A. C., & Müllen, K. (2005). The chemistry of organic nanomaterials. *Angewandte Chemie International Edition*, *44*(35), 5592–5629.

Gunaratne, S. A. (2001). Paper, printing and the printing press: A horizontally integrative macrohistory analysis. *Gazette (Leiden, Netherlands)*, *63*(6), 459–479.

Habibi, Y., Lucia, L. A., & Rojas, O. J. (2010). Cellulose nanocrystals: Chemistry, self-assembly, and applications. *Chemical Reviews*, *110*(6), 3479–3500.

Hayashi, N., Kondo, T., & Ishihara, M. (2005). Enzymatically produced nano-ordered short elements containing cellulose Iβ crystalline domains. *Carbohydrate Polymers*, *61*(2), 191–197.

Henriksson, M., Berglund, L. A., Isaksson, P., Lindstrom, T., & Nishino, T. (2008). Cellulose nanopaper structures of high toughness. *Biomacromolecules*, *9*(6), 1579–1585.

Herrick, F. W., Casebier, R. L., Hamilton, J. K., & Sandberg, K. R. (1983). Microfibrillated cellulose: Morphology and accessibility. *Journal of Applied Polymer Science*, *37*, 2.

Hou, D., Li, T., Chen, X., He, S., Dai, J., Mofid, S. A., ... Hu, L. (2019). Hydrophobic nanostructured wood membrane for thermally efficient distillation. *Science Advances*, *5*(8), eaaw3203.

Hon, D. N. S. (1994). Cellulose: A random walk along its historical path. *Cellulose*, *1*, 1–25.

Hu, J., Zhang, Q., & Lee, D. J. (2018). Kraft lignin biorefinery: A perspective. *Bioresource Technology*, *247*, 1181–1183.

Hu, L., Liu, N., Eskilsson, M., Zheng, G., McDonough, J., Wagberg, L., et al. (2013). Silicon-conductive nanopaper for Li-ion batteries. *Nano Energy*, *2*(1), 138–145.

Huang, J., Zhu, H., Chen, Y., Preston, C., Rohrbach, K., Cumings, J., & Hu, L. (2013). Highly transparent and flexible nanopaper transistors. *ACS Applied Nanomaterial Forums*, 7(3), 2106–2113.

Huang, P., Wu, M., Kuga, S., Wang, D., Wu, D., & Huang, Y. (2012). One-step dispersion of cellulose nanofibers by mechanochemical esterification in an organic solvent. *ChemSusChem*, *5*(12), 2319–2322.

Jiang, F., & Hsieh, Y. L. (2014). Super water absorbing and shape memory nanocellulose aerogels from TEMPO-oxidized cellulose nanofibrils via cyclic freezing–thawing. *Journal of Materials Chemistry A*, *2*(2), 350–359.

Kamel, S. (2007). Nanotechnology and its applications in lignocellulosic composites, a mini review. *Express Polymer Letters*, *1*(9), 546–575.

Kaniewska, M., & Trojanowicz, M. (2011). Chiral biosensors and immunosensors. *Biosensors: Emerging Materials and Applications*, 99.

Kashef M. and Sabouni A., 2010. Nanotechnology and the building industry. In: *Proceedings of the international conference on Nanotechnology: Fundamentals and applications*, Ottawa, Ontario, Canada.

Khodari, M., Bilitewski, U., & Basry, A. A. H. (2015). Screen-printed electrodes for amperometric determination of iodide. *Electroanalysis*, *27*(2), 281–284.

Kim, D. Y., & Steckl, A. J. (2010). Electrowetting on paper for electronic paper display. *ACS Applied Materials & Interfaces*, *2*(11), 3318–3323.

Kim, K. S., Zhao, Y., Jang, H., Lee, S. Y., Kim, J. M., Kim, K. S., ... Hong, B. H. (2009). Large-scale pattern growth of graphene films for stretchable transparent electrodes. *Nature*, *457*(7230), 706–710.

Kim, P. J., Kim, K., & Pol, V. G. (2019). A comparative study of cellulose derived structured carbons on the electrochemical behavior of lithium metal-based batteries. *Energy Storage Materials*, *19*, 179–185.

Klemm, D., Heublein, B., Fink, H. P., & Bohn, A. (2005). Cellulose: Fascinating biopolymer and sustainable raw material. *Angewandte Chemie International Edition*, *44*(22), 3358–3393.

Klemm, D., Kramer, F., Moritz, S., Lindström, T., Ankerfors, M., Gray, D., & Dorris, A. (2009). Cellulose nanocrystals: Chemistry, self-Assembly, and applications. *Angewandte Chemie International Edition*, *50*(24), 5438–5466.

Lafuma, A., & Quéré, D. (2003). Superhydrophobic states. *Nature Materials*, *2*(7), 457–460.

Li, M., Rui, J., Liu, D., Su, F., Li, Z., Qiao, H., ... Guo, M. (2020). Liquid transport in fibrillar channels of ion-associated cellular nanowood foams. *ACS Applied Materials & Interfaces*, *12*(52), 58212–58222.

Li, T., Song, J., Zhao, X., Yang, Z., Pastel, G., Xu, S., ... Jiang, F. (2018). Anisotropic, lightweight, strong, and super thermally insulating nanowood with naturally aligned nanocellulose. *Science Advances*, *4*(3), eaar3724.

Liu, S. Q., Hu, F. T., Liu, C. B., Chen, F., Wu, Z. Y., Liang, Z. Q., . . . Chen, Z. G. (2013). Graphene sheet-starch platform based on the groove recognition for the sensitive and highly selective determination of iodide in seafood samples. *Biosensors and Bioelectronics*, *47*, 396–401.

Marchessault, R. H., Morehead, F. F., & Walter, N. M. (1959). Liquid crystal systems from fibrillar polysaccharides. *Nature*, *184*(4686), 632–633.

Martins, R., Barquinha, P., Pereira, L., Correia, N., Gonçalves, G., Ferreira, I., & Fortunato, E. (2008). Write-erase and read paper memory transistor. *Applied Physics Letters*, *93*(20), 203501.

Mazzeo, A. D., Kalb, W. B., Chan, L., Killian, M. G., Bloch, J. F., Mazzeo, B. A., & Whitesides, G. M. (2012). BA 15 Mazzeo and GM Whitesides. *Advanced Materials*, *24*, 2850–2856.

Miwa, M., Nakajima, A., Fujishima, A., Hashimoto, K., & Watanabe, T. (2000). Effects of the surface roughness on sliding angles of water droplets on superhydrophobic surfaces. *Langmuir: the ACS Journal of Surfaces and Colloids*, *16*(13), 5754–5760.

Mohieldin, S. D., et al. (2011). Nanotechnology in pulp and paper industries: A review. *Key Engineering Materials*, *471*, 251–256.

Morales-Narváez, E., Golmohammadi, H., Naghdi, T., Yousefi, H., Kostiv, U., Horak, D., . . . Merkoçi, A. (2015). Nanopaper as an optical sensing platform. *ACS Nano*, *9*(7), 7296–7305.

Nakajima, A., Fujishima, A., Hashimoto, K., & Watanabe, T. (1999). Preparation of transparent superhydrophobic boehmite and silica films by sublimation of aluminum acetylacetonate. *Advanced Materials*, *11*(16), 1365–1368.

Nechyporchuk, O., Belgacem, M. N., & Bras, J. (2016). Production of cellulose nanofibrils: A review of recent advances. *Industrial Crops and Products*, *93*, 2–25.

Politano, A., Marino, A. R., Campi, D., Farías, D., Miranda, R., & Chiarello, G. (2012). Elastic properties of a macroscopic graphene sample from phonon dispersion measurements. *Carbon*, *50*(13), 4903–4910.

Pourreza, N., Golmohammadi, H., Naghdi, T., & Yousefi, H. (2015). Green in-situ synthesized silver nanoparticles embedded in bacterial cellulose nanopaper as a bionanocomposite plasmonic sensor. *Biosensors and Bioelectronics*, *74*, 353–359.

Puurunen, K., & Vasara, P. (2007). Opportunities for utilising nanotechnology in reaching near-zero emissions in the paper industry. *Journal of Cleaner Production*, *15*(13–14), 1287–1294.

Ramakrishna, S., Fujihara, K., Teo, W. E., Yong, T., Ma, Z., & Ramaseshan, R. (2006). Electrospun nanofibers: solving global issues. *Materials Today*, *9*(3), 40–50.

Ranby, B. G. (1949). Aqueous colloidal solutions of cellulose micelles. *Acta Chemica Scandinavica*, *3*(5), 649–650.

Rånby, B. G., & Ribi, E. D. (1950). Über den feinbau der zellulose. *Experientia*, *6*(1), 12–14.

Saito, T., & Isogai, A. (2004). TEMPO-mediated oxidation of native cellulose. The effect of oxidation conditions on chemical and crystal structures of the water-insoluble fractions. *Biomacromolecules*, *5*(5), 1983–1989.

Saito, T., Kimura, S., Nishiyama, Y., & Isogai, A. (2007). Cellulose nanofibers prepared by TEMPO-mediated oxidation of native cellulose. *Biomacromolecules*, *8*(8), 2485–2491.

Saleh, T. A. (2015). Isotherm, kinetic, and thermodynamic studies on Hg (II) adsorption from aqueous solution by silica-multiwall carbon nanotubes. *Environmental Science and Pollution Research*, *22*(21), 16721–16731.

Saleh, T. A., & Baig, N. (2019). Efficient chemical etching procedure for the generation of superhydrophobic surfaces for separation of oil from water. *Progress in Organic Coatings*, *133*, 27–32.

Sehaqui, H., Liu, A., Zhou, Q., & Berglund, L. A. (2010). Fast preparation procedure for large, flat cellulose and cellulose/inorganic nanopaper structures. *Biomacromolecules*, *11*(9), 2195–2198.

Sehaqui, H., Zhou, Q., Ikkala, O., & Berglund, L. A. (2011). Strong and tough cellulose nanopaper with high specific surface area and porosity. *Biomacromolecules*, *12*(10), 3638–3644.

Semmler, J., Bley, K., Klupp Taylor, R. N., Stingl, M., & Vogel. (2019). Particulate coatings with optimized haze properties. *Advanced Functional Materials*, *29*(4), 1806025.

Sharma, A., Thakur, M., Bhattacharya, M., Mandal, T., & Goswami, S. (2019). Commercial application of cellulose nano-composites—A review. *Biotechnology Reports*, *21*, e00316.

Shema, A. I., Balarabe, M. K., & Alfa, M. T. (2018). Energy efficiency in residential building using nano composite material. *International Journal of Civil Engineering and Technology*, *9*(3), 853–864.

Siegel, A. C., Phillips, S. T., Dickey, M. D., Lu, N., Suo, Z., & Whitesides, G. M. (2010). Foldable printed circuit boards on paper substrates. *Advanced Functional Materials*, *20*(1), 28–35.

Sjostrom, E. (2013). *Wood chemistry. Fundamentals and application* (2nd ed.). London: Academic Press.

Smook, G. (2016). *Handbook for pulp & paper technologists (The Smook Book)* (4th ed.). TAPPI Press.

Takacs, E., Wojnarovits, L., Földváry, C., Hargittai, P., Borsa, J., & Sajo, I. (2000). Effect of combined gamma-irradiation and alkali treatment on cotton–cellulose. *Radiation Physics and Chemistry*, *57*(3–6), 399–403.

Turbak, A. F., Snyder, F. W., & Sandberg, K. R. (1983). Microfibrillated cellulose, a new cellulose product: Properties, uses, and commercial potential. *Journal of Applied Polymer Science*, *37*(9), 815–827.

Wang, Q., Yao, Q., Liu, J., Sun, J., Zhu, Q., & Chen, H. (2019). Processing nanocellulose to bulk materials: A review. *Cellulose*, *26*(13), 7585–7617.

Wegner, T. H., & Jones, E. P. (2009). A fundamental review of the relationships between nanotechnology and lignocellulosic biomass. *The Nanoscience and Technology of Renewable Biomaterials*, *1*, 1–41.

Wicklein, B., Kocjan, A., Salazar-Alvarez, G., Carosio, F., Camino, G., Antonietti, M., & Bergström, L. (2015). Thermally insulating and fire-retardant lightweight anisotropic foams based on nanocellulose and graphene oxide. *Nature nanotechnology*, *10*(3), 277–283.

Yao, W., Bae, K. J., Jung, M. Y., & Cho, Y. R. (2017). Transparent, conductive, and superhydrophobic nanocomposite coatings on polymer substrate. *Journal of Colloid and Interface Science*, *506*, 429–436.

Zheng, Y., Lin, H. M., & Tsao, G. T. (1998). Pretreatment for cellulose hydrolysis by carbon dioxide explosion. *Biotechnology Progress*, *14*(6), 890–896.

Zhou, T., Wang, J. W., Huang, M., An, R., Tan, H., Wei, H., … He, J. (2019). Breathable nanowood biofilms as guiding layer for green on-skin electronics. *Small (Weinheim an der Bergstrasse, Germany)*, *15*(31), 1901079.

Zhu, H., Luo, W., Ciesielski, P. N., Fang, Z., Zhu, J. Y., Henriksson, G., … Hu, L. (2016). Wood-derived materials for green electronics, biological devices, and energy applications. *Chemical Reviews*, *116*(16), 9305–9374.

Zhu, Z., Xiao, G., Chen, J., & Fu, S. (2020). Wood nanotechnology: A more promising solution toward energy issues: A mini-review. *Cellulose*, *27*, 8513–8526.

Zor, E., Morales-Narváez, E., Alpaydin, S., Bingol, H., Ersoz, M., & Merkoçi, A. (2017). Graphene-based hybrid for enantioselective sensing applications. *Biosensors and Bioelectronics*, *87*, 410–416.

CHAPTER 7

Pulp and paper industry-based pollutants, and their adverse impacts

Komal Rizwan[1], Tahir Rasheed[2], Muhammad Bilal[3] and Hafiz M.N. Iqbal[4]
[1]Department of Chemistry, University of Sahiwal, Sahiwal, Pakistan
[2]Interdisciplinary Research Center for Advanced Materials, King Fahd University of Petroleum and Minerals, Dhahran, Saudi Arabia
[3]School of Life Science and Food Engineering, Huaiyin Institute of Technology, Huai'an, P.R. China
[4]Tecnologico de Monterrey, School of Engineering and Sciences, Monterrey, Mexico

7.1 Introduction

Paper is produced at a large scale to meet the needs of human beings but this industry releases a huge amount of organic (chlorocatechols, hexadecanoic acids, methoxy phenols, octacosane, chlorosyringols, β-sitosterol trimethylsilyl ether, tetrachlorocatechol, chloroguaiacols, terpenes, methanol, decalone, benzoic acid), inorganic (Zn, Mg, Fe, Cu, etc.), gaseous (carbon dioxide, hydrogen sulfides, methyl mercaptan, chlorine dioxide, etc.), and solid wastes (sludge) contaminants in the environment (Pola, Collado, Oulego, Calvo, & Díaz, 2021). The paper production process consumes a high amount of energy and water (Singh & Chandra, 2019; Veluchamy & Kalamdhad, 2017). Recycled waste paper is utilized as raw material for newspaper production. Globally, paper is produced at a large scale on an annual basis (G. Singh, Kaur, Khatri, & Arya, 2019), and a huge amount of toxic pollutants (various gases, solids, and liquids) are discharged from these industries (Fig. 7.1) (Gopal, Sivaram, & Barik, 2019; Singh & Chandra, 2019). Paper mills have been declared among toxic industries. The wastewater released from these industries is damaging the ecosystem, including our aquatic system. (Haq & Raj, 2020; Ram, Rani, Gebru, & Abrha, 2020).

Various processes, including deinking, pulping, and washing are primary sources to generate solid waste materials. The degree of waste generated depends on various factors such as raw material employed, grade of paper, and processing techniques. Generally, the solid waste material is wet and possesses a potent quantity of ash (Monte, Fuente, Blanco, & Negro, 2009). Various gases such as carbon dioxide, hydrogen sulfides (H_2S), methyl mercaptan, and sulfur-based molecules are discharged in the air from solid waste material. These discharged gases and solid waste material are responsible for causing respiratory issues, cardiac disorder, eye and skin irritation, along with some other complications including nausea and headache (Singh & Chandra, 2019). The generation of paper includes pulping, bleaching, and finishing

Nanotechnology in Paper and Wood Engineering
DOI: https://doi.org/10.1016/B978-0-323-85835-9.00005-2

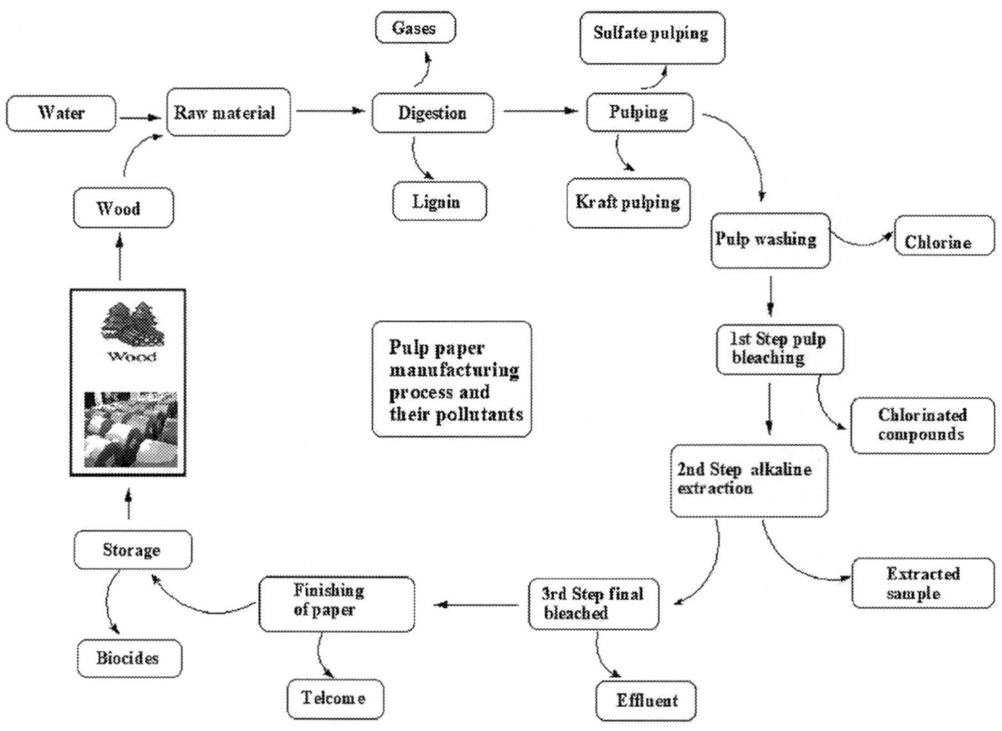

Figure 7.1 Schematic of paper manufacturing process and generation of various organic pollutants in each step of paper making (Singh & Chandra, 2019).

processes. The removal of noncellulosic materials (lignin) is called bleaching. The employment of chlorine-based molecules results in the generation of organo-chlorine molecules in the bleaching process. These compounds are discharged into the water system as effluents that are measured as adsorbable organic halogens (AOXs). These molecules, including bleaching contaminants, are responsible for causing skin, respiratory, and different chronic diseases. Mutagenic disorders and reproductive diseases in water-living bodies and humans are also caused by these effluents (Dwivedi, Vivekanand, Pareek, Sharma, & Singh, 2010). Currently, industries are employing eco-friendly elemental chlorine-free bleaching agents (peroxy acetic acid, hydrogen peroxide, sodium per-carbonate, and ozone, etc.) to produce highly white and bright kind of paper. Enzymes are potent and efficient to reduce the implementation of bleaching agents (Raj, Kumar, Singh, & Prakash, 2018). The utilization of prebleaching enzyme (xylanase) has been reported widely and commercially employed in various countries including Europe, North America, and Asia. The released contaminants from paper industries may be organic and inorganic and these also include natural molecules as resins, salts, organic-based halides, lignins, phenols, and cellulose-based compounds (Abhishek, Dwivedi, Tandan, & Kumar, 2017; Raj et al., 2018). Still, more research and studies are required to focus on the toxicology of these effluents. The accumulation, recalcitrance, and bio-magnification of harmful effluents of water require consideration for mitigation in animals, humans, and aquatic life (González, Sarria, & Sánchez, 2010). The effluents' effect has been considered many times but still, studies are needed to access the considerable ways for mitigation. The release of eco-friendly components into the aquatic system is under consideration by researchers (Olaniran & Igbinosa, 2011). Char generation from the waste effluents (lignins, cellulose, and hemi-cellulose) plays an important part in agriculture and wastewater treatment (Yaashikaa, Kumar, Varjani, & Saravanan, 2019). Traditional technologies are employed to diminish effluents from paper and pulp industries, but the presence of pollutants in processed effluents needs upgradation of techniques. Microbial fuel cell, bleaching technique, anaerobic digestion, and transformation routes are important techniques among all.

7.2 Waste effluents from the pulp and paper industry

Paper mills are the main source to generate huge waste worldwide. Due to the high amount of waste generation and contamination, the paper mill industries are placed in the category of 17 high polluting industries in India (Sonkar, Kumar, Dutt, & Kumar, 2019) and worldwide at 6th rank (Virendra, Purnima, Sanjay, Anil, & Rita, 2014). During the process of paper pulp production and finishing, the effluents of paper mills are produced. Features of the waste produced depend on the raw material used and various processes: bleaching, washing of pulp, and wood.

7.3 Pollutants from pulp and paper industry: categories and characteristics

Worldwide, the pulp and paper industries have been placed at 6th position in the list of dangerous, polluting industries, and discharge of their processed and unprocessed waste materials in water is the main source to cause environmental pollution (Vishal Kumar, Marín-Navarro, & Shukla, 2016). The disposal of pollutants in the ecosystem causes harmful diseases. The generated waste material from the paper industry is characterized into main four categories, including wastewater, solid, gaseous, and particle wastes (Sonkar et al., 2019). Wastewater consists of suspended-solids (fibers, dirt, pigments, and bark), dissolved inorganic materials, and bleaching materials. Toxic compounds, dyes, alcohols, turpentine, sizing agents, and thermal loads are also the main part of discharged effluents (Hooda, Bhardwaj, & Singh, 2015; Hooda, Bhardwaj, & Singh, 2018). Sludge is produced after primary and secondary wastewater treatment. Barks, fibers, ash from coal boilers, and lime sludge are also effluents produced during mills processing (Wang, Hung, Lo, & Yapijakis, 2005). In paper and pulp industries, air contamination is caused by the fabrication of sulfur-based volatile molecules (during sulfite pulping process), also some gases including dimethyl-sulfide, hydrogen sulfide, and methyl mercaptan etc. These gases are released during various processes (digester, evaporator, and recovery unit) (Pokhrel & Viraraghavan, 2004). The pollutants from paper and pulp mills mostly exist in the form of complexes. Mainly chromium, cadmium, and manganese are emitted in the environment from paper mills. And in nonmetallic elements, mostly calcium, sulfates, sodium, and chlorine-based compounds are released from paper mills (Chandra & Singh, 2012). These are produced during pulping, bleaching, and washing processes.

7.4 Adverse health impacts of pulp and paper industry pollutants

The waste generated in paper mills can be recycled and reused in safe manners, and this reduces the landfill. The prohibitory laws are implemented in various countries regarding the release of waste materials at landfill areas. Already existing rules what are those, enlist some and high increased taxes forced to develop novel methods for management of paper industry wastes (United States EPA, 1995). While keeping in view the volume, the sludge is the main effluent produced after wastewater management, and its production rate is variable in industries (Canmet, 2005). The discharged liquid effluents keep great biological and chemical oxygen demands, sodium, toxic heavy metals, and chlorine-based derivatives, and these are out of the acceptable limit of waste regulations guidelines (Elliott & Mahmood, 2005). Halides-based organic compounds are produced from bleaching effluents. These waste materials possess harmful contaminants, and these are released into the water system, directly responsible for polluting the aquatic system. Such a direct release of effluents to the aquatic system is harmful to aquatic life and may cause mutagenesis. Some common examples of harmful effects of pollutants from the pulp and paper industry are shown in Fig. 7.2. While some molecules in waste materials show difficulty in biodegradation and they gather in food chain and they consist of woody materials, cellulose fibers, lignin-based materials, acetic acid, saccharinic acid, and formic acids. Phytotoxicity and chromosomal issues are generated in onions with discharged effluents from the paper mill industry (Yadav & Chandra, 2018). Different research groups proposed the existence of various carcinogenic components in effluents (Monge-Corella et al., 2008). Methyl mercaptan is also the main effluent generated in the paper mill industry, and this is responsible for damaging the electron transport system by preventing cytochrome oxidase in humans (Singh & Chandra, 2019). Harmful effects of copper metal on aquatic life (catfish and trouts) in the existence of high pH and calcium in paper mill wastewater have been reported (Elliott & Mahmood, 2005). Heavy metals discharged from paper mills are reported to be accumulated in organs of various living species (Lindholm-Lehto, Knuutinen, Ahkola, & Herve, 2015). A proper irrigation system is absent, so farmers utilize polluted water sources for irrigation purposes, which causes heavy metals,

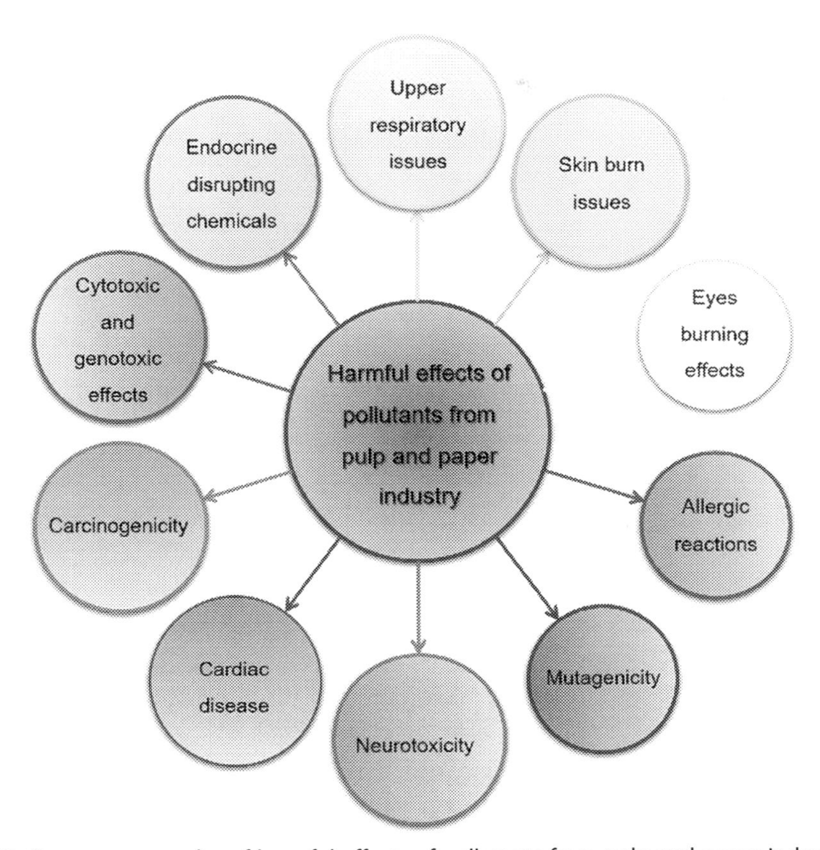

Figure 7.2 Common examples of harmful effects of pollutants from pulp and paper industry.

including various other contaminants in the agriculture system. According to reports, almost 4247 personnel were investigated, and among them, 380 cancer cases were determined, and most were suffering from ovarian cancer (Soskolne & Sieswerd, 2010). Scientists reported that workers of paper mill industries have a great risk of lungs, ovarian, breast, prostate, and stomach cancer (Monge-Corella et al., 2008). The use of chlorine-based compounds may form organo chlorine-based molecules, and these compounds, after discharge in water, are measured as Adsorbable Organic Halogens (Krigstin & Sain, 2006). These compounds also cause respiratory and other chronic disorders (Singh & Chandra, 2019).

7.5 Environmental implications regarding pulp and paper industry waste

Rules regarding the environment are reviewed on a regular basis. For maintaining the air and water quality, the parameters and limits of toxic chemicals are given by the United States environmental protection agency worldwide. For the management of

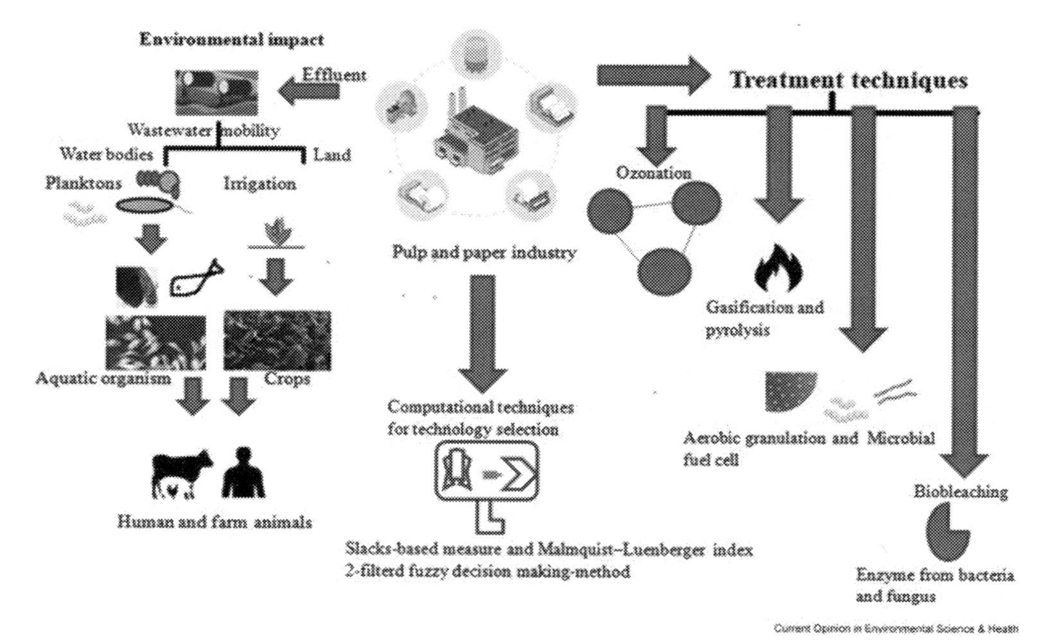

Figure 7.3 Representation of pulp and paper wastes effect on the ecosystem and various treatment techniques (Gupta, Liu, & Shukla, 2019). *Adopted from Gupta, G.K., Liu, H., & Shukla, P. (2019). Pulp and paper industry—based pollutants, their health hazards and environmental risks. Current Opinion in Environmental Science & Health, 12, 48—56.*

waste materials, the industries have their own system as well under certain relevant rules of the state (S. Gupta, Saksena, & Baris, 2019). Experiments were conducted on the fishes of Baikal lake; the paper mill producing paper for a long time exhibited agreement with environmental norms and suitable waste treatment measures (Kotelevtsev, Stepanova, Lindström-Seppä, & Hänninen, 2000). There are various toxic compounds like dichoroguicol, trichlorophenol, dichlorophenol, tetrachloroguicol, and pentachlorophenol in the paper industry wastewater that may lead to a decrease in the survival of animals (Virendra et al., 2014). Dangerous effects of wastewater generated in the pulp and paper industry, along with different treatment techniques, have been presented in Fig. 7.3.

The waste material from the paper industry impacts the food chain. The water bodies polluted with industrial wastewater become unsuitable for the growth of aquatic life, including fishes, microbes, and planktons. This is also responsible to destroy the economy of the fish market directly. Invasion of mutagens (present into wastewater) in zooplanktons, other aquatic life (seals, fishes, roach, etc.) existing near the paper industry has been reported (Stepanova et al., 2000). Paper industry wastewater also has genotoxic effects (Bhat, Cui, Li, & Vig, 2019). Bacterial communities existing in wastewater as *Fusibacter* and *Desulfobulbus* are used as pollution indicators because contaminants from the paper industry are responsible for altering the microbes present in wastewater. Harmful materials accumulation becomes toxic for consumers and causes different diseases as cancer, skin irritations, and lung problems (Guo, Zhao, Lu, Jia, & Sun, 2016). Paper industry waste material contains high amount of various metals as chromium (Cr) and Cadmium (Cd), which cause damage to the liver, including other health issues (Borah, Singh, Rangan, Karak, & Mitra, 2018). Wastewater having heavy metals employed for irrigation purposes causes phyto-toxicity (Muchuweti et al., 2006). Thus producers and consumers are affected subsequently, and bad effects are imposed on ecosystem functioning (Vashi, Iorhemen, & Tay, 2018). Paper industry wastes have adverse effects on soil microbes in terms of toxicity. Ethanolic extract of waste showed reduced luminescence of *Vibrio fischeri*, which revealed the toxicity of waste (Fraser, O'Halloran, & Van Den Heuvel, 2009). Fertilizers obtained from pulp waste liquor showed toxic effects on earthworms (Song et al., 2018). Production of malondialdehyde and reactive oxidation species (ROS) increased by using fertilizers. The enhancement of free radicals and various antioxidant enzymes causes damages to DNA (Song et al., 2018). Effluents released from paper mills are released in the aquatic system, and water characteristics are badly affected by contaminants present in wastewater. Long-term effects of pulp and paper industry effluents such as the ecological toxicity and reduced recovery of caddis larvae gills have been reported (Ratia, Vuori, & Oikari, 2012). Resin acids and retene existing in effluents exhibit geno-toxicity to sea bass (Gravato & Santos, 2002). Endocrine-disrupting compounds are also included in a major class of contaminants that also exist in paper mill wastewater. In various performed assays; the wastewater from the paper mill was found mutagenic and genotoxic (comet assay) (Balabanič, Filipič, Klemenčič, & Žegura, 2017).

7.6 Techniques for wastewater treatment

The paper and pulp industry generates a massive amount of waste materials, so there is a dire need to find suitable ways to manage future challenges and safely dump the waste materials. Technology development provides easy ways to provide waste to different industries for their usage as secondary materials. This may lead to a decrease in waste materials, and industries shall become efficient to reduce environmental risks generated by waste materials. To reduce different waste materials (organic and inorganic) from pulp and paper wastewater, various techniques have been employed. For the treatment of wastewater from different industries, Advanced oxidation processes have been employed (Cesaro, Belgiorno, Siciliano, & Guida, 2019). Advance oxidation techniques help in the reduction of the toxic effects of waste materials. Optimized Fenton process with the aid of response surface method reduced the COD (V Kumar, Suraj, & Ghosh, 2019). Fenton process is widely employed because it is cost-effective and can be carried out in normal solar irradiation. Lab-scale efficacy of the Fenton method has been exhibited in the removal of total organic carbon (TOC) and COD (Xu, Wang, & Hao, 2007). The electro-Fenton process is better in comparison to the conventional Fenton method because of hydrogen peroxide generation. With this process, greater removal of TOCs is achieved. Some other AOPs as Ozonation, electro-coagulation, and sonication play an important role in the removal of harmful effluents from waste. There is a substantial effect on wastewater through biological and ozonation treatment (Bajpai, 2017). For the ozonation process, ozone is used, and it fastens the treatment process of waste because of free radicals' generation. This process helps in the removal of lignin and also reduces COD and BOD. During effluents treatment, the efficacy of ozonation is affected by wastewater characteristics (Mainardis, Buttazzoni, De Bortoli, Mion, & Goi, 2020). The sonication process helps manage waste, but inorganic ions existing in waste materials decrease the efficacy of sonication. This process alone cannot significantly treat the waste but greatly enhances the efficacy in combination with the Fenton process. COD removal was observed by utilizing sonication along with Fenton and photolysis, respectively (Al-Bsoul et al., 2020; Gogate, Thanekar, & Oke, 2020). The electrocoagulation process is also efficient for the removal of humic acid present in waste effluents (Kamali et al., 2019). This approach can also effectively mitigate COD, NH_3, TSS, and color (Izadi, Hosseini, Darzi, Bidhendi, & Shariati, 2018). Black liquor gasification with carbon was conducted with 24% efficacy which showed that apparatus clogged due to the existence of lignin in liquor (Purkarová, Ciahotný, Šváb, Skoblia, & Beňo, 2018). During this process, H_2S (hydrogen sulfide) gas was produced, which is toxic for humans.

By utilizing steel mill waste materials, ferric aluminum sulfate chloride (polymeric in nature) was prepared, serving as a composite coagulant along with polyacrylamide and diminished chroma and COD. At a large industrial scale, this approach worked in an excellent manner (Yang, Li, Zhang, Wen, & Ni, 2019). Various oxidation-based processes for efficient removal of toxic effluents from pulp and paper industry wastes have been described in Table 7.1. Aerobic granulars have great tolerance to toxicity and higher shock resistance and are considered efficient to treat wastewater containing high contents of BOD and COD (Morais, Silva, & Borges, 2016). Aerobic granular sludge efficiently eradicates tannins and lignins from the wastewater (Vashi, Iorhemen, & Tay, 2019). For the removal of chromophoric molecules from waste materials of the paper industry, the biosorbents technique is also emerging. Bio-sorbents prepared by usage of industrial wastes offer economical and convenient ways for treatment (Kakkar, Malik, & Gupta, 2018). To find the best technology for a specific industry, computational tools are advantageous. They aid to decrease waste discharge based on processing technology, employed raw materials, and industry infrastructure. A 2-filtered fuzzy decision-making strategy was developed, and it provided a promising practice to assess and choose technology to exclude losses for industries (Akhundzadeh & Shirazi, 2017). To evaluate and regulate

Table 7.1 Various oxidation-based processes for efficient removal of toxic effluents from pulp and paper industry wastes.

Sr. #	Treatment processes	Optimized conditions	Electrodes/catalysts/process	Removal efficacy from waste (%)				References
				COD	Color	Turbidity	Miscellaneous	
1	Electrocoagulation	Time approx. = 25 min	Iron	60	95	–	Lignin/tannin > 70%	Wagle, Lin, Nawaz, and Shipley (2020)
		pH = 4.0, Time = 10 min	Aluminum	–	–	–	Humic acid 93%	Barhoumi et al. (2019)
		Time = 40 min, pH = 7.0	SS-304	82	99	–	TOC 79%, TDS 43%	D. Kumar and Sharma (2019)
		Time = 60 min pH = 8 @ 20 V	Iron	70	–	65	TDS 51%, TOC 68%	Hugar and Marol (2020)
		Time = 60 min pH = 7 @ 10 V	Aluminum and iron	79.5%	98.5	–	TSS 83.4%, NH_3 85.3%	Izadi et al. (2018)
2	Photo-electrocatalytic oxidation	Time = 37 min, pH = 4.4		–	–	–	TOC 93% decrease	Singh and Garg (2019)
		Time = 12 hrs	Au/TiO_2NTs	63.5	–	–	84% degradation of 4-Chloroguaiacol, 44.4% TOC	Rajput, Changotra, Sangal, and Dhir (2021)
		Time = 60 min, Current density = 0.75 mA cm^{-2} PMS (peroxy-monosulfate) = 6.0 mM	Combined Coagulation ($FeCl_3$) and electroactivated-PMS process	53	–	–	–	Jaafarzadeh, Ghanbari, and Alvandi (2017)
		Time = 72.5 min, under UV/LED	TiO_2/Polystyrene composite	89.0			Lignin 93.97%	Haghighi, Rahmani, Kariminejad, and Sene (2019)
		Pressure 5 g/L, @ room temp, Time = 24 hrs, 12 hrs solar irradiation	Copper oxide nanomaterials	–	–	–	84% TOC	Kamali, Khalaj, Costa, Capela, and Aminabhavi (2020)
		Time = 6 hrs	Ag_2O (1.4 wt.%) /ZnO	66	41	–	–	Marques, Ferrari-Lima, Slusarski-Santana, ad Fernandes-Machado (2017)
		Time = 90 min, pH = 8 & H_2O_2 200 mg/L	UV/H_2O_2	85.4	92.6	2.7	Total phenol 94.5	Neves et al. (2020)
		Time = 37 min, pH = 4.4 @ H_2O_2 dose = 6.8 mM, $[H_2O_2]$: $[Fe^{2+}]$ = 65	–	–	–	–	TOC 93%, 90% dechlorinations (PCP, 4-CP, 2,4,6-TCP, 2,4-DCP)	Singh and Garg (2019)

(*Continued*)

Table 7.1 (Continued)

Sr. #	Treatment processes	Optimized conditions	Electrodes/catalysts/process	Removal efficacy from waste (%)				References
				COD	Color	Turbidity	Miscellaneous	
3	Ultrasound	2:5 (nZVI: H_2O_2), Time = 30 min	–	–	–	–	55% adsorbable organohalogens (AOX)	Erhardt et al. (2021)
4	Fenton-based oxidation processes	pH = 3.5, Time = 60 min	–	87.3	95.9	95.86	Removal of 93.59% NH_3 nitrogen, Total nitrogen 51.73%, Total phosphorus 84.75	Xing et al. (2020)
		Time = 15 min, pH = 5.36	–	91.2	99.9	–	Removal of TOC 78.5%	Buthiyappan and Raman (2019)
		Time = 90 min, pH = 8, Fe^{+2} 5 mg/L & H_2O_2 200 mg/L	–	90.4	93.4	10.5	Total phenols 88.2%	Neves et al. (2020)
		Temp. = 60 °C and Fe^{2+}Conc. = 0.4 mM	–	88.01	–	–	–	Abbas and Abbas (2019)
		pH = 3.33, Fe (III) 1000 mg/L H_2O_2 dosage = 14.55 mM. Time = 60 min	Combined aerobic moving bed biofilm reactor (MBBR) & Fenton process	53.73%	–	–	–	Brink, Sheridan, and Harding (2018)
		Time = 60 min, pH = 7	–	77	–	–	–	Zhou, Kang, Zhou, Zhong, and Xing (2018)
		Under UV irradiation	$ZnAl_2O_4/BiPO_4/H_2O_2$	73.98	84.54	–	–	Tian et al. (2020)
		Time = 10 min, pH = 2, Temp 60 °C	–	–	80%	–	90% adsorbable organohalogens (AOX)	Ribeiro, Marques, Portugal, and Nunes (2020)
		Time = 30 min, pH = 3, Dosage of Fe^{2+} 3 mM, H_2O_2=6 mM	–	92.1%	90.3	–	–	Abedinzadeh, Shariat, Monavari, and Pendashteh (2018)
		Time 60 min, Fe^{2+}/H_2O_2=0.32 mM	–	63	–	–	–	Brink, Sheridan, and Harding (2017)
5	Ozonation	Ozone dosage around 590–610 mg O_3/L	Ozonation after biological treatment	81	–	–	BOD 63%, TSS 33%	Mainardis et al. (2020)
			Coupled thermophilic biofilm-based systems (TBSs) with ozonation	90.48	87.89	–	–	Tang et al. (2020)

environmental effects, the Slacks-based measure and the Malmquist-Luenberger index are significant parameters. Malmquist-Luenberger index may provide technical modification and effective change controlled by adverse outputs. Slacks-based measure narrates eco-efficacy (Yu, Shi, Wang, Chang, & Cheng, 2016).

7.7 Waste to value aspects

There are some valuable strategies that can be employed for the significant use of paper industry waste. On-site conversion of waste material into useful products is great for development Effluents obtained from the pulp and paper industry can be properly utilized and dispose of through valorization techniques. Clinker generation, ethanol, and sorbents production are included in valorization techniques. They can be useful for producing new sources (Gupta & Shukla, 2020) is one of the beneficial methods to use paper industry waste. The waste obtained from pulp contains various oxides of calcium, aluminum, silica, and iron, which are prerequisite for clicker production. Lime mud, ash, and sludge are also the main waste materials for cement formation. Effluents obtained from pulp are calcined at a higher temperature, so it is useful for building blocks formation. To match the strength of building materials, 10% effluents may be mixed with clay (Azevedo et al., 2019). Production of nano-silica from the ash of the boiler is also another approach for the utilization of pulp effluents. By precipitation method, nano-silica was prepared and further mixed with sludge for the production of cement clinker. Cement produced using nano-silica was found efficient for constructing building materials (Vashistha, Singh, Dutt, & Kumar, 2019). Bricks and tiles can also be produced by utilizing paper industry-based wastes (dos Santos, Cabrelon, de Sousa Trichês, & Quinteiro, 2019).

Sludge obtained from the paper industry is constituted of organic and inorganic contaminants and it affects the environment. Anaerobic digestion of sludge was carried out with two different methods: incineration and second incineration, along with pyrolysis to compare the results (Mohammadi et al., 2019). Biochar was produced in anaerobic digestion and pyrolysis. Biochar is promising for the sustainability of the ecosystem and for forest productivity especially in those countries where sludge is produced at a large scale. Reduction in toxicity was observed after Biochar formation because of effluents discharge. This happens because of the decrease in metal mobility in biochar production as compared to parent effluents (Liu, Luo, & Shukla, 2020). Biochar produced from waste is also used as an adsorbent to remove effluents from the soil, and it can improve the quality of sales (Fakayode, Aboagarib, Zhou, & Ma, 2020). Biochar is produced from the carbonization of sludge and it can be used to reduce the heavy metals leaching risk and maintains the micro-environment of soil (Chen et al., 2020). Biochar formed from lignin can be employed as carbon black.

Scientists also produced nano-biochar from industrial waste and employed it as a filler of styrene rubber and it enhanced the strength of rubber (Jiang et al., 2020).

Lignin from sludge can also be excluded by the utilization of anaerobic digestion. Thus, lignin can be degraded and can be transformed into some other beneficial products (Khan & Ahring, 2019). Anaerobic digestion is also helpful for biogas production (Lopes, Silva, Rosa, & de Ávila Rodrigues, 2018). In the paper industry, bleaching is done at a large scale for white and bright paper production, and bleaching by using enzymes may help decrease contaminants. At a large scale in the paper industry, the two enzymes xylanase and laccase are employed to carry the bio-bleaching process. Enzymes usage reduces the consumption of chlorine and strength; the quality of the product is enhanced. The pulp can also be used efficiently as an adsorbent. After bleaching of pulp through enzymes, it can be mixed with carbon to generate adsorbent. Various toxic drugs (sulfamethoxazole, carbamazepine) and other effluents from wastewater can be removed through these adsorbents (Oliveira, Calisto, Santos, Otero, & Esteves, 2018). For wastewater treatment and energy production, the microbial fuel cell is a new strategy in use. Microbes can be used to generate hydrogen from wastewater, and this may economically provide benefits to industries. There are few advantages associated with the microbial fuel cell approach like electrical energy production, eco-friendly, economical, more environmental sustainability, and elimination of ventilation methodology (Neto, Reginatto, & De Andrade, 2018). For the reduction of waste materials, the generation of useful chemicals is also an emerging trend. Lignin is present in high quantity in the waste of paper industry and it can be depolymerized to form monomers of phenols, acids, aldehyde, and various aromatics (Sun, Fridrich, de Santi, Elangovan, & Barta, 2018). Various carboxylic acids can be synthesized from lignin by opting Fenton phenomena. Wet oxidation of lignin was found significant in depolymerization phenomena. Irmak and colleagues reported the production of carboxylic acid derivatives from lignin at various temperatures and pressures (Irmak, Kang, & Wilkins, 2020).

7.8 Conclusion

The paper industry generates a huge amount of waste for every ton of generated paper. The generated waste material causes serious effects on both aquatic and terrestrial life. The discharged effluents pose serious health issues, from skin diseases to mutagenic disorders. Although rules are made by governing concerned bodies to keep the environment sustainable but compliance with rules by industries is a serious issue. Advanced techniques are needed to reduce pressure on the industrial sector. The prepared coagulants and aerobic granules from the steel-mill wastes include new technologies to be focused on as they are eco-friendly and economical. Electro-coagulation and adsorbent-based techniques for wastewater treatment are novel established areas during the last decades. Electricity production through a microbial fuel cell and remediation with the help of efficient micro-organisms may be a great step for the generation of energy and the disposal of waste materials. Only employment of technology cannot be potent for development, but there is a need to install artificial intelligence programs and computational study to help the industrial sector find suitable techniques to mitigate waste materials load on the environment. Employment of computational tools would help in energy efficacy and decrease paper industry waste production by employing proper models at every processing step. There is a need to develop more efficient and sophisticated models to incorporate industry-based technology.

Acknowledgment

Consejo Nacional de Ciencia y Tecnología (MX) is thankfully acknowledged for partially supporting this work under Sistema Nacional de Investigadores (SNI) program awarded to Hafiz M.N. Iqbal (CVU: 735340).

Conflict of interests

The author(s) declare no conflicting interests.

References

Abbas, Z. I., & Abbas, A. S. (2019). Oxidative degradation of phenolic wastewater by electro-fenton process using MnO2-graphite electrode. *Journal of Environmental Chemical Engineering*, 7(3), 103108.

Abedinzadeh, N., Shariat, M., Monavari, S. M., & Pendashteh, A. (2018). Evaluation of color and COD removal by Fenton from biologically (SBR) pre-treated pulp and paper wastewater. *Process Safety and Environmental Protection*, *116*, 82–91. Available from https://doi.org/10.1016/j.psep.2018.01.015.

Abhishek, A., Dwivedi, A., Tandan, N., & Kumar, U. (2017). Comparative bacterial degradation and detoxification of model and kraft lignin from pulp paper wastewater and its metabolites. *Applied Water Science*, 7(2), 757–767.

Akhundzadeh, M., & Shirazi, B. (2017). Technology selection and evaluation in Iran's pulp and paper industry using 2-filterd fuzzy decision making method. *Journal of Cleaner Production*, *142*, 3028–3043.

Al-Bsoul, A., Al-Shannag, M., Tawalbeh, M., Al-Taani, A. A., Lafi, W. K., Al-Othman, A., & Alsheyab, M. (2020). Optimal conditions for olive mill wastewater treatment using ultrasound and advanced oxidation processes. *Science of the Total Environment*, *700*, 134576.

Azevedo, A., Marvila, T., Fernandes, W. J., Alexandre, J., Xavier, G., Zanelato, E., & Mendes, B. (2019). Assessing the potential of sludge generated by the pulp and paper industry in assembling locking blocks. *Journal of Building Engineering*, *23*, 334–340.

Bajpai, P. (2017). *Pulp and paper industry: Emerging waste water treatment technologies*. Elsevier.

Balabanič, D., Filipič, M., Klemenčič, A. K., & Žegura, B. (2017). Raw and biologically treated paper mill wastewater effluents and the recipient surface waters: Cytotoxic and genotoxic activity and the presence of endocrine disrupting compounds. *Science of the Total Environment*, *574*, 78–89.

Barhoumi, A., Ncib, S., Chibani, A., Brahmi, K., Bouguerra, W., & Elaloui, E. (2019). High-rate humic acid removal from cellulose and paper industry wastewater by combining electrocoagulation process with adsorption onto granular activated carbon. *Industrial Crops and Products*, *140*, 111715.

Bhat, S. A., Cui, G., Li, F., & Vig, A. P. (2019). Biomonitoring of genotoxicity of industrial wastes using plant bioassays. *Bioresource Technology Reports*, *6*, 207–216.

Borah, P., Singh, P., Rangan, L., Karak, T., & Mitra, S. (2018). Mobility, bioavailability and ecological risk assessment of cadmium and chromium in soils contaminated by paper mill wastes. *Groundwater for Sustainable Development, 6*, 189–199.

Brink, A., Sheridan, C., & Harding, K. (2018). Combined biological and advance oxidation processes for paper and pulp effluent treatment. *South African Journal of Chemical Engineering, 25*, 116–122. Available from https://doi.org/10.1016/j.sajce.2018.04.002.

Brink, A., Sheridan, C. M., & Harding, K. G. (2017). The Fenton oxidation of biologically treated paper and pulp mill effluents: A performance and kinetic study. *Process Safety and Environmental Protection, 107*, 206–215. Available from https://doi.org/10.1016/j.psep.2017.02.011.

Buthiyappan, A., & Raman, A. A. A. (2019). Energy intensified integrated advanced oxidation technology for the treatment of recalcitrant industrial wastewater. *Journal of Cleaner Production, 206*, 1025–1040.

Canmet, E. (2005). Pulp and paper sludge to energy—Preliminary Assessment of Technologies. *Natural Resources Canada, CANMET Energy. Report, 34*(1), 152, *0173e479*.

Cesaro, A., Belgiorno, V., Siciliano, A., & Guida, M. (2019). The sustainable recovery of the organic fraction of municipal solid waste by integrated ozonation and anaerobic digestion. *Resources, Conservation and Recycling, 141*, 390–397.

Chandra, R., & Singh, R. (2012). Decolourisation and detoxification of rayon grade pulp paper mill effluent by mixed bacterial culture isolated from pulp paper mill effluent polluted site. *Biochemical Engineering Journal, 61*, 49–58.

Chen, C., Liu, G., An, Q., Lin, L., Shang, Y., & Wan, C. (2020). From wasted sludge to valuable biochar by low temperature hydrothermal carbonization treatment: Insight into the surface characteristics. *Journal of Cleaner Production, 263*, 121600.

Gupta, G. K., Liu, H., & Shukla, P. (2019). Pulp and paper industry—based pollutants, their health hazards and environmental risks. *Current Opinion in Environmental Science & Health, 12*, 48–56.

dos Santos, V. R., Cabrelon, M. D., de Sousa Trichês, E., & Quinteiro, E. (2019). Green liquor dregs and slaker grits residues characterization of a pulp and paper mill for future application on ceramic products. *Journal of Cleaner Production, 240*, 118220.

Dwivedi, P., Vivekanand, V., Pareek, N., Sharma, A., & Singh, R. P. (2010). Bleach enhancement of mixed wood pulp by xylanase—laccase concoction derived through co-culture strategy. *Applied Biochemistry and Biotechnology, 160*(1), 255–268.

Elliott, A., & Mahmood, T. (2005). Survey benchmarks generation, management of solid residues. *Pulp Pap, 79*(12), 49–55.

Erhardt, C. S., Basegio, T. M., Capela, I., Rodríguez, A. L., Machado, Ê. L., López, D. A. R., & Bergmann, C. P. (2021). AOX degradation of the pulp and paper industry bleaching wastewater using nZVI in two different agitation processes. *Environmental Technology & Innovation, 22*, 101420.

Fakayode, O. A., Aboagarib, E. A. A., Zhou, C., & Ma, H. (2020). Co-pyrolysis of lignocellulosic and macroalgae biomasses for the production of biochar—A review. *Bioresource Technology, 297*, 122408.

Fraser, D. S., O'Halloran, K., & Van Den Heuvel, M. R. (2009). Toxicity of pulp and paper solid organic waste constituents to soil organisms. *Chemosphere, 74*(5), 660–668.

Gogate, P., Thanekar, P., & Oke, A. (2020). Strategies to improve biological oxidation of real wastewater using cavitation based pre-treatment approaches. *Ultrasonics Sonochemistry, 64*, 105016.

González, L. F., Sarria, V., & Sánchez, O. F. (2010). Degradation of chlorophenols by sequential biological-advanced oxidative process using Trametes pubescens and TiO_2/UV. *Bioresource Technology, 101*(10), 3493–3499.

Gopal, P., Sivaram, N., & Barik, D. (2019). *Paper industry wastes and energy generation from wastes. In* Energy from Toxic Organic Waste for Heat and Power Generation (pp. 83–97). Elsevier.

Gravato, C., & Santos, M. (2002). Juvenile sea bass liver biotransformation and erythrocytic genotoxic responses to pulp mill contaminants. *Ecotoxicology and Environmental Safety, 53*(1), 104–112.

Guo, J., Zhao, L., Lu, W., Jia, H., & Sun, Y. (2016). Bacterial communities in water and sediment shaped by paper mill pollution and indicated bacterial taxa in sediment in Daling River. *Ecological Indicators, 60*, 766–773.

Gupta, G. K., & Shukla, P. (2020). Insights into the resources generation from pulp and paper industry wastes: Challenges, perspectives and innovations. *Bioresource Technology, 297*, 122496.

Gupta, S., Saksena, S., & Baris, O. F. (2019). Environmental enforcement and compliance in developing countries: Evidence from India. *World Development*, *117*, 313–327.

Haghighi, M., Rahmani, F., Kariminejad, F., & Sene, R. A. (2019). Photodegradation of lignin from pulp and paper mill effluent using TiO2/PS composite under UV-LED radiation: Optimization, toxicity assessment and reusability study. *Process Safety and Environmental Protection*, *122*, 48–57.

Haq, I., & Raj, A. (2020). *Pulp and paper mill wastewater: ecotoxicological effects and bioremediation approaches for environmental safety. In* Bioremediation of industrial waste for environmental safety (pp. 333–356). Springer.

Hooda, R., Bhardwaj, N. K., & Singh, P. (2015). Screening and identification of ligninolytic bacteria for the treatment of pulp and paper mill effluent. *Water, Air, & Soil Pollution*, *226*(9), 1–11.

Hooda, R., Bhardwaj, N. K., & Singh, P. (2018). Brevibacillus parabrevis MTCC 12105: a potential bacterium for pulp and paper effluent degradation. *World Journal of Microbiology and Biotechnology*, *34*(2), 1–10.

Hugar, G. M., & Marol, C. K. (2020). Feasibility of electro coagulation using iron electrodes in treating paper industry wastewater. *Sustainable Water Resources Management*, *6*(4), 1–10.

Irmak, S., Kang, J., & Wilkins, M. (2020). Depolymerization of lignin by wet air oxidation. *Bioresource Technology Reports*, *9*, 100377. Available from https://doi.org/10.1016/j.biteb.2019.100377.

Izadi, A., Hosseini, M., Darzi, G. N., Bidhendi, G. N., & Shariati, F. P. (2018). Treatment of paper-recycling wastewater by electrocoagulation using aluminum and iron electrodes. *Journal of Environmental Health Science and Engineering*, *16*(2), 257–264.

Jaafarzadeh, N., Ghanbari, F., & Alvandi, M. (2017). Integration of coagulation and electro-activated HSO5 − to treat pulp and paper wastewater. *Sustainable Environment Research*, *27*(5), 223–229. Available from https://doi.org/10.1016/j.serj.2017.06.001.

Jiang, C., Bo, J., Xiao, X., Zhang, S., Wang, Z., Yan, G., & He, H. (2020). Converting waste lignin into nano-biochar as a renewable substitute of carbon black for reinforcing styrene-butadiene rubber. *Waste management*, *102*, 732–742.

Kakkar, S., Malik, A., & Gupta, S. (2018). Treatment of pulp and paper mill effluent using low cost adsorbents: An overview. *Journal of Applied and Natural Science*, *10*(2), 695–704.

Kamali, M., Alavi-Borazjani, S. A., Khodaparast, Z., Khalaj, M., Jahanshahi, A., Costa, E., & Capela, I. (2019). Additive and additive-free treatment technologies for pulp and paper mill effluents: Advances, challenges and opportunities. *Water Resources and Industry*, *21*, 100109.

Kamali, M., Khalaj, M., Costa, M. E. V., Capela, I., & Aminabhavi, T. M. (2020). Optimization of kraft black liquor treatment using ultrasonically synthesized mesoporous tenorite nanomaterials assisted by Taguchi design. *Chemical Engineering Journal*, *401*, 126040.

Khan, M. U., & Ahring, B. K. (2019). Lignin degradation under anaerobic digestion: Influence of lignin modifications-A review. *Biomass and Bioenergy*, *128*, 105325.

Kotelevtsev, S., Stepanova, L., Lindström-Seppä, P., & Hänninen, O. (2000). Monooxygenase activities in the liver of fish living in Lake Baikal and after treatment with waste waters from the Baikalsk Pulp and Paper Mill. *Aquatic Ecosystem Health & Management*, *3*(2), 271–276.

Krigstin, S., & Sain, M. (2006). Characterization and potential utilization of recycled paper mill sludge. *Pulp Paper Canada*, *107*(5), 29–32.

Kumar, D., & Sharma, C. (2019). Remediation of pulp and paper industry effluent using electrocoagulation process. *Journal of Water Resource and Protection*, *11*(3), 296.

Kumar, V., Marín-Navarro, J., & Shukla, P. (2016). Thermostable microbial xylanases for pulp and paper industries: trends, applications and further perspectives. *World Journal of Microbiology and Biotechnology*, *32*(2), 34.

Kumar, V., Suraj, P., & Ghosh, P. (2019). *Optimization of COD removal by advanced oxidation process through response surface methodology from pulp & paper industry wastewater.*

Lindholm-Lehto, P. C., Knuutinen, J. S., Ahkola, H. S., & Herve, S. H. (2015). Refractory organic pollutants and toxicity in pulp and paper mill wastewaters. *Environmental Science and Pollution Research*, *22*(9), 6473–6499.

Liu, H., Luo, J., & Shukla, P. (2020). Effluents detoxification from pulp and paper industry using microbial engineering and advanced oxidation techniques. *Journal of Hazardous Materials*, *398*, 122998.

Lopes, A. D. C. P., Silva, C. M., Rosa, A. P., & de Ávila Rodrigues, F. (2018). Biogas production from thermophilic anaerobic digestion of kraft pulp mill sludge. *Renewable Energy*, *124*, 40–49.

Mainardis, M., Buttazzoni, M., De Bortoli, N., Mion, M., & Goi, D. (2020). Evaluation of ozonation applicability to pulp and paper streams for a sustainable wastewater treatment. *Journal of cleaner production*, *258*, 120781.

Marques, R. G., Ferrari-Lima, A. M., Slusarski-Santana, V., & Fernandes-Machado, N. R. C. (2017). Ag2O and Fe2O3 modified oxides on the photocatalytic treatment of pulp and paper wastewater. *Journal of Environmental Management*, *195*, 242–248. Available from https://doi.org/10.1016/j.jenvman.2016.08.034.

Mohammadi, A., Sandberg, M., Venkatesh, G., Eskandari, S., Dalgaard, T., Joseph, S., & Granström, K. (2019). Environmental performance of end-of-life handling alternatives for paper-and-pulp-mill sludge: Using digestate as a source of energy or for biochar production. *Energy*, *182*, 594–605.

Monge-Corella, S., García-Pérez, J., Aragonés, N., Pollán, M., Pérez-Gómez, B., & López-Abente, G. (2008). Lung cancer mortality in towns near paper, pulp and board industries in Spain: A point source pollution study. *BMC Public Health*, *8*(1), 1–11.

Monte, M. C., Fuente, E., Blanco, A., & Negro, C. (2009). Waste management from pulp and paper production in the European Union. *Waste Management*, *29*(1), 293–308.

Morais, I. L. H., Silva, C. M., & Borges, C. P. (2016). Aerobic granular sludge to treat paper mill effluent: organic matter removal and sludge filterability. *Desalination and Water Treatment*, *57*(18), 8119–8126.

Muchuweti, M., Birkett, J., Chinyanga, E., Zvauya, R., Scrimshaw, M. D., & Lester, J. (2006). Heavy metal content of vegetables irrigated with mixtures of wastewater and sewage sludge in Zimbabwe: Implications for human health. *Agriculture, Ecosystems & Environment*, *112*(1), 41–48.

Neto, S. A., Reginatto, V., & De Andrade, A. R. (2018). *Microbial fuel cells and wastewater treatment In* Electrochemical water and wastewater treatment (pp. 305–331). Elsevier.

Neves, L. C., de Souza, J. B., de Souza Vidal, C. M., Herbert, L. T., de Souza, K. V., Martins, K. G., & Young, B. J. (2020). Phytotoxicity indexes and removal of color, COD, phenols and ISA from pulp and paper mill wastewater post-treated by UV/H2O2 and photo-Fenton. *Ecotoxicology and Environmental Safety*, *202*, 110939.

Olaniran, A. O., & Igbinosa, E. O. (2011). Chlorophenols and other related derivatives of environmental concern: properties, distribution and microbial degradation processes. *Chemosphere*, *83*(10), 1297–1306.

Oliveira, G., Calisto, V., Santos, S. M., Otero, M., & Esteves, V. I. (2018). Paper pulp-based adsorbents for the removal of pharmaceuticals from wastewater: A novel approach towards diversification. *Science of the Total Environment*, *631*, 1018–1028.

Pokhrel, D., & Viraraghavan, T. (2004). Treatment of pulp and paper mill wastewater—a review. *Science of the Total Environment*, *333*(1–3), 37–58.

Pola, L., Collado, S., Oulego, P., Calvo, P. Á., & Díaz, M. (2021). Characterisation of the wet oxidation of black liquor for its integration in Kraft paper mills. *Chemical Engineering Journal*, *405*, 126610.

Purkarová, E., Ciahotný, K., Šváb, M., Skoblia, S., & Beňo, Z. (2018). Supercritical water gasification of wastes from the paper industry. *The Journal of Supercritical Fluids*, *135*, 130–136.

Raj, A., Kumar, S., Singh, S. K., & Prakash, J. (2018). Production and purification of xylanase from alkaliphilic Bacillus licheniformis and its pretreatment of eucalyptus kraft pulp. *Biocatalysis and Agricultural Biotechnology*, *15*, 199–209.

Rajput, H., Changotra, R., Sangal, V. K., & Dhir, A. (2021). Photoelectrocatalytic treatment of recalcitrant compounds and bleach stage pulp and paper mill effluent using Au-TiO2 nanotube electrode. *Chemical Engineering Journal*, *408*, 127287.

Ram, C., Rani, P., Gebru, K. A., & Abrha, M. G. M. (2020). Pulp and paper industry wastewater treatment: use of microbes and their enzymes. *Physical Sciences Reviews*, *5*(10).

Ratia, H., Vuori, K.-M., & Oikari, A. (2012). Caddis larvae (Trichoptera, Hydropsychidae) indicate delaying recovery of a watercourse polluted by pulp and paper industry. *Ecological Indicators*, *15*(1), 217–226.

Ribeiro, J. P., Marques, C. C., Portugal, I., & Nunes, M. I. (2020). Fenton processes for AOX removal from a kraft pulp bleaching industrial wastewater: Optimisation of operating conditions and cost assessment. *Journal of Environmental Chemical Engineering*, *8*(4), 104032.

Singh, A. K., & Chandra, R. (2019). Pollutants released from the pulp paper industry: Aquatic toxicity and their health hazards. *Aquatic toxicology*, *211*, 202–216.

Singh, G., Kaur, S., Khatri, M., & Arya, S. K. (2019). Biobleaching for pulp and paper industry in India: Emerging enzyme technology. *Biocatalysis and Agricultural Biotechnology*, *17*, 558–565.

Singh, S., & Garg, A. (2019). Performance of photo-catalytic oxidation for degradation of chlorophenols: Optimization of reaction parameters and quantification of transformed oxidized products. *Journal of Hazardous Materials*, *361*, 73–84. Available from https://doi.org/10.1016/j.jhazmat.2018.08.055.

Song, P., Ping, L., Gao, J., Li, X., Zhu, M., & Wang, J. (2018). Ecotoxicological effects of fertilizers made from pulping waste liquor on earthworm *Eisenia fetida*. *Ecotoxicology and Environmental Safety*, *166*, 237–241.

Sonkar, M., Kumar, M., Dutt, D., & Kumar, V. (2019). Treatment of pulp and paper mill effluent by a novel bacterium *Bacillus* sp. IITRDVM-5 through a sequential batch process. *Biocatalysis and Agricultural Biotechnology*, *20*, 101232.

Soskolne, C. L., & Sieswerd, L. E. (2010). Cancer risk associated with pulp and paper mills: a review of occupational and community epidemiology. *Chronic Diseases and Injuries in Canada*, 29.

Stepanova, L., Lindström-Seppä, P., Hänninen, O., Kotelevtsev, S., Glaser, V., Novikov, C., & Beim, A. (2000). Lake Baikal: biomonitoring of pulp and paper mill waste water. *Aquatic Ecosystem Health & Management*, *3*(2), 259–269.

Sun, Z., Fridrich, B. I., de Santi, A., Elangovan, S., & Barta, K. (2018). Bright Side of Lignin Depolymerization: Toward New Platform Chemicals. *Chemical Reviews*, *118*, 614–678.

Tang, Z., Lin, Z., Wang, Y., Zhao, P., Kuang, F., & Zhou, J. (2020). Coupling of thermophilic biofilm-based systems and ozonation for enhanced organics removal from high-temperature pulping wastewater: Performance, microbial communities, and pollutant transformations. *Science of the Total Environment*, *714*, 136802. Available from https://doi.org/10.1016/j.scitotenv.2020.136802.

Tian, Q., Ran, M., Fang, G., Ding, L., Pan, A., Shen, K., & Deng, Y. (2020). ZnAl2O4/BiPO4 composites as a heterogeneous catalyst for photo-Fenton treatment of textile and pulping wastewater. *Separation and Purification Technology*, *239*, 116574. Available from https://doi.org/10.1016/j.seppur.2020.116574.

United States EPA S. (1995). *EPA office of compliance sector notebook project: Profile of pulp and paper industry*. EPA/310-R-95–015, Washington, DC 20460, USA.

Vashi, H., Iorhemen, O., & Tay, J. (2018). Degradation of industrial tannin and lignin from pulp mill effluent by aerobic granular sludge technology. *Journal of Water Process Engineering*, *26*, 38–45.

Vashi, H., Iorhemen, O., & Tay, J. (2019). Extensive studies on the treatment of pulp mill wastewater using aerobic granular sludge (AGS) technology. *Chemical Engineering Journal*, *359*, 1175–1194.

Vashistha, P., Singh, S., Dutt, D., & Kumar, V. (2019). Sustainable utilization of paper mill solid wastes via synthesis of nano silica for production of belite based clinker. *Journal of Cleaner Production*, *224*, 557–565.

Veluchamy, C., & Kalamdhad, A. S. (2017). Influence of pretreatment techniques on anaerobic digestion of pulp and paper mill sludge: a review. *Bioresource Technology*, *245*, 1206–1219.

Virendra, K., Purnima, D., Sanjay, N., Anil, K., & Rita, K. (2014). Biological approach for the treatment of pulp and paper industry effluent in sequence batch reactor. *Journal of Bioremediation and Biodegradation*, *5*(3).

Wagle, D., Lin, C.-J., Nawaz, T., & Shipley, H. J. (2020). Evaluation and optimization of electrocoagulation for treating Kraft paper mill wastewater. *Journal of Environmental Chemical Engineering*, *8*(1), 103595. Available from https://doi.org/10.1016/j.jece.2019.103595.

Wang, L. K., Hung, Y.-T., Lo, H. H., & Yapijakis, C. (2005). *Waste treatment in the process industries*. CRC Press.

Xing, L., Kong, M., Xie, X., Sun, J., Wei, D., & Li, A. (2020). Feasibility and safety of papermaking wastewater in using as ecological water supplement after advanced treatment by fluidized-bed Fenton coupled with large-scale constructed wetland. *Science of the Total Environment*, *699*, 134369.

Xu, M., Wang, Q., & Hao, Y. (2007). Removal of organic carbon from wastepaper pulp effluent by lab-scale solar photo-Fenton process. *Journal of Hazardous Materials*, *148*(1–2), 103–109.

Yaashikaa, P., Kumar, P. S., Varjani, S. J., & Saravanan, A. (2019). Advances in production and application of biochar from lignocellulosic feedstocks for remediation of environmental pollutants. *Bioresource Technology*, *292*, 122030.

Yadav, S., & Chandra, R. (2018). Detection and assessment of the phytotoxicity of residual organic pollutants in sediment contaminated with pulp and paper mill effluent. *Environmental Monitoring and Assessment*, *190*(10), 1–15.

Yang, S., Li, W., Zhang, H., Wen, Y., & Ni, Y. (2019). Treatment of paper mill wastewater using a composite inorganic coagulant prepared from steel mill waste pickling liquor. *Separation and Purification Technology*, *209*, 238–245.

Yu, C., Shi, L., Wang, Y., Chang, Y., & Cheng, B. (2016). The eco-efficiency of pulp and paper industry in China: an assessment based on slacks-based measure and Malmquist–Luenberger index. *Journal of Cleaner Production*, *127*, 511–521.

Zhou, H., Kang, L., Zhou, M., Zhong, Z., & Xing, W. (2018). Membrane enhanced COD degradation of pulp wastewater using Cu_2O/H_2O_2 heterogeneous Fenton process. *Chinese Journal of Chemical Engineering*, *26*(9), 1896–1903.

Further reading

Dixit, M., Gupta, G. K., & Shukla, P. (2020). Insights into the resources generation from pulp and paper industry wastes: Challenges, perspectives and innovations. *Bioresource Technology*, *297*, 122496.

PART II

Applications

CHAPTER 8

Pharmaceutical applications of nanocellulose

Shweta Mishra[1] **and Anil M. Pethe**[2]
[1]Shobhaben Pratapbhai Patel School of Pharmacy and Technology Management, SVKM's NMIMS (Deemed-to-be-University), Mumbai, India
[2]Datta Meghe College of Pharmacy, Datta Meghe Institute of Medical Sciences, Deemed-to-be-University, Sawangi (Meghe), Wardha, India

8.1 Introduction

Nowadays starch has got more attention. Due to its biodegradability, biosafety, and sustainability, cellulose has been the subject of strenuous research (Villanova et al., 2011). It has been used by humans since incident time (Yang et al., 2008). Method of papermaking was first developed by china. Although higher plants are more important for the production of pulp and paper, they could be obtained from various sources, including plants, bacterial sources, and marine sources (Glass et al., 2012). Nanocellulose is defined as the nanoscale structure product or extract obtained from aboriginal cellulose. It is mainly crystalline with at least one dimension on a nanoscale are known as Nanocellulose (Lejeune & Deprez, 2010). It shows both the characteristics as it is made up of two words nano & cellulose. It possesses all the properties of both the terms being nanostructure it possesses various properties such as optical properties, rheological and high-specific surface area as well as cellulose properties like mechanical strength, good potential for chemical modification, and biocompatibility. Generally, the family of nanocellulose is classified into three subtypes:

1. Cellulose nanocrystals (CNC)
2. Bacterial cellulose (BC)
3. Nanofibrillated cellulose (NFC)

8.2 Methods of preparation

There are so many methods for the preparation of nanocellulose. It can be prepared by biomaterials using various chemicals or mechanical means (Achaby et al., 2017). Cellulose fibers which are obtained after chemical treatment showed different characteristics including crystal structure, morphology, aspect ratio, degree of crystallinity, and surface chemistry, which could be varied. Acid hydrolysis is used for removing the

Nanotechnology in Paper and Wood Engineering
DOI: https://doi.org/10.1016/B978-0-323-85835-9.00012-X

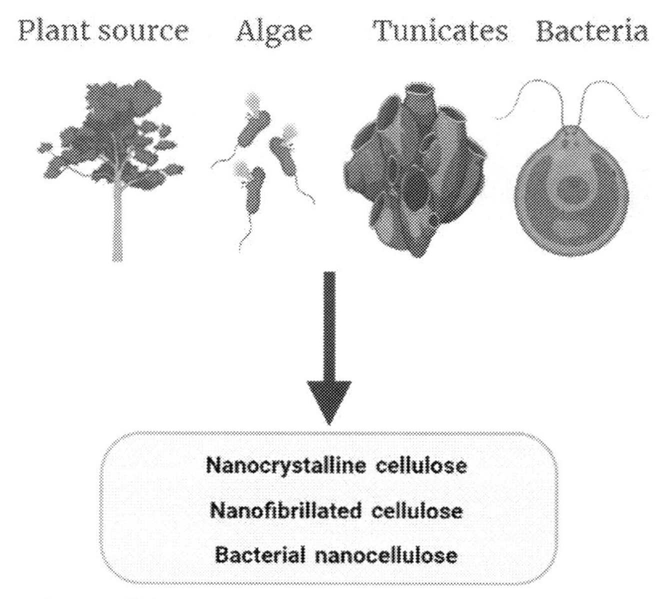

Figure 8.1 Sources of nanocellulose preparation.

amorphous region while keeping the crystalline region which forms a rod-like structure. Intra and intermolecular hydrogen bonds are used for forming the extended network. Nanocrystalline cellulose (NCC) is explored for various kinds of drug delivery, as a drug excipient, and in tissue engineering for various purposes. NFC is used for nanocomposite preparation due to its good mechanical strength and BC is generally used for tissue engineering. Common methods for the preparation of NCC are explained in Fig. 8.1.

8.2.1 Acid hydrolysis for nanocellulose preparation

CNCs are generally prepared by acid hydrolysis technique (Johar et al., 2012a).

8.2.1.1 Step I (Alkali treatment)

The alkali treatment is used for the purification of cellulose. The sample is usually treated with an alkali solution refluxed for 2 h, filtered, and washed by using distilled water.

8.2.1.2 Step II (Bleaching process)

Bleaching is performed by adding acidic acid buffer solution, distilled water, and aqueous chlorite at reflux for 4 h (at 100°C–130°C on silicon oil bath). Then after cooled and filtered by using distilled water. This treatment is repeated several times (Fig. 8.2).

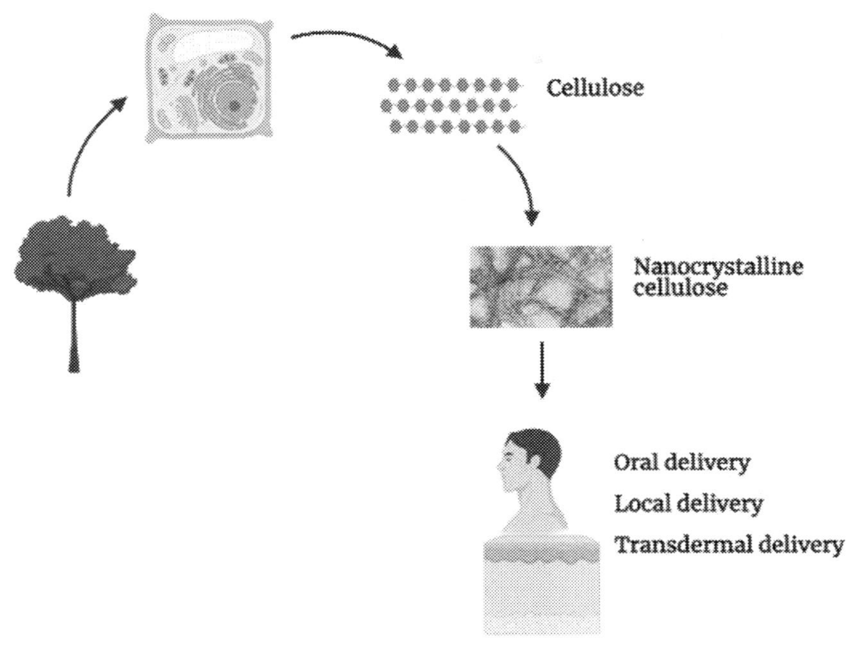

Figure 8.2 Preparation and use of nanocellulose in drug delivery.

8.2.1.3 Step III (Hydrolysis treatment)

After bleaching, hydrolysis is performed by using 10 mol L^{-1} of sulfuric acid at 50°C temperature under continuous stirring for 40 min. It is centrifuged at 10,000 rpm at 10°C for 10 min. Then the suspension is dialyzed using distilled water until its pH reaches 5–6. Final suspension is then sonicated for 30 min.

Hydrochloric acid and sulfuric acid have been widely used because it successfully degrades cellulose fibers (Shankar et al., 2018). They react with hydroxyl group via esterification process allowing grafting of the anionic sulfate ester group. This anionic group promotes NCC dispersion in water and formed a negative electrostatic layer around NCCs (Johar et al., 2012b). This process promotes hydrolytic cleavage of glycosidic bonds and cleaves the amorphous region of cellulose by penetrating it. There are some alternative techniques to acid hydrolysis such as enzymatic hydrolysis, TEMPO oxidation, and treatment with ionic liquids (Klemm et al., 2011).

8.3 Application of NCC

Nanocellulose is widely used for different drug delivery systems. Being a biocompatible material, cellulose is extensively subjected to modification as it has ample opportunity for surface functionalization. Various attempts have been made for enhancing cellulose properties.

NCCs have been extensively used for drug delivery because of their excellent colloidal stability. Various strategies for different kinds of drug release have been studied and performed by renowned researchers. KohHann Suk and coworkers have worked on the encapsulation of cow urine in nanocrystalline cellulose. They have taken various concentrations of urine to analyze minimum inhibitory concentration and then encapsulated it by the microemulsion method. The microbial growth of encapsulated CNPs was tested by following the preliminary test with disk diffusion and turbidimetric assay. After this they are characterized by X-ray photoelectron spectroscopy, Fourier transforms infrared spectroscopy, X-ray powder diffraction, atomic force microscopy, and scanning electron microscopy. On *B. subtilis* agar plate, encapsulated urine and normal urine both showed excellent results. The positive outcome of this is that it is not inhibiting *Escherichia coli* at a normal concentration that is usually found in human beings which involves the synthesis of vitamin B and K. Cow urine alone and encapsulated cow urine had positive results against *A. Niger* but encapsulated had greater efficacy against it (Peng, Dhar, Liu, & Tam, 2011).

Marina Lima de Fontes and coworkers worked on the anticancer activity of doxorubicin bacterial nanocellulose. They prepared BC with carboxy methylcellulose which adheres to BC microfibrils in a discontinuous mode and interferes with their aggregation. CMC interfere in the steric hindrance of BC by altering the hydrogen bond between microfibrils. BC/CMC biocomposites were successfully obtained by using soluble cellulose derivative HS culture medium is modified in situ and prepared BC/CMC nanocomposites successfully. This cellulose material acts as coating material and forms "coating islands," which strongly influenced biomedical applications (Fontes et al., 2017). Gram-negative bacteria are used in the synthesis of BC such as *glucnocetabacter*, *agrobacteriase*, *aerobactor*, and *achromobaceter*.

"Click" chemistry is used for synthesizing Poly ethyl ethylene phosphate modified CNCs by grafting process. It is a combination of ring-opening polymerization and Cu I catalyzed azide—alkyne cycloaddition. Doxorubicin had encapsulated by electrostatic interaction into poly (ethyl ethylene phosphate) modified CNCs. This carrier could be used for delivering anticancer molecules electrostatic and releasing them into the tumor environment (Wang, He, Zhang, Tam, & Ni, 2015).

Hydrogels are hydrophilic polymer networks that may absorb from 10%—20% (an arbitrary lower limit) up to thousands of times of their dry weight in water. Hydrogels may be chemically stable or they may degrade and eventually disintegrate and dissolve (Hoffman, 2012). The main drawback of hydrogels is their poor mechanical property. There are different nanocomposites prepared by Jinping Zhou and coworkers which were further tested for cytotoxicity and inflammatory reaction on the mice having liver cancer. Results showed that nanocomposite hydrogels have great potentials for delivering drugs (You et al., 2016). Dufresne A. and coworkers prepared a biocompatible double-membrane hydrogel having cationic cellulose and anionic alginate in its structure. It had one external and the internal membrane made up of neat alginate and cationic CNCs respectively. Cationic CNCs enhanced the stability through electrostatic interaction between anionic and cationic CNCs. It could be used for codelivery of two drugs (Lin, Gèze, Wouessidjewe, Huang, & Dufresne, 2016).

Gautier M.A. Ndong Ntoutoume and coworkers used the ionic association technique for preparing cyclodextrin CNCs. CD/CNCs complex is further used for encapsulation of curcumin. Then it is compared with normal curcumin and it showed greater cytotoxicity against cancer cell lines (Ndong Ntoutoume et al., 2016). Natural cotton wool is used for the preparation of CNCs and functionalization is done by using disulfide bond linked poly (2-(dimethyl amino) ethyl methacrylate) (PDMAEMA). PDMAEMA chains can be easily cleaved from CNCs. CNC−SS PD showed suppressing activity on cancer-growing cells. It demonstrated that for effective gene/drug delivery redox responsive polycation is the preferable method (Hu et al., 2015). Fu-JianXu and coworkers used the control polymerization technique for gold CNCs particle conjugate with heterogeneous brush-coated polymer. CNCs were proposed via controllable polymerization techniques. CNCs were individually grafted by PDMAEMA and poly (poly (ethylene glycol) ethyl ether methacrylate) brushes (Zhang et al., 2010).

Maren Roman and coworker's targeted chemotherapeutic agents to folate receptor-positive cancer cells by synthesizing folic acid conjugate CNCs. Elongated nanoparticles are different and have a specific advantage over spherical nanoparticles. This work showed that folic acid covalently attaching to CNCs was the only way to target CNCs to folate receptor-positive mammalian cells (Dong, Cho, Lee, & Roman, 2014).

Nanoparticle aggregation is the main reason for the decreasing performance of non-materials. Stabilization of unstable nanoparticles with water-dispersible and biocompatible nanoparticles is a promising strategy and alternative for use of capping agents. Dopamine is used for modification of CNCs and followed by in situ generation and AgNPs fixed on CNC surface through reducing Ag ion by polydopamine coated CNC. The result showed increased stability of AgNPs and a fourfold increase in antibacterial activity against *E. coli* and *Bacillus subtilis* (Shi et al., 2015). Xingyu Jiang and coworkers prepared different sizes of AgNPs by using redox reaction between $Ag+$ and glucose at room temperature. It develops a colorimetric and nonenzymatic assay for glucose detection with a lower limit of 0.116 μm. These NCC-assisted AgNPs showed enhanced antibacterial activity for both gram-positive and negative bacteria. It can also help in various areas like clinical diagnosis and biomanufacturing (Wang et al., 2016). Kam Chiu Tam and coworkers worked on a green approach and used it for fabricating CNC@polyrhodamine core sheath NPs based in aqueous solution. A high amount of polyrhodanine coating could be obtained by varying the ratio of CNCs, monomer, and concentration of oxidant Fe (I). The result showed antimicrobial properties against *E. coli* and *B. subtilis*. it has also given an idea that core sheath materials can be used as a potential candidate for antimicrobial application (Tang et al., 2015a).

Reza A. Ghiladi and coworkers synthesized and characterized cationic Porphyrin surface-modified CNCs. Cu (I) and 1, 3-dipolar cycloaddition occurs between azide groups on the cellulosic surface which fixed Porphyrin to the surface of CNCs. The resulting CNC-Por complex is characterized by different parameters and it showed excellent efficacy toward the photodynamic inactivation of *Mycobacterium Smegmatis* and *Staphylococcus Aureus*, only slight activity against *E. coli* (Feese, Sadeghifar, Gracz, Argyropoulos, & Ghiladi, 2011). Daniele Oliveira de Castro and coworkers worked on surface functionalization, they esterified the surface of CNCs by using nontoxic resin acids i.e. rosin. The result showed etherification proceed from the surface of CNCs and antimicrobial activity of rosin grafted modified and neat CNCs have strong antibacterial activity against gram-positive and negative bacteria (De Castro, Bras, Gandini, & Belgacem, 2016). Mauri A. Kostiainen and coworkers worked and prepared cationic poly brush modified CNCs which were prepared from intrinsically anionic CNCs by surface-initiated atom-transfer radical polymerization of poly-(N,N-dimethyl amino ethyl methacrylate) from the crystal surface. It is further quaternized with methyl iodide. These results show the feasibility of the modified CNCs in virus binding and concentrating, and opens a new way for them to be used as transduction enhancers for viral delivery applications (Rosilo et al., 2014).

Dilip Kumar pal and coworkers worked and prepared Gliclazide loaded methyl cellulose mucoadhesive microcapsules by using a central composite design. This formulation showed prolonged systemic absorption to maintain lower blood glucose levels and improved patient compliance (Pal & Nayak, 2011) only cellulose Mohammad L. Hassan and coworkers prepared oxidized CNCs by incorporating them into chitosan nanoparticles to control the release rate of Repaglinide. Hydrogen bonding retard the release of drug by producing sustain or control release profile (Abo-Elseoud et al., 2018).

Microcrystalline cellulose is the established and the most important drug excipient in the preparation of solid oral dosage form. Ruzica Kolakovic and coworkers worked on Spray-Dried Cellulose Nanofibers as Novel Tablet Excipients. They prepared nanofibrils cellulose and compared it with commercial MCC. Nanofibrils cellulose has a better ability to pack and lower powder porosity than commercial MCC. That material has good flow property and addition to commercial MCC improves flow ability to the mixture. This study shows that spray-dried CNF can be used as an excipient for tablet production (Kolakovic et al., 2011).

BC and NFC are mainly explored for transdermal drug delivery system (TDDS) due to its absorbing and imbibing property which helps in faster wound healing. Its mechanical and physical stability BC is used for dermal preparations. Wei et al. prepared a BC film by immersing a lyophilized BC film in a benzalkonium chloride (BZK) solution and then lyophilizing the wet film. The result showed high water-absorbing capacity and sustained antibacterial activity (for 24 h), and could therefore be used as a wound dressing film. Not only BNC (Bacterial Nanocellulose) but also NFC has shown a high potential application in TDDS. As an example, Sarkar et al. prepared NFC/chitosan transdermal film for the delivery of ketorolec tromethamine in which the NFC acts as an elegant carrier. Their findings showed sustained release profiles of the drug from matrices 40% of the drug was released in 10 h by the addition of 1 wt.% NFC in the formulation. They concluded that NFC could be a potential candidate for controlled TDDS. Nanocellulose is used for various drug deliveries some of them are listed in Table 8.1.

BC, Bacterial cellulose; *CNC*, cellulose nanocrystal; *NCC*, nanocrystalline cellulose.

Table 8.1 Nanocrystalline cellulose in different drug delivery.

Sr. no.	Type of cellulose	Application	References
Drug delivery			
1	Polymer-modified cellulose nanocrystals	For controlled release of doxorubicin	Wang et al. (2015)
2	Cationic nanocrystalline cellulose	Subcutaneous and sustained delivery of doxorubicin	You et al. (2016)
3	Cationic CNC with anionic alginate	For codelivery of drugs	Lin et al. (2016)
4	Cyclodextrin/cellulose nanocrystals complexes	Antiproliferative effect of curcumin on cell lines of colorectal and prostatic cancer	Ndong Ntoutoume et al. (2016)
5	Cotton cellulose nanocrystals	For gene/drug delivery	Hu et al. (2015)
6	Heterogeneous polymer-modified cellulose nanocrystals conjugated with gold nanoparticles	For multifunctional therapy	Hu et al. (2016)
7	Folic acid-modified CNC	Targeted delivery in cancer treatment	Dong et al. (2014)
8	Silver cellulose polydopamine coated CNCs	Antibacterial activity against Gram-positive and negative bacteria	Shi et al. (2015)
9	NCC-based silver nanoparticles	For clinical diagnosis, biomanufacturing, and antibacterial action	Wang et al. (2016)

(*Continued*)

Table 8.1 (Continued)

Sr. no.	Type of cellulose	Application	References
10	Polyrhodanine-coated CNCs	Antimicrobial activity against *Escherichia coli* and *Bacillus subtilis*	Tang et al. (2015b)
11	Porphyrin-cellulose Nanocrystals	For use in photo bactericidal	Feese et al. (2011)
12	Surface grafted cellulose nanocrystals	Strong antibacterial activity especially against Gram-negative bacteria	De Castro et al. (2016)
13	Cationic polymer modified cellulose	As transduction enhancers for viral delivery	Rosilo et al. (2014)
14	Polyrhodanine-coated cellulose nanocrystals	Antibacterial activity against *E. coli* and *B. subtilis*	Tang et al. (2015a)
15	Alginate—methyl cellulose mucoadhesive microcapsules	For controlling release of gliclazide	Pal and Nayak (2011)
16	Chitosan cellulose nanocrystals nanocomposite	For controlling release of repaglinide	Abo-Elseoud et al. (2018)
Transdermal drug delivery system			
17	BC membrane	Berberine sulfate	Li et al. (2019) and Huang et al. (2013)
18	BNC (Bacterial Nanocellulose) hydrogel	Povidone iodine	Fu, Zhou, Zhang, and Yang (2013a)
19	BNC (Bacterial Nanocellulose) hydrogel	Polyhexanide	Fu, Zhou, Zhang, and Yang (2013b)
20	Zein/BNC (Bacterial Nanocellulose) nanocomposite	Silymarine	Maneerung, Tokura, and Rujiravanit (2008)
21	Octenidine	BNC (Bacterial Nanocellulose) polaxmar gel	Singla et al. (2017)
22	Ibuprofen	Magnetic NCC hydrogel	Tritt-Goc, Jankowska, Pogorzelec-Glaser, Pankiewicz, and èawniczak (2018)
23	Curcumin	NCC film	Ahrem et al. (2014)
24	Bovine serum albumin	NCC hydrogel	Martínez Ávila et al. (2014)
25	Bovine serum albumin	NCC cyclodextrin hydrogel	Zamarioli, Martins, Carvalho, and Freitas (2015)

8.4 Conclusion

This chapter summarizes the basic introduction, preparation, and uses of nanocellulose. It has been shown that during the last 15 years an increasing number of researches in the area of cellulose nanocomposite have been developed. It is used in various fields in biomedical starting from drug delivery to treatment. Cellulose is an attractive material because it is cheap, light in weight, and has an unlimited supply. Supply could be easily met from both land and sea sources. The next century will see a plethora of innovations in sustainable nanomaterial as we are moving towards the integration of these materials by replacing nonrenewable materials with those that are sustainable and green. It can be the engine that drives this innovation and offer solutions that provide a balance between the environment and the needs of consumers.

References

Abo-Elseoud, W. S., Hassan, M. L., Sabaa, M. W., Basha, M., Hassan, E. A., & Fadel, S. M. (2018). Chitosan nanoparticles/cellulose nanocrystals nanocomposites as a carrier system for the controlled release of repaglinide. *International Journal of Biological Macromolecules*, *111*, 604–613. Available from https://doi.org/10.1016/j.ijbiomac.2018.01.044.

AchabyEl, M., Kassab, Z., Aboulkas, A., & Gaillard, C. (2017). Reuse of Red Algae Waste for the Production of Cellulose Nanocrystals and its Application in Polymer Nanocomposites. *International Journal of Biological Macromolecules*. Available from https://doi.org/10.1016/j.ijbiomac.2017.08.067.

Ahrem, H., Pretzel, D., Endres, M., Conrad, D., Courseau, J., Müller, H., ... Kinne, R. W. (2014). Laser-structured bacterial nanocellulose hydrogels support ingrowth and differentiation of chondrocytes and show potential as cartilage implants. *Acta Biomaterialia*, *10*, 1341–1353. Available from https://doi.org/10.1016/j.actbio.2013.12.004.

De Castro, D. O., Bras, J., Gandini, A., & Belgacem, N. (2016). Surface grafting of cellulose nanocrystals with natural antimicrobial rosin mixture using a green process. *Carbohydrate Polymers*, *137*, 1–8. Available from https://doi.org/10.1016/j.carbpol.2015.09.101.

Dong, S., Cho, H. J., Lee, Y. W., & Roman, M. (2014). Synthesis and cellular uptake of folic acid-conjugated cellulose nanocrystals for cancer targeting. *Biomacromolecules*, *15*, 1560–1567. Available from https://doi.org/10.1021/bm401593n.

Feese, E., Sadeghifar, H., Gracz, H. S., Argyropoulos, D. S., & Ghiladi, R. A. (2011). Photobactericidal porphyrin-cellulose nanocrystals: synthesis, characterization, and antimicrobial properties. *Biomacromolecules*, *12*, 3528–3539. Available from https://doi.org/10.1021/bm200718s.

Fontes, M. L. D., Meneguin, A. B., Tercjak, A., Gutierrez, J., Stringhetti, B., Cury, F., & Barud, H. S. (2017). Effect of in situ modification of bacterial cellulose with carboxymethylcellulose on its nano/microstructure and methotrexate release properties. *Carbohydrate Polymers*. Available from https://doi.org/10.1016/j.carbpol.2017.09.061.

Fu, L., Zhou, P., Zhang, S., & Yang, G. (2013a). Evaluation of bacterial nanocellulose-based uniform wound dressing for large area skin transplantation. *Materials Science and Engineering C*, *33*, 2995–3000. Available from https://doi.org/10.1016/j.msec.2013.03.026.

Fu, L., Zhou, P., Zhang, S., & Yang, G. (2013b). Evaluation of bacterial nanocellulose-based uniform wound dressing for large area skin transplantation. *Materials Science and Engineering: C*, *33*, 2995–3000. Available from https://doi.org/10.1016/J.MSEC.2013.03.026.

Glass, D. C., Moritsugu, K., Cheng, X., & Smith, J. C. (2012). REACH coarse-grained simulation of a cellulose fiber. *Biomacromolecules*, *13*, 2634–2644. Available from https://doi.org/10.1021/bm300460f.

Hoffman, A. S. (2012). Hydrogels for biomedical applications. *Advanced Drug Delivery Reviews*. Available from https://doi.org/10.1016/j.addr.2012.09.010.

Hu, H., Hou, X.-J., Wang, X.-C., Nie, J.-J., Cai, Q., & Xu, F.-J. (2016). Gold nanoparticle-conjugated heterogeneous polymer brush-wrapped cellulose nanocrystals prepared by combining different controllable polymerization techniques for theranostic applications. *Polymer Chemistry*, 7, 3107–3116. Available from https://doi.org/10.1039/C6PY00251J.

Hu, H., Yuan, W., Liu, F. S., Cheng, G., Xu, F. J., & Ma, J. (2016). Redox-responsive polycation-functionalized cotton cellulose nanocrystals for effective cancer treatment. *ACS Applied Materials and Interfaces*, 7, 8942–8951. Available from https://doi.org/10.1021/acsami.5b02432.

Huang, L., Chen, X., Nguyen, T. X., Tang, H., Zhang, L., & Yang, G. (2016). Nano-cellulose 3D-networks as controlled-release drug carriers. *Journal of Materials Chemistry B*, *1*, 2976–2984. Available from https://doi.org/10.1039/c3tb20149j.

Johar, N., Ahmad, I., & Dufresne, A. (2012). Extraction, preparation and characterization of cellulose fibres and nanocrystals from rice husk. *Industrial Crops and Products*, *37*, 93–99. Available from https://doi.org/10.1016/j.indcrop.2011.12.016.

Klemm, D., Kramer, F., Moritz, S., Lindström, T., Ankerfors, M., Gray, D., & Dorris, A. (2011). Nanocelluloses: A new family of nature-based materials. *Angewandte Chemie - International Edition*, *50*, 5438–5466. Available from https://doi.org/10.1002/anie.201001273.

Kolakovic, R., Peltonen, L., Laaksonen, T., Putkisto, K., Laukkanen, A., & Hirvonen, J. (2011). Spray-dried cellulose nanofibers as novel tablet excipient. *AAPS PharmSciTech*, *12*, 1366–1373. Available from https://doi.org/10.1208/s12249-011-9705-z.

Lejeune, A., & Deprez, T. (2010). *Cellulose : structure and properties, derivatives and industrial uses*. Nova Science Publishers.

Li, J., Yang, Z., Liu, H., Qiu, M., Zhang, T., Li, W., & Tan, S. (2019). ADS-J1 disaggregates semen-derived amyloid fibrils. *Biochemical Journal*, *476*, 1021–1035. Available from https://doi.org/10.1042/BCJ20180886.

Lin, N., Gèze, A., Wouessidjewe, D., Huang, J., & Dufresne, A. (2016). Biocompatible double-membrane hydrogels from cationic cellulose nanocrystals and anionic alginate as complexing drugs codelivery. *ACS Applied Materials & Interfaces*, *8*, 6880–6889. Available from https://doi.org/10.1021/acsami.6b00555.

Maneerung, T., Tokura, S., & Rujiravanit, R. (2008). Impregnation of silver nanoparticles into bacterial cellulose for antimicrobial wound dressing. *Carbohydrate Polymers*, *72*, 43–51. Available from https://doi.org/10.1016/j.carbpol.2007.07.025.

Martínez Ávila, H., Schwarz, S., Feldmann, E. M., Mantas, A., Von Bomhard, A., Gatenholm, P., & Rotter, N. (2014). Biocompatibility evaluation of densified bacterial nanocellulose hydrogel as an implant material for auricular cartilage regeneration. *Applied Microbiology and Biotechnology*, *98*, 7423–7435. Available from https://doi.org/10.1007/s00253-014-5819-z.

Ndong Ntoutoume, G. M. A., Granet, R., Mbakidi, J. P., Brégier, F., Léger, D. Y., Fidanzi-Dugas, C., & Sol, V. (2016). Development of curcumin–cyclodextrin/cellulose nanocrystals complexes: New anticancer drug delivery systems. *Bioorganic & Medicinal Chemistry Letters*, *26*, 941–945. Available from https://doi.org/10.1016/J.BMCL.2015.12.060.

Pal, D., & Nayak, A. K. (2011). Development, optimization, and anti-diabetic activity of gliclazide-loaded alginate–methyl cellulose mucoadhesive microcapsules. *AAPS PharmSciTech*, *12*, 1431–1441. Available from https://doi.org/10.1208/s12249-011-9709-8.

Peng, B. L., Dhar, N., Liu, H. L., & Tam, K. C. (2011). Chemistry and applications of nanocrystalline cellulose and its derivatives: A nanotechnology perspective. *Canadian Journal of Chemical Engineering*, *89*, 1191–1206. Available from https://doi.org/10.1002/cjce.20554.

Rosilo, H., McKee, J. R., Kontturi, E., Koho, T., Hytönen, V. P., Ikkala, O., & Kostiainen, M. A. (2014). Cationic polymer brush-modified cellulose nanocrystals for high-affinity virus binding. *Nanoscale*, *6*, 11871–11881. Available from https://doi.org/10.1039/C4NR03584D.

Shankar, S., Oun, A. A., & Rhim, J. W. (2018). Preparation of antimicrobial hybrid nano-materials using regenerated cellulose and metallic nanoparticles. *International Journal of Biological Macromolecules*, *107*, 17–27. Available from https://doi.org/10.1016/j.ijbiomac.2017.08.129.

Shi, Z., Tang, J., Chen, L., Yan, C., Tanvir, S., Anderson, W. A., & Tam, K. C. (2015). Enhanced colloidal stability and antibacterial performance of silver nanoparticles/cellulose nanocrystal hybrids. *Journal of Materials Chemistry B, 3*, 603–611. Available from https://doi.org/10.1039/C4TB01647E.

Singla, R., Soni, S., Kulurkar, P. M., Kumari, A., Mahesh, S., Patial, V., & Yadav, S. K. (2017). In situ functionalized nanobiocomposites dressings of bamboo cellulose nanocrystals and silver nanoparticles for accelerated wound healing. *Carbohydrate Polymers, 155*, 152–162. Available from https://doi.org/10.1016/j.carbpol.2016.08.065.

Tang, J., Song, Y., Tanvir, S., Anderson, W. A., Berry, R. M., & Tam, K. C. (2015a). Polyrhodanine coated cellulose nanocrystals: A sustainable antimicrobial agent. *ACS Sustainable Chemistry and Engineering, 3*, 1801–1809. Available from https://doi.org/10.1021/acssuschemeng.5b00380.

Tang, J., Song, Y., Tanvir, S., Anderson, W. A., Berry, R. M., & Tam, K. C. (2015b). Polyrhodanine coated cellulose nanocrystals: A sustainable antimicrobial agent. *ACS Sustainable Chemistry & Engineering, 3*, 1801–1809. Available from https://doi.org/10.1021/acssuschemeng.5b00380.

Tritt-Goc, J., Jankowska, I., Pogorzelec-Glaser, K., Pankiewicz, R., & èawniczak, P. (2018). Imidazole-doped nanocrystalline cellulose solid proton conductor: synthesis, thermal properties, and conductivity. *Cellulose, 25*, 281–291. Available from https://doi.org/10.1007/s10570-017-1555-8.

Villanova, J. C. O., Ayres, E., Carvalho, S. M., Patrício, P. S., Pereira, F. V., & Oréfice, R. L. (2011). Pharmaceutical acrylic beads obtained by suspension polymerization containing cellulose nanowhiskers as excipient for drug delivery. *European Journal of Pharmaceutical Sciences, 42*, 406–415. Available from https://doi.org/10.1016/j.ejps.2011.01.005.

Wang, H., He, J., Zhang, M., Tam, K. C., & Ni, P. (2015). A new pathway towards polymer modified cellulose nanocrystals via a "grafting onto" process for drug delivery. *Polymer Chemistry, 6*, 4206–4209. Available from https://doi.org/10.1039/C5PY00466G.

Wang, S., Sun, J., Jia, Y., Yang, L., Wang, N., Xianyu, Y., & Jiang, X. (2016). Nanocrystalline cellulose-assisted generation of silver nanoparticles for nonenzymatic glucose detection and antibacterial agent. *Biomacromolecules, 17*, 2472–2478. Available from https://doi.org/10.1021/acs.biomac.6b00642.

Yang, B., Jiang, Y., Zhao, M., Shi, J., & Wang, L. (2008). Effects of ultrasonic extraction on the physical and chemical properties of polysaccharides from longan fruit pericarp. *Polymer Degradation and Stability, 93*, 268–272. Available from https://doi.org/10.1016/j.polymdegradstab.2007.09.007.

You, J., Cao, J., Zhao, Y., Zhang, L., Zhou, J., & Chen, Y. (2016). Improved mechanical properties and sustained release behavior of cationic cellulose nanocrystals reinforeced cationic cellulose injectable hydrogels. *Biomacromolecules, 17*, 2839–2848. Available from https://doi.org/10.1021/acs.biomac.6b00646.

Zamarioli, C. M., Martins, R. M., Carvalho, E. C., & Freitas, L. A. P. (2018). Nanoparticles containing curcuminoids (Curcuma longa): Development of topical delivery formulation. *Brazilian Journal of Pharmacognosy, 25*, 53–60. Available from https://doi.org/10.1016/j.bjp.2014.11.010.

Zhang, T., Wang, W., Zhang, D., Zhang, X., Ma, Y., Zhou, Y., & Qi, L. (2010). Biotemplated synthesis of gold nanoparticle-bacteria cellulose nanofiber nanocomposites and their application in biosensing. *Advanced Functional Materials, 20*, 1152–1160. Available from https://doi.org/10.1002/adfm.200902104.

CHAPTER 9

Nano-biodegradation of plastic materials

Alcides Lopes Leão[1], Ivana Cesarino[1], Milena Chanes de Souza[1], Otavio Augusto Titton Dias[2] and Mohammad Jawaid[3]

[1]Department of Bioprocesses and Biotechnology, School of Agriculture (FCA), São Paulo State University (UNESP), Botucatu, São Paulo, Brazil

[2]Centre for Biocomposites and Biomaterials Processing, John H. Daniels Faculty of Architecture, Landscape, and Design, University of Toronto, Toronto, ON, Canada

[3]Universiti Putra Malaysia, Serdang, Malaysia

9.1 Introduction

Nanotechnology has brought many benefits to materials science, as it is possible to adapt material structures at extremely small scales to achieve specific properties (Rajak, 2018). Using nanotechnology, materials can be made effectively lighter, stronger, more durable, more reactive, more like a sieve, or better electrical conductors, among many other characteristics (Crawford, 2016). The incorporation of these nanomaterials into polymers, for example, can improve the mechanical properties and add new functionalities to the materials, for example, drug release and capture of specific substances (Boisseau & Loubaton, 2011; National Nanotechnology Initiative, 2021). However, the end of the useful life of polymer-nanoparticle composites is still in early days in terms of environmental impact, and the life cycle assessment should be conducted to compare the pros and cons of every material (Balaguer, Aliaga, Fito, & Hortal, 2016). In addition to applications for use in polymeric materials, nanoparticles also have a broad range of potential applications, as will be described in the next topics.

This chapter investigates the influence of the nanoparticles on the thermal degradation process of polymer composites, since it is clear that nanotechnology brings better physical and mechanical properties for the polymer composites and specifically for the bioplastics. As reported by Borgna (2017), the addition of nanoscale particles mainly in the packaging industry aims to optimize and improve the mechanical and barrier properties which typically exhibit poorer performance when compared to the conventional fossil based plastics. The advantage of the utilization of nanotechnology is that the amount of filler or reinforcement applied is very low, which reduces some eventual problem that arises compared to conventional inorganic fillers such as density, changes in the surface and appearance, reduction in the processability, etc. Just as a comparison for conventional fillers (fiberglass or calcium carbonate), the load is around 15 wt.% while for nanoscale only 5–6 wt.% is enough (Liu et al., 2019). Leão et al. (2012)

Nanotechnology in Paper and Wood Engineering
DOI: https://doi.org/10.1016/B978-0-323-85835-9.00008-8

reported that the addition of a low amount of nanoscale fillers into polymers leads to improvement in mechanical and physical properties.

9.2 Applications

Among the actual list of applications, few can be listed: packaging, materials, coating, therapy techniques, diagnostics, complex drug delivery systems, food science, electronics, energy, water and air treatments, sensors, information technology, homeland security, transportation, energy, food safety, ballistic applications, and environmental science (Murty, Shankar, Raj, Rath, & Murday, 2013). Many everyday commercial products are currently on the market and in daily use that rely on nanoscale materials and processes, such as antibacterial sanitary surfaces through self-cleaning construction materials. Nanotechnology is present in many medical aspects including medical tools, needleless vaccines, vaccine scaffold, etc. (Contera, La Bernardino de Serna, & Tetley, 2020). For the optical applications, films, antireflective, scratch resistant and electrically conductive, are the most important in this area (Fig. 9.1) (Bharmoria & Ventura, 2019).

In the textile sector, nanotechnology has been used for developing biocides, multifunctional cotton fabrics, surface modifiers, nonodor fabrics, color changing agents, wrinkling, staining, mimetism, ballistic application, etc., based on its high modulus (Gavrilenko & Wang, 2021; Sawhney et al., 2008). Even military applications are under development, with fabrics that change colors, with monitoring capacities and energy generation (Sonawane, Patil, & Sonawane, 2018).

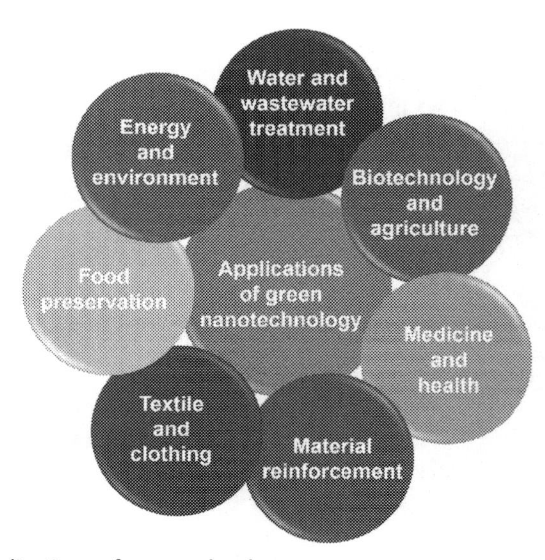

Figure 9.1 Current applications of nanotechnology.

In the field of transportation, the main aspect of the nanoreinforcement is the lightweight of the materials that are important in cars, trucks, airplanes, boats, and spacecraft. In the field of sports few can be listed such as baseball bats, tennis rackets, bicycles, motorcycle helmets, automobile parts, luggage, and power tool housings. For example, the combination of light weight and conductivity makes them ideal for applications such as wearable flexible strain sensors, electromagnetic shielding, and thermal management (Shalaby & Saad, 2020). Modern applications of nanoengineered materials in automotive industry include high-power rechargeable battery systems, thermoelectric materials for temperature control, tires with lower rolling resistance, high-efficiency sensors and electronics, catalyst converter, smart solar panels, and fuel shells and additives for cleaner exhaust and extended range (Mangaraj, Das, & Malla, 2020).

Nanotechnology has opened up new avenues of research in computing and electronics and led to the fabrication of more portable systems that enabled the storage of larger amounts of information to be seen in ultra-high-definition displays and televisions, with sharp colors and low energy consumption. These screens can be flexible, bendable, foldable, rollable, and stretchable. Nanotechnology also can be used in high efficiency solar panels, with nanostructured solar cells that could be made in flexible rolls and paintable (Leão et al., 2012; National Nanotechnology Initiative, 2021).

There are also many ways that nanotechnology can help detect and clean up environmental contaminants. That is the case of low-cost environmental sensors for emerging contaminants allowing monitoring treatment of impurities in water and wastewater (Ejeian et al., 2018).

The reduction of a commercial jet aircraft weight by 20% has been reported to reduce its fuel consumption by 15%. The use of advanced nanomaterials with twice the strength of conventional composites would reduce the gross weight of a launch vehicle by as much as 63% according to NASA studies (National Nanotechnology Initiative, 2021).

Recently, nanocellulose-reinforced composites have drawn great attention due to their lightweight. The versatility of nanocellulose and its surface-tailoring capabilities have been demonstrated as a potential material for application in a wide array of industrial sectors including electronics, construction, packaging, food, energy, health care (e.g., fully transparent sun blockers), automotive, and defense (Kargarzadeh et al., 2017; Saba & Jawaid, 2017).

Nanotechnology although is considered one of the powered drive for the world's technology, still is under consideration the possible hazardous impacts over the human health or even the environment, opening the demand for a new area called nanotoxicology (Bai, Sabouni, & Husseini, 2018).

9.3 Nanocellulose

Cellulose has drawn significant attention from many researchers as it is a sustainable and renewable macromolecule. That is a good motivation to study cellulose-based materials mainly in nanoscale as nanomaterials with practically new applications. Nanocellulose is classified as cellulose nanocrystals (CNC), cellulose nanofiber (CNF), and bacterial nanocellulose. These nanostructured cellulose materials possess different properties like particle size, morphology, and crystallinity induced by the different extraction methods and different sources (Leão et al., 2012).

Nanocellulose is a term that refers to cellulose fibrils with at least one dimension of up to 100 nm in size (Yang et al., 2018). It can be obtained from several lignocellulosic sources including coconut shell, cassava bagasse, eucalyptus wood, cotton linter, bamboo, banana leaf, grass, corn, rice, etc. (Pandey, Ahn, Lee, Mohanty, & Misra, 2010; Santana, do Rosário, Pola, & Otoni, 2017). In addition, they can also be isolated from other varieties of cellulosic resources, such as animals (e.g., tunicates), bacteria, and algae. Cellulose is the most abundant raw material on the planet, and, in principle, they can be extracted from almost any cellulosic material using different procedures (Thakur, Thakur, & Prasanth, 2014). The most common forms of nanofibers extraction are physical (e.g., mechanical refining), biological (treatment with specific bacteria and enzymes), or chemical methods (treatment with acids and alkalis), or a combination of these methods, as shown in Fig. 9.2 (Zhang,

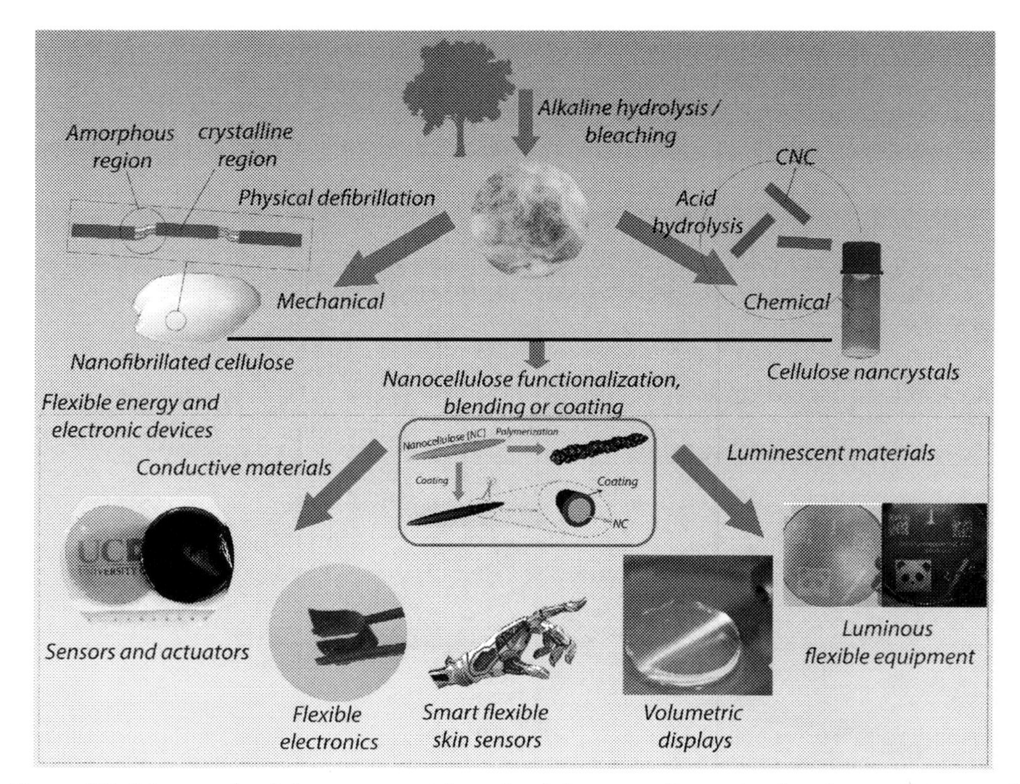

Figure 9.2 Scheme of cellulose nanocrystals and cellulose nanofiber extraction from cellulose, and their surface functionalization to be used in flexible energy and electronic materials (Dias et al., 2020).

Barhoum, Xiaoqing, Li, & Samyn, 2019; Zhong & Peng, 2017). About 1 billion tons of pulp is produced in several factories each year. Additionally, cellulose is renewable, biodegradable, inexpensive, thermally stable, lightweight, and has many other good properties (Kafy et al., 2017).

Nanocellulose can be obtained in the form of nanofibers (CNF), nanocrystal (CNC), or nanowhisker, has been reported as an interesting sustainable and tunable template for the production of high-performance materials (Dias et al., 2020; Thakur et al., 2014). CNFs and CNCs are different in shape, size, and composition (Xu et al., 2013; Xu et al., 2014).

Due to several factors such as environmental concern, sustainable products, bioeconomy, circular economy are opening new markets for nanocellulose. The market for nanocellulose is projected to increase from US$297 million in 2020 to US$783 million by 2025, at a compound annual growth rate of 21.3% (Grand View Research, 2020). The major nanocellulose market players are Fiberlean technologies (United Kingdom), Borregard (Norway), Nippon Paper Industries (Japan), Celluforce INC (Canada), Kruger INC (Canada), Stora Enso (Finland), Rise Innventia (Sweden), American Process Inc. (United States), FPInnovations (Canada), UPM-Kymmene Oyj (Finland), Melodea (Israel), Cellucomp (Scotland), Blue Goose Refineries (Canada), Oji Holdings Corporation (Japan), VTT (Finland), and Sappi (South Africa) (Markets & Markets, 2020). In Brazil, it is reported that Suzano, a Brazilian pulp and paper company will start a nanocellulose plant in Aracruz, at an estimated price of US$8000 per ton and a capacity of 2000 per month.

9.3.1 Cellulose nanofibers

CNF usually have the dimensions of several micrometers and 5–20 nm wide, and is considered as one of the most attractive bionanomaterial in the world due to its good mechanical properties, such as Young's high modulus ($\sim$140 GPa), high strength (few GPa), low density, and anisotropic physical characteristics. In the highly crystalline CNFs, the elastic modulus of the crystalline domain of nanocellulose is 150 and 50 GPa for longitudinal and transverse directions, respectively (Thakur, Thakur, & Kessler, 2017).

Cellulose fibers are composed of cellulose chain molecules with nanofibers, as shown in Fig. 9.3 with its hierarchical structure. These nanofibers or elementary fibrils are 2–20 nm in diameter and a few micrometers in length (Zhang et al., 2019), as can be observed in Fig. 9.4 and Fig. 9.5. Even with all the listed properties, such as high crystallinity, mechanical properties, surface area, aspect ratio, and surface charge, which can be tuned as a tailor-made process, Kafy et al. (2017) still wrote that CNFs were not competitive for structural compounds due to the fact that they are too short.

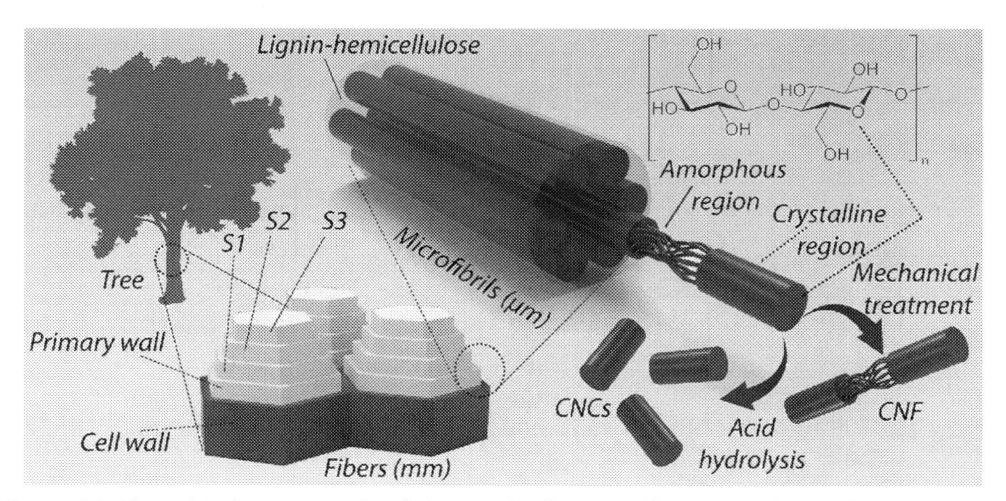

Figure 9.3 Hierarchical structure of cellulose and schematic illustration of crystalline and amorphous regions in cellulose fiber (Dias et al., 2020). *CNC*, Cellulose nanocrystals; *CNF*, cellulose nanofiber.

Figure 9.4 Cellulose nanofibers (Kouzegaran, 2020).

9.3.2 Cellulose nanocrystals

CNC is a rigid rod with dimensions varying with a diameter of 3–20 nm and length from 100 nm to microns. As observed in Fig. 9.6 and Fig. 9.7 CNCs are obtained

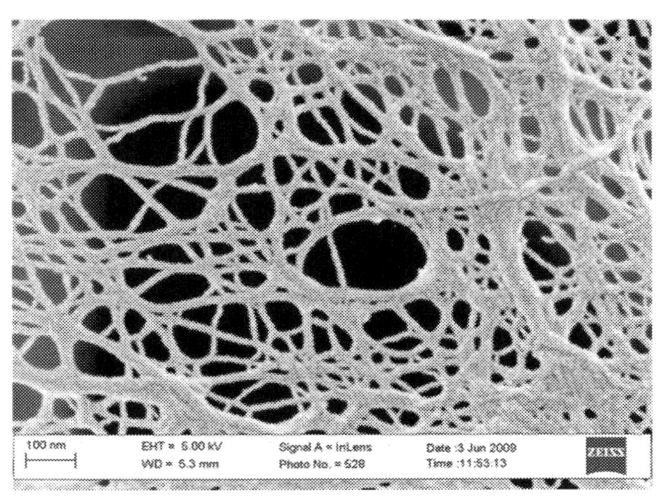

Figure 9.5 Scanning electron microscopy image of cellulose nanofibers produced by mechanical grinding for 1 h (Tsuzuki, Zhang, Rana, Liu, & Wang, 2010).

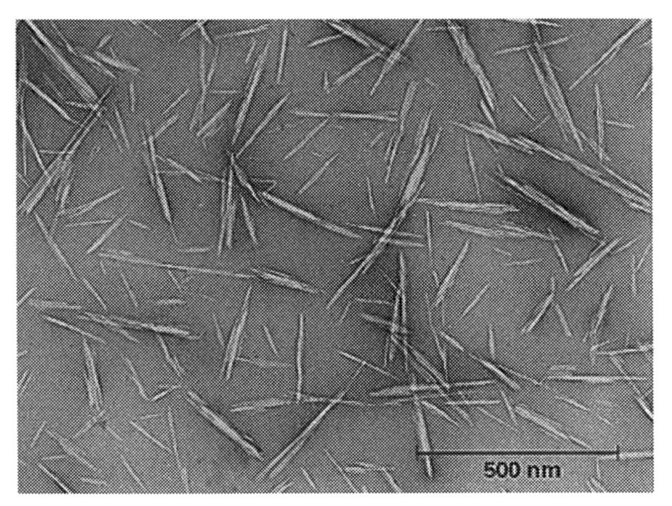

Figure 9.6 Transmission electron micrograph of cellulose nanocrystals from ramie fibers. Compared to microfibrillated cellulose, the fibers are shorter and not interconnected to each other (Menezes & Siqueira, 2009).

through selective acid hydrolysis of amorphous domains of cellulose, where crystalline regions are maintained. As main characteristics, CNCs have high aspect ratio, high surface area, high mechanical strength, renewability, light weight, wide availability, low cost, ability to form hydrogen bonds, hydrophobic, and a liquid crystalline nature. The CNC is considered an outstanding material in which its modulus has approximately the theoretical strength of a single fiber, means that it is stronger than steel and

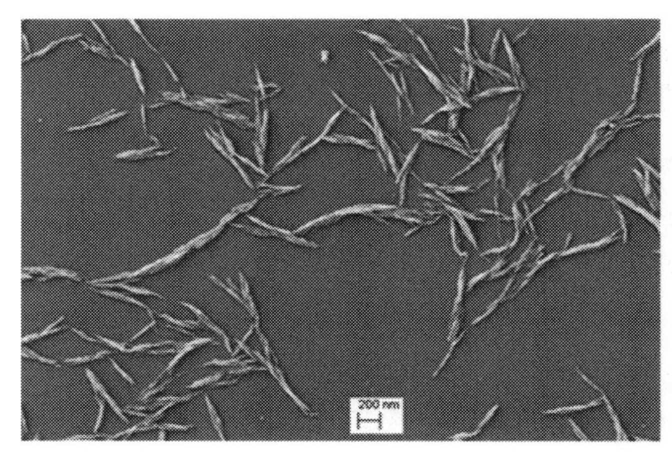

Figure 9.7 Aqueous suspension of nanocrystalline cellulose (Nanocrystalcell, 2021).

similar to Kevlar (Eichhorn, 2011; Menezes & Siqueira, 2009; Moon, Martini, Nairn, Simonsen, & Youngblood, 2011; Shankaran, 2018).

9.4 Degradability

The world of plastics started around 1950, with a global exponential growth since then, resulting in a worldwide production of 400 MT year^{-1} with an estimate of reaching 600 MT year^{-1} by the year 2030, with a lot of this ending up in oceans around the globe (Chamas et al., 2020). Therefore, a sustainable end-life management plan for plastic has been extensively investigated. One of the main reasons for the plastic's dramatic growth is related to its durability, but it directly affects its end-of-life (Narancic, Cerrone, Beagan, & O'Connor, 2020). This brings the demand for better knowledge for the mechanism of degradation of those polymers in such a way that a public policy can manage to reduce its impact over the environment. From the large and still growing list of polymers in use around the globe for endless applications, one of the most popular is the polyolefins (Stelmachowski, 2010). It is a very long-lasting polymer in the environment due to its high molecular weight, hydrophobicity, lack of functional groups that would allow to be attacked by enzymes, microorganism (decay fungi, molds, stain fungi, and marine borers), and environmental agents, such as water, ultraviolet (UV) radiations (Newborough, Highgate, & Vaughan, 2002; Sowmya, Ramalingappa, Krishnappa, & Thippeswamy, 2014). The problem is that the additives including antioxidants stabilizers, anti-UV, etc., used to extend the useful life of the plastics, result in more durability after its usage. Unfortunately, only 18% of the plastics postconsumers are recycled and 24% processed in waste to energy plants. The remaining 58% is inappropriately dumped into landfill sites (Chamas et al., 2020).

Polymer degradation can be a combination of physical and chemical agents, such as heat that occurs due to the macromolecular degradation and cross-linking. The reaction occurs in two types: chain-end degradation and random degradation. The chain ends are the starting point and monomers are lost, being called depolymerization. The other is a random breakdown that results in a sharp reduction in the molecular weight, although slow compared to the depolymerization (Singh, Ruj, Sadhukhan, & Gupta, 2019). Table 9.1 shows the main degrading agents and their types of degradation, data compilation made by literature review (Glaser, 2019; Gu, 2003; Holländer & Thome, 2004; Kiedrowski et al., 2020; Kost, Leong, & Langer, 1989; Middleton et al., 2013; Pathak & Navneet, 2017; Pospíšil, Nešpůrek, & Pilař, 2008; Weidner, 2012; Weikart & Yasuda, 2000; Yousif & Haddad, 2013; Zakrevskii, Sudar, Zaopo, & Dubitsky, 2003)

The degradation rates of the different plastics vary depending upon their composition, gramature, shape, size, additives, and exposition conditions. The degradation route has two pathways: abiotic and biotic. The mechanisms that degrade the plastics can be physical (microwaves, cracking, embrittlement, abrasion, and flaking) or chemical (at the molecular level, related to oxidation or hydrolysis). Sometimes the chemical degradation is related to radiant energy, that results in bond breaking in the material of the light-sensitive parts under light exposure. Several environmental factors affect the rate of degradation including the moisture, temperature, light, and even microorganisms (Chamas et al., 2020).

Table 9.1 Degrading agents and type of degradation.

Degrading agents	Types of degradation
Light (ultraviolet, visible)	Photochemical degradation
X-ray, gamma-ray, etc.	High energy radiation-induced degradation
Laser light	Ablative photo degradation
Electrical field	Electrical aging
Plasma	Corrosive degradation, etching
Microorganism	Biodegradation
Enzymes	Biocorrosion
Stress forces	Mechanical degradation
Ultrasound	Ultrasonic degradation
Chemicals (acid, base, etc.)	Chemical degradation
Heat	Thermal degradation and/or decomposition
Oxygen, ozone	Oxidative degradation (oxidation), ozonolysis
Heat and oxygen	Thermoxidative degradation and/or decomposition combustion
Light and oxygen	Photo-oxidation

9.4.1 Degradation

The degradation is divided in two main groups, where degradation consists in reducing the polymer into small pieces, visible or not, under specific conditions, and in absence of microorganisms. On the other hand, biodegradation requires the presence of microorganisms. Degradation occurs through losses in the polymer chain and is considered an irreversible phenomenon resulting in deterioration of the properties of the polymer (Kumar & Maiti, 2016). Several factors can act in synergism. Several processes are responsible for the degradation in the polymer (Bhuvaneswari, 2018; Kumar & Maiti, 2016), as listed below:

1. Thermal degradation is initiated via heat and temperature. Thermal oxidation occurs in the presence of oxygen and temperature. And thermal hydrolysis occurs in the presence of water and temperature.
2. Chemical degradation includes hydrolytic degradation caused in the presence of water, and etching takes place via acid, alkali, salt, or reactive gases.
3. Mechanical degradation includes physical wear, including cracking and erosion, abrasive forces, stress, and fatigue.
4. Photo degradation is caused by electromagnetic radiation;
5. High energy radiation includes X-ray, alpha, beta, and gamma rays; and photo-oxidation is caused by the presence of oxygen and UV and visible light.
6. Biological degradation is mediated by biological agents, including enzymatic degradation (lipase, proteinase) and microbial degradation caused by bacteria or fungi.
7. High-energy radiation degradation.
8. Ultrasonic wave degradation.

Polymer degradation is defined by changes in the properties of a polymer, including tensile strength, color, shape, and molecular weight, occurring by one or more environmental factors, such as heat, light, chemicals, or any other applied force (Sowmya, Ramalingappa, Krishnappa, & Thippeswamy, 2015). These changes in the chemical and/or physical structure of the polymer chain diminish the molecular weight of the polymer. These reactions are considered positive in the case of an intentional biodegradation or when one wants to reduce the molecular weight of a polymer. The degradation is influenced by the polymer structure. Aromatic functionality degrades easier by UV radiation, in contrast to hydrocarbon-based units that are more sensitive to thermal degradation. So, this hydrocarbon-based units do not recommend utilization at high temperature. (Lu, Solis-Ramos, Yi, & Kumosa, 2018; Yousif & Haddad, 2013).

The degradation of polymers to form smaller molecules may proceed by random scission, that is in the case of polyethylene (PE), or specific scission. For example, the PE at temperatures above 450°C, results in a mix of hydrocarbon derivatives. Nevertheless, sometimes the degradation is not considered a problem but understanding this process is important in its degradation, reuse, and recyclability (Trache et al., 2020; Weidner, 2012).

Aiming to evaluate the degradation rates and uniformize the literature data available are important to have a common denominator either for the different polymers, conditions, or pathways. A metric is discussed, the specific surface degradation rate (SSDR), which covers a very wide range, with some of the variability arising due to degradation studies conducted in different environments. It has been reported that SSDRs for high density polyethylene (HDPE) in the marine environment varies from almost 0 to 11 μm year^{-1}. For the same environment conditions, an extrapolation for half-lives goes from 58 years for bottles to 1200 years for pipes. Interestingly the SSDRs for HDPE and polylactic acid (PLA) are at the same range in the marine environment, considering that PLA degrades much faster in land (Chamas et al., 2020).

9.4.2 Biodegradation

Polymers have a long history since the prehistoric era. Data around 1000 BC were found about the use of a polymer by the Chinese. This polymer, a varnish extracted from a tree called *Rhus vernicflua*, was used in making vests and housing material. The aim of this varnish was to increase the furniture durability, since it was a waterproof coating; used later in 1950s. The man-made polymers had a strong development in the 20th century, with the development of polymeric products based on organic molecules which were nondegradable, opening the doors for the biodegradable polymers (Feldman, 2008).

The biodegradable polymers debuted in the 1980s. By definition, it is a process capable of decomposing polymeric materials into carbon dioxide, methane, water, inorganic compounds, or biomass with the help of some microorganisms. Biodegradation generally begins with fragmentation followed by mineralization by microorganisms and can be divided into aerobic and anaerobic degradations, which can be seen in Fig. 9.8, and are expressed by the equation:

$$\text{Polymer} + CO_2 \rightarrow CH_4 + H_2O + \text{biomass} + \text{residue(s)}$$

Biodegradability is affected by the origin of the polymer, chemical structure, and the exposition environment. Therefore the materials may be divided into three groups: the mixtures of synthetic polymers and substances which are easily digestible by microorganisms (interalia, modified natural polymers, natural polymers such as starch and cellulose); synthetic polymers with groups that are susceptible to hydrolytic microbials; and the biopolymers from bacterial sources, for example, polyhydroxybutyrate (PHB) (Leja & Lewandowicz, 2010; Siracusa, Rocculi, Romani, & Rosa, 2008). Table 9.2 shows the list of microorganisms that are responsible for most of the biodegradation processes for polymers.

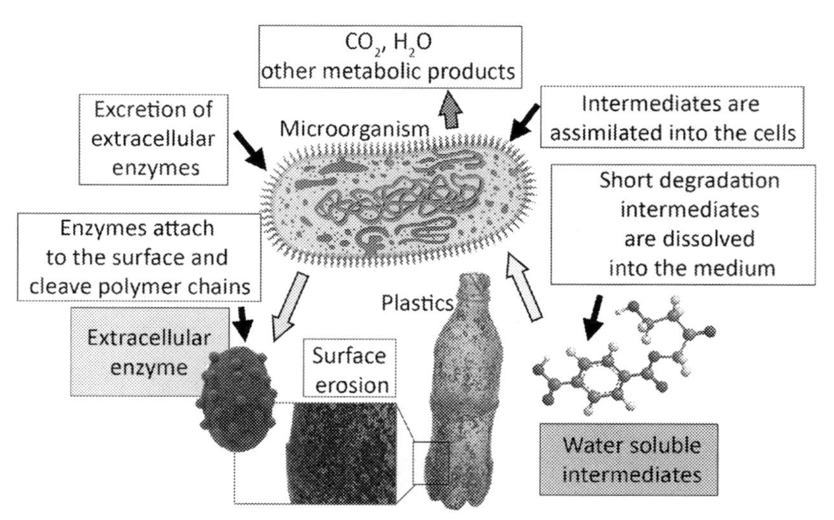

Figure 9.8 General mechanism of plastic degradation under aerobic degradation (Adapted from Müller, 2005).

Table 9.2 List of potential organisms to degrade different types of plastics.

Plastics	Microorganisms
PE	*Brevibacillus borstelensis*; *Rhodococcus ruber* *Penicillium simplicissimum*; YK *Comamonas acidovorans*; TB-35 *Curvularia senegalensis*
PU	*Aureobasidium pullulans*; *Cladosporium* sp.; *Fusarium solani*; *Pseudomonas chlororaphis*; *Pseudomonas putida* AJ
PVC	*Ochrobactrum* TD; *Pseudomonas fluorescens* B-22 *Aspergillus niger van Tieghem* F-1119
PlasticizedPVC	*Aureobasidium pullulans*
BTA Copolyester	*Thermomonospora fusca*

Notes: *PE*, Polyethylene; *PU*, polyurethane; *PVC*, polyvinyl chloride; *BTA Copolyester*, aliphatic-aromatic copolyester.
Source: Adapted from Bhuvaneswari, G.H. (2018). Degradability of polymers. Recycling of Polyurethane Foams (pp. 29–44). Elsevier.

9.5 Nonbiodegradable polymers

The most used plastics worldwide are not biodegradable and are the ones listed by numbers: PE, polypropylene (PP), polystyrene, polyvinyl chloride (PVC), and polyethylene terephthalate (PET) with an expressive growth rate of 9%, from 1950 to 2008 (Chanprateep, 2010). The use of microbial and enzymatic processes of biodegradation can be the way to reduce the impact of such plastics in the environment (Bhatia, 2013).

Table 9.3 Worldwide production of plastic resins in 2019 and its market share based on commercial value.

Resin	Recycling code	Market share (%)	Market share (US$ billion)
Polyethylene terephthalate[a]	1	8.99	23.9
High density polyethylene[a]	2	32.32	85.19
Polyvinyl chloride	3	4.33	11.5
Low density polyethylene[a]	4	12.79	34.0
Polypropylene[a]	5	15.20	40.41
Polystyrene	6	3.65	9.7
Polyethylene furanoate[a]	—	0.00	Incipient
Polytrimethylene terephthalate[a]	—	0.56	1.5
Poly(butylene-adipate-co-terephthalate)[a], **	—	0.21	0.55
Polybutylene succinate[a],[b]	—	0.04	0.11
Poliácido láctico[a],[b]	—	0.20	0.54
Polyamide[a]	—	10.77	28.63
Polihidroxialcanoatos[a],[b]	—	0.02	0.06
Polycaprolactone[b]	—	0.07	0.19
Acrylonitrile butadiene styrene	—	10.65	28.3
Starch Blends[a],[b]	—	0.19	0.5
Other Biodegradables[b]	—	—	Incipient

[a]Can be or are biologically derived.
[b]Biodegradable.

The biopolymers still represent a small amount of the global resin production in quantity, but in value represents more, since they are more expensive than its counterparts. The worldwide polymers production can be seen in Table 9.3.

The disposal of nondegradable plastics has triggered the development of biodegradable plastics, but still some properties are under investigation such as brittleness, poor insulation, and limitations as gas barriers. The use of nanoparticles or nanocomposites represent a new way toward biodegradation and sustainable applications (Kumar & Maiti, 2016).

The presence of micro- and nanoplastics is found in all biomass. Bratovcic (2017) carried out a study submitting plastics for three weeks of UV irradiation, generating particles at nanoscale. Successful degradation of low density polyethylene (LDPE) was reported with the addition of polypyrrole/TiO_2 (PPy/TiO_2) nanocomposite as photocatalyst.

9.6 Bioplastics

For almost any conventional plastic material or corresponding application, there is a bioplastic or biopolymer alternative. Bioplastics have the same properties as conventional plastics; however, they are bio-based, biodegradable, or both. Bioplastics can be classified into three different groups, see Fig. 9.9. Nonbiodegradable plastics can be from a renewable raw material source or partially from renewable sources, such as PE, PP, or PET (so-called drop-ins) and bio-based technical performance polymers, such as polytrimethylene terephthalate or thermoplastic polyester elastomers. It is worth mentioning that not all bioplastics are biodegradable, that is, it is not enough to leave it on the ground for it to decompose. The decomposition of some bioplastics lasts as long as that of plastics from petroleum, about 500 years. PLA and polyhydroxyalkanoates or polybutylene succinate, for example, are plastics from renewable and biodegradable raw material sources. There are also plastics that are sources of nonrenewable raw materials, derived from fossil resources, and that are biodegradable, such as poly butylene-adipate-co-terephthalate (European Bioplastics, 2021).

Currently, bioplastics represent 1% of the 368 million tons of plastic produced annually. However, that number is growing (Fig. 9.10), because these bioplastics, such as PE, PP, PET, or PVC with biological or partially biological basis, are technically

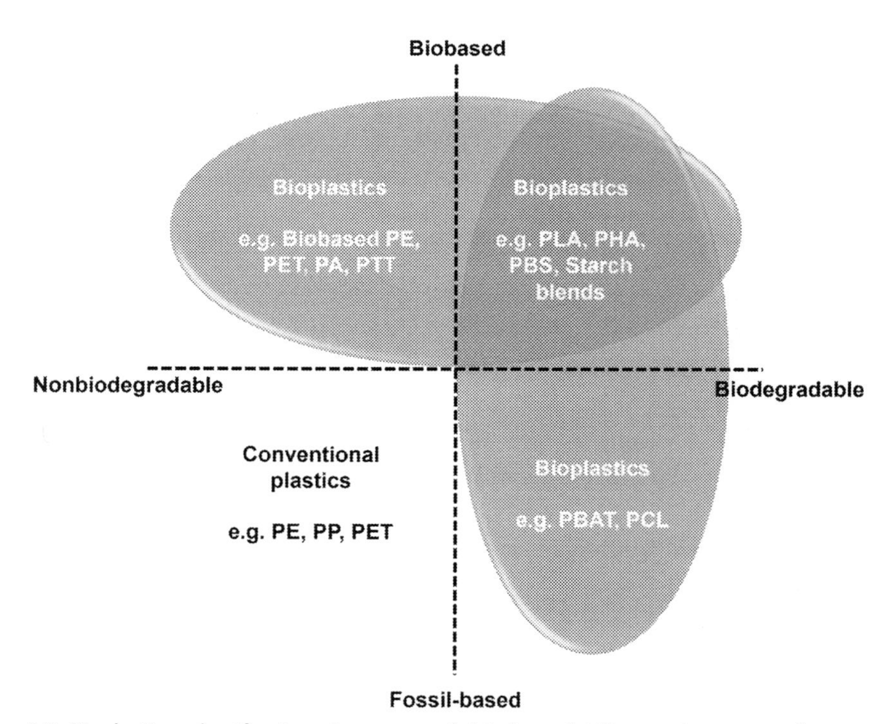

Figure 9.9 Bioplastics classification in terms of biodegradability and source of raw material (Adapted from European Bioplastics, 2021). *PA*, Polyamide; *PBAT*, poly butylene-adipate-co-terephthalate; *PBS*, polybutylene succinate; *PCL*, polycaprolactone; *PE*, polyethylene; *PEF*, polyethylene furanoate; *PET*, Polyethylene terephthalate; *PHA*, polyhydroxyalkanoates; *PLA*, polylactic acid; *PP*, polypropylene; *PTT*, polytrimethylene terephthalate.

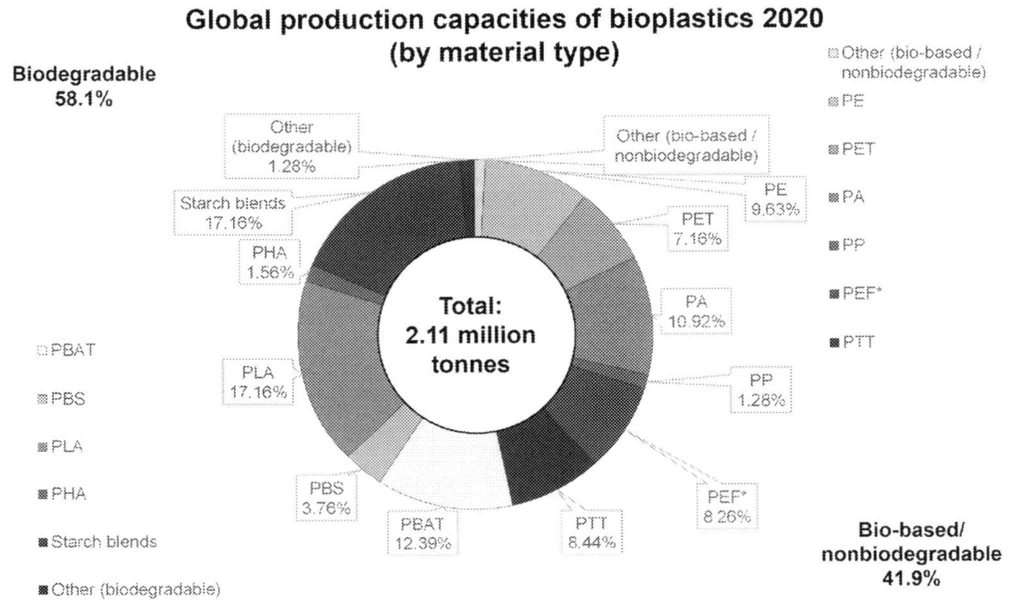

Figure 9.10 Global production of bioplastics 2020 (Adapted from European Bioplastics, 2021). *PA*, Polyamide; *PBAT*, poly butylene-adipate-co-terephthalate; *PBS*, polybutylene succinate; *PE*, polyethylene; *PEF*, polyethylene furanoate; *PET*, Polyethylene terephthalate; *PHA*, polyhydroxyalkanoates; *PLA*, polylactic acid; *PP*, polypropylene; *PTT*, polytrimethylene terephthalate.

equivalent to their fossil counterparts. In addition, bioplastics help reduce a product's carbon footprint. Additionally, along with the growing variety of bioplastic materials, properties such as flexibility, durability, printability, transparency, barrier, heat resistance, gloss, and more have been significantly improved (Gökçe, 2018).

Since the biopolymers can offer innovative properties and be sustainable for the bioeconomy, they are a promising alternative to replace the fossil-based plastics in the market (European Bioplastics, 2021). The investigation of nanocomposites and their applications are divided into rigid packaging materials and improvement of plastics films (Borgna, 2017).

9.7 Biodegradable polymers

Biodegradable polymers are a class of polymers that were developed and are under constant investigation as a replacement to man-made plastic, since they are sustainable and have lower carbon footprint. The biodegradable polymers represent a small amount, around 1%, of the world's total plastic production, but is growing by 20%–30% more than fossil-based polymers, still not able to meet the world's demand (Madhavan Nampoothiri, Nair, & John, 2010). In theory, all polymers are biodegradable because they degrade over time, which can take decades. Nonetheless, only that can quickly degrade in a few months to two years are considered biodegradable polymers (Sadi, 2010).

The blends of composites with different polymers, including natural polymers is a way to tune the biodegradability, of these materials or increasing or reducing its rate. Leão, Correa, and Souza (2008) used PHB matrix in blends with natural fibers aiming to improve its mechanical properties, cost reduction, and tune the biodegradation rate. In the case of other blends with PHB, cellulose, or wood more studies are needed mainly in the field of degradation tests (Letcher, 2020).

Balaguer et al. (2016) investigated the biodegradation and environmental impact of PLA composite film embedded with inorganic nanoscale particles such as montmorillonite, calcium carbonate, and silicon dioxide.

Another study carried out by Deetuam, Samthong, Choksriwichit, and Somwangthanaroj (2020) showed that PLA films completely disintegrated into visually indistinguishable residues after 6–7 weeks of incubation in a composting environment. It was also observed that nanomaterials did not significantly reduce the level of biodegradability of PLA or increase the ecotoxicity in plants.

Several natural fibers were used as reinforcement in a PLA matrix at different levels (10 and 20 wt.%) with an improvement in its degradability. This fact was explained by the cellulose hydroxyl groups that interact with the polyester groups, favoring its hydrolysis (Kumar & Maiti, 2016).

9.8 Effect of nanocellulose on biodegradability

Nanoparticles are listed for more than a 100 applications, as discussed before, but a more recent application is its ability to favor the plastics disintegration (Bhatia, Girdhar, Tiwari, & Nayarisseri, 2014; Pathak & Navneet, 2017).

Since the plastics are not reactive and, in most cases, have a noticeable resistance against microbial attack, they are long lasting material in the environment. LDPE, for example, can be found everywhere as a pollutant and contamination agent due its inappropriate disposal and nondegradation properties. Nanoparticles influence the enhancement of polymer degradation (Pandey et al., 2015).

Recently, a positive effect of the nanoscale reinforcements over the growth of LDPE degrading microorganisms was observed, concluding that the nanoparticles helped the biodegradation process (Bhatia et al., 2014; Mohanan, Montazer, Sharma, & Levin, 2020).

The potential interest in biodegradable plastics is derived from the ecofriendly elimination via microbial action which converts the plastics into smaller entities resulting in a more sustainable system. One of the great advantages of nanoparticles is the ability to adjust the rate of biodegradation based on the required specification.

Therefore chemical, physical, and biological properties of the biodegradable polymers can be modified and controlled for sustainable applications in medicine and other areas (Kumar & Maiti, 2016).

9.9 Conclusions

Considering the current state-of-the-art of nanoparticle-filled polymers, the addition of nanosize components in the polymer matrix makes possible to influence the degradation and tune biodegradation rate of the composite. The filler size and morphology is an important factor. However, the effect of different nanoparticles embedded within plastics in terms of biodegradability is not fully clear.

The development of nanocomposites is a way to overcome some of the disadvantages of the conventional biopolymers specially for packaging applications, including permeability to water and gases, improve the mechanical properties, brittleness, reduce the weight, etc.

The utilization of nanoscale reinforcements or fillers due to specific characteristics such as higher specific surface area results in better mechanical properties. This occurs because the nanoparticles act as nucleating agents that at the end can control the biodegradation rate of the composite. As an example, polylactide layered silicate nanocomposites had a better biodegradation when compared to a neat polymer matrix.

Even though recent breakthroughs have emerged in this field, more research in nanotechnology is needed.

References

Bai, R. G., Sabouni, R., & Husseini, G. (2018). Green nanotechnology—A road map to safer nanomaterials. *Applications of nanomaterials* (pp. 133–159). Elsevier. Available from http://doi.org/10.1016/B978-0-08-101971-9.00006-5.

Balaguer, M. P., Aliaga, C., Fito, C., & Hortal, M. (2016). Compostability assessment of nano-reinforced poly(lactic acid) films. *Waste Management (New York, N.Y.)*, *48*, 143–155. Available from https://doi.org/10.1016/j.wasman.2015.10.030.

Bharmoria, P., & Ventura, S. P. M. (2019). Optical applications of nanomaterials. In A. H. Bhat, I. Khan, M. Jawaid, F. O. Suliman, H. Al-Lawati, & S. M. Al-Kindy (Eds.), *Nanomaterials for healthcare, energy and environment. Advanced structured materials* (pp. 1–29). Singapore: Springer Singapore. Available from http://doi.org/10.1007/978-981-13-9833-9_1.

Bhatia, M. (2013). Implicating nanoparticles as potential biodegradation enhancers: A review. *Journal of Nanomedicine & Nanotechnology*, *04*(04). Available from https://doi.org/10.4172/2157-7439.1000175.

Bhatia, M., Girdhar, A., Tiwari, A., & Nayarisseri, A. (2014). Implications of a novel *Pseudomonas* species on low density polyethylene biodegradation: An in vitro to in silico approach. *SpringerPlus*, *3*, 497. Available from https://doi.org/10.1186/2193-1801-3-497.

Bhuvaneswari, G.,H. (2018). Degradability of polymers. *Recycling of polyurethane foams*, 29–44. Available from https://doi.org/10.1016/B978-0-323-51133-9.00003-6, Elsevier.

Boisseau, P., & Loubaton, B. (2011). Nanomedicine, nanotechnology in medicine. *Comptes Rendus Physique*, *12*(7), 620–636. Available from https://doi.org/10.1016/j.crhy.2011.06.001.

Borgna, Ilaria. (2017) Nanoparticles and bioplastics: a winning pair for packaging. <https://www.kosmeticaworld.com/2017/04/01/nanoparticles-bioplastics-winning-pair-packaging/> Accessed 18.01.21.

Bratovcic, A. (2017). Degradation of micro- and nano-plastics by photocatalytic methods. *Journal of Nanoscience and Nanotechnology Applications*, *3*(3), 1–9. Available from https://doi.org/10.18875/2577-7920.3.304.

Chamas, A., Moon, H., Zheng, J., Qiu, Y., Tabassum, T., Jang, J. H., ... Suh, S. (2020). Degradation rates of plastics in the environment. *ACS Sustainable Chemistry & Engineering*, *8*(9), 3494–3511. Available from https://doi.org/10.1021/acssuschemeng.9b06635.

Chanprateep, S. (2010). Current trends in biodegradable polyhydroxyalkanoates. *Journal of Bioscience and Bioengineering*, *110*(6), 621–632. Available from https://doi.org/10.1016/j.jbiosc.2010.07.014.

Contera, S., La Bernardino de Serna, J., & Tetley, T. D. (2020). Biotechnology, nanotechnology and medicine. *Emerging topics in life sciences [Online]*, *4*(6), 551–554.

Crawford, Mark. (2016) 10 Ways Nanotechnology Impacts Our Lives. <https://www.asme.org/topics-resources/content/10-ways-nanotechnology-impacts-lives> Accessed 9.02.21.

Deetuam, C., Samthong, C., Choksriwichit, S., & Somwangthanaroj, A. (2020). Isothermal cold crystallization kinetics and properties of thermoformed poly(lactic acid) composites: Effects of talc, calcium carbonate, cassava starch and silane coupling agents. *Iran Polymer Journal*, *29*(2), 103–116. Available from https://doi.org/10.1007/s13726-019-00778-4.

Dias, O. A. T., Konar, S., Leão, A. L., Yang, W., Tjong, J., & Sain, M. (2020). Current state of applications of nanocellulose in flexible energy and electronic devices. *Frontiers in Chemistry*, *8*, 420. Available from https://doi.org/10.3389/fchem.2020.00420.

Eichhorn, S. J. (2011). Cellulose nanowhiskers: Promising materials for advanced applications. *Soft Matter*, 7(2), 303–315. Available from https://doi.org/10.1039/C0SM00142B.

Ejeian, F., Etedali, P., Mansouri-Tehrani, H.-A., Soozanipour, A., Low, Z.-X., Asadnia, M., ... Razmjou, A. (2018). Biosensors for wastewater monitoring: A review. *Biosensors & Bioelectronics*, *118*, 66–79. Available from https://doi.org/10.1016/j.bios.2018.07.019.

European Bioplastics Bioplastic materials. 2021. <https://www.european-bioplastics.org/bioplastics/materials/> Accessed 10.02.21.

Feldman, D. (2008). Polymer history. *Designed Monomers and Polymers*, *11*(1), 1–15. Available from https://doi.org/10.1163/156855508X292383.

Gavrilenko, O., & Wang, X. (2021). Functionalizing nanofibrous materials for textile applications. *Electrospun polymers and composites* (pp. 471–512). Elsevier. Available from http://doi.org/10.1016/B978-0-12-819611-3.00016-9.

Glaser, J. A. (2019). Biological degradation of polymers in the environment. In A. Gomiero (Ed.), *Plastics in the environment*. IntechOpen. Available from http://doi.org/10.5772/intechopen.85124.

Gökçe, E. (2018). Rethinking sustainability: A research on starch based bioplastic. *Journal of Sustainable Construction Materials and Technologies*, *3*(3), 249–260. Available from https://doi.org/10.29187/jscmt.2018.28.

Gu, J.-D. (2003). Microbiological deterioration and degradation of synthetic polymeric materials: Recent research advances. *International Biodeterioration & Biodegradation*, *52*(2), 69–91.

Grand View Research. Nanocellulose market size, share & trends analysis report by products, by application, regional outlook, competitive strategies, and segment forecasts, 2019 To 2025. **2020**. <https://www.grandviewresearch.com/industry-analysis/nanocellulose-market> Accessed 25.01.21.

Holländer, A., & Thome, J. (2004). Degradation and stability of plama polymers. In H. Biederman (Ed.), *Plasma polymer films* (pp. 247–277). Published By Imperial College Press And Distributed By World Scientific Publishing Co. Available from http://doi.org/10.1142/9781860945380_0007.

Kafy, A., Kim, H. C., Zhai, L., Kim, J. W., Van Hai Kang, T. J., & Kim, J. (2017). Cellulose long fibers fabricated from cellulose nanofibers and its strong and tough characteristics. *Scientific Reports*, 7(1), 17683. Available from https://doi.org/10.1038/s41598-017-17713-3.

Kargarzadeh, H., Mariano, M., Huang, J., Lin, N., Ahmad, I., Dufresne, A., & Thomas, S. (2017). Recent developments on nanocellulose reinforced polymer nanocomposites: A review. *Polymer*, *132*(2016), 368–393. Available from https://doi.org/10.1016/j.polymer.2017.09.043.

Kiedrowski, K., Jakobs, F., Kielhorn, J., Johannes, H.-H., Kowalsky, W., Kracht, D., ... Ristau, D. (2020). Laser-induced degradation and damage morphology in polymer optical fibers. In C.-A.

Bunge, K. Kalli, & P. Peterka (Eds.), *Micro-structured and specialty optical fibres VI* (p. 3). SPIE. Available from http://doi.org/10.1117/12.2553999.

Kost, J., Leong, K., & Langer, R. (1989). Ultrasound-enhanced polymer degradation and release of incorporated substances. *Proceedings of the National Academy of Sciences of the United States of America, 86*(20), 7663–7666.

Kouzegaran, Vahid Javan. Cellulose Nanofibers. <https://nanografi.com/blog/cellulose-nanofibers/> Accessed 10.02.21.

Kumar, S., & Maiti, P. (2016). Controlled biodegradation of polymers using nanoparticles and its application. *RSC Advances, 6*(72), 67449–67480. Available from https://doi.org/10.1039/C6RA08641A.

Leão, A. L., Cherian, B. M., Souza, S. F., De Sain, M., Narine, S., Caldeira, M. S., & Toledo, M. A. S. (2012). Use of primary sludge from pulp and paper mills for nanocomposites. *Molecular Crystals and Liquid Crystals, 556*(1), 254–263. Available from https://doi.org/10.1080/15421406.2012.635974.

Leão, A.L., Correa, C.A., Souza, S.F. Extrusion of PET/LDPE, PHB and PP blended with natural fibers for profiles and injection molded applications. In: Proceedings of *10th international conference on progress in wood & biofibres plastic composites, 2008.*

Leja, K., & Lewandowicz, G. (2010). Polymers biodegradation and biodegradable polymers – A review. *Polish Journal of Environmental Studies*, 255–266. [Online]. <https://www.researchgate.net/publication/230793131_Polymer_Biodegradation_and_Biodegradable_Polymers_-_a_Review>.

Letcher, T. (2020). *Plastic waste and recycling: Environmental impact, societal issues*. Elsevier Academic Press.

Lu, T., Solis-Ramos, E., Yi, Y., & Kumosa, M. (2018). UV degradation model for polymers and polymer matrix composites. *Polymer Degradation and Stability, 154*(4), 203–210. Available from https://doi.org/10.1016/j.polymdegradstab.2018.06.004.

Madhavan Nampoothiri, K., Nair, N. R., & John, R. P. (2010). An overview of the recent developments in polylactide (PLA) research. *Bioresource Technology, 101*(22), 8493–8501. Available from https://doi.org/10.1016/j.biortech.2010.05.092.

Mangaraj, A., Das, A. K., & Malla, B. (2020). Nanoparticles used in construction and other industries: A review. *EasyChair Preprint*, 1–12. [Online]. <https://easychair.org/publications/preprint_download/nDHq>.

Markets and Markets. *Nanocellulose market by type (MFC & NFC, CNC/NCC, and others), application (pulp & paper, composites, biomedical & pharmaceutical, electronics & sensors, and others), region (Europe, North America, APAC, and rest of world) - Global forecast to 2025*. 2020. <https://www.marketsandmarkets.com/Market-Reports/nano-cellulose-market-56392090.html> (Accessed 25.01.21).

Menezes, A., J., Siqueira, G., Curvelo, A. A. S, & Dufresne, A. (2009). Extrusion and characterization of functionalized cellulose whiskers reinforced polyethylene nanocomposites. *Polymer, 50*(19), 4552–4563. Available from https://doi.org/10.1016/j.polymer.2009.07.038.

Middleton, J., Burks, B., Wells, T., Setters, A. M., Jasiuk, I., Predecki, P., … Kumosa, M. (2013). The effect of ozone on polymer degradation in polymer core composite conductors. *Polymer Degradation and Stability, 98*(1), 436–445. Available from https://doi.org/10.1016/j.polymdegradstab.2012.08.018.

Mohanan, N., Montazer, Z., Sharma, P. K., & Levin, D. B. (2020). Microbial and enzymatic degradation of synthetic plastics. *Frontiers in Microbiology, 11*, 580709. Available from https://doi.org/10.3389/fmicb.2020.580709.

Moon, R. J., Martini, A., Nairn, J., Simonsen, J., & Youngblood, J. (2011). Cellulose nanomaterials review: Structure, properties and nanocomposites. *Chemical Society Reviews, 40*(7), 3941–3994. Available from https://doi.org/10.1039/C0CS00108B.

Müller, R. -J. (2005). Biodegradability of polymers: Regulations and methods for testing. In A. Steinbüchel (Ed.), *In* Biopolymers online (p. 49). Wiley. Available from http://doi.org/10.1002/3527600035.BPOLA012.

Murty, B. S., Shankar, P., Raj, B., Rath, B. B., & Murday, J. (2013). Applications of nanomaterials. In B. S. Murty, P. Shankar, B. Raj, B. B. Rath, & J. Murday (Eds.), *In* Textbook of nanoscience and nanotechnology (pp. 107–148). Berlin, Heidelberg: Springer Berlin Heidelberg. Available from http://doi.org/10.1007/978-3-642-28030-6_4.

Nanocrystalcell. Nanocrystalline cellulose (NCC). 2021. <https://www.nanocrystacell.eu/> Accessed 10.02.21.

Narancic, T., Cerrone, F., Beagan, N., & O'Connor, K. E. (2020). Recent advances in bioplastics: Application and biodegradation. *Polymers, 12*(4). Available from https://doi.org/10.3390/polym12040920.

National Nanotechnology Initiative. Benefits and applications: Official website of the United States. <https://www.nano.gov/you/nanotechnology-benefits> Accessed 18.01.21.

Newborough, M., Highgate, D., & Vaughan, P. (2002). Thermal depolymerisation of scrap polymers. *Applied Thermal Engineering*, *22*(17), 1875–1883. Available from https://doi.org/10.1016/S1359-4311(02)00115-1.

Pandey, J. K., Ahn, S. H., Lee, C. S., Mohanty, A. K., & Misra, M. (2010). Recent advances in the application of natural fiber based composites. *Macromolecular Materials and Engineering*, *295*(11), 975–989. Available from https://doi.org/10.1002/mame.201000095.

Pandey, P., Swati., Harshita., Manimita., Shraddha., Yadav, M., & Tiwari, A. (2015). Nanoparticles accelerated in-vitro biodegradation of LDPE: A review. *Pelagia Research Library*, *6*, 17–22. [Online]. <https://www.imedpub.com/articles/nanoparticles-accelerated-invitro-biodegradation-of-ldpe-a-review.pdf>.

Pathak, V. M., & Navneet. (2017). Review on the current status of polymer degradation: A microbial approach. *Bioresources and Bioprocessing*, *4*(1), 451. Available from https://doi.org/10.1186/s40643-017-0145-9.

Pospíšil, J., Nešpůrek, S., & Pilař, J. (2008). Impact of photosensitized oxidation and singlet oxygen on degradation of stabilized polymers. *Polymer Degradation and Stability*, *93*(9), 1681–1688. Available from https://doi.org/10.1016/j.polymdegradstab.2008.06.004.

Rajak, A. (2018). Nanotechnology and its application. *Journal of Nanomedicine & Nanotechnology*, *09*(03). Available from https://doi.org/10.4172/2157-7439.1000502.

Saba, N., & Jawaid, M. (2017). Recent advances in nanocellulose-based polymer nanocomposites. *Cellulose-reinforced nanofibre composites* (pp. 89–112). Elsevier. Available from http://doi.org/10.1016/B978-0-08-100957-4.00004-8.

Sadi, R.K. Estudo da compatibilização e da degradação de blendas polipropileno/poli (3-hidroxibutirato) (PP/PHB), 2010.

Santana, J. S., do Rosário, J. M., Pola, C. C., Otoni, C. G., Fátima Ferreira Soares, N. de, Camilloto, G. P., & Cruz, R. S. (2017). Cassava starch-based nanocomposites reinforced with cellulose nanofibers extracted from sisal. *Journal of Applied Polymer Science*, *134*(12), 201. Available from https://doi.org/10.1002/app.44637.

Sawhney, A. P. S., Condon, B., Singh, K. V., Pang, S. S., Li, G., & Hui, D. (2008). Modern applications of nanotechnology in textiles. *Textile Research Journal*, *78*(8), 731–739. Available from https://doi.org/10.1177/0040517508091066.

Shalaby, M. N., & Saad, M. M. (2020). Advanced material engineering and nanotechnology for improving sports performance and equipment. *International Journal of Psychosocial Rehabilitation*, 2314–2322. <https://www.researchgate.net/publication/341670625_Advanced_Material_Engineering_and_Nanotechnology_for_Improving_Sports_Performance_and_Equipment> Accessed 9.02.21.

Shankaran, D. R. (2018). Cellulose nanocrystals for health care applications. *Applications of nanomaterials* (pp. 415–459). Elsevier. Available from http://doi.org/10.1016/B978-0-08-101971-9.00015-6.

Singh, R. K., Ruj, B., Sadhukhan, A. K., & Gupta, P. (2019). Thermal degradation of waste plastics under non-sweeping atmosphere: Part 1: Effect of temperature, product optimization, and degradation mechanism. *Journal of Environmental Management*, *239*, 395–406. Available from https://doi.org/10.1016/j.jenvman.2019.03.067.

Siracusa, V., Rocculi, P., Romani, S., & Rosa, M. D. (2008). Biodegradable polymers for food packaging: A review. *Trends in Food Science & Technology*, *19*(12), 634–643. Available from https://doi.org/10.1016/j.tifs.2008.07.003.

Sonawane, G. H., Patil, S. P., & Sonawane, S. H. (2018). Nanocomposites and its applications. *Applications of Nanomaterials* (pp. 1–22). Elsevier. Available from http://doi.org/10.1016/B978-0-08-101971-9.00001-6.

Sowmya, H. V., Ramalingappa., Krishnappa, M., & Thippeswamy, B. (2014). Degradation of polyethylene by *Trichoderma harzianum*—SEM, FTIR, and NMR analyses. *Environmental Monitoring and Assessment*, *186*(10), 6577–6586. Available from https://doi.org/10.1007/s10661-014-3875-6.

Sowmya, H. V., Ramalingappa., Krishnappa, M., & Thippeswamy, B. (2015). Degradation of polyethylene by *Penicillium simplicissimum* isolated from local dumpsite of Shivamogga district. *Environment, Development and Sustainability*, *17*(4), 731–745. Available from https://doi.org/10.1007/s10668-014-9571-4.

Stelmachowski, M. (2010). Thermal conversion of waste polyolefins to the mixture of hydrocarbons in the reactor with molten metal bed. *Energy Conversion and Management*, *51*(10), 2016–2024. Available from https://doi.org/10.1016/j.enconman.2010.02.035.

Thakur, M. K., Thakur, V. K., & Prasanth, R. (2014). Nanocellulose-based polymer nanocomposites: An introduction. In V. K. Thakur (Ed.), *Nanocellulose polymer nanocomposites* (pp. 1–15). Hoboken, NJ, USA: John Wiley & Sons, Inc. Available from http://doi.org/10.1002/9781118872246.ch1.

Thakur, V. K., Thakur, M. K., & Kessler, M. R. (Eds.), (2017). *Handbook of composites from renewable materials*. Hoboken, NJ, USA: John Wiley & Sons, Inc.

Trache, D., Tarchoun, A. F., Derradji, M., Hamidon, T. S., Masruchin, N., Brosse, N., & Hussin, M. H. (2020). Nanocellulose: From fundamentals to advanced applications. *Frontiers in Chemistry*, *8*, 392. Available from https://doi.org/10.3389/fchem.2020.00392.

Tsuzuki, T., Zhang, L., Rana, R., Liu, Q., Wang, X. Production of green nanomaterials. In: *Proceedings of 2010 international conference on nanoscience and nanotechnology; IEEE, 2010–2010* (pp 150–153). Availabel from https://doi.org/10.1109/ICONN.2010.6045222.

Weidner, S. M. (2012). Mass spectrometry. *Polymer science*. A comprehensive reference (pp. 93–109). Elsevier. Available from http://doi.org/10.1016/B978-0-444-53349-4.00023-6.

Weikart, C. M., & Yasuda, H. K. (2000). Modification, degradation, and stability of polymeric surfaces treated with reactive plasmas. *Journal of Polymer Science Part A Polymer Chemistry*, *38*(17), 3028–3042, 10.1002/1099-0518(20000901)38:17 < 3028:AID-POLA30 > 3.0.CO;2-B.

Xu, X., Liu, F., Jiang, L., Zhu, J. Y., Haagenson, D., & Wiesenborn, D. P. (2013). Cellulose nanocrystals vs. cellulose nanofibrils: A comparative study on their microstructures and effects as polymer reinforcing agents. *ACS Applied Materials & Interfaces*, *5*(8), 2999–3009. Available from https://doi.org/10.1021/am302624t.

Xu, X., Wang, H., Jiang, L., Wang, X., Payne, S. A., Zhu, J. Y., & Li, R. (2014). Comparison between cellulose nanocrystal and cellulose nanofibril reinforced poly(ethylene oxide) nanofibers and their novel Shish-Kebab-like crystalline structures. *Macromolecules*, *47*(10), 3409–3416. Available from https://doi.org/10.1021/ma402627j.

Yang, S., Xie, Q., Liu, X., Wu, M., Wang, S., & Song, X. (2018). Acetylation improves thermal stability and transmittance in FOLED substrates based on nanocellulose films. *RSC Advances*, *8*(7), 3619–3625. Available from https://doi.org/10.1039/C7RA11134G.

Yousif, E., & Haddad, R. (2013). Photodegradation and photostabilization of polymers, especially polystyrene: Review. *SpringerPlus*, *2*, 398. Available from https://doi.org/10.1186/2193-1801-2-398.

Zakrevskii, V. A., Sudar, N. T., Zaopo, A., & Dubitsky, Y. A. (2003). Mechanism of electrical degradation and breakdown of insulating polymers. *Journal of Applied Polymer Science*, *93*(4), 2135–2139. Available from https://doi.org/10.1063/1.1531820.

Zhang, K., Barhoum, A., Xiaoqing, C., Li, H., & Samyn, P. (2019). Cellulose nanofibers: Fabrication and surface functionalization techniques. In A. Barhoum, M. Bechelany, & A. S. H. Makhlouf (Eds.), *Handbook of Nanofibers* (pp. 409–449). Cham: Springer International Publishing. Available from http://doi.org/10.1007/978-3-319-53655-2_58.

Zhong, L., & Peng, X. Biorenewable (2017). Nanofiber and nanocrystal: Renewable nanomaterials for constructing novel nanocomposites. In V. K. Thakur, M. K. Thakur, & M. R. Kessler (Eds.), *In* Handbook of composites from renewable materials (pp. 155–226). Hoboken, NJ, USA: John Wiley & Sons, Inc. Available from http://doi.org/10.1002/9781119441632.ch130.

CHAPTER 10

Production of microfibrillated cellulose fibers and their application in polymeric composites

Ming Liu, Katrin Greta Hoffmann, Thomas Geiger and Gustav Nyström
Cellulose and Wood Materials Laboratory, Swiss Federal Laboratories for Materials Science and Technology (Empa), Dübendorf, Switzerland

10.1 Microfibrillated cellulose fiber production

10.1.1 Microstructure of microfibrillated cellulose

Cellulose is the most abundant, biodegradable, and renewable biopolymer on earth with low density (1.5 g cm^{-3}). Microfibrillated cellulose (MFC) fibers, as one of the most attractive cellulose fibers exhibiting extraordinarily high specific surface area (>150 $m^2\ g^{-1}$ BET), high aspect ratio (>1000), high strength (1−3 GPa), and high stiffness (100−130 GPa), are potential reinforcements to substitute conventional or synthetic fillers for polymer-reinforced (nano)composites (Zimmermann, Pöhler, & Geiger, 2004).

Compared to other forms of nanocellulose, MFC exhibits a unique interpenetrated web-like network structure in the micrometer size (up to 100 μm), and a diameter of an individual fiber in nanometer size (<100 nm) (Fig. 10.1). It is usually obtained by defibrillating cellulose through mechanical disintegration under high mechanical shear forces from wood pulp (Chen, Zhu, & Tong, 2016), plant fibers, and agricultural residues (e.g., sugar beetroots, sweet potato residue, and potato pulp) (Chinga-Carrasco, 2011; Goussé, Chanzy, Cerrada, & Fleury, 2004; Nguyen et al., 2013; Osong, Norgren, & Engstrand, 2016; Singh, Kaushik, & Ahuja, 2016; Tonoli et al., 2012; Zuluaga, Putaux, Restrepo, Mondragon, & Gañán, 2007).

10.1.2 Chemical composition of microfibrillated cellulose

MFC, mainly produced from wood pulp via mechanical grinding or disintegration (Khalil et al., 2014), has the same chemical composition as the starting pulp materials. The wood pulp is primarily composed of 60%−85% cellulose, 10%−30% hemicellulose, and 0%−10% lignin depending on the pulping process and whether the pulp is bleached or not (Table 10.1). In general, unbleached pulp has a high lignin content up to 30% of its weight, while bleached pulp is found to have a lignin content of <2% (Table 10.1).

Nanotechnology in Paper and Wood Engineering
DOI: https://doi.org/10.1016/B978-0-323-85835-9.00003-9

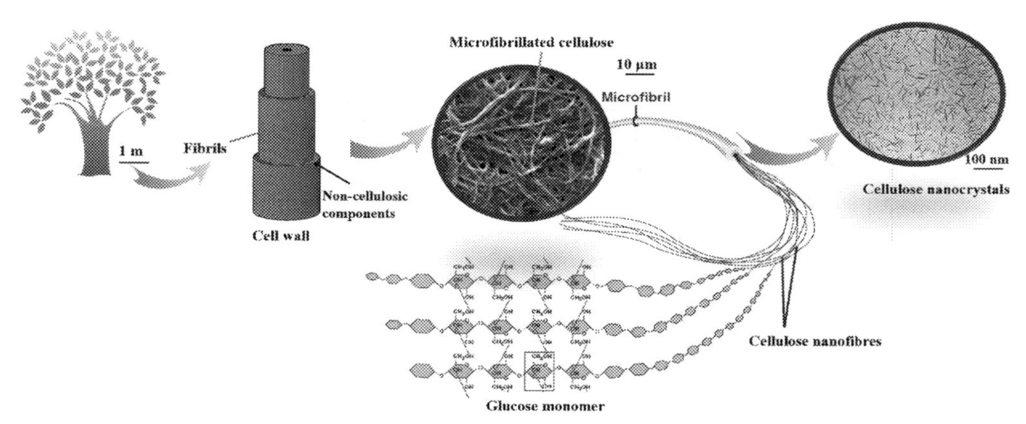

Figure 10.1 Schematic of the macro and microstructure of microfibrillated cellulose.

Other nonwood plant fibers and agro-residues are also potential resources for MFC production as they generally have low lignin content and high pectin content (Table 10.2), which makes the extraction and defibrillation process easier and more efficient. Those nonwood plant fibers include hemp, flax, and kenaf fibers, etc., and the agro-residues involve all types of food residues such as sugar beet, carrot, potato, soybean residues, orange, banana peels, wheat straw, and coffee pulp (Table 10.2). This demonstrates an interesting new route for the use of agro byproducts or food residues in high-performance and sustainable materials and biocomposites.

In comparison with acid-digested cellulose fibers, namely cellulose nanocrystals, MFC typically exhibits a few percent, up to 30%, hemicellulosic polysaccharides in its composition (Tables 10.1 and 10.2). The crosslinked network of hemicellulosic matrices interacting with cellulose nanofibers is usually considered as the main integrity of the wood and plant cell walls providing cell wall strength and stiffness (El Awad Azrak, Costakis, Moon, Schueneman, & Youngblood, 2020; Goussé, Chanzy, Cerrada, & Fleury, 2004). Therefore, the presence of hemicellulosic polymers in the MFC will ultimately offer unique properties for the final fiber-filled composites.

10.1.3 Techniques for microfibrillated cellulose fiber production

MFC derived from wood pulp is currently commercially available from different suppliers such as Exilva of Borregaard AS (Norway), Valida of Sappi Ltd. (Sappi Europe, Belgium), Celova of Weidmann AG (Switzerland), and Celish of Daicel Miraizu Ltd. MFC derived from agro-residues is also commercially available such as Curran, produced by Cellucomp Ltd., from waste streams of root vegetables. MFC has demonstrated the high potential of offering exceptional mechanical and rheological properties for various applications, such as painting, coating, personal care, home care, paper, food, concrete, and composite materials.

Table 10.1 The chemical composition of wood pulps.

Pulp type	Weight percentage (%)				References
	Cellulose	Hemicellulose	Lignin	Extractives	
Bleached softwood	79.2 (0.2)	20.0 (0.1)	0.8 (0.1)	0 (0)	Spence et al. (2010)
Bleached hardwood	78.0 (0.2)	20.3 (0.1)	1.3 (0.1)	0.5 (0.1)	Spence et al. (2010)
Unbleached hardwood	78.0 (0.5)	19.3 (0.1)	2.4 (0.4)	0.3 (0.2)	Spence et al. (2010)
Unbleached softwood	69.0 (2.5)	22.0 (0.7)	8.8 (1.8)	0.2 (0.1)	Spence et al. (2010)
Unbleached softwood high lignin	65.2 (0.8)	20.1 (0.1)	13.8 (0.7)	0.8 (0.6)	Spence et al. (2010)
Thermomechanical pulp	37.7 (0.6)	29.2 (0.1)	31.2 (0.5)	1.9 (0.0)	Spence et al. (2010)
Unbleached birch kraft pulp	74.2	21.8	5.6	0.6	Saastamoinen et al. (2012)
Bleached birch kraft pulp	74.7	26.6	ND	ND	Saastamoinen et al. (2012)
Unbleached kraft pulp	84.1 (0.4)	10.29	12.9 (0.1)	1.1 (0.3)	da Silva, Mocchiutti, Zanuttini, and Ramos (2006)
Unbleached pulp	57.1	11.2	27.4	NA	Jackson and Line (1997)
Unbleached kraft softwood[a]	65.0 (1.9)	10.3 (0.5)	13.5 (0.7)	NA	Shawky Sol, Sakran She, Ahmad, and Abdel-Atty (2017)
Unbleached kraft hardwood[a]	67.7 (2.8)	12.4 (0.6)	17.6 (1.2)	NA	Shawky Sol et al. (2017)
Unbleached kraft hardwood[a]	71.2 (2.6)	13.1 (0.5)	16.3 (0.6)	NA	Shawky Sol et al. (2017)
Unbleached kraft hardwood[a]	70.4 (2.2)	11.0 (0.4)	12.2 (0.6)	NA	Shawky Sol et al. (2017)

Notes: Values shown in parentheses present standard deviations.
NA, Not available; *ND*, not detected.
[a]Unbleached kraft pulp was obtained from different tree species.

Table 10.2 The chemical composition of microfibrillated cellulose fibers extracted from plant pulp and food residues.

Resources	Weight percentage (%)					References
	Cellulose	Pectin	Lignin	Hemicellulose	Protein	
Orange peel[a]	52.3 (1.7)	24.0 (0.4)	1.5 (1.7)	NA	5.5 (0.4)	de Melo E. M. (2018)
Orange peel[a]	57.0 (1.7)	9.8 (0.4)	3.7 (1.7)	NA	5.3 (0.4)	de Melo E. M. (2018)
Orange peel[a]	61.5 (1.7)	6.6 (0.4)	3.7 (1.7)	NA	4.5 (0.4)	de Melo E. M. (2018)
Orange peel[a]	68.9 (1.7)	2.9 (0.4)	6.9 (1.7)	NA	2.6 (0.4)	de Melo E. M. (2018)
Orange peel[a]	68.7 (1.7)	1.5 (0.4)	9.4 (1.7)	NA	1.9 (0.4)	de Melo E. M. (2018)
Orange peel[a]	68.1 (1.7)	1.4 (0.4)	11.4 (1.7)	NA	1.7 (0.4)	de Melo E. M. (2018)
Potato	76.0	1.8	/	NA	NA	Dufresne, Dupeyre, and Vignon (2000)
Sugar beet pulp	14.4	74.2	3.0	NA	NA	Hietala, Sain, and Oksman (2017)
Soybean	64.3	NA	1.1	NA	16.4	Li, Wang, Hou, and Li (2018)
Carrot	70		4.2	23	NA	Berglund, Noël, Aitomäki, Öman, and Oksman (2016)
Coffee pulp[a]	19–26	NA	18–30	24–45	NA	Bekalo and Reinhardt (2010)
Coffee pulp[a]	40–49	NA	33–35	25–32	NA	Bekalo and Reinhardt (2010)
Wheat straw	97.6	NA	1.7	5.5	NA	Liu et al. (2017)
Kenaf[a]	82.6 (0.9)	NA	1.8 (0.4)	11.8 (0.4)	NA	Jonoobi, Harun, Shakeri, Misra, and Oksmand (2009)
Kenaf[a]	92.8 (0.5)	NA	0.5 (0.3)	4.7 (0.7)	NA	Jonoobi et al. (2009)
Hemp	85.6 (0.6)	1.5 (0.1)	NA	4.1 (0.3)	NA	Liu et al. (2016); Rana and Gupta (2020)

Notes: *NA*, Not available.
[a]The samples were derived from different pretreatment processes.

Cellulose fibers primarily present in the cell walls of wood, plant, or food residues, are interlocked by the heterogeneous polysaccharides of hemicellulose (e.g., xylan, xyloglucan, arabinoxylan) providing the strength to the cell walls (Liu et al., 2016; Liu et al., 2017). The interlocked network of microfibrils and hemicellulose is further embedded in a matrix of pectic and aromatic substances, and other components (e.g., proteins) (Li, Du, Wang, & Yu, 2014; Liu et al., 2017; 2017; Neagu, 2005; Roach et al., 2011; Salmén, 2004). It is thus necessary to remove the noncellulosic components (e.g., hemicellulose and lignin) that bind the cellulose fibers together to produce more separated fibers. Many techniques including chemical and enzymatic treatments in combination with or without other mechanical/physical treatments (e.g., refining, steam explosion, blending, or milling) are reported to be effective to extract cellulose fibers from wood, plant, and agro-residues (Fig. 10.2 and Table 10.3) (Liu et al., 2016; Schultz-Jensen et al., 2013).

Wood pulp, a lignocellulosic fibrous material, is the main resource for the production of MFC. It is usually prepared by separating cellulose fibers from wood chips chemically with different chemicals (acid, base, and bleaching agents) accompanied by mechanical treatments. Several different pulping processes available can be used to extract the wood fibers, which may be classified into two general groups, mechanical and chemical pulping processes.

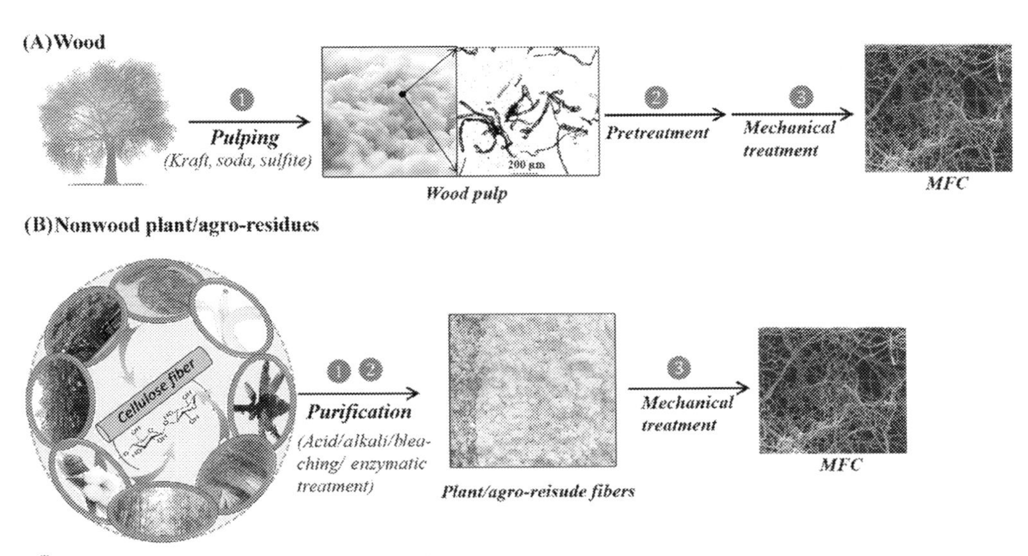

Figure 10.2 Schematic of microfibrillated cellulose (MFC) production from (A) wood and (B) nonwood plant and agro-residues.

Table 10.3 Different processing methods for the production of microfibrillated cellulose fibers.

Raw materials	Pretreatment/purification	Mechanical defibrillation	Posttreatment	References
Bleached kraft pulp	No	Refining (disk refiner)	No	Tonoli et al. (2012)
Bleached kraft pulp	No	Knife milling	Sonication	Tonoli et al. (2012)
Carrot residues	Bleaching	Ultrafine grinding	No	Berglund et al. (2016)
Soybean residues	Petroleum ether extraction/acid hydrolysis	Homogenization (HP-4 L high-pressure homogenizer)	No	Li et al. (2018)
Potato pulp	Alkali/bleaching	Blending/homogenization (Manton-Gaulin laboratory homogenizer)	No	Dufresne et al. (2000)
Kraft pulp	Refining (Valley beater)	Homogenization (15 MR two-stage Manton-Gaulin homogenizer)	No	Spence et al. (2010)
Wheat straw	Steam explosion/alkali/bleaching/acid + ultrasonication	Homogenization (Fluko FA25 high shear homogenizer)	No	Kaushik and Singh (2011)
Dissolved pulp	No	Grinding (MKCA6-3, Masuko grinder)	No	Iwamoto, Nakagaito, and Yano (2007)
Argan shells	Hammer milling/ethanol extraction/alkali/bleaching	Homogenization (NanoVater, NV200)	No	Bouhoute et al. (2020)
Hemp fiber	Alkali/acid/alkali	Grinding	No	Rana and Gupta (2020)
Wood powder	Ethanol-benzene extraction/bleaching/alkali treatment	Grinding (MKCA6-3 Masuko grinder)	No	Iwamoto, Abe, and Yano (2008)
Bamboo fiber	Steam explosion/alkali treatment	Grinding (MKCA6-3, Masuko grinder)	No	Vu, Sinh, Choi, and Pham (2017)
Softwood sulfite pulp	Mechanical beating/enzymatic treatment	Homogenization (microfluidizer M-110EH)	No	Mikkonen et al. (2012); Saini, Belgacem, Missoum, and Bras (2015)
Milled beech wood	Delignification/autoclaving in ethanol-water	Grinding (MKCA6−2, Masuko grinder) + homogenization (APV 1000)	No	Herzele, Veigel, Liebner, Zimmermann, and Gindl-altmutter (2016)

Bleached softwood kraft pulp	Refining/enzymatic treatment	Beating and refining	No	Tian et al. (2017)
Bleached spruce sulfite pulp	No	Homogenization (Gaulin M12)	No	Jarnthong, Wang, Wang, Wang, & Li, 2015; Stenstad, Andresen, Tanem, and Stenius (2008)
Date palm fruit stalk	Alkali/bleaching/enzymatic treatment	High-shear mixing/grinding (MKCA6-2, Masuko grinder)	No	Jonoobi, Harun, Shakeri, Misra, & Oksman, 2009
Bleached kraft birch pulp	No	Grinding (Masuko grinder)	No	Karlsson, Larsson, Pettersson, & Wågberg, 2020
Sugar beet pulp	Alkali/bleaching or alkali treatment alone	Blending (Ultra Turrax IKA)/ homogenization (APV1000)	No	Pinkl, Veigel, Colson, and Gindl-altmutter (2017)
Kraft pulp	Refining (Claflin conical refiner)	Homogenization (10.51 homogenizer)	No	Khalighi, Berger, & Ersoy, 2020
Kraft hardwood pulp	Beating or none	Homogenization	No	Spence, Venditti, Rojas, Habibi, and Pawlak (2011)
Kraft hardwood pulp	Beating	Microfluidizer	No	Spence et al. (2011)
Soybean straw	Chemical (alkali/bleaching) + enzymatic treatment	Turrax blending and sonication	No	Martelli-Tosi, Torricillas, Martins, de Assis, and Tapia-Blácido (2016)
Kraft pulp	Alkali/bleaching/refining/enzymatic pretreatments (endo-1,4-xylanase)	Mechanical refining and homogenizing	No	Tozluoglu et al. (2018)

(*Continued*)

Table 10.3 (Continued)

Raw materials	Pretreatment/purification	Mechanical defibrillation	Posttreatment	References
Bleached kraft pulp	Enzymatic treatment (endoglucanase, lytic polysaccharide monooxygenase, endoxylanase)	Sonication	No	Hu, Tian, Renneckar, and Saddler (2018)
Kraft pulp	Two sequential enzymatic treatments (xylanase, mannanase, or combined with cellulase) or further combined with mechanical pretreatments, e.g., shredding, beating, and steam explosion	Refining, defibrillation, beating, friction, grinding, high shear defibrillation, homogenization (e.g., microfluidizer), or other known mechanical fiber treatments	Friction	Heiskanen et al. (2011)
Kraft pulp	Ball milling followed by enzymatic treatment (endoglucanase, beta-1,4-glucosidase, cellobiohydrolases)	Sonication	No	Squinca et al. (2020)

The mechanical pulping process refers to the process to produce wood pulp by a mechanical grinding process. The chemical pulping process is the process to produce wood pulp by the use of different chemicals. It can be further classed as sulfite, sulfate (Kraft), and soda pulping processes according to chemicals applied in the pulping process.

Analogous to the pulping process applied for wood fiber separation, similar processes were developed to extract cellulose-rich fibers from nonwood resources by using chemical, mechanical, and enzymatic processes (Fig. 10.2 and Table 10.3). For the chemical treatments, acids (e.g., sulfuric acid and hydrochloric acid) and alkali (e.g., sodium hydroxide and potassium hydroxide) are usually used at the initial stage to remove hemicellulose from the raw materials, and different bleaching agents such as sodium chlorite ($NaClO_2$) and peroxides (H_2O_2) are further used to remove lignin. In comparison to the chemical treatment, enzymatic treatment provides a much milder alternative to remove the noncellulosic polymers more efficiently and specifically. Monocomponent enzymes with monoactivity of β-1,4-endoglucanase, α-1,4-endo-polygalacturonase, β-1,4-endo-xylanase, and β-1,4-endo-mannanase, or enzyme cocktails with multiple activities thereof are commonly used to remove pectin and hemicellulose polysaccharides. Oxidase enzymes such as laccases and peroxidases are applied for lignin removal to produce cellulose-rich fine fibers.

The cellulose-rich fibers extracted from wood or agro-residues need to be further exposed to one principal mechanical treatment (e.g., grinding and homogenization) or a serial of different mechanical treatments (blending or milling combined with grinding, homogenization, and extrusion) to defibrillate the fibers to produce the MFC fibers (Table 10.3). It normally starts with cellulose-rich fibers suspended in an aqueous medium (dry matter content ≤ 2 wt.%) under high shearing forces so that microfibrils are separated and formed. Upon the formation of microfibrils, the water molecules fill the gaps among the formed fibrils to stabilize the whole network through interacting with the fibrils by hydrogen bonding, forming a paste-like viscous suspension with solid content usually <2 wt.%. However, this mechanical process is very energy consuming. Therefore pretreatment is necessary to reduce the energy consumption in the subsequent mechanical processing to make the production of the MFC fibers more sustainable and energy efficient (Heiskanen, Backfolk, Vehviläinen, Kamppuri, & Nousiainen, 2011; Kaushik & Singh, 2011; Liu et al., 2017; Squinca, Bilatto, Badino, & Farinas, 2020; Tozluoglu, Poyraz, & Candan, 2018).

Various pretreatment techniques were applied including steam explosion, beating, grinding, and dispersing, all aiming to open up the network structure of the cell walls. This provides improved accessibility for chemicals or enzymes in the subsequent treatment toward the noncellulosic components that bind the fibrils together at their interfaces. The strategy of applying pretreatment preceding the mechanical treatment is thereby especially important not only to improve the energy efficiency of the MFC production but also to avoid fibril breakage and damage during the mechanical defibrillation process.

Furthermore, the pH of the aqueous solution during the mechanical treatment was found to affect the efficiency of the defibrillation process. It is preferably above 10, as it was shown to increase the efficiency of the mechanical treatment of cellulose fibers, and thus to reduce the energy consumption for MFC production (Heiskanen et al., 2011). This could probably be attributed to the increased swelling capacity of the cellulose fibers and charge repulsion at the pH of 10 (Chang, He, Zhou, & Zhang, 2011; Karlsson, Larsson, Pettersson, & Wågberg, 2020).

10.2 Microfibrillated cellulose application in polymeric composites

Polymeric nanocomposites based on MFC have attracted a lot of interest in various fields, for example, packaging, personal protective and sports equipment, automotive, elastomeric compounds, construction insulation, and flexible electronics, as there is an increasing demand for innovative materials from renewable resources for high-performance applications. MFC fibers with a high surface area and aspect ratio (>1000), a low density ($1.5\ g\ cm^{-3}$), high strength ($1-3$ GPa), and stiffness ($100-130$ GPa) are potential reinforcements to substitute synthetic fibers, such as glass fibers, for load-bearing structural composite materials (Lee et al., 2014).

In principle, any type of polymer can be used to fabricate cellulose-filled polymeric biocomposites. Based on the resource, these polymers can be categorized into natural polymers (e.g., xyloglucan, xylan, galactomannan, starch, chitosan, etc.) and synthetic polymers. In this chapter, the natural polymers, or biopolymers, refer to polymers extracted from plants such as polysaccharides, starch, and proteins. For simplicity, natural rubber is not included in this category. The synthetic polymers are generally categorized into thermosets (e.g., epoxy resins, unsaturated polyester resins, polyurethanes, etc.), thermoplastics (e.g., polyethylene (PE), polypropylene (PP), poly(lactic acid) (PLA), polyglycolide, polyesters, etc.), and elastomers (e.g., natural rubber, styrene-butadiene rubber, polybutadiene rubber, ethylene propylene diene monomer rubber, etc.) (Fig. 10.3). The process to incorporate MFC in the above-mentioned polymers is primarily dependent on the type of polymer selected and the form of MFC used (e.g., MFC suspension or dried MFC).

Most of natural polymers are hydrophilic with good compatibility with MFC. Therefore, the MFC/natural polymers biocomposites can be readily fabricated using compression molding, casting methods, or drying in an oven or under air after homogenizing MFC and the polymers with a mixer, kneader, or extruder (Hu, Tian, Renneckar, & Saddler, 2018; Lee, Aitomäki, Berglund, Oksman, & Bismarck, 2014; Lems, Winklehner, Hansmann, Gindl-Altmutter, & Veigel, 2019; Lendvai, Karger-Kocsis, Kmetty, & Drakopoulos, 2016; Lengowski et al., 2020). Usually, there are no covalent bonds present in the resulting MFC/

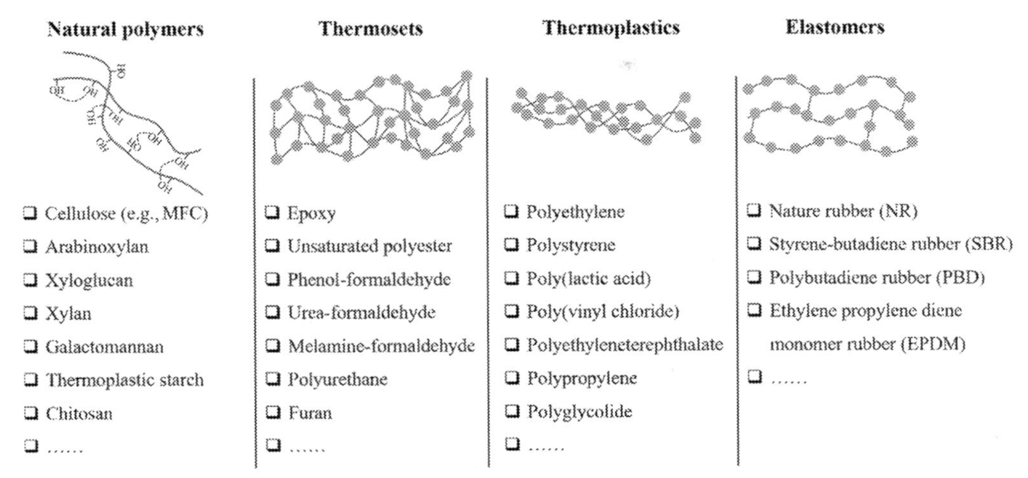

Figure 10.3 The commonly used polymers for biocomposite fabrication in combination with microfibrillated cellulose (MFC).

biopolymer composites. Instead, their integrity is provided through hydrogen bonds among MFC and between reinforcement and matrix (Fig. 10.3).

Thermosets are usually present in liquid form. The fabrication of MFC/thermoset composites can be done through a set of sequential steps including solvent exchange, mixing with resins, solvent evaporation, and curing (Vu et al., 2017). The solvent exchange step is tedious and time consuming, while it is in particular necessary for several resins, for example, epoxies, which can potentially react with the water from the MFC aqueous suspension. For resins not sensitive to water, the solvent exchange can principally be skipped. However, before curing, the water needs to be removed completely in any case, since it can interfere with the curing reactions between the resin and hardener.

Thermoplastics are one of the most widely used polymer classes in our daily life. The MFC/thermoplastic composites are thereby intensively studied topics. The incorporation of MFC fibers is not only aimed at reinforcing the thermoplastics for desired mechanical properties but also at reducing the consumption of petroleum-based polymers as a direct consequence. The manufacturing of MFC/thermoplastic composites is usually done through extrusion, internal mixing, compression molding, or injection molding, with dried and hydrophobized MFC fibers. How well the MFC fillers are dispersed in the composite is primarily dependent on the surface properties of polymers, fillers, and the mixing process. Consequently, all those factors in turn affect the mechanical properties of the final composites.

MFC is also a promising candidate to replace conventional reinforcements (e.g., carbon black and silica) in elastomeric compounds to obtain a final product with high sustainability and light weight. Depending on the type of elastomer, three different processing techniques are adapted to introduce MFC-based fillers into the final elastomeric compound. When latex emulsions (e.g., natural rubber latex) were used as the matrix, the MFC-reinforced latex composites were often fabricated by premixing MFC aqueous suspension with the latex aqueous suspension together with crosslinking compounds (e.g., sulfur and accelerators). This was followed by prevulcanization in an aqueous solution under heating and stirring. The final rubber compounds can be obtained by casting the mixture on a glass plate followed by drying it in an oven or at ambient temperature (Ghanadpour, Carosio, Larsson, & Wågberg, 2015; M. Liu et al., 2017; Q. Liu et al., 2017; Lu, Askeland, & Drzal, 2008; Lucenius, Valle-Delgado, Parikka, & Österberg, 2019).

However, the most current commercial elastomer compounds (e.g., tires and rubber rings) are manufactured through a melting-based process where a good dispersion of fillers and chemicals in the matrix is mainly achieved through mechanical shearing and mixing (Sae-Oui et al., 2017; Sattayanurak et al., 2019). Thereafter, the green compounds are cured under high temperature and high pressure in corresponding molds. To adapt to this conventional rubber compound manufacturing line, MFC in dried powder form or MFC-containing master batches should be first developed. In Section 10.9, recent efforts to obtain hydrophobic MFC powder or MFC-containing master batches will be discussed. Moreover, it is interesting to explore the potential of those anisotropic fibers in tailoring the trade-off relationship among rolling resistance, wet grip, and abrasion to develop more energy-saving and more sustainable elastomer compounds.

10.2.1 Microfibrillated cellulose in natural polymers

When water is removed from aqueous MFC suspensions, the cellulose microfibrils tend to form strong, irreversible aggregates via hydrogen bonding (Silva et al., 2021). This process is also known as hornification from wood pulps in papermaking (Chen et al., 2018; Fernandes Diniz, Gil, & Castro, 2004; Salmén & Stevanic, 2018; Weise, 1998). Here, pore closures and aggregation of cellulose microfibrils in the cell wall of the wood fibers lead to a reduced surface area and a decrease in water retention value (Chen et al., 2018; Salmén & Stevanic, 2018; Weise, 1998). With increased defibrillation of pulp to MFC, the smaller fibril diameter and larger surface area enable more contact points between the individual cellulose fibrils and therefore stronger interactions will form among fibrils. The fine fibrils can fill voids and a strong and stable network is created (Lee, Aitomäki, Berglund, Oksman, & Bismarck, 2014). The MFC's tendency to aggregate can be used to form films, nanopapers, and all-cellulose composite boards, where MFC acts simultaneously as binder and reinforcement. Moreover, it can be applied as an environmentally friendly binder in paper laminates and particleboards. Strong interactions can be also formed with other biopolymers, such as hemicellulose polysaccharides, starch, and lignin.

10.2.1.1 Pure MFC films/nanopapers

The terms MFC "film" and MFC "nanopaper" are used synonymously in the literature (Lavoine, Desloges, Dufresne, & Bras, 2012), and describe dense films composed of randomly orientated cellulose microfibrils with a thickness ranging from a few microns to more than 100 μm (Aulin, Gällstedt, & Lindström, 2010; Minelli et al., 2010; Plackett et al., 2010). These MFC films can be formed from highly diluted MFC suspensions (solid content smaller than 1 wt.%) via different processing routes such as casting evaporation (Chen, Zhu, & Tong, 2016; Jackson & Line, 1997; Missio et al., 2018; Mnich et al., 2020; Schultz-Jensen et al., 2013), vacuum filtration (Neagu, Gamstedt, & Berthold, 2006; Nguyen, 2013), and pressurized filtration (Österberg et al., 2013). Further drying and densification can be achieved through oven drying (Neagu, Gamstedt, & Berthold, 2006; Nguyen, 2013; Schultz-Jensen et al., 2013) or hot-pressing (Österberg et al., 2013; Schultz-Jensen et al., 2013).

Especially evaporation of the aqueous medium at ambient conditions or oven drying at low temperatures after filtration is very time consuming. Here, film production takes 12 h to several days (Chen, Zhu, & Tong, 2016; Jackson & Line, 1997; Missio et al., 2018; Mnich et al., 2020; Neagu, Gamstedt, & Berthold, 2006; Nguyen et al., 2013; Schultz-Jensen et al., 2013). Therefore, considerable effort was made to reduce processing times. Österberg et al. (2013) developed a method to manufacture MFC films in only 1−2.5 h by pressurized filtration at 2.5 bar combined with hot-pressing at 100 °C. Another fast method to prepare high-quality MFC films was presented by Schultz-Jensen et al. (2013) using a sheet former (Rapid-Köthen) and a sheet dryer. The first large-scale continuous production of MFC films on a support, for example a polymer film, was patented by VTT & Aalto University Foundation (2013).

Generally, the properties of the MFC films are influenced by the composition, crystallinity, and defibrillation degree of the MFC used. The nanopapers are transparent to translucent, often additionally possessing a high haze, and show low thermal expansion coefficients (Aulin et al., 2010; Minelli et al., 2010; Operamolla, 2019; Plackett et al., 2010). The high crystallinity of the cellulose and the entangled network structure of the microfibrils lead to interesting barrier properties (Kumar et al., 2014; Lavoine et al., 2012). The exceptionally low oxygen transmission of unmodified MFC films at dry conditions is the most prominent. Österberg et al. (2013), for example, reported an oxygen transmission rate below $0.6\ cm^3\ \mu m\ m^{-2}\ d^{-1}\ kPa^{-1}$ at relative humidity smaller than 65% for 120 μm thick films. Typical for hydrophilic membranes, oxygen permeability sharply increases over 70% relative humidity and water vapor barrier properties are relatively poor (Aulin et al., 2010; Kumar et al., 2014; Minelli et al., 2010; Österberg et al., 2013; Plackett et al., 2010). Contrasting, pure MFC films have excellent grease resistance against a variety of oils (Aulin et al., 2010; Kumar et al., 2014).

Related to the cellulose source and the processing, the reported mechanical properties of pure MFC films vary broadly (Siró & Plackett, 2010). The films can show high stiffness and strength in combination with large strain-to-failure. This results in higher toughness compared to normal wood-pulp paper (Neagu, Gamstedt, & Berthold, 2006). Very high tensile properties were, for example, obtained for the nanopaper formed by the Rapid-Köthen process. It has a tensile modulus of 13.4 GPa, a tensile strength of 232 MPa, and strain-to-failure of 5% (Schultz-Jensen et al., 2013). The wet strength is around 10% of the dry strength for unmodified films (Österberg et al., 2013).

These physical properties, the renewable and compostable character, and the up-scalable production make MFC films very interesting for packaging applications, as separator film for batteries, as a substrate for printed electronics, displays, and solar cells, and biomedical applications (Chen, Zhu, & Tong, 2016; Missio, 2018; Nordqvist et al., 2007; Schultz-Jensen et al., 2013). To tailor challenging properties, as for example, the hydrophilic nature, or to introduce new properties to the films, a large variety of chemical modifications and combinations with noncellulosic materials are discussed in the literature (Ghanadpour, Carosio, Larsson, & Wågberg, 2015; Minelli et al., 2010; Sehaqui, Zimmermann, & Tingaut, 2014; Shen & Feng, 2018; Yan et al., 2014).

10.2.1.2 Pure MFC boards

The idea to grind or mill plant material in watery suspension until it forms a hornlike bulk substance after drying for the fabrication of household goods dates back to a patent in the year 1921 (Schönbeck, 1921). More recently, the commercial material Zelfo was established, using refined microfiber pulp for casting and molding of cellulosic bulk material. Its density ranges from $0.3-1.5\ g\ cm^{-3}$ with a maximum flexural modulus and strength of 9.4 GPa and 95 MPa (Svoboda et al., 2000). Finally, Lee, Aitomäki, Berglund, Oksman, & Bismarck, (2014) introduced high-performance boards purely made out of homogenized bleached softwood kraft pulp. Compared to the production of thin MFC films, the challenge of removing large quantities of water from the MFC suspensions (water content $\geq$ 98 wt.%) in an efficient way is even greater for MFC boards with a height of a few millimeters. First, the solid content of the MFC suspensions is increased to around 10 wt.% by filtration or centrifugation (Lee, Aitomäki, Berglund, Oksman, & Bismarck, 2014; Qiang, Patel, & Manas-Zloczower, 2020; Rana & Gupta, 2020; Roach et al., 2011). Then, solid plates are formed by cold or hot pressing in porous molds under pressures ranging from 1 to 120 MPa. Often, a long dewatering stage is followed by a shorter hot pressing stage between 140 °C–150 °C (Lee, Aitomäki, Berglund, Oksman, & Bismarck, 2014; Roach et al., 2011). The pressing process can be complemented by a drying step in a (vacuum) oven (Lee, Aitomäki, Berglund, Oksman, & Bismarck, 2014; Qiang, Patel, & Manas-Zloczower, 2020; Rana

& Gupta, 2020). Lee, Aitomäki, Berglund, Oksman, & Bismarck, 2014), for example, dewatered 10 wt.% MFC slurry gradually under 5 MPa at room temperature to a solid content of 50 wt.%. The moist plates were further dried in an oven at 105 °C followed by a 30 min hot-pressing step at 100 MPa and 150 °C.

The dense cellulose boards are very stiff with a flexible modulus up to 17 GPa (Arévalo & Peijs, 2016). The oven-dried samples of Lee, Aitomäki, Berglund, Oksman, and Bismarck, (2014) showed a flexible modulus of 16 GPa and flexible strength of 250 MPa. Tensile properties were reported by Yousefi et al. (2018) at 60% relative humidity with a tensile modulus of 3.2 GPa and tensile strength of 85 MPa. It is important to note here that the humidity during conditioning and testing greatly alters the mechanical properties. The key to produce boards with high mechanical performance is the enhancement of surface area through the fibrillation treatment. An increased degree of fibrillation results in higher stiffness and strength, although degradation through the treatment should be avoided (Herzele, Veigel, Liebner, Zimmermann, & Gindl-altmutter, 2016; Lee, Aitomäki, Berglund, Oksman, & Bismarck, 2014; Roach et al., 2011). Even though MFC boards are water sensitive, they do not lose their integrity in a wet state. Typically, water uptake after immersion in water stabilizes over time at around 40–50 wt.%, and good dimensional recovery is observed (Arévalo & Peijs, 2016; Tonoli et al., 2019; Yousefi et al., 2018).

Due to their exceptional mechanical properties combined with fully biobased, renewable origin, and biodegradability, MFC boards are interesting candidates to replace petrochemical structural and nonstructural materials in the area of transport and building applications, consumer electronics, household goods, and sporting equipment. Current drawbacks include the long processing times, water sensitivity, and the tendency to shrinkage and warping during drying. Especially the latter is still an unsolved problem for three-dimensional (3D) molding (Rol et al., 2020). Batch-wise processing needs to be replaced by industrial scalable continuous processing to gain commercial relevance (Arévalo & Peijs, 2016).

Several recent developments address these challenges. In contrast to traditional fibrillation techniques, it is also possible to produce good quality MFC with a twin-screw extrusion process in combination with enzymatic or chemical pretreatment (Rol et al., 2017; Rol, Billot, Bolloli, Beneventi, & Bras, 2020). The process yields MFC with a solid content of 20%–25%. It is further possible to continuously form MFC sheets by single-screw extrusion from MFC pastes with such high solid content (El Awad Azrak, Costakis, Moon, Schueneman, & Youngblood, 2020). El Awad Azrak et al. (2020) obtained defect-free MFC sheets with an output rate of over 7 kg h^{-1} (wet) after adding carboxymethyl cellulose (9 wt.% of the final dry product) as a processing agent. The sheets can be stacked and consolidated in a hot-press or by calendering.

10.2.1.3 MFC in other natural oligomers and polymers

MFC shows high potential to form pure MFC-based films and boards, thanks to their high specific surface area and strong interactions among fibrils via intra- and intermolecular hydrogen bonds. However, the processing of pure MFC products cannot be adapted to current industrial processing, for example extrusion and injection molding, due to the no flowability of pure MFC. The introduction of other natural polymers may probably address this issue by improving the flowability of the MFC. Therefore, the application of MFC in other natural oligomers (e.g., lignin, xylan, arabinoxylan, xyloglucan, tannin, and other polyphenol-containing moieties) and polymers (e.g., chitin, gelatin, starch, thermoplastic starch, and some proteins) (Table 10.4) has received a lot of attention as biocomposite materials.

MFC shows a good reinforcing effect in those natural oligomers and polymers. It enables a significant improvement in tensile properties of the resulting materials (Table 10.4). The high affinity between MFC and natural polymers may even result in a further increase in tensile properties and tensile toughness compared with the pure MFC films (Goussé, Chanzy, Cerrada, & Fleury, 2004; Lems, Winklehner, Hansmann, Gindl-Altmutter, & Veigel, 2019). Moreover, the hybrid materials showed new features such as antioxidant properties, surface hydrophobicity, enhanced

Table 10.4 Examples of natural oligomers and polymers used for microfibrillated cellulose-reinforced films/boards.

Types of matrices	Method	Improvement	References
Arabinoxylan	Solution casting	Elasticity and tensile properties	Stevanic et al. (2012)
Galactoglucomannan	Filtration/hot pressing	Tensile strength and toughness	Goussé, Chanzy, Cerrada, & Fleury, 2004
Xylan	Solution casting	Tensile properties and hydrophobicity	Iwamoto, Yamamoto, Lee, & Endo, 2014
Starch	Solution casting	Tensile properties	Dufresne and Vignon (1998)
Starch/hydroxypropyl starch	Solution casting	Tensile properties, air permeability	Lengowski et al. (2020)
Chitosan	Solution casting	Wet properties	Nordqvist et al. (2007)
Tannin	Solution casting	Air permeability, antioxidant properties	Missio et al. (2018)
Soy protein and clove essential oil	Solution casting	Tensile properties, air, and vapor permeability	Ortiz, Salgado, Dufresne, and Mauri (2018)

mechanical properties under high humidity, air and vapor permeability, etc. (Table 10.4).

The incorporation of MFC in those natural oligomers and polymers was usually done by solution casting followed by drying either in an oven or under controlled ambient temperature and humidity. This approach is favored by the hydrophilic character of both MFC and the natural oligomers and polymers, and the presence of a high amount of water in MFC suspension. As an exception, the successful fabrication of MFC/thermoplastic starch composites through an extrusion process was demonstrated (Hietala, Mathew, & Oksman, 2013). This successful case showed the potential and possibility of using MFC in thermoplastic starch composite at a large scale. For the other natural polymers, new methods with high scalability that are compatible with the currently available industrial molding processes will be highly appreciated.

Moreover, there are some interesting crosslinkable phenolic molecules derived from plant and agro-residues that can be used to fabricate novel biocomposite materials (Mnich et al., 2020). Those crosslinkable phenolic molecules can often be found in four biopolymer classes lignin, extensins, glucuronoarabinoxylan, and side chains of rhamnogalacturonan-I (Mnich et al., 2020). These biopolymers can be used to form crosslinked firm hydrogels under the catalysis of enzymes, which can be potential matrices for MFC-reinforced composite materials. The hydrogels are endowed with antioxidant and antimicrobial activities due to the presence of phenolic molecules. This approach was already studied for hydrogels (Li, Du, Wang, & Yu, 2014), fiber cell composites (Liu et al., 2017), and cellulose nanocrystal-based composites (Paës, von Schantz, & Ohlin, 2015). It is therefore worth investigating the use of the same approach with MFC as reinforcing agents to achieve improved mechanical properties and other desired features.

10.2.2 Microfibrillated cellulose in thermoplastics

Melt compounding for incorporation of MFC into thermoplastics is mostly performed in an internal mixer or a single- or twin-screw extruder. Most well-known and commonly used thermoplastics can be used to manufacture MFC/thermoplastics composite including PLA, PP, PE, polycarbonate, and poly (methyl methacrylate) (Table 10.5). The MFC is normally prepared in an aqueous solution by mechanical treatment forming a gel-like suspension. The high water content in the suspension (>90 wt.%) together with the hydrophilic nature of the MFC are the main restrictions for the successful use of MFC in MFC/thermoplastic composites.

Ishikura and Yano (2020) presented a straightforward method to fabricate MFC/polyester nanocomposites starting from the MFC water suspension. A premixed polyester fiber (1.2 deniers with a length of 5 mm) and MFC suspension was first dewatered by filtration and compression followed by drying in the oven. The dried samples

Table 10.5 Commonly used thermoplastics and molding conditions used for microfibrillated cellulose-based composites

Thermoplastics	Molding	Molding conditions	References
Polyester	Compression molding	200 °C	Ishikura and Yano (2020)
Thermoplastic starch	Twin-screw extrusion and compression molding	Extruder: 100 °C–120 °C Compression molding: 130 °C	Lendvai, Karger-Kocsis, Kmetty, and Drakopoulos (2016)
Polylactic acid	Mixing and compression molding	Mixing: 160 °C, 20 min; Compression molding: 220 °C	Meng et al. (2018)
Polypropylene	Compression molding and injection molding	Compression molding: 180 °C, 10 MPa, 5 min; Injection molding: 190 °C	(Iwamoto, Yamamoto, Lee, & Endo, 2014)
Polypropylene/ Polyethylene	Twin-screw extrusion	Extruder: 210 °C	Palange et al. (2019)
Polycarbonate	Compression molding	Compression molding: 210 °C, 1 MPa, 1 min	Panthapulakkal and Sain (2012)
Poly(vinylidene fluoride)	Dimethylformamide solution casting	NA	Poothanari, Michaud, Damjanovic, and Leterrier (2021)
Poly(methyl methacrylate)	Dimethylformamide solution casting	NA	Islam et al. (2016)

Notes: NA, Not available.

were then hot-pressed to obtain the final MFC/polyester composites. Similarly, MFC/polyester composites with 3D structures were also successfully designed and fabricated. An increase in tensile properties (e.g., tensile strength and Young's modulus) was observed, in particular for the composites with high MFC content (i.e., above 50 wt.%). In this study, the MFC was unmodified, so the hydrophilic surface of MFC may contribute to the less pronounced increase in the tensile properties with increasing MFC content as the authors observed.

PLA is another type of thermoplastic polyester usually produced by fermentation of renewable agricultural residues followed by polymerization. PLA is usually regarded as a green polymer as it is biodegradable in industrial composting plants under controlled conditions. Therefore MFC/PLA composites are receiving special attention to produce 100% biodegradable materials. Meng et al. (2018) proposed a method using a Brabender Intelli-Torque Plasti-Corder Prep-Mixer to prepare MFC/PLA composites with freeze-dried and chopped MFC fibers, and the addition of 2–5 wt.% epoxidized soybean oil to improve the dispersion of MFC fibers in the resulting composites. It was revealed that the addition of the epoxidized soybean oil (2–5 wt.%) in the MFC/PLA composites could significantly improve the ductility and toughness of the composites, but a sharp drop in strain and toughness, as well as the other mechanical properties, was found with a further increase in the loading of the epoxidized soybean oil to 10 wt.%. In principle, the epoxidized soybean oil works as a compatibilizer in the composite to enhance the coherence between the MFC fillers and the PLA matrix. However, the properties went down when too much compatibilizer was added. Though desired toughness and strength of the MFC/PLA composites were achieved, the preparation of the dried MFC using a freeze drier followed by further mechanical grinding will limit the successful use of the MFC at scale.

To further improve the surface properties of MFC, an attempt to use coated MFC with polyoxyethylene(10)nonylphenyl ether (PNE) to prepare MFC/PP composites was made by Iwamoto et al., (2014) using a twin-screw extruder. The PNE-coated MFC fibers were prepared by mixing MFC and PNE in aqueous suspension followed by freeze-drying. Before mixing the coated fibers with polymer in the extruder, the PNE-coated MFC fibers were first dispersed in toluene and mixed with predissolved PP and maleic anhydride grafted PP in toluene followed by evaporation of the toluene on a hot plate and further drying in a vacuum oven. It was observed that the dispersion of the MFC fillers was improved by coating with the PNE compatibilizer and the use of the maleic anhydride grafted PP. The PP composites containing 10 wt.% PNE-coated MFC showed a significant increase in Young's modulus from 497 to 722 MPa and yield strength from 21 to 32 MPa compared with the neat PP. However, the complex sample preparation method using freeze-drying and the organic solvent toluene is still far from solving the above-mentioned barriers that limit the successful application of MFC in large quantities.

More attempts to further improve the hydrophobicity of the MFC fillers by using chemical modifications were made. One method proposed by Palange, Johns, Scurr, Phipps, and Eichhorn (2019), as an example, was to coat the surface of MFC fillers in an aqueous solution with tannic acid, which exists in both phenol and quinone form in the solution. Besides, a basic condition is favorable for the formation of the quinone form from its phenol form. The quinone form of tannic acid can further react with amines through amidation reactions. In this study, Palange et al. (2019) applied both octadecylamine and hexylamine to further react with tannic acid on the surface of the MFC. In this way, a hydrophobic form of MFC fillers was obtained from undried MFC slurry with tannic acid and different amines. The study also proved the modification to be successful to improve the dispersion of the MFC fibers in PP and PE polymers. An increase in the modulus was noticed when the loading of the MFC fillers was greater than 2 wt.%. Despite positive mechanical properties were obtained by incorporating the modified MFC fillers in the polymers, the presence of some aggregates was still noticed, which was assumed to impair the mechanical performance and durability of the composites.

To avoid using wet MFC suspensions, dried MFC could make it easier to prepare MFC/thermoplastic composites. Dried MFC films were studied as reinforcing agents in polycarbonate matrix. The dried films were placed between polycarbonate sheets, and the composites were simply fabricated by pressing the stacked films by compression molding (Panthapulakkal & Sain, 2012). The modulus of the polycarbonate was increased by about 100% by incorporating only 18% of MFC fillers. An increase of 30% in tensile strength was observed by the authors for the composites with the same fiber content. Despite these positive results, large deviations in the tensile properties occurred and no clear increase in tensile properties along with the increase in the fiber content was achieved. This might be attributed to the high hydrophilicity of the unmodified MFC sheets that could eventually weaken the interface between the MFC and the polycarbonate.

The hydrophilicity of the MFC fibers and high water content of the MFC suspension are still the main barriers for the successful application of MFC fibers in thermoplastic composite materials, especially for high-performance structural components at an industrial scale. The high water content not only increases the additional cost to remove the water, but also makes the surface hydrophobization through chemical reactions (e.g., nucleophilic acyl substitution, nucleophilic addition, and nucleophilic substitution) less efficient as the water molecules can compete with the cellulose-hydroxyl groups toward the reactions. Many attempts have been made, some of which were discussed in detail above, to manufacture MFC/thermoplastics composites by using freeze-dried MFC powder, dried MFC films, coated MFC powder with or without post modification to improve the dispersion of MFC fillers in the resulting thermoplastic composites.

Continuous efforts are still required to further address these two barriers, especially for upscaling of the MFC surface functionalization and fabrication of the MFC-based composites at scale. The surface hydrophobization, in general, is necessary for most nonpolar polymers to give good compatibility and coherence. The reported methods are not yet efficient enough to give the required hydrophobicity or are too complicated with the use of large quantities of (toxic) organic chemicals. To make the manufacturing process acceptable from both economic and environmental aspects, more simple and more efficient techniques are the key to address those issues.

10.2.3 Microfibrillated cellulose in thermosets

MFC fibers have also attracted significant interest as an alternative reinforcement in thermoset-based composite materials due to the unique characteristics of thermosets. The liquid state of the thermosets as monomers before curing, some being even compatible with waterborne systems, makes them easier to mix with MFC fillers in practice, even with MFC aqueous suspension. Both dried MFC in powder, foam, or aerogel form and wet MFC dispersed in either organic solvent or water were already studied to introduce MFC into a thermoset.

To skip a predrying step, some studies successfully demonstrated the fabrication of composites through a solvent-exchange process or suspension polymerization approaches (Fig. 10.4 and Table 10.6).

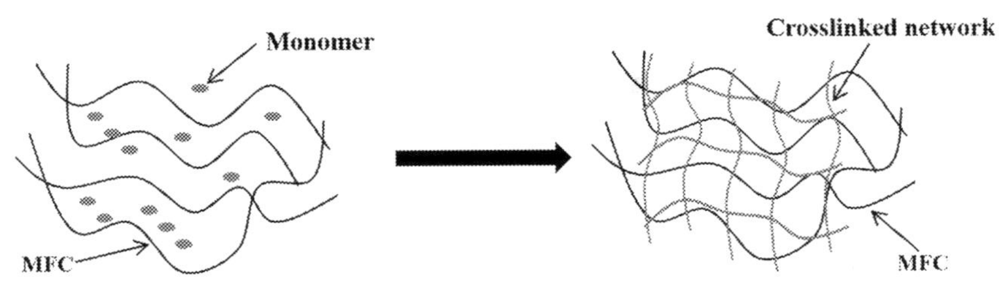

Figure 10.4 Schematic of microfibrillated cellulose (MFC)/thermoset composite fabrication (before curing—left; after curing—right).

Table 10.6 Commonly used thermosets and conditions used for the fabrication of microfibrillated cellulose (MFC)-based composites.

Thermoset	Molding/mixing system	Curing conditions	References
Polyester	Suspension polymerization	103 °C, 2 h	Yan et al. (2016)
Epoxy	MFC in acetone	40 °C, 12 h followed by 120 °C, 2 h	Lu et al. (2008)
Epoxy	Freeze-dried MFC	80 °C, 4 h followed by 120 °C, 4 h	Qiang et al. (2020)
Epoxy	Ice-templating and freeze-dried MFC	60 °C, 1 h followed by 80 °C, 1 h	Nissilä et al. (2019)
Poly(furfuryl alcohol)	In-situ polymerization in MFC aqueous slurry followed by freeze-drying	80 °C, 24 h	Lems, Winklehner, Hansmann, Gindl-Altmutter, and Veigel (2019)
Unsaturated Polyester	MFC in acetone	90 °C, 2 h followed by 140 °C, 3 h	Ansari, Skrifvars, and Berglund (2015)
Waterborne polyurethane	MFC aqueous suspension	NA	Hormaiztegui et al. (2016)
Melamine-formaldehyde	Dried MFC films	160 °C, 10 min	Henriksson and Berglund (2007)
Phenol-formaldehyde	Freeze-dried MFC	160 °C, 30 min	Yan et al. (2016)

Notes: NA, Not available.

Among those studies, many researchers have paid attention to epoxy-based systems. Various strategies were tested to prepare MFC/epoxy composites including (1) freeze-dried MFC impregnated in epoxy/hardener mixture under vacuum followed by curing (Qiang, Patel, & Manas-Zloczower, 2020); (2) solvent-exchanged MFC in acetone suspension mixed with epoxy/hardener followed by evaporation and then curing (Lu, Askeland, & Drzal, 2008); and (3) epoxy/hardener infused in MFC-based aerogel prepared via vacuum infusion molding (Nissilä, Hietala, & Oksman, 2019).

In the study by Qiang et al. (2020), the MFC/epoxy composites were fabricated from freeze-dried MFC foam. Epichlorohydrin was applied as fiber modifier as well as crosslinker between MFC fillers and epoxy matrix. The modification showed a positive impact on the hydrophobization of the surface with limited fiber agglomeration without impairing the fiber properties, for example, crystallinity of the fibers (Qiang et al., 2020).

Lu et al. (2008) attempted to hydrophobize the surface of MFC before mixing with the epoxy monomers with different silanes including 3-aminopropyltriethoxysilane, 3-glycidoxypropyltrimethoxysilane, and a titanate coupling agent (Lica 38) in acetone solvent. Silanization can be carried out in a simple way, and it is also more tolerant with water or moisture in the samples and the environment. Successful silanization of MFC, confirmed using Fourier-transform infrared spectroscopy, X-ray photoelectron spectroscopy, and contact angle measurements is capable of altering the surface character of MFC from hydrophilic to hydrophobic. As a result, better and strong adhesion between the MFC fillers and the epoxy polymer matrix was observed for the silane-treated fibers, which eventually resulted in better mechanical properties of the composite materials (Lu et al., 2008). With a 5 wt.% addition of MFC fillers, the storage modulus at 130 °C was measured to increase from 9.7 MPa for the neat epoxy resins to 37.3 MPa for the composites with unmodified MFC fillers, and then further to 65.6 MPa for the samples with 3-aminopropyltriethoxysilane modified MFC fillers. Desired mechanical properties were obtained through a simple modification with organosilanes, but the sample preparation using a solvent exchange method from water to acetone could limit the application of MFC in the epoxy system at large scales.

As summarized in Table 10.6, similar approaches with freeze-dried MFC powder/foam or solvent-exchanged MFC were applied for other thermosetting systems, such as unsaturated polyester and phenol-formaldehyde systems. Similarly, a limited improvement was achieved for the composites made from unmodified MFC, which is mainly due to the weak interfacial properties between hydrophilic MFC fillers and hydrophobic polymers. Besides, the two main issues including dewatering and surface functionalization have not yet been fully addressed toward the application of MFC at a large scale.

Some efforts, which addressed those barriers and demonstrated new possibilities for manufacturing MFC/thermosets composites in more simple ways, will be discussed more specifically below. One of the examples is the work done by Yan et al. (2016), who proposed a new route toward embedding MFC fillers in a nonpolar thermoset matrix without using any organic solvent. It uses the unique features of lignin present in the MFC (Yan et al., 2016). The amphiphilic surface of lignin in MFC is capable of stabilizing an emulsion of unsaturated polyester monomers in an aqueous solution. Upon polymerization of the resin in the aqueous solution, the formed thermoset microspheres are embedded in the MFC network through hydrophobic lignin moieties. A mechanically stable porous network structure is then formed after conventional drying in an oven. It can further be used as a master batch after milling for the composite fabrication. Despite requiring a high content of lignin residues, the process is rather simple compared to the other above-mentioned methods.

Another example is the preparation of a waterborne polyurethane/MFC composite in an aqueous solution (Hormaiztegui et al., 2016). The synthesis of the waterborne biobased polyurethane was carried out from a reaction between a polycaprolactone diol or a biobased macrodiol derived from castor oil and isophorone diisocyanate. 2,2-Bis(hydroxymethyl)propionic acid was introduced into the synthesis to produce the ionic centers needed to formulate the aqueous polyurethane suspensions. The composites were fabricated by mixing the synthesized waterborne polyurethane and MFC aqueous suspension, followed by drying in an oven. The two examples are worth mentioning as both demonstrated novel strategies to minimize water-related problems.

10.2.4 Microfibrillated cellulose in elastomers

Elastomers owe their viscoelastic properties with weak intermolecular interactions sharing similarities with both thermosets (requiring vulcanization) and thermoplastics (flowing ability above glass transition temperature). In the rubber industry, especially in tire application, the currently used fillers include carbon black and silica. Carbon black is derived from fossil fuel, and silica from minerals. MFC fibers with low density and high sustainability are of high potential to be used as new reinforcing fillers in the elastomer compounds.

Many studies focused on fabricating MFC-rubber compounds based on latex aqueous emulsions (Table 10.7). The fabrication starts with premixing the MFC suspension with rubber latex aqueous suspension (e.g., natural rubber latex suspension) together with crosslinking curatives (e.g., sulfur or peroxides and accelerators). It is followed by prevulcanization in the aqueous solution under heating and stirring. The final rubber compounds can be prepared by casting the mixture on a glass plate followed by drying in an oven or at ambient temperature (Abraham et al., 2013; 2013; JaKato et al., 2015; Jarnthong et al., 2015). Overall, good reinforcing efficiency with improved tensile properties and dynamic mechanical properties were found, which is presumable due to the formation of a three-dimensional MFC network interlocking with the rubber matrix polymer network (Abraham et al., 2013).

As discussed above, the limitations including weak interfacial interactions between MFC and hydrophobic polymers and dewatering are still valid in this section. Some

Table 10.7 Commonly used types of rubber and microfibrillated cellulose (MFC) for the fabrication of MFC-based compounds.

Rubber	Modification	Mixing process	References
Styrene–butadiene copolymer rubber	Esterification (palmitoyl chloride and 3,3′-dithiopropionic acid chloride)	Melt-mixing process (internal mixer)	Fumagalli et al. (2018)
High ammonia natural rubber latex	Unmodified/freeze-dried MFC	Solution casting	Jarnthong et al. (2015)
High ammonia natural rubber latex	Esterification (stearic acid chloride)	Coagulation (unmodified MFC); solution casting followed by roll milling for the modified MFC	Kato et al. (2015)
Natural rubber latex	Unmodified/MFC suspension	Solution casting	Abraham et al. (2013); Liu et al., 2017
Natural rubber latex	Sodium alginate coated MFC	Solution casting	Supanakorn et al. (2021)
Natural rubber latex	Thermoplastic starch-MFC mixture	Extrusion compounding	Drakopoulos et al. (2017)

attempts were made to address those limitations by tailoring the surface of the MFC to make it more hydrophobic. One approach, using aqueous carboxylated MFC obtained by 2,2,6,6-tetramethylpiperidine-1-oxy radical oxidation mixed with tetra-alkylammonium ($^{+}NR_4$) counter ions to improve the interfacial interactions with rubber latex was proposed (Fukui et al., 2018). It resulted in a significant improvement in storage modulus by up to 55 times at 25 °C, and in Young's modulus by up to 12 times. Similar results were also observed for oleic acid modified MFC in natural rubber by mixing the modified MFC with rubber in toluene solution (Kato et al., 2015). Although positive results were reported, the approach of using rubber latex or dissolved rubber in solvents is limited to laboratory scales, since this method is not compatible with the current rubber compound manufacturing lines that mainly depend on melt-mixing processing.

Until now, there are very limited studies dedicated to filled rubber compounds prepared from MFC fillers based on melt-mixing processing. One study was published in 2018 to fabricate MFC-based elastomer compounds with chemically modified freeze-dried MFC and a diene elastomer matrix via melt-mixing processing (Fumagalli et al., 2018). The freeze-dried MFC was modified with both palmitoyl chloride to obtain a nonreactive hydrophobic surface, and with 3,3′-dithiopropionic acid chloride to introduce reactive groups (i.e., disulfide bonds) that can covalently bind the nanocellulose to the dienic matrix in a subsequent vulcanization process using sulfur as the curing agent. A significant increase in tensile strength was observed from 17 MPa for samples with unmodified MFC to 37 MPa with palmitoyl chloride modified MFC, and further to 47 MPa with 3,3′-dithiopropionic acid chloride modified MFC, when 17 vol.% MFC was loaded in the rubber compounds. This work demonstrates the high potential of the MFC-based fillers in elastomer compounds to replace conventional fillers, silica and carbon black. However, further optimization of the mechanical properties of the compounds and deeper analysis of the structure-properties relationship in the MFC-based compounds will still be required. Also, the freeze-drying process before surface functionalization of the MFC limits the possibilities for upscaling.

10.3 Future perspectives

The use of MFC for biocomposites has received considerable interest in recent years. As discussed throughout this review, high water content ($\geq$90 wt.%) and high surface hydrophilicity are the two biggest barriers to the application of MFC in hydrophobic polymers. The conventional fiber surface modification techniques (e.g., nucleophilic acyl substitution, nucleophilic addition, and nucleophilic substitution) for surface modification of dried cellulose fibers cannot be applied directly to the MFC suspension. Before any chemical modification, the priority is to get rid of the water from the MFC aqueous suspension. The widely used dewatering methods include freeze-drying and solvent exchange via centrifugation or filtration, which are either not scalable or consume large quantities of organic solvents leading to high costs of the final products.

Continuous efforts are still in demand to further address those above-mentioned issues. One route is to come up with new solutions to deal with the large quantities of water in scalable and economically efficient methods. Moreover, new and more efficient modifiers are urgent to be identified or synthesized to block the interaction among fibrils while drying and to ideally form covalent bonds on the surface of the dried cellulose fibrils as coupling agents. Only when the two limitations are addressed properly, the application of MFC-based fillers in high-performance polymeric composites can be considerably boosted.

Another interesting topic is to use MFC alone or together with other natural polymers such as starch, hemicelluloses, chitins, polyphenols, and amino acid-based oligomers or polymers. Most of those natural materials are hydrophilic leading to better compatibility and high coherence with MFC. This avoids the necessity for premodification of the MFC at the very first step. For these applications, dewatering of MFC suspension, pure or mixed with other natural polymers, can be done in scalable and conventional drying methods (e.g., hot pressing, oven drying). Eventually, a postmodification can be more easily performed on the dried products if necessary. This opens a new door for the successful use of MFC for biobased composite materials.

References

Abraham, E., et al. (2013). Physicomechanical properties of nanocomposites based on cellulose nanofibre and natural rubber latex. *Cellulose*, *20*(1), 417–427. Available from https://doi.org/10.1007/s10570-012-9830-1.

Abraham, E., Thomas, M. S., John, C., Pothen, L. A., Shoseyov, O., & Thomas, S. (2013). Green nanocomposites of natural rubber nanocellulose: Membrane transport, rheological and thermal degradation characterisations. *Industrial Crops and Products*, *51*, 415–424. Available from https://doi.org/10.1016/j.indcrop.2013.09.022.

Andresen, M., Stenstad, P., Møretrø, T., Langsrud, S., & Syverud, K. (2007). Nonleaching antimicrobial films prepared from surface-modified microfibrillated cellulose. *Biomacromolecules*, *8*, 2149–2155. Available from https://doi.org/10.1021/bm070304e.

Ansari, F., Skrifvars, M., & Berglund, L. (2015). Nanostructured biocomposites based on unsaturated polyester resin and a cellulose nanofiber network. *Composites Science and Technology*, *117*, 298–306. Available from https://doi.org/10.1016/j.compscitech.2015.07.004.

Arévalo, R., & Peijs, T. (2016). Binderless all-cellulose fibreboard from microfibrillated lignocellulosic natural fibres. *Composites Part A Applied Science and Manufacturing*, *83*, 38–46. Available from https://doi.org/10.1016/j.compositesa.2015.11.027.

Aulin, C., Gällstedt, M., & Lindström, T. (2010). Oxygen and oil barrier properties of microfibrillated cellulose films and coatings. *Cellulose*, *17*(3), 559–574. Available from https://doi.org/10.1007/s10570-009-9393-y.

Bekalo, S. A., & Reinhardt, H. W. (2010). Fibers of coffee husk and hulls for the production of particleboard. *Materials and Structures*, *43*(8), 1049–1060. Available from https://doi.org/10.1617/s11527-009-9565-0.

Berglund, L., Noël, M., Aitomäki, Y., Öman, T., & Oksman, K. (2016). Production potential of cellulose nanofibers from industrial residues: Efficiency and nanofiber characteristics. *Industrial Crops and Products*, *92*, 84–92. Available from https://doi.org/10.1016/j.indcrop.2016.08.003.

Bouhoute, M., et al. (2020). Microfibrillated cellulose from *Argania spinosa* shells as sustainable solid particles for O/W Pickering emulsions. *Carbohydrate Polymers*, *251*(6), 116990. Available from https://doi.org/10.1016/j.carbpol.2020.116990.

Chang, C., He, M., Zhou, J., & Zhang, L. (2011). Swelling behaviors of pH- and salt-responsive cellulose-based hydrogels. *Macromolecules*, *44*(6), 1642–1648. Available from https://doi.org/10.1021/ma102801f.

Chen, Y. M., Jiang, Y., Wan, J. Q., Wu, Q. T., Wei, Z. B., & Ma, Y. W. (2018). Effects of wet-pressing induced fiber hornification on hydrogen bonds of cellulose and on properties of eucalyptus paper sheets. *Holzforschung*, *72*(10), 829–837. Available from https://doi.org/10.1515/hf-2017-0214.

Chen, N., Zhu, J. Y., & Tong, Z. (2016). Fabrication of microfibrillated cellulose gel from waste pulp sludge via mild maceration combined with mechanical shearing. *Cellulose*, *23*, 2573–2583. Available from https://doi.org/10.1007/s10570-016-0959-1.

Chinga-Carrasco, G. (2011). Cellulose fibres, nanofibrils and microfibrils: The morphological sequence of MFC components from a plant physiology and fibre technology point of view. *Nanoscale Research Letters*, *6*(1), 417. Available from https://doi.org/10.1186/1556-276X-6-417.

de Melo E. M., Microfibrillated cellulose and high-value chemicals from orange peel residues (Doctoral thesis), Department of chemistry, University of York, United Kingdom. 2018. <https://etheses.whiterose.ac.uk/23195/1/Microfibrillated%20cellulose%20and%20high-value%20chemicals%20from%20orange%20peel%20residues_Eduardo%20Melo>

Drakopoulos, S. X., Karger-Kocsis, J., Kmetty, L., Lendvai., & Psarras, G. C. (2017). Thermoplastic starch modified with microfibrillated cellulose and natural rubber latex: A broadband dielectric spectroscopy study. *Carbohydrate Polymers*, *157*, 711–718. Available from https://doi.org/10.1016/j.carbpol.2016.10.036.

Dufresne, A., Dupeyre, D., & Vignon, M. R. (2000). Cellulose microfibrils from potato tuber cells: Processing and characterization of starch-cellulose microfibril composites. *Journal of Applied Polymer Science*, *76*(14), 2080–2092. Available from https://doi.org/10.1002/(SICI)1097-4628(20000628)76:14<2080:AID-APP12>3.0.CO;2-U.

Dufresne, A., & Vignon, M. R. (1998). Improvement of starch film performances using cellulose microfibrils. *Macromolecules*, *31*(8), 2693–2696. Available from https://doi.org/10.1021/ma971532b.

El Awad Azrak, S. M., Costakis, W. J., Moon, R. J., Schueneman, G. T., & Youngblood, J. P. (2020). Continuous processing of cellulose nanofibril sheets through conventional single-screw extrusion. *ACS Applied Polymer Materials*, *2*(8), 3365–3377. Available from https://doi.org/10.1021/acsapm.0c00477.

Fernandes Diniz, J. M. B., Gil, M. H., & Castro, J. A. A. M. (2004). Hornification - Its origin and interpretation in wood pulps. *Wood Science and Technology*, *37*(6), 489–494. Available from https://doi.org/10.1007/s00226-003-0216-2.

Fiorote, J. A., Freire, A. P., de, D., Rodrigues, S., Martins, M. A., Andreani, L., & Valadares, L. F. (2019). Preparation of composites from natural rubber and oil palm empty fruit bunch cellulose: Effect of cellulose morphology on properties. *BioResources*, *14*(2), 3168–3181. Available from https://doi.org/10.15376/biores.14.2.3168-3181.

Fukui, S., Ito, T., Saito, T., Noguchi, T., & Isogai, A. (2018). Counterion design of TEMPO-nanocellulose used as filler to improve properties of hydrogenated acrylonitrile-butadiene matrix. *Composites Science and Technology*, *167*, 339–345. Available from https://doi.org/10.1016/j.compscitech.2018.08.023.

Fumagalli, M., et al. (2018). Rubber materials from elastomers and nanocellulose powders: Filler dispersion and mechanical reinforcement. *Soft Matter*, *14*(14), 2638–2648. Available from https://doi.org/10.1039/c8sm00210j.

Ghanadpour, M., Carosio, F., Larsson, P. T., & Wågberg, L. (2015). Phosphorylated cellulose nanofibrils: A renewable nanomaterial for the preparation of intrinsically flame-retardant materials. *Biomacromolecules*, *16*(10), 3399–3410. Available from https://doi.org/10.1021/acs.biomac.5b01117.

Gordobil, O., Egüés, I., Urruzola, I., & Labidi, J. (2014). Xylan-cellulose films: Improvement of hydrophobicity, thermal and mechanical properties. *Carbohydrate Polymers*, *112*, 56–62. Available from https://doi.org/10.1016/j.carbpol.2014.05.060.

Goussé, C., Chanzy, H., Cerrada, M. L., & Fleury, E. (2004). Surface silylation of cellulose microfibrils: Preparation and rheological properties. *Polymer*, *45*(5), 1569–1575. Available from https://doi.org/10.1016/j.polymer.2003.12.028.

Hansen, N. M. L., Blomfeldt, T. O. J., Hedenqvist, M. S., & Plackett, D. V. (2012). Properties of plasticized composite films prepared from nanofibrillated cellulose and birch wood xylan. *Cellulose*, *19*(6), 2015–2031. Available from https://doi.org/10.1007/s10570-012-9764-7.

Hassan, M. L., Bras, J., Hassan, E. A., Silard, C., & Mauret, E. (2014). Enzyme-assisted isolation of microfibrillated cellulose from date palm fruit stalks. *Industrial Crops and Products*, *55*, 102–108. Available from https://doi.org/10.1016/j.indcrop.2014.01.055.

Heiskanen, I., Backfolk, K., Vehviläinen, M., Kamppuri, T., & Nousiainen, P. (2021). *Process for producing microfibrillated cellulose*, WO 2011/004301 A1.

Henriksson, M., & Berglund, L. A. (2007). Structure and properties of cellulose nanocomposite films containing melamine formaldehyde. *Journal of Applied Polymer Science*, *106*, 2817–2824. Available from https://doi.org/10.1002/app.

Henriksson, M., Berglund, L. A., Isaksson, P., Lindström, T., & Nishino, T. (2008). Cellulose nanopaper structures of high toughness. *Biomacromolecules*, *9*(6), 1579–1585. Available from https://doi.org/10.1021/bm800038n.

Herzele, S., Veigel, S., Liebner, F., Zimmermann, T., & Gindl-altmutter, W. (2016). Reinforcement of polycaprolactone with microfibrillated lignocellulose. *Industrial Crops and Products*, *93*, 302–308. Available from https://doi.org/10.1016/j.indcrop.2015.12.051.

Hietala, M., Mathew, A. P., & Oksman, K. (2013). Bionanocomposites of thermoplastic starch and cellulose nanofibers manufactured using twin-screw extrusion. *European Polymer Journal*, *49*(4), 950–956. Available from https://doi.org/10.1016/j.eurpolymj.2012.10.016.

Hietala, M., Sain, S., & Oksman, K. (2017). Highly redispersible sugar beet nanofibers as reinforcement in bionanocomposites. *Cellulose*, *24*(5), 2177–2189. Available from https://doi.org/10.1007/s10570-017-1245-6.

Hormaiztegui, M. E. V., Mucci, V. L., Santamaria-Echart, A., Corcuera, M. Á., Eceiza, A., & Aranguren, M. I. (2016). Waterborne polyurethane nanocomposites based on vegetable oil and microfibrillated cellulose. *Journal of Applied Polymer Science*, *133*(47), 1–12. Available from https://doi.org/10.1002/app.44207.

Hu, J., Tian, D., Renneckar, S., & Saddler, J. N. (2018). Enzyme mediated nanofibrillation of cellulose by the synergistic actions of an endoglucanase, lytic polysaccharide monooxygenase (LPMO) and xylanase. *Scientific Reports*, *8*(1), 4–11. Available from https://doi.org/10.1038/s41598-018-21016-6.

Ishikura, Y., & Yano, H. (2020). Microfibrillated-cellulose-reinforced polyester nanocomposites prepared by filtration and hot pressing: Bending properties and three-dimensional formability. *Journal of Applied Polymer Science*, *137*(33), 1–7. Available from https://doi.org/10.1002/app.48192.

Islam, M. T., Montarsolo, A., Zoccola, M., Canetti, M., Cacciamani, A., & Bertini, F. (2016). Effect of individualized cellulose fibrils on properties of poly(methyl methacrylate) composites. *Journal of Macromolecular Science Part B*, *55*(9), 867–883. Available from https://doi.org/10.1080/00222348.2016.1207589.

Iwamoto, S., Abe, K., & Yano, H. (2008). The effect of hemicelluloses on wood pulp nanofibrillation and nanofiber network characteristics. *Biomacromolecules*, *9*(3), 1022–1026. Available from https://doi.org/10.1021/bm701157n.

Iwamoto, S., Nakagaito, A. N., & Yano, H. (2007). Nano-fibrillation of pulp fibers for the processing of transparent nanocomposites. *Applied Physics A*, *89*(2), 461–466. Available from https://doi.org/10.1007/s00339-007-4175-6.

Iwamoto, S., Yamamoto, S., Lee, S. H., & Endo, T. (2014). Mechanical properties of polypropylene composites reinforced by surface-coated microfibrillated cellulose. *Composites Part A: Applied Science and Manufacturing*, *59*, 26–29. Available from https://doi.org/10.1016/j.compositesa.2013.12.011.

Jackson, M. J., & Line, M. A. (1997). Organic composition of a pulp and paper mill sludge determined by FTIR, 13C CP MAS NMR, and chemical extraction techniques. *Journal of Agricultural and Food Chemistry*, *45*(6), 2354–2358. Available from https://doi.org/10.1021/jf960946l.

Jarnthong, M., Wang, F., Wang, X. Y., Wang, R., Li, J. H., et al. (2015). Preparation and properties of biocomposite based on natural rubber and bagasse nanocellulose. *MATEC Web of Conferences*, *26*, 1–4. Available from https://doi.org/10.1051/matecconf/20152601005.

Jonoobi, M., Harun, J., Shakeri, A., Misra, M., & Oksman, K. (2009). Chemical composition, crystallinity, and thermal degradation of bleached and unbleached kenaf bast (Hibiscus cannabinus) pulp and nanofibers. *BioResources*, *4*(2), 626–639. Available from https://doi.org/10.15376/biores.4.2.626-639.

Karlsson, R. M. P., Larsson, P. T., Pettersson, T., & Wågberg, L. (2020). Swelling of cellulose-based fibrillar and polymeric networks driven by ion-induced osmotic pressure. *Langmuir*, *36*(41), 12261–12271. Available from https://doi.org/10.1021/acs.langmuir.0c02051.

Kato, H., Nakatsubo, F., Abe, K., & Yano, H. (2015). Crosslinking via sulfur vulcanization of natural rubber and cellulose nanofibers incorporating unsaturated fatty acids. *RSC Advances*, *5*(38), 29814–29819. Available from https://doi.org/10.1039/c4ra14867c.

Kaushik, A., & Singh, M. (2011). Isolation and characterization of cellulose nanofibrils from wheat straw using steam explosion coupled with high shear homogenization. *Carbohydrate Research*, *346*(1), 76–85. Available from https://doi.org/10.1016/j.carres.2010.10.020.

Khalighi, S., Berger, R. G., & Ersoy, F. (2020). Cross-linking of wheat bran arabinoxylan by fungal laccases yields firm gels. *Processes*, *8(1)*(36). Available from https://doi.org/10.3390/pr8010036.

Khalil, H. P. S. A., et al. (2014). Production and modification of nanofibrillated cellulose using various mechanical processes: A review. *Carbohydrate Polymers*, *99*, 649–665. Available from https://doi.org/10.1016/j.carbpol.2013.08.069.

Kumar, V., et al. (2014). Comparison of nano- and microfibrillated cellulose films. *Cellulose*, *21*(5), 3443–3456. Available from https://doi.org/10.1007/s10570-014-0357-5.

Lauri, J., Koponen, A., Haavisto, S., & Czajkowski, J. (2017). Analysis of rheology and wall depletion of microfibrillated cellulose suspension using optical coherence tomography. *Cellulose*, *24*(11), 4715–4728. Available from https://doi.org/10.1007/s10570-017-1493-5.

Lavoine, N., Desloges, I., Dufresne, A., & Bras, J. (2012). Microfibrillated cellulose - Its barrier properties and applications in cellulosic materials: A review. *Carbohydrate Polymers*, *90*(2), 735–764. Available from https://doi.org/10.1016/j.carbpol.2012.05.026.

Lee, K. Y., Aitomäki, Y., Berglund, L. A., Oksman, K., & Bismarck, A. (2014). On the use of nanocellulose as reinforcement in polymer matrix composites. *Composites Science and Technology*, *105*, 15–27. Available from http://doi.org/10.1016/j.compscitech.2014.08.032.

Lems, E. M., Winklehner, S., Hansmann, C., Gindl-Altmutter, W., & Veigel, S. (2019). Reinforcing effect of poly(furfuryl alcohol) in cellulose-based porous materials. *Cellulose*, *26*(7), 4431–4444. Available from https://doi.org/10.1007/s10570-019-02348-6.

Lendvai, L., Karger-Kocsis, J., Kmetty, Á., & Drakopoulos, S. X. (2016). Production and characterization of microfibrillated cellulose-reinforced thermoplastic starch composites. *Journal of Applied Polymer Science*, *133*(2). Available from https://doi.org/10.1002/app.42971.

Lengowski, E. C., et al. (2020). Different degree of fibrillation: Strategy to reduce permeability in nanocellulose-starch films. *Cellulose*, *27*(18), 10855–10872. Available from https://doi.org/10.1007/s10570-020-03232-4.

Liu, M., Baum, A., Odermatt, J., Berger, J., Yu, L., Zeuner, B., ... Meyer, A. S. (2017). Oxidation of lignin in hemp fibers by laccase: Effects on mechanical properties of hemp fibers and unidirectional fiber/epoxy composites. *Composites Part A: Applied Science and Manufacturing*, *95*, 377–387. Available from https://doi.org/10.1016/j.compositesa.2017.01.026.

Liu, M., Meyer, A. S., Fernando, D., Silva, D. A. S., Geoffrey, D., & Thygesen, A. (2016). Effect of pectin and hemicellulose removal from hemp fibres on the mechanical properties of unidirectional hemp epoxy composites. *Composites Part A: Applied Science and Manufacturing*, *90*, 724–735. Available from https://doi.org/10.1016/j.compositesa.2016.08.037.

Li, X., Du, G., Wang, S., & Yu, G. (2014). Physical and mechanical characterization of fiber cell wall in castor (*Ricinus communis* L.) stalk. *Bioresources*, *9*(1), 1596–1605. Available from https://ojs.cnr.ncsu.edu/index.php/BioRes/article/view/4969.

Liu, C., Zhang, Y., Wang, S., Meng, Y., & Hosseinaei, O. (2014). Micromechanical properties of the interphase in cellulose nanofiber-reinforced phenol formaldehyde bondlines. *BioResources*, *9*(3), 5529–5541. Available from https://doi.org/10.15376/biores.9.3.5529-5541.

Liu, M., Silva, D. A. S., Fernando, D., Meyer, A. S., Madsen, B., Daniel, G., & Thygesen, A. (2016). Controlled retting of hemp fibres: Effect of hydrothermal pre-treatment and enzymatic retting on the mechanical properties of unidirectional hemp/epoxy composites. *Composites Part A: Applied Science and Manufacturing*, *88*, 253–262. Available from https://doi.org/10.1016/j.compositesa.2016.06.003.

Liu, Q., Lu, Y., Jacquet, N., Ouyang, C., He, W., Yan, W., ... Richel, A. (2017). Isolation of high-purity cellulose nanofibers from wheat straw through the combined environmentally friendly methods of steam explosion, microwave-assisted hydrolysis, and microfluidization. *ACS Sustainable Chemistry & Engineering*, *5*(7), 6183–6191. Available from https://doi.org/10.1021/acssuschemeng.7b01108.

Li, P., Wang, Y., Hou, Q., & Li, X. (2018). Isolation and characterization of microfibrillated cellulose from agro-industrial soybean residue (Okara). *BioResources*, *13*(4), 7944–7956. Available from https://doi.org/10.15376/biores.13.4.7944-7956.

Lucenius, J., Valle-Delgado, J. J., Parikka, K., & Österberg, M. (2019). Understanding hemicellulose-cellulose interactions in cellulose nanofibril-based composites. *Journal of Colloid and Interface Science*, *555*, 104–114. Available from https://doi.org/10.1016/j.jcis.2019.07.053.

Lu, J., Askeland, P., & Drzal, L. T. (2008). Surface modification of microfibrillated cellulose for epoxy composite applications. *Polymer*, *49*. Available from https://doi.org/10.1016/j.polymer.2008.01.028.

Martelli-Tosi, M., Torricillas, M. D. S., Martins, M. A., de Assis, O. B. G., & Tapia-Blácido, D. R. (2016). Using commercial enzymes to produce cellulose nanofibers from soybean straw. *Journal of Nanomaterials*, *2016*. Available from https://doi.org/10.1155/2016/8106814.

Meng, X., et al. (2018). Toughening of nanocelluose/PLA composites via bio-epoxy interaction: Mechanistic study. *Materials & Design*, *139*(2017), 188–197. Available from https://doi.org/10.1016/j.matdes.2017.11.012.

Mikkonen, K. S., et al. (2012). Arabinoxylan structure affects the reinforcement of films by microfibrillated cellulose. *Cellulose*, *19*, 467–480. Available from https://doi.org/10.1007/s10570-012-9655-y.

Minelli, M., et al. (2010). Investigation of mass transport properties of microfibrillated cellulose (MFC) films. *Journal of Membrane Science*, *358*(1–2), 67–75. Available from https://doi.org/10.1016/j.memsci.2010.04.030.

Missio, A. L., et al. (2018). Nanocellulose-tannin films: From trees to sustainable active packaging. *Journal of Cleaner Production*, *184*, 143–151. Available from https://doi.org/10.1016/j.jclepro.2018.02.205.

Mnich, E., et al. (2020). Phenolic cross-links: building and de-constructing the plant cell wall. *Natural Product Reports*, *37*, 919–961. Available from https://doi.org/10.1039/c9np00028c.

Neagu, R. C., Gamstedt, E. K., & Berthold, F. (2006). Stiffness contribution of various wood fibers to composite materials. *Journal of Composite Materials*, *40*(8), 663–699. Available from https://doi.org/10.1177/0021998305055276.

Nguyen, H. D., et al. (2013). A novel method for preparing microfibrillated cellulose from bamboo fibers. *Advances in Natural Sciences: Nanoscience and Nanotechnology*, *4*, 1–9. Available from https://doi.org/10.1088/2043-6262/4/1/015016.

Nissilä, T., Hietala, M., & Oksman, K. (2019). A method for preparing epoxy-cellulose nanofiber composites with an oriented structure. *Composites Part A Applied Science and Manufacturing*, *125*, 105515. Available from https://doi.org/10.1016/j.compositesa.2019.105515.

Nobuta, K., et al. (2016). Characterization of cellulose nanofiber sheets from different refining processes. *Cellulose*, *23*(1), 403–414. Available from https://doi.org/10.1007/s10570-015-0792-y.

Nordqvist, D., Idermark, J., Hedenqvist, M. S., Gällstedt, M., Ankerfors, M., & Lindström, T. (2007). Enhancement of the wet properties of transparent chitosan-acetic-acid-salt films using microfibrillated cellulose. *Biomacromolecules*, *8*(8), 2398–2403. Available from https://doi.org/10.1021/bm070246x.

Oinonen, P., Areskogh, D., & Henriksson, G. (2013). Enzyme catalyzed cross-linking of spruce galactoglucomannan improves its applicability in barrier films. *Carbohydrate Polymers*, *95*(2), 690–696. Available from https://doi.org/10.1016/j.carbpol.2013.03.016.

Operamolla, A. (2019). Recent advances on renewable and biodegradable cellulose nanopaper substrates for transparent light-harvesting devices: Interaction with humid environment. *International Journal of Photoenergy*, *2019*. Available from https://doi.org/10.1155/2019/3057929.

Ortiz, C. M., Salgado, P. R., Dufresne, A., & Mauri, A. N. (2018). Microfibrillated cellulose addition improved the physicochemical and bioactive properties of biodegradable films based on soy protein and clove essential oil. *Food Hydrocolloids*, *79*, 416–427. Available from https://doi.org/10.1016/j.foodhyd.2018.01.011.

Osong, S. H., Norgren, S., & Engstrand, P. (2016). Processing of wood-based microfibrillated cellulose and nanofibrillated cellulose, and applications relating to papermaking: A review. *Cellulose*, *23*(1), 93–123. Available from https://doi.org/10.1007/s10570-015-0798-5.

Österberg, M., et al. (2013). A fast method to produce strong NFC films as a platform for barrier and functional materials. *ACS Applied Materials & Interfaces*, *5*(11), 4640–4647. Available from https://doi.org/10.1021/am401046x.

Paës, G., von Schantz, L., & Ohlin, M. (2015). Bioinspired assemblies of plant cell wall polymers unravel the affinity properties of carbohydrate-binding modules. *Soft Matter*, *11*(33), 6586–6594. Available from https://doi.org/10.1039/c5sm01157d.

Palange, C., Johns, M. A., Scurr, D. J., Phipps, J. S., & Eichhorn, S. J. (2019). The effect of the dispersion of microfibrillated cellulose on the mechanical properties of melt-compounded polypropylene–polyethylene copolymer. *Cellulose*, *26*(18), 9645–9659. Available from https://doi.org/10.1007/s10570-019-02756-8.

Panthapulakkal, S., & Sain, M. (2012). Preparation and characterization of cellulose nanofibril films from wood fibre and their thermoplastic polycarbonate composites. *International Journal of Polymer Science*, *2012*. Available from https://doi.org/10.1155/2012/381342.

Pinkl, S., Veigel, S., Colson, J., & Gindl-altmutter, W. (2017). Nanopaper properties and adhesive performance of microfibrillated cellulose from different (Ligno-)cellulose raw materials. *Polymers (Basel).*, *9*, 3–5. Available from https://doi.org/10.3390/polym9080326.

Plackett, D. (2010). Microfibrillated cellulose and new nanocomposite materials: a review, pp. 459–494. Available from https://doi.org/10.1007/s10570–010–-9405-y.

Plackett, D., et al. (2010). Physical properties and morphology of films prepared from microfibrillated cellulose and microfibrillated cellulose in combination with amylopectin. *Journal of Applied Polymer Science*, *117*(6), 3601–3609. Available from https://doi.org/10.1002/app.32254.

Poothanari, M. A., Michaud, V., Damjanovic, D., & Leterrier, Y. (2021). Surface modified microfibrillated cellulose-poly(vinylidene fluoride) composites: Beta-phase formation, viscoelastic and dielectric performance. *Polymer International*, 0–2. Available from https://doi.org/10.1002/pi.6202.

Qiang, Y., Patel, A., & Manas-Zloczower, I. (2020). Enhancing microfibrillated cellulose reinforcing efficiency in epoxy composites by graphene oxide crosslinking. *Cellulose*, *27*(4), 2211–2224. Available from https://doi.org/10.1007/s10570-019-02916-w.

Rana, S. S., & Gupta, M. K. (2020). Isolation of nanocellulose from hemp (Cannabis sativa) fibers by chemo-mechanical method and its characterization. *Polymer Composites*, *41*, 5257–5268. Available from https://doi.org/10.1002/pc.25791.

Roach, M. J., et al. (2011). Development of cellulosic secondary walls in flax fibers requires beta-galactosidase. *Plant Physiology*, *156*(3), 1351–1363. Available from https://doi.org/10.1104/pp.111.172676.

Rodionova, G., Lenes, M., Eriksen, Ø., & Gregersen, Ø. (2011). Surface chemical modification of microfibrillated cellulose: Improvement of barrier properties for packaging applications. *Cellulose*, *18*, 127–134. Available from https://doi.org/10.1007/s10570-010-9474-y.

Rol, F., et al. (2017). Pilot-scale twin screw extrusion and chemical pretreatment as an energy-efficient method for the production of nanofibrillated cellulose at high solid content. *ACS Sustainable Chemistry & Engineering*, *5*(8), 6524–6531. Available from https://doi.org/10.1021/acssuschemeng.7b00630.

Rol, F., Billot, M., Bolloli, M., Beneventi, D., & Bras, J. (2020). Production of 100% cellulose nanofibril objects using the molded cellulose process: A feasibility study. *Industrial & Engineering Chemistry Research*, *59*(16), 7670–7679. Available from https://doi.org/10.1021/acs.iecr.9b06127.

Rol, F., Vergnes, B., El Kissi, N., & Bras, J. (2020). Nanocellulose production by twin-screw extrusion: Simulation of the screw profile to increase the productivity. *ACS Sustainable Chemistry & Engineering*, *8*(1), 50–59. Available from https://doi.org/10.1021/acssuschemeng.9b01913.

Saastamoinen, P., et al. (2012). Laccase aided modification of nanofibrillated cellulose with dodecyl gallate. *BioResources*, 7(4), 5749–5770. Available from https://doi.org/10.15376/biores.7.4.5749-5770.

Sae-Oui, P., Suchiva, K., Sirisinha, C., Intiya, W., Yodjun, P., & Thepsuwan, U. (2017). Effects of blend ratio and SBR type on properties of carbon black-filled and silica-filled SBR/BR tire tread compounds. *Advances in Materials Science and Engineering*, *2017*. Available from https://doi.org/10.1155/2017/2476101.

Saini, S., Belgacem, M. N., Missoum, K., & Bras, J. (2015). Natural active molecule chemical grafting on the surface of microfibrillated cellulose for fabrication of contact active antimicrobial surfaces. *Industrial Crops and Products*, *78*, 82–90. Available from https://doi.org/10.1016/j.indcrop.2015.10.022.

Salmén, L. (2004). Micromechanical understanding of the cell-wall structure. *Comptes Rendus Biologies*, *327*(9–10), 873–880. Available from https://doi.org/10.1016/j.crvi.2004.03.010.

Salmén, L., & Stevanic, J. S. (2018). Effect of drying conditions on cellulose microfibril aggregation and "hornification". *Cellulose*, *25*(11), 6333–6344. Available from https://doi.org/10.1007/s10570-018-2039-1.

Sattayanurak, S., Noordermeer, J. W. M., Sahakaro, K., Kaewsakul, W., Dierkes, W. K., & Blume, A. (2019). Silica-reinforced natural rubber: Synergistic effects by addition of small amounts of secondary fillers to silica-reinforced natural rubber tire tread compounds. *Advances in Materials Science and Engineering*, *2019*. Available from https://doi.org/10.1155/2019/5891051.

Schönbeck, M. (1921). Verfahren zur Herstellung einer bearbeitungsfähigen Masse aus organischen Rohstoffen. *Reichspatentamt Dtsch. Reich, Patentschrift Nr.*, *334494*, 1921.

Schultz-Jensen, N., et al. (2013). Pretreatment of the macroalgae *Chaetomorpha linum* for the production of bioethanol - Comparison of five pretreatment technologies. *Bioresource Technology*, *140*, 36–42. Available from https://doi.org/10.1016/j.biortech.2013.04.060.

Sehaqui, H., Liu, A., Zhou, Q., & Berglund, L. A. (2010). Fast preparation procedure for large, flat cellulose and cellulose/inorganic nanopaper structures. *Biomacromolecules*, *11*(9), 2195–2198. Available from https://doi.org/10.1021/bm100490s.

Sehaqui, H., Zimmermann, T., & Tingaut, P. (2014). Hydrophobic cellulose nanopaper through a mild esterification procedure. *Cellulose*, *21*, 367–382. Available from https://doi.org/10.1007/s10570-013-0110-5.

Shawky Sol, A., Sakran She, M., Ahmad, F., & Abdel-Atty, M. (2017). Evaluation of paper pulp and paper making characteristics produced from different African woody trees grown in Egypt. *Research Journal of Forestry*, *11*(2), 19–27. Available from https://doi.org/10.3923/rjf.2017.19.27.

Shen, Z., & Feng, J. (2018). Highly thermally conductive composite films based on nanofibrillated cellulose in situ coated with a small amount of silver nanoparticles. *ACS Applied Materials & Interfaces*, *10*, 24193–24200. Available from https://doi.org/10.1021/acsami.8b07249.

Silva, L. E., et al. (2021). Redispersion and structural change evaluation of dried microfibrillated cellulose. *Carbohydrate Polymers*, *252*(2020). Available from https://doi.org/10.1016/j.carbpol.2020.117165.

da Silva, T. A., Mocchiutti, P., Zanuttini, M. A., & Ramos, L. P. (2006). Chemical characterization of pulp components in unbleached softwood kraft fibers recycled with the assistance of a laccase/HBT system. *BioResources*, *2*(4), 616–629. Available from https://doi.org/10.15376/biores.2.4.616-629.

Singh, M., Kaushik, A., & Ahuja, D. (2016). Surface functionalization of nanofibrillated cellulose extracted from wheat straw: Effect of process parameters. *Carbohydrate Polymers*, *150*, 48–56. Available from https://doi.org/10.1016/j.carbpol.2016.04.109.

Siró, I., & Plackett, D. (2010). Microfibrillated cellulose and new nanocomposite materials: A review. *Cellulose*, *17*, 459–494. Available from https://doi.org/10.1007/s10570-010-9405-y.

Spence, K. L., Venditti, R. A., Rojas, O. J., Habibi, Y., & Pawlak, J. J. (2010). The effect of chemical composition on microfibrillar cellulose films from wood pulps: Water interactions and physical properties for packaging applications. *Cellulose*, *17*(4), 835–848. Available from https://doi.org/10.1007/s10570-010-9424-8.

Spence, K. L., Venditti, R. A., Rojas, O. J., Habibi, Y., & Pawlak, J. J. (2011). A comparative study of energy consumption and physical properties of microfibrillated cellulose produced by different processing methods. *Cellulose*, *18*, 1097–1111. Available from https://doi.org/10.1007/s10570-011-9533-z.

Squinca, P., Bilatto, S., Badino, A. C., & Farinas, C. S. (2020). Nanocellulose production in future biorefineries: An integrated approach using tailor-made enzymes. *ACS Sustainable Chemistry & Engineering*, *8*(5), 2277–2286. Available from https://doi.org/10.1021/acssuschemeng.9b06790.

Stenstad, P., Andresen, M., Tanem, B. S., & Stenius, P. (2008). Chemical surface modifications of microfibrillated cellulose. *Cellulose*, *15*, 35–45. Available from https://doi.org/10.1007/s10570-007-9143-y.

Stevanic, J. S., Bergström, E. M., Gatenholm, P., Berglund, L., & Salmén, L. (2012). Arabinoxylan/nanofibrillated cellulose composite films. *Journal of Materials Science*, *47*(18), 6724–6732. Available from https://doi.org/10.1007/s10853-012-6615-8.

Supanakorn, G., Varatkowpairote, N., Taokaew, S., & Phisalaphong, M. (2021). Alginate as dispersing agent for compounding natural rubber with high loading microfibrillated cellulose. *Polymers (Basel).*, *13*(3), 1–11. Available from https://doi.org/10.3390/polym13030468.

Svoboda, M. A., Lang, R. W., Bramsteidl, R., Ernegg, M., & Stadlbauer, W. (2000). Zelfo—An engineering material fully based on renewable resources. *Molecular Crystals and Liquid Crystals Science and Technology. Section A. Molecular Crystals and Liquid Crystals*, *353*, 47–58. Available from https://doi.org/10.1080/10587250008025647.

Tian, X., et al. (2017). Enzyme-assisted mechanical production of microfibrillated cellulose from Northern bleached softwood kraft pulp. *Cellulose*, *24*(9), 3929–3942. Available from https://doi.org/10.1007/s10570-017-1382-y.

Tonoli, G. H. D., et al. (2012). Cellulose micro nanofibres from Eucalyptus kraft pulp: Preparation and properties. *Carbohydrate Polymers*, *89*(1), 80–88. Available from http://doi.org/10.1016/j.carbpol.2012.02.052.

Tonoli, G. H. D., et al. (2019). Cellulose sheets made from micro/nanofibrillated fibers of bamboo, jute and eucalyptus cellulose pulps. *Cellulose Chemistry and Technology*, *53*(3–4), 291–305. Available from https://doi.org/10.35812/CelluloseChemTechnol.2019.53.29.

Tozluoglu, A., Poyraz, B., & Candan, Z. (2018). Examining the efficiency of mechanic/enzymatic pretreatments in micro/nanofibrillated cellulose production. *Maderas: Ciencia y Tecnologia*, *20*(1), 67–84. Available from https://doi.org/10.4067/S0718-221X2018005001601.

VTT and Aalto University Foundation, *Method for the preparation of NFC films on supports*, WO 2013/060934 A2, 2013.

Vu, C. M., Sinh, L. H., Choi, H. J., & Pham, T. D. (2017). Effect of micro/nano white bamboo fibrils on physical characteristics of epoxy resin reinforced composites. *Cellulose*, *24*(12), 5475–5486. Available from https://doi.org/10.1007/s10570-017-1503-7.

Weise, U. (1998). Hornification - Mechanisms and terminology. *Paperi ja Puu/Paper Timber*, *80*(2), 110–115.

Yan, Y., et al. (2016). Microfibrillated lignocellulose enables the suspension-polymerisation of unsaturated polyester resin for novel composite applications. *Polymers*, *8*(7), 1–11. Available from https://doi.org/10.3390/polym8070255.

Yano, H., & Nakahara, S. (2004). Bio-composites produced from plant microfiber bundles with a nanometer unit web-like network. *Journal of Materials Science*, *39*(5), 1635–1638. Available from https://doi.org/10.1023/B:JMSC.0000016162.43897.0a.

Yan, C., et al. (2014). Highly stretchable piezoresistive graphene–nanocellulose nanopaper for strain sensors. *Advanced Materials*, *26*(13), 2022–2027. Available from https://doi.org/10.1002/adma.201304742.

Yousefi, H., Azad, S., Mashkour, M., & Khazaeian, A. (2018). Cellulose nanofiber board. *Carbohydrate Polymers*, *187*, 133–139. Available from https://doi.org/10.1016/j.carbpol.2018.01.081.

Zimmermann, T., Pöhler, E., & Geiger, T. (2004). Cellulose fibrils for polymer reinforcement. *Advanced Engineering Materials*, *6*(9), 754–761. Available from https://doi.org/10.1002/adem.200400097.

Zuluaga, R., Putaux, J.-L., Restrepo, A., Mondragon, I., & Gañán, P. (2007). Cellulose microfibrils from banana farming residues: isolation and characterization. *Cellulose*, *14*, 585–592. Available from https://doi.org/10.1007/s10570-007-9118-z.

CHAPTER 11

Nanotechnology: application and potentials for heterogeneous catalysis

Yaser Dahman, Nishanth Ignatius, Anthony Poblete, Aleksa Krunic, Peter Ma, Nishil Gosalia, Tayyub Ali and Yaser Dahman
Department of Chemical Engineering, Ryerson University, Toronto, ON, Canada

11.1 Introduction

Nanotechnology has tremendous potential, it has the capability to change the face of the medical industry, create ultra-small electronic devices, lightning-fast telecommunications, and change the way of life forever. Nanotechnology development is aimed at being woven in every aspect of product development due to the vast amount of advantages it possesses. Within industrial process, nanotechnology is helping to increase the efficiency and cost effectiveness of integral processes. This is predominant when analyzing process that incorporate a catalyst to facilitate reactions.

Nanotechnology can be defined as the "engineering with atomic precision" (Ramsden, 2011). Nanotechnology is interested in the development, engineering, and fabrication of technology at the nanoscale. The nanoscale range is between 1 and 100 nm, as such, nanoscale materials cannot be seed with the naked eye. This makes nanotechnology a challenging field to master, with major hurdles that need to be overtaken before being commercialized successfully. Nevertheless, nanotechnology is an exciting new field that can influence the world in a big way.

This study will focus on the recent developments of nanotechnology in heterogeneous catalysis reactions. Catalysis is the increased rate of a chemical reaction due to the presence of a catalyst. A catalyst has the ability to increase the rate of reaction without itself being consumed or physically changed. More specifically, heterogeneous catalysis is when the reactants and the catalysts differ in phase. Typically, the reactants are in liquid or gaseous phase while the catalyst itself is solid. Catalysts are used in 90% of chemical processes worldwide. Some examples of where catalysts are used include the production of ammonia, catalytic cracking of gas oil, reforming of naphtha, and production of sulfuric acid.

Current research is focused on integrating nanotechnology with catalysts. This is mainly done through the utilization of nanoparticles spread on the surface of the catalyst and supported on nanosupports. The motivation for development of this process is that it can enable

Nanotechnology in Paper and Wood Engineering
DOI: https://doi.org/10.1016/B978-0-323-85835-9.00013-1

more cost-effective, environmentally friendly and efficient processes. Nanocatalysts can improve catalytic properties while reducing the amount of catalyst material that is needed. This is beneficial as not only does nanotechnology allows more profitability but can also increase efficiency. When breaking particles down to nanoscale, the properties change as compared to the bulk substance. This enables researchers to investigate alternative catalysts elements that could be more economically viable then its counterpart. This study will focus on recent developments when it comes to nanotechnology and catalysis. Furthermore, relevant applications and how nanotechnology has helped to secure a vivid future will be discussed.

11.2 Dehalogenation and hydrogenation reactions

Many of the world's largest industries such as food, petrochemical and pharmaceutical depend on large scale processes and chemical reactions. Dehalogenation and hydrogenation are only two of the many important reactions that are used to manufacture some of the world's most popular products like vegetable oil while aiding tremendously in the production of fuels.

Hydrogenation is a chemical reaction of adding hydrogen atoms to carbon—carbon multiple bonds (Carlson, 2004). Additionally, for hydrogenation to occur at a practical rate a catalyst must be present. Typically, the goal of hydrogenation is to break any existing carbon—carbon double or triple bonds by saturating the compound with carbon and hydrogen. Common catalysts consist of a heavy metal in finely divided form. Aside from its use in the production of fats and oils and gasoline, biological hydrogenation also occurs in the human body. In this instance the reaction is much more complex as hydrogen is not found in the body and instead requires a carrier molecule while an enzyme acts as the catalyst.

Dehalogenation, often referred to as chemical dehalogenation is described as a chemical process to remove halogens (usually chlorine) from a chemical contaminant rendering it less hazardous (Environmental Protection Agency, 1996). Dehalogenation can be applied to common halogenated contaminants including polychlorinated biphenyls, pesticides, and compounds found in swimming pool chemicals and textile production. The two most common versions of chemical dehalogenation used in processes are glycolate dehalogenation and base-catalyzed decomposition process. The former involves the use of the chemical reagent APEG which consists of an alkali metal hydroxide and a polyethylene glycol, while the latter process utilizes a sodium bicarbonate and mixes it with the contaminated substance (EPA, 1996).

A common theme in both dehalogenation and hydrogenation is the presence of a third party. In the case of dehalogenation (Fig. 11.1) this is the APEG or sodium

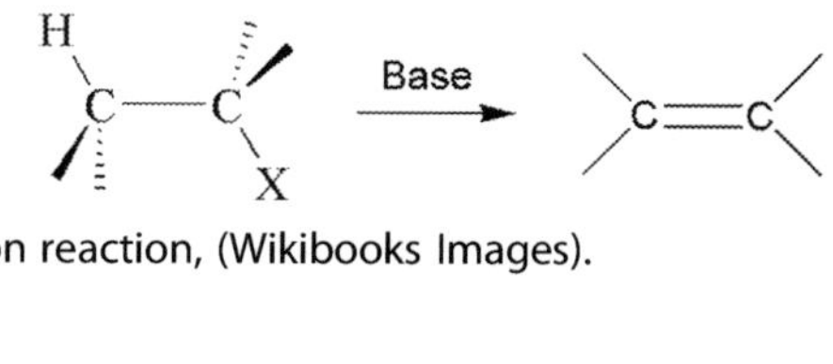

H
C—C
X
Base
C=C

Figure 11.1 Dehalogenation reaction, (Wikibooks Images).

bicarbonate and in hydrogenation the heavy metal catalyst or enzyme acts as the third party. Nanotechnology has the potential to play a large role in the growth and applications of these processes because of the impact that the nanoscale has on catalysts, more specifically in this case, heterogeneous catalysts.

11.2.1 Catalytic application of biogenic platinum nanoparticles for hydrogenation of cinnamaldehyde to cinnamyl alcohol

In a study performed by Zheng et al. biogenic platinum nanoparticles (PtNPs) were used in the hydrogenation of cinnamaldehyde to cinnamyl alcohol. PtNPs have been demonstrated to exhibit high catalytic activity, owing to their high surface area and consequently, well-defined PtNPs of particular shapes and sizes have been synthesized through various methods (Chen & Holt-Hindle, 2010). Currently cinnamyl alcohol is produced via two methods: one being the saponification storax oil and cinnamon oil among others using hot NaOH solution and the other being the reduction of cinnamaldehyde by benzyl alcohol aluminum or aluminum isopropoxide. The aforementioned methods are both environmentally unfriendly and costly, producing cinnamyl alcohol through the hydrogenation of cinnamaldehyde, however, overcomes these disadvantages (Zhang, Liao, Xu, & Yu, 2000). An efficient catalyst for hydrogenation is crucial and heterogeneous catalysts have been of great interest because of their separability and recyclability (Merlo, Machado, Vetere, Faria, & Casella, 2010). A platinum supported by titania (Pt/TiO_2) catalyst has been known to exhibit excellent catalytic activity due to the strong interaction between Pt and TiO_2 with high reducibility (Colmenares et al., 2011). The study by Zheng et al. utilized PtNPs which were biosynthesized by reducing aqueous Na_2PtCl_4 with Cacumen Pladycladi extract as the reducing agent (Fig. 11.2).

The use of biosynthesized PtNPs as opposed to a standard Pt loading on pure TiO_2 showed a 600% increase in selectivity to cinnamyl alcohol. Increasing Pt loading from 0.5% to 1%, increased the conversion of cinnamaldehyde more than triple and the rates of conversion and selectivity increased as well. Increasing further to 2% Pt leads to a slight increase in conversion to cinnamaldehyde but a decrease in selectivity. Cobalt doping also had an effect on the conversion of the reaction. As the Pt/TiO_2 nanocatalyst was doped at low concentrations (0.5%−1.5%) selectivity toward cinnamyl alcohol as well as conversion of cinnamaldehyde increased. Similar to the increase in Pt to 2%, an increase in doping to 2% resulted in a decrease in conversion.

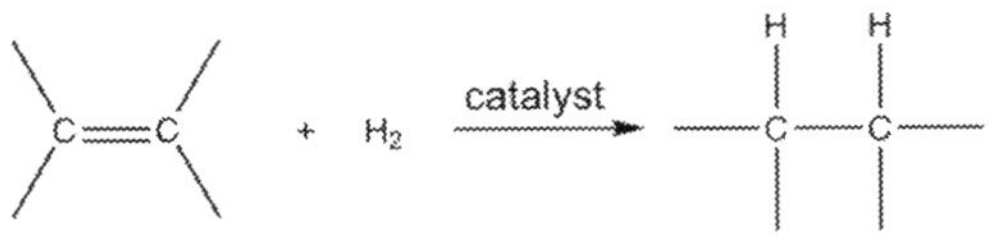

$$\mathrm{C{=}C} + \mathrm{H_2} \xrightarrow{\text{catalyst}} \mathrm{HC{-}CH}$$

Figure 11.2 Hydrogenation reaction, alkene to alkane (Wikibooks Images).

Durability tests on the catalyst showed that following three cycles conversion remained at a very high rate with selectivity toward cinnamyl alcohol remaining unchanged.

The results obtained through the study demonstrated that at appropriate concentrations the use of biogenic PtNPs supported on TiO_2 could be an efficient catalyst in the hydrogenation reaction. Similarly, appropriate levels of cobalt doping showed that the catalyst can be even more efficient. Finally, the durability of the nano-based catalyst was apparent in the way conversion held up after three cycles, and had minimal decrease following the fifth cycle owing to PtNPs potential as a catalyst in the catalytic hydrogenation of cinnamaldehyde.

11.2.2 Excellent catalytic properties over nanocomposite catalysts for selective hydrogenation of halnitrobenzenes

The synthesis of organic dyes, perfumes, herbicides, pesticides, and medicines requires aromatic haloamines as an organic intermediate. Production of these amines is accomplished primarily by the selective hydrogenation of the appropriate halonitro compound over metal catalysts. An issue however has been the hydrodehalogenation of the haloamine product over most of the metal catalysts used (Wang, Biradar, Duncan, & Asefa, 2010). Studies conducted by Liang et al., have shown the use of nanocomposite catalysts to improve catalytic properties in the hydrogenation of halonitrobenzenes (Fig. 11.3).

Through the use of a partially reduced Pt/γ-Fe_2O_3 nanocomposite catalyst for selective dehydrogenation 2,4-dinitrochlorobenzene and iodonoitrobenzene, the

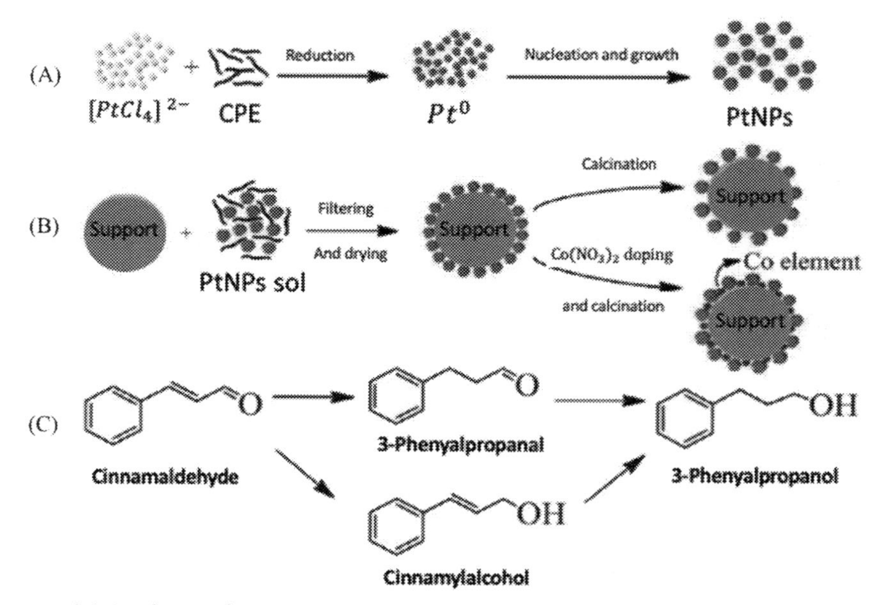

Figure 11.3 (A) Synthesis of PtNPs. (B) Preparation of platinum and cobalt–platinum catalysts. (C) Hydrogenation of cinnamaldehyde. *CPE*, Cacumen Pladycladi extract; *PtNPs*, platinum nanoparticles.

reactions of interest were accomplished with high rates and superior selectivity versus chlorophenylenediamines (CPDA) and iodoanilines (IAN) of which both are products of dehalogenation. Over the nanocomposite catalyst of use, the hydrodehalogenation of CPDA and IAN was fully inhibited for the first time (Liang, Wang, Wang, Liu, & Liu, 2008). It was also found that for the first time cobalt could hardly be chemisorbed on the Pt nanoparticles deposited on the partially reduced γ-Fe_2O_3 nanoparticles; this discovery implied a weak tendency of back donation of the d-electron from the nanoparticles to the π^* antibonding orbitals of the adsorbed molecules (Liang, Wang, Wang, Liu, & Liu, 2008). The weak tendency of back donation plays a major role in suppressing the hydrodehalogenation of haloanilines in the hydrogenation of halonitrobenzenes over the nanocomposite catalyst.

11.2.3 An efficient and reusable heterogeneous catalyst for dehalogenation reaction

The widespread applications of aromatic and aliphatic halogenated compounds have raised environmental and health issues with regards to their proper disposal. Past methods of disposal included ignition however this resulted in a large amount of greenhouse gases. A study by Patra, Dutta, and Bhaumik (2012), used an iron containing, highly ordered, mesoporous material with high surface area, good pore wall stability, and good mechanical, thermal, and chemical stability in a dehalogenation reaction with the intent of finding a more economical and environmentally friendly method to deal with the disposal of halogenated compounds. Patra et al. (2012) were able to develop a new highly ordered mesoporous $TiO_2-Fe_2O_3$ mixed oxide as a heterogeneous and reusable catalyst for the dehalogenation of various aryl halides under mild conditions.

Transmission electron microscopic (TEM) images of the mixed oxide sample showed that pores are present throughout the specimen, arranged in a honeycomb-like hexagonal array. The mesopores were regularly ordered with a cavity diameter of approximately 3.1 nm (Patra et al., 2012). Further investigation revealed that the sample was composed of nanoparticles approximately 40 nm in size which are self-assembled in order to form large aggregated particles with dimensions in the range of 120–200 nm with Fe, Ti, and O atoms being distributed uniformly on the material. The porous framework of the $TiO_2-Fe_2O_3$ mixed oxide allowed very good catalytic activity to be exhibited because it facilitates interaction between the active site of the catalyst and the reactant molecule. This is highlighted even more so when contrasted with the same reaction performed using a homogeneous catalyst ($FeCl_3$) which showed very poor conversion due to only the surface Fe taking part in the reaction (Patra et al., 2012) (Fig. 11.4).

Following the successful synthesis of $TiO_2-Fe_2O_3$ mixed oxide via hydrothermal sol–gel synthetic approach, a hexagonally ordered porous structure was obtained. The nanoparticles that composed the catalyst allowed for high surface area, good pre wall stability, and the presence of reactive Fe metal sites to activate the carbon–halogen

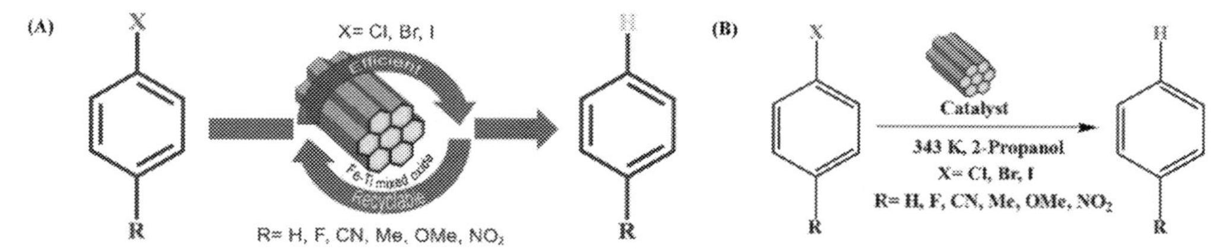

Figure 11.4 (A) Mesoporous catalyst in dehalogenation. (B) Dehalogenation using ordered iron–titanium mixed-oxide materials.

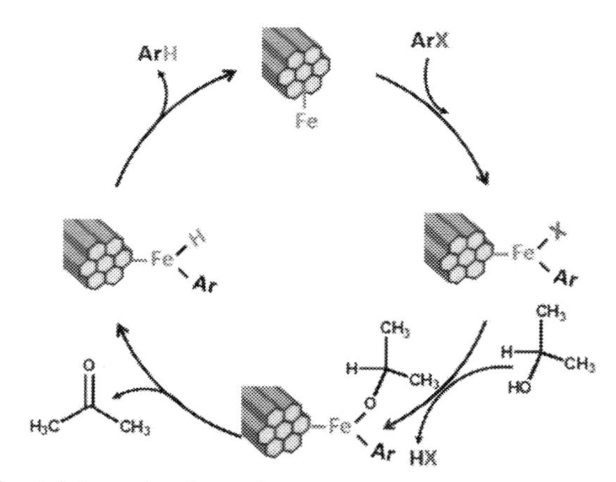

Figure 11.5 Proposed catalytic cycle of reaction.

bond cleavage. Excellent catalytic activity was shown when using the mixed-oxide material in the dehalogenation of aryl iodides, bromides, and chlorides tolerating -F, -CN, $-CH_3$, $-OCH_3$, and $-NO_2$ groups in the presence of 2-propanol as solvent (Patra et al., 2012). The heterogeneous nature of the mesoporous catalyst allowed it to remain in a separate solid phase in the reaction mixture. A benefit of this was the easy recovery of the catalyst through simple filtration and furthermore the catalyst was reused efficiently (Fig. 11.5).

With regard to the environmental issues, the use of mesoporous mixed-oxide catalysts can significantly improve environmental cleanup by simply converting the hazardous halogens into much less hazardous compounds while retaining the catalyst for reuse and thus waste reduction.

11.2.4 Looking to the future

The future of nanotechnology in catalysis is focused on five areas: active sites for metals, active sites for oxides, environment around active sites, reaction engineering, and catalysis for nanotechnology (Kung & Kung, 2004). Reaction engineering is relevant to hydrogenation and dehalogenation discussed above, because it can be possible to design catalyst systems where a movable segment is present, which is controlled by a hinge to bring up a blocking group or retrieve it from an active site, thereby turning the catalyst off and on (Kung & Kung, 2004). This mechanism would allow hydrogenation and dehalogenation reactions to be controlled to a higher degree potentially allowing for improved efficiency and selectivity amongst other benefits. Challenges remain in nanotechnology and catalysts such as the need to control active sites and binding sites in atoms; however, the ultimate goal continues to focus on the design of custom catalysts to achieve perfect selectivity in a catalytic reaction such as hydrogenation or dehalogenation.

11.3 Hydrosilylation reactions

Hydrosilylation, also referred to as hydrosilylation is an addition reaction between organic and inorganic silicon hydrides across unsaturated bonds (carbon—carbon double or triple bonds) as shown in Fig. 11.6. The most common bonds that are looked at are C—C, C—O, and C—N. This particular reaction was first observed in 1947 by Leo Sommer and it involved the reaction of trichlorosilane and 1-octene while acetyl peroxide was present.

Now what makes these catalysts so desirable in heterogeneous catalysis is that they not only increase the efficiency and turnover rates but also help to improve the regioselectivity. That is the preference of a certain product in favor of all other possible results as shown in Fig. 11.7, HCl reacts with propene and gives two possibilities (1-chloropropane and 2-chloropropane), where in this case 1-chloropropane is the major product.

As a result, this reaction has become important in the preparation of organosilicon compounds found in laboratory settings and in the industry (Noll, 1968). In this section, various catalysts that are used as well as what is produced by the reaction, its applications in nanotechnology, along with a better understanding of why it is superior

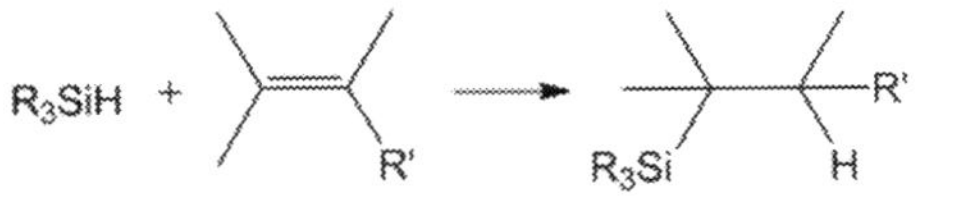

Figure 11.6 Hydrosilylation is the addition of an H—Si bond across a double bond.

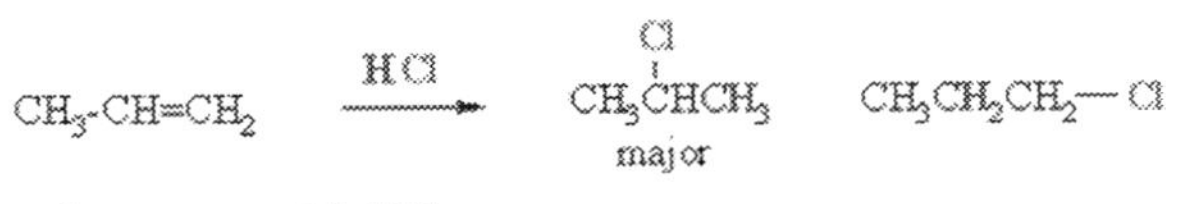

Figure 11.7 Reaction of propene with HCl.

to other methods, and finally possible obstacles that may be encountered in its use will be discussed.

11.3.1 Advancement over the years: platinum-based catalysts

As aforementioned organosilicon compounds are of major importance because they are the building blocks of all silicones and sol–gel silicates (Rappoport & Apeloig, 2001). By performing the hydrosilylation reaction of alkynes, vinylsilanes (copolymer plastics) are produced and are found to be nontoxic, very stable reagents, and adapt to various conditions. Over the last 60–70 years the use of platinum-based catalysts has been studied and its use is still being reviewed to this date.

Chloroplatinic acid (H_2PtCl_6) was used as a selective and very efficient alkene in hydrosilation catalysis dating back to 1957 (Speier, Webster, & Barnes, 1957). It allowed for high conversion of hydrosilylated products in a short period of time and also created very little side products. With ideal conditions, low amounts of catalyst were needed to achieve a full reaction, and as a result platinum-based hydrosilation immediately became an important reaction in the preparation of organosilicon materials for academic and industrial uses.

A report by Karstedt (1973) explained the significance of a compound that was soluble in silicone oil (coined the Karstedt catalyst shown in Fig. 11.8), this developed into one of the most widely used catalysts in the synthesis of organosilicon compounds. Although this process was done using homogeneous catalysis its study allowed for the development of a more desirable heterogeneous approach.

11.3.2 Recent breakthroughs in platinum catalysts

In the previous section there was mention of homogeneous catalysts being used and the reason for this was that they also showed high reaction activity and solubility in silicon products, making them industrially useful and important. However, the downfall was that separating and recycling of these homogeneous catalysts from certain solutions after the reaction was complete were almost impossible. This meant that the catalysts could only be used once and with platinum being an expensive species, not being able to recover and reuse it, made its use unappealing. Trace amounts of platinum could

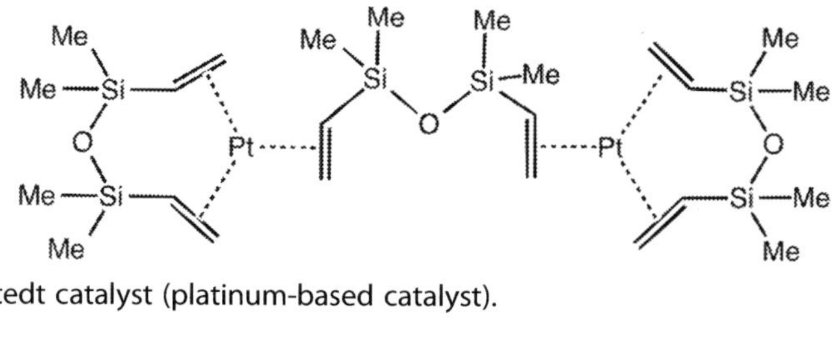

Figure 11.8 Karstedt catalyst (platinum-based catalyst).

also be found in the end product and for many applications this would be deemed unacceptable.

Therefore in the past 10 years, much effort has been put into finding a more efficient and recyclable solid catalyst alternative. PtNPs were heterogenized in an attempt to solve this problem of efficiency and recyclability, but most of these attempts proved to be ineffective, as catalytic activity were reported to be significantly less than using the method of homogeneous catalysts. The first survey of catalysts for hydrosilation in a heterogeneous environment was published by Fiedorow and Wawrzynczak (2006). This was then followed by Ying, Jiajian, Jiayun, Guoqiao, and Xiaonian (2011) explaining the various methods on how to prepare heterogeneous catalysts for hydrosilation. Many new supported platinum-based catalysts were found that could be utilized in the hydrosilation of alkynes, as a result of these reports.

Although few solid catalysts are found to be recyclable, there is still ongoing research being conducted to find a way to tackle this issue to improve stability and selectivity. Another problem that must be confronted is the possibility of the catalysts creating undesirable metal particle sintering. The goal is to eventually develop catalytic materials that are better suited for practical, every-day applications by being able to overcome these obstacles.

11.3.3 Heterogeneous versus homogeneous catalysts in hydrosilylation: nanotechnology applications

The switch to heterogeneous catalysts from homogeneous catalysts for use in hydrosilation has gained much momentum in recent years due to the fact that they are found to be a less expensive and safer alternative. Generally (with heterogeneous catalysts), a lot less amount of catalyst is required to achieve a full reaction and thus this reduces economic and environmental costs. Silicon, a product of hydrosilation reactions, is used for many applications, one being in silicone breast implants. Having even the slightest amounts of platinum leaking from these implants can potential lead to contamination of the nearby pectoral tissues, posing an immediate threat to human health (Arepalli, Bezabeh, & Brown, 2002). Again, using heterogeneous catalysts combined with nanochemistry may help to avoid this problem of leaching and also increase recyclability of the catalyst. Using solid nanocatalysts with entrapped platinum also allows for a highly controllable reaction and a much higher product yield.

11.3.4 Platinum-supported nanoparticles

An example of one of the first heterogeneous platinum-catalyzed hydrosilation alkyne reactions was reported by (Alonso et al., 2011). This involved the use of Pt/TiO_2 catalyst to undergo the reaction shown in Fig. 11.9.

The catalyst was then recovered by filtration and reused, this was repeated but eventually catalytic activity decreased substantially and results were deemed to be unsatisfactory.

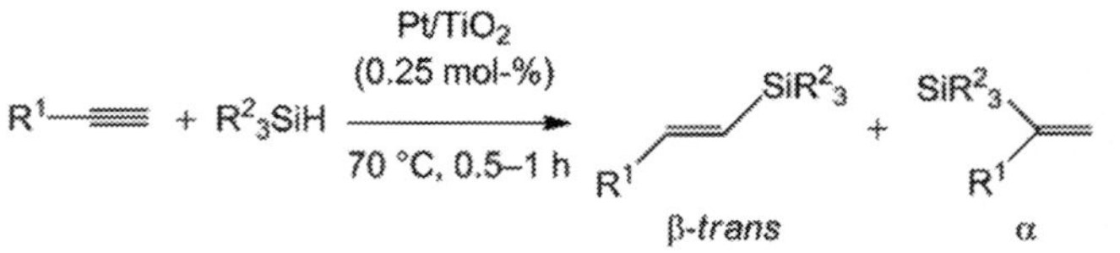

Pt/TiO_2 (0.25 mol-%)
R^1— + R^2_3SiH → 70 °C, 0.5–1 h
SiR^2_3 + SiR^2_3
R^1 R^1
β-*trans* α

Figure 11.9 Platinum-catalyzed hydrosilation of alkynes using platinum-supported by titania (Pt/TiO_2) catalyst.

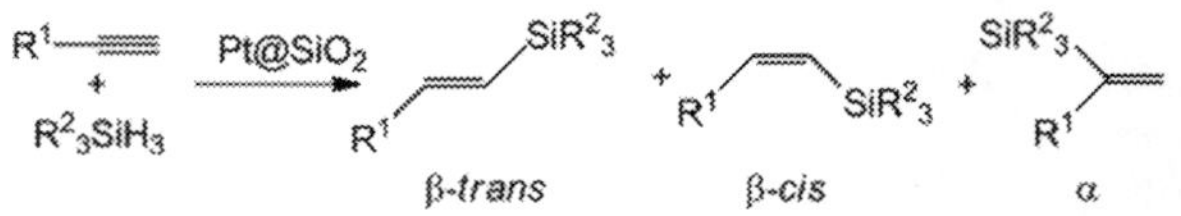

R^1— + $R^2_3SiH_3$ → $Pt@SiO_2$
SiR^2_3 + SiR^2_3 + SiR^2_3
R^1 R^1 R^1
β-*trans* β-*cis* α
at-

Figure 11.10 Hydrosilation of alk-1-enes with silanes using sol—gel platinum catalysts ($Pt@SiO_2$).

A study by Jiménez, Martínez-Rosales, & Cervantes (2003) showed the hydrosilation of alk-1-enes with silanes brought about by sol—gel entrapped platinum catalysts ($Pt@SiO_2$), this is shown in Fig. 11.10.

Again, the yields were found to be unsatisfactory in comparison to homogeneous catalysis, as well as having poor selectivity. Also, after each reaction the medium lost catalytic activity and therefore required repeated activation, this in turn meant loss of catalyst in each step.

However, in the most recent study by Zhang, Wang, Cheng, Li, and Zhang (2012) the use of Pt^0 nanoparticles along with meth-branched polymer (HTD-2) allowed for increased catalytic activity similar to traditional homogeneous reactions by use of Karstedt catalysts. In the hydrosilation of styrene with 1,1,3,3-tetramethyldisiloxane the polymer and Pt^0 catalysts achieved remarkably high conversions in comparison to Speier and Karstedt catalysts used homogeneously as shown in Fig. 11.11.

11.3.5 Leach-proof and sinter-proof catalysts

It is evident that supported metal nanoparticles (SMNPs) are becoming very important to the field of organic chemistry due to their use as catalysts for a wide range of chemical reactions (Astruc, Lu, & Ruiz Aranzaes, 2005). The atomic structures of naked nanoparticles contain many unsaturated sites that allow for the adsorbing of reactants and catalyzes their conversion. However, from current literature reviews the problem that many heterogeneous catalysts (acting as reservoirs for SMNPs) faces is the leaching into solutions in order to catalyze reactions. This makes it undesirable for industrial uses.

In terms of sinter-proofing catalysts, work done by (Jia & Schüth, 2011) using metal nanoparticles with tuned shape, size, and composition that are then embedded by porous support shells has shown promising results. Since sintering is the result of metal particles moving on support surfaces, keeping the nanoparticles in the inner parts

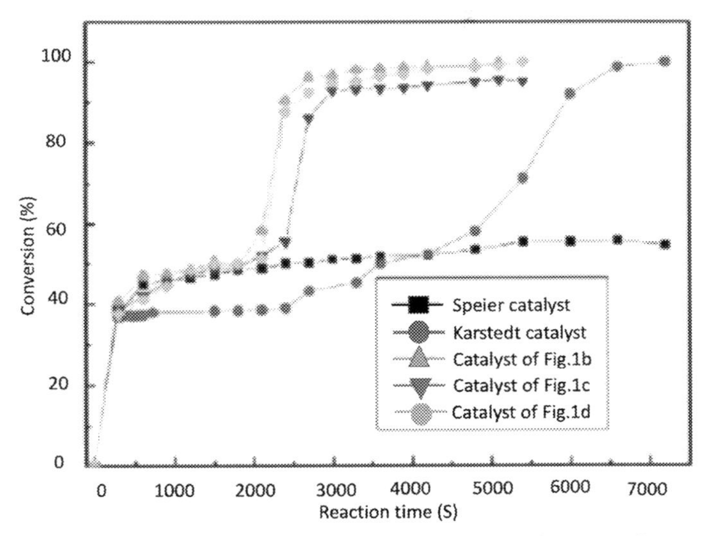

Figure 11.11 Comparison of conversion (%) vs Reaction time (S). Where, catalysts at different mass ratio of HTD-2, THF, and Pt: (Fig1b) HTD-2:THF:Pt = 1: 4: 0.17; (Fig1c) HTD-2:THF:Pt = 1: 4.5: 0.17; (Fig1d) HTD-2:THF:Pt = 1: 4: 0.15.

of porous solids would allow for as little agglomeration as possible and allows the catalyst to be recycled.

Continued work to improve these two characteristics in heterogeneous catalysts will allow the industry to move forward and replace the current use of homogeneous processes.

11.3.6 A look into the future of heterogeneous catalysts in hydrosilylation

The continued work in nanochemistry has allowed for examples of the first leach-proof and recyclable heterogeneous catalysts in hydrosilation reactions. Not only does this provide a safer and cheaper option to traditional hydrosilation methods, but also opens up a pathway to more selective hydrosilation processes and products free of contaminants (leach proof). Solid nanocatalysts based on PtNPs will have a continued use in the silicon industry to allow for controllable reaction processes and higher product yields (Geisberger, Baumann, & Daniels, 2006). It is now time to replace the homogeneous hydrosilation with heterogeneous processes based on the fast-growing field of nanotechnology.

11.4 C–C coupling reactions

C–C coupling reaction stands for carbon–carbon coupling reaction, and generally refers to reactions in which two fragments of hydrocarbon are coupled in presence of a catalyst (usually metal). An example of this type of reaction would be when the reaction between an organometallic compound of the type R—M (where R is the organic fragment and M is the main group center) with an organic halide of the type R'-X, which forms the product R—R' with a new carbon—carbon bond (Crabtree, 2001). The two main types of coupling reactions known are heterocoupling and homocoupling. Heterocoupling is done with two different partners, that is, of alkene with an alkyl halide to give an alkene that is substituted. Homocoupling reactions couple two partners that are identical, that is, the coupling of two acetylideo which results in the formation of a dialkyne. This type of reaction is known as Glaser coupling (Hartwig, 2010).

The process of C—C coupling usually start off by the oxidative addition of an organic halide to a catalyst. This is then followed by the second partner undergoing transmetalation. Transmetalation is usually referred to organometallic reactions which involve ligands transfer from one metal to another metal. Transmetalation here places both coupling partners on one metal center, and gets rid of (eliminates) the functional groups. The next and last step in this process is reductive elimination. The two coupling fragments undergo reductive elimination and regenerate the catalyst along with giving the organic product. Due to the fact that unsaturated organic groups add readily, they are known to couple more easily as opposed to others.

Operating conditions for C—C coupling reactions require the use of reagents that are vulnerable to oxygen and water. It is however not acceptable to assume that all coupling reactions require rejection of water. Coupling reactions using palladium as catalyst can be performed in aqueous solutions by making use sulfonated phosphines (water soluble).

A general mechanism of the reaction taking place in C—C coupling is shown in Fig. 11.12.

The most common types of C—C coupling reactions (Fig. 11.13) include but are not limited to the following:

1. Negishi cross coupling, occurs between organohalide and organozinc compounds.
2. Suzuki cross coupling, occurs between aryle halides and boronic acids.
3. Stille cross coupling, occurs between organohalides and organotin compounds.
4. Heck cross coupling, occurs between alkenes and aryle halides.
5. Sonogashira cross coupling, occurs between aryl halides and alkynes.
6. Wurtz cross coupling, occurs between two alkyl halides and sodium.

11.4.1 Catalysts

Catalysts are often defined as substances that alter the rate of a reaction without themselves undergoing any permanent chemical change. They play a vital role in carbon—carbon coupling reactions. Various substances including copper, cobalt, platinum, and iron may be used depending on the type of reaction and the desired product. The most common and the most effective catalyst, however, is palladium. Palladium, a rare, expensive, and a

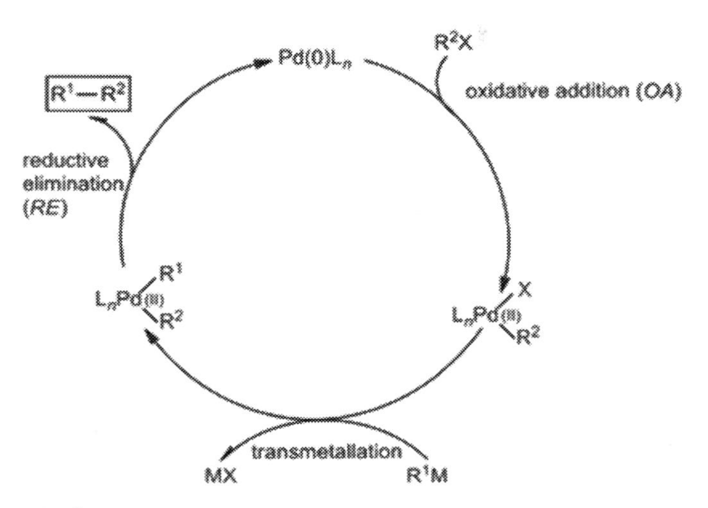

Pd(0)Ln
R2X
oxidative addition (OA)
R1—R2
reductive elimination (RE)
LnPd(II) R1 R2
LnPd(II) X R2
transmetallation
MX
R1M
R1, R2 = C groups. X = I, Br, Cl, OTf, etc. M = metal countercation.
M = Zn, Zr, Al (Negishi), B (Suzuki), Mg (Kumada), Sn (Stille), Si (Hiyama)

Figure 11.12 C—Ccoupling mechanism (Xu, Kim, Wei, & Negishi, 2014).

Heck
Sonogashira
ArSnBu3
Stille
ArB(OH)2
Suzuki
Hal
R'MgBr
Kumada
R'ZnBr
Negishi
Amines
HN R1 R2

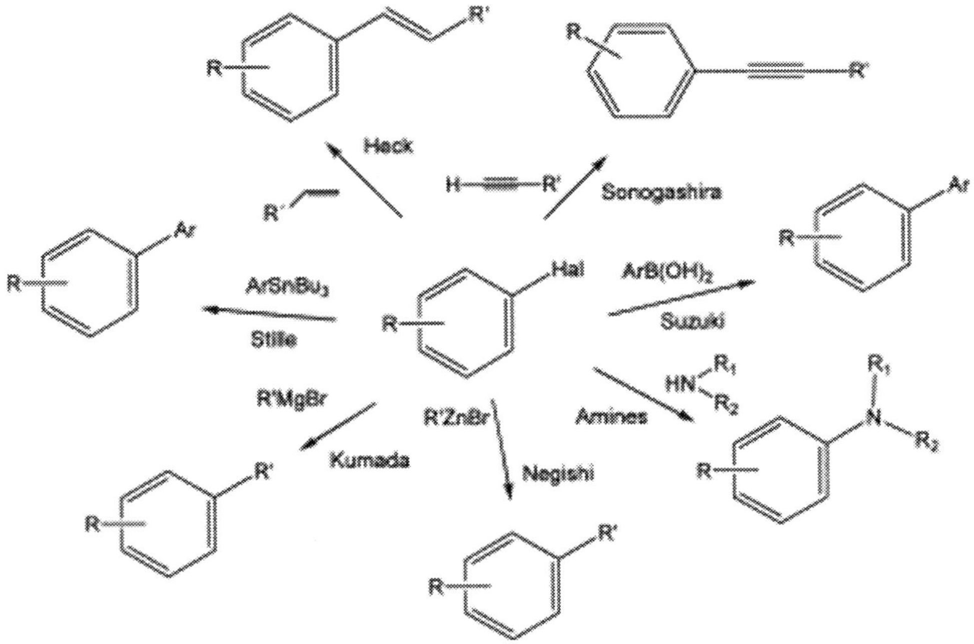

Figure 11.13 C—C coupling reactions using palladium (Herrmann, Öfele, Preysing, & Schneider, 2003).

noble metal makes a robust catalyst, and is known not only for its high functional group tolerance, but also because of low organo-palladium sensitivity toward air and water.

11.4.2 Nanoparticles as catalysts

Over the past decade the use of transition metal nanoparticles (palladium nanoparticles mainly) for C−C coupling reactions has drawn much attention due to their important applications including Heck, Sonogashira, and Suzuki reactions. The catalytic formation of C−C bonds along with hydrogenation and reduction of double bonds plays a crucial role in organic synthesis. Using catalysts can be expensive, and harmful to the environment, and this calls for new catalytic systems that can replace existing homogeneous ones. Metal nanoclusters behave differing properties from the classical colloids, and exhibit a number of advantages. They have a smaller size range from 1 to 10 nm (in diameter), narrower size dispersion, have reproducible synthesis, and a large surface area. In organic and aqueous solvents, they are known to be isolable and redissolvable. Transition metal nanoparticles also have better catalytic properties, that is, they are more active and usually exhibit reproducible activities and display higher selectivites (Balanta, Godard, & Claver, 2011). One of the main advantages of the use of nanoparticles as catalysts is they have shown potential to be recovered and reused. Reutilization of catalysts can prove to be extremely beneficial not only due to reduced expenses, but also because of a small environmental footprint. Metal nanoparticles used for catalytic application are produced by two main methods. The first being reduction of metal salts, and the second being organometallic complex decomposition (Widegren & Finke, 2003). In the synthesis of metal nanoparticles, it is important to remember that a protective agent must be used in order to ensure that they remain stable and do not form bulk metal. This can be done by electrostatic stabilization and steric stabilization. The first one is usually known to be of use in aqueous media, and the second one is used in aqueous and organic solvents. It is imperative that the right stabilizing agent be used without which attaining the desired size for nanoparticles may be compromised, therefore not achieving catalysts of high activity.

11.4.3 Use of nanoparticles in Heck reaction

Palladium nanoparticles were first used in this type of reaction in 1996 by Beller et al. by making use of colloids. Use of these systems showed accurate and good results for the Heck arylation of styrene by activated aryl bromides (low catalyst loading), and provided limited activity for deactivated aryl bromides and chlorides. Palladium nanoparticles prepared by electrochemical reduction (stabilized by propylene carbonate) proved to be successful in converting activated aryl bromides in high yields along with activated chlorobenzene (Reetz & Lohmer, 1996). It was later discovered by Reetz & Lohmer (1996) that using a mixture of [PdX_2] salt alongside tetra-arylphosphonium salt activated aryl chlorides, and is used in the production of a predecessor of Naproxen in the industry.

Further research in this field led to the discovery of formation of palladium nanoparticles of size 1.6 nm while studying the Heck reaction between iodobenzene and ethyl acrylate in N-methyl pyrrolodinone (Reetz & Lohmer, 1996). Use of palladium nanoparticles for Heck reaction, has therefore proven to be successful, and in future has the potential to make this process more efficient and friendlier.

11.4.4 Use of nanoparticles in Sonogashira reaction

Palladium nanoparticles can be stabilized by polyvinylpyrrolidine to act as catalysts for the reaction of aryl iodides and bromides with phenylacetylene (Sonogashira reaction) (Wang et al., 2010). Through further research in Sonogashira reactions, it was reported that palladium nanoparticles of one-pot acted as efficient catalysts for the reaction between arylacetylenes and 2-iodophenols. This reaction took place in water, and there was no ligand or cocatalyst (copper) present. In this particular study, the role of the stabilizer was played by amine, since the formation of palladium nanoparticles was accompanied by amine (Saha, Dey, & Ranu, 2010). Another area of study where Sonogashira reaction is completed with the help of nanoparticles, is the use of palladium nanoparticles that have the support of graphite oxide along with graphenes that are derived chemically. These catalysts can also assist in the Suzuki–Miyaura reaction (activated aryle chlorides) and the Heck coupling (activated aryl bromides) (Balanta et al., 2011).

Catalyst reuse and reutilization has been an area of interest for scientists, since it could lead to significant cost savings in addition to being sustainable to the environment. Mixed metal palladium cobalt nanospheres (these are hollow and are bi-metallic) are known to act as effective and reusable catalysts in aqueous Sonogashira reactions (Li, Zhu, Liu, Xie, & Li, 2010). When selecting and using a catalyst, one of the main concerns is its activity. Reutilization of catalysts often reduces the activity with successive uses. These mixed metal nano-spheres have reported to be recycled up to seven times without significantly losing their activity, which is due to the fact that they hollow, and therefore provide a greater number of palladium reaction active sites.

11.4.5 Use of nanoparticles in the Stille reaction

Palladium nanoparticles, known to have catalytic properties, are efficiently reactive for the Stille reaction. Experimental results showed that considerable amount of product was yielded using a fairly low quantity of catalyst, that is, Pd-4-capped palladium nanoparticles with nontraditional aqueous solvent conditions at room temperature (Pacardo et al., 2009). Through research, it has been reported that the underlying principle that guides these catalytic reactions using palladium nanoparticles could be an atom leaching mechanism. This reaction is driven by the particles, which act as catalytic reservoirs with active metal species (Astruc, 2007). The first step, the oxidative addition step is carried out by palladium atoms being abstracted from nanoparticles. The oxidative addition is reported to occur at two separate sites, that is, Pd^0 (free) atoms and at the surface of nanoparticles. The catalytic cycle may therefore occur in solution, giving the final product and regenerating Pd^0 (Pacardo & Knecht, 2013). Until the reaction goes to completion, these atoms can be recycled (by the catalytic process). In this reaction, palladium nanoparticles have been reported to demonstrate a wide range of reactivity when reacted with aryl halides. Palladium nanoparticles, therefore play a vital role in Stille reaction by acting as catalysts and pushing the reaction to completion.

Nanotechnology, therefore play a vital role in C–C coupling reactions. They are known to have numerous advantages over the tradition system which includes the use of colloids as catalyst. Reutilization is one of the main benefits of using metal nanoparticles. Over time, it has been proven that aryl iodides and bromides can efficiently and effectively be used as substrates; however, recent studies and researches have also shown that palladium nanoparticles can be used as catalysts in activating aryl chlorides, which is significantly more challenging. During the production of palladium nanoparticles, it is imperative to use a stabilizing agent, since this helps not only in preventing the formation of palladium black which is less active, but also minimizing leaching from the surface of molecular species. The use of nanotechnology in C–C coupling reactions is vast and is still growing. Research groups all over the world are showing more an increased interest in this field, and the future holds a strong potential for this method to change the face of coupling reactions by making them less expensive, more environment friendly, less toxic, faster and more rapid, more easily accessible, and the ability to be reutilized.

11.5 Fuel cell technology

Nanotechnology continues to shape the future of all technological aspects in the world. One application where nanotechnology can help excel the efficiency of a system is in fuel cells. Fuel cells have been gaining traction due to the decline of the health of the environment and the growing greenhouse gas emissions. Fuel cells have the potential to replace fossil fuels in many applications without emitting any harmful emissions. This could set the future foundation for fuel cells to be a sustainable energy source. Currently, the most common technology is the form of a proton exchange membrane fuel cell (PEMFC). Although other fuel cell technologies do exist, PEMFC is the most common due to their abundance of advantages. PEMFCs can operate at temperatures less than 100°C that allow for rapid start-up, easier to manufacture compared to other fuel cells, compact and lightweight, and respond to loading variability's extremely quickly (Buchanan, 2015). They consist of heterogeneous catalysts at the

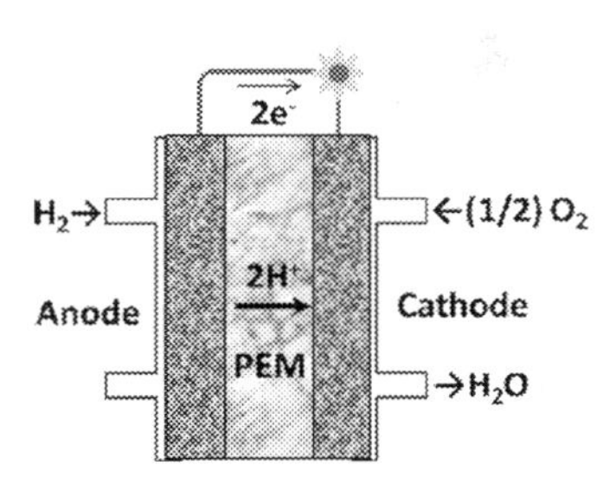

Figure 11.14 Schematic representation of a hydrogen fuel cell reaction (Mazumder, Lee, & Sun, 2010). *PEM*, Proton exchange membrane.

anode and cathode that help facilitate the reactions; this is where nanotechnology is influencing fuel cell development in a big way.

Typically, there are two main types of fuel cells, one that use hydrogen as the fuel source and the second that uses methanol. A fuel cell works by converting the chemical energy contained in these fuels into useful electrical energy. Fuel cells function by oxidizing the fuel at the anode with the use of a catalysts and then again reducing oxygen at the cathode with another catalyst, the schematic representation of this process can be seen in Fig. 11.14. The positive ions flow through the proton exchange membrane and crossover to the cathode. On the other hand, the proton exchange membrane is impermeable to electrons, therefore they must find another path to the cathode, and this is the basis for capturing the electrical energy. In the case of a hydrogen fuel cell, water and thermal energy are the only reactants that are present within this reaction. The corresponding chemical equation is: $H_2 + 0.5O_2H_2O$. Similarly, for a methanol fuel cell the corresponding equation is: $CH_3OH + 1.5O_2CO_2 + 2H_2O$.

11.6 Platinum catalysts

There lie certain limitations with regard to the fuel cell technology that hold back the potential of these cells being successful in the mass market. The main obstacles in commercialization are catalysts' costs and reliability (Mazumder et al., 2010). Primarily, platinum has been used as the main catalyst ever since the inception of the fuel cell idea. Platinum is a precious metal making it scarce and expensive, based on a loading of 5 kW g^{-1} the platinum catalyst alone accounts for 50% of the total cost of the fuel cell. Platinum is also readily poisoned by the presence of CO and other contaminants. Lastly, when platinum is present in acidic fuel cell reactions the active surface area is reduced and the performance deteriorates (Mazumder et al., 2010). In oxygen reduction reactions the surface of the platinum has a tendency to be covered by the electrolyte layer. This causes the number of active sites to dramatically decline, thus the efficiency is reduced. Nanotechnology is working to reduce costs and increase the efficiency of fuel stacks by refining the catalyst technology through the implementation of nanoparticles as part of the design.

With respect to increasing efficiency and reducing costs, the research has mainly shifted in three ways. First, there has been relentless effort to optimize the existing platinum catalysts by the means of reducing particle size to the nonorange, controlling morphology, optimizing structure, and increasing uniformity. Second, development of bimetallic catalyst that further enhances catalytic property while still reducing costs, two nanoparticles of interest are usually platinum and some other alloy. The final development is the shift to search for a platinum-free catalyst alternative by using nanoparticles as the new research base (Zhu, Kim, Tsao, Zhang, & Yang, 2015).

11.6.1 Platinum nanoparticles

The development of mono dispersed PtNPs with controlled sizes and shapes was one of the first developments that was brought to light. Through various research efforts, it was determined that the optimum size for a platinum catalyst is around 3 nm. This value ensures that the maximum mass activity is attained. Decreasing the size beyond 3 nm does not seem to have any significant effect on performance. Another important aspect of nanoparticle platinum catalysts is the morphology and structure of the dispersed material. It was determined that a (1 0 0) structure is the most active (Zhu et al., 2015). Not only do PtNPs have the ability to increase activity and efficiency, they can also improve catalytic properties which were degraded with the use of bulk platinum. Examples of these phenomena were demonstrated by the work conducted by Kim and Jhi (2011). Within their research paper, they proved that when PtNPs were deposited on nitrogen-doped grapheme, the catalyst could tolerate carbon monoxide much more effectively than that of platinum in its bulk form. Advancements in using PtNPs as catalysts propel the fuel cell technology closer to being widely implemented. They have proven to increase efficiency and because of this, less platinum has to be used, thus driving down costs and making the technology more economically viable.

11.6.2 Alternative catalysts material

With the use of nanoparticles there has been a push to use bimetallic catalysts that consist of platinum and some other metal in order to achieve favorable catalytic properties. Binary and ternary platinum alloys exhibit increased activity as compared to just a pure platinum catalyst. Examples of binary alloys include platinum with iron, cobalt, nickel, or copper. When analyzing a nanoplatinum—copper binary material and a nanoplatinum—copper—nickel ternary material, it was determined that both mass and specific activity was increased 2—5 times as compared to pure platinum (Zhou & Zhang, 2015). Among the various binary alloys, a platinum—nickel nanoparticle catalyst appeared to have the highest oxygen reduction reaction activity with outstanding durability. Although platinum-alloy nanocatlysts have the opportunity to take over as the catalysts of choice, the price of platinum still remains a glaring issue. As such, there has been a push to create nonplatinum nanocatalysts that would effectively take the place of the platinum counterpart.

Nanopalladium-based catalysts have been gaining traction in an effort to deviate from the conventional platinum-based catalyst. Based on the extensive research, palladium catalysts seem to have fruitful future in fuel cell applications. It has been shown that nanostructure palladium particles exhibit high elector catalytic activities in oxidation of small molecules and are more resilient to carbon monoxide poisoning (Wei & Chen, 2012). The TEM images for 4.5-nm palladium nanoparticles can be seen in Fig. 11.15. Alternatively, research in to gold-based nanoparticles is also being conducted. Gold differs from regular palladium and platinum as it has been found to be highly active for oxygen reduction reactions in alkaline electrolytes, as opposed to platinum and palladium which are used for electrolyte fuel cells. It was determined that the (1 0 0) gold plane was the most active, this confines to the results that were discussed earlier within this report. With respect to gold nanoparticles it was uncovered that catalysis was still size dependent, with the smaller particles being more active (Mazumder et al., 2010). The TEM images for 3-nm gold particles can be seen in Fig. 11.15.

Other non-PtNPs that show great promise are iron–nitrogen and cobalt–nitrogen-based catalyst. Future work for nanoparticles in fuel cell catalysts includes trying to optimize the technology and creating a more efficient production route that can be commercialized simply.

11.6.3 Supporting materials

Concurrently to the research conducting in finding a nanomaterial that will effectively take the place of bulk platinum, research has been underway on finding a support

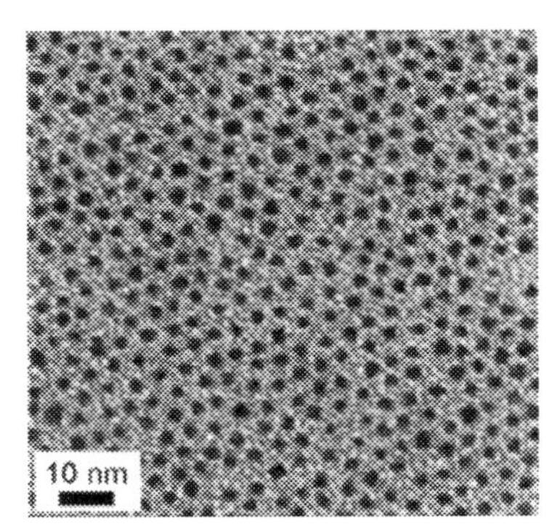

Figure 11.15 Left: Transmission electron microscopic (TEM) images for 4.5-nm palladium particles (Mazumder et al., 2010). Right: TEM images for 3-nm gold particles (Mazumder et al., 2010).

material for the nanoparticles that will favorably increase catalyst properties. The metal catalysts are usually deposited on supports that exhibit high electrical conductivity as well as stability. This has primarily been various carbon materials, these include carbon nanotubes, carbon nanofibers, grapheme, fullerene, and tungsten carbide (Wei & Chen, 2012). One novel application of a nanosupport is carbon nanodots. Carbon nanodots possess unique electronic, optical, and thermal properties. The work conducted by Wei and Chen (2012) dispersed palladium nanoparticles on carbon nanodots using the facile green method. The analysis of the result indicated that this hybrid material provided high electrical conductivity and the active surface was not blocked for organic molecular fuel oxidation, this makes them promising for alkaline fuel cell applications. Graphene oxide supports have also been investigated as supports for bimetallic nanoparticle catalyst. Gupta, Yola, Atar, Üstündağ, and Solak (2014) within their research they concluded that bimetallic nanoparticles that were supported on graphene oxide demonstrated enhanced electrochemical efficiency for methanol oxidation–reduction reaction reactions with regard to diffusion efficiency, oxidation potential and peak current. Silver–gold nanoparticles exhibited the maximum stimulation when deposited on graphene oxide supports. This result can be seen in Fig. 11.16 graph, which compared the scan rate versus the peak current. It can be seen that silver–gold nanoparticles deposited on graphene oxide supports result in a greater peak current than other nanocomposites tested.

Nanotechnology has allowed for specific control of catalysts' properties. Most of the nanoparticles discussed above were monodispersed on supports. This means that the size and structure of each particle were strictly controlled. This allows for decreased variability in catalysts' performance and more efficient reactions. Nanotechnology has the ability to precisely control structure properties. This was shown by Song et al. (2010), where the effect of pore morphology of mesoporous carbons on electrocatalytic activity on PtNP for fuel cell reactions was studied. The study was conducted by comparing supports of ordered mesoporous carbon (CMK-3) and disordered wormhole-like mesoporous carbon (WMC) materials. The corresponding structures can be seen in Fig. 11.17.

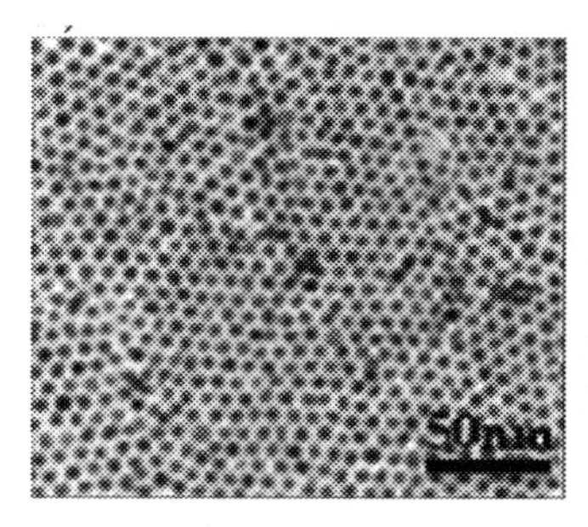

Figure 11.16 Scan rate versus peak current of graphene oxide supported materials (Gupta et al., 2014).

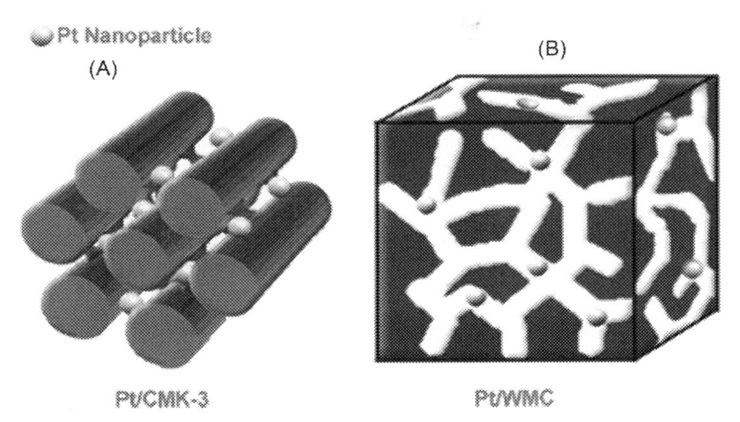

Figure 11.17 (A) Ordered mesoporous carbon (CMK-3) support morphology. (B) Wormhole-like mesoporous carbon (WMC) support morphology.

In the case of CMK-3 carbon supports, its structural properties allow for high degree of order. Furthermore, this facilitated for good three-dimensional interconnection nanospacing of the carbon nanorods. These properties allowed for superior mass transfer and the electrolyte can access the PtNPs more readily.

In the case of the WMC supports, the poor morphology of the structure and low connectivity of the wormhole like mesopores actually decrease the mass transfer abilities. This work is beneficial because when designing porous materials, it is now known that the pore morphology should be given a great amount of detail and be designed so that it contains performance-enhancing properties.

11.6.4 Fuel cell outlook

Nanotechnology has helped evolve fuel cell technology to new heights. Without the research, commitment and money that has been invested into further developing nanocatalysts, the efficiency and cost effectiveness of this technology today would not exist. Fuel cells have an immense amount of potential; this was first evident when former president Bush pumped US$1.2 billion into hydrogen fuel cell development. However, this funding was cut by president Obama as he viewed that this technology still needs refinement and is still in its infancy before implementation. Nevertheless, the United States Department of Energy has invested interest in the technology. They set targets for efficiency of the fuel cell catalysts, in 2014 researchers exceeded 2017 targets. This was primarily due to advancements in bimetallic nanocatlysts. This proves the fact that nanotechnology is helping to shape the face of fuel cell technology forever.

11.7 Heavy oil technology

Pollution and environmental concerns owing to the increase in fossil fuel production and relating to conventional oil recovery methods are the main reasons for the extensive research being devoted to the development of greener methods for producing heavy oil, as well as, continuing the increase in biodiesel fuel production. In this section of the study, solid catalyst-based methods for improving both, the petroleum-based fuel and the biodiesel fuel production, while decreasing the negative environmental impacts associated with them will be discussed. Additionally, the advantages of using heterogeneous solid catalysts in place of homogeneous base catalysts will be addressed.

11.7.1 Heavy oil recovery methods

Enhanced oil recovery methods such as steam-assisted gravity drainage (SAGD), although considerably effective in their recovery percentages, offer a great deal of environmental drawbacks. In addition to the large amounts of natural gas and freshwater required during the heavy oil extraction, the recovery technique is known for producing extremely high levels of greenhouse gas emissions. Newer, alternative oil recovery methods that have come to be known as toe to heel air injection and vapor extraction (VAPEX) have significantly grown in popularity over recent years due to their potential for equally high oil recovery rates, in addition to being more environmentally friendly. The customarily non-thermal VAPEX method relies on molecular diffusion of a vaporized solvent such as propane, in order to decrease oil viscosity and enhance its ability to flow. Similar to SAGD, it also incorporates the use of two horizontally parallel wells, with the main and most significant difference being that SAGD is a thermal-based method that relies on generating hot steam for heating the heavy oil and bitumen found in the operational reservoirs.

11.7.2 Nanotechnology application

Pourabdollah, Zarringhalam Moghaddam, Kharrat, and Mokhtari (2011) demonstrated in their research the potential advantages of incorporating the use of montmorillonite (MMT) nanoclays in the improvement of the VAPEX process. MMT is an exceptionally popular heterogeneous catalyst due to various reasons such as cost, environmental friendliness, as well as swelling capabilities that allow it to retain foreign molecular substances within its interlayers (Kumar, Dhkashinamoorthy, & Pitchumani, 2014). Asphaltenes are a brownish-black substance found in crude oil that is responsible for its low the American Petroleum Institute (API) gravity. Aside from increasing the thickness of the oil, the presence of asphaltenes also decreases oil refinability and makes extraction more difficult and expensive. Fig. 11.18 illustrates how the MMT nanocatalysts can absorb asphaltenes found in heavy oil and bitumen within its layers.

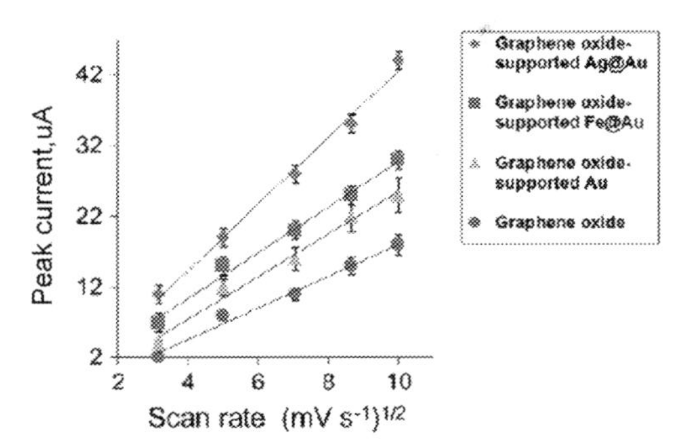

Figure 11.18 Entrapment of asphaltenes within montmorillonite sublayers.

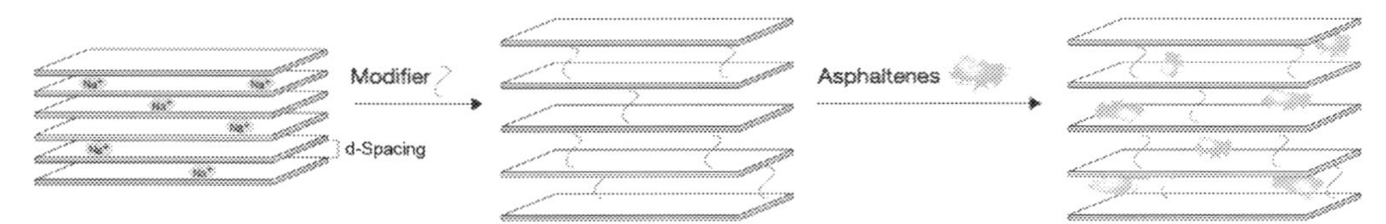

Figure 11.19 Asphaltene retention in (A) vapor extraction (VAPEX) cell without montmorillonites (MMTs) versus (B) VAPEX cell with MMTs.

Pourabdollah et al. (2011) experimental set-up included the usage of two cells: one packed with standard oil beads without the MMT nanocatalysts and the other with glass beads including the MMT nanocatalysts. This was done to be able to carry out a comparative analysis between the two and determine precisely the effectiveness of the MMT heterogeneous catalyst in removing asphaltenes from viscous oil. Additionally, propane was used as the solvent due to its capability of being more effective in de-asphalting than other hydrocarbons in lower pressure experimental conditions. As can be seen in Fig. 11.19, the MMT-assisted VAPEX cell succeeded in retaining a much higher level of asphaltenes. Due to the negative charge encouraged by the MMT crystal lattice, the adsorption of the heavy oil components including asphaltenes, metals, cation, and sulfur atoms was made possible. Furthermore, an additional advantage of using MMT nanocatalysts is that they can be effectively washed in order to avoid the catalyst degradation that would result from the accumulation of too much asphaltene content (Pourabdollah et al., 2011).

Fig. 11.19 also shows that the experimental results indicate an up to 22% increase in asphaltene content removal from the oil with the incorporation of MMT nanoparticles in the VAPEX cell. This positive outcome confirms that with the incorporation of heterogeneous MMT nanocatalysts the produced oil will be less viscous, more pure, and easier to recover. Additionally, this means that less refining processes will be required to produce high-quality fractions of oil. Therefore less natural gas, as well as, freshwater sources will need to be utilized and this in turn will promote a healthier

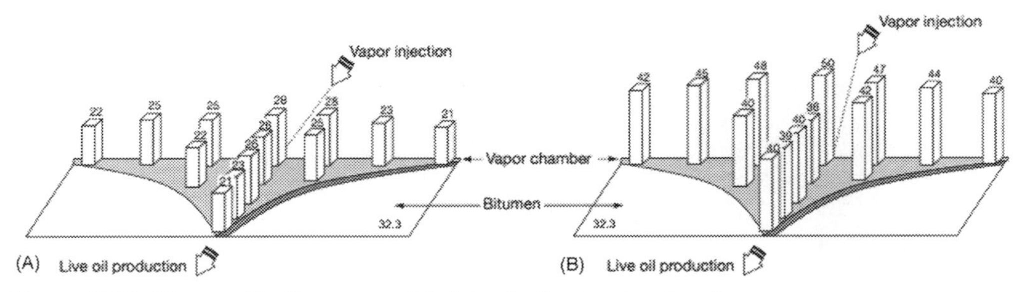

Figure 11.20 Montmorillonite nanoparticle introduction to glass beads.

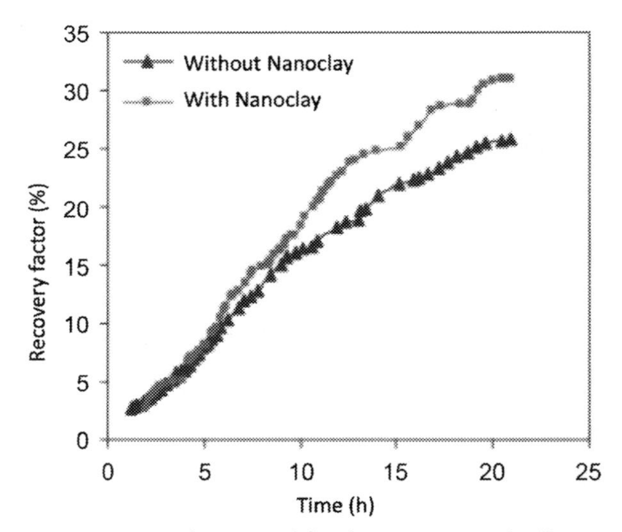

Figure 11.21 Recovery percentages determined for the experimental cells.

and greener environment than otherwise would be possible with more conventional methods that do not incorporate heterogeneous solid catalysts. Fig. 11.20 illustrates how the MMT nanoparticles were mixed with the glass beads used in the experiment.

Pourabdollah et al. (2011) also elucidated graphically the increase in recovery factor that resulted from the improved VAPEX nanoincorporated cell over the conventional VAPEX cell. Fig. 11.21 shows an overall increase in recovery percentages by 30 ($\pm$ 4)%. This, combined with the improved dissolution ability of the solvent, owing to the heterogeneous nanocatalysts, also demonstrates that a higher recovery percentage can be achieved even with the injection of less vaporized solvent. Therefore in addition to improved oil production rates, the usage of MMT nanocatalysts reduces production costs by minimizing the amount of solvent required.

As mentioned previously, the need for greener and renewable fuel sources stems from the world's increasing demand for energy. Currently, the majority of biodiesel production utilizes alkaline homogeneous catalysts (Romero, Martinez, & Natividad, 2011). Since homogeneous catalysts are of the same phase as the reaction mixture they are placed in, this creates some significant drawbacks. Two of the main disadvantages of using homogeneous catalysts in biodiesel production include higher wastewater pollution as well as higher purification costs because there is not an effective method for removing the catalyst. Subsequently, this means that the catalyst cannot be reused. Although homogeneous catalysts due offer advantages such as handling and competitive prices, the biodiesel industry's interest is shifting toward the application of heterogeneous catalysts. This shift toward heterogeneous catalysts would offer benefits such as lower energy consumptions, reduced costs, easier catalyst recovery, superior catalytic activity, and more environmentally friendly nature (Akia, Yazdani, Motaee, Han, & Arandiyan, 2014; Thanh, Okitsu, Boi, & Maeda, 2012).

11.8 Supercritical water gasification

Gasification is an important industrial process that involves the conversion of biomass into hydrogen, carbon monoxide, and carbon dioxide, when combined are commonly referred to as syngas. This syngas is considered a renewable energy source since it is produced from organic based materials and it plays a vital role in the efforts to reduce the world's dependency on nonrenewable fossil fuels (Akia et al., 2014; Demirbas, 2011). However, gasification has a drawback. It cannot process wet biomass, unless it is dried. Furthermore, the dying of the biomass is both time consuming and costly. For this reason, conventional gasification by itself cannot be used to convert wet biomass substances that could potentially be converted to precious fuel sources (Boukis, Galla, Muller, & Dinjus, 2007).

However, hydrothermal gasification with the incorporation of heterogeneous catalysts is able to form desirable compounds such as hydrogen from wet biomass ($>70\%$ water content) (Azadi, Farnood, 2010; Boukis et al., 2007). Therefore as an alternative to classic gasification, considerable research has been done on the sub and supercritical water gasification (SWCG) of biomass. Low temperature (374°C–550°C) as opposed to high-temperature (550°C–700°C) SWCG processes, although not always necessary, can benefit from the incorporation of a solid catalyst to help overcome energy barriers experienced at these lower temperatures Additionally, at these lower temperatures, the unwanted formation of char is produced if nanocatalysts are not incorporated in the reaction mixture. So far, activated carbon, transition metals, and oxides have been the main catalysts used in the SWCG of biomass (Azadi & Farnood, 2010).

Activated carbon is able to offer excellent stability and a high degree of metal dispersion. For this reason, it is commonly used in hydrogenation reactions. Additionally, it is known as the most cost-effective heterogeneous catalyst, however, its incorporation in high-temperature SWCG has been deemed not effective in improving the rate of gasification or the syngas yield. Heterogeneous catalysts, such as, supported nickel and supported ruthenium catalysts have also been used in SWCG. The latter showing higher reactivity, metal dispersion, better sintering resistance, and oxidation resistance. However, the presence of sulfur atoms is known to be able to degrade the ruthenium catalyst and render it inactive (Azadi & Farnood, 2010). Therefore much more research is needed to find more suitable catalysts that will improve the gasification process and the conversion of wet biomass to a viable energy source that can be used in place of conventional fuel sources.

11.9 Magnetic nanoparticles

11.9.1 Nanoscale magnetic stirring bars for heterogeneous catalysis

The reaction rate and a reduction in energy consumption can be significantly enhanced through effective mixing. Effective mixing allows mass transfer to arise in chemical reactions, particularly in heterogeneous catalysis. To acquire optimal mixing, magnetic stirring bars are often utilized. Conventional stirring methods are not applicable to nanoscale reactors, such as microdroplets and micelles, which are essential for lab-on-chip applications and microliter biological assessments (Yang et al., 2015). Conventional stirring methods cannot be used because the stirring bars are much larger than the nanoscale reactor. Consequently, it is necessary to design stirring bars on a nanoscale level that are proficient at rotating under external impact. Magnetic nanoparticles are placed into firm nanochains through the process of magnetic induction and external coating through induced self-assembly.

This section will be analyzing a two-step synthesis method toward creating nanometer magnetic stirring bars that comprise palladium nanoparticles for heterogeneous catalysis. One-dimensional nanochains were prepared by using Fe_3O_4 as building blocks and is referred to as Fe_3O_4-NC (Deng et al., 2005). By utilizing ultrasound and Fe_3O_4's paramagnetic response, a cross linked polymer (cyclotriphosphazene-co-4,4'-sulfonyldiphenol) was coated as the shell. This allowed stability and the new Fe_3O_4 nanochains are referred to as Fe_3O_4-NC-PZS (Zhou, Meng, Feng, Zhang, & Lu, 2010). The coating provided extra functioning sites to place the palladium nanoparticles onto the surface of the Fe_3O_4-NC-PZS, in an attempt to promote catalytic activity (Chen, Cui, Niu, Jiang, & Song, 2010). The synthesized Fe_3O_4-NC-PZS-Pd efficiently stirred the microdroplets and was capable of accelerating mass transfer by presenting better catalytic activity compared to a palladium catalyst.

Fig. 11.22 is a visual representation of the reaction.

Under ultrasonic irradiation, Fe_3O_4 nanoparticles were placed into nanochains and demonstrated constant morphology. The presence of Fe_3O_4 nanoparticles allowed a

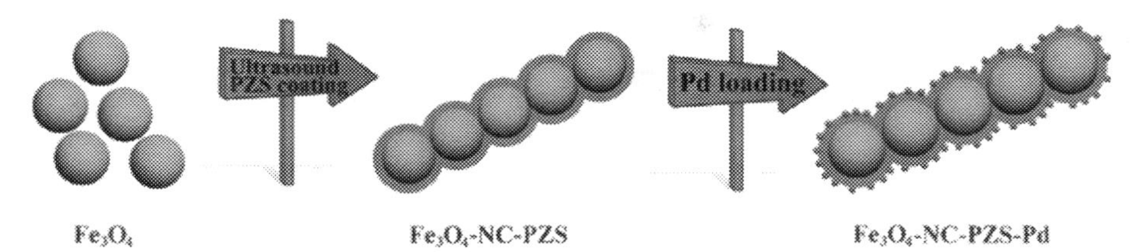

Figure 11.22 Preparation of Fe_3O_4-NC-PZS-Pd.

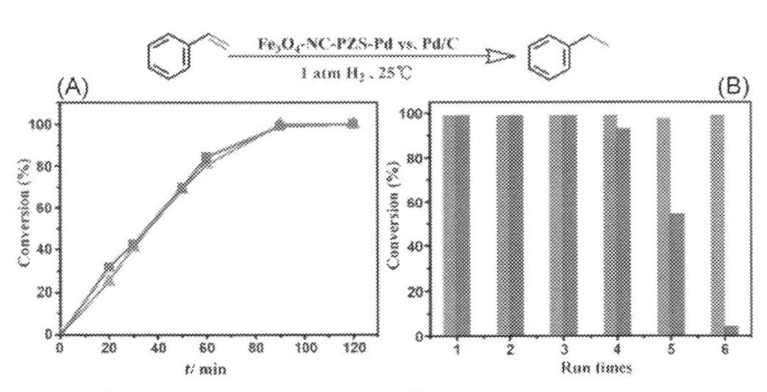

Figure 11.23 (A) Conversion per cent versus time. (B) Conversion per cent versus run times.

good magnetic response and could be used not only as a stirrer, but also a catalyst. The properties of Fe_3O_4-NCPZS-Pd were tested by observing the hydrogenation of styrene. Two properties were observed: the dispersivity of the palladium nanoparticles was better and the Fe_3O_4 nanoparticles gave better mixing effects. These two properties resulted in improved heterogeneous catalytic activity.

Fig. 11.23 shows: (1) conversion versus reaction time for styrene hydrogenation with Fe_3O_4-NC-PZS-Pd in green and palladium catalyst in red; and (2) recycling of Fe_3O_4-NC-PZS-Pd and commercial palladium catalyst.

From observing the graphs, it is noted that there was no reduction of activity in Fe_3O_4-NC-PZS-Pd after six cycles. A decrease was observed for the Pd/C after three runs in graph (B). Once the reaction was completed, the system was analyzed using plasma atom emission spectroscopy and no palladium was discovered. After multiple catalytic cycles, Fe_3O_4-NC-PZS-Pd preserved its morphology while the nanoparticles were still on the surface. The results presented excellent catalytic activity and stability. This occurs due to several factors such as: the nitrogen atoms present in PZS which stops aggregation of nanoparticles. The PZS layer also guards the Fe_3O_4 and its interior core from corrosion. After the reaction is complete, the Fe_3O_4-NC-PZS-Pd can be recovered with a magnet and has a recovery of almost 100% (Yang et al., 2015). The palladium catalyst must be recovered through centrifugation because it is dispersed throughout the solution. A test was also conducted to see whether the Fe_3O_4-NC-PZS-Pd was resistant to corrosion when placed in acidic solutions like nitric acid. It

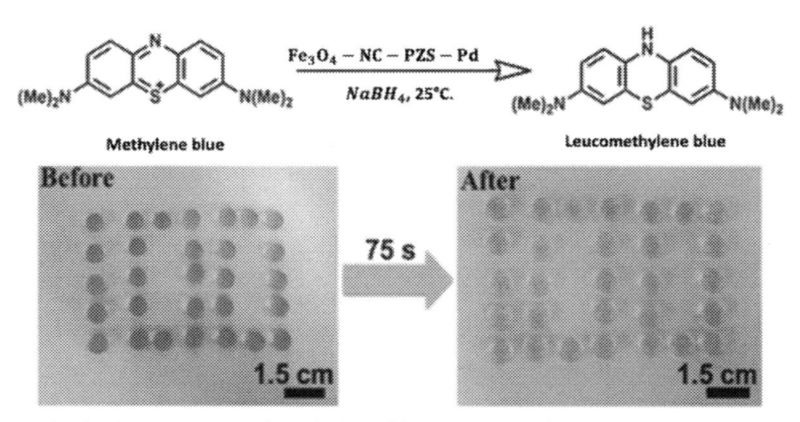

Figure 11.24 The hydrogenations of methylene blue in arrays of microdroplets.

was found that the PZS layer was unaffected for over 48 h and formed protection whereas other samples without the PZS completely dissolved (Yang et al., 2015).

The hydrogenations of methylene blue in arrays of microdroplets were used to test the catalytic activity of Fe_3O_4-NC-PZS-Pd. The change in color of the methylene blue solution is an indicator of the reaction progress. It was observed that the microdroplets went from blue to colorless in 75 s after the addition of Fe_3O_4-NC-PZS-Pd. The palladium catalyst had a much lower reaction rate due to worse mixing in the microscopic system. Fig. 11.24 is a visual representation of the process.

In summary, using nanoscale heterogeneous catalysis in magnetic stirring bars proved to be extremely effective. This was proven by examining Fe_3O_4-NC-PZS-Pd in solution versus a palladium catalyst. It was noted that there was an increase in mass transfer, catalytic activity and stability.

11.9.2 Nanoscale magnetic catalyst for biodiesel production

As conventional oil supplies constantly diminish and pollution emissions continuously increase, new studies are being conducted for an alternative renewable fuel source. Biodiesel, defined as fatty acid methyl esters, has attracted great publicity as a potential source of energy (Demirbas, 2011). Biodiesel is usually produced by reacting vegetable oil/animal fats with methanol in the presence of a catalyst. The products of this reaction are methyl ester and glycerin. This process frequently occurs within a homogeneous catalyst but encountered too many problems such as: high costs and complications involving the separation process. Purification is also time consuming and produces lots of waste water. Researchers began looking into heterogeneous catalysts as a substitute for the transesterification process to overcome these challenges. It was discovered that nano-magnetic heterogeneous catalysts are a better choice because of high recovery from magnetic separation. Ying and Chen (2007) have stabilized cells of *Bacillus subtilis* with magnetic particles, particularly Fe_3O_4 for the transesterification of Chinese tallow with methanol. The magnetic nanoparticles show advantages in easy separation, high resistance to saponification, good rigidity, favorable acid resistance, catalytic activity, and are easily reusable (Wen, Wang, Lu, Hy, & Han, 2010).

The nano-magnetic solid catalyst was prepared with 5 g of Fe_3O_4 and 100 g of MOs (MgO, CaO, and SrO). The mixture was then mixed with an aqueous solution of KF and dried at 105°C for 24 h. Afterwards, the material was heated to a temperature range from 300°C to 800°C in a muffle furnace. The final catalyst is then stored in a desiccator. The catalytic activities of several catalysts were evaluated in the transesterification of Stillingia oil with methanol. $KF/CaO-Fe_3O_4$ was carried out at 65°C for 3 h with 25 g of Stillingia oil, 13.4 mL of methanol, and 1 g of magnetic nanocatalyst. This was compared to CaO, KF/CaO, $KF/MgO\text{-}Fe_3O_4$, and $KF/SrO\text{-}Fe_3O_4$. The results are summarized based on surface area ($m^2\ g^{-1}$), pore diameter (nm) and yield (%) for each catalyst in Table 11.1.

It was found that the catalytic activity of $KF/CaO-Fe_3O_4$ was the highest (Ying & Chen, 2007). It was also noted that the catalytic activity is dependent on the surface area of the catalyst. The magnetic core did not have any impact on the catalytic activity of the catalyst and allowed for easy recovery. The stability tests for the nano-magnetic catalysts were conducted by dissolving the catalysts in methanol. Once the catalyst was removed from the methanol, the concentration of calcium, magnesium, and strontium was analyzed by using atomic absorption spectrometry. The concentration respectively founded to be 30, 24, and 83 ppm. The presence of a crystal phase improved the stability of the catalysts. $KF/CaO-Fe_3O_4$ was then tested for reusability through experimentation by carrying out reaction cycles. After three hours of transesterification, the nanocatalysts were separated by a permanent magnet. It was washed with anhydrous methanol and dried at 105°C (Hu, Guan, Wang, & Han, 2011). The catalyst that was recovered was used again in a second reaction under the same reaction conditions. After running 14 trials, it was found that the catalyst maintained its catalytic activity even after being reused. It was only after 16 trials that the catalytic activity began to decrease. The rate of transesterification declines gradually in the

Table 11.1 Catalysts' yields at 600°C.

Catalyst	Surface area ($m^2\ g^{-1}$)	Pore diameter (nm)	Yield (%)
CaO	4.6	8.6	85.9
KF/CaO	19.2	39.4	96.8
$KF/CaO-Fe_3O_4$	20.8	42.0	95.0
$KF/MgO-Fe_3O_4$	4.1	13.7	83.6
$KF/SrO-Fe_3O_4$	5.9	21.3	87.6

Stillingia oil and methanol solution. The reason for the occurrence is the decline in components throughout the reaction and separation process. The yield obtained from biodiesel reached the fresh catalyst level. It is noted that after reusing the catalyst 14 times, the recovery of the catalyst is 90%. After 20 times of reusing the catalyst, the recovery was still above 84% indicating that magnetic separation is successful. The decrease in recovery occurs for two main reasons consisting of the dissolution of catalysts, and the decline of catalyst in the midsection of the collection operation. Through experimentation, it was established that the nano-magnetic catalyst $KF/CaO-Fe_3O_4$ demonstrated good durability and high recovery (Hu et al., 2011).

In summary, the nano-magnetic solid catalyst $KF/CaO-Fe_3O_4$ was developed through the impregnation method. The catalyst has a unique porous structure and its average particle diameter size is 50 nm. This catalyst can be reused 14 times without facing deterioration with a recovery of 90%. The yield of biodiesel is 95% and the reaction process takes three hours. Conclusively, nano-magnetic catalyst could be the future with a good prospect of application and development.

11.10 Conclusion

Significant research efforts and investments have been made into nanotechnology over the past few years, consequently the technology has advanced at a rapid pace. This advancement has allowed many industries to reap the benefits. Industries affected include food, petroleum, pharmaceuticals, and renewable energy amongst others.

Heterogeneous catalysts play a large role in each of the aforementioned industries and the effectiveness of the catalysts is affected by reactions. Dehalogenation, hydrogenation, and C−C coupling reactions all used nanotechnology to improve catalyst performance by increasing factors such as efficiency and reusability due to their increased surface area and ability to be more targeted. This was further proven in hydrosilylation reaction and C−C coupling reactions. Within C−C coupling reactions, recent literature that implemented nanoparticles in the Heck, Sonogawshira, and Stille reactions were analyzed. It was determined that yet again these nanocatalysts could indeed increase efficiency of the process.

More tangible applications were also analyzed, such as the application of nanotechnology in fuel cells. It was found that recent developments in nanoparticles as catalysts has progressed fuel cell research immensely. The nanoparticles alleviate the limitations found in the traditional platinum catalyst and also can increase efficiency and lessen costs. Next, the applications and potentials for nanotechnology and oil recovery were analyzed. In an effort to improve VAPEX recovery, the effect of MMT nanoclay was researched. In doing so, many VAPEX properties were enhanced such as asphaltene content removal. Lastly, magnetic nanoparticles were investigated with magnetic stirrers to increase the effective stirring and the use of magnetic nanoparticles in biodiesel production.

This chapter has highlighted many potentials of nanotechnology in heterogonous catalysis. The research conducted has concluded that nanotechnology is actively influencing process in which catalysts exist. The outlook for the future of this technology seems positive, if optimized, process will become significantly more efficient and cost effective.

References

Akia, M., Yazdani, F., Motaee, E., Han, D., & Arandiyan, H. (2014). A review on conversion of biomass to biofuel by nanocatalsysts. *Biofuel Research Journal, 1*, 16−25.

Alonso, F., Buitrago, R., Moglie, Y., Ruiz-Martínez, J., Sepúlveda-Escribano, A., & Yus, M. (2011). Hydrosilylation of alkynes catalysed by platinum on titania. *Journal of Organometallic Chemistry, 696*, 368−372.

Arepalli, S. R., Bezabeh, S., & Brown, S. L. (2002). Allergic reaction to platinum in silicone breast implants. *Journal of Long-Term Effects of Medical Implants, 12*, 299−306.

Astruc, D., Lu, F., & Ruiz Aranzaes, J. (2005). Nanoparticles as recyclable catalysts: the frontier between homogeneous and heterogeneous catalysis. *Angewandte Chemie International Edition, 44*, 7852−7872.

Astruc, D. (2007). Palladium nanoparticles as efficient green homogeneous and heterogeneous carbon-carbon coupling precatalysts: a unifying view. *Inorganic Chemistry, 46*, 1884−1894.

Azadi, P., & Farnood, R. (2010). Review of heterogeneous catalysts from sub- and supercritical water gasification of biomass and wastes. *International Journal of Hydrogen Energy*, 9529−9541. Available from https://doi.org/10.1016/j.ijhydene.2011.05.081.

Balanta, A., Godard, C., & Claver, C. (2011). Pd nanoparticles for C−C coupling reactions. *Chemical Society Reviews, 40*(10), 4973. Available from https://doi.org/10.1039/c1cs15195a.

Boukis, N., Galla, U., Muller, H., & Dinjus, E., (2007). Biomass gasification in superctitical water. Experimental progress achieved with the Verana Pilot Plant. In: *15th European biomass conference & exhibition*. 1013−1016.

Buchanan, F. (2015). PEM fuel cells: Theory, performance and applications. Nova Publishers.

Carlson, G. L. (2004). Hydrogenation. In (3rd ed.K. L. Lerner, & B. W. Lerner (Eds.), *The Gale encyclopedia of science* (3Detroit: Gale. <http://go.galegroup.com.ezproxy.lib.ryerson.ca/ps/i.do?id = GALE%7CCX3418501177&v = 2.1&u = rpu_main&it = r&p = GVRL&sw = w&asid = d0ac62fd931220f5839bf3cd6f41cd71>.

Chen, A. C., & Holt-Hindle, P. (2010). Platinum-based nanostructured materials: Synthesis, properties, and applications. *Chemical Reviews, 110*, 3767−3804.

Chen, Z., Cui, Z.-M., Niu, F., Jiang, L., & Song, W.-G. (2010). Pd nanoparticles in silica hollow spheres with mesoporous walls: a nanoreactor with extremely high activity. *Chem. Commun., 46*, 6524−6526.

Colmenares, J. C., Magdziarz, A., Aramendia, M. A., Marinas, A., Marinas, J. M., Urbano, F. J., & Navio, J. A. (2011). Influence of the strong metal support interaction effect (SMSI) of Pt/TiO(2) and Pd/TiO (2) systems in the photocatalytic biohydrogen production from glucose solution. *Catalysis Communication, 16*, 1−6.

Crabtree, R. H. (2001). *The organometallic chemistry of the transition metals*. New York: John Wiley & Sons.

Demirbas, A. (2011). Competitive liquid biofuels from biomass. *Applied Energy, 88*(1), 17−28. Available from https://doi.org/10.1016/j.apenergy.2010.07.016.

Deng, H., Li, X., Peng, Q., Wang, X., Chen, J., & Li, Y. (2005). Monodisperse magnetic single-crystal ferrite microspheres. *Angewandte Chemie International Edition, 44*, 2782−2785, Angew. Chem. 2005, 117, 2842−2845.

Environmental Protection Agency (EPA). (1996). *A citizen's guide to chemical dehalogenation*. Environmental Protection Agency, Washington, D.C., United States.

Fiedorow, R., & Wawrzynczak, A. (2006). Catalysts for hydrosilylation, in heterogeneous systems. In B. Marciniec (Ed.), *Education in advanced chemistry* (10, pp. 327–344). Poznan Poland: Wydawnictwo Poznanskie.

Geisberger G., F. Baumann, A. Daniels Wacker Chemie, United States Patent 7,145,028, 2006.

Gupta, V. K., Yola, M. L., Atar, N., Üstündağ, Z., & Solak, A. O. (2014). Electrochemical studies on graphene oxide-supported metallic and bimetallic nanoparticles for fuel cell applications. *Journal of Molecular Liquids*, *191*, 172–176. Available from https://doi.org/10.1016/j.molliq.2013.12.014.

Hartwig, J. F. (2010). *Organotransition Metal Chemistry, from Bonding to Catalysis*. New York: University Science Books, ISBN 1–891389-53-X.

Herrmann, W. A., Öfele, K., Preysing, D. V., & Schneider, S. K. (2003). Phospha-palladacycles and N-heterocyclic carbene palladium complexes: Efficient catalysts for CC-coupling reactions. *Journal of Organometallic Chemistry*, *687*(2), 229–248. Available from https://doi.org/10.1016/j.jorganchem.2003.07.028.

Hu, S., Guan, Y., Wang, Y., & Han, H. (2011). Nano-magnetic catalyst for biodiesel production. *Applied Energy*, *1*(88), 2685–2690.

Jia, C.-J., & Schüth, F. (2011). Colloidal metal nanoparticles as a component of designed catalyst. *Physical Chemistry Chemical Physics: PCCP*, *13*, 2457–2487.

Karstedt B.D. General Electric, United States 3775452, 1973.

Jiménez, R., Martínez-Rosales, J. M., & Cervantes, J. (2003). The activity of Pt/SiO2 catalysts obtained by the sol-gel method in the hydrosilylation of 1-alkynes. *Canadian Journal of Chemistry*, *81*, 1370–1375.

Kim, G., & Jhi, S. (2011). Carbon monoxide-tolerant platinum nanoparticle catalysts on defect-engineered graphene. *ACS Nano*, *5*(2), 805.

Kumar, B. S., Dhkashinamoorthy, A., & Pitchumani, K. (2014). K10 Montmorillonite clays as environmentally benign catalysts for organic reactions. *Catalysis Sciencie & Technology*, *8*(4), 2378–2396. Available from https://doi.org/10.1039/C4CY00112E.

Kung, M. C., & Kung, H. H. (2004). Nanotechnology: Applications and potentials for heterogeneous catalysis. *Catalysis Today*, *97*(4), 219–224. Available from https://doi.org/10.1016/j.cattod.2004.07.055.

Li, H., Zhu, Z., Liu, J., Xie, S., & Li, H. (2010). Hollow palladium–cobalt bimetallic nanospheres as an efficient and reusable catalyst for Sonogashira-type reactions. *Journal of Materials Chemistry*, *20*, 4366.

Liang, M., Wang, X., Wang, Y., Liu, H., & Liu, H. (2008). Excellent catalytic properties over nanocomposite catalysts for selective hydrogenation of halonitrobenzenes. *Journal of Catalysis*, *255*(2), 335–342. Available from https://doi.org/10.1016/j.jcat.2008.02.025.

Mazumder, V., Lee, Y., & Sun, S. (2010). Recent development of active nanoparticle catalysts for fuel cell reactions. *Advanced Functional Materials*, *20*(8), 1224–1231. Available from https://doi.org/10.1002/adfm.200902293.

Merlo, A. B., Machado, B. F., Vetere, V., Faria, J. L., & Casella, M. L. (2010). PtSn/SiO(2) catalysts prepared by surface controlled reactions for the selective hydrogenation of cinnamaldehyde. *Applied Catalysis A: General*, *383*, 43–49.

Noll, W. (1968). *Chemistry and technology of silicones*. New York: Academic Press.

Pacardo, D. B., Sethi, M., Jones, S. E., Naik, R. R., & Knecht, M. R. (2009). Biomimetic synthesis of Pd nanocatalysts for the Stille coupling reaction. *ACS Nano*, *3*, 1288–1296.

Pacardo, D. B., & Knecht, M. R. (2013). Exploring the mechanism of Stille C–C coupling viapeptide-capped Pd nanoparticles results in low temperature reagent selectivity. *Catalysis Science Technolology*, *3* (3), 745–753. Available from https://doi.org/10.1039/c2cy20636f.

Patra, A. K., Dutta, A., & Bhaumik, A. (2012). Highly ordered mesoporous TiO2-Fe2O3 mixed oxide synthesized by sol-gel pathway: An efficient and reusable heterogeneous catalyst for dehalogenation reaction. *ACS Applied Materials & Interfaces*, *4*(9), 5022.

Pourabdollah, K., Zarringhalam Moghaddam, A., Kharrat, R., & Mokhtari, B. (2011). Improvement of heavy oil recovery in the VAPEX process using montmorillonite nanoclays. *Oil & Gas Science and Technology – Revue d'IFP Energies Nouvelles*, *66*(6), 1005–1016. Available from https://doi.org/10.2516/ogst/2011109.

Ramsden, J. (2011). *Nanotechnology: An introduction*. MA, USA: William Andrew.

Reetz, M. T., & Lohmer, G. (1996). Propylene carbonate stabilized nanostructured palladium clusters as catalysts in Heck reactions. *Chemical Communication*, 1921.

Romero, R., Martinez, S. L., & Natividad, R. (2011). Biodiesel production by using heterogeneous catalysts. In M. Manzanera (Ed.), *Alternative fuel* (pp. 372–379). InTech, ISBN: 97809530307-372-9.

Saha, D., Dey, R., & Ranu, B. C. (2010). A simple and efficient one-pot synthesis of substituted benzo[b]furans by Sonogashira coupling–5-endo-dig cyclization catalyzed by palladium nanoparticles in water under ligand- and copper-free aerobic conditions. *European Journal of Organic Chemistry*, 6067.

Song, S., Liang, Y., Li, Z., Wang, Y., Fu, R., Wu, D., & Tsiakaras, P. (2010). Effect of pore morphology of mesoporous carbons on the electrocatalytic activity of pt nanoparticles for fuel cell reactions. *Applied Catalysis B, Environmental*, *98*(3), 132–137. Available from https://doi.org/10.1016/j.apcatb.2010.05.021.

Speier, J. L., Webster, J. A., & Barnes, C. H. (1957). The addition of silicon hydrides to olefinic double bonds. Part II. The use of group VIII metal catalysts. *Journal of the American Chemical Society*, *79*, 974–979.

Thanh, L. T., Okitsu, K., Boi, L. V., & Maeda, Y. (2012). Catalytic technologies for biodiesel fuel production and utilization of glycerol: A review. *Catalysts*, *2*, 191–222. Available from https://doi.org/10.3390/catal2010191.

Wang, Y., Biradar, A. V., Duncan, C. T., & Asefa, T. (2010). Silica nanosphere-supported shaped Pdnanoparticles encapsulated with nanoporous silica shell: Efficient and recyclable nanocatalysts. *Journal of Materials Chemistry*, *20*, 7834.

Wei, W., & Chen, W. (2012). "Naked" pd nanoparticles supported on carbon nanodots as efficient anode catalysts for methanol oxidation in alkaline fuel cells. *Journal of Power Sources*, *204*, 85–88. Available from https://doi.org/10.1016/j.jpowsour.2012.01.032.

Wen, L. B., Wang, Y., Lu, D. L., Hy, S. Y., & Han, S. Y. (2010). Preparation of KF/CaO nanocatalysts and its application in biodiesel production from Chinese tallow seed oil. *Fuel*, *89*, 2267–2271.

Widegren, J. A., & Finke, R. G. (2003). A review of the problem of distinguishing true homogeneous catalysis from soluble or other metal-particle heterogeneous catalysis under reducing conditions. *Journal of Molecular Catalysis A: Chemical*, *191*, 187.

Xu, S., Kim, E. H., Wei, A., & Negishi, E. (2014). Pd- and Ni-catalyzed cross-coupling reactions in the synthesis of organic electronic materials. *Science and Technology of Advanced Materials*, *15*(4), 044201. Available from https://doi.org/10.1088/1468-6996/15/4/044201.

Yang, S., Cao, C., Sun, Y., Huang, P., Wei, F., & Song, W. (2015). Nanoscale magnetic stirring bars for heterogeneous catalysis in microscopic system. *Heterogeneous Catalysis*, 2661–2664. Available from https://doi.org/10.1002/anie.201410360.

Ying, B., Jiajian, P., Jiayun, L., Guoqiao, L., & Xiaonian, L. (2011). Preparation methods and application of heterogeneous catalysts for hydrosilylation. *Progress in Chemistry*, *23*, 2466–2477.

Ying, M., & Chen, G. Y. (2007). Study on the production of biodesial by magnetic cell biocatylyst based on lipase-producing Bacillus subtilis. *Applied Biochemistry and Biotechnology*, *137*, 793–803.

Zhang, D., Wang, J., Cheng, X., Li, T., & Zhang, A. (2012). Synthesis of heterogeneous shape-controllable nano-hyperbranched polymer/Pt(0) catalyst with high catalytic activity in hydrosilylation. *Macromolecular Research*, *20*, 549–551.

Zhang, Y. K., Liao, S. J., Xu, Y., & Yu, D. R. (2000). Catalytic selective hydrogenation of cinnamaldehyde to hydrocinnamaldehyde. *Applied Catalysis A—General*, *192*, 247–251.

Zhou, J., Meng, L., Feng, X., Zhang, X., & Lu, Q. (2010). One-pot synthesis of highly magnetically sensitive nanochains coated with a highly cross-linked and biocompatible polymer. *Angewandte Chemie International Edition*, *49*, 8476–8479.

Zhou, Y., & Zhang, D. (2015). Nano PtCu binary and PtCuAg ternary alloy catalysts for oxygen reduction reaction in proton exchange membrane fuel cells. *Journal of Power Sources*, *278*, 396–403. Available from https://doi.org/10.1016/j.jpowsour.2014.12.08.

Zhu, F., Kim, J., Tsao, K., Zhang, J., & Yang, H. (2015). Recent development in the preparation of nanoparticles as fuel cell catalysts. *Current Opinion in Chemical Engineering*, *8*, 89–97. Available from https://doi.org/10.1016/j.coche.2015.03.005.

CHAPTER 12

Lignin removal from pulp and paper industry waste streams and its application

Vivek Yadav[1], Adarsh Kumar[2], Muhammad Bilal[3], Tuan Anh Nguyen[4] and Hafiz M.N. Iqbal[5]
[1]State Key Laboratory of Crop Stress Biology in Arid Areas, College of Horticulture, Northwest A&F University, Yangling, P.R. China
[2]Department of Environmental Microbiology, School for Environmental Science, Babasaheb Bhimrao Ambedkar University (A Central University), Lucknow, India
[3]School of Life Science and Food Engineering, Huaiyin Institute of Technology, Huai'an, P.R. China
[4]Institute for Tropical Technology, Vietnam Academy of Science and Technology, Hanoi, Vietnam
[5]Tecnologico de Monterrey, School of Engineering and Sciences, Monterrey, Mexico

12.1 Introduction

Lignin is an important source of bioenergy on planet earth and is utilized extensively for value-added products and bioenergy sources (Fig. 12.1). Lignocellulosic biomass is composed of hemicellulose, cellulose, and lignin monomers (Cho & Park, 2018; Sagues et al., 2018). The broad availability and low economic values have increased the interest of researchers in its use to generate better values and more sustainable use (Bilal et al., 2020; Bilal & Iqbal, 2020; Ferreira et al., 2020). Lignocellulosic biomass is the most abundant and considerably cheaper than crude oil. The major constituents of these substances are cellulose (35%–50%), hemicellulose (20%–35%), and lignin (10%–25%) (Fatma et al., 2018). Lignocellulosic biomass mainly includes agriculture residues, waste from agro-industries. Lignocellulose is the most abundant biorenewable material in the world. The majority of lignocellulosic biomass is derived from nontree plant species. Lignocellulose is the most abundant biorenewable material on the planet earth. In comparison to animal manure, lignocellulosic biomass is characterized by a low buffer capacity and high C/N ratios. The traditional use of lignocellulosic biomass is for composting and bulking agents. Basically, lignin was considered as a low-value biowaste, and no importance was given to these polymers for industrial uses (Luo & Abu-Omar, 2017; Ragauskas et al., 2014). Nevertheless, the traditional approach has evolved, and it has been utilized for value addition and energy perspective (Agrawal et al., 2014; Ľudmila, Michal, Andrea, & Aleš, 2015; Luo & Abu-Omar, 2017; Ragauskas et al., 2014). Pulp and paper are considered as one of the largest industrial commodities in the world. These industries have many influences on global greenery share.

Nanotechnology in Paper and Wood Engineering
DOI: https://doi.org/10.1016/B978-0-323-85835-9.00019-2

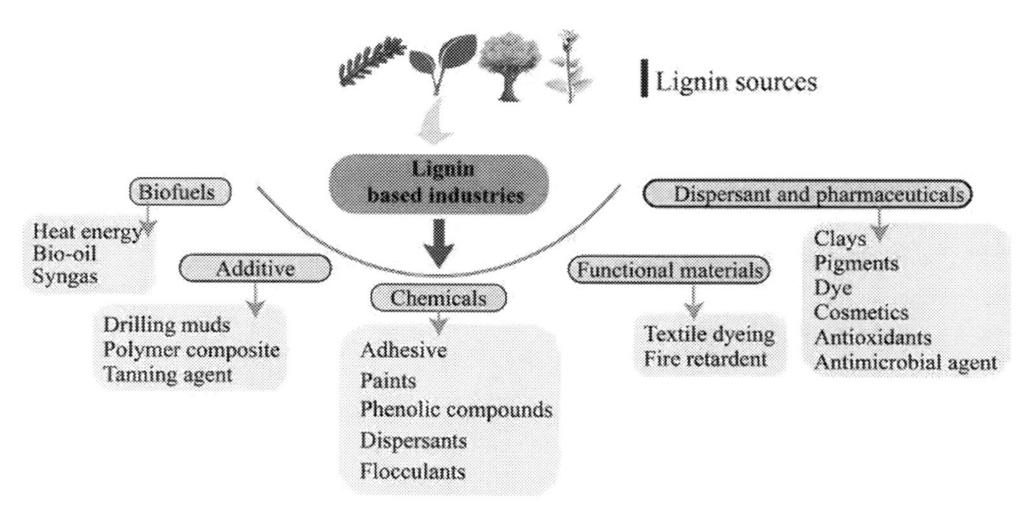

Figure 12.1 Lignin-based industries and application in different value-added products.

With life advancing, the share of pulp and paper industries continues to increase. The effluent and wastewater generated from these industries include organic and inorganic contaminants that mostly originate from tannins, lignins, resins, and chlorine compounds (Zainith et al., 2019). These industries are considered to be among the biggest producers of lignin biopolymers. In total, these industries generate between 50 and 70 mt of lignin per year. Some researchers assumed this could hit the mark of 225 mt $year^{-1}$ by 2030. (Luo & Abu-Omar, 2017; Mandlekar et al., 2018). Pulp and paper industries are complex structure industries with various mills, products, and processes. The paper and pulp mills are majorly divided into three types including pulp mills, recyclable paper processing, and hybrid mills. Pulp and paper industries require a huge number of resources including water and wood, and generates a huge quantity of waste materials. In general, 905.8 million m3 of water is needed in the paper and pulp industries, and 695.7 million m3 of wastewater is generated annually (Elango et al., 2017). The basic major steps in paper industries are pulping followed by bleaching. In the initial step, pulping is the major source of pollutant generation. Different methods and technologies evolve with time, and manufacturing and processing steps are much more advanced than ever. Therefore the waste varies significantly in terms of waste type and quantity, and water consumption is not the only problem, but solid wastes, including sludge generated from wastewater and air emissions, are major threatening concerns. The lignin-related compounds and lignin are a part of the effluent waste generated by these industries. Waste effluents containing substances derived from lignin, disturb aquatic life and damage the natural ecosystem. (Haq & Raj, 2020; Raj et al., 2014). The present wastewater treatment techniques are not efficient in terms of labor requirement and cost and they also hold residual effects (Singhal &

Thakur, 2009). In order to address the problems of wastewater from paper and pulp industries, different methods, including biological, chemical, and physical treatment methods, have been implicated (Fu & Viraraghavan, 2001; Pokhrel & Viraraghavan, 2004; Ragunathan & Swaminathan, 2004; Santos et al., 2002; Taseli, & Gokcay, 1999; Tezel et al., 2001). Different treatment methods have positive and negative sides. For instance, biological methods are not efficient for cellulose, long-chain polymers, and cyclic group of lignans. Similarly, biological methods are not useful for color removal. In general, in biochemical methods, less biodegradability index indicates that the specific wastewater cannot be treated well by biochemical methods (Khansorthong & Hunsom, 2009). Activated sludge has many shortcomings, including poor settling characteristics, sensitivity to shock, toxicity, and the inability of biodegrading poorly biodegradable substances such as lignin. Lignin removal is specific to the use of purified enzymes, crude enzymes, and whole culture preparations with various fungal strains (Srivastava et al., 2005; Munir et al., 2015; Bilal & Asgher, 2016; Xia et al., 2019; Li et al., 2019a, b; Fernandes et al., 2020). However, the fungal strains are less efficient due to their ability to adopt adverse environmental conditions. Notably, some bacterial strains showed the potential to produce unique sets of enzymes that can utilize low molecular lignans and produce economically important compounds (Raj et al., 2007a, b; Chandra et al., 2007).

12.2 Lignin: biosynthesis to utilization

Lignin constitutes the second most profusely available polymer on the earth after cellulose. It is an encrusting material in which the cellulose microfibrils are embedded, and typically 15%–40% of tissue mass in terrestrial plants is lignin. It is a natural complex phenolic polymer that exists in all vascular plants. All terrestrial plants contain lignin polymers, and some aquatic organisms contain either lignin or lignin-like compounds. The biosynthesis method of lignin has attracted a strong interest, as lignin is the major contributing factor to the recalcitration of biomass feedstocks (Bilal & Iqbal, 2020; Studer et al., 2011).

12.2.1 Nature of lignin

Naturally, lignin is a strong fiber that is not easily decomposable. Relating to strong bonds, it also seems to have abundant hydrogen bonds. It is linked in a complex and diverse manner to carbohydrate (hemicellulose) in wood. It is the major structural feature of the cell wall of plants. Basically, it facilitates water transport, reinforces the cell wall, and acts as a defense barrier for biotic pathogens. Lignin is a generic term that is used to represent a large group of aromatic polymers. High complicity in the nature of lignin provides enough evidence that it could be a part of various functions in plants. Many functions are still unknown and basic characters of lignans monomers and unite are discovered. The main building blocks of lignin are syringyl alcohol, coniferyl alcohol, and coumaryl alcohol units derived from the polymerization of the hydroxycinnamyl alcohols, p-coumaryl, coniferyl, and sinapyl alcohol, respectively (Boerjan et al., 2003).

12.2.2 Overview of lignin: biosynthesis and distribution

The biosynthesis process of lignin begins from the general phenylpropanoid pathway. This pathway also includes the shikimate intermediates (Fig. 12.2). There are two main stages in the development of lignin: monolignol biosynthesis and monolignol polymerization by free-radical coupling. Tyrosine and phenylalanine act as an initial substrate (Barros et al., 2016). Ammonia lyase activity converts phenylalanine cinnamic and p-coumaric acids, respectively. Further, after reducing carboxylic acid, the biosynthesis of hydroxycinnamyl alcohols instigates (Vanholme et al., 2013). The reduction of carboxylic acid moiety proceeds with reductase enzymes. Finally, the cinnamic acid is converted to the basic building blocks of lignin. Altogether, guaiacyl (G), syringyl (S), and p-hydroxyphenyl (H) units are considered lignin units (Lu et al., 2015). These

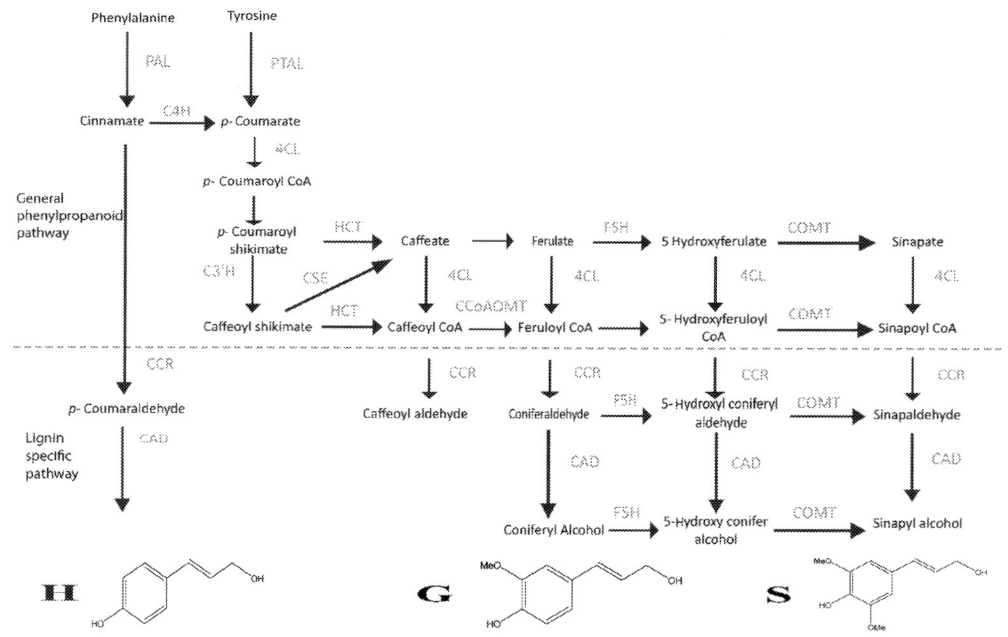

Figure 12.2 Overview of lignin biosynthesis by the phenylpropanoid pathway. Enzymes and intermediates are shown in a different color. *CAD*, Caffeoyl-CoAO methyltransferase; *CCoAOMT*, cinnamic acid 4-hydroxylase; *CCR*, cinnamoyl-CoA reductase; *C3′H*, p-coumaroyl shikimate 3′-hydroxylase; *C4H*, (hydroxy)cinnamyl alcohol dehydrogenase; *4CL*, 4-hydroxycinnamoyl-CoA ligase; *COMT*, caffeic acid/5-hydroxyferulic acid O-methyltransferase; *CSE*, caffeoyl shikimate esterase; *F5H*, coniferaldehyde/ferulate 5-hydroxylase; *HCT*, hydroxycinnamoyl-CoA: shikimate/quinate hydroxycinnamoyl transferase; *PAL*, phenylalanine ammonia-lyase; *PTAL*, phenylalanine tyrosine ammonia-lyase.

basic lignin units are synthesized in the cytoplasm and transferred to cell wall units. Lignin composition and amount varied greatly from extraction source and were influenced by many different factors, including environmental and developmental factors. Moreover, a vast difference has been discovered in lignin composition and amount between the plant of the same species. For instance, the monolignol ratio (S/G) ranged from 1.0 to 3.0, and a fraction of lignin ranged from 15.7% to 27.9% in the cell wall of *Populus trichocarpa* (Studer et al., 2011). Similarly, cellulose, hemicellulose, and lignin are not equally dispersed within the cell walls of the same species and different plant species. The arrangement and quantity of these components of the plant cell wall vary depending on the organisms, tissues, and maturity of the plant cell wall.

12.2.3 Sources of lignin waste generation

Lignin is derived from lignocellulose biomass, and it represents nearly 30% of the lignocellulose biomass. Lignin is a three-dimensional polymer produced from the phenylpropanoid pathway. Apart from mechanical strength to the secondary cell wall, it works as a glue that provides compressive binding to plant tissue and individual fiber.

12.2.4 Industrial sources of lignin

12.2.4.1 Kraft lignin

This specific type of lignin is produced during the process of sulfate (kraft) cooking. It is an important lignin biowaste from pulp industries. The volume of kraft lignin is 85% of the total world lignin production (Tejado et al., 2007). The kraft lignin is different from native lignin in terms of phenolic hydroxyl groups. Annually, a significant part of such lignin is used for energy production, and a small share of 2% is used for value-added products (Lora, 2008). Kraft lignin has found many value-added products, but most of them are small-scale production units. The high value products from kraft lignin could be pesticides, fertilizer, carbon fiber, thermoplastic polymers, binders, and resins (Christopher, 2012).

12.2.4.2 Soda lignin

Soda lignin is a byproduct of soda-anthraquinone pulping processes. Basically, soda lignin is produced from nonwoods fibers, straws, and flax. This product is mostly utilized in developing countries of Asia and South America, as a source for the production of paper and other hardwood products (Lora, 2008). The production of such industries is minimal due to the availability of annual feedstock. Soda lignin is a sulfur-free product and closer to natural lignin in comparison to kraft lignin.

12.2.4.3 Organosolv lignins

The lignin produced from organosolv pulping process is considered as an organosolv lignin. These are low in molecular weight and high chemical purity. Organosolv lignins are poor in water solubility due to their hydrophobic nature. Chemically these are very pure forms of lignin with variable chemical reactive chains. Based on this, these lignins are suitable to use as a substitute for kraft and soda lignin for value-added products in many industries. The basic utilization is the formulation of inks.

12.2.4.4 Lignosulphonates/sulfite lignin

Lignin, which is extracted by the sulfite process, is called sulfite lignin or lignosulphonates. Sulfite process is a popular and common process used in many industries and is based on neutral or acidic treatment with different sulfates or bisulfates. Sulfite lignin is water soluble in nature with several charged groups. Chemically this type of lignins are higher in molecular weight with a high amount of ash content. Sulfite lignin is used to produce the number of value-added products such as colloidal suspensions, dispersing agents, animal feeds (Lora, 2008), particleboards (Jin et al., 2015), surfactants, and cement additives (Grierson et al., 2005; Ansari & Pawlik, 2007).

12.3 Techniques for lignin removal

It has been earlier described that lignin is found in different conformations and has different removal processes. Wastewater contaminants of paper industries mainly consist of organic pollutants and lignin and its derivative compounds in recalcitrant in nature. The lignins were responsible for the brownish color of pulp—paper mill wastewater. The pulp—paper mill wastewater color reduction is also indicated the removal of lignin component. Lignin is not only the problem, but it has a broad range of applications in different industries like textile, biofuels, chemicals, food, pharmaceutical, biochemical, and cosmetics (Asgher et al., 2018; Bilal et al., 2018; Luo et al., 2020). However, lignin removal from the paper industry is not a target for its applications for value-added products (Cotana et al., 2014). The different techniques for lignin removal and its recovery have been developed.

12.3.1 Physicochemical processes

Several methods are offered for the lignin removal from agro-waste and paper mill wastewater, that is, adsorption, coagulation—precipitation, membrane technologies, advanced oxidation process, and ozonation, etc. However, its industrial application is also implemented in several industries.

12.3.1.1 Coagulation and precipitation

The separation of suspended particles and organic pollutants from the solution through flocs formation for the settlement of the particles is known as coagulation—precipitation. The organic particles and coagulants may create interactive forces at the controlled temperature, pH, and dosage of coagulants during the coagulation method. There are four steps in the completion of the coagulation process, that is, enmeshment, adsorption, neutralization of charge, and precipitation. As two significant stages of charged neutralization and sweep flocculation, the coagulation process can be simplified even further.

The adsorption and enmeshment on cationic metal hydroxides for the impurity's removal will take place during flocculation/coagulation for the precipitation of organic contaminants (Duan & Gregory, 2003). The charge neutralization process takes place when positive- and negative-charged particles clouded with counter ionic particles. The pH range for charge neutralization, 4–5.5, is the most desirable (Chang et al., 1993). The addition of coagulants contributes to colloid destabilization. A higher counter-ion valence, greater destabilizing effect, and lower dose are needed for the coagulation process. The colloid particles are absorbed on the surface of the coagulant and then destabilized and precipitated. Researchers have tried many coagulants for lignin removal. The aluminum and inorganic iron salt are utilized as a primary coagulant for wastewater treatment. Garg et al. (2010) also examined that ammonium alum, aluminum sulfate and ferrous sulfate were used as the best coagulant. According to Jian-Ping et al. (2011) aluminum chloride, modified natural polymer, starch-g-PAM-g-PDMC [polyacrylamide and poly(2-methacryloyloxyethyl)trimethyl ammonium chloride] are also used as a coagulant for lignin removals. Moreover, coagulation using poly aluminum chloride and copper sulfate was examined by Kumaret al. (2011). Therefore oxotitanium sulfate and aluminum sulfate mixture as a coagulant were more effective for lignin recovery (Chernoberezhskii et al., 2002). The electrochemical method is technically as well as economically feasible for the wastewater treatment and bioremediation process of industrial wastewater on large scale. In this process, the soluble anode is inserted inside the wastewater and dissolution of metal hydroxide takes place and forms flocks to separate. The electrocoagulation technique may be considered for the decolorization of pulp and paper industry wastewater (Table 12.1).

12.3.1.2 Adsorption

The adsorption method is very effective for color and organic pollutants removal from the wastewater of different industries. Different absorbents are also used for the color and lignin removal from wastewater of pulp and paper mill and these adsorbents are activated carbon, silica, coal ash, etc. (Kamali & Khodaparast, 2015). Color from wastewater has been removed up to 90% through activated carbon, fuller soil, charcoal, and coal ash (Murthy et al., 1991). The same results were also reported by Shawwa et al. (2001), on the use of activated coke for the color and chemical oxygen demand (COD) removal, up to 90%, of the bleached paper mill wastewater through adsorption methods. Lignin removal was also recorded through the adsorption process by the use of blast furnace dust (80.4%) and slag (61%) (Das & Patnaik, 2000). The 41.38% color and 60.87% COD

Table 12.1 Various coagulants were used for lignin removal.

Coagulants	Chemical formula	Coagulation mechanism	Workable pH range	Hydrolysis product charge	Reported removal of studied parameter	References
Aluminum sulfate	$Al_2(SO)_3 \cdot 18H_2O$	Charged neutralization and sweep flocculation	2.0–7.0	Positive	Kraft lignin recovery 80%	Chernoberezhskii et al. (2002)
Aluminum chloride	$Al(Cl)_3$	Charged neutralization and sweep flocculation	2.0–8.0	Positive	Color removal 90%	Jian-Ping et al. (2011)
Copper sulfate	$CuSO_4$	Sweep	2.0–8.0	Positive	Color removal 76% and COD removal 74%	Kumar et al. (2011)
Commercial alum	$KAl(SO_4)_2 \cdot 12H_2O$	Sweep	4.5–7.0	Positive	Color removal 90% and COD removal 63%	Garg et al. (2010)
Electrocoagulation	N-A	Sweep	4.0–7.0	Positive	Lignin removal (Al and Fe anode) 80%	Garg et al. (2010)
Ferrous sulfate	$FeSO_4$	Sweep	4.0–7.0	Positive	COD removal 60%	Garg et al. (2010)
Mixture of oxotitanium sulfate and aluminum sulfate	$TiOSO_4$; $2H_2O$ $Al_2(SO_4)_3$; $18H_2O$	Parallel heterocoagulation	2.0–8.0	Multicomponent solution of both positive and negative charges	Kraft lignin recovery 90%	Chernoberezhskii et al. (2002)
Oxotitanium sulfate	$TiOSO_4$; $2H_2O$	Heterocoagulationmechanism	2.0–4.0	Negative	Virtually complete kraft lignin recovery	Chernoberezhskii et al. (2002)
Polyaluminum chloride	PAC	Charged neutralization and sweep flocculation	2.0–8.0	Positive	Color removal 90% and COD removal 80%	Kumar et al. (2011)

Notes: *COD*, Chemical oxygen demand.

were also removed by the use of polyaluminum silicate chloride (400 mg L^{-1}), and cheap adsorbent bentonite (450 mg L^{-1}) through tertiary adsorption coagulation treatment (Xilei et al., 2010).

12.3.1.3 Membrane technologies

Membrane techniques is most effective and extensively used for the treatment of various industrial wastewater recently. But it is very difficult to use this system on a wide scale because of technological limitations and high costs (Greenlee et al., 2010). Therefore by using pretreatment methods, the efficiency of this method can be increased. The removal of color, biochemical oxygen demand (BOD), total dissolved solids, and COD through the membrane electrochemical reactor from paper mill wastewater has been also reported by Chanworrawoot and Hunsom (2012). Gönder et al. (2012) reported that COD (89%), sulfate (97%), total hardness (83%), coefficient of spectral absorption (95%), and conductivity (50%) removed from the paper mill wastewater water by the technique ultrafiltration membranes at pH 10.

12.3.1.4 Ozonation

The ozonation techniques are used for disinfection of the wastewater with a broad range of applications in wastewater treatment. But, according to Yamamoto (2001) ozonation process was used for the decolorization and detoxification of color, COD, and various contaminants from the industrial wastewater. Moreover, photocatalysis and ozonation methods were effective for the treatment of COD, total organic carbon, color, and toxicity of the pulp–paper industry effluents (Torrades et al., 2001; Yeber et al., 1999). If the high dose of ozone for 15 min were utilized for the removal of color, it decreased up to 95%–97% (Sevimli & Sarikaya, 2002). Furthermore, the ozonation technique for the pulp–paper mill wastewater treatment improved BOD, COD, and lignin by 40%, 11%, and 46% respectively, but it was insufficient for the color removal (Ruas et al., 2007). The degradation and removal of lignin depends on the ozone dose during the ozonization phase of the solution of alkali lignin (Michniewicz et al., 2012).

12.3.1.5 Advanced oxidation processes

AOP technique is successful for the lignin removal from the pulp–paper industry wastewater (Abedinzadeh et al., 2018). This process is also known for the detoxification of the organic substances from the wastewater and also effective for the nonbiodegradable compounds which are not degraded by microorganisms properly (Merayo et al., 2013). The most important things of this technique are the formation of most reactive oxidizing species, that is, hydroxyl radicals which are nonselective by nature, also come under the mechanism of AOPs (Al-Rasheed, 2005).

12.3.2 Removal of lignin by biological means

The biological removal of the lignin from paper mill wastewater is a valid alternative source for the elimination of the chemical treatment process. Different ligninocellulytic microorganisms (actinomycetes, bacteria, and fungi) secrete ligninolytic enzymes, that is, laccase, lignin peroxidase, and manganese peroxidase which are involved in the detoxification and biodegradation of lignin (Bilal et al., 2017a,b,c; Kumar & Chandra, 2020; Kumar et al., 2020a; Singh et al., 2020a,b). Some potential bacterial species were used for the delignification of paper mill wastewater such as lignin degradation of 70%–80% (*Pseudomonas putida and Acinetobacter calcoaceticus*), COD removal (*Aeromonas formicans* and *Bacillus* sp.), etc. (Brown & Chang, 2014; Gupta et al., 2001; Raj et al., 2007; Singh et al., 2020b). The most efficient lignin degradation by a white-rot fungus has been reported due to their capability of secreting ligninolytic enzymes also received our attention (Asgher et al., 2016a,b,c; Leonowicz et al., 1999). This fungus uses lignin as a carbon source and degrades it. Moreover, several *Basidiomycetes* fungal species are also reported for lignin degradation by their extracellular ligninolytic enzymes which require low-molecular-weight mediators (Kumar & Chandra, 2020; Kumar et al., 2020a,b). This is advanced biological technique for the removal of lignin by the use of *Aspergillus foetidus, Phanerochaete chrysosporium* and *Trametes versicolor* (Fu & Viraraghavan, 2001). Furthermore, the fungus *P. chrysosporium* is well-known for the potential of degrading and detoxifying xenobiotic contaminants present in wastewater (Kumar & Chandra, 2020). Basidiomycete *Bjerkandera adusta* is mostly found in Europe. This fungus can degrade xenobiotic compounds and also capable of lignin degradation of paper mill wastewater (Sodaneath et al., 2017). It has been already reported that *B. adusta* has the ability of lignin biomineralization of soil and decolorization of industrial dye due to its lacquer and manganese peroxidase activity, but not to date at the industrial stage (Anastasi et al., 2010; Wang et al., 2002) (Table 12.2).

12.4 Gainful utilization of lignin

Cho and Park (2018) recently focused on the use of lignin for its industrial application and production of various value-added products. Most of the bacterial species have the potential of secretion of ligninocellulytic enzymes that convert lignin into bioplastic and polyhydroxyalkanoates (PHAs), that is, *Bacillus megaterium*, *Cupriavidus basilensis* B-8, *Ralstonia eutropha* H16, and *P. putida* KT2440 (Gasser et al., 2014; Shi et al., 2017). Moreover, PHAs are produced by microorganisms which have a broad range of application such as biomedical materials, films, biocompatible drug delivery, coatings, and organic and inorganic coatings (Brodin et al., 2017). The use of lignin by *P. putida* to convert lignin into PHA and the mechanism of conversion were also investigated (Salvachúa et al., 2015). Several strains of *Rhodococcus* have been well studied for their

Table 12.2 Summary of biological removal of lignin.

Microorganisms	Lignin reduction (%)	References
Bacteria		
Bacillus amyloliquefaciens SL-7	28.55	Mei et al. (2020)
Bacillus velezensis	40.39	Verma et al. (2020)
Arthrobacter sp. C2	40.1	Jiang et al. (2019)
Pseudomonas fluorescens DSM 50,090 and *Rodococcus. opacus* DSM1069	80	Ravi et al. (2019)
Serratia liquefaciens	72	Haq et al. (2017)
Paenibacillus sp. strain LD-1	54	Raj et al. (2014)
Citrobacter freundii (FJ581026) and *Citrobacter* sp. (FJ581023)	71	Chandra and Bharagava (2013)
Bacillus cereus (ITRC-S6) and *Serratia marcescens* (ITRC-S7)	30–42	Chandra et al. (2007)
Bacillus sp.	37	Raj et al. (2007)
Bacillus subtilis and *Micrococcus luteus*	97	Tyagi et al. (2014)
C. freundii and *S. marcescens*	65	Abhishek et al. (2015)
Fungus		
Trametes villosa	35.68	Silva et al. (2014)
Trametes versicolor	15	Plácido et al. (2013)
Ganoderma lucidum	39.6	Asgher et al. (2013)
Pycnoporus cinnabarinus	40	Camarero et al. (2004)
Shizophyllum commune	47.5–63.6	Asgher et al. (2016c)
Pleurotus sapidus	53.1	Asgher et al. (2016b)
Pleurotus eryngii	39.15–56.9	Asgher et al. (2016a)
Merulius aureus and *Fusariumsam bucinum*	79	Malaviya and Rathore (2007)
Gliocladium virens	52	Kamali and Khodaparast (2015)
Emericella nidulans var. nidulans	37	Singhal and Thakur (2009)
Phanerochaete chrysosporium MTCC No. 787	71	Chopra and Singh (2012)
Aspergillus fumigates MTCC No. 3377	51	Chopra and Singh (2012)
Aspergillus flavus strain F10	39–61	Barapatre and Jha (2016)
Cunninghamella echinulata FR3	30	Xie et al. (2015)
Pleurotus ostreatus	37.7–46.5	Li et al. (2019)
Echinodontiumtaxodii	66.7–73.3	Shi et al. (2013)

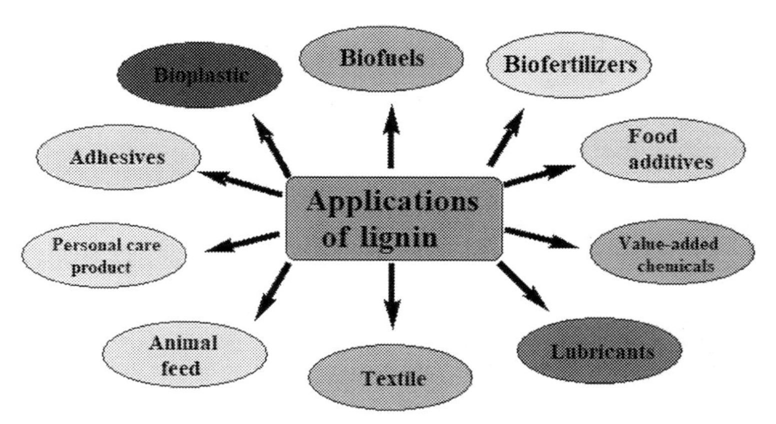

Figure 12.3 Applications of lignin in different areas.

metabolic pathways, β-*ketoadipate* pathway of lignin compound for lipid synthesis and bioaccumulation by lignin compounds (He et al., 2017). Different studies have revealed the ability of *Rodococcus opacus* to use lignin compounds to produce triglycerides (Wells et al., 2015). The microorganism may catabolites the aromatic compounds and sources of carbon into its metabolite such as pyruvate from different matrices. The synthesis of value-added products from the microorganisms by their primary or secondary metabolic activities produced, for example, cis, cis-muconate, which acts as a functional chemical substitute of microbes (Rodriguez et al., 2017). The cis, cis-muconate was genetically modified by *P. putida* from the lignin and *P. putida* utilized aromatic compounds and it contained very little biomass. Vardon et al. (2016), developed cis-cis-muconate accumulating *P. putida* KT2440 which could assimilate with a higher yield of 0.7 g L^{-1} cis-cis-muconate. This is a metabolite in the metabolism of catechol which is an intermediate in β-ketoadipate pathway, when accumulated blocks the consecutive conversion route (Xie et al., 2016). In addition, pyruvate is intermediate that can be converted into value-added products such as lactate, alcohols, amino acids, and terpenoids in different metabolic pathways. Therefore for advancement and improved application possibilities, the exceptional adaptability of bacterial ligninolytic enzymes to function under a wide variety of circumstances is advantageous (Fig. 12.3).

12.5 Conclusion

Lignocellulosic ethanol has been identified as a green alternative for meeting energy needs. Paper and pulp industries are an important source of various contaminants and a significant amount of wastewater depending on the type of processes used in industries. The pulp and paper industry has the benefit of having byproducts that can be used for renewable energy resources. Lignin is a vital waste that has excellent value. Wastewater is highly contaminated with lignin and lignin derivatives. Lignin valorization could be a better aspect of generating income from paper and pulp industries. Lignin depolymerization is a highly beneficial strategy. Despite its dynamic nature, the use of advanced depolymerization methods in the presence of catalysts/biocatalysts is beneficial. Here, it is significant to study the various processes/methods involved in lignin extraction and valorization. Over the past several decades, advances have fostered the engineering of metabolic pathways in fungal and bacterial strains to make them more productive. The plethora of available research on biological lignin extraction methods continues to illustrate evidence of its importance and efficiency. Some desirable mechanisms are still fragmentary, and more research is required to bring things down to the threshold. A metabolic-pathway technique could be used as a critical tool in making the lignin extraction method more effective.

References

Abedinzadeh, N., Shariat, M., Monavari, S. M., & Pendashteh, A. (2018). Evaluation of color and COD removal by Fenton from biologically (SBR) pre-treated pulp and paper wastewater. *Process Safety and Environmental Protection*, *116*, 82–91. Available from https://doi.org/10.1016/j.psep.2018.01.015.

Abhishek, A., Dwivedi, A., Tandan, N., & Kumar, U. (2015). Comparative bacterial degradation and detoxification of model and Kraft lignin from pulp paper wastewater and its metabolites. *Applied Water Science*, 7(2). Available from https://doi.org/10.1007/s13201-015-0288-9.

Agrawal, A., Kaushik, N., & Biswas, S. (2014). Derivatives and applications of lignin—an insight. *The SciTech Journal*, *1*(7), 30–36.

Al-Rasheed, R.A., 2005. Water treatment by heterogeneous photocatalysis an overview. *Presented at 4th SWCC acquired experience symposium*, Jeddah, Saudi Arabia.

Anastasi, A., Spina, F., Prigione, V., Tigini, V., Giansanti, P., & Varese, G. C. (2010). Scale-up of a bioprocess for textile wastewater treatment using *Bjerkandera adusta*. *Bioresource Technology*, *101*, 3067–3075.

Ansari, A., & Pawlik, M. (2007). Floatability of chalcopyrite and molybdenite in the presence of lignosulfonates. Part II. Hallimond tube flotation. *Minerals Engineering*, *20*(6), 609–616.

Asgher, M., Ahmad, Z., & Iqbal, H. M. N. (2013). Alkali and enzymatic delignification of sugarcane bagasse to expose cellulose polymers for saccharification and bio-ethanol production. *Industrial Crops and Products*, *44*, 488–495.

Asgher, M., Khan, S. W., & Bilal, Md (2016a). Optimization of lignocellulolytic enzyme production by *Pleurotus eryngii* WC 888 utilizing agro-industrial residues and bio-ethanol production. *Romanian Biotechnological Letters*, *21*(1), 11133.

Asgher, M., Ijaz, A., & Bilal, M. (2016b). Lignocellulose-degrading enzyme production by *Pleurotus sapidus* WC 529 and its application in lignin degradation/Lignoselüloz-çözücü enzim üretiminde *Pleurotus sapidus* WC 529 ve lignin parçalanmasındaki uygulamaları. *Turkish Journal of Biochemistry*, *41*(1), 26–36.

Asgher, M., Wahab, A., Bilal, M., & Iqbal, H. M. N. (2016c). Lignocellulose degradation and production of lignin modifying enzymes by *Schizophyllum commune* IBL-06 in solid-state fermentation. *Biocatalysis and Agricultural Biotechnology*, *6*, 195–201.

Asgher, M., Wahab, A., Bilal, M., & Iqbal, H. M. (2018). Delignification of lignocellulose biomasses by alginate—chitosan immobilized laccase produced from *Trametes versicolor* IBL-04. *Waste and Biomass Valorization*, *9*(11), 2071–2079.

Barapatre, A., & Jha, H. (2016). Decolourization and biological treatment of pulp and paper mill effluent by lignin-degrading fungus *Aspergillus flavus* Strain F10. *International Journal of Current Microbiology and Applied Sciences*, *5*, 19–32.

Barros, J., Serrani-Yarce, J. C., Chen, F., Baxter, D., Venables, B. J., & Dixon, R. A. (2016). Role of bifunctional ammonia-lyase in grass cell wall biosynthesis. *Nature Plants*, *2*(6), 1–9.

Bilal, M., & Asgher, M. (2016). Biodegradation of agrowastes by lignocellulolytic activity of an oyster mushroom, *Pleurotus sapidus*. *Journal of the National Science Foundation of Sri Lanka*, *44*, 4.

Bilal, M., Asgher, M., Iqbal, H. M., Hu, H., & Zhang, X. (2017a). Delignification and fruit juice clarification properties of alginate-chitosan-immobilized ligninolytic cocktail. *LWT- Food, Science and Technology*, *80*, 348–354.

Bilal, M., Asgher, M., Iqbal, H. M., Hu, H., & Zhang, X. (2017b). Biotransformation of lignocellulosic materials into value-added products—A review. *International Journal of Biological Macromolecules*, *98*, 447–458.

Bilal, M., Asgher, M., Iqbal, H. M., & Ramzan, M. (2017c). Enhanced bio-ethanol production from old newspapers waste through alkali and enzymatic delignification. *Waste and Biomass Valorization*, *8*(7), 2271–2281.

Bilal, M., & Iqbal, H. M. (2020). Ligninolytic enzymes mediated ligninolysis: an untapped biocatalytic potential to deconstruct lignocellulosic molecules in a sustainable manner. *Catalysis Letters*, *150*(2), 524–543.

Bilal, M., Nawaz, M. Z., Iqbal, H., Hou, J., Mahboob, S., Al-Ghanim, K. A., & Cheng, H. (2018). Engineering ligninolytic consortium for bioconversion of lignocelluloses to ethanol and chemicals. *Protein and Peptide Letters*, *25*(2), 108–119.

Bilal, M., Wang, Z., Cui, J., Ferreira, L. F. R., Bharagava, R. N., & Iqbal, H. M. (2020). Environmental impact of lignocellulosic wastes and their effective exploitation as smart carriers–A drive towards greener and eco-friendlier biocatalytic systems. *Science of the Total Environment*, *722*, 137903.

Boerjan, W., Ralph, J., Baucher, M. (2003). Lignin Biosynthesis. *Annual Review of Plant Biology*, *54*(1), 519–546. Available from https://doi.org/10.1146/annurev.arplant.54.031902.134938.

Brodin, M., Vallejos, M., Opedal, M. T., Area, M. C., & Chinga-Carrasco, G. (2017). Lignocellulosics as sustainable resources for production of bioplastics–a review. *Journal of Cleaner Production*, *162*, 646–664.

Brown, M. E., & Chang, M. C. (2014). Exploring bacterial lignin degradation. *Current Opinion in Chemical Biology*, *19*, 1–7.

Camarero, S., Garcıa, O., Vidal, T., Colom, J., del Rıo, J. C., Gutiérrez, A., ... Martınez, Á. T. (2004). Efficient bleaching of non-wood high-quality paper pulp using laccase-mediator system. *Enzyme and Microbial Technology*, *35*(2–3), 113–120.

Chandra, R., & Bharagava, R. N. (2013). Bacterial degradation of synthetic and kraft lignin by axenic and mixed culture and their methabolic products. *Journal of Environmental Biology*, *34*, 991–999.

Chandra, R., Raj, A., Purohit, H. J., & Kapley, A. (2007). Characterization and optimization of three potential aerobic bacterial strains for kraft lignin degradation from pulp paper waste. *Chemosphere*, *67*, 839–846.

Chang, Q., Fu, J. Y., & Li, Z. L. (1993). *Principles of Flocculation*. Lanzhou, China: Lanzhou University Press.

Chanworrawoot, K., & Hunsom, M. (2012). Treatment of wastewater from pulp and paper mill industry by electrochemical methods in membrane reactor. *Journal of Environmental Management*, *113*, 399–406.

Chernoberezhskii, Y. M., Dyagileva, A. B., Atanesyan, A. A., & Leshchenko, T. V. (2002). Influence of the kraft lignin concentration on the efficiency of its coagulation recovery from aqueous electrolyte solutions. *Russian Journal of Applied Chemistry*, 7, 1166–1169.

Cho, H. U., & Park, J. M. (2018). Biodiesel production by various oleaginous microorganisms from organic wastes. *Bioresource Technology*, *256*, 502–508.

Chopra, A. K., & Singh, P. P. (2012). Removal of color, COD and lignin from pulp and paper mill effluent by *Phanerochaete chrysosporium* and *Aspergillus fumigates*. *Journal of Chemical and Pharmaceutical Research*, *4*, 4522–4532.

Christopher, L. P. (2012). Integrated forest biorefineries: Current state and development potential. In L. P. Christopher (Ed.), *Integrated forest biorefineries: Challenges and opportunities* (pp. 1–66). Cambridge: Royal Society of Chemistry.

Cotana, F., Cavalaglio, G., Nicolini, A., Gelosia, M., Coccia, V., Petrozzi, A., Brinchi, L., 2014. Lignin as co-product of second generation bioethanol production from lignocellulosic biomass, *Energy Procedia*, *45*, 52–60.

Das, C. P., & Patnaik, L. N. (2000). Removal of lignin by industrial solid wastes. *Practice Periodical of Hazardous, Toxic, Radioactive Waste Management*, *4*, 156–161.

Duan, J., & Gregory, J. (2003). Coagulation by hydrolyzing metal salts. *Advances in Colloid Inter-face Science*, *100–102*, 475–502.

Elango, B., Rajendran, P., & Bornmann, L. (2017). A scientometric analysis of international collaboration and growth of literature at the macro level. *Malaysian Journal of Library & Information Science*, *20*, 41–50.

Fatma, S., Hameed, A., Noman, M., Ahmed, T., Shahid, M., Tariq, M., ... Tabassum, R. (2018). Lignocellulosic biomass: A sustainable bioenergy source for the future. *Protein and Peptide Letters*, *25*(2), 148–163. Available from https://doi.org/10.2174/0929866525666180122144504, PMID:. Available from 29359659.

Fernandes, C. D., Nascimento, V. R. S., Meneses, D. B., Vilar, D. S., Torres, N. H., Leite, M. S., ... Ferreira, L. F. R. (2020). Fungal biosynthesis of lignin-modifying enzymes from pulp wash and *Luffa cylindrica* for azo dye RB5 biodecolorization using modeling by response surface methodology and artificial neural network. *Journal of Hazardous Materials*, *399*, 123094.

Ferreira, L. F. R., Torres, N. H., de Armas, R. D., Fernandes, C. D., da Silva Vilar, D., Aguiar, M. M., ... Bharagava, R. N. (2020). Fungal lignin-modifying enzymes induced by vinasse mycodegradation and its relationship with oxidative stress. *Biocatalysis and Agricultural Biotechnology*, *27*, 101691.

Fu, Y., & Viraraghavan, T. (2001). Fungal decolorization of dye wastewaters: A review. *Bioresource Technology*, *79*(3), 251–262.

Garg, A., Mishra, I. M., & Chand, S. (2010). Effectiveness of coagulation and acid precipitation processes for the pre-treatment of diluted black liquor. *Journal of Hazardous Materials*, *180*, 158–164.

Gasser, E., Ballmann, P., Dröge, S., Bohn, J., & König, H. (2014). Microbial production of biopolymers from the renewable resource wheat straw. *Journal of Applied Microbiology*, *117*, 1035–1044. Available from https://doi.org/10.1111/jam.12581.

Gönder, Z. B., Arayici, S., & Barlas, H. (2012). Treatment of pulp and paper mill wastewater using ultra-filtration process: Optimization of the fouling and rejections. *Industrial & Engineering Chemistry Research*, *51*, 6184–6195.

Greenlee, L. F., Testa, F., Lawler, D. F., Freeman, B. D., & Moulin, P. (2010). Effect of antiscalants on precipitation of an RO concentrate: Metals precipitated and particle characteristics for several water compositions. *Water Research*, *44*, 2672–2684.

Grierson, D.E., Safi, M., Xu, L., & Liu, Y. (2005). Simplified methods for progressive-collapse analysis of buildings. In *Proceedings of the Structures congress 2005: Metropolis and beyond* (pp. 1–8). New York, NY.

Gupta, V. K., Minocha, A. K., & Jain, N. (2001). Batch and continuous studies on treatment of pulp mill wastewater by *Aeromonas formicans*. *Journal of Chemical Technology and Biotechnology (Oxford, Oxfordshire: 1986)*, *76*, 547–552.

Haq, I., Kumar, S., Raj, A., Lohani, M., & Satyanarayana, G. N. V. (2017). Genotoxicity assessment of pulp and paper mill effluent before and after bacterial degradation using *Allium cepa* test. *Chemosphere*, *169*, 642–650.

Haq, I., & Raj, A. (2020). Pulp and paper mill wastewater: Ecotoxicological effects and bioremediation approaches for environmental safety. In R. N. Bharagava, & G. Saxena (Eds.), *Bioremediation of industrial waste for environmental safety: Volume II: Biological agents and methods for industrial waste management* (pp. 333–356). Singapore: Springer.

He, Y., Li, X., Xue, X., Swita, M. S., Schmidt, A. J., & Yang, B. (2017). Biological conversion of the aqueous wastes from hydrothermal liquefaction of algae and pine wood by *Rhodococci*. *Bioresource Technology*, *224*, 457–464.

Jiang, C., Cheng, Y., Zang, H., Chen, X., Wang, Y., Zhang, Y., ... Li, C. (2019). Insight into biodegradation of lignin and the associated degradation pathway by psychrotrophic *Arthrobacter* sp. C2 from the cold region of China. *Cellulose*. Available from https://doi.org/10.1007/s10570-019-02858-3.

Jian-Ping, W., Chen, Y.-Z., Wang, Y., Yuan, S.-J., & Yu, H.-Q. (2011). Optimization of the coagulation-flocculation process for pulp mill wastewater treatment using a combination of uniform design and response surface methodology. *Water Research*, *45*, 5633–5640.

Jin, M., Slininger, P. J., Dien, B. S., Waghmode, S., Moser, B. R., Orjuela, A., ... Balan, V. (2015). Microbial lipid-based lignocellulosic biorefinery: Feasibility and challenges. *Trends in Biotechnology*, *33*, 43–54.

Kamali, M., & Khodaparast, Z. (2015). Review on recent developments on pulp and paper mill wastewater treatment. *Ecotoxicology and Environmental Safety*, *114*, 326–342.

Khansorthong, S., & Hunsom, M. (2009). Remediation of wastewater from pulp and paper mill industry by the electrochemical technique. *Chemical Engineering Journal*, *151*(1–3), 228–234.

Kumar, A., & Chandra, R. (2020). Ligninolytic enzymes and its mechanisms for degradation of lignocellulosic waste in environment. *Heliyon*, *6*, e03170.

Kumar, A., Singh, A. K., Ahmad, S., & Chandra, R. (2020a). Optimization of laccase production by *Bacillus* sp. strain AKRC01 in presence of agro-waste as effective substrate using response surface methodology. *Journal of Pure and Applied Microbiology*, *14*(1), 351–362.

Kumar, A., Singh, A. K., & Chandra, R. (2020b). Comparative analysis of residual organic pollutants from bleached and unbleached paper mill wastewater and their toxicity on *Phaseolus aureus* and *Tubifex tubifex*. *Urban Water Journal*, *17*(10), 860–870.

Kumar, P., Teng, T. T., Chand, S., & Wasewar, K. L. (2011). Treatment of paper and pulp mill effluent by coagulation. *International Journal of Civil and Environmental Engineering*, *3*, 3.

Leonowicz, A., Matuszewska, A., Luterek, J., Ziegenhagen, D., Wojtaś-Wasilewska, M., Cho, N. S., & Rogalski, J. (1999). Biodegradation of lignin by white rot fungi. *Fungal Genetics and Biology: FG & B*, *27*, 175–185.

Li, F., Ma, F., Zhao, H., Zhang, S., Wang, L., Zhang, X., & Yu, H. (2019). A lytic polysaccharide monooxygenase from a white-rot fungus drives the degradation of lignin by a versatile peroxidase. *Applied and Environmental Microbiology*, *85*, e02803–e02818. Available from https://doi.org/10.1128/AEM.02803-18.

Li, X., Xia, J., Zhu, X., Bilal, M., Tan, Z., & Shi, H. (2019). Construction and characterization of bifunctional cellulases: Caldicellulosiruptor-sourced endoglucanase, CBM, and exoglucanase for efficient degradation of lignocellulose. *Biochemical Engineering Journal*, *151*, 107363.

Lora, J. (2008). Industrial commercial lignins: Sources, properties and applications. In M. N. Belgacem, & A. Gandini (Eds.), *Monomers. polymers and composites from renewable resources* (pp. 225–241). Amsterdam: Elsevier.

Lu, F., Karlen, S. D., Regner, M., Kim, H., Ralph, S. A., Sun, R. C., … Ralph, J. (2015). Naturally p-hydroxybenzoylated lignins in palms. *BioEnergy Research*, *8*(3), 934–952.

Ľudmila, H., Michal, J., Andrea, Š., & Aleš, H. (2015). Lignin, potential products and their market value. *Wood Research*, *60*(6), 973–986.

Luo, H., & Abu-Omar, M. M. (2017a). *Chemicals from lignin. Encyclopedia of Sustainable Technologies* (pp. 573–585). Amsterdam, The Netherlands: *Elsevier*.

Luo, H., Zheng, P., Bilal, M., Xie, F., Zeng, Q., Zhu, C., … Wang, Z. (2020). Efficient bio-butanol production from lignocellulosic waste by elucidating the mechanisms of *Clostridium acetobutylicum* response to phenolic inhibitors. *Science of the Total Environment*, *710*, 136399.

Malaviya, P., & Rathore, V. S. (2007). Bioremediation of pulp and paper mill effluent by a novel fungal consortium isolated from polluted soil. *Bioresource Technology*, *98*, 3647–3651.

Mandlekar, N., Cayla, A., Rault, F., Giraud, S., Salaün, F., Malucelli, G., & Guan, J.-P. (2018). *An overview on the use of lignin and its derivatives in fire retardant polymer systems. Lignin-trends and applications*. IntechOpen.

Mei, J., Shen, X., Gang, L., Xu, H., Wu, F., & Sheng, L. (2020). A novel lignin degradation bacteria-*Bacillus amyloliquefaciens* SL-7 used to degrade straw lignin efficiently. *Bioresource Technology*, *310*, 123445.

Merayo, N., Hermosilla, D., Blanco, L., Cortijo, L., & Blanco, A. (2013). Assessing the application of advanced oxidation processes, and their combination with biological treatment, to effluents from pulp and paper industry. *Journal of Hazardous Materials*, *262*, 420–427. Available from https://doi.org/10.1016/j.jhazmat.2013.09.005.

Michniewicz, M., Stufka-Olczyk, J., & Milczarek, A. (2012). Ozone degradation of lignin; its impact upon the subsequent biodegradation. *Fibres & Textiles in Eastern Europe*, *20*, 191–196.

Munir, N., Asgher, M., Tahir, I. M., Riaz, M., Bilal, M., & Shah, S. A. (2015). Utilization of agro-wastes for production of ligninolytic enzymes in liquid state fermentation by *Phanerochaete chrysosporium*-IBL-03. *IJCBS*, *7*, 9–14.

Murthy, B. S. A., Sihorwala, T. A., Tilwankar, H. V., & Killedar, D. J. (1991). Removal of colour from pulp and paper mill effluents by sorption technique- A case study. *Indian Journal of Environmental Protection*, *11*, 360.

Plácido, J., Imam, T., & Capareda, S. (2013). Evaluation of ligninolytic enzymes, ultrasonication and liquid hot water as pretreatments for bioethanol production from cotton gin trash. *Bioresource Technology*, *139*, 203–208.

Pokhrel, D., & Viraraghavan, T. (2004). Treatment of pulp and paper mill wastewater—A review. *Science of the Total Environment*, *333*(1–3), 37–58.

Ragauskas, A. J., Beckham, G. T., Biddy, M. J., Chandra, R., Chen, F., Davis, M. F., . . . Keller, M. (2014). Lignin valorization: Improving lignin processing in the biorefinery. *Science (New York, N.Y.)*, *344*, 1246843. Available from https://doi.org/10.1126/science.1246843.

Ragunathan, R., & Swaminathan, K. (2004). Biological treatment of a pulp and paper industry effluent by *Pleurotus* spp. *World Journal of Microbiology and Biotechnology*, *20*(4), 389–393.

Raj, A., Kumar, S., Haq, I., & Singh, S. K. (2014). Bioremediation and toxicity reduction in pulp and paper mill effluent by newly isolated ligninolytic *Paenibacillus* sp. *Ecological Engineering*, *71*, 355–362.

Raj, A., Reddy, M. M., Chandra, R., Purohit, H. J., & Kapley, A. (2007a). Biodegradation of kraft-lignin by *Bacillus* sp. isolated from sludge of pulp and paper mill. *Biodegradation*, *18*, 783–792.

Raj, A., Reddy, M. K., & Chandra, R. (2007b). Identification of low molecular weight aromatic compounds by gas chromatography–mass spectrometry (GC–MS) from kraft lignin degradation by three *Bacillus* sp. *Int. Biodeterior. Biodegrad.*, *59*, 292–296.

Ravi, K., Abdelaziz, O. Y., Nöbel, M., García-Hidalgo, J., Gorwa-Grauslund, M. F., Hulteberg, C. P., & Lidén, G. (2019). Bacterial conversion of depolymerized kraft lignin. *Biotechnology for Biofuels*, *12*, 56. Available from https://doi.org/10.1186/s13068-019-1397-8.

Rodriguez, A., Salvachúa, D., Katahira, R., Black, B. A., Cleveland, N. S., Reed, M., . . . Gladden, J. M. (2017). Base-catalyzed depolymerization of solid lignin-rich streams enables microbial conversion. *ACS Sustainable Chemistry & Engineering*, *5*, 8171–8180.

Ruas, D. B., Mounteer, A. H., Lopes, A. C., Gomes, B. L., Brandao, F. D., & Girondoll, L. M. (2007). Combined chemical biological treatment of bleached eucalypt kraft pulp mill effluent. *Water Science & Technology*, *55*(6), 143–150.

Sagues, W. J., Bao, H., Nemenyi, J. L., & Tong, Z. (2018). Lignin-first approach to biorefining: Utilizing Fenton's reagent and supercritical ethanol for the production of phenolic and sugars. *ACS Sustainable Chemistry & Engineering*, *6*, 4958–4965.

Salvachúa, D., Karp, E. M., Nimlos, C. T., Vardon, D. R., & Beckham, G. T. (2015). Towards lignin consolidated bioprocessing: simultaneous lignin depolymerization and product generation by bacteria. *Green Chemistry: An International Journal and Green Chemistry Resource: GC*, *17*, 4951–4967.

Santos, A. Z., Tavares, C. R. G., & Gomes-da-Costa, S. M. (2002). Treatment of the effluent from a kraft bleach plant with the white-rot fungus *Pleurotus ostreatoroseus* Sing. *Brazilian Journal of Chemical Engineering*, *19*(4), 371–375.

Sevimli, M. F., & Sarikaya, H. Z. (2002). Ozone treatment of textile effluents and dyes: Effect of applied ozone dose, pH and dye concentration. *Journal of Chemical Technology and Biotechnology (Oxford, Oxfordshire: 1986)*, *77*, 842–850.

Shawwa, A. R., Smith, D. W., & Sego, D. C. (2001). Color and chlorinated organics removal from pulp wastewater using activated petroleum coke. *Water Research*, *35*, 745–749.

Shi, Y., Chai, L., Tang, C., Yang, Z., Zheng, Y., Chen, Y., et al. (2013). Biochemical investigation of kraft lignin degradation by *Pandoraea* sp. B-6 isolated from bamboo slips. *Bioprocess and Biosystems Engineering*, *36*, 1957–1965.

Shi, Y., Yan, X., Li, Q., Wang, X., Xie, S., Chai, L., et al. (2017). Directed bioconversion of kraft lignin to polyhydroxyalkanoate by *Cupriavidus basilensis* B-8 without any pretreatment. *Process Biochemistry*, *52*, 238–242. Available from https://doi.org/10.1016/j.procbio.2016.10.004.

Silva, M. L. C., de Souza, V. B., da Silva Santos, V., Kamida, H. M., de Vasconcellos-Neto, J. R. T., Góes-Neto, A., & Koblitz, M. G. B. (2014). Production of manganese peroxidase by *Trametes villosa* on unexpensive substrate and its application in the removal of lignin from agricultural wastes. *Advances in Bioscience and Biotechnology*, *5*(14), 1067.

Singh, A. K., Kumar, A., & Chandra, R. (2020a). Detection of refractory organic pollutants from pulp paper mill effluent and their toxicity on *Triticum aestivum; Brassica campestris* and *Tubifex-tubifex*. *Journal of Experimental Biology and Agricultural Sciences*, *8*(5), 663–675.

Singh, A. K., Kumar, A., & Chandra, R. (2020b). Residual organic pollutants detected from pulp and paper industry wastewater and their toxicity on *Triticum aestivum* and *Tubifex-tubifex* worms. *Materials Today: Proceedings*. Available from https://doi.org/10.1016/j.matpr.2020.10.862.

Singhal, A., & Thakur, I. S. (2009). Decolourization and detoxification of pulp and papermill effluent by *Emericella nidulans* var. nidulans. *Journal of Hazardous Materials*, *171*, 619–625.

Sodaneath, H., Lee, J. I., Yang, S. O., Jung, H., Ryu, H. W., & Cho, K. S. (2017). Decolorization of textile dyes in an air-lift bioreactor inoculated with *Bjerkandera adusta* OBR105. *Journal of Environmental Science and Health Part A*, *52*, 1099–1111.

Srivastava, V. C., Mall, I. D., & Mishra, I. M. (2005). Treatment of pulp and paper mill wastewaters with poly aluminium chloride and bagasse fly ash. *Colloids and Surfaces A: Physicochemical and Engineering Aspects*, *260*(1–3), 17–28.

Studer, M. H., DeMartini, J. D., Davis, M. F., Sykes, R. W., Davison, B., Keller, M., . . . Wyman, C. E. (2011). Lignin content in natural Populus variants affects sugar release. *Proceedings of the National Academy of Sciences*, *108*(15), 6300–6305.

Taseli, B. K., & Gokcay, C. F. (1999). Biological treatment of paper pulping effluents by using a fungal reactor. *Water science and technology*, *40*(11–12), 93–99.

Tejado, A., Pena, C., Labidi, J., Echeverria, J. M., & Mondragon, I. I. (2007). Physicochemical characterization of lignins from different sources for use in phenol–formaldehyde resin synthesis. *Bioresource Technology*, *98*, 1655–1663.

Tezel, U., Guven, E., Erguder, T. H., & Demirer, G. N. (2001). Sequential (anaerobic/aerobic) biological treatment of Dalaman SEKA pulp and paper industry effluent. *Waste Management*, *21*(8), 717–724.

Torrades, F., Pera,l, J., Perez, M., Domenech, X., Hortal, J. A. G., & Riva, M. C. (2001). Removal of organic contaminants in bleached kraft effluents using heterogeneous photocatalysis and ozone. *Tappi Journal*, *84*, 63.

Tyagi, S., Kumar, V., Singh, J., Teotia, P., Bisht, S., & Sharma, S. (2014). Bioremediation of pulp and paper mill effluent by dominant aboriginal microbes and their consortium. *International Journal of Environmental Research*, *8*, 561–568.

Vanholme, R., Cesarino, I., Rataj, K., Xiao, Y., Sundin, L., Goeminne, G., . . . Boerjan, W. (2013). Caffeoyl shikimate esterase (CSE) is an enzyme in the lignin biosynthetic pathway in *Arabidopsis*. *Science (New York, N.Y.)*, *341*(6150), 1103–1106.

Vardon, D. R., Rorrer, N. A., Salvachúa, D., Settle, A. E., Johnson, C. W., Menart, M. J., . . . Beckham, G. T. (2016). Cis, cis-Muconic acid: Separation and catalysis to bio-adipic acid for nylon-6,6 polymerization. *Green Chemistry: an International Journal and Green Chemistry Resource: GC*, *18*, 3397–3413.

Verma, M., Ekka, A., Mohapatra, T., & Ghosh, P. (2020). Optimization of kraft lignin decolorization and degradation by bacterial strain *Bacillus velezensis* using response surface methodology. *Journal of Environmental Chemical Engineering*, *8*(5), 104270.

Wang, Y., Vazquez-Duhalt, R., & Pickard, M. A. (2002). Purification, characterization, and chemical modification of manganese peroxidase from *Bjerkandera adusta* UAMH 8258. *Current Microbiology*, *45*, 77–87.

Wells, T., Wei, Z., & Ragauskas, A. (2015). Bioconversion of lignocellulosic pretreatment effluent via oleaginous *Rhodococcus opacus* DSM 1069. *Biomass Bioenergy*, *72*, 200–205.

Xia, J., Yu, Y., Chen, H., Zhou, J., Tan, Z., He, S., . . . Li, X. (2019). Improved lignocellulose degradation efficiency by fusion of β-glucosidase, exoglucanase, and carbohydrate-binding module from *Caldicellulosiruptor saccharolyticus*. *BioResources*, *14*(3), 6767–6780.

Xie, S., Qin, X., Cheng, Y., et al. (2015). Simultaneous conversion of all cell wall components by an oleaginous fungus without chemi-physical pre- treatment. *Green Chemistry: an International Journal and Green Chemistry Resource: GC*, *17*, 1657–1667.

Xie, S., Ragauskas, A. J., & Yuan, J. S. (2016). Lignin conversion: Opportunities and challenges for the integrated biorefinery. *Industrial Biotechnology.*, *12*, 161–167.

Xilei, D., Tingzhi, L., Weijiang, D., & Huiren, H. (2010). Adsorption and coagulation tertiary treatment of pulp and paper mills wastewater. *Biomedical Engineering (ICBBE)*.

Yamamoto, S. (2001). Ozone treatment of bleached kraft pulp and waste paper. *Japan Tappi Journal*, *55*, 90−97.

Yeber, M. C., Rodriquez, J., Freer, J., Baeza, J., Duran, N., & Mansilla, H. D. (1999). Advanced oxidation of a pulp mill bleaching wastewater. *Chemosphere*, *39*, 1679−1688.

Zainith, S., Purchase, D., Saratale, G. D., Ferreira, L. F. R., Bilal, M., & Bharagava, R. N. (2019). Isolation and characterization of lignin-degrading bacterium *Bacillus aryabhattai* from pulp and paper mill wastewater and evaluation of its lignin-degrading potential. *3 Biotech*, *9*(3), 1−11.

Further reading

Bajwa, D. S., Pourhashem, G., Ullah, A. H., & Bajwa, S. G. (2019). A concise review of current lignin production, applications, products and their environmental impact. *Industrial Crops and Products*, *139*, 111526.

Christopher, L. P., et al. (2014). Lignin biodegradation with laccasemediator systems. *Frontiers in Energy Research*, *2*, 12.

Faison, B. D., & Kirk, T. K. (1985). Factors involved in the regulation of a ligninase activity in *Phanerochaete chrysosporium*. *Applied and Environmental Microbiology Journal*, *49*, 299−304.

Yadav, V., Wang, Z., Wei, C., Amo, A., Ahmed, B., Yang, X., & Zhang, X. (2020). Phenylpropanoid pathway engineering: An emerging approach towards plant defense. *Pathogens*, *9*(4), 312. Available from https://doi.org/10.3390/pathogens9040312. PMID: 32340374; PMCID: PMC7238016.

CHAPTER 13

Nanotechnology in packaging of food and drugs

Marzieh Badiei[1], Nilofar Asim[2], Nurul Asma Samsudin[3], Nowshad Amin[3] and Kamaruzzaman Sopian[2]
[1]Independent Researcher, Mashhad, Iran
[2]Solar Energy Research Institute, Universiti Kebangsaan Malaysia, Bangi, Malaysia
[3]Institute of Sustainable Energy, Universiti Tenaga Nasional (@The National Energy University), Kajang, Malaysia

13.1 Introduction

Nanotechnology is a powerful tool that significantly contributes to the development of smart packaging materials for foods and pharmaceuticals. Nanotechnology has been used for developing light but strong materials with improved biodegradable properties for the fabrication of sensors or indicators for consumer information (Mihindukulasuriya & Lim, 2014). Moreover, nanotechnology has led to the development of materials with improved mechanical, barrier, and antimicrobial properties, extending the shelf life of foods and drugs during storage. Packaging materials should be oxygen (O_2)-, moisture-, and heat-resistant to control the microbial degradation of foods and drugs. In addition, the migration of packaging components from the packaging to the drugs and foods should be traced and controlled (Jildeh & Matouq, 2020).

Polymer nanocomposites are mixtures of polymer as matrix with inorganic or organic fillers. Fillers with different particle shapes and a nanometric range in dimension (nanoparticles or NPs), produce polymer nanocomposites. Materials made up of NPs possess a relatively high surface-to-volume ratio that causes high surface activity (Fu et al., 2019). The nanomaterial can improve the properties of the final nanocomposite. Accordingly, nanocomposites possess improved barrier properties and mechanical strength that provide advanced structures for food- and drug-packaging applications. New bio-based polymer composites have been used to develop edible and biodegradable films. Using biopolymers in packaging materials improves the biodegradability and biocompatibility of the film material, making it recyclable with low waste discharge to the environment (Wróblewska-Krepsztul et al., 2018). However, the poor performance of edible and biodegradable polymers, such as poor gas and moisture barrier, low mechanical strength, and low heat stability, has limited their industrial application.

Nanotechnology in Paper and Wood Engineering
DOI: https://doi.org/10.1016/B978-0-323-85835-9.00009-X

Several nanocomposites have been fabricated by adding NPs as reinforcement agents to polymers. Reinforcing compounds improve the physical features of packaging materials, mainly the mechanical and barrier properties. Fillers with high aspect ratio and specific surface provide good reinforcing agents and barrier property against the diffusion of undesirable gas molecules, such as O_2, carbon dioxide (CO_2), water vapor, and ultraviolet (UV) radiation (Wang et al., 2019a). Moreover, the dimensional stability, heat resistance, and stiffness of packaging materials increase. Consequently, the overall barrier properties of a nanocomposite against vapors, gases, and UV improve (López de Dicastillo et al., 2020).

In addition to reinforcing NPs, bioactive nanomaterials can be incorporated in the biopolymer matrix to act as antimicrobial and antioxidant agents and for UV absorbance (Al-Tayyar et al., 2020). They are responsible for providing active properties to protect and preserve the freshness of foods or drugs through intrinsic and/or extrinsic factors (Sharma et al., 2020a). The preparation of cellulose-based composites and their application in antibacterial activity have been reported (Li et al., 2018). Metal and metal oxide NPs (Hoseinnejad et al., 2018) and essential oils (Sharma et al., 2020b) are the most popular antimicrobial and antioxidant agents. These nanomaterials decrease the demand for preservative materials and extend the shelf life of foods or drugs. Antimicrobial nanomaterials can also be incorporated into capsulation materials to preserve flavor and aroma and increase the quality and shelf life of foods and drugs (Froiio et al., 2019).

Nanomaterials, mainly as nanosensors or indicators, could be used as a sensitive part of intelligent packaging systems to provide feedback to consumers about the actual condition and freshness of packaged foods and drugs (Warriner et al., 2014). Nanosensors and indicators detect pathogens and spoiling agents, such as O_2 and humidity, in the package environment (Yousefi et al., 2019). Meanwhile, active packaging places emphasis on the development of packaging materials to preserve the environmental conditions of foods, and intelligent packaging uses smart functionalities, like sensing, detection, and communication, to provide information about the quality of foods and drugs. Fig. 13.1 shows the categories of active and intelligent packaging systems and their properties.

Nanocellulose (NC) mainly includes cellulose nanocrystals (CNCs) and nanofibrillated cellulose (NFC), and has been used to produce nanocomposites for food- and drug-packaging applications (Yu et al., 2017). Cellulose has a long history of application in food packaging in several forms, such as paper, board, cellulose acetate, methylcellulose (MC), hydroxypropyl cellulose, hydroxypropyl methylcellulose, and carboxymethylcellulose (CMC) (Shaghaleh et al., 2018). NC (Vilarinho et al., 2018) offers distinctive characteristics, such as high specific surface area, and possesses a high concentration of active functional groups. These are unique features for surface modification that develop the mechanical properties of nanocomposites. NC has low O_2

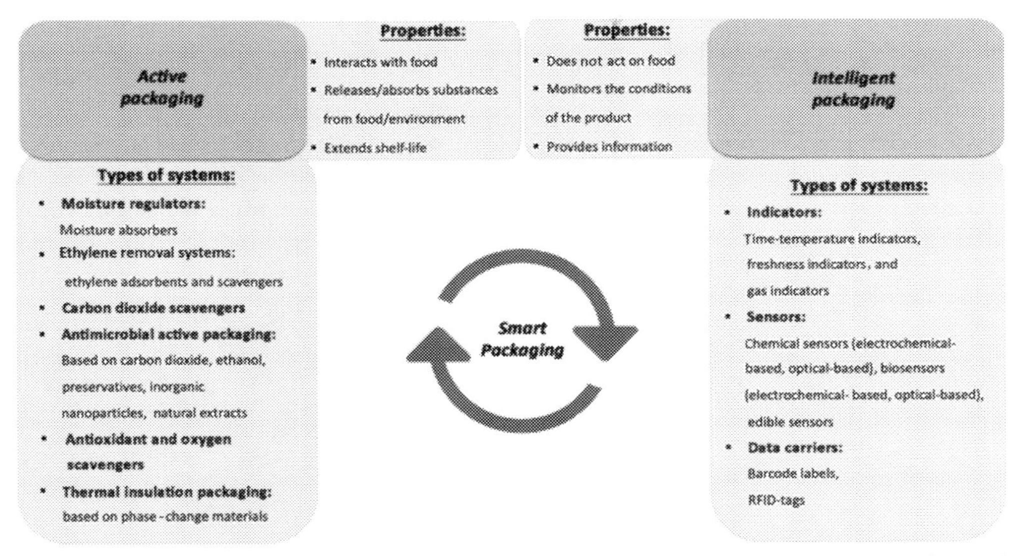

Figure 13.1 Active and intelligent packaging systems and their properties (Drago et al., 2020). *RFID*, Flexible radio frequency identification.

permeability, with poor moisture barrier ability compared to petroleum-based polymers. NFC is a good O_2 barrier and provides flexibility to the polymer matrix. Meanwhile, nanocrystalline cellulose provides significant tensile strength and a large surface area. For example, an antimicrobial cellulose paper for food and drug packaging was prepared through an impregnation method that allows Zn^{2+} to enter the fibers of a cellulose paper substrate (Li et al., 2021).

To address these weaknesses, nanocomposite technology has been used as an option to enhance the overall properties of composites (Ferrer et al., 2017). This chapter presents the common applications of nanotechnology in the packaging industry, with emphasis on NC as the reinforcement agent of packaging materials for food and drug products.

13.2 Nanocellulose for reinforcement of nanocomposites

The possibility of improving the functional properties of polymers for food and drug packaging using NPs has resulted in the development of a wide range of nanocomposites. The reinforcement technique using nanomaterials is used as a promising approach for improving the performance properties of packaging materials. Good moisture and water barrier properties and mechanical and thermal stability, which are required of packaging materials, are obtained from reinforcing the polymer matrix with NPs (Lopes et al., 2018). These NPs provide many active zones inside the polymeric matrix structure, enhancing the reinforcement effects.

Cellulose nanomaterials, mainly cellulose nanofibers (CNFs), have been used as a reinforcement agent in different synthetic polymers and biopolymers and have led to the development of cellulose-reinforced polymers. The utilization of CNCs (Li et al., 2020) and cellulose nanowhiskers (Cao et al., 2020) as reinforcement agents for nanocomposite synthesis has been reported, but NC fibers (NCFs) have a higher aspect ratio and more amorphous regions in their molecular structure, providing better mechanical strength and flexibility and a higher optical clarity in nanocomposites that fill the gaps of packaging materials (Xu et al., 2013). The combination of these key properties is important for their envisioned application as food- and drug-packaging materials. Fig. 13.2 schematically illustrates the impact of CNC reinforcement on barrier properties, mainly by reducing the pores and defects in the polymer matrix.

CNFs as organic nanoreinforcement agents are environment friendly and widely available, with low cost in manufacturing. Accordingly, they are inherently appropriate as a reinforcement agent for fabrication of food- and drug-packaging nanocomposites (Abdul Rashid et al., 2018; Bhatnagar & Sain, 2005). Cellulose-reinforced nanocomposites take advantage of improved gas barrier properties. A NC with a high aspect ratio offers a permeation path for gases, like water vapor, O_2, and CO_2, while decreasing the transfer rate of the gases through the package (Khosravi et al., 2020).

According to a study (Mani & Umapathy, 2020), reinforcing a film of pectin and fenugreek gum with microfibrillated cellulose (0.01 wt.%/wt.%) improves the mechanical and thermal stability of nanocomposites.

In another study (Moreirinha et al., 2020), the arabinoxylans biopolymer was reinforced with microfibrillated cellulose to prepare homogeneous and translucent films with high thermal and mechanical stability. Refined carrageenan was reinforced with NCFs to prepare a film with enhanced thermal stability and moisture barrier property and increased hydrophobic surface, which is suitable for food packaging (Sedayu et al., 2020). The incorporation of maximum 7.5 wt.%/wt.% of bacterial cellulose (BC)

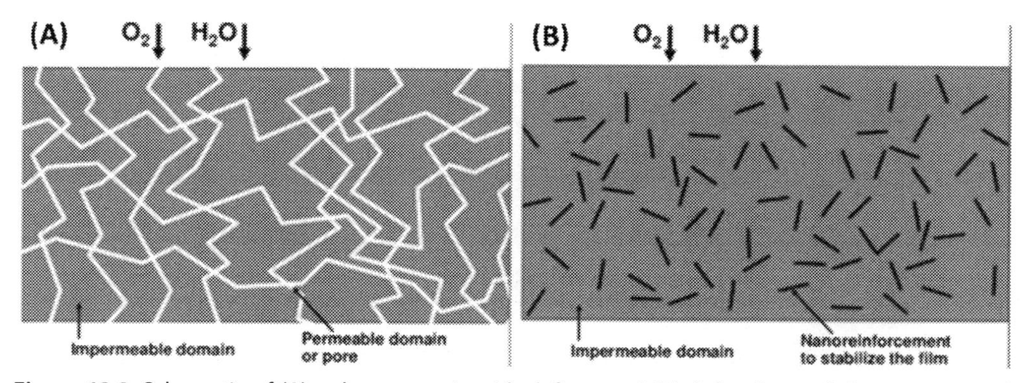

Figure 13.2 Schematic of (A) polymer matrix with defects and (B) defect-free cellulose nanocrystal-reinforced polymer matrix with improved barrier properties (Hubbe et al., 2017).

nanowhiskers into a polymeric blend reduced the water vapor permeability and increased the mechanical strength and the elastic modulus (Haghighi et al., 2020).

In another study (Cao et al., 2020), the effect of carboxylated CNC whisker as the reinforcement agent of cassia-gum-based edible film for oil packaging was investigated by analyzing the mechanical, thermal, and barrier stabilities of the film. The incorporation of the reinforcement agent reduced the light transmittance and improved the oil permeability and the mechanical stability.

Li et al. (2020) reported the fabrication of nanocomposites with a homogenous microstructure of CNC-grafted rosin (CNC-R) hybrids through the solution casting method. The CNC-R nanohybrid, as the reinforcing agent, was well-dispersed in the poly (3-hydroxybutyrate-co-3-hydroxyvalerate) matrix to improve the mechanical and thermal stability of the nanocomposite film for application as an alternative packaging material.

The overall performance of packaging materials, mainly the mechanical, thermal, and barrier properties, can be improved by adding NPs as reinforcement agents.

13.3 Active packaging

Active packaging has emerged as a promising packaging technology to provide satisfactory solutions for the demands of manufacturers and consumers. Antimicrobial packaging systems are categorized based on their mechanisms of actions.

One type of packaging system is the active releasing system in which the bioactive compound, such as the antimicrobial and antioxidant agent, is released into the headspace of the packaging to interact with the surface of the product. Such interactions are projected to improve the shelf life and safety of packaged foods or drugs. Antimicrobial agents are incorporated into the polymer matrix to prepare antimicrobial films that can gradually release them in a specific amount to the food or drug surface to perform their biocide actions. The controlled release of the antimicrobial material into the packaging needs an internal stimuli, mainly O_2 content, relative humidity, pH, or enzymatic activity, as the driving force (Ho et al., 2011). They primarily decrease the O_2 and moisture content of the internal environment to minimize the enzymatic deteriorative reactions of aerobic bacteria (Al-Tayyar et al., 2020). Fig. 13.3 schematically illustrates different ways of incorporating and releasing antimicrobial agents in packaging systems.

For example, a multicomponent active film packaging material was developed, consisting of titanium dioxide (TiO_2) and rosemary oil, as strong antibacterial agents in a CNF/whey protein polymer matrix (Alizadeh-Sani et al., 2020). The active packaging increased the shelf life of lamb meat up to 15 days. The antimicrobial agent is slowly released from the nanocomposite materials into the packaging environment, leading to the strong antimicrobial activity of the packaging material. Rosemary as an antimicrobial agent contains phenolic compounds that can interact with enzymes and

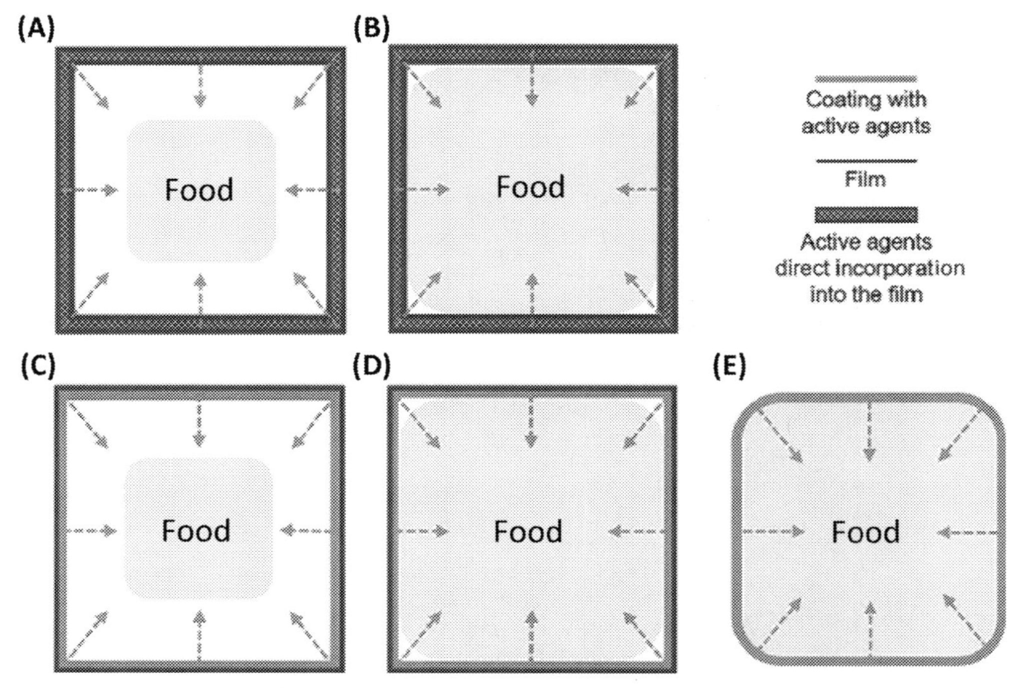

Figure 13.3 Schematic of incorporation and release of antimicrobial agents from packaging material. (A) Direct incorporation into packaging material and gradual diffusion into the headspace. (B) Direct incorporation into packaging material and gradual diffusion with direct contact. (C) Surface coating on the inner layer of packaging material and gradual diffusion into the headspace. (D) Surface coating on the inner layer of packaging material and gradual diffusion with direct contact. € Edible film and gradual diffusion with direct contact (Halonen et al., 2020).

disrupt the mechanism of microbes. TiO_2 and rosemary oil can delay the harmful oxidation process of meat and lipid oxidation during refrigerated storage. Some studies used encapsulated essential oils as an antimicrobial agent to protect fruits, vegetables, and meats while alleviating the negative sensory effects of essential oils. Bioactive materials in the form of solid particles or liquid droplets can be incorporated into nanocapsules or microcapsules to provide a barrier between the packaged product and the antimicrobial agent and control its release (Becerril et al., 2020). NPs have also been widely used to encapsulate antimicrobial substances (Fig. 13.4).

Tsai et al. (2018) prepared silymarin (SMN)–zein NPs that were absorbed onto BC nanofibers to form NPs/nanofibers nanocomposite. The SMN–zein NPs improved the release of SMN from the packaging film for the protection of salmon muscle from lipid oxidation. The nanoencapsulation of NPs in a multilayer packaging system may control the release of NPs into the packaging (Prakash et al., 2018). The concept of nanotechnology was used in fabricating multilayer packaging materials to

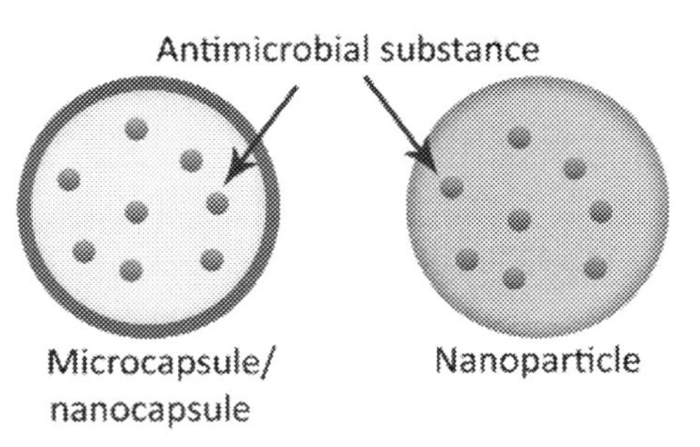

Figure 13.4 Nanoparticle and nanocapsule loaded with antimicrobial substance (Becerril et al., 2020).

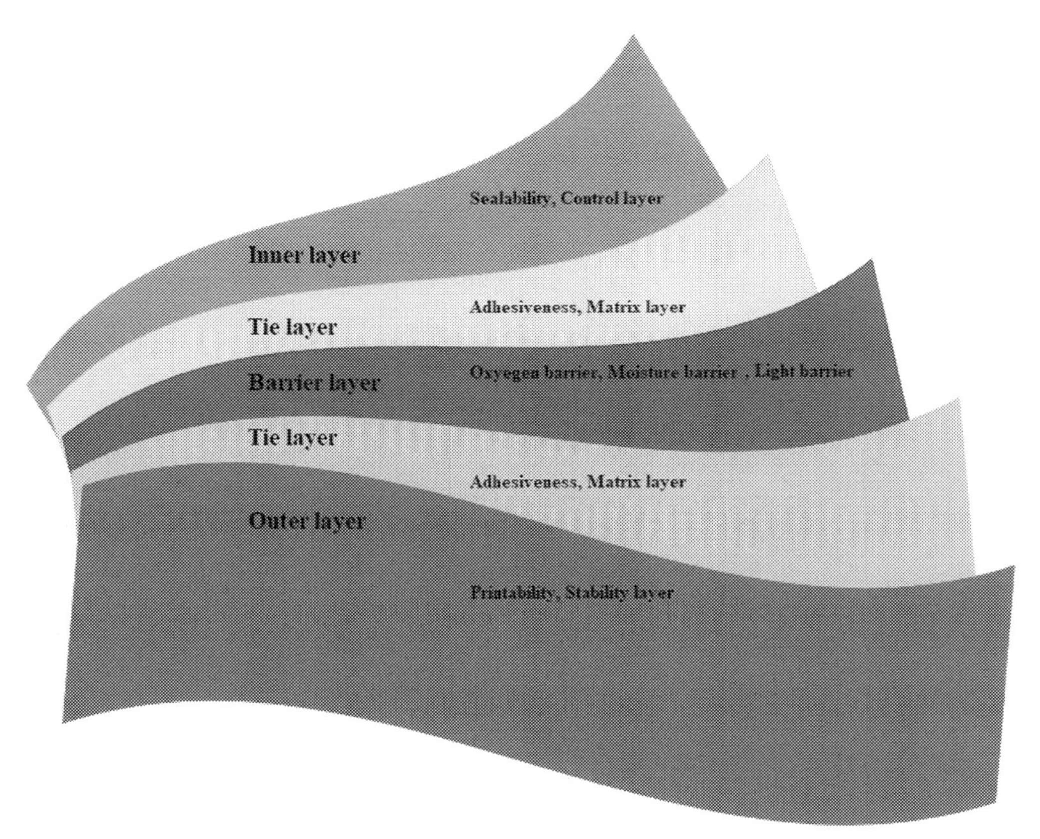

Figure 13.5 Each component of a multilayered packaging material carries significant functions to the overall system.

improve the functional characteristics. Multilayer packaging consists of several polymer layers as one packaging material (Fig. 13.5).

The antimicrobial NP embedded in one layer migrates to the surface and interacts with the pathogens. For example, an antimicrobial multilayer packaging material was developed from polypropylene films coated with CMC and containing essential oil as an antimicrobial agent (Honarvar et al., 2017). In another study, a NC-based multilayer coating film was fabricated using CNF/CNC to obtain an effective barrier against oil and grease (Tyagi et al., 2019).

Active scavenging system is another type of packaging system in which NPs are purposely incorporated into the polymer matrix to prepare a carrier suitable for integration into the packaging material. The active scavenging packaging system involves the removal of detrimental gases, including moisture, CO_2, O_2, and ethylene, from the packaging headspace or surrounding. The functional groups in the polymer and antimicrobial agent help immobilize the NP. Several NPs, mainly metals, metal oxides, and carbon nanotubes, have been mostly used to prepare antimicrobial packaging materials. For example, TiO_2 NPs have been widely used to produce O_2 scavenger films. TiO_2 is an inert, nontoxic, odorless, and cheap material with a high microbial inhibition activity against a wide variety of bacteria and fungi. TiO_2 can be photoinduced by UV radiation, leading to the promotion of its electrons' photocatalytic reactions. These electrons can transfer to the O_2 in air to produce photo-generated holes with a strong oxidation potential for reaction with compounds (Fig. 13.6).

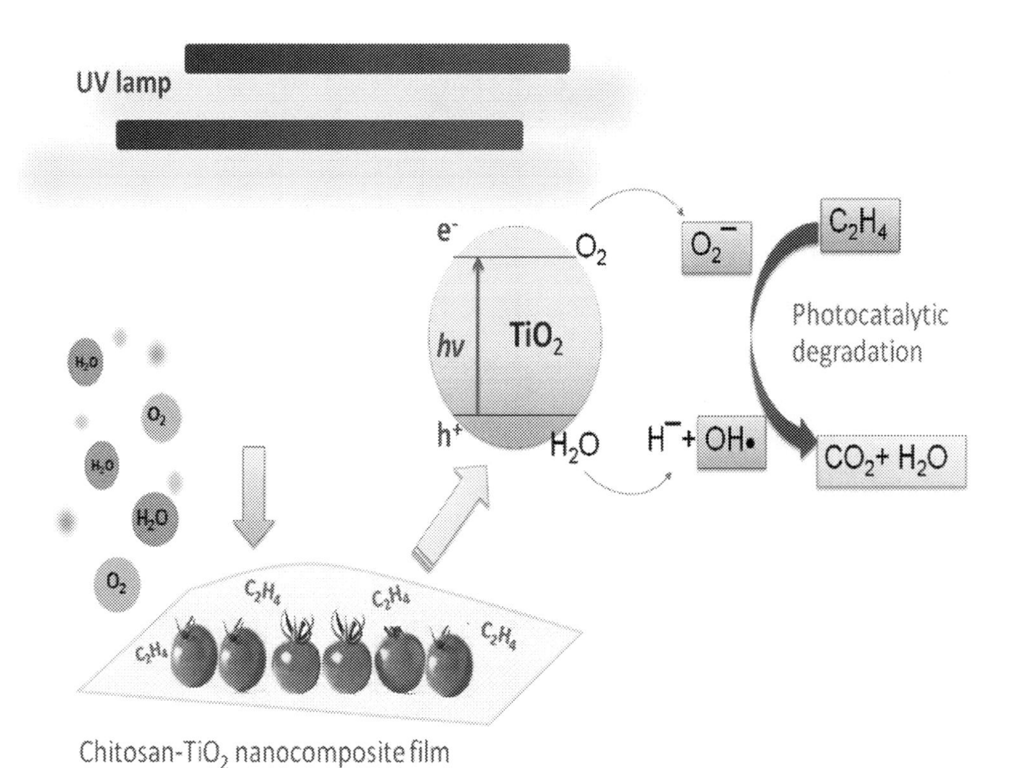

Figure 13.6 Mechanism of titanium dioxide (TiO_2) as oxygen scavenger in packaging of tomato fruit (Kaewklin et al., 2018).

Accordingly, the catalytic reactions of TiO_2 with an unlimited scavenger capacity have been exploited for removing O_2, water vapor, and ethylene vapors (Mihindukulasuriya & Lim, 2014). Owing to the presence of many -OH groups on the cellulose surface, TiO_2 can form hydrogen bonds with cellulose, so TiO_2 is compatible with polysaccharides-based edible films. The cellulose/metal oxide composite forms a chemically stable hybrid composite, which exhibits an antimicrobial property and is evaluated as a packaging material. In fact, TiO_2 NPs can improve the mechanical, thermal, antimicrobial, and water barrier properties of polysaccharide-based edible films (Xie & Hung, 2018). However, few applications of cellulose/TiO_2 nanocomposite films for antimicrobial purposes are reported in the literature.

As an example, Arularasu et al., 2020 synthesized a CNF/TiO_2 nanocomposite and proved its effective antibacterial activity, with promising application in food packaging. The nanocomposite structure has numerous pores, and $Ti4+$ ions can readily penetrate the cellulose polymer matrix through its pores to form a coordination bond between the $Ti4+$ ions and the hydroxyl group of cellulose. The potential mechanism proposed for the antibacterial activity of a nanocomposite material is the photocatalytic reaction, which could generate different reactive O_2 species (ROS). Salama and Aziz (2020) fabricated a composite edible coating of CMC blended with chitosan biguanidine hydrochloride (ChBg). The incorporation of TiO_2 NPs (CMC/ChBg/nTiO_2) improved the UV light shield, antibacterial properties, and thermal, mechanical, and barrier properties of polysaccharide-based edible films. Accordingly, the coating increased the shelf life of green bell peppers. TiO_2 NPs have been used in many studies as an antimicrobial agent in liquid suspension under UV radiation. For example, Xie and Hung (2019) prepared a TiO_2-embedded cellulose acetate film and evaluated the effectiveness of photodisinfection in direct contact with contaminated liquid solutions.

Moreover, zinc oxide (ZnO) (Roy & Rhim, 2020), silicon oxide (SiO_2) (Liu et al., 2020), and silver (Ag) NPs (Chen et al., 2020) are among the most studied nontoxic strong antibacterial agents for their ability as photocatalytic disinfecting agents. For example, a new formulation composed of CMC/CNC-immobilized Ag NPs was developed as an antibacterial paper coating (He et al., 2020). The coated papers as a packaging material extended the shelf life of strawberries up to seven days. ZnO NPs were used in fabricating active nanocomposite films from soy protein isolate, which was reinforced with different CNC-containing agents, including CNC, CNC/ZnONP, and CNC@ZnONP hybrids (Xiao et al., 2020). The incorporation of reinforcing agents improved the mechanical strength and water vapor and O_2 barrier properties of films. Incorporating ZnO NPs in the protein matrix and generating ROS inhibited the growth of foodborne pathogens in pork preservation. Meanwhile, the CNC@ZnONP nanohybrids prepared using an in situ growth method prevented the migration of zinc from the nanocomposite film into the food system due to the electrostatic attraction between CNC and the ZnO NPs. Moreover, an antimicrobial cellulose paper for food and drug packaging was prepared using ZnO NPs on the paper surface through a papermaking process (Li et al., 2021).

13.4 Intelligent packaging

The term intelligent packaging refers to packaging systems with extended functions used for foods and pharmaceuticals. The intelligent functions include sensing for detection, factor tracing, and communication (to provide information and notification of possible problems). The intelligent functions are responsible for the extended shelf life by ensuring safety and quality. Food packaging with these functionalities helps detect the freshness status of packaged food or displays information about the environment inside the package for the convenience of suppliers and consumers (Yousefi et al., 2019).

Indicators and sensors are the two main sensing technologies used in intelligent packaging (Wang et al., 2019b). Sensors and indicators use sensing and traceability techniques and quality indicators to detect various chemical changes or microbial growth. These technologies measure gases, such as O_2, CO_2, ethylene, and ammonia (NH_3), volatile organic compounds, humidity, microbial growth, and temperature changes. Indicators monitor the information on product status or the surrounding environmental factors of the packaging itself in real time. Sensors are more complex than indicators. Sensors generate a signal for detecting a physical or chemical property. Indicators display color change or an increase in color intensity as immediate visual and qualitative information (Wang et al., 2019b). Colorimetric indicators are the most common sensors for qualitative information, but they cannot provide quantitative information about the packaged products or collective measurement and time data. The two common colorimetric indicators are time—temperature and gas indicators, which send notifications on the real-time status of packaged foods (Bumbudsanpharoke & Ko, 2019). Other potential packaging sensors are electrical or flexible radio frequency identification (RFID) sensors (Bibi et al., 2017). The RFID tag is usually attached on the packaging film of the tracked item. The emphasis is on the intelligent films based on natural and biodegradable polymers and their key applications in the food- and drug-packaging systems. Paper composed of cellulosic fibers is often used as a carrier substrate for other materials, such as electrodes, to detect gases (Barandun et al., 2019; Kim et al., 2014; Listyarini et al., 2018).

Nanotechnology has created excellent opportunities in smart packaging, mostly through the development of physical and chemical nanosensors and indicators to provide the real-time status of food freshness (Dobrucka, 2020). Nanosensors with diagnostic and indicator functions are usually integrated into packaging materials to monitor and indicate the quality of products during storage. Surface coatings that use NPs can sense or detect the surrounding environment. Nanosensors respond to environmental imbalance due to the change in O_2 level, relative humidity, or temperature that accelerates reactions for the degradation of food or drug and decreases the shelf life. Thereafter, NPs used in packaging materials as reactive particles, sense different decomposition parameters and monitor through different mechanisms, including color

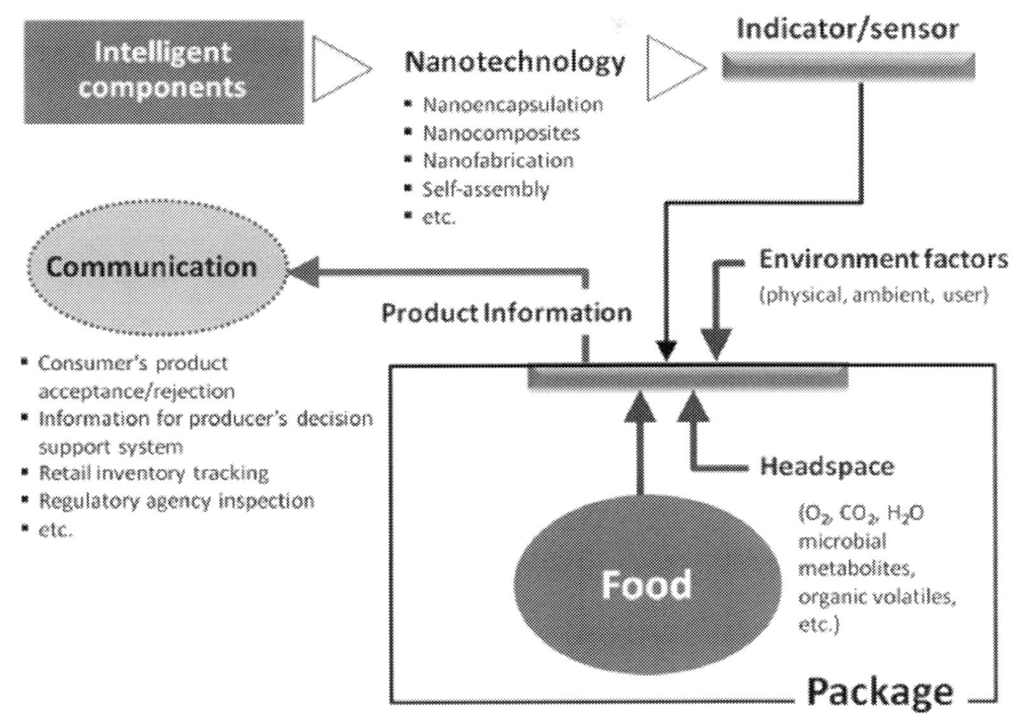

Figure 13.7 Contribution of nanotechnology in intelligent packaging concept (Mihindukulasuriya & Lim, 2014).

change, electrical readout, or fluorescence in real time (Fig. 13.7). Visible indicators monitor the micropores or sealing defects breach in packaging systems that expose products to chemical and microbial contaminants.

13.4.1 Gas indicator/sensor

Metabolites are produced in the headspace of packaging during the storage period, thereby causing the microbial spoilage of foods or drugs. For example, acidic CO_2 develops during the metabolism of microorganisms or the deterioration of proteins, producing alkaline volatile nitrogen compounds (Lu et al., 2020). In a gas indicator, metabolites are the main quality indicating factors, which are detected through the color change from a chemical or enzymatic reaction, indicating that gas indicators are commonly chemical sensors. Several gas indicators have been developed to monitor the presence and concentration of CO_2, N_2, O_2, water vapor, ethanol, NH_3, hydrogen sulfide, and the volatile organic compounds inside a package to indicate the freshness or spoilage of food. Various types of indicators/sensors that highlight the presence of gases, which can be used as package or printed labels on the packaging film, have emerged (Mohammadian et al., 2020).

An NH_3 sensor based on hydroxyapatite NPs within a matrix of cellulose nanofibrils has been reported. The sensing ability of the films, which was studied in the presence of NH_3 vapor by two probe electrodes, exhibited the detection limit of NH_3 at a concentration as low as 5 ppm (Narwade et al., 2019). Colorimetric pH-sensitive indicators have been widely used for detecting CO_2 to monitor the freshness of food through visual color change (Zhang & Lim, 2018). For example, Saliu and Della Pergola (2018) developed a pH-sensitive indicator of CO_2 based on anthocyanin/Îμ-polylysine in a food grade cellulose matrix. When acidic CO_2 is released, a reversible reaction with Îμ-polylysine occurs and leads to the dramatic decrease of the pH, which can be highlighted by the color variation of the anthocyanin dye. The indicator has been successfully used in monitoring the quality of poultry meat. Guo et al. (2020) developed a colorimetric barcode made up of 20 different types of porous nanocomposites, which comprise of dye-loaded chitosan NPs embedded on cellulose acetate. Chitosan and CA are biodegradable and nontoxic, making them biocompatible and environment-friendly options for screening food freshness. Each of the 20 dyes responds within a different pH range, covering the detection range of gases emitted from rotting meats. When exposed to different gases with varying concentrations, each bar of the barcode will display a different color or range of coloration. Highly sensitive NH_3 sensors fabricated from polymerized polydiacetylenes stabilize with CNCs in the chitosan matrix (Nguyen et al., 2019). Barandun et al. (2019) developed paper-based electrical gas sensors because these gases are soluble in water at room temperature, such as CO_2 and NH_3. The dissolved gas in the thin film of water adsorbed on the cellulose fibers within paper generates electrical conductivity. The electrical output is converted to a digital signal for consumers. This sensor can be implemented into tags on food packaging and has been successfully tested for meat products, such as fish and poultry. Fig. 13.8 shows a schematic of the sensing mechanism for alkaline gas. The sensors could be integrated into RFID tags and read by an NFC-enabled smartphone.

13.4.2 Time–temperature indicators/sensors

Temperature is a crucial factor that plays an important role in ensuring the quality and safety of perishable food products. Time–temperature indicators respond to changes in temperature resulting from a chemical, enzymatical, microbiological, or mechanical process during distribution, transport, and storage (Pandian et al., 2020). These indicators monitor irreversible changes in physical characteristics, usually visual color change, color movement, or shape deformation, in response to temperature exposed to the product (Muller & Schmid, 2019). These indicators are part of an important trend in smart packaging for the direct improvement of the quality control of food packaging. Indeed, these indicators can evaluate the changes in packaged food products, such as exposure to lower or higher temperature than the recommended temperature, to

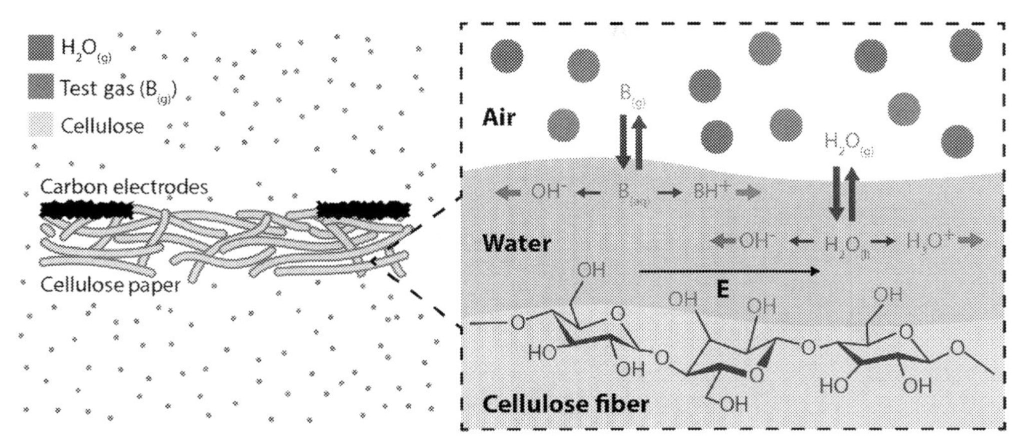

Figure 13.8 Sensing mechanism of alkaline gas. Dissolved alkaline gas ($B_{(aq)}$) generates cations (BH^+) and hydroxide anions ($OH^{\bar{a}}$). Production of other ions is traced and measured by two carbon electrodes (Barandun et al., 2019).

predict their freshness. These sensors are especially important for packaging chilled and frozen foods (Mohan & Ravishankar, 2019). For example, a colorimetric temperature indicator was developed based on the melting point of a transparent solvent and color change of a dye compound in the system (Lorite et al., 2017). Solvent and dye powder are separated using a cellulose dialysis membrane. The indicator used an electrically conductive film of carbon nanotube, which is a part of a RFID reader, to provide visible information. Over the melting point, solvent flows to develop visible color change when the dye compound is reached.

13.5 Conclusion

The exploitation of nanomaterials in food and drug packaging with the innovation of traditional packaging technology has greatly helped to provide the requirements of the market, including fresh and high-quality products. Moreover, consumers are now leaning toward the use of natural materials, such as biopolymers, including NC materials, in the food and drug industry. Hence, developing multifunctional active packaging materials that alleviate the negative environmental impacts of packaging has become possible. In addition, nanomaterials and nanosensors that guarantee the protection of food and drug products against environmental damage by revealing a colorimetric, chemical, or electric sign in real time have attracted huge attention. Indeed, these improvements can be optimized by using nanomaterials in packaging systems. Nanocomposites incorporate some essential properties into packaging materials to enhance their functionalities.

However, whether the NPs incorporated into active and intelligent packaging systems are safe in terms of penetration into food and drug products must be considered. The second issue regarding the industrial use of nanotechnology in packaging materials is the costs associated with novel technologies and machinery. Future studies should consider main factors, mainly cost and the multifunctionality of packaging for commercial application. To overcome these issues, new strategies must be developed to alleviate these problems and compete with the current technology in terms of efficiency and global competitiveness.

References

Abdul Rashid, E. S., Muhd Julkapli, N., & Yehye, W. A. (2018). Nanocellulose reinforced as green agent in polymer matrix composites applications. *Polymers for Advanced Technologies*, *29*, 1531–1546.

Alizadeh-Sani, M., Mohammadian, E., & Mcclements, D. J. (2020). Eco-friendly active packaging consisting of nanostructured biopolymer matrix reinforced with TiO2 and essential oil: Application for preservation of refrigerated meat. *Food Chemistry*, 126782.

Al-Tayyar, N. A., Youssef, A. M., & Al-Hindi, R. (2020). Antimicrobial food packaging based on sustainable bio-based materials for reducing foodborne pathogens: A review. *Food Chemistry*, *310*, 125915.

Arularasu, M., Harb, M., & Sundaram, R. (2020). Synthesis and characterization of cellulose/TiO2 nanocomposite: Evaluation of in vitro antibacterial and in silico molecular docking studies. *Carbohydrate Polymers*, *249*, 116868.

Barandun, G., Soprani, M., Naficy, S., Grell, M., Kasimatis, M., Chiu, K. L., ... GüDer, F. (2019). Cellulose fibers enable near-zero-cost electrical sensing of water-soluble gases. *ACS Sensors*, *4*, 1662–1669.

Becerril, R., NerÃn, C., & Silva, F. (2020). Encapsulation systems for antimicrobial food packaging components: An update. *Molecules*, *25*, 1134.

Bhatnagar, A., & Sain, M. (2005). Processing of cellulose nanofiber-reinforced composites. *Journal of Reinforced Plastics and Composites*, *24*, 1259–1268.

Bibi, F., Guillaume, C., Gontard, N., & Sorli, B. (2017). A review: RFID technology having sensing aptitudes for food industry and their contribution to tracking and monitoring of food products. *Trends in Food Science & Technology*, *62*, 91–103.

Bumbudsanpharoke, N., & Ko, S. (2019). Nanomaterial-based optical indicators: Promise, opportunities, and challenges in the development of colorimetric systems for intelligent packaging. *Nano Research*, *12*, 489–500.

Cao, L., Ge, T., Meng, F., Xu, S., Li, J., & Wang, L. (2020). An edible oil packaging film with improved barrier properties and heat sealability from cassia gum incorporating carboxylated cellulose nano crystal whisker. *Food Hydrocolloids*, *98*, 105251.

Chen, Q.-Y., Xiao, S.-L., Shi, S. Q., & Cai, L.-P. (2020). A One-pot synthesis and characterization of antibacterial silver nanoparticle–cellulose film. *Polymers*, *12*, 440.

Dobrucka, R. (2020). Application of nanotechnology in food packaging. *Journal of Microbiology, Biotechnology and Food Sciences*, *2020*, 353–359.

Drago, E., Campardelli, R., Pettinato, M., & Perego, P. (2020). Innovations in smart packaging concepts for food: An extensive review. *Foods*, *9*, 1628.

Ferrer, A., Pal, L., & Hubbe, M. (2017). Nanocellulose in packaging: Advances in barrier layer technologies. *Industrial Crops and Products*, *95*, 574–582.

Froiio, F., Mosaddik, A., Morshed, M. T., Paolino, D., Fessi, H., & Elaissari, A. (2019). Edible polymers for essential oils encapsulation: Application in food preservation. *Industrial & Engineering Chemistry Research*, *58*, 20932–20945.

Fu, S., Sun, Z., Huang, P., Li, Y., & Hu, N. (2019). Some basic aspects of polymer nanocomposites: A critical review. *Nano Materials Science*, *1*, 2–30.

Guo, L., Wang, T., Wu, Z., Wang, J., Wang, M., Cui, Z., . . . Chen, X. (2020). Portable food-freshness prediction platform based on colorimetric barcode combinatorics and deep convolutional neural networks. *Advanced Materials*, *32*, 2004805.

Haghighi, H., Gullo, M., La China, S., Pfeifer, F., Siesler, H. W., Licciardello, F., & Pulvirenti, A. (2020). Characterization of bio-nanocomposite films based on gelatin/polyvinyl alcohol blend reinforced with bacterial cellulose nanowhiskers for food packaging applications. *Food Hydrocolloids*, *113*, 106454.

Halonen, N. J., Palvölgyi, P. S., Bassani, A., Fiorentini, C., Nair, R., Spigno, G., & Kordas, K. (2020). Bio-based smart materials for food packaging and sensors—A review. *Frontiers in Materials*, 7, 82.

He, Y., Li, H., Fei, X., & Peng, L. (2020). Carboxymethyl cellulose/cellulose nanocrystals immobilized silver nanoparticles as an effective coating to improve barrier and antibacterial properties of paper for food packaging applications. *Carbohydrate Polymers*, *252*, 117156.

Ho, B. T., Joyce, D. C., & Bhandari, B. R. (2011). Release kinetics of ethylene gas from ethylene—α-cyclodextrin inclusion complexes. *Food Chemistry*, *129*, 259—266.

Honarvar, Z., Farhoodi, M., Khani, M. R., Mohammadi, A., Shokri, B., Ferdowsi, R., & Shojaee-Aliabadi, S. (2017). Application of cold plasma to develop carboxymethyl cellulose-coated polypropylene films containing essential oil. *Carbohydrate Polymers*, *176*, 1—10.

Hoseinnejad, M., Jafari, S. M., & Katouzian, I. (2018). Inorganic and metal nanoparticles and their antimicrobial activity in food packaging applications. *Critical Reviews in Microbiology*, *44*, 161—181.

Hubbe, M. A., Ferrer, A., Tyagi, P., Yin, Y., Salas, C., Pal, L., & Rojas, O. J. (2017). Nanocellulose in thin films, coatings, and plies for packaging applications: A review. *BioResources*, *12*, 2143—2233.

Jildeh, N. B., & Matouq, M. (2020). Nanotechnology in packing materials for food and drug stuff opportunities. *Journal of Environmental Chemical Engineering*, *8*, 104338.

Kaewklin, P., Siripatrawan, U., Suwanagul, A., & Lee, Y. S. (2018). Active packaging from chitosan-titanium dioxide nanocomposite film for prolonging storage life of tomato fruit. *International Journal of Biological Macromolecules*, *112*, 523—529.

Khosravi, A., Fereidoon, A., Khorasani, M. M., Naderi, G., Ganjali, M. R., Zarrintaj, P., . . . Gutierrez, T. J. (2020). Soft and hard sections from cellulose-reinforced poly (lactic acid)-based food packaging films: A critical review. *Food Packaging and Shelf Life*, *23*, 100429.

Kim, J.-H., Mun, S., Ko, H.-U., Yun, G.-Y., & Kim, J. (2014). Disposable chemical sensors and biosensors made on cellulose paper. *Nanotechnology*, *25*, 092001.

Li, F., Abdalkarim, S. Y. H., Yu, H.-Y., Zhu, J., Zhou, Y., & Guan, Y. (2020). Bifunctional reinforcement of green biopolymer packaging nanocomposites with natural cellulose nanocrystal—Rosin hybrids. *ACS Applied Bio Materials*, *3*, 1944—1954.

Li, J., Cha, R., Mou, K., Zhao, X., Long, K., Luo, H., . . . Jiang, X. (2018). Nanocellulose-based antibacterial materials. *Advanced Healthcare Materials*, 7, 1800334.

Li, M., Feng, Q., Liu, H., Wu, Y., & Wang, Z. (2021). In situ growth of nano-ZnO/GQDs on cellulose paper for dual repelling function against water and bacteria. *Materials Letters*, *283*, 128838.

Listyarini, A., Sholihah, W. & Imawan, C. A paper-based colorimetric indicator label using natural dye for monitoring shrimp spoilage. IOP Conference Series: Materials Science and Engineering, 2018. 367, 012045.

Liu, J., Chen, P., Qin, D., Jia, S., Jia, C., Li, L., . . . Shao, Z. (2020). Nanocomposites membranes from cellulose nanofibers, SiO2 and carboxymethyl cellulose with improved properties. *Carbohydrate Polymers*, *233*, 115818.

Lopes, T., Bufalino, L., JÃonior, M., Tonoli, G., & Mendes, L. (2018). Eucalyptus wood nanofibrils as reinforcement of carrageenan and starch biopolymers for improvement of physical properties. *Journal of Tropical Forest Science*, *30*, 292—303.

López de Dicastillo, C., Velásquez, E., Rojas, A., Guarda, A., & Galotto, M. J. (2020). The use of nanoadditives within recycled polymers for food packaging: Properties, recyclability, and safety. *Comprehensive Reviews in Food Science and Food Safety*, *19*, 1760—1776.

Lorite, G. S., Selkälä, T., Sipola, T., Palenzuela, J., Jubete, E., viÃ ± uales, A., . . . Uusitalo, S. (2017). Novel, smart and RFID assisted critical temperature indicator for supply chain monitoring. *Journal of Food Engineering*, *193*, 20—28.

Lu, P., Yang, Y., Liu, R., Liu, X., Ma, J., Wu, M., & Wang, S. (2020). Preparation of sugarcane bagasse nanocellulose hydrogel as a colourimetric freshness indicator for intelligent food packaging. *Carbohydrate Polymers*, *249*, 116831.

Mani, J., & Umapathy, M. J. (2020). Preparation and properties of microfibrillated cellulose einforced pectin/fenugreek gum biocomposite. *New Journal of Chemistry*, *44*, 18792–18802.

Mihindukulasuriya, S., & Lim, L.-T. (2014). Nanotechnology development in food packaging: A review. *Trends in Food Science & Technology*, *40*, 149–167.

Mohammadian, E., Alizadeh-Sani, M., & Jafari, S. M. (2020). Smart monitoring of gas/temperature changes within food packaging based on natural colorants. *Comprehensive Reviews in Food Science and Food Safety*, *19*, 2885–2931.

Mohan, C. & Ravishankar, C. 2019. Active and intelligent packaging systems-Application in seafood. *World Journal of Aquaculture Research & Development*, *1*, 10–16. <http://drs.cift.res.in/handle/123456789/4396>.

Moreirinha, C., Vilela, C., Silva, N. H., Pinto, R. R., Almeida, A., Rocha, M. A. M., ... Freire, C. S. (2020). Antioxidant and antimicrobial films based on brewers spent grain arabinoxylans, nanocellulose and feruloylated compounds for active packaging. *Food Hydrocolloids*, *108*, 105836.

Muller, P., & Schmid, M. (2019). Intelligent packaging in the food sector: A brief overview. *Foods*, *8*, 16.

Narwade, V. N., Anjum, S. R., Kokol, V., & Khairnar, R. S. (2019). Ammonia-sensing ability of differently structured hydroxyapatite blended cellulose nanofibril composite films. *Cellulose*, *26*, 3325–3337.

Nguyen, L. H., Naficy, S., Mcconchie, R., Dehghani, F., & Chandrawati, R. (2019). Polydiacetylene-based sensors to detect food spoilage at low temperatures. *Journal of Materials Chemistry C*, *7*, 1919–1926.

Pandian, A. T., Chaturvedi, S., & Chakraborty, S. (2020). Applications of enzymatic time—temperature indicator (TTI) devices in quality monitoring and shelf-life estimation of food products during storage. *Journal of Food Measurement and Characterization*, 1–18.

Prakash, B., Kujur, A., Yadav, A., Kumar, A., Singh, P. P., & Dubey, N. (2018). Nanoencapsulation: An efficient technology to boost the antimicrobial potential of plant essential oils in food system. *Food Control*, *89*, 1–11.

Roy, S., & Rhim, J.-W. (2020). Carboxymethyl cellulose-based antioxidant and antimicrobial active packaging film incorporated with curcumin and zinc oxide. *International Journal of Biological Macromolecules*, *148*, 666–676.

Salama, H. E., & Aziz, M. S. A. (2020). Optimized carboxymethyl cellulose and guanidinylated chitosan enriched with titanium oxide nanoparticles of improved UV-barrier properties for the active packaging of green bell pepper. *International Journal of Biological Macromolecules*, *165*, 1187–1197.

Saliu, F., & Della Pergola, R. (2018). Carbon dioxide colorimetric indicators for food packaging application: Applicability of anthocyanin and poly-lysine mixtures. *Sensors and Actuators B: Chemical*, *258*, 1117–1124.

Sedayu, B. B., Cran, M. J., & Bigger, S. W. (2020). Reinforcement of refined and semi-refined carrageenan film with nanocellulose. *Polymers*, *12*, 1145.

Shaghaleh, H., Xu, X., & Wang, S. (2018). Current progress in production of biopolymeric materials based on cellulose, cellulose nanofibers, and cellulose derivatives. *RSC Advances*, *8*, 825–842.

Sharma, R., Jafari, S. M., & Sharma, S. (2020a). Antimicrobial bio-nanocomposites and their potential applications in food packaging. *Food Control*, *112*, 107086.

Sharma, S., Barkauskaite, S., Jaiswal, A. K., & Jaiswal, S. (2020b). Essential oils as additives in active food packaging. *Food Chemistry*, *343*, 128403.

Tsai, Y.-H., Yang, Y.-N., Ho, Y.-C., Tsai, M.-L., & Mi, F.-L. (2018). Drug release and antioxidant/antibacterial activities of silymarin-zein nanoparticle/bacterial cellulose nanofiber composite films. *Carbohydrate polymers*, *180*, 286–296.

Tyagi, P., Lucia, L. A., Hubbe, M. A., & Pal, L. (2019). Nanocellulose-based multilayer barrier coatings for gas, oil, and grease resistance. *Carbohydrate Polymers*, *206*, 281–288.

Vilarinho, F., Sanches Silva, A., Vaz, M. F., & Farinha, J. P. (2018). Nanocellulose in green food packaging. *Critical Reviews in Food Science and Nutrition*, *58*, 1526–1537.

Wang, J., Liu, X., Jin, T., He, H., & Liu, L. (2019a). Preparation of nanocellulose and its potential in reinforced composites: a review. *Journal of Biomaterials Science, Polymer Edition*, *30*, 919–946.

Wang, L., Wu, Z., & Cao, C. (2019b). Technologies and fabrication of intelligent packaging for perishable products. *Applied Sciences*, *9*, 4858.

Warriner, K., Reddy, S. M., Namvar, A., & Neethirajan, S. (2014). Developments in nanoparticles for use in biosensors to assess food safety and quality. *Trends in Food Science & Technology*, *40*, 183–199.

Wróblewska-Krepsztul, J., Rydzkowski, T., Borowski, G., Szczypiński, M., Klepka, T., & Thakur, V. K. (2018). Recent progress in biodegradable polymers and nanocomposite-based packaging materials for sustainable environment. *International Journal of Polymer Analysis and Characterization*, *23*, 383–395.

Xiao, Y., Liu, Y., Kang, S., Wang, K., & Xu, H. (2020). Development and evaluation of soy protein isolate-based antibacterial nanocomposite films containing cellulose nanocrystals and zinc oxide nanoparticles. *Food Hydrocolloids*, *106*, 105898.

Xie, J., & Hung, Y.-C. (2018). UV-A activated TiO2 embedded biodegradable polymer film for antimicrobial food packaging application. *LWT*, *96*, 307–314.

Xie, J., & Hung, Y.-C. (2019). Methodology to evaluate the antimicrobial effectiveness of UV-activated TiO2 nanoparticle-embedded cellulose acetate film. *Food Control*, *106*, 106690.

Xu, X., Liu, F., Jiang, L., Zhu, J., Haagenson, D., & Wiesenborn, D. P. (2013). Cellulose nanocrystals vs. cellulose nanofibrils: A comparative study on their microstructures and effects as polymer reinforcing agents. *ACS Applied Materials & Interfaces*, *5*, 2999–3009.

Yousefi, H., Su, H.-M., Imani, S. M., Alkhaldi, K., M. Filipe, C. D., & Didar, T. F. (2019). Intelligent food packaging: A review of smart sensing technologies for monitoring food quality. *ACS Sensors*, *4*, 808–821.

Yu, H.-Y., Zhang, H., Song, M.-L., Zhou, Y., Yao, J., & Ni, Q.-Q. (2017). From cellulose nanospheres, nanorods to nanofibers: Various aspect ratio induced nucleation/reinforcing effects on polylactic acid for robust-barrier food packaging. *ACS Applied Materials & Interfaces*, *9*, 43920–43938.

Zhang, Y., & Lim, L.-T. (2018). Colorimetric array indicator for NH3 and CO2 detection. *Sensors and Actuators B: Chemical*, *255*, 3216–3226.

CHAPTER 14

Enzyme cocktail: a greener approach for biobleaching in paper and pulp industry

Adarsh Kumar[1], Prerna[1], Muhammad Bilal[2], Ajay Kumar Singh[1], Sheel Ratna[1], K.R. Talluri Rameshwari[3], Ishtiaq Ahmed[4] and Hafiz M.N. Iqbal[5]

[1]Department of Environmental Microbiology, School for Environmental Science, Babasaheb Bhimrao Ambedkar University (A Central University), Lucknow, India
[2]School of Life Science and Food Engineering, Huaiyin Institute of Technology, Huai'an, P.R. China.
[3]Division of Microbiology, Department of Water & Health, Faculty of Life Sciences, JSS Academy of Higher Education and Research, Mysuru, India.
[4]School of Medical Science, Menzies Health Institute Queensland, Griffith University, Southport, QLD, Australia.
[5]Tecnologico de Monterrey, School of Engineering and Sciences, Monterrey, Mexico.

14.1 Introduction

Pulp and paper industries are globally dominated by Asian, East Latin American, Northern European, North American, and Australian companies. Total paper production will reach around 490 million tonnes by 2020, increasing 24.0% over the previous decade. According to current global paper production statistics, China is the largest producer, followed by the United States. India's market share is around 3.7%, and it is steadily increasing at a rate of 7.0% per year (Balda, Sharma, Capalash, & Sharma, 2019; Kaur, Bhardwaj, & Lohchab, 2017). Pulp and paper industries are, without a doubt, a significant and ever-growing part of the global economy. However, it must continue to encounter cost-cutting initiatives and environmental protection agency pressures to reduce and release hazardous chemicals without sacrificing paper quality. Furthermore, as people have become more aware of the importance of maintaining environmental protection standards, market demand for chlorine-free bleached paper has grown significantly in recent years. As a result, the pulp and paper industry will benefit from sustainable paper production using greener technologies.

Alternatives to chlorine-based pulp bleaching have been used, including ozone, hydrogen peroxide, oxygen delignification, raw pulp selection, and enzyme-assisted technologies (N. Bhardwaj, Kumar, & Verma, 2019b; N.K. Bhardwaj, Kaur, Chaudhry, Sharma, & Arya, 2019a; A.K. Singh, Kumar, & Chandra, 2020a, 2020b; G. Singh, Kaur, Khatri, & Arya, 2019; Tripathi, Bhardwaj, & Ghatak, 2019). Oxygen delignification plants are becoming increasingly popular as a pulp pretreatment option, but they have several drawbacks, including high installation costs, elevated temperature (80°C−120°C), highly alkaline conditions (pH > 10), and high pressure (600−800 kPa) for achieving a

Nanotechnology in Paper and Wood Engineering
DOI: https://doi.org/10.1016/B978-0-323-85835-9.00007-6

rational reaction efficiency. The recovery cycle is also increased, raising the time and energy required for paper production. Hydrogen peroxide is a costly substitute for industrial use. Ozone treatment has decent effects, but it degrades the fibers, resulting in final paper products losing strength. Depending on the raw material, industries use a variety of pulps, including softwood, hardwood, and agricultural waste such as cereal straw, wheat straw, and sugarcane bagasse. Wood-based virgin pulp dominates the paper industry, and it is expected to account for half of all industrial wood harvest by 2050. However, due to the rising demand for paper and declining forest cover areas around the world, the use of agropulp in the paper industry is increasing. Agricultural waste can be more cost effective than wood-based pulp due to its local availability and unique properties. It would also be environmentally sustainable because it would be less dependent on forests (Gonzalo et al., 2017). According to industry experts, by 2025, the use of agro-waste such as wheat straw and bagasse for sustainable paper productions would have doubled from its current capacity of 21 million tonnes, owing to the need for less water and electricity, lower production costs, the use of fewer chemicals as well as the availability of an alternative to a virgin wood pulp (Gonzalo et al., 2017).

When opposed to traditional pulp bleaching methods, which are expensive and cause pollution, enzyme-based bleaching has the advantage of being environmentally friendly. The worldwide demand for enzymes in the pulp and paper industry is expected to rise from US$73 million in 2019 to US$95 million in 2024. Various lignocellulosic enzymes, such as xylanases and laccases, have been extensively investigated in the laboratory to substitute chemical bleaching procedures with elemental chlorine-free, cost-effective, and environmentally friendly biobleaching activities. On the other hand, laccases have a significant disadvantage in that they only work in the presence of a mediator in pulp biobleaching, which is impractical for large-scale use due to their high cost (Singh & Arya, 2019). In the pulp and paper industry, xylanases are frequently used to increase pulp value by depolymerizing xylan, which is bonded to lignin in plant cell walls. At the laboratory scale, bacterial and fungal xylanases have been used to biobleach various pulp forms (Kumar, Kumar, Chhabra, & Shukla, 2019). In various industrial processes, fungal xylanase has 76 main drawbacks, including thermal instability, narrow acidic pH range, mass, and higher production costs, and thus bacterial xylanase has become a hardened competitor. *Bacillus* strains are preferred choices for the pulp and paper industry because of superior yields of thermo-alkali stable hemicellulolytic enzymes (Yardimci & Cekmecelioglu, 2018). This chapter mainly spotlights the uses of microbial enzymes as a greener and ecofriendly approach in the paper and pulp industry for biobleaching purposes.

14.2 Microbial enzyme applications in biobleaching

Generally, the removal of lignin from pulp—paper waste via chemical or gases/steam is called bleaching, but biobleaching or prebleaching is the term used for lignin removal via different microbial enzymes cocktails such as laccases, xylanases, cellulase, protease, amylase, manganese peroxidases (MnPs), lignin peroxidases (LiPs), and lipases (Gupta, Kapoor, & Shukla, 2020; Kumar, Thakur, Singh, & Shah, 2020; Singh et al., 2019).

14.2.1 Laccases

Laccase (EC 1.10.3.2, benzenediol: O_2 oxidoreductases) is a multicopper enzyme that removes lignin using low molar mass chemical mediators (Kumar & Chandra, 2020). Laccases can oxidize a wide variety of inorganic and organic substrates in both mild and extreme conditions, and they can be found in bacteria, fungi, plants, and insects (Kumar, Singh, Ahmad, & Chandra, 2020a; Singh, Kaur, Puri, & Sharma, 2015). Nathan, Rani, Gunaseeli, and Kannan (2018) tested the biobleaching potential of laccase produced by *Fusarium equiseti* VKF-2 isolated from mangrove. The laccase facilitated the maximum hexenuronic acid release with Kappa number reduction. There was a Δ brightness of 15% following 4 hour of enzymatic treatment. Finally, they recommended laccase treatment as a green and feasible bioleaching process for newspaper waste. Sharma et al. (2018), the optimized condition through response surface methodology for laccase production under solid-state fermentation (SSF). Laccase production from *Ganoderma lucidum* RCK 2011 was optimised using response surface methods under SSF conditions, resulting in an eightfold increase over the unoptimized media. Remarkably, the biobleaching potentials of both in situ and in vitro SSF laccases were similar, resulting in a 25% reduction in chlorine dioxide required to achieve the same pulp brightness as the pulp treated chemically (Sharma et al., 2018). Therefore they recommended in situ SSF laccase as an economical and ecofriendly treatment approaches for paper industries. Boruah, Sarmah, Das, and Goswami (2019) isolated laccase producing *Kocuria* sp. PSB-1 for bamboo pulp bleaching. The reported isolated PSBS-1 strain significantly enhanced ISO brightness up to 80.03 ± 1.68% after laccase, mediator, peroxidase, and ozone bleaching. Further X-ray diffraction, Fourier transform infrared spectroscopy, and scanning electron microscopy analyses supported significant changes in lignin spectra, cellulose crystallinity index, and morphology, such as the appearance of cracks, grooves, and crevices due to ligninolytic activity. The laccase (lac) genes of *Madurella mycetomatis* were cloned and heterologously expressed in the methyl-trophic host *Pichia pastoris* by Tulek et al. (2021). They reported recombinant laccase (MmLac) significantly enhanced the whiteness in textile bleaching.

14.2.2 Lignin and manganese peroxidases (heme peroxidase)

The uses of ligninolytic enzymes in the bleaching process in the paper industry were summarized by Chaurasia and Bhardwaj (2019). Saleem, Khurshid, and Ahmed (2018) applied laccase, xylanase, and MnP individually and sequentially for biobleaching of oven-dried paper. These findings suggested that xylanase give better kappa number reduction and improvement in brightness compared to laccase and MnP.

Hamedi, Vaez Fakhri, and Mahdavi (2020) isolated *Streptomyces rutgersensis* UTMC 2445 from lignocellulosic rich soil and screened for xylanase production and bleaching. At specific conditions (pH 5–6, temperature 30°C, time 6 h), biobleaching increased the brightness of mechanical paper pulp compared to control. Further, after 60 min, enzymatic treatment significantly reduced the consumption of chemical and achieved up to 55% higher brightness.

14.2.3 Cellulase

Cellulose degradation requires a variety of cellulase combination. Cellulase classified into various types, namely endoglucanase (E.C. 3.2.1.4) and exoglucanase (E.C. 3.2.1.91), which cleaves internally and at free ends of cellulose and breaks cellulose subsequently into disaccharides. During the biobleaching of paper–pulp waste applied without carbohydrate-binding sites cellulose (Kumar & Rani, 2019). In comparison to previous research, the enzyme allowed for a greater drop in Kappa number and Hexenuronic acid (Hex A). For varying treatment times, a brightness of approximately 10% was reached with both cellulase and xylanase (Kumar et al., 2018). Beside of above tear strength of recycled paper.

14.2.4 Xylanase

Xylanase (3.2.1.8) is found in abundance in nature, and it is produced in large quantities by bacteria, fungi, yeast, and algae (N. Bhardwaj, Kumar, & Verma, 2019b; N.K. Bhardwaj, Kaur, Chaudhry, Sharma, & Arya, 2019a). Xylanase is widely applied for bioconversion of lignocellulosic materials and bleaching kraft pulps to increase pulp brightness (Kmetzki, Henn, Moraes, Silva, & Kadowaki, 2019). Xylanases are hydrolytic enzymes that cleave the complex plant cell wall polysaccharide xylan's β-1,4 backbones. Xylan is the most common hemicellulosic component in both soft and hard foods. After cellulose, it is the most abundant renewable polysaccharide. Xylanases and related debranching enzymes are produced by actinomycetes, bacteria, yeast, and fungi and are responsible for hemicellulose hydrolysis (Walia, Guleria, Mehta, Chauhan, & Parkash, 2017). Sharma et al., 2017 studied the advantage of ultrafiltered xylanase–pectinase over crude enzyme for the first time. They assessed the effect of an ultrafiltered xylanase–pectinase mixture produced by *Bacillus pumilus* AJK on agro waste-based media (Nagar & Gupta, 2020). At optimized conditions (pH: 8.5, temperature: 55°C, time: 2 h), ultrafiltered-enzyme reduced kappa number 15.2% at xylanase–pectinase (4.0–0.8 IU) per gram of pulp and reduction of Cl_2-ClO_2 consumption (30%–28.86%), biological oxygen demand (18.13%), and chemical oxygen demand (21.66%). Kraft pulp biobleaching through xylanase produced by *B. pumilus* SV-205 under SSF on wheat brawn was studied by Nagar and Gupta (2020). At optimized conditions such as enzyme dose (12.5 IU g^{-1}), pH 10.0, and incubation period 120 min at 60°C temperature xylanase pretreatment enhanced brightness by 1.05 point with 0.8 points reduction in kappa number. Patel and Dudhagara (2020) worked on optimized xylanase production and biobleaching of rice straw pulp from hot spring isolated *Bacillus tequilensis* strain UD-3. *B. tequilensis* UD-3 produced 8.54 IU mL^{-1} xylanase at optimized conditions of pH 8.0, temperature 50°C, and boosted at 1.0 mM concentration of metal ions such as Mn^{+2}, Fe^{+3}, Cu^{+4}, and Zn^{+4} (Patel & Dudhagara, 2020). Further, Zn^{+4} concoction with xylanase reduced kappa number (17.24 ± 0.02) after Xyl + Zn bleaching. A significant reduction was also seen in reducing sugar (50%), phenol (29.19%), and lignin compounds (35.86% and 40.48%) during the Xyl + Zn treatment of pulp samples compared to the only xylanase.

Xylanase and laccases are renewable green, highly specific and cost-effective bioleaching agents that work individually or in a cocktail to improve the quality of paper (A. Sharma, Balda, Gupta, Capalash, & Sharma, 2020; D. Sharma, Chaudhary, Kaur, & Arya, 2020). Further, Sridevi, Narasimha, and Devi (2019) reported xylanase from *Penicillium* sp. reduced kappa number (23.1−20.8) with increased brightness (40.1−39.2) from the paper pulp after treatment. Angural et al. (2020) applied a cocktail of laccase (L), xylanase (X), and mannanase (M) from thermo-alkali stable *B. tequilensis* LXM 55 produced pulp biobleaching. Enzymes cocktail led to 49.35% reduction in kappa number and enhancement in brightness to 11.59% and whiteness to 4.11%. They reported that no mediator was required for the application of laccase. Further, *Bacillus* sp. NG-27 and *Bacillus nealsonii* PN-11 produce xylanase and mannanase, respectively (Angural et al., 2021). The cocktail enzyme (X + M) at optimized pH 8.0 and 65°C temperature within 1 h increased brightness and whiteness up to 11% and 75%, respectively, with a 45.64% reduction in kappa (Angural et al., 2021).

14.2.5 Lipases

Lipases (E.C. 3.1.1.3) are a class of carboxylesterase that catalyzes the hydrolysis of long-chain acylglycerols to glycerol, free fatty acids, and mono- and diglycerides. Its various industrial applications predict that by 2020, the global market will raise to about $590.5 million (Ktata et al., 2020). Lipase-producing bacteria *Bacillus, Pseudomonas, Burkholderia,* and fungi, such as *Aspergillum, Rhizopus, Penicillium,* and *Candida* are reported in the previous literature (Singh & Mukhopadhyay, 2012). Nathan and Rani (2020) studied the role of lipase in enzyme cocktail for deinking paper pulp. They found at optimized conditions, lipase activity was 12.36 times higher (105.12 U g^{-1}), and brightness increased up to 14.5% by cellulase, xylanase, and lipase enzymatic cocktail. *Bacillus halodurans* enzyme cocktail (xylanase, 15 U; pectinase, 2 U; -amylase, 2.5 U; protease, 2 U; and lipase, 1.8 U) was used to biobleaching of unbleached and oxygen delignification based pulp for 90 minutes at optimized environmental conditions such as temperature (65°C) and pH (9.0−9.5), followed by reduced chemical treatment. In the case of unbleached agropulp, cocktail pretreatment increased tensile strength (23.55%), burst factor (20.3%), and tear factor (3.17%) and reduced kappa number (19.5%) of pulp with the effect on the brightness.

14.2.6 Protease

Proteases (EC 3.4.21.112) are a large family of enzymes that catalyze the hydrolysis of peptide bonds in proteins and polypeptides. Proteolytic enzymes produced by commercially available microorganisms are typically a mix of endopeptidases (proteinases) and exopeptidases. Microbial proteolytic enzymes are available from a variety of fungi and bacteria. With a few exceptions, most fungal proteases can tolerate and act efficiently over a broad pH range (~4.0−8.0), while bacterial proteases work best over a narrow range (nearly pH 7−8). For centuries, fungal protease has been used in a variety of industrial applications, including the pulp and paper industry, the food industry, and so forth. Soybeans or other grains are steamed and inoculated with *Aspergillus flavus-oryzae* or *Aspergillus tamarii* spores in these applications. The koji is covered with brine after maximum enzyme production has occurred, and enzymatic digestion is allowed to occur. This method of producing soy sauce is also used sparingly in this country. There is no attempt to separate the enzymes from the producing organisms in these applications. Enzymatic processing has several advantages over acid or alkaline protein hydrolysis, including the use of basic machinery and the absence of amino acid destruction or racemization.

14.2.7 Amylase

α-amylase (EC 3.2.1.1) enzyme catalyze the hydrolysis of 1,4-glycosidic linkages in starch to generated simple compounds like maltotriose, glucose, and maltose (Gupta, Gigras, Mohapatra, Goswami, & Chauhan, 2003; Kumar, Singh, & Chandra, 2020b). The main benefit of using microorganisms to make amylases is the low cost of bulk production and the ease with which microbes can be modified to produce enzymes with desired properties. α-Amylase comes from a variety of fungi, yeasts, and bacteria. In the industrial sector, enzymes derived from fungi and bacteria have dominated (Gupta et al., 2003). α-Amylases may be used in a variety of industrial processes, including cooking, fermentation, textiles, paper, detergents, and pharmaceuticals as well as fine-chemical industries. However, due to advances in biotechnology, amylases are now used in a variety of fields including clinical, medicinal, textile, food, brewing, and analytical chemistry as well as starch saccharification (Kandra, 2003; Kumar, Singh, & Chandra, 2020b).

14.3 Pulp and papermaking processes

Pulping is the method of converting agricultural waste or wood into a versatile fiber that can be used to produce paper. Many pulping processes may be used, depending on the final products. The most common technique, that is, kraft pulping is a hot alkaline sulfide digestion of wood that removes the majority of the lignin and other residues, but the strong cellulosic fiber is stabled. Moreover, writing and printing paper qualities are also made with an acid sulfite pulping process. Furthermore, mechanical, chemo-thermomechanical, and thermomechanical pulping processes have much higher production of fibers, but with lower strength and optical properties. Newsprint or magazine stock can be made with such fibers. If the end usage includes white paper, the fiber is washed and then bleached after pulping. After that, additional manipulations such as scaling, filler addition, and color addition can be made to create the final paper product.

Recycled fibers are generated from the recovered postconsumer paper is a major fiber source in addition to primary fiber. The basic steps in recycling paper are screening, bleaching, pulping, cleaning, flotation, and removal of other contaminants. Each of these steps may benefit from the use of enzymes. The remainder of this analysis will go into how enzymes are used in these processes, as well as how enzymes can be modified to best meet the needs of the process.

14.3.1 Use of enzyme in pulping

Several enzymes were used in pulping process can increase fiber yield, reduce energy requirements for further refining, or alter the fiber in specific ways. Cellulases have been used in a variety of ways in the paper industry. Pretreatments with enzymes such as cellulase, pectinase, and hemicellulase have been shown to improve the kraft pulping of sycamore chips and other pulp sources (Moldes & Vidal, 2008). This enzyme combination resulted in improved pulp delignification and lower bleaching chemical costs without compromising the paper's strength. As a result of the high cost of enzymes and concerns about the effectiveness of a large enzyme assisting low molecular weight pulping chemicals, cellulases have been challenged as a way to improve the kraft process (Kumar, Singh, Ahmad, & Chandra, 2020a; Woolridge, 2014). Cellulase testing was conducted on mechanical pulps. In this case, the brightness and energy required for refining with a crude cellulase on *Radiata pine* versus the energy required after a cellobiohydrolase treatment on spruce mechanical pulps yield conflicting results (Woolridge, 2014). Laccase and protease, for example, have been shown to reduce the amount of energy needed for mechanical pulping (Singh, Capalash, Kaur, Puri, & Sharma, 2016).

Laccase treatment has the added benefit of improving fiber bonding, which improves the paper's strength (Singh et al., 2016). For decades, enzymes have been used in the refining of virgin fibers. Cellulases and xylanases were used to treat kraft pulp, and both enzymes decreased the amount of energy needed for further refining (Collins, Gerday, & Feller, 2005). Cellulases must be used with caution so that the strength of the fibers is not weakened (Collins et al., 2005). On unbleached kraft pulps, xylanase treatments are more successful than on bleached kraft pulps (Bajpai, 1999). The use of enzymes can also increase the yield of thermomechanical pulp. The use of an acetyl esterase to deesterify the soluble O-acetyl galactoglucomannans of Norway spruce resulted in the galactoglucomannans precipitating onto the fiber and the fiber yield (Marques, Pala, Alves, Amarak-Collaco, & Gama, 2003).

Dissolving pulps are made from cellulose that is bright and has a consistent molecular weight. These pulps are processed to produce soluble reactive carbohydrate chains, which are then extruded into fibers or films. The treatment reduced the viscosity and chain length of an eucalyptus and acacia pulp while increasing its reactivity by endoglucanase (Singh et al., 2016). Moreover, at the same protein dosage, endoglucanases were more effective at hydrolyzing the pulp than cellobiohydrolases (Marques et al., 2003). Cellulases were used in a two-step process in which alkali-insoluble content was extracted and then recombined with the first extraction to produce a more dissolving pulp, alkali soluble, than pulp treated directly with enzymes. Although, the fibers from the two-step processes had significantly lower solubility when compared at the same degree of hydrolysis (Walia et al., 2017).

14.3.2 Enzyme use in bleaching

The pulp and paper mills are well-known for contributing significantly to environmental pollution due to their heavy use of chemicals and energy. Different chemicals are used during the bleaching process, including H_2SO_4, Cl_2, ClO_2, NaOH, and H_2O_2. In bleached effluents, these chemicals combine with lignin and carbohydrates to produce a significant amount of contaminants. The pulp and paper industry has been forced by environmental demand to reduce pollutant generation in the bleaching section (Singh et al., 2020a, 2020b). Several enzymes have emerged as simple, cost-effective, and environmentally friendly of pulp bleaching alternatives. Biobleaching, also known as prebleaching, is the process of pretreating pulp with enzymes. Biobleaching is achieved using several microbial enzymes, that is, laccases, xylanases, MnPs, pectinases, and LiPs. In general, xylanase has an indirect influence on the final pulp brightness. It eliminates bleaching agent reactants as well as barriers to the bleaching process. The oxidative enzymes were attacked on lignin components. Laccases react with lignin-derived compounds using oxygen, whereas LiPs and MnPs use hydrogen peroxide to breakdown lignin in kraft pulp, for increasing brightness of pulp produced by the *Tinea versicolor* and *Pseudomonas chrysosporium* (Kumar & Chandra, 2020).

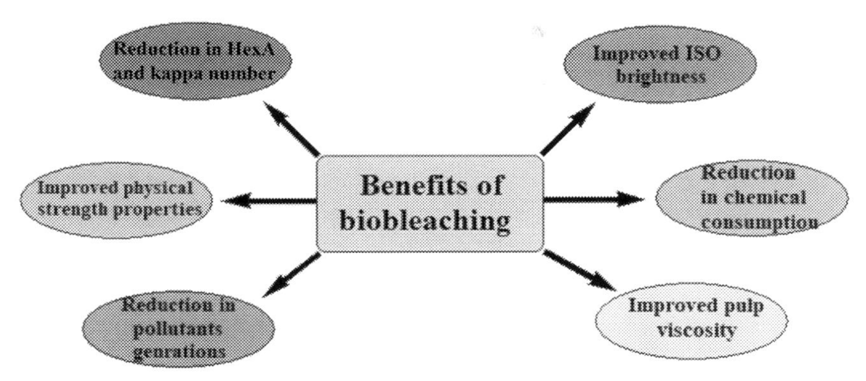

Figure 14.1 Schematic of benefits of biobleaching. *HexA*, hexeneuronic acid.

MnPs and LiPs also require moderate hydrogen peroxide levels for activity, but when it is present at high amount they are inactivated. Both peroxidase enzymes were mostly produced by white-rot fungus, and have been reported by several researchers. The uses of these enzymes were limited due to their high cost of hydrogen peroxide requirements as well as sensitivity to hydrogen peroxide. Although enzymes are widely available, they are still susceptible to hydrogen peroxide. The hydrogen peroxide needed could be provided by continuous low-level hydrogen peroxide addition or a hydrogen peroxide-producing device (glucose oxidase and glucose). The use of biobleaching pulp reduces the use of chlorinated chemicals and forms adsorbable organic halides while increasing the pulp's consistency. The kappa number of pulps is reduced by enzyme pretreatment, and ISO brightness is significantly improved. Biobleaching of pulp resulted in improved physical strength properties and pulp viscosity (Fig. 14.1).

Laccases work directly to act on lignin and can be extracted from pulp, while xylanases help enhance delignification by making pulp more vulnerable to bleaching chemicals. They have different delignification mechanisms (Virk, Sharma, & Capalash, 2012). As a result, enzyme-based, environmentally sustainable processes could be more efficient than recent methods (Kapoor, Kapoor, & Kuhad, 2007). Biobleaching of mixed wood pulp was tested using a mixed-enzyme preparation of laccase and xylanase. Cocultivating mutant *Pleurotus ostreatus* MTCC 1804 and *Penicillium oxalicum* SAU (E)-3.510 yielded the enzymes by the SSF process. The use of a mixed enzyme preparation with xylanase bleaching pulp, on the other hand, resulted in a substantial reduction in kappa number and an increase in brightness. A mixed enzyme solution (xylanase:laccase, 22:1) resulted in enhanced delignification of pulp when bleaching was done at 10% pulp consistency (55°C, pH 9.0) for 3 h (Dwivedi, Vivekanand, Pareek, Sharma, & Singh, 2009). The cost effectiveness of producing xylanase and laccase up to 10 kg substrate level in elemental chlorine-free bleaching of eucalyptus kraft pulp was investigated. The ClO_2 savings were greater at laboratory scale with the potentially relevant xylanase (25%) followed by laccase (15%) compared to the pulp prebleached separately. When used at pilot scale (50 kg pulp), the sequential enzyme treatment improved pulp properties (50% reduced postcolor number, 15.71% increased tear index) and reduced organo-chlorinated compound levels in bleach effluents (34%) (Sharma et al., 2014).

14.3.3 Enzyme use in modifications and fiber recycling

Paper recycling's main goals are to remove ink and other contaminants while maintaining the fibers optical and strength properties. Enzymes may aid in dewatering (drainage) and contaminant removal and improve the bond between recycled fibers. Sheet formation is hindered by secondary fiber drainage resistance, which slows the paper machine's activity and increases drying energy (Nathan, Rani, Gunaseeli, & Kannan, 2014). Enzyme treatments that assist fiber separation in the washing and flotation processes can also help with deinking and contaminant removal as shown in Fig. 14.2.

Many previous researchers had used lignocellulolytic enzymes for pulp modification. Cellulase, xylanase, and laccase were used to alter pulp in our tests (Nathan et al., 2018). During the recycling process, the optimum pulp modification was found to be 5% pulp paper treated with 10% (vol./wt.) crude cellulase enzyme provided by Trichoderma *viride* VKF3 (Nathan et al., 2018). Laccase derived from *F. equiseti* VKF2 could decrease lignin content to the maximum level when combined with a 5% pulp consistency and a 15% enzyme dosage (Nathan et al., 2018). It was discovered that a low cellulase enzyme dosage from *E. coli* SD5 was suitable for pulp modification (Nathan et al., 2018).

14.3.4 Refining and drainage

Endoglucanases used properly will increase the drainage rates of recycled fibers beyond what polymer addition can achieve (Lammi & Heikkurinen, 1997). efining is a mechanical treatment of pulp that modifies that structure of the fibers in order to achieve

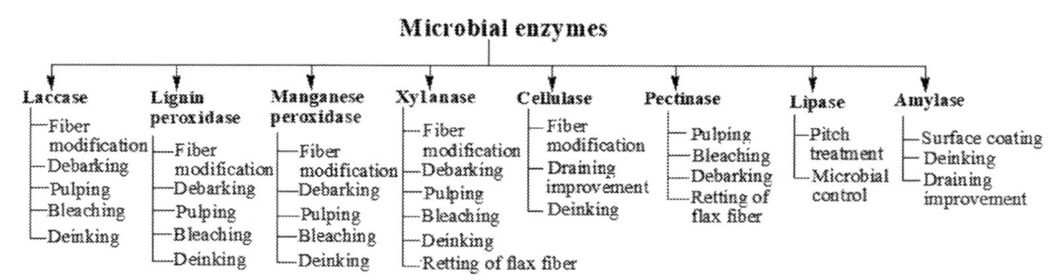

Figure 14.2 Application of various enzymes used in the paper and pulp industries and their functions.

desired paper making properties. While, the uses of enzymes at the same dose level before refining and impact on energy savings and physical strength achieved of pulp. With a suitable pH, temperature, and consistency, as well as proper mixing, treatments can be done in little as 15 min. The preparation and activity of the enzyme determine the enzyme dosage and treatment period (Brandberg & Kulachenko, 2017). After refining, adding enzymes increases freeness, allowing the paper machine to operate very efficiently (Genco, 1999). This effect may be caused by the removal of cellulose microfibrils from paper fiber surfaces (Brodin & Eriksen, 2014). Endoglucanases that target amorphous cellulose often cause rapid loss of fiber strength, so treatment doses must be kept low in both the cases (Brodin & Eriksen, 2014). Pala, Mota, and Gama (2004) looked into using cellulose-binding domains (CBDs) isolated from cellulase preparations after proteolytic digestion to prevent losing strength. CBDs improved both drainage rates and paper strength properties at low doses, but the benefits of strength indices were less pronounced at higher dose rates. Pala et al. (2004) proposed that the increased microfibrilation of the fiber surface was responsible for the improved strength. Moreover, the increased drainage was also responsible for the residual hydrolytic activity, as CBDs preparations with detectable reducing sugar release increased drainage more than those without detectable activity.

Furthermore, fiber determines the efficacy of enzyme treatment. Chemically treated fibers are much more sensitive to cellulase action than mechanical fibers. Cellulases in low doses can increase hand-sheet density and coarseness while reducing strength. When it comes to kraft fiber, these treatments are the most effective (Bajpai, 1999). Alkaline cellulase activity can be beneficial because bond strength restoration is most important with secondary fibers, and recycled fibers are also filled with calcium carbonate (Bajpai, 1999). However, no reports of highly active alkaline cellulases have been found. In fiber recycling mill treatment water, starches used in sizings can build up and hinder drainage. Alpha-amylase can help with drainage in such instances, possibly by lowering the viscosity of the backwater (Singh et al., 2016). Amylase treatments have been shown to increase the drainage of recycled paper pulp, allowing the paper machines to run faster.

Therefore the key recycled paper sources are old mixed newspapers, office waste, and old corrugated containers. Several coatings, adhesives, glues, and binders found in the recycled paper can cause stickies to form in the pulp slurry (Ballinas-Casarrubias et al., 2020). Stickies that clump together to form larger particles cause problems in the papermaking process and must be removed. An unknown esterase was discovered to reduce the size and number of stickies in recycled paper sources (Ballinas-Casarrubias et al., 2020). Mostly cellulase and hemicellulase enzymes have been used as additives to increase the drainage of old corrugated container liner boards and pulps, allowing for the use of less polymeric additives to strengthen the paper (Bajpai, 1999). Therefore some anionic surfactants were commonly used during recycling of papers, which can cause problems when cellulases and xylanases are used (Bajpai, 1999).

Cellulase and xylanase activity is increased when cationic or nonionic surfactants are used (Bajpai, 1999; Jurasek & Paice, 1992).

14.3.5 Microbial enzyme-assisted deinking specific

Enzymes are used as a replacement for polluting technologies to overcome their disadvantages (Srinivasan & Rele, 1999). Enzymatic deinking has sparked a lot of interest in addressing these issues due to its high efficiency and low environmental impact (Pala, Mota, & Gama, 2006; Thomas, 1994). In the bleaching process, enzyme-based biobleaching could remove the need for chlorine and chlorine compounds (Raghukumar, Usha, Gaud, & Mishra, 2004). White-rot fungi were used to degrade residual lignin in the pulp using ligninolytic enzymes like laccase, lignin, and manganese peroxide as well as hemicellulolytic enzymes like xylanase (Casimir-Schenkel, Davis, Fiechter, Gygin, & Murray, 1995).

14.3.5.1 Enzymatic deinking and paper characteristics

Paper and pulp undergo morphological, physical, and chemical changes as a result of their enzymatic treatment. The degree of these changes varies depending on the enzymes used. The effects of various enzymes on their morphological and chemical properties are discussed in the following section. Furthermore, deinking is addressed, which is determined by a change in brightness level.

14.3.5.2 Advantages of biodeinking

The following are the key benefits of biological deinking:

- Low-toxic effluents from an environmentally safe process.
- With the chemical deinking technique, you can achieve a comparable brightness.
- Enzymatic deinking makes use of simple to make microbial enzymes.
- The technique is economically valuable.
- Energy demand is reduced.
- Many physical properties, such as burst index and tensile strength, are enhanced as a result.

14.3.5.3 Challenges of biological deinking

The main points that come to mind when addressing biological deinking complexities are the outweighing benefit and ease of chemical deinking methods. Chemical deinking is still used in many industries, although it is detrimental to the environment. The major challenges of biological deinking are:

1. Enzyme-based deinking is dependent on the enzyme's efficiency under different environmental conditions such as optimum activity can be at mesophilic conditions, and lower in deinking efficiency at psychrophilic and thermophilic conditions.
2. In the paper and pulp industries, the enzyme must tolerate inhibitory effects from various additives.

3. The enzyme production process must be easy, scalable, repeatable, and economically viable.
4. Pure enzymes are expensive, which could lift the cost of enzymatic deinking. As a result, the partially purified or crude enzyme should be able to conduct the deinking process on its own.
5. Enzymatic deinking could result in higher recycled paper quality, both in terms of brightness and mechanical properties.
6. The enzyme source should be nonpathogenic, and the enzyme itself should not cause any harm even if it is revealed to the handlers.
7. In the deinking process, enzyme cocktails are more successful than individual enzymes. When using enzyme combinations, however, the mixture does not have any antagonistic effects.
8. Enzymes should not be blocked by catabolite repression or other mechanisms. Sugar moieties would be released into the pulp solution during cellulose and xylan degradation, which could contribute to catabolite feedback mechanisms.
9. Unlike other industrial applications, using immobilized enzyme in the paper and pulp industries can be difficult, reducing the chances of the enzyme being extracted and reused.
10. During enzymatic deinking, the organic load would be higher, which could result in a high biological oxygen demand. Before it can be released, this effluent must be properly treated.

Future researchers in the field of enzymatic deinking would need to fix these issues to make enzymatic deinking economically feasible and implementable. These issues must be resolved to make the enzymatic deinking process fully environmentally friendly.

14.3.5.4 Future directions in deinking research

Enzyme-based deinking has a promising future ahead of it. In today's world, waste management is a major concern, and successful recycling of paper waste can help to minimize a significant amount of solid waste. This would contribute to a reduction in greenhouse gas emissions as well as the threat posed by solid waste. Enzyme-assisted deinking is less harmful to the environment and easier to do. Biological deinking will make paper recycling a fully environmentally friendly operation. As previously mentioned, there are several barriers to implementing enzyme-based deinking that must be tackled in future research. The main considerations for the commercialization of this technology are safety and viability. Researchers should continue to find research gaps and work to close them to make the enzymatic deinking process an effective paper recycling process.

14.3.6 Removal of pitch

Pitch is the parts of plant wood composed resin acids, fatty acids, esters of fatty acids, waxes, sterols, and glycerol of other fats which are soluble in methylene. It is approximately 10% of total wood weight component and causes a toxic problem in human beings. Biological reduction of pitch using enzymes is a very effective method. Lipase enzymes are mostly used in pitch removal. Very less commercialization of lipases for removal of pitch is available. This method is very helpful for pitch control, which reduced problems associated with wastewater (Kumar, Singh, & Chandra, 2020b; Singh, Kaur, Khatri, & Arya, 2019). It may also help for the pitch removal from the machinery, to control the bacterial pathogen from the pulp in paper industries, and useful for cleaning of the paper machines. There are many advantages of this method in the paper industry, that is, reducing bleached chemicals, quality improvement, ecofriendly technology, nontoxic, reduced organic load, and cost effective. It may also play a very important role in the color reduction of effluent and control wood loss (Ekman & Holmbom, 2000). However, natural process of degradation may take longer time, but it may influence the degradation process. During chemical pulping process, sulfite is used, which increases resins concentrations in the effluent and highly toxic to aquatic and terrestrial ecosystem. But this method may control the toxic properties of resins present in the effluent of pulp paper industry. Hence, this method is very effective in recent time due to its ecofriendly technique to wastewater pitch removal and control pollution in the environment.

14.3.7 Removal of slime

The broad definition of the slime included problems associated with bacterial and fungal growth in the pulp and paper mill. The moisture condition of the paper mill is favorable for the production and growth of slime in the paper industry. There is the presence of starch, sugar, wood extractives, and phosphorus, nitrogenous compounds, and some other additives act as favorable nutrient sources for slime growth (Singh et al., 2019; Singh et al., 2020a). The requirement of aeration in the paper plants is important, but it may also be affected due to the growth of aerobic bacteria. The paper machine is the best culture medium for the growth of slime due to the maintenance of pH control, temperature, and other parameters as per the requirement of pulp production. It may also help in slime growth and healthy life. Slime is the previous name of the microbial origin within the deposition of organic matter in the paper industry. Practically, paper mill worked with sterilized system (Ullah, 2011). Microbes contaminate the paper mills, and different slime compounds are not characterized yet. The strategy of the biocides is to destroy the targeted source of slime present at the contaminated site. However, in some cases, slime compounds that are specific and highly spread are characterized and follow the removal strategy. Alpha-2,6-linked polymer of fructose forms film of slime. Several *Pseudomonas* and *Bacillus* bacteria, especially these, grow in around paper machines and secret these enzymes (A. Kumar, Singh, & Chandra, 2020b; V. Kumar, Thakur, Singh, & Shah, 2020). When it comes to fine paper, the amount of inhibiting compounds is higher. The enzyme levan hydrolase can hydrolyze this polymer to low-molecular-weight polymers that are water soluble, thereby cleaning the slime out of the system. Henkel Corp. commercial levan hydrolase, under the name EDC-1, is supplied as a product (Morristown, United States). There is no effect of the enzyme on cellulose, and not toxic for paper (Kiuru, 2011). The enzyme is normally applied at the paper machine's headbox, although it has also been added at the dryer discharge in some situations. The enzyme works best at pH 5.0 and is successful at pH 4.0−8.0. Grace Dearborn's research is still focused on preventing biofilm formation in water circuits of the paper machine. The extracellular polysaccharides present in biofilm are specified by combination of enzymes or by specific enzymes (Blanco, 2003). Dearborn had to create a system for simulating biofilm formation as well as analytical methods for determining biofilm components initially. Darazyme is the product of this family. However, in acidic and alkaline condition trial of wood free printing, wood containing, and writing grades, paper making permitted eliminating microbiocide from the paper mill toxic waste material.

14.3.8 Removal of shives

The small and bundle like structures of the fiber present during the wood digestion process, which are not divided into individual forms during the pulping process, are known as shives. These are looking like a splinter and are dark in color with respect to the rest of the splinters. It is the pulp's one of the most relevant parameters for determining the quality of a product. Mostly shive is the best count for the quality of kraft bleached pulp. It has been discovered that Shivex, a new enzyme formulation, can be used through bleaching, can improve the quality of shive removal. Mills will increase the yield of brown stock by treating it with Shivex. The shive elimination during the bleaching process is up to 55% (Sanchez-Echeverri et al., 2021).

This allows the mill to reduce its real shive count and increases its margin safety against the shives depending on the shive level in incoming brown stock and required shive level of bleaching pulp. The rise in shive removal is followed by a boost in pulp bleaching quality. As a result, mills will minimize chlorine consumption in bleach plants without losing shive counts (Coroller et al., 2013). It is as a multicomponent protein, a mixture in which xylanases are primary, but the extent to which the enzyme eliminates shives is irrelevant to the xylanase activity or effectiveness of bleach boosting.

14.3.9 Debarking

All wood processing starts with the removal of the bark. This phase necessitates a considerable amount of energy. Comprehensive debarking is needed for high-quality chemical and mechanical pulp because a small amount of bark residue may darken the product. Full debarking and high energy demand resulted in raw material loss because of the prolonged treatment process in the mechanical drum. The cambium, which is made up of just one layer of cells, forms the boundary between bark and wood (Singh et al., 2020b).

Inside, the living cell contains xylem plant cells and outside, phloem cells found in the stem. High and low pectin content and lignin are typical characteristics of the cambium in every wood species studied. The content of pectins in cambium cells varies among the wood species but may be as high as 40% of the dry weight. The phloem also includes a lot of pectic and hemicellulosic compounds (Lefeuvre et al., 2015; Singh et al., 2020a). Pectinases have been identified as key enzyme of the whole process; xylanases also involved because of their higher hemicellulosic content in the cambium phloem. After pretreatment with pectinolytic enzymes, the energy needed for debarking was found to be reduced by up to 80%. One of the most complicated aspects of enzymatic debarking is the lack of enzyme penetration in entire cambium logs (Singh et al., 2019).

14.3.10 Retting of flax fibers

The hemp and flax like plant fiber have been processed with enzymes. The binding material of plant tissues has been removed by enzymes secreted by microorganisms in situ is currently affecting fiber liberation (Terzopoulou et al., 2015). This phase is thought to be regulated by pectinases, but xylanases may also be involved. Treatment with artificial enzyme mixtures in place of slow natural retting may become as new fiber liberation technology (Peponi, Biagiotti, Torre, Kenny, & Mondragon, 2008).

14.3.11 Reduction of vessel picking

Recently, use of tropical hardwoods like eucalyptus and bamboo for pulp production has globally increased. Since such tree's growth is very fast, and use of chips and pulps can be used in a number of ways. The vessel elements of tropical hardwoods are, however, large and hard, and they do not fibrillate during normal beating. As a result, they stick up out of the surface of the paper (Lee, Sapuan, Lee, & Hassan, 2016). The vessels are ripped out during the printing process, leaving voids. The value of tropical hardwood pulps is decreased as a result of this trait. Increased beating can improve vessel fibrillation and stability over time, but it can also cause poor drainage. Honshu Paper Co. filed a patent in industrial uses of cellulases to increase flexibility and reliability of hardwood vessels in the industry. The use of enzymes decreased vessel picking by 85%. Smoothness and tensile strength also improved at the same time. The amount of time it takes for water to drain has also increased (Agnihotri, Dutt, & Tyagi, 2010).

14.3.12 Cellulose-binding domains

Many researchers have recognized that the use of CBDs in the field of biotechnology has enormous potential. If term "cellulose-binding domain" is used for the key word in a computerized search for patent, there are more than 150 results returned. On a large scale, cellulose is an excellent matrix of affinity purification. (Mahjoub, Yatim, Mohd Sam, & Hashemi, 2014). The physical properties of this chemically inert matrix are excellent, and it has a weak affinity for nonspecific protein binding. It comes in different shapes and sizes and is pharmaceutically healthy and reasonably priced (Hubbe, Ayoub, Daystar, Venditti, & Pawlak, 2013). CBDs have been found to be useful for the physical and chemical modification of composite materials, as well as modified materials production with highly improved quality and properties, according to current research. CBDs may be used for the alteration of polysaccharide materials via in vitro or in-vivo agro-biotechnology (Moudood, Rahman, Ochsner, Islam, & Francucci, 2019). On cellulose-containing fabrics, CBDs cause nonhydrolytic fiber disruption. The future applications of the "CBD technology" include changes in individual cell and alter the whole organism (Akil et al., 2011). The nutrition and texture values of crops and the product made by them can be altered by promoter and trafficking signals of these genes' expression.

14.4 Modifying enzymes to attain activity under specific conditions

Routine work is going on specific enzyme cloning of DNA sequences of organisms. Different new xylanases, cellulases enzymes, have been modified and cloned based on their activities and characterized. The production of these enzymes is without using any natural host (Akin et al., 2004). Often these enzymes can be produced without ever having to deal with the natural host. With the plethora of genomic sequences being discerned, many new enzymes might be available by searching the newsequences for regions that code for specific enzymes. This directed approach has provided new potential enzymes for use. In polymerase chain reaction, experiments were conducted to generate these types of enzymes (Zhao et al., 2017). This method has been used to create new enzymes from unidentified species found in the environment. The 2,5-diketo-D-gluconic acid reductase cloning unknown organism is an example of this method. The resulting these enzymes have significantly high catalytic properties than previously discovered from other species. Although there may be some limitations on what might be made by recombinant organisms, there are many enzymatic targets that might benefit from this approach. All of the cellular functions of thermophilic organisms include active enzymes (Adamsen, Akin, & Rigsby, 2002). These enzymes must be thermostable due to the existence of the environment in which these species live. From these DNA species, several enzymes have been cloned. Thermophilic bacterial enzymes can withstand high temperatures, but it will operate more slowly at low temperatures. Thus, a thermophilic enzyme should be sought when a process has to remain hot or requires the use of solvents, but many other alternatives exist when the process does not have to be heated and is at a nondenaturing pH. However, if the fiber must be explicitly heated for the enzyme to be successful, use of thermophilic enzymes coasted higher for the given treatment process. When the crystalline phase of an enzyme has been determined, some precise planned adjustments can be made, and the reaction in the protein's behavior can be predicted (George, Mussone, & Bressler, 2014). The Bacillus *circulans* enzyme xylanase has been improved due to disulfide bonds during site-specific mutation for the thermostability. However, thermostability was improved as a result of the planned improvements. Just one of the eight improvements resulted in an improvement in the optimum activity at optimum temperature. Site-specific modifications were also made to *P. chrysosporium* MnP, allowing the enzyme to withstand higher hydrogen peroxide concentrations than the wild form. The increased resistance was partially compensated by a decline in the enzyme's hydrogen peroxide affinity (Singh & Chandra, 2019).

Awareness of the protein's sequence and some understanding of how changes can be imparted are needed for site-specific changes. The *Streptomyces lividans* sequence of xylanase was compared with thermostable xylanase of the same family, providing a roadmap for improving the *S. lividans* enzyme's thermostability. The different variants of thermostable *S. lividans* xylanases were discovered (Luo et al., 2009). The presence of thermostabilizing domains has been discovered. The 29 amino acids of xylanase A of *Thermomonospora fusca* gave *Trichoderma reesei* xylanase II and *S. lividans* xylanase B improved thermostability. Further, random fragmentation of gene shuffling of *S. lividans* and *T. fusca* genes resulted in improved temperature optimization in the *S. lividans* xylanase B as variants (Joo, Pack, Kim, & Yoo, 2011). The modified *T. reesei* enzyme's thermostability and thermal activity have been integrated by BioBrite HB60C, which represents enhanced activity in the elemental chlorine-free bleaching process. Other xylanase 10 family domains have thermostabilizing properties. *Thermomonospora alba*, Cellulomonas *fimi*, *Streptomyces olivaceoviridis*, *Neocallimastix frontalis*, and *Fibrobacter succinogene* have examined their all domains (Bedarkar et al., 1992). These findings suggest that these domains are often transferred their functions into the new chimeric protein's functions. It may have the ability to bind cellulose of plants and some other proteins and change substrate-binding domains and also alter the domains of the active sites, among other things. Proteins with different characteristics can be generated by rearranging modules or domains. The combination of cellulase activity with (1-3,1-4)-beta-glucanase activity is an example of the modified enzyme. However, unique linkages present in barley glucan were degraded by this multienzyme (Moreau, Shareck,

Kluepfel, & Morosoli, 1994). The unique linkages present in barley glucan were degraded by this multienzyme (Suzuki, Hatagaki, & Oda, 1991).

The enzyme design of in silico could lead to further advancements in this field. These types of modifications necessitate in-depth structural awareness of an enzyme's active site. On the other hand, nature's modifications are finite and operate within the framework of what was already present and operating (Shoichet, Baase, Kuroki, & Matthews, 1995). The bond angles, principles of protein folding, hydrogen bonding solvation, hydrophobic cores, and some other forces which provided different possibilities that cannot be tested in an enzyme designed by using in silico methodology. This technology can already be used to build thermophilic protein variants. The limit of temperature for enzyme activity may be high at 200°C (Kumar, Singh, Ahmad, & Chandra, 2020a). The challenge is to narrow down these potential structural variations to those that can be easily checked.

14.5 Environmental and manufacturing benefits

Biobleaching is a new and hot discovery due to the mediator of laccase for kraft and pulp bleaching. As a result, the environmentally sustainable chlorine-free (totally chlorine free) and elemental chlorine-free bleaching techniques are needed to reduce cellulose, hemicellulose, and lignin contents from the dissolving tank of pulp to maintain the high brightness of pulp and also improve the toxicity profile of effluent and also reduced the content of absorbable organic halogen (Singh & Chandra, 2019). Biological pulp prebleaching methods using xylanase allow for the selective removal of 20% of xylan enzyme in pulp saves 25% chlorinated chemicals. Alternatively, lignolytic enzymes of white-rot fungi can be used to bleach pulp, allowing for chemical savings and establishing a chlorine-free bleaching method (Hynning, 1996). The advantages of biobleaching are reduced adsorbable organic halogen, reduced consumption of bleaching chemical, improved pulp, improved brightness, paper quality, pollution load reduction, and effluent toxicity reduction (Kangas, Jappinen, & Savolainen, 1984).

The paper industries focus on biological substitutes and use very less in paper manufacturing to lower capital, costs, and reduce toxic environmental effects in the ecosystem. One recent application of biological treatments process has been to reduce refining energy usage in mechanical pulping methods. Certain fungal treatments have been shown to accomplish this goal without causing damage to the resulting fiber, and likely with higher quality fiber as a result (Keith, 1976). Wood chips have also been successfully pretreated for chemical pulping processes. The aims of this application are more uniform delignification, increased yield, or reduced chemical consumption. Chips treatment with cellulose and hemicellulose enzymes is still in its early stages of development. After 20 days, pretreatment of hard wood chips with *P. chrysosporium* improves kraft pulp yield, but the effect is more pronounced after 30 days (Kim et al., 2016).

14.6 Innovation and implementation

White-rot fungi are the most effective for delignification due to production of ligninolytic extracellular oxidative enzymes. The white-rot fungi have properties to degrade and delignify the content of pulp paper mill effluent and it also have the ability to extend the brightness of the pulp by their enzymatic activities. The enzyme MnP is thought to be the most essential and useful for kraft and pulp biobleaching process (Chaurasia, 2019). The pretreatment for the sulfite pulp by using *Aureobasidium pullulans* xylanases increases the brightness, alkali volubility, which is an important characteristic of producing viscose rayon grade pulp. The toxicity of xylanases is to reduce the hemicellulose content of pulp in dissolving tank (Chen, Stemple, Kumar, & Wei, 2016). The biobleaching process by using white-rot fungus (*Ceriporiopsis subvermispora)*, could highly increase the brightness, of pulp and also affected the cellulosic content of the paper industry (Choinowski, Blodig, Winterhalter, & Piontek, 1999).

In the pulp and paper industry, the use of the environmentally friendly processes is becoming more widespread, and biotechnological approaches and processes are at the forefront of new research for ecofriendly environment. The xylanase prebleaching of pulp is a biotechnology application in Indian paper industries (Singh et al., 2020a). In India, extensive research and development work on biobleaching by using enzymes in pulp bleaching and material preparation is popular nowadays. As a result of increased pressure to reduce the content of organic chlorine bonded compounds present in effluent, mainly paper industries are becoming involved in this biobleaching and biotreatment methods and have begun mill trials also (Kumar, Singh, & Chandra, 2020b; Singh, Kumar, & Chandra, 2020b). The commercialization of freeness monitoring and biomechanical pulping would help to further optimize these processes while lowering the costs of enzymes and fungi. Enzyme technology can decrease the energy and capital costs and improves the qualities of degraded fiber furnishes and decrease the toxic environmental effect of paper manufacturing.

14.7 Conclusion

The role of cocktail enzymes in biobleaching processing is addressed in this chapter. Enzyme-based ecofriendly and greener biobleaching technologies are essential and productive, if applied judiciously by the pulp and paper industry. These biotechnologies are intended to use fewer bleaching chemicals, which are toxic and pollute the environment, especially after paper mill effluents are discharged into rivers and ponds. Less chemical use means less water is used to wash bleaching chemicals from the bleached pulp, which is particularly important in developing countries where water scarcity is a major issue. Nevertheless, due to elevated manufacture costs, research related to enzyme biosynthesis and deployment in biobleaching is restricted to lab scale. Large-scale application of enzymes in pulp bleaching is still in the infancy stage. This highlights the necessity for effective and low-cost enzyme production and application technologies in the pulp industry.

Acknowledgments

The financial assistance as UGC-NON fellowship from University Grant Commission, New Delhi, to Adarsh Kumar, PhD scholar is highly acknowledged.

References

Adamsen, A. P. S., Akin, D. E., & Rigsby, L. L. (2002). Chelating agents and enzyme retting of flax. *Textile Research Journal*, *72*(4), 296–302.

Agnihotri, S., Dutt, D., & Tyagi, C. H. (2010). Complete characterization of bagasse of early species of *Saccharum officinerum*-CO 89003 for pulp and paper making. *BioResources*, *5*(2), 1197–1214. Available from https://doi.org/10.15376/biores.7.4.5247-5257.

Akil, H. M., Omar, M. F., Mazuki, A. A. M., Safiee, S., Ishak, Z. A. M., & Abu Bakar, A. (2011). Kenaf fiber reinforced composites: A review. *Materials & Design*, *32*(8–9), 4107–4121.

Akin, D. E., Henriksson, G., Evans, J. D., Adamsen, A. P. S., Foulk, J. A., & Dodd, R. B. (2004). Progress in enzyme-retting of flax. *Journal of Natural Fibers*, *1*(1), 21–47.

Angural, S., Bala, I., Kumar, A., Kumar, D., Jassal, S., & Gupta, N. (2021). Bleach enhancement of mixed wood pulp by mixture of thermo-alkali-stable xylanase and mannanase derived through co-culturing of Alkalophilic *Bacillus* sp. NG-27 and *Bacillus nealsonii* PN-11. *Heliyon*, 7(1), e05673. Available from https://doi.org/10.1016/j.heliyon.2020.e05673.

Angural, S., Kumar, A., Kumar, D., Warmoota, R., Sondhi, S., & Gupta, N. (2020). Lignolytic and hemicellulolytic enzyme cocktail production from *Bacillus tequilensis* LXM 55 and its application in pulp biobleaching. *Bioprocess and Biosystems Engineering*, *43*(12), 2219–2229. Available from https://doi.org/10.1007/s00449-020-02407-4.

Bajpai, P. (1999). Application of enzymes in the pulp and paper industry. *Biotechnology Progress*, *15*, 147–157.

Balda, S., Sharma, A., Capalash, N., & Sharma, P. (2019). *Microbial enzymes for eco-friendly recycling of waste-paper by deinking. Microbes for sustainable development and bioremediation* (pp. 43–54). CRC Press. Available from https://doi.org/10.1201/9780429275876-2.

Ballinas-Casarrubias, L., González-Sánchez, G., Eguiarte-Franco, S., Siqueiros-Cendón, T., Flores-Gallardo, S., Duarte Villa, E., ... Rascón-Cruz, Q. (2020). Chemical characterization and enzymatic control of stickies in kraft paper production. *Polymers*, *12*(1), 245. Available from https://doi.org/10.3390/polym12010245.

Bedarkar, S., Gilkes, N. R., Kilburn, D. G., Kwan, E., Rose, D. R., Miller, R. C., Jr, ... Withers, S. G. (1992). Crystallization and preliminary X-ray diffraction analysis of the catalytic domain of Cex, an exo–1,4-glucanase and -1,4-xylanase from the bacterium *Cellulomonas fimi*. *Applied Microbiology and Biotechnology*, *228*, 693–695.

Bhardwaj, N., Kumar, B., & Verma, P. (2019b). A detailed overview of xylanases: An emerging biomolecule for current and future prospective. *Bioresources and Bioprocessing*, *6*(1), 1–36. Available from https://doi.org/10.1186/s40643-019-0276-2.

Bhardwaj, N. K., Kaur, D., Chaudhry, S., Sharma, M., & Arya, S. (2019a). Approaches for converting sugarcane trash, a promising agro residue, into pulp and paper using soda pulping and elemental chlorine-free bleaching. *Journal of Cleaner Production.*, *217*, 225–233. Available from https://doi.org/10.1016/j.jclepro.2019.01.223.

Microbiology in papermaking. In A. Blanco (Ed.), *Recent research developments in applied microbiology and biotechnology (Vol. 1)*. Kerala: Research Signpost.

Boruah, P., Sarmah, P., Das, P. K., & Goswami, T. (2019). Exploring the lignolytic potential of a new laccase producing strain *Kocuria* sp. PBS-1 and its application in bamboo pulp bleaching. *International Biodeterioration & Biodegradation*, *143*, 104726.

Brandberg, A., & Kulachenko, A. (2017). The effect of geometry changes on the mechanical stiffness of fiber/fiber bonds. In W. Batchelor, & D. Söderberg (Eds.), *Fundamentals of papermaking fibres, advances in paper science and technology* (pp. 683–719). Lancashire: The Pulp and Paper Fundamental Research Society.

Brodin, F. W., & Eriksen, O. (2014). Preparation of individualised lignocellulose microfibrils based on thermomechanical pulp and their effect on paper properties. *Nordic Pulp and Paper Research Journal*, *30*(3), 443–451.

Chaurasia, B. (2019). Biological pretreatment of lignocellulosic biomass (Water hyacinth) with different fungus for enzymatic hydrolysis and bio-ethanol production resource: Advantages, future work and prospects. *Acta Scientific Agriculture*, *3*(5).

Chaurasia, S. K., & Bhardwaj, N. K. (2019). Biobleaching-An ecofriendly and environmental benign pulp bleaching technique: A review. *Journal of Carbohydrate Chemistry*, *38*(2), 87–108. Available from https://doi.org/10.1080/07328303.2019.1581888.

Chen, Y., Stemple, B., Kumar, M., & Wei, N. (2016). Cell surface display fungal laccase as a renewable biocatalyst for degradation of persistent micropollutants bisphenol A and sulfamethoxazole. *Environmental Science & Technology*, *50*, 8799–8808.

Choinowski, T., Blodig, W., Winterhalter, K. H., & Piontek, K. (1999). The crystal structure of lignin peroxidase at 1.70 Å resolution reveals a hydroxy group on the Cβ of tryptophan 171: A novel radical site formed during the redox cycle. *Journal of Molecular Biology*, *286*, 809–827.

Collins, T., Gerday, C., & Feller, G. (2005). Xylanases, xylanase families and extremophilic xylanases. *FEMS Microbiology Reviews*, *29*, 3–23.

Coroller, G., Lefeuvre, A., Le Duigou, A., Bourmaud, A., Ausias, G., Gaudry, T., et al. (2013). Effect offlaxfibres individualisation on tensile failure offlax/epoxy unidirectionalcomposite. *Composites Part A: Applied Science and Manufacturing*, *51*, 62–70. Available from https://doi.org/10.1016/j.compositesa.2013.03.018.

Dwivedi, P., Vivekanand, V., Pareek, N., Sharma, A., & Singh, R. P. (2009). Bleach enhancement of mixed wood pulp by xylanase-laccase concoction derived through co-culture strategy. *Applied Biochemistry and Biotechnology*, *160*, 55–68.

Ekman, R., & Holmbom, B. (2000). The chemistry of wood resin. In E. Back, & L. Allen (Eds.), *Pitch control, wood resin andderesination* (pp. 37–76). Atlanta: Tappi Press.

Casimir-Schenkel, J., Davis, S., Fiechter, A., Gygin, B., Murray, E., et al., 1995. *Pulp bleaching with thermostable xylanase of Thermomonospora fusca*. United States patent No. 5407827, Apr. 18, 1995. Available from https://patents.google.com/patent/US5407827A/en.

Genco, J.M., 1999. Fundamental processes in stock preparation and refining. In: *Proceedings of Tappi pulping conference* (pp. 57–96).

George, M., Mussone, P. G., & Bressler, D. C. (2014). Surface and thermal characterization of natural fibres treated with enzymes. *Industrial Crops and Products*, *53*, 365–373.

Gonzalo, A., Bimbela, F., Sánchez, J. L., Labidi, J., Marín, F., & Arauzo, J. (2017). Evaluation of different agricultural residues as raw materials for pulp and paper production using a semichemical process. *Journal of Cleaner Production.*, *156*, 184–193. Available from https://doi.org/10.1016/j.jclepro.2017.04.036.

Gupta, R., Gigras, P., Mohapatra, H., Goswami, V. K., & Chauhan, B. (2003). Microbial α-amylases: A biotechnological perspective. *Process Biochemistry*, *38*, 1599–1616.

Gupta, G. K., Kapoor, R. K., & Shukla, P. (2020). *Advanced techniques for enzymatic and chemical bleaching for pulp and paper industries. Microbial Enzymes and Biotechniques* (pp. 43–56). Singapore: Springer. Available from https://doi.org/10.1007/978-981-15-6895-43.

Hamedi, J., Vaez Fakhri, A., & Mahdavi, S. (2020). Biobleaching of mechanical paper pulp using *Streptomyces rutgersensis* UTMC 2445 isolated from a lignocellulose-rich soil. *Journal of Applied Microbiology*, *128*(1), 161–170. Available from https://doi.org/10.1111/jam.14489.

Hubbe, M. A., Ayoub, A., Daystar, J. S., Venditti, R. A., & Pawlak, J. J. (2013). Enhanced absorbent products incorporating cellulose and its derivatives: A review. *BioResources*, *8*(4), 6556–6629. Available from https://doi.org/10.15376/biores.8.4.6556-6629.

Hynning, P. A. (1996). Separation, identification, and quantification of compounds of industrial effluents with bioconcentration potential. *Water Research*, *30*, 1103–1108.

Joo, J. C., Pack, S. P., Kim, Y. H., & Yoo, Y. J. (2011). Thermostabilization of *Bacillus circulans* xylanase: Computational optimization of unstable residues based on thermal fluctuation analysis. *Journal of Biotechnology*, *151*, 56–65. Available from https://doi.org/10.1016/j.jbiotec.2010.10.002.

Jurasek, L., Paice, M., 1992. *Proceedings of the international symposium on pollution prevention in the manufacture of pulp & paper-opportunities and barriers*, Washington, DC.

Kandra, L. (2003). α-Amylases of medical and industrial importance. *Journal of Molecular Structure (Theochem)*, *666–667*, 487–498.

Kangas, J., Jappinen, P., & Savolainen, H. (1984). Exposure to hydrogen sulfide, mercaptans and sulfur dioxide in pulp industry. *American Industrial Hygiene Association Journal*, *45*(12), 787–790.

Kapoor, M., Kapoor, R. K., & Kuhad, R. C. (2007). Differential and synergistic effects of xylanase and laccase mediator system (LMS) in bleaching of soda and waste pulps. *Journal of Applied Microbiology*, *103*, 305–317.

Kaur, D., Bhardwaj, N. K., & Lohchab, R. K. (2017). Prospects of rice straw 439 as a raw material for paper making. *Waste Management (New York, N.Y.)*, *60*, 127–139.

Keith, L. H. (1976). Identification of organic compounds in unbleached treated kraft paper mill wastewaters. *Environmental Science & Technology*, *10*, 555–564.

Kim, S., Thiessen, P. A., Bolton, E. E., Chen, J., Fu, G., Gindulyte, A., ... Bryant, S. H. (2016). PubChem substance and compound databases. *Nucleic Acids Research*, *44*(D1), D1202–D1213.

Kiuru, J., Unigrafia, Oy, Helsinki, Finland 2011. *Interactions of chemical variations and biocide performance at paper machines* (Doctoral dissertations). Aalto University Publication Series. Available from http://urn.fi/URN:ISBN:978-952-60-4455-2.

Kmetzki, A. C., Henn, C., Moraes, S. S., Silva, N. F., & Kadowaki, M. K. (2019). Physicochemical characteristics of fungal xylanases and their potential for biobleaching of kraft and non-wood pulps. *Annual Research & Review in Biology*, 1–7. Available from https://doi.org/10.9734/arrb/2019/v34i430160.

Ktata, A., Karray, A., Mnif, I., & Bezzine, S. (2020). Enhancement of *Aeribacillus pallidus* strain VP3 lipase catalytic activity through optimization of medium composition using Box-Behnken design and its application in detergent formulations. *Environ Sci Pollut Res Int*, *27*(11), 12755–12766. doi:10.1007/s11356-020-07853-x.

Kumar, A., & Chandra, R. (2020). Ligninolytic enzymes and its mechanisms for degradation of lignocellulosic waste in environment. *Heliyon*, *6*, e03170.

Kumar, A., Singh, A. K., Ahmad, S., & Chandra, R. (2020a). Optimization of laccase production by *Bacillus* sp. strain AKRC01 in presence of agro-waste as effective substrate using response surface methodology. *Journal of Pure & Applied Microbiology*, *14*(1), 351–362.

Kumar, A., Singh, A. K., & Chandra, R. (2020b). Comparative analysis of residual organic pollutants from bleached and unbleached paper mill wastewater and their toxicity on *Phaseolus aureus* and *Tubifex tubifex*. *Urban Water Journal*, *17*(10), 860–870.

Kumar, N. V., & Rani, M. E. (2019). *Microbial enzymes in paper and pulp industries for bioleaching application. Research trends of microbiology* (pp. 1–11). MedDocs eBooks.

Kumar, V., Kumar, A., Chhabra, D., & Shukla, P. (2019). Improved biobleaching of mixed hardwood pulp and process optimization using novel GA-ANN and GA-ANFIS hybrid statistical tools. *Bioresource Technology*, *271*, 274–282. Available from https://doi.org/10.1016/j.biortech.2018.09.115.

Kumar, V., Thakur, I. S., Singh, A. K., & Shah, M. P. (2020). Application of metagenomics in remediation of contaminated sites and environmental restoration. *Emerging technologies in environmental bioremediation* (2020, pp. 197–232). Elsevier, ISBN 9780128198605.

Lammi, T., & Heikkurinen, A. (1997). Changes in fibre wall structure during defibration. *Fundamentals of papermaking materials, vol 1 of transactions of the 11th fundamental research symposium* (pp. 641–662). Cambridge: Mechanical Engineering Publications Limited.

Lee, C. H., Sapuan, S. M., Lee, J. H., & Hassan, M. R. (2016). Melt volume flow rate and melt flow rate of kenaf fibre reinforced Floreon/magnesium hydroxide biocomposites. *Springer Plus*, *5*, 1680.

Lefeuvre, A., Le Duigou, A., Bourmaud, A., Kervoelen, A., Morvan, C., & Baley, C. (2015). Analysis of the role of the main constitutive polysaccharides in the flaxfibre mechanical behaviour. *Industrial Crops and Products*, *76*, 1039–1048. Available from https://doi.org/10.1016/j.indcrop.2015.07.062.

Luo, H., Li, J., Yang, J., Wang, H., Yang, Y., Huang, H., ... Yao, B. (2009). A thermophilic and acid-stable family-10 xylanase from the acidophilic fungus *Bispora* sp. MEY-1. *Extremophiles: Life Under Extreme Conditions*, *13*, 849–857. Available from https://doi.org/10.1007/s00792-009-0272-0.

Mahjoub, R., Yatim, J. M., Mohd Sam, A. R., & Hashemi, S. H. (2014). Tensile properties of kenaf fiber due to various conditions of chemical fiber surface modifications. *Construction and Building Materials*, *55*, 103–113, 2014.

Marques, S., Pala, H., Alves, M. T., Amarak-Collaco., Gama, F. M., et al. (2003). Characterisation and application of glycanases secreted by *Aspergillus terreus* CCMI 498 and *Trichoderma viride* CCMI 84 for enzymatic deinking of mixed office wastepaper. *Journal of Biotechnology*, *100*, 209–219.

Moldes, D., & Vidal, T. (2008). Laccase–HBT bleaching of eucalyptus kraft pulp: Influence of the operating conditions. *Bioresource Technology*, *99*, 8565–8570.

Moreau, A., Shareck, F., Kluepfel, D., & Morosoli, R. (1994). Increase in catalytic activity and thermostability of the xylanase A of *Streptomyces lividans* 1326 by site-specific mutagenesis. *Enzyme and Microbial Technology*, *16*, 420–424. Available from https://doi.org/10.1016/0141-0229(94)90158-9.

Moudood, A., Rahman, A., Ochsner, A., Islam, M., & Francucci, G. (2019). Flax fiber and its composites: An overview of water and moisture absorption impact on their performance. *Journal of Reinforced Plastics and Composites*, *38*(7), 323–339.

Nagar, S., & Gupta, V. K. (2020). Hyper production and eco-friendly bleaching of kraft pulp by xylanase from *Bacillus pumilus* SV-205 using agro waste material. *Waste and Biomass Valorization*, *12*, 1–13. Available from https://doi.org/10.1007/s12649-020-01258-0.

Nathan, V. K., & Rani, M. E. (2020). A cleaner process of deinking waste paper pulp using *Pseudomonas mendocina* ED9 lipase supplemented enzyme cocktail. *Environmental Science and Pollution Research*, *27* (29), 36498–36509. Available from https://doi.org/10.1007/s11356-020-09641-z.

Nathan, V.K., Rani, M.E., Gunaseeli, R., Kannan, N.D., 2014. Potential of xylanase from *Trichoderma viride* VKF3 in waste paper pulp characteristics modification. In: *Proceeding of international conference on chemical, environmental and biological sciences (CEBS 2014)* (pp. 54–60), Kuala Lumpur, Malaysia.

Nathan, V. K., Rani, M. E., Gunaseeli, R., & Kannan, N. D. (2018). Paper pulp modification and deinking efficiency of cellulase-xylanase complex from *Escherichia coli* SD5. *International Journal of Biological Macromolecules*, *111*, 289–295.

Pala, H., Mota, M., & Gama, F. M. (2004). Enzymatic vs chemical deinking of non-impact ink printed paper. *Journal of Biotechnology*, *108*, 79–89.

Pala, H., Mota, M., & Gama, F. M. (2006). Factors influencing MOW deinking: Laboratory scale studied. *Enzyme and Microbial Technology*, *38*, 81–87.

Patel, K., & Dudhagara, P. (2020). Optimization of xylanase production by *Bacillus tequilensis* strain UD-3 using economical agricultural substrate and its application in rice straw pulp bleaching. *Biocatalysis and Agricultural Biotechnology*, *30*, 101846. Available from https://doi.org/10.1016/j.bcab.2020.101846.

Peponi, L., Biagiotti, J., Torre, L., Kenny, J. M., & Mondragon, I. (2008). Statistical analysis of the mechanical properties of natural fibers and their composite materials. I. Natural fibers. *Polymer Composites*, *29*(3), 313–320.

Raghukumar, C., Usha, M., Gaud, V. R., & Mishra, R. (2004). Xyalanse of marine fungi of potential use for biobleaching of paper pulp. *Journal of Industrial Microbiology & Biotechnology*, *31*, 433–441.

Saleem, R., Khurshid, M., & Ahmed, S. (2018). Laccases, manganese peroxidases and xylanases used for the bio-bleaching of paper pulp: An environmental friendly approach. *Protein and Peptide Letters*, *25* (2), 180–186. Available from https://doi.org/10.2174/0929866525666180122100133.

Sanchez-Echeverri, L. A., Ganjian, E., Medina-Perilla, J. A., Quintana, G. C., Sanchez-Toro, J. H., & Tyrer, M. (2021). Mechanical refining combined with chemical treatment for the processing of Bamboo fibres to produce efficient cement composites. *Construction and Building Materials*, *269*, 121232. Available from https://doi.org/10.1016/j.conbuildmat.2020.121232, ISSN 0950–0618.

Sharma, A., Balda, S., Gupta, N., Capalash, N., & Sharma, P. (2020). Enzyme cocktail: An opportunity for greener agro-pulp biobleaching in paper industry. *Journal of Cleaner Production*, *271*, 122573. Available from https://doi.org/10.1016/j.jclepro.2020.122573.

Sharma, A., Jain, K. K., Srivastava, A., et al. (2018). Potential of in situ SSF laccase produced from *Ganoderma lucidum* RCK 2011 in biobleaching of paper pulp. *Bioprocess Biosyst Eng*, *42*, 367–377. Available from https://doi.org/10.1007/s00449-018-2041-x.

Sharma, A., Thakur, V. V., Shrivastava, A., et al. (2014). Xylanase and laccase based enzymatic kraft pulp bleaching reduces adsorbable organic halogen (AOX) in bleach effluents: a pilot scale study. *Bioresour Technol*, *169*, 96–102. Available from https://doi.org/10.1016/j.biortech.2014.06.066.

Sharma, D., Agrawal, S., Yadav, R. D., & Mahajan, R. (2017). Improved efficacy of ultrafiltered xylanase–pectinase concoction in biobleaching of plywood waste soda pulp. *3 Biotech*, 7(1), 2. Available from https://doi.org/10.1007/s13205-017-0614-z.

Sharma, D., Chaudhary, R., Kaur, J., & Arya, S. K. (2020). Greener approach for pulp and paper industry by xylanase and laccase. *Biocatalysis and Agricultural Biotechnology*, 101604. Available from https://doi.org/10.1016/j.bcab.2020.101604.

Shoichet, B. K., Baase, W. A., Kuroki, R., & Matthews, B. W. (1995). A relationship between protein stability and protein function. *Proceedings of the National Academy of Science*, *92*, 452–456. Available from https://doi.org/10.1073/pnas.92.2.452.

Singh, A. K., & Mukhopadhyay, M. (2012). Overview of fungal lipase: a review. *Applied Biochemistry and Biotechnology*, *166*(2), 486–520. Available from https://doi.org/10.1007/s12010-011-9444-3.

Singh, A. K., & Chandra, R. (2019). Pollutants released from the pulp paper industry: Aquatic toxicity and their health hazards. *Aquatic Toxicology*, *211*, 202–216.

Singh, A. K., Kumar, A., & Chandra, R. (2020a). Detection of refractory organic pollutants from pulp paper mill effluent and their toxicity on *Triticum aestivum*; *Brassica campestris* and *Tubifex-tubifex*. *Journal of Experimental Biology and Agricultural Sciences*, *8*(5), 663–675.

Singh, A. K., Kumar, A., & Chandra, R. (2020b). Residual organic pollutants detected from pulp and paper industry wastewater and their toxicity on *Triticum aestivum* and *Tubifex-tubifex* worms. *Materials Today: Proceedings*. Available from https://doi.org/10.1016/j.matpr.2020.10.862.

Singh, G., & Arya, S. K. (2019). Utility of laccase in pulp and paper industry: A progressive step towards the green technology. *International Journal of Biological Macromolecules*. Available from https://doi.org/10.1016/j.ijbiomac.2019.05.168.

Singh, G., Capalash, N., Kaur, K., Puri, S., & Sharma, P. (2016). Enzymes: Applications in pulp and paper industry. In G. Dhillon, Singh, & S. Kaur (Eds.), *Agro-industrial wastes as feedstock for enzyme production: Apply and exploit the emerging and valuable use options of waste biomass* (pp. 157–172). Academic Press.

Singh, G., Kaur, K., Puri, S., & Sharma, P. (2015). Critical factors affecting laccase-mediated biobleaching of pulp in paper industry. *Applied Microbiology and Biotechnology*, *99*(1), 155–164. Available from https://doi.org/10.1007/s00253-014-6219-0.

Singh, G., Kaur, S., Khatri, M., & Arya, S. K. (2019). Biobleaching for pulp and paper industry in India: Emerging enzyme technology. *Biocatalysis and Agricultural Biotechnology*, *17*, 558–565. Available from https://doi.org/10.1016/j.bcab.2019.01.019.

Sridevi, A., Narasimha, G., & Devi, P. S. (2019). Production of xylanase by *Penicillium* sp. And its biobleaching efficiency in paper and pulp industry. *International Journal of Pharmaceutical Sciences and Research*, *10*(3), 1307–1311. Available from https://doi.org/10.13040/IJPSR.0975-8232.10(3)0.1307-11.

Srinivasan, M. C., & Rele, M. V. (1999). Microbial xylanases for paper industry. *Current Science*, 77, 137–142.

Suzuki, Y., Hatagaki, K., & Oda, H. (1991). A hyperthermostable pullulanase produced by an extreme thermophile, *Bacillus flavocaldarius* KP 1228, and evidence for the proline theory of increasing protein thermostability. *Applied Microbiology and Biotechnology*, *34*, 707–714.

Terzopoulou, Z. N., Papageorgiou, G. Z., Papadopoulou, E., Athanassiadou, E., Alexopoulou, E., & Bikiaris, D. N. (2015). Green composites prepared from aliphatic polyesters and bast fibers. *Industrial Crops and Products*, *68*, 60–79.

Thomas, W. J. (1994). Comparison of enzyme-enhanced with conventional deinking of xerographic and laser-printed paper. *Taapi Journal*, 77, 173–179.

Tripathi, S. K., Bhardwaj, N. K., & Ghatak, H. R. (2019). Improvement in selectivity of ozone bleaching using DTPA as carbohydrate protector for wheat straw pulp. *Nordic Pulp and Paper Research Journal*, *34*(3), 271–279. Available from https://doi.org/10.1515/npprj-2018-0035.

Tulek, A., Karataş, E., Çakar, M. M., Aydın, D., Yılmazcan, Ö., & Binay, B. (2021). Optimisation of the production and bleaching process for a new laccase from *Madurella mycetomatis*, expressed in *Pichia pastoris*: From secretion to yielding prominent. *Molecular Biotechnology*, *63*(1), 24–39. Available from https://doi.org/10.1007/s12033-020-00281-9.

Ullah, S., 2011. *Biocides in papermaking chemistry* (Master's thesis). University of Jyvaskyla, Finland.

Virk, A. P., Sharma, P., & Capalash, N. (2012). Use of laccase in pulp and paper industry. *Biotechnology Progress*, *28*, 21–32.

Walia, A., Guleria, S., Mehta, P., Chauhan, A., & Parkash, J. (2017). Microbial xylanases and their industrial application in pulp and paper biobleaching: A review. *3 Biotech*, 7(1), 11. Available from https://doi.org/10.1007/s13205-016-0584-6.

Woolridge, E. M. (2014). Mixed enzyme systems for delignification of lignocellulosic biomass. *Catalysts*, *4*, 1–35.

Yardimci, G. O., & Cekmecelioglu, D. (2018). Assessment and optimization of xylanase production using co-cultures of *Bacillus subtilis* and *Kluyveromyces marxianus*. *3 Biotech*, *8(7)*. Available from https://doi.org/10.1007/s13205-018-1315-y.

Zhao, D., Liu, P., Pan, C., Du, R., Ping, W., & Ge, J. (2017). Flax retting by degumming composite enzyme produced by *Bacillus licheniformis* HDYM-04 and effect on fiber properties. *Ee Journal of Ee Textile Institute*, *108*(4), 507–510.

Kumar, V.N., Rani, M.E., Gunaseeli, R., Kannan, N.D. (2018). Paper pulp modification and deinking efficiency of cellulase-xylanase complex from *Escherichia coli* SD5. Int J Biol Macromol. May;111:289-295. doi: 10.1016/j.ijbiomac.2017.12.126.

CHAPTER 15

Electrospun cellulose composite nanofibers and their biotechnological applications

Sumeet Malik[1], Adnan Khan[1], Nisar Ali[2], Farman Ali[3], Abbas Rahdar[4], Sikandar I. Mulla[5], Tuan Anh Nguyen[6] and Muhammad Bilal[7]

[1]Institute of Chemical Sciences, University of Peshawar, Peshawar, Pakistan
[2]Key Laboratory for Palygorskite Science and Applied Technology of Jiangsu Province, National & Local Joint Engineering Research Centre for Deep Utilization Technology of Rock-salt Resource, Faculty of Chemical Engineering, Huaiyin Institute of Technology, Huai'an, P.R. China
[3]Department of Chemistry, Hazara University, Dhodial, Pakistan
[4]Department of Physics, Faculty of Science, University of Zabol, Zabol, Iran
[5]Department of Biochemistry, School of Applied Sciences, REVA University, Bangalore, India
[6]Institute for Tropical Technology, Vietnam Academy of Science and Technology, Hanoi, Vietnam
[7]School of Life Science and Food Engineering, Huaiyin Institute of Technology, Huai'an, P.R. China

15.1 Introduction

In the last few decades, the exploitation of naturally abundant sources that are renewable and biodegradable as well has grabbed the attention of most scientists (Ali et al., 2021; Nawaz, Khan, Ali, Ali, & Bilal, 2020). The reason to focus on such raw materials is that they are easily available, environmentally friendly, and sustainable resources (Ali, Bilal, Khan, Ali, Yang, et al., 2020; Ali, Uddin, et al., 2020). Cellulose, one of such advantageous materials, has the greatest abundance in nature (Ali, Ahmad, et al., 2020; Eslahi, Mahmoodi, Mahmoudi, Zandi, & Simchi, 2020). The supreme properties of cellulose include its ease of extraction from its source, biocompatible, biodegradable, and nontoxic nature. Its greatest abundance is attributed to the fact of it being the backbone of all plant fibers. So, plants could be considered as a rich source of cellulose (Ali, Khan, Malik, et al., 2020; Khan et al., 2020; Moohan et al., 2020). The morphological insight of cellulose shows that it is a linear condensation polymer, which consists of repeating units of D-anhydroglucopyranose joined together by β-1,4-glycosidic linkages. Based on its structural features, cellulose could also be considered as 1,4-β,D glucan (Ali, Khan, Bilal, et al., 2020; Tayeb, Amini, Ghasemi, & Tajvidi, 2018). The cellulose tends to have a supermolecular structure with adjacent chain units oriented at an angle of 180 degrees to each other. These repeating units are the anhydro cellobiose unit and the number of repeating units per molecule is half a degree of polymerization, which is equal to 14,000 in natural cellulose (Ali, Khan,

Nanotechnology in Paper and Wood Engineering
DOI: https://doi.org/10.1016/B978-0-323-85835-9.00016-7

Nawaz, et al., 2020; Aziz et al., 2020; Wsoo, Shahir, Bohari, Nayan, & Abd Razak, 2020). The cellulose has a structure that could be considered as slender rod-like crystalline microfibrils. The solid crystalline structure may have a highly ordered crystalline region as well as low ordered regions called the amorphous region. The promising properties associated with cellulose like high mechanical strength, renewability, lightweight, optical transparency, etc. have made them a competent candidate in various applications (Ilyas, Sapuan, Sanyang, Ishak, & Zainudin, 2018; Sartaj et al., 2020) (Fig. 15.1). The properties of the cellulose-based materials could further be enhanced by shrinking them from micron to nanoscale (10×10^{-3} to 100×10^{-3} μm). The nanorange entities may have properties like small sizes and high surface-to-volume ratios, flexible surface chemistry, stiffness, and tensile strength (Ali, Naz, Shah, Khan, & Nawaz, 2020; Zhu et al., 2020). These properties greatly depend upon the preparatory and fabrication methods of polymeric nanofibers. These methods include self-assembly (Naeem, Alfred, Lv, Zhou, & Wei, 2018), drawing (Wang et al., 2017), template synthesis (Tamahkar, Bakhshpour, & Denizli, 2019), phase separation (Ali, Bilal, Khan, Ali, & Iqbal, 2020; Halim et al., 2019), electrospinning (Ashraf et al., 2020), rotary jet spinning (Barhoum et al., 2019), multilayer coextrusion (Cheng et al., 2019), etc. These techniques have greatly been utilized but they have certain demerits associated with them, which limit their uses (Ali, Azeem, et al., 2020; Ali, Bilal, Nazir, et al., 2020). The most commonly faced difficulties while considering these methods include uncontrollable orientation, discontinuity in the process, the uncontrollable diameter of the fiber produced, use of limited polymer, etc. (Ali, Khan, Malik, et al., 2020; Dong et al., 2012). These processes usually produce single long length

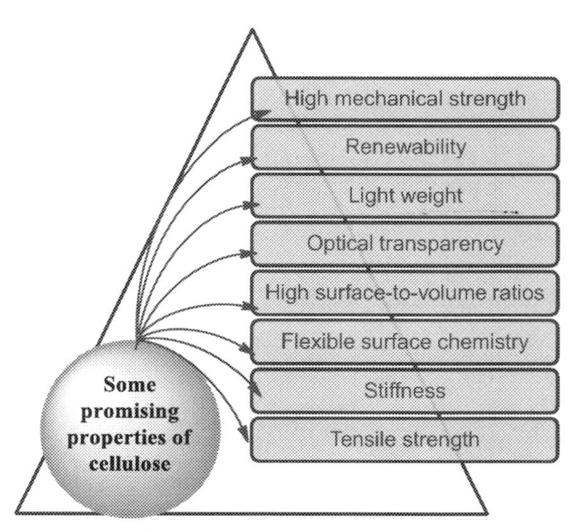

Figure 15.1 Promising properties of cellulose.

nanofibers rather than nanosize ones. The longer time it takes to produce nanofibers is yet another anomaly. Based on these reasons, electrospinning has become a greatly utilized technique for the fabrication of polymers. Electrospinning is a simple and efficient technique for fabricating nanofibers in submicron diameters (nanorange) using polymer in melt form or solutions (Joshi et al., 2015; Khan, Ali, et al., 2019). This chapter covers the properties and applications of electrospun cellulose nanofibers.

15.2 Electrospinning

Electrospinning, also termed electrostatic fiber spinning, is a fine and versatile technique that produces nanofibers at the nano-to-microfibers level (Khalil, Davoudpour, Bhat, Rosamah, & Tahir, 2015; Khan, Shah, Mehmood, Ali, & Khan, 2019). The electrospinning works on the principle of applying an electrical voltage against the electrodes, which brings about a spinning motion of the solution (Fig. 15.2). This spinning motion results in the production of a stretched and charged polymer jet, which on evaporation of the solvent turns into solid fiber (Jordan, Viswanath, Kim, Pokorski, & Korley, 2016; Khan, Khalil, & Khan, 2019). The morphological features of the obtained fiber depend upon the operating parameters during the process of electrospinning, such as the applied voltage, the concentration of the solution, viscosity, surface tension, distance between the needle and target, etc. (Aboamera, Mohamed, Salama, Osman, & Khattab, 2019; Ali et al., 2018). The required pore size, available surface area, porosity, and crystallinity of the fibers could be controlled through thermal annealing (He et al., 2014; Sohni et al., 2018). The electrospinning depends on a balance setup between the surface tension of the polymer solution and the charge repulsion force applied by the external field. The polymer solution

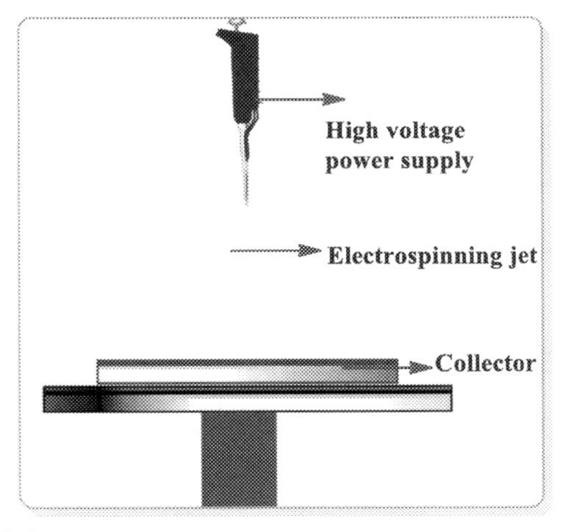

Figure 15.2 Schematic of electrospinning process.

becomes a conical-shaped droplet under the influence of external electric force being applied, called the Taylor cone (Saeed et al., 2018; Shi et al., 2012). The threshold point is achieved, when the repulsive force provided by the external electric field overcomes the surface tension of the polymer solution; a charged jet is obtained from the solution reservoir (Rojas, Montero, & Habibi, 2009; Shah, Din, Khan, & Shah, 2018). The electrostatic force draws the charged jet towards the collector, through alternating or direct current, resulting in the evaporation of the solvent and fabrication of polymer fibers on the collector. The shapes of the nanofibers could be controlled by bringing about some changes in the electrospinning setup (Ponnamma, Parangusan, Tanvir, & AlMa'adeed, 2019). The different structures of nanofibers that could be obtained include aligned (Cai et al., 2016), core-shell (Li, Yang, Yu, Du, & Yang, 2018), hollow (Zanjani, Saner Okan, Menceloglu, & Yildiz, 2015), multilayer (Mouro, Fangueiro, & Gouveia, 2020), etc. The viscosity of the solution and the ultimate polymer chain entanglement are greatly affected by the molecular weight. If the molecular weight is high, highly viscous solution will result in the formation of a continuous polymer jet instead of beads or droplets. Hence, the viscosity, surface tension of the solution, and charge density greatly affect the formation of the beads (Wan, Wang, Yang, Guo, & Yin, 2016). An optimum concentration is also necessary to bring about a proper droplet size. A more high concentration makes the polymer solution highly viscous, while a low concentration makes droplets before reaching the collector (Yang et al., 2019). The electrical conductivity is another key factor in the electrospinning process. The greater the conductivity, the smaller will be the diameter of the fibers formed. The solvent used must be carefully chosen because the vapor pressure of the solvent affects the evaporation rate (Salami et al., 2017). Other operating factors like temperature, humidity, pressure, etc. also affect the final product.

15.3 Electrospinning of cellulose composite nanofibers

Due to the relatively large abundance, renewability, and biodegradability, cellulose is greatly being used for the production of electrospun fibers. However, natural cellulose-based fibers through electrospinning are very difficult to design (Wang, Kong, & Ziegler, 2019) (Fig. 15.3). The reason is the insolubility of cellulose in water as well as many other organic solvents. This problem could be overcome by incorporating other materials, that is, metals, polymers, etc. forming composites. These composites could easily be treated through electrospinning to produce electrospun cellulose composite nanofibers (Kargarzadeh et al., 2017). Some of the solvents have been reported that have the tendency to dissolve cellulose. These include N-methylmorpholine-N-oxide /H_2O (monohydrate-NMMO), lithium chloride (LiCl)/dimethylacetamide (DMAc), ionic liquids, trifluoroacetic acid, etc. The cellulose could be dissolved in monohydrate-NMMO, but tertiary amine oxide solvents are not true solvent systems and must be dissolved in

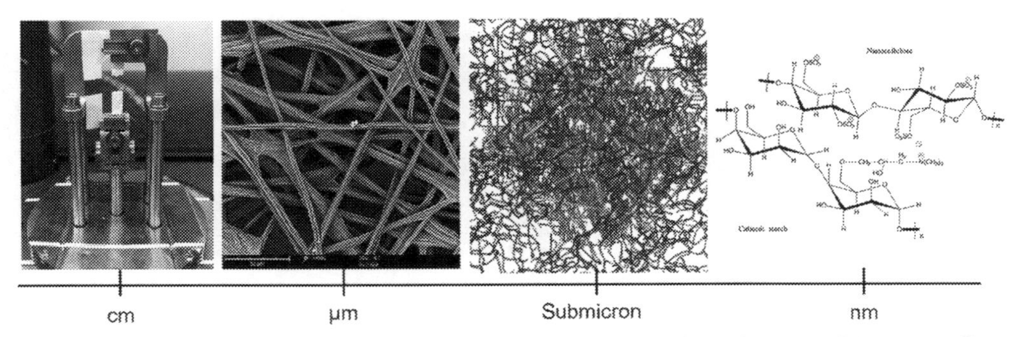

Figure 15.3 Electrospinning process for the fabrication process of starch/nanocellulose nanofibers (Wang et al., 2019).

water or other organic solvents like dimethyl sulfoxide (Pasaoglu & Koyuncu, 2020). Alongside, the hydration of NMMO by two or more water molecules may cause the loss of its water dissolving ability. LiCl/DMAc could also be used as a solvent system for the electrospinning of cellulose, but the preparation of LiCl/DMAc solution is a hard task. Similarly, some ionic liquids like 1-ethyl-3-methylimidazolium chloride, 1-butyl-3-methylimidazolium chloride, and 1-ethyl-3-methylimidazolium acetate have also been used for dissolving cellulose as well as other biopolymers (Hsieh, 2018). Due to the said issues associated with the dissolving of cellulose into solvents, attention has been diverted to the incorporation of cellulose with other components forming electrospun cellulose composite nanofibers, which have found applications in various fields. The electrospinning aids the production of cellulose-based composite nanofibers having a delicate, complicated framework that has the capability to be used for various useful applications (Gabr et al., 2014). The obtained electrospun cellulose nanocomposite fibers tend to have increased mechanical properties like Young's modulus and elongation at break. A great deal of research has been performed using cellulose and cellulose derivatives for the production of electrospun three-dimensional nanocomposites. The most commonly incorporated groups with cellulose include chitosan (CS), lignin, synthetic fibers, poly(ethylene-co-vinyl alcohol), poly methyl methacrylate, polystyrene, polyethylene oxide, etc. (Hatch et al., 2019).

15.4 Applications of electrospun cellulose composite nanofibers

The electrospun cellulose composite nanofibers are being utilized in many fields these days due to their cost-effective and renewable properties. The electrospun cellulose composite nanofiber has combined the useful properties of cellulose as well as the benefits of the electrospinning method. Some of the major fields making use of these nanofibers are discussed in detail as follows.

15.4.1 Electrospun cellulose composite nanofibers as sensors

Among the most unique properties of electrospun cellulose composite nanofibers, some of the most noticeable ones are their highly porous surface with small pore sizes and high adsorption capacity. This property makes them very well suited for usage as sensors. Devarayan and Kim (2015) prepared a pH sensor using electrospun cellulose nanofibers. First, the red cabbage extract was immobilized on the surface of electrospun cellulose nanofibers through adsorption. In the next step, crosslinking of the electrospun cellulose nanofibers was performed using bifunctional diisocyanate. The prepared Red cabbage cellulose nonwoven fabrics (RC/Cs-ESNW) nanofibers were used as pH sensors against different pH at the simulated conditions. The sensor showed different colors at each pH and showed stability at varying temperatures and over a long period of time. The RC/Cs-ESNW nanofibers could successfully be used for health monitoring purposes. Jia, Yu, Zhang, Dong, and Li (2016) prepared a novel sensor for accurate ammonia detection. In order to design this ammonia sensor, polyethyleneimine (PEI) and graphene oxide (GO) were assembled onto the surface of electrospun cellulose acetate (CA) nanofibers, following the electrostatic layer-by-layer self-assembly technique. The prepared CA/PEI/GO-based quartz crystal microbalance sensor presented a satisfactory sensitivity response toward ammonia. The reversibility and high selectivity of the prepared sensor toward the target proved its promising utilization for ammonia detection. Teodoro, Shimizu, Scagion, and Correa (2019) performed lead detection by developing a tri composite sensor. An e-tongue based sensor unit was designed by incorporating ternary nanocomposites using polyamide (PA6) electrospun nanofibers, cellulose nanowhiskers (CNWs), and silver nanoparticles together. The electrospun nanofibers provided the chance to detect analytes, even present in very low concentration, due to their high surface area and highly porous nature. The obtained sensor provided satisfactory results for the detection of lead at even the lowest concentration of $10\,\text{nmol}\,L^{-1}$. Teodoro, Migliorini, Facure, and Correa (2019) also developed electrospun nanofibers-based electrochemical sensor for the detection of mercury. The sensor was prepared by the combination of CNWs/reduced GO and PA6 electrospun nanofibers. The (PA6/CNW:rGO) hybrid nanocomposite showed an excellent efficiency toward the detection of Hg(II) even at the lowest concentrations. The specificity of the sensor was confirmed through interference studies. The sensors also showed a high efficiency toward the real environmental sample. Some of the cellulose nanofibers used as a sensors are shown in Table 15.1.

15.4.2 Electrospun cellulose composite nanofibers in drug delivery

Drug delivery is yet another field that is making use of the electrospun cellulose composite nanofibers. The highly flexible nature and hold over the drug release kinetics help in controlling the drug release. Also, a variety of drug loading methods could be used by making use of different electrospinning methods to obtain high encapsulation and loading capacity as shown in Table 15.2. Wang et al. (2011) prepared electrospun

Table 15.1 Electrospun cellulose composite nanofibers-based sensors.

Electrospun cellulose composite nanofibers-based sensors	Counterpart	Target	Reference
Cellulose/TiO_2/PANI composite nanofibers	TiO_2/PANI	Ammonia	Pang et al. (2016)
PVP fibers/36 degrees $LiTaO_3$ surface acoustic wave sensor	$LiTaO_3$	Hydrogen	He et al. (2010)
Hybrid fillers of carbon nanotubes/cellulose nanocrystal into electrospun polyurethane membranes (TPU/CNT–CNC)	CNTs	Strain sensing	Zhu, Zhou, Liu, and Fu (2019)
Cellulose acetate electrospun fibers decorated with rhodamine B-functionalized core-shell ferrous nanoparticles (γ-Fe_2O_3/SiO_2/RhB NPs-functionalized CA nanocomposite fibers)	γ-Fe_2O_3/SiO_2/RhB	Ammonia	Petropoulou et al. (2020)
Electrospun cellulose acetate fibers modified with antibody specific to 25-hydroxy vitamin D-3 and bovine serum albumin (BSA/AB-25OHD_3/CAEF/RCP)	Antibody specific to 25-OHD_3 and BSA	25-OHD_3	Chauhan and Solanki (2019)
Cellulose nanocrystal-reinforced polymethylmethacrylate(PMMA) fiber	PMMA	Quartz tuning fork sensor	Kim, Park, and Jeon (2020)
Nanofiber membranes of cellulose acetate/multiwalled carbon nanotubes (MWCNTs)/PVP (CA/MWCNTs/PVP)	MWCNTs/PVP	Ascorbic acid	Zhai et al. (2015)
Electrospun nanofibers and conductive magnetic nanoparticles (MNPs)	MNPs	*Escherichia coli* O157:H7	Luo et al. (2011)
Encapsulating enzyme into metal-organic framework (MOF) during in situ growth on cellulose acetate nanofibers	ZIF-8 MOF	Glucose	Li et al. (2020)
Chitosan/poly(vinyl alcohol) blend nanofibers	Chitosan	Pirimiphos-methyl	El-Moghazy et al. (2016)

Notes: *PANI*, Polyaniline; *PVP*, polyvinylpyrrolidone.

Table 15.2 Electrospun cellulose composite nanofibers in drug delivery.

Electrospun cellulose composite nanofibers	Counterpart	Drug used	Reference
Cellulose acetate (CA) composite nanofibers membrane	–	Tetracycline Hydrochloride	Hu et al. (2018)
Zein/ethyl cellulose-based nanofibers	Corn protein zein	Indomethacin	Lu, Wang, Li, Qiu, and Wei (2017)
Functionalized cellulose nanocrystal-poly[2-(dimethylamino)ethyl methacrylate] (CNC-*g*-PDMAEMA) reinforced poly(3-hydroxybutyrate-co-3-hydroxyvalerate) electrospun composite membranes	PDMAEMA nanoparticles	Tetracycline hydrochloride (TCH)	Chen et al. (2020)
Cellulose nanocrystal incorporated with polycaprolactone solution	Polycaprolactone	TCH	Hivechi, Bahrami, and Siegel (2019)
TCH-loaded poly(lactic acid)/polyvinylpyrrolidone (PVP)/TCH-multiwall carbon nanotubes (MWCNTs) composite fibrous mats	MWCNTs	TCH	Bulbul, Eskitoros-Togay, Demirtas-Korkmaz, and Dilsiz (2019)
MIL-125-NH_2/TiO_2 nanocomposite incorporated with polyacrylonitrile/cellulose acetate (MIL: Materials Institute Lavoisier)	Metal-organic framework	Doxorubicin HCl and MCF-7 (MCF-7: Michigan Cancer Foundation-7)	Bahmani et al. (2020)
Core-shell composite micro-/nano-fibers of polyurethane (PU) and cellulose acetate phthalate (CAP)	PU	Intravaginal drug delivery	Hua et al. (2016)
Coaxial electrospun composite of carboxymethyl cellulose (TCMC) and poly (ethylene oxide) (PEO)	PEO	TCH	Esmaeili and Haseli (2017a)
Sodium carboxymethyl cellulose with methyl acrylate (TCMC) and PEO	Methyl acrylate (TCMC) and PEO	TCH	Esmaeili and Haseli (2017b)
PVP/CA/glycerin/garlic hybrid composite	PVP	In vitro antibacterial drug	Edikresnha, Suciati, Munir, and Khairurrijal (2019)

composite nanofibers that contained nanoparticles via fabrication following the one-step, single-nozzle electrospinning technique. The nanoparticles were completely incorporated inside the composite nanofibers and formed a core-sheath structure. The prepared nanoparticles contained dual drugs, which were then fabricated with the electrospun nanofibers. The model drugs used were rhodamine B and naproxen. The results showed that a controlled and programmed release of the model drugs could be obtained confirming the efficiency of the electrospun composite nanofibers in the field of drug delivery. Cheng et al. (2017) prepared a composite of cellulose nanocrystals (CNCs) with those of the electrospun nanofibrous membranes of poly(3-hydroxybutyrate-co-3-hydroxyvalerate) (PHBV). The incorporation of the CNCs with that of the PHBV enhanced the hydrophilic nature of the composite nanofibrous membranes. The finally obtained PHBV/CNC composite membranes were then used for the controlled drug release of tetracycline hydrochloride. The drug loading efficiency was highly enhanced for up to 98% and a drug release of 86% was observed in 540 h. Cui et al. (2018) developed the drug-loaded polyvinyl alcohol (PVA)/CS composite nanofibers and used them for controlled drug delivery purposes. First, the PVA/CS composite nanofibers were prepared by loading the model drug, ampicillin sodium, following the electrospinning procedure (Fig. 15.4). The cross-linking of the composite was also performed using glutaraldehyde as the crosslinker and isopropyl alcohol as the solvent. The resultant crosslinked PVA/CS composite nanofibers were obtained with fine morphology and hydrophobic nature.

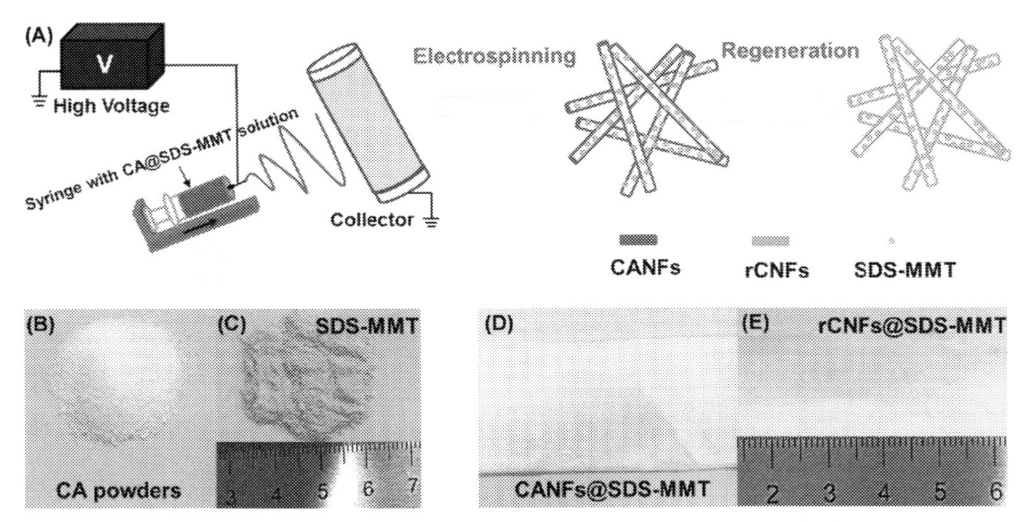

Figure 15.4 (A) Synthesis of cellulose@SDS-MMT nanofiber mats (B–E) Photo of cellulose acetate (CA) powders, SDS-MMT, CANFs@SDS-MMT, and rCNFs@SDS-MMT (Cai et al., 2017). *rCNFs*, Reinforced cationic cellulose nanofibers; *SDS-MMT*, sodium dodecyl sulfonate-modified montmorillonite; *CANFs*, cellulose acetate nanofibers.

The controlled release of ampicillin sodium was evaluated successfully and the kinetics followed the Korsmeyer-Peppas model. Yan, White, Yu, and Zhao (2014) studied the preparation of CA and polyvinylpyrrolidone (PVP) nanocomposite as a drug delivery system for ferulic acid as the model drug. The composite was prepared via a modified coaxial electrospinning procedure using N,N-dimethylacetamide, an organic solvent, as the sheath fluid. The obtained FA/PVP-loaded CA nanofibers exhibited great efficiency for the sustained drug release and transdermal drug delivery.

15.4.3 Electrospun cellulose composite nanofibers in environmental remediation

The electrospun cellulose composite nanofibers have obtained great applications in the field of wastewater remediation. The removal of contaminants from wastewater bodies has been a milestone for researchers due to the ever-growing pollution. Cellulose-based composites have given a convenient solution for the removal of contaminants based on their robust properties shown in Table 15.3. Phan, Lee, Huang, Mukai, and Kim (2019) prepared a composite of CS/cellulose nanofibers, through the electrospinning method. The mixture of CS and CA was treated in the co-solvent system using trifluoroacetic/acetic acid followed by treating with Na_2CO_3. The Na_2CO_3 causes the CS neutralization and CA deacetylation to convert it to cellulose. The obtained CS/CL nanocomposite had great adsorptive properties. The nanocomposite was then used for the removal of metal ions. The maximum adsorption capacity obtained by the CS/CL nanofiber composite for As(V), Pb(II), and Cu(II) was 39.4, 57.3, and 112.6 mg g^{-1}, respectively. Cai et al. (2017) prepared cellulose-based composite nanofibers and used them as adsorbents for the removal of Cr(VI) ions from the wastewater bodies. The nanofibers composite was prepared through electrospinning of CA through organically modified montmorillonite (MMT) and then performing the deacetylation. The obtained composite nanofiber mats of cellulose@SDS-MMT possessed an excellent adsorption capacity for the removal of Cr(VI) ions. Hamad et al. (2020) prepared an organic/inorganic hybrid of pure CA nanofibers and it's impregnated with hydroxyapatite (CA/HAp) nanocomposite fibers by electrospinning. The resultant (CA/HAp) nanocomposite fibers were used as efficient adsorbents for the removal of Pb(II) and Fe(III) ions (Fig. 15.5).

The results showed that high removal efficiencies of 99.7% and 95.46% were obtained for lead and iron, respectively. Gebru and Das (2017) prepared electrospun CA/TiO_2 and used them for the removal of toxic metal ions. The prepared CA/TiO_2 had a smooth morphological surface with high porosity and surface area. The cellulose composite nanofibers exhibited excellent efficiency for the adsorptive removal of Pb(II) and Cu(II) with an adsorption capacities of 25 and 23 mg g^{-1}, respectively.

Table 15.3 Electrospun cellulose composite nanofibers in environmental remediation.

Electrospun cellulose composite nanofibers	Counterpart	Target	References
Novel NH_2-functionalized cellulose acetate (CA)/silica composite nanofibrous membranes	Silica	Cr(VI)	Taha, Wu, Wang, and Li (2012)
CA/P(DMDAAC-AM) composite nanofibrous membrane	P(DMDAAC-AM)	Acid black 172	Xu, Peng, Zhang, Wang, and Lou (2020)
Surface functionalized cellulose acetate/graphene oxide composite nanofibers	Graphene oxide	Indigo carmine	Aboamera, Mohamed, Salama, Osman, and Khattab (2018)
Electrospun cellulose-graphene nanocarbon fibers	Graphene	Hg(II)	Bhalara, Balasubramanian, and Banerjee (2015)
$H_4SiW_{12}O_{40}$ (SiW_{12})/CA composite nanofibrous membrane	$H_4SiW_{12}O_{40}$ (SiW_{12})	Tetracycline and methyl orange	Li et al. (2017)
Beta-cyclodextrin (β-CD) functionalized CA nanofibers	β-CD	Phenanthrene	Celebioglu, Demirci, and Uyar (2014)
Electrospun montmorillonite-impregnated CA nanofiber membranes	Montmorillonite	Ciprofloxacin	Das, Barui, and Adak (2020)
ZnO/cellulose nanocomposite	ZnO	Methylene blue	Lefatshe, Muiva, and Kebaabetswe (2017)
Cationized cellulose nanofibers	3-Chloro-2-hydroxypropyl trimethylammonium chloride	SO_4^{2-}	Muqeet et al. (2017)
Cationic cellulose nanofibers	Positively charged quaternary ammonium groups	SO_4^{2-}	Sehaqui et al. (2016)

Notes: CA, cellulose acetate; *P(DMDAAC-AM)*, poly(dimethyldiallylammonium chloride-acrylamide).

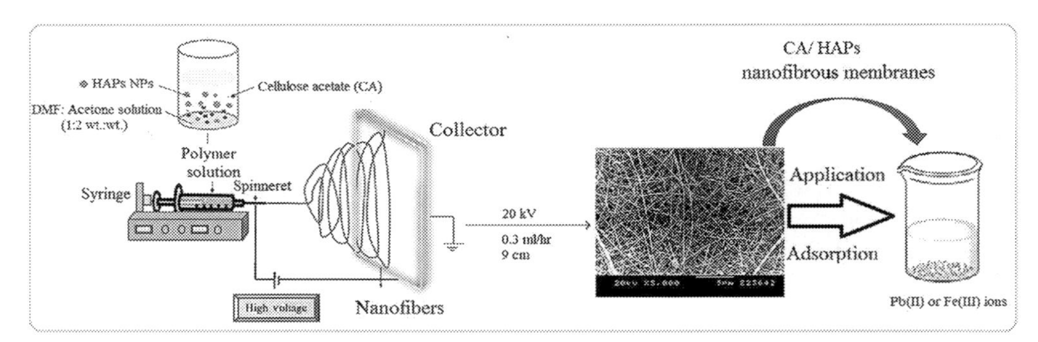

Figure 15.5 Schematic of the electrospinning process for the preparation CA/Hap (Hamad et al., 2020). *CA*, Cellulose acetate; *HAp*, hydroxyapatite; *NPs*, nanoparticles.

15.4.4 Electrospun cellulose composite nanofibers in tissue engineering

One of the issues related to organs/tissues treatment in humans is the difficulty to design proper scaffolds/synthetic materials that can be used as an alternative of the target tissue or organ. In this context, attention has been diverted toward the usage of electrospun cellulose composite nanofibers as potent materials for the purposes of cell proliferation and growth (Table 15.4). A brief view into the tissue engineering field shows that the electrospun cellulose composite nanofibers have greatly been used as porous membranes for skin, in blood vessels and nerve regeneration, and as threefold scaffolds for bone and cartilage regenerations. Ao et al. (2017) have evaluated the usage of electrospun cellulose/nano-HA nanocomposite nanofibers for the bone tissue engineering purpose. In the first step, native cotton cellulose and HA ($Ca_{10}(PO_4)_6(OH)_2$) were fabricated together to obtain the nanocomposite. The solvent system used was the LiCl/DMAc that had the tendency to dissolve cotton cellulose without causing any degradation or side reaction with cellulose. The obtained nanocomposite showed an excellent strength and biocompatibility making it a useful candidate for the bone tissue engineering applications. Abudula et al. (2019) prepared electrospun composite scaffolds by incorporating a matrix of polylactic acid (PLA) and polybutylene succinate (PBS) with those of the cellulose nanofibrils. The prepared electrospun cellulose nanofibers reinforced PLA/PBS composite scaffolds were used in vascular tissue regeneration. Various properties of the composite nanofibers like biodegradation, wettability, and protein adsorption capacity were studied and found satisfactory. The developed composite was thus found potent for the vascular tissue engineering applications. Thunberg et al. (2015) presented a report on the synthesis of conductive polypyrrole onto the electrospun cellulose nanofibers. The obtained scaffold was evaluated for its activity in neural tissue engineering. First, the cellulose nanofibers were electrospun through CA followed by surface modification using in situ pyrrole polymerization. The modification of cellulose nanofibers caused an increase in their conductivity for

Table 15.4 Electrospun cellulose composite nanofibers in tissue engineering.

Electrospun cellulose composite nanofibers	Counterpart	Target	References
Electrospun fibrous bionanocomposite scaffolds reinforced with cellulose nanocrystals (CNCs) fabricated with maleic anhydride grafted poly (lactic acid)	Maleic anhydride grafted poly(lactic acid)	Bone tissue engineering	Zhou et al. (2013)
Three-dimensional bionanocomposite poly (butylene succinate)/ CNC scaffolds	Poly(butylene succinate)	Tissue engineering	Huang et al. (2018)
Poly(lactic acid)/CNC composite scaffolds	Poly(lactic acid)	Bone tissue regeneration	Patel, Dutta, Hexiu, Ganguly, and Lim (2020)
Aligned electrospun cellulose/CNCs nanocomposite nanofibers loaded with bone morphogenic protein-2 (BMP-2)	Protein-2 (BMP-2)	Bone tissue engineering	Zhang et al. (2019)
Cellulose acetate/gelatin/ nanohydroxyapatite (CA/ Gel/nHA) nanocomposite mats	Gelatin/ nanohydroxyapatite	Wound dressing	Samadian et al. (2018)
Electrospun cellulose nanofibers	–	Human dental follicle cells	He et al. (2015)
Chitosan/nanocrystals cellulose-graft-poly(N-vinylcaprolactam) nanofibers	Chitosan	Skin tissue engineering	Ghorbani, Nezhad-Mokhtari, Sohrabi, and Roshangar (2020)
Cellulose nanowhiskers embedded and aligned with CA propionate matrix	–	Vascular tissue engineering	Pooyan, Kim, Jacob, Tannenbaum, and Garmestani (2013)
Electrospun three-dimensional CA/lactic acid nanofibers	Lactic acid	Bone tissue engineering	Lee et al. (2020)
Chitosan–polyethylene oxide matrix reinforced with CNCs	Chitosan	Wound dressing	Naseri, Mathew, Girandon, Fröhlich, and Oksman (2015)

up to 10^5 times. The electrospun cellulose/polypyrrole nanofibers scaffolds provide a better cell adhesion of SH-SY_5Y human neuroblastoma cells altering their morphology to more neuron-like phenotype. Zhang et al. (2015) prepared poly(ethylene glycol) (PEG)-grafted CNCs and incorporated them with PLA following the electrospinning technique. The obtained electrospun PLA/CNC-g-PEG nanocomposite scaffolds were for the tissue engineering purposes in human mesenchymal stem cells. The detailed analysis of the PLA/CNC-g-PEG nanocomposite scaffolds showed that they exhibited high mechanical strength, cell viability, and proliferation cell count and showed great applications in bone tissue engineering.

15.5 Conclusion

Cellulose's ability to form hydrogen bonding provides it a high mechanical strength. The cellulose-based nanocomposites tend to have high surface area and small sizes, great stability, etc. The electrospinning technique offers great advantages and ease of preparation for the design of cellulose-based nanocomposite fibers. The electrospun cellulose composite nanofibers have greatly been used in various fields including environmental remediation for the removal of contaminants, biomedical fields, biotechnology, etc. There is still a slot left to be filled in the exploitation of electrospun cellulose composite nanofibers due to the insolubility of cellulose in conventional solvents. But a variety of solvents are still available that can dissolve cellulose properly increasing its applicability toward different fields.

Conflict of interests

The authors declare no conflict of interest.

References

Aboamera, N. M., Mohamed, A., Salama, A., Osman, T. A., & Khattab, A. (2018). An effective removal of organic dyes using surface functionalized cellulose acetate/graphene oxide composite nanofibers. *Cellulose*, *25*(7), 4155–4166.

Aboamera, N. M., Mohamed, A., Salama, A., Osman, T. A., & Khattab, A. (2019). Characterization and mechanical properties of electrospun cellulose acetate/graphene oxide composite nanofibers. *Mechanics of Advanced Materials and Structures*, *26*(9), 765–769.

Abudula, T., Saeed, U., Memic, A., Gauthaman, K., Hussain, M. A., & Al-Turaif, H. (2019). Electrospun cellulose nano fibril reinforced PLA/PBS composite scaffold for vascular tissue engineering. *Journal of Polymer Research*, *26*(5), 110.

Ali, N., Ahmad, S., Khan, A., Khan, S., Bilal, M., Ud Din, S., . . . Khan, H. (2020). Selenide-chitosan as high-performance nanophotocatalyst for accelerated degradation of pollutants. *Chemistry—An Asian Journal*, *15*(17), 2660–2673.

Ali, N., Azeem, S., Khan, A., Khan, H., Kamal, T., & Asiri, A. M. (2020). Experimental studies on removal of arsenites from industrial effluents using tridodecylamine supported liquid membrane. *Environmental Science and Pollution Research*, *27*(11), 1–12.

Ali, N., Bilal, M., Khan, A., Ali, F., & Iqbal, H. M. (2020). Design, engineering and analytical perspectives of membrane materials with smart surfaces for efficient oil/water separation. *TrAC Trends in Analytical Chemistry*, *127*, 115902.

Ali, N., Bilal, M., Khan, A., Ali, F., Yang, Y., Khan, M., ... Iqbal, H. M. (2020). Dynamics of oil-water interface demulsification using multifunctional magnetic hybrid and assembly materials. *Journal of Molecular Liquids*, *312*, 113434.

Ali, N., Bilal, M., Khan, A., Ali, F., Yang, Y., Malik, S., ... Iqbal, H. M. (2021). Deployment of metal-organic frameworks as robust materials for sustainable catalysis and remediation of pollutants in environmental settings. *Chemosphere*, *272*, 129605.

Ali, N., Bilal, M., Nazir, M. S., Khan, A., Ali, F., & Iqbal, H. M. (2020). Thermochemical and electrochemical aspects of carbon dioxide methanation: A sustainable approach to generate fuel via waste to energy theme. *Science of The Total Environment*, *712*, 136482.

Ali, N., Kamal, T., Ul-Islam, M., Khan, A., Shah, S. J., & Zada, A. (2018). Chitosan-coated cotton cloth supported copper nanoparticles for toxic dye reduction. *International Journal of Biological Macromolecules*, *111*, 832–838.

Ali, N., Khan, A., Bilal, M., Malik, S., Badshah, S., & Iqbal, H. (2020). Chitosan-based bio-composite modified with thiocarbamate moiety for decontamination of cations from the aqueous media. *Molecules*, *25*(1), 226.

Ali, N., Khan, A., Malik, S., Badshah, S., Bilal, M., & Iqbal, H. M. (2020). Chitosan-based green sorbent material for cations removal from an aqueous environment. *Journal of Environmental Chemical Engineering*, *8*(5), 104064.

Ali, N., Khan, A., Nawaz, S., Bilal, M., Malik, S., Badshah, S., & Iqbal, H. M. (2020). Characterization and deployment of surface-engineered chitosan-triethylenetetramine nanocomposite hybrid nano-adsorbent for divalent cations decontamination. *International Journal of Biological Macromolecules*, *152*, 663–671.

Ali, N., Naz, N., Shah, Z., Khan, A., & Nawaz, R. (2020). Selective transportation of molybdenum from model and ore through poly inclusion membrane. *Bulletin of the Chemical Society of Ethiopia*, *34*(1), 93–104.

Ali, N., Uddin, S., Khan, A., Khan, S., Khan, S., Ali, N., ... Bilal, M. (2020). Regenerable chitosan-bismuth cobalt selenide hybrid microspheres for mitigation of organic pollutants in an aqueous environment. *International Journal of Biological Macromolecules*, *161*, 1305–1317.

Ao, C., Niu, Y., Zhang, X., He, X., Zhang, W., & Lu, C. (2017). Fabrication and characterization of electrospun cellulose/nano-hydroxyapatite nanofibers for bone tissue engineering. *International Journal of Biological Macromolecules*, *97*, 568–573.

Ashraf, R., Sofi, H. S., Akram, T., Rather, H. A., Abdal-hay, A., Shabir, N., ... Sheikh, F. A. (2020). Fabrication of multifunctional cellulose/TiO_2/Ag composite nanofibers scaffold with antibacterial and bioactivity properties for future tissue engineering applications. *Journal of Biomedical Materials Research Part A*, *108*(4), 947–962.

Aziz, A., Ali, N., Khan, A., Bilal, M., Malik, S., Ali, N., & Khan, H. (2020). Chitosan-zinc sulfide nanoparticles, characterization and their photocatalytic degradation efficiency for azo dyes. *International Journal of Biological Macromolecules*.

Bahmani, E., Zonouzi, H. S., Koushkbaghi, S., Hafshejani, F. K., Chimeh, A. F., & Irani, M. (2020). Electrospun polyacrylonitrile/cellulose acetate/MIL-125/TiO_2 composite nanofibers as an efficient photocatalyst and anticancer drug delivery system. *Cellulose*, *27*(17), 10029–10045.

Barhoum, A., Pal, K., Rahier, H., Uludag, H., Kim, I. S., & Bechelany, M. (2019). Nanofibers as new-generation materials: From spinning and nano-spinning fabrication techniques to emerging applications. *Applied Materials Today*, *17*, 1–35.

Bhalara, P. D., Balasubramanian, K., & Banerjee, B. S. (2015). Spider—web textured electrospun composite of graphene for sorption of Hg (II) ions. *Materials Focus*, *4*(2), 154–163.

Bulbul, Y. E., Eskitoros-Togay, Ş. M., Demirtas-Korkmaz, F., & Dilsiz, N. (2019). Multi-walled carbon nanotube-incorporating electrospun composite fibrous mats for controlled drug release profile. *International Journal of Pharmaceutics*, *568*, 118513.

Cai, J., Chen, J., Zhang, Q., Lei, M., He, J., Xiao, A., ... Xiong, H. (2016). Well-aligned cellulose nanofiber-reinforced polyvinyl alcohol composite film: Mechanical and optical properties. *Carbohydrate Polymers*, *140*, 238−245.

Cai, J., Lei, M., Zhang, Q., He, J. R., Chen, T., Liu, S., ... Fei, P. (2017). Electrospun composite nanofiber mats of cellulose@ organically modified montmorillonite for heavy metal ion removal: Design, characterization, evaluation of absorption performance. *Composites Part A: Applied Science and Manufacturing*, *92*, 10−16.

Celebioglu, A., Demirci, S., & Uyar, T. (2014). Cyclodextrin-grafted electrospun cellulose acetate nanofibers via "Click" reaction for removal of phenanthrene. *Applied Surface Science*, *305*, 581−588.

Chauhan, D., & Solanki, P. R. (2019). Hydrophilic and insoluble electrospun cellulose acetate fiber-based biosensing platform for 25-hydroxy vitamin-D3 detection. *ACS Applied Polymer Materials*, *1*(7), 1613−1623.

Chen, Y., Abdalkarim, S. Y. H., Yu, H. Y., Li, Y., Xu, J., Marek, J., ... Tam, K. C. (2020). Double stimuli-responsive cellulose nanocrystals reinforced electrospun PHBV composites membrane for intelligent drug release. *International Journal of Biological Macromolecules*.

Cheng, J., Li, H., Cao, Z., Wu, D., Liu, C., & Pu, H. (2019). Nanolayer coextrusion: An efficient and environmentally friendly micro/nanofiber fabrication technique. *Materials Science and Engineering: C*, *95*, 292−301.

Cheng, M., Qin, Z., Hu, S., Dong, S., Ren, Z., & Yu, H. (2017). Achieving long-term sustained drug delivery for electrospun biopolyester nanofibrous membranes by introducing cellulose nanocrystals. *ACS Biomaterials Science & Engineering*, *3*(8), 1666−1676.

Cui, Z., Zheng, Z., Lin, L., Si, J., Wang, Q., Peng, X., & Chen, W. (2018). Electrospinning and crosslinking of polyvinyl alcohol/chitosan composite nanofiber for transdermal drug delivery. *Advances in Polymer Technology*, *37*(6), 1917−1928.

Das, S., Barui, A., & Adak, A. (2020). Montmorillonite impregnated electrospun cellulose acetate nanofiber sorptive membrane for ciprofloxacin removal from wastewater. *Journal of Water Process Engineering*, *37*, 101497.

Devarayan, K., & Kim, B. S. (2015). Reversible and universal pH sensing cellulose nanofibers for health monitor. *Sensors and Actuators B: Chemical*, *209*, 281−286.

Dong, H., Strawhecker, K. E., Snyder, J. F., Orlicki, J. A., Reiner, R. S., & Rudie, A. W. (2012). Cellulose nanocrystals as a reinforcing material for electrospun poly (methyl methacrylate) fibers: Formation, properties and nanomechanical characterization. *Carbohydrate Polymers*, *87*(4), 2488−2495.

Edikresnha, D., Suciati, T., Munir, M. M., & Khairurrijal, K. (2019). Polyvinylpyrrolidone/cellulose acetate electrospun composite nanofibres loaded by glycerine and garlic extract with in vitro antibacterial activity and release behaviour test. *RSC Advances*, *9*(45), 26351−26363.

El-Moghazy, A. Y., Soliman, E. A., Ibrahim, H. Z., Marty, J. L., Istamboulie, G., & Noguer, T. (2016). Biosensor based on electrospun blended chitosan-poly (vinyl alcohol) nanofibrous enzymatically sensitized membranes for pirimiphos-methyl detection in olive oil. *Talanta*, *155*, 258−264.

Eslahi, N., Mahmoodi, A., Mahmoudi, N., Zandi, N., & Simchi, A. (2020). Processing and properties of nanofibrous bacterial cellulose-containing polymer composites: A review of recent advances for biomedical applications. *Polymer Reviews*, *60*(1), 144−170.

Esmaeili, A., & Haseli, M. (2017a). Electrospinning of thermoplastic carboxymethyl cellulose/poly (ethylene oxide) nanofibers for use in drug-release systems. *Materials Science and Engineering: C*, *77*, 1117−1127.

Esmaeili, A., & Haseli, M. (2017b). Optimization, synthesis, and characterization of coaxial electrospun sodium carboxymethyl cellulose-graft-methyl acrylate/poly (ethylene oxide) nanofibers for potential drug-delivery applications. *Carbohydrate Polymers*, *173*, 645−653.

Gabr, M. H., Phong, N. T., Okubo, K., Uzawa, K., Kimpara, I., & Fujii, T. (2014). Thermal and mechanical properties of electrospun nano-celullose reinforced epoxy nanocomposites. *Polymer Testing*, *37*, 51−58.

Gebru, K. A., & Das, C. (2017). Removal of Pb (II) and Cu (II) ions from wastewater using composite electrospun cellulose acetate/titanium oxide (TiO2) adsorbent. *Journal of Water Process Engineering*, *16*, 1−13.

Ghorbani, M., Nezhad-Mokhtari, P., Sohrabi, H., & Roshangar, L. (2020). Electrospun chitosan/nano-crystalline cellulose-graft-poly (N-vinylcaprolactam) nanofibers as the reinforced scaffold for tissue engineering. *Journal of Materials Science*, *55*(5), 2176–2185.

Halim, A., Xu, Y., Lin, K. H., Kobayashi, M., Kajiyama, M., & Enomae, T. (2019). Fabrication of cellulose nanofiber-deposited cellulose sponge as an oil-water separation membrane. *Separation and Purification Technology*, *224*, 322–331.

Hamad, A. A., Hassouna, M. S., Shalaby, T. I., Elkady, M. F., Abd Elkawi, M. A., & Hamad, H. A. (2020). Electrospun cellulose acetate nanofiber incorporated with hydroxyapatite for removal of heavy metals. *International Journal of Biological Macromolecules*, *151*, 1299–1313.

Hatch, K. M., Hlavatá, J., Paulett, K., Liavitskaya, T., Vyazovkin, S., & Stanishevsky, A. V. (2019). Nanocrystalline cellulose/polyvinylpyrrolidone fibrous composites prepared by electrospinning and thermal crosslinking. *International Journal of Polymer Science*, *2019*.

He, X., Arsat, R., Sadek, A. Z., Wlodarski, W., Kalantar-Zadeh, K., & Li, J. (2010). Electrospun PVP fibers and gas sensing properties of PVP/36 YX LiTaO3 SAW device. *Sensors and Actuators B: Chemical*, *145*(2), 674–679.

He, X., Cheng, L., Zhang, X., Xiao, Q., Zhang, W., & Lu, C. (2015). Tissue engineering scaffolds electrospun from cotton cellulose. *Carbohydrate Polymers*, *115*, 485–493.

He, X., Xiao, Q., Lu, C., Wang, Y., Zhang, X., Zhao, J., ... Deng, Y. (2014). Uniaxially aligned electrospun all-cellulose nanocomposite nanofibers reinforced with cellulose nanocrystals: Scaffold for tissue engineering. *Biomacromolecules*, *15*(2), 618–627.

Hivechi, A., Bahrami, S. H., & Siegel, R. A. (2019). Drug release and biodegradability of electrospun cellulose nanocrystal reinforced polycaprolactone. *Materials Science and Engineering: C*, *94*, 929–937.

Hsieh, Y. L. (2018). Cellulose nanofibers: Electrospinning and nanocellulose self-assemblies. *Advanced Green Composites*, 67–96.

Hu, S., Qin, Z., Cheng, M., Chen, Y., Liu, J., & Zhang, Y. (2018). Improved properties and drug delivery behaviors of electrospun cellulose acetate nanofibrous membranes by introducing carboxylated cellulose nanocrystals. *Cellulose*, *25*(3), 1883–1898.

Hua, D., Liu, Z., Wang, F., Gao, B., Chen, F., Zhang, Q., ... Huang, C. (2016). pH responsive polyurethane (core) and cellulose acetate phthalate (shell) electrospun fibers for intravaginal drug delivery. *Carbohydrate Polymers*, *151*, 1240–1244.

Huang, A., Peng, X., Geng, L., Zhang, L., Huang, K., Chen, B., ... Kuang, T. (2018). Electrospun poly (butylene succinate)/cellulose nanocrystals bio-nanocomposite scaffolds for tissue engineering: Preparation, characterization and in vitro evaluation. *Polymer Testing*, *71*, 101–109.

Ilyas, R. A., Sapuan, S. M., Sanyang, M. L., Ishak, M. R., & Zainudin, E. S. (2018). Nanocrystalline cellulose as reinforcement for polymeric matrix nanocomposites and its potential applications: A review. *Current Analytical Chemistry*, *14*(3), 203–225.

Jia, Y., Yu, H., Zhang, Y., Dong, F., & Li, Z. (2016). Cellulose acetate nanofibers coated layer-by-layer with polyethylenimine and graphene oxide on a quartz crystal microbalance for use as a highly sensitive ammonia sensor. *Colloids and Surfaces B: Biointerfaces*, *148*, 263–269.

Jordan, A. M., Viswanath, V., Kim, S. E., Pokorski, J. K., & Korley, L. T. (2016). Processing and surface modification of polymer nanofibers for biological scaffolds: A review. *Journal of Materials Chemistry B*, *4*(36), 5958–5974.

Joshi, M. K., Tiwari, A. P., Pant, H. R., Shrestha, B. K., Kim, H. J., Park, C. H., & Kim, C. S. (2015). In situ generation of cellulose nanocrystals in polycaprolactone nanofibers: Effects on crystallinity, mechanical strength, biocompatibility, and biomimetic mineralization. *ACS Applied Materials & Interfaces*, 7(35), 19672–19683.

Kargarzadeh, H., Mariano, M., Huang, J., Lin, N., Ahmad, I., Dufresne, A., & Thomas, S. (2017). Recent developments on nanocellulose reinforced polymer nanocomposites: A review. *Polymer*, *132*, 368–393.

Khalil, H. A., Davoudpour, Y., Bhat, A. H., Rosamah, E., & Tahir, P. M. (2015). *Electrospun cellulose composite nanofibers. Handbook of polymer nanocomposites. Processing, performance and application* (pp. 191–227). Berlin, Heidelberg: Springer.

Khan, A., Ali, N., Bilal, M., Malik, S., Badshah, S., & Iqbal, H. (2019). Engineering functionalized chitosan-based sorbent material: characterization and sorption of toxic elements. *Applied Sciences*, *9*(23), 5138.

Khan, A., Shah, S. J., Mehmood, K., Ali, N., & Khan, H. (2019). Synthesis of potent chitosan beads a suitable alternative for textile dye reduction in sunlight. *Journal of Materials Science: Materials in Electronics*, *30*(1), 406–414.

Khan, H., Gul, K., Ara, B., Khan, A., Ali, N., Ali, N., & Bilal, M. (2020). Adsorptive removal of acrylic acid from the aqueous environment using raw and chemically modified alumina: Batch adsorption, kinetic, equilibrium and thermodynamic studies. *Journal of Environmental Chemical Engineering*, *8*(4), 103927.

Khan, H., Khalil, A. K., & Khan, A. (2019). Photocatalytic degradation of alizarin yellow in aqueous medium and real samples using chitosan conjugated tin magnetic nanocomposites. *Journal of Materials Science: Materials in Electronics*, *30*(24), 21332–21342.

Kim, W., Park, E., & Jeon, S. (2020). Performance enhancement of a quartz tuning fork sensor using a cellulose nanocrystal-reinforced nanoporous polymer fiber. *Sensors*, *20*(2), 437.

Lee, J., Moon, J. Y., Lee, J. C., Hwang, T. I., Park, C. H., & Kim, C. S. (2020). Simple conversion of 3D electrospun nanofibrous cellulose acetate into a mechanically robust nanocomposite cellulose/calcium scaffold. *Carbohydrate Polymers*, *253*, 117191.

Lefatshe, K., Muiva, C. M., & Kebaabetswe, L. P. (2017). Extraction of nanocellulose and in-situ casting of ZnO/cellulose nanocomposite with enhanced photocatalytic and antibacterial activity. *Carbohydrate Polymers*, *164*, 301–308.

Li, J. J., Yang, Y. Y., Yu, D. G., Du, Q., & Yang, X. L. (2018). Fast dissolving drug delivery membrane based on the ultra-thin shell of electrospun core-shell nanofibers. *European Journal of Pharmaceutical Sciences*, *122*, 195–204.

Li, W., Li, T., Li, G., An, L., Li, F., & Zhang, Z. (2017). Electrospun H4SiW12O40/cellulose acetate composite nanofibrous membrane for photocatalytic degradation of tetracycline and methyl orange with different mechanism. *Carbohydrate Polymers*, *168*, 153–162.

Li, X., Feng, Q., Lu, K., Huang, J., Zhang, Y., Hou, Y., ... Wei, Q. (2020). Encapsulating enzyme into metal-organic framework during in-situ growth on cellulose acetate nanofibers as self-powered glucose biosensor. *Biosensors and Bioelectronics*, *171*, 112690.

Lu, H., Wang, Q., Li, G., Qiu, Y., & Wei, Q. (2017). Electrospun water-stable zein/ethyl cellulose composite nanofiber and its drug release properties. *Materials Science and Engineering: C*, *74*, 86–93.

Luo, Y., Nartker, S., Wiederoder, M., Miller, H., Hochhalter, D., Drzal, L. T., & Alocilja, E. C. (2011). Novel biosensor based on electrospun nanofiber and magnetic nanoparticles for the detection of E. coli O157: H7. *IEEE Transactions on Nanotechnology*, *11*(4), 676–681.

Moohan, J., Stewart, S. A., Espinosa, E., Rosal, A., Rodríguez, A., Larrañeta, E., ... Domínguez-Robles, J. (2020). Cellulose nanofibers and other biopolymers for biomedical applications. A review. *Applied Sciences*, *10*(1), 65.

Mouro, C., Fangueiro, R., & Gouveia, I. C. (2020). Preparation and characterization of electrospun double-layered nanocomposites membranes as a carrier for *Centella asiatica* (L.). *Polymers*, *12*(11), 2653.

Muqeet, M., Malik, H., Mahar, R. B., Ahmed, F., Khatri, Z., & Carlson, K. (2017). Cationization of cellulose nanofibers for the removal of sulfate ions from aqueous solutions. *Industrial & Engineering Chemistry Research*, *56*(47), 14078–14088.

Naeem, M. A., Alfred, M., Lv, P., Zhou, H., & Wei, Q. (2018). Three-dimensional bacterial cellulose-electrospun membrane hybrid structures fabricated through in-situ self-assembly. *Cellulose*, *25*(12), 6823–6830.

Naseri, N., Mathew, A. P., Girandon, L., Fröhlich, M., & Oksman, K. (2015). Porous electrospun nanocomposite mats based on chitosan–cellulose nanocrystals for wound dressing: Effect of surface characteristics of nanocrystals. *Cellulose*, *22*(1), 521–534.

Nawaz, A., Khan, A., Ali, N., Ali, N., & Bilal, M. (2020). Fabrication and characterization of new ternary ferrites-chitosan nanocomposite for solar-light driven photocatalytic degradation of a model textile dye. *Environmental Technology & Innovation*, *20*, 101079.

Pang, Z., Yang, Z., Chen, Y., Zhang, J., Wang, Q., Huang, F., & Wei, Q. (2016). A room temperature ammonia gas sensor based on cellulose/TiO2/PANI composite nanofibers. *Colloids and Surfaces A: Physicochemical and Engineering Aspects*, *494*, 248–255.

Pasaoglu, M. E., & Koyuncu, I. (2020). Substitution of petroleum-based polymeric materials used in the electrospinning process with nanocellulose: A review and future outlook. *Chemosphere*, *269*, 128710.

Patel, D. K., Dutta, S. D., Hexiu, J., Ganguly, K., & Lim, K. T. (2020). Bioactive electrospun nanocomposite scaffolds of poly (lactic acid)/cellulose nanocrystals for bone tissue engineering. *International Journal of Biological Macromolecules*, *162*, 1429–1441.

Petropoulou, A., Kralj, S., Karagiorgis, X., Savva, I., Loizides, E., Panagi, M., ... Riziotis, C. (2020). Multifunctional gas and pH fluorescent sensors based on cellulose acetate electrospun fibers decorated with rhodamine B-functionalised core-shell ferrous nanoparticles. *Scientific Reports*, *10*(1), 1–14.

Phan, D. N., Lee, H., Huang, B., Mukai, Y., & Kim, I. S. (2019). Fabrication of electrospun chitosan/cellulose nanofibers having adsorption property with enhanced mechanical property. *Cellulose*, *26*(3), 1781–1793.

Ponnamma, D., Parangusan, H., Tanvir, A., & AlMa'adeed, M. A. A. (2019). Smart and robust electrospun fabrics of piezoelectric polymer nanocomposite for self-powering electronic textiles. *Materials & Design*, *184*, 108176.

Pooyan, P., Kim, I. T., Jacob, K. I., Tannenbaum, R., & Garmestani, H. (2013). Design of a cellulose-based nanocomposite as a potential polymeric scaffold in tissue engineering. *Polymer*, *54*(8), 2105–2114.

Rojas, O. J., Montero, G. A., & Habibi, Y. (2009). Electrospun nanocomposites from polystyrene loaded with cellulose nanowhiskers. *Journal of Applied Polymer Science*, *113*(2), 927–935.

Saeed, K., Sadiq, M., Khan, I., Ullah, S., Ali, N., & Khan, A. (2018). Synthesis, characterization, and photocatalytic application of Pd/ZrO 2 and Pt/ZrO 2. *Applied Water Science*, *8*(2), 60.

Salami, M. A., Kaveian, F., Rafienia, M., Saber-Samandari, S., Khandan, A., & Naeimi, M. (2017). Electrospun polycaprolactone/lignin-based nanocomposite as a novel tissue scaffold for biomedical applications. *Journal of Medical Signals and Sensors*, 7(4), 228.

Samadian, H., Salehi, M., Farzamfar, S., Vaez, A., Ehterami, A., Sahrapeyma, H., ... Ghorbani, S. (2018). In vitro and in vivo evaluation of electrospun cellulose acetate/gelatin/hydroxyapatite nanocomposite mats for wound dressing applications. *Artificial Cells, Nanomedicine, and Biotechnology*, *46* (sup1), 964–974.

Sartaj, S., Ali, N., Khan, A., Malik, S., Bilal, M., Khan, M., ... Khan, S. (2020). Performance evaluation of photolytic and electrochemical oxidation processes for enhanced degradation of food dyes laden wastewater. *Water Science and Technology*, *81*(5), 971–984.

Sehaqui, H., Mautner, A., de Larraya, U. P., Pfenninger, N., Tingaut, P., & Zimmermann, T. (2016). Cationic cellulose nanofibers from waste pulp residues and their nitrate, fluoride, sulphate and phosphate adsorption properties. *Carbohydrate Polymers*, *135*, 334–340.

Shah, S., Din, S., Khan, A., & Shah, S. A. (2018). Green synthesis and antioxidant study of silver nanoparticles of root extract of *Sageretiathea* and its role in oxidation protection technology. *Journal of Polymers and the Environment*, *26*(6), 2323–2332.

Shi, Q., Zhou, C., Yue, Y., Guo, W., Wu, Y., & Wu, Q. (2012). Mechanical properties and in vitro degradation of electrospun bio-nanocomposite mats from PLA and cellulose nanocrystals. *Carbohydrate Polymers*, *90*(1), 301–308.

Sohni, S., Gul, K., Ahmad, F., Ahmad, I., Khan, A., Khan, N., & Bahadar Khan, S. (2018). Highly efficient removal of acid red-17 and bromophenol blue dyes from industrial wastewater using graphene oxide functionalized magnetic chitosan composite. *Polymer Composites*, *39*(9), 3317–3328.

Taha, A. A., Wu, Y. N., Wang, H., & Li, F. (2012). Preparation and application of functionalized cellulose acetate/silica composite nanofibrous membrane via electrospinning for Cr (VI) ion removal from aqueous solution. *Journal of Environmental Management*, *112*, 10–16.

Tamahkar, E., Bakhshpour, M., & Denizli, A. (2019). Molecularly imprinted composite bacterial cellulose nanofibers for antibiotic release. *Journal of Biomaterials Science, Polymer Edition*, *30*(6), 450–461.

Tayeb, A. H., Amini, E., Ghasemi, S., & Tajvidi, M. (2018). Cellulose nanomaterials—Binding properties and applications: A review. *Molecules*, *23*(10), 2684.

Teodoro, K. B., Migliorini, F. L., Facure, M. H., & Correa, D. S. (2019). Conductive electrospun nanofibers containing cellulose nanowhiskers and reduced graphene oxide for the electrochemical detection of mercury (II). *Carbohydrate Polymers*, *207*, 747–754.

Teodoro, K. B., Shimizu, F. M., Scagion, V. P., & Correa, D. S. (2019). Ternary nanocomposites based on cellulose nanowhiskers, silver nanoparticles and electrospun nanofibers: Use in an electronic tongue for heavy metal detection. *Sensors and Actuators B: Chemical*, *290*, 387–395.

Thunberg, J., Kalogeropoulos, T., Kuzmenko, V., Hägg, D., Johannesson, S., Westman, G., & Gatenholm, P. (2015). In situ synthesis of conductive polypyrrole on electrospun cellulose nanofibers: Scaffold for neural tissue engineering. *Cellulose*, *22*(3), 1459–1467.

Wan, Z., Wang, L., Yang, X., Guo, J., & Yin, S. (2016). Enhanced water resistance properties of bacterial cellulose multilayer films by incorporating interlayers of electrospun zein fibers. *Food Hydrocolloids*, *61*, 269–276.

Wang, H., Kong, L., & Ziegler, G. R. (2019). Fabrication of starch-nanocellulose composite fibers by electrospinning. *Food Hydrocolloids*, *90*, 90–98.

Wang, S., Jiang, F., Xu, X., Kuang, Y., Fu, K., Hitz, E., & Hu, L. (2017). Super-strong, super-stiff macrofibers with aligned, long bacterial cellulose nanofibers. *Advanced Materials*, *29*(35), 1702498.

Wang, Y., Qiao, W., Wang, B., Zhang, Y., Shao, P., & Yin, T. (2011). Electrospun composite nanofibers containing nanoparticles for the programmable release of dual drugs. *Polymer Journal*, *43*(5), 478–483.

Wsoo, M. A., Shahir, S., Bohari, S. P. M., Nayan, N. H. M., & Abd Razak, S. I. (2020). A review on the properties of electrospun cellulose acetate and its application in drug delivery systems: A new perspective. *Carbohydrate Research*, 107978.

Xu, Q., Peng, J., Zhang, W., Wang, X., & Lou, T. (2020). Electrospun cellulose acetate/P (DMDAAC-AM) nanofibrous membranes for dye adsorption. *Journal of Applied Polymer Science*, *137*(15), 48565.

Yan, J., White, K., Yu, D. G., & Zhao, X. Y. (2014). Sustained-release multiple-component cellulose acetate nanofibers fabricated using a modified coaxial electrospinning process. *Journal of Materials Science*, *49*(2), 538–547.

Yang, Y., Li, W., Yu, D. G., Wang, G., Williams, G. R., & Zhang, Z. (2019). Tunable drug release from nanofibers coated with blank cellulose acetate layers fabricated using tri-axial electrospinning. *Carbohydrate Polymers*, *203*, 228–237.

Zanjani, J. S. M., Saner Okan, B., Menceloglu, Y. Z., & Yildiz, M. (2015). Design and fabrication of multi-walled hollow nanofibers by triaxial electrospinning as reinforcing agents in nanocomposites. *Journal of Reinforced Plastics and Composites*, *34*(16), 1273–1286.

Zhai, Y., Wang, D., Liu, H., Zeng, Y., Yin, Z., & Li, L. (2015). Electrochemical molecular imprinted sensors based on electrospun nanofiber and determination of ascorbic acid. *Analytical Sciences*, *31*(8), 793–798.

Zhang, C., Salick, M. R., Cordie, T. M., Ellingham, T., Dan, Y., & Turng, L. S. (2015). Incorporation of poly (ethylene glycol) grafted cellulose nanocrystals in poly (lactic acid) electrospun nanocomposite fibers as potential scaffolds for bone tissue engineering. *Materials Science and Engineering: C*, *49*, 463–471.

Zhang, X., Wang, C., Liao, M., Dai, L., Tang, Y., Zhang, H., ... Ji, P. (2019). Aligned electrospun cellulose scaffolds coated with rhBMP-2 for both in vitro and in vivo bone tissue engineering. *Carbohydrate Polymers*, *213*, 27–38.

Zhou, C., Shi, Q., Guo, W., Terrell, L., Qureshi, A. T., Hayes, D. J., & Wu, Q. (2013). Electrospun bio-nanocomposite scaffolds for bone tissue engineering by cellulose nanocrystals reinforcing maleic anhydride grafted PLA. *ACS Applied Materials & Interfaces*, *5*(9), 3847–3854.

Zhu, L., Zhou, X., Liu, Y., & Fu, Q. (2019). Highly sensitive, ultrastretchable strain sensors prepared by pumping hybrid fillers of carbon nanotubes/cellulose nanocrystal into electrospun polyurethane membranes. *ACS Applied Materials & Interfaces*, *11*(13), 12968–12977.

Zhu, Q., Yao, Q., Sun, J., Chen, H., Xu, W., Liu, J., & Wang, Q. (2020). Stimuli induced cellulose nanomaterials alignment and its emerging applications: A review. *Carbohydrate Polymers*, *230*, a115609.

CHAPTER 16

Treatment of pulp and paper industry waste effluents and contaminants

Adnan Khan[1], Sumeet Malik[1], Nisar Ali[2], Muhammad Bilal[3], Farooq Sher[4], Vineet Kumar[5], Luiz Fernando Romanholo Ferreira[6,7] and Hafiz M.N. Iqbal[8]

[1]Institute of Chemical Sciences, University of Peshawar, Peshawar, Pakistan
[2]Key Laboratory for Palygorskite Science and Applied Technology of Jiangsu Province, National & Local Joint Engineering Research Centre for Deep Utilization Technology of Rock-salt Resource, Faculty of Chemical Engineering, Huaiyin Institute of Technology, Huai'an, P.R. China
[3]School of Life Science and Food Engineering, Huaiyin Institute of Technology, Huai'an, P.R. China
[4]Department of Engineering, School of Science and Technology, Nottingham Trent University, Nottingham, United Kingdom
[5]Department of Botany, School of Life Science, Guru Ghasidas Vishwavidyalaya (A Central University), Bilaspur, India
[6]Graduate Program in Process Engineering, Tiradentes University (UNIT), Aracaju, Brazil
[7]Institute of Technology and Research (ITP), Tiradentes University (UNIT), Aracaju, Brazil
[8]Tecnologico de Monterrey, School of Engineering and Sciences, Monterrey, Mexico

16.1 Introduction

Paper and pulp are one of the largest industries across the world, which use wood as raw material to produce pulp, paper, and other cellulose-based products (Ali et al., 2021; Khan et al., 2021; Novais, Carvalheiras, Senff, & Labrincha, 2018). This is considered one of the most important industries due to the large-scale production of paper and paper-based products because of its vast needs. Owing to the largest production of paper and pulp industry, its byproduct generation is also massive in scale (Fig. 16.1). This industrial sector is majorly exploiting the energy and water resources for their operation (Nawaz, Khan, Ali, Ali, & Bilal, 2020; Sun, Wang, Shi, & Klemeš, 2018; Yang et al., 2021). The principle of the paper and pulp industry is to process the cellulose present in the wood to obtain pulp, which is further used in paper production. Some other sources of cellulose include jute, sisal, flax, cereal straw, reeds, etc. are also utilized for the manufacture of paper and pulp products (Ali, Uddin, Khan, et al., 2020; Ali, Ahmad, Khan, et al. 2020; Mohammadi et al., 2019). The major raw products used in paper production are only wood or wastes from industrial effluents (cellulosic source) and processed to produce a paper of different quality.

Despite the high production of the paper and pulp industry, the demand is constantly rising (Ali, Bilal, Khan, et al., 2020; Khan et al., 2020; Saeli, Tobaldi, Seabra, & Labrincha, 2019). A survey performed, showed that the world production of paper has leveled up to more than 390 million tons, as per Food and Agriculture Organization Corporate Statistical Database. There is still a chance of an enormous rise in paper production due to

Nanotechnology in Paper and Wood Engineering
DOI: https://doi.org/10.1016/B978-0-323-85835-9.00018-0

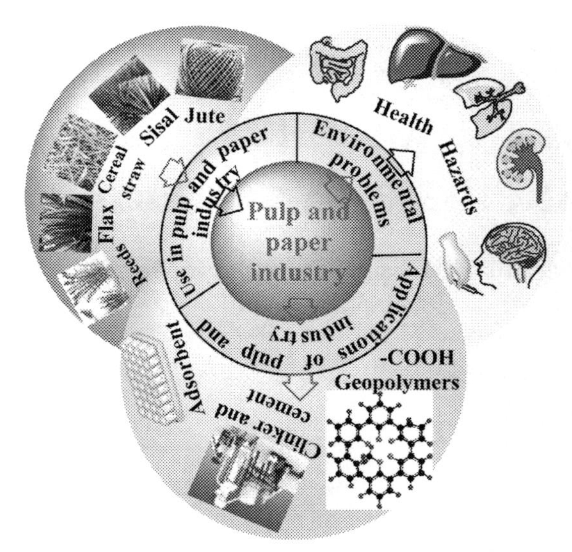

Figure 16.1 Resources, wastes, and challenges of pulp and paper industry.

its excessively increasing need (Ali, Khan, Malik, et al., 2020; Chakraborty et al., 2019). Another source that is extensively employed for running the paper and pulp industry is water consumption. The paper production operations require a high amount of water as up to 5−100 m^3/1t of paper produced (Ali, Khan, Bilal, et al., 2020; de Azevedo et al., 2019). The range of water consumption is based on different factors like the quality of the paper, the extent of water re-usage, substrate level, and so on. In some of the developing countries, the water consumption is very high because the recycling of wastewater is extremely poor (Ali, Khan, Nawaz, et al., 2020; Toczyłowska-Mamińska, 2017). This lower recycling ratio of the wastewater compared to its consumption at high levels is causing massive damage to the environment (Ali, Naz, Shah, Khan, & Nawaz, 2020; Ma et al., 2019). While in developed countries, this wastewater is recycled in the industrial mills, and advanced treatment processes are also being employed to purify water and manage hazardous waste (Azevedo et al., 2019).

A rough estimate shows that about 3bln m^3 of wastewater obtained from the paper and pulp industry needs to be managed. Another major aspect of the operations and milling of the paper and pulp industry is deforestation (Branco, Serafim, & Xavier, 2019; Gupta & Shukla, 2020). A number of forests are destroyed, and trees are cut down to meet the wood needs of this industry. Massive deforestation ultimately affects the climatic and weather changes contributing to global warming and the greenhouse effect (Aziz et al., 2020; van Ewijk, Stegemann, & Ekins, 2021). Hence, the establishment of the paper and pulp industries also has a negative impact on the environment in climatic aspects. The presented chapter covers the environmental impact of hazards coming out of the paper and pulp industry and their treatment by different techniques.

16.2 Processing of paper and pulp industry

As mentioned earlier, the major raw material used by the pulp and paper industry is wood, as the cellulosic source. Some other industrial residues, like cotton liners, bagasse, which are the sugar cane residues, are also used as raw products (Mazhar et al., 2019; Sartaj et al., 2020; Fig. 16.2). The wood used as raw material in the process of paper manufacturing can be split up and composed of four components, such as cellulose, hemicelluloses, lignin, and other extractives. Cellulose makes up the major portion of the wood (about 50%) and is the constituent of interest also. The hemicelluloses play the role of binding among the paper fibers (Kumar & Christopher, 2017). Lignin has a complex organic structure, and it renders strength to the wood fibers and is burned during the pulping process of papermaking. While the extractives like waxes and resins, and so on, are also removed during the paper manufacture. Overall, the paper and pulp mill's industrial operation consists of converting the raw products into pulp followed by the conversion of pulp into paper (Ali, Bilal, Nazir, et al., 2020; Sharma, Tripathi, & Chandra, 2021a,b). There are several steps involved in the manufacture of the paper, which can be split into raw material preparation (debarking, chipping, and conveying), pulping (chemical, mechanical, and recycled paper), chemical recovery (evaporation, recovery boiler, recausticizing, and calcination), bleaching, stock preparation, dewatering, pressing/drying,

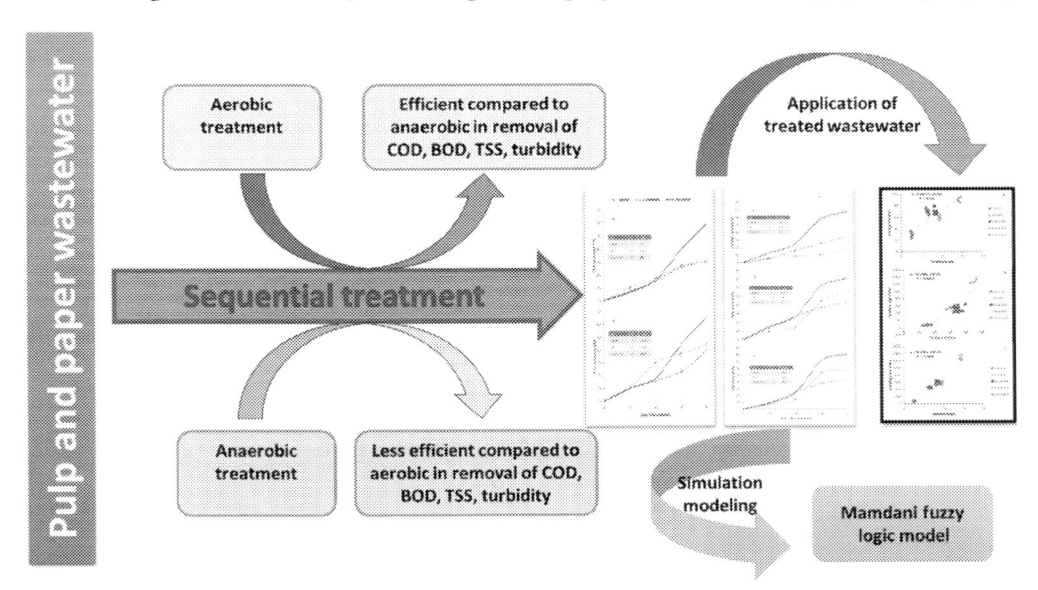

Figure 16.2. Sequential processing of pulp and paper wastewater (Mazhar et al., 2019). *BOD*, Biological oxygen demand; *COD*, chemical oxygen demand; *TSS*, total suspended solids. *Reproduced with permission from reference Sadat Mazhar, Allah Ditta, Laura Bulgariu, Iftikhar Ahmad, Munir Ahmed, Ata Allah Nadiri. (2019). Sequential treatment of paper and pulp industrial wastewater: Prediction of water quality parameters by Mamdani Fuzzy Logic model and phytotoxicity assessment, Chemosphere, Volume 227, Pages 256–268. https://doi.org/10.1016/j.chemosphere.2019.04.022. Copyright 2019 Elsevier.*

and finishing (Ali, Azeem, Khan, et al., 2020; Gonzalo et al., 2017). The operation commences with the raw material preparation by debarking and chipping the woods.

The debarked wood is then processed in the form of chips to reduce the size and obtain a proper shape homogenously to avoid any inconvenience. The chips are then screened to remove the odd-sized or shaped chips along with the removal of sawdust (Ali, Bilal, Khan, Ali, & Iqbal, 2020; Simão, Hotza, Raupp-Pereira, Labrincha, & Montedo, 2018). The processing of the chips is performed at optimized conditions to avoid high energy consumption. The perfectly sized chips are then leveled up to pulping. The pulping consists of the separation of lignin from the wood. Different types of pulping procedures are carried out to get pulp fibers from the wood, which can further be processed (Ali, Khan, Malik, et al., 2020; Simão et al., 2017). Different pulping procedures include chemical pulping (kraft sulfate pulp and sulfite pulp), semichemical pulping (cold-caustic process and neutral sulfite process), mechanical pulping (stone groundwood, refiner mechanical, thermomechanical, and chemimechanical). These types of pulping processes have extensively been utilized, but the most commonly employed method is the Kraft pulping procedure. In the case of the kraft pulping process, the alkaline chemicals (sodium hydroxide and sodium sulfide) are added to the raw material under high temperature and pressure and subjected to cooking (Khan, Ali, et al., 2019; Yang et al., 2019). The cooked products come out in the form of disintegrated fibers or pulp. The obtained pulp has a characteristic dark brown color. The obtained pulp is cleaned, screened, and fractionated further before subjecting to bleaching. The bleaching process is indeed an important step as it renders brightness to the paper for further usage. The bleaching step uses some alkali, like sodium hydroxide, to remove the alkali-soluble lignin (Ali et al., 2019; Li et al., 2018). The focus is given to the chlorine-free chemicals to be used in the bleaching process. The recovery of some chemicals is also performed to utilize the worn-out chemicals further, avoiding the contamination rate. The finally obtained pulp is then converted to paper by the deposition of the cellulosic fibers onto the moving fabric with the water's removal. The left-out water is removed through pressing and drying (Didone et al., 2017; Khan, Shah, Mehmood, Ali, & Khan, 2019).

16.3 Types of pollutants and their characteristics

The detailed discussion about the operations of the paper and pulp industry shows a high-scale demand. Alongside the enormous level of production by this industry, the effluents production is also very high. The paper and pulp industry utilizes a number of resources and volumes of water for its proper working (Brink, Sheridan, & Harding, 2018; Khan, Khalil, & Khan, 2019). At the same time, emissions of pollutants are causing hazardous impacts on the environment and living organisms. Among the highly running industries, the paper and pulp industry come 6th regarding effluents production and environment polluting industry. Most of the harmful effluents are produced during the pulping and bleaching section of the procedures. These steps make use of a number of different chemicals, which become the cause of the effluents production (Ali, Kamal, et al., 2018; Usman, Ma, Wasif Zafar, Haseeb, & Ashraf, 2019). These chemicals are usually not properly processed or unprocessed when emitted to the environment, leading to detrimental effects. The effluents emerging from the paper and pulp industry are either emitted directly in the air or come out along with the wastewater, ultimately becoming an unwanted part of the environment. Based on the nature of the effluents, they are divided into the following categories.

16.3.1 Gaseous effluents emissions into the air

The large-scale airborne pollution in the form of gaseous emissions is caused by the paper and pulp industry. The steps of paper production like chipping (cooking and storage), pulping (pulp washing), bleaching (using chemical agents), evaporation, recovery, storage, and so on, produce gaseous contaminants at each stage (Elakkiya & Niju, 2020; Sohni et al., 2018). These contaminants significantly pollute the air causing ill effects. The major effluents in the gaseous form emitted from these processes include reduced sulfur compounds like methyl mercaptan, hydrogen sulfide, dimethyl sulfide, particulate emissions of nitrogen oxide and sulfur compounds, and so on. Other air-borne emissions include steam, fly ash due to energy and fuel burning, coal burning, elevated temperatures, and so on (Ali, Ismail, et al., 2018; Zainith, Chowdhary, & Bharagava, 2019; Fig. 16.3).

16.3.2 Solid wastes emitted in the wastewater

Another major category of pollutants obtained from the paper and pulp industry is solid wastes. The solid wastes obtained from the paper and pulp industry are produced mainly by wood bark processing. The unprocessed wood bark, left-outs of the recovery boiler, screening of the chips, salts used in the processing of the wood, sawdust, dregs, and grits from the causticizing process, and so on (Lindholm-Lehto, Knuutinen, Ahkola, & Herve, 2015; Wahid et al., 2017). The solid products are obtained through different steps of paper and pulp processing. Apart from these solid wastes, other heavy metals like chromium (Cr), manganese (Mn), cadmium (Cd) iron (Fe), and nonmetals like calcium, magnesium, sodium, sulfate, chlorine, and so on. These metals and nonmetallic elements also have toxic nature depending upon their presence and paper stiffness. The sludge produced during the paper and pulp production process also comes in the form of unwanted pollutants into the environment (Shah, ud Din, Khan, & Shah, 2018; Sharma, Sarma, & Devi, 2020).

16.3.3 Liquid wastes emitted as wastewater

The water used in the different operations of paper and pulp production becomes highly polluted due to the addition of chemicals. These pollutants' containing water is

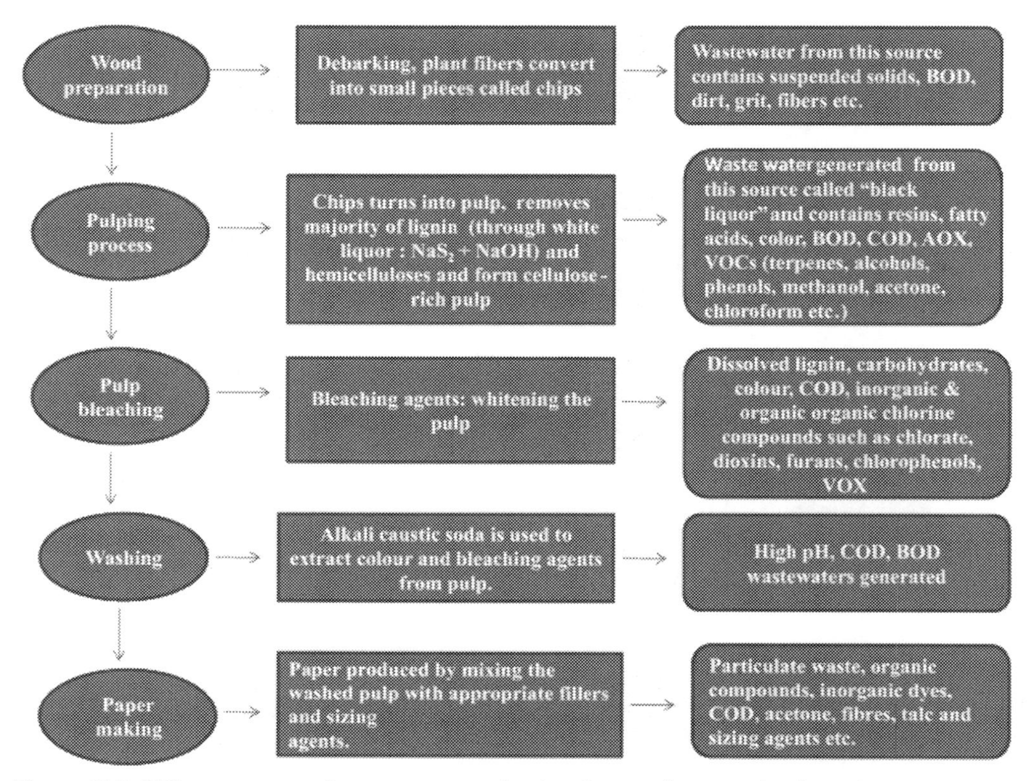

Figure 16.3 Different stages of wastewater production from pollutants of pulp and paper industry (Zainith et al., 2019). *AOX*, Organic halides; *BOD*, biological oxygen demand; *COD*, chemical oxygen demand; *VOCs*, volatile organic compounds; *VOX*, volatile halogenated compounds. *Reproduced with permission from reference Zainith, S., Chowdhary, P., & Bharagava, R. N. (2019). Recent advances in physico-chemical and biological techniques for the management of pulp and paper mill waste. Emerging and Eco-Friendly Approaches for Waste Management, 271–297. Copyright 2019 Springer.*

discharged out of the industry in the form of wastewater, which highly pollutes the environment (CB & Babu, 2018; Saeed et al., 2018). The wastewater obtained from the paper and pulp industry is generally thought to be high in suspended particles, biological oxygen demand (BOD), chemical oxygen demand (COD), phosphorus, nitrogen, acetic acid, formic acid, and so on. Other organic substances like organic halides, chlorinated organic compounds including furans and dioxins are also a part of wastewater.

16.4 Environmental impact of effluents

The environmental impacts of wastes coming out of the paper and pulp industries cannot be overlooked. The wastewater management projects are working on massive scales to minimize the hazardous effects of the effluents obtained from different industries (Beckline, Yujun, Eric, & Kato, 2016; Neelofar et al., 2017; Fig. 16.4). But

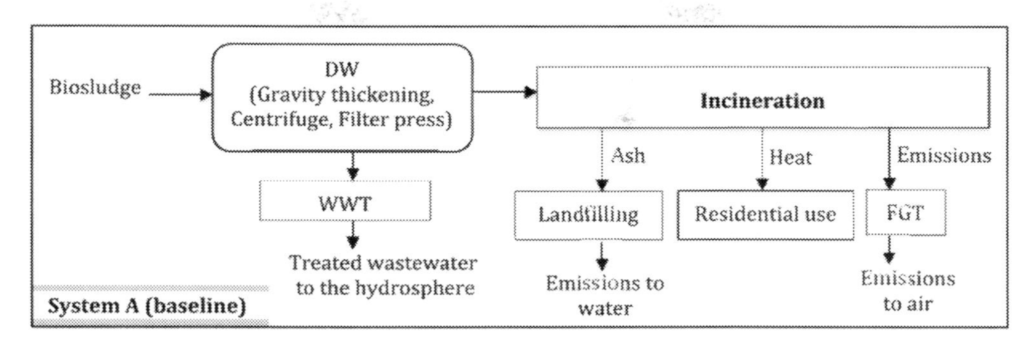

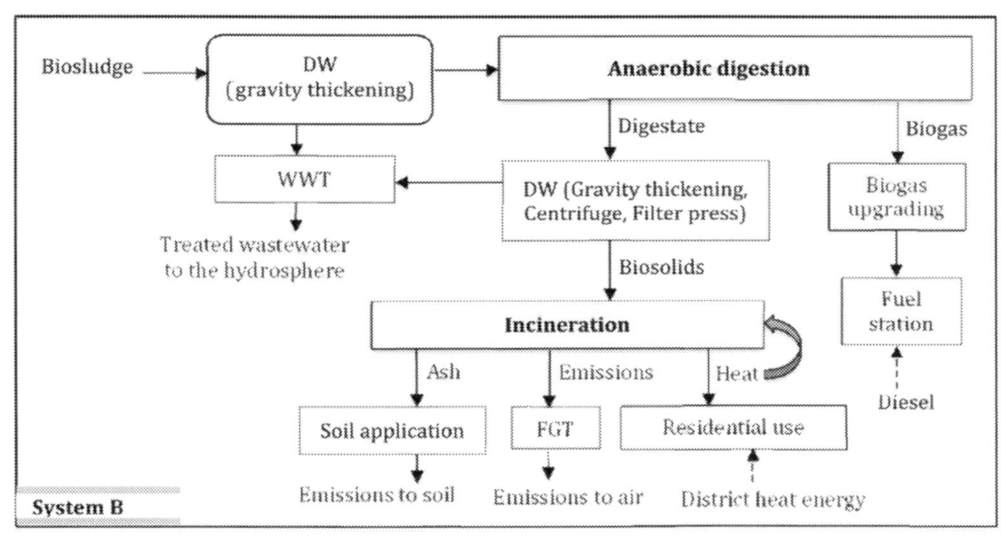

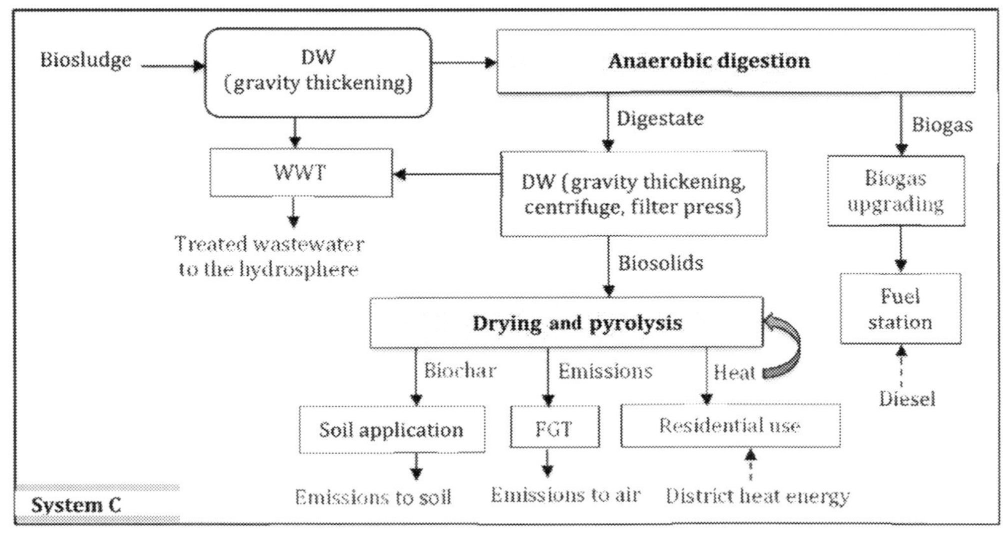

Figure 16.4 Schematic life cycle of (a) incineration, (b) anaerobic digestion (AD)-incineration, and (c) AD-pyrolysis. *DW*, Dewatering; *FGT*, flue gas treatment; *WWT*, wastewater treatment (Mohammadi et al., 2019).

the sludge and wastewater produced from the paper and pulp industry are still deteriorating the environment. The effluents coming out of the paper and pulp industry become a part of the environment causing serious threats to living organisms. The effluents become a part of the food chain, ultimately reaching out to humans, animals, and plants. The environmentally active compounds are a major cause of surface water body pollution (Khan et al., 2017; Söderholm, Bergquist, & Söderholm, 2019). Most of these contaminants have mutagenic and carcinogenic nature, which makes their presence questionable. Such pollutants cause various diseases and ailments in human beings, animals, plants (terrestrial and aquatic). One of the common problems caused by effluents is deoxygenation.

The accumulation of the fiber and wood chips causes a blanket's deposition on the sediment surface, thus inhibiting the oxygen supply to the benthic fauna, causing harmful effects (Khan, Badshah, & Airoldi, 2015; Mohammadi et al., 2019). The air emission of gases from the paper and pulp industry's operation leads to increased production of the greenhouse effect, smog, acid rain, and so on. These disastrous effects are ruining the environment. Similarly, the discharges of these industries into the water bodies make their way to the food chains and enter the human bodies causing serious ailments like lung diseases, allergies, kidney failures, and so on. The growth of plants and animals is also affected by such contaminants (Hussain et al., 2016; Khan, Badshah, & Airoldi, 2011; Ramírez-Malule, Quiñones-Murillo, & Manotas-Duque, 2020). In short, the paper and pulp industry is one of the major contributors to rising world pollution.

16.5 Treatment of paper and pulp industry contaminants

The gist of the whole discussion is that the paper and pulp industry has harmful effects on the environment. In order to minimize the harmful effects being caused by the paper and pulp industry, various steps have been taken (Deeba, Pruthi, & Negi, 2018; Khan, Khalil, Khan, Saeed, & Ali, 2016; Khan, Khan, et al., 2016). One such methodology is to use the primary treatment methods to prevent environmental hazards caused by paper and pulp industry effluents (Kong, Hasanbeigi, & Price, 2016; Rahim, Ullah, Khan, & Haris, 2016). In the primary treatment process, the suspended solids and waste products (chips, bark, gravel, chemicals, etc.) are removed from the wastewater by screening and using settling tanks (Ávila, Reyes, Bayona, & García, 2013; Khan, Wahid, Ali, Badshah, & Airoldi, 2015). Another form of wastewater treatment is the secondary treatment process. In the secondary treatment process, different types of bioreactors are used for the treatment of the wastewater like aerated lagoons (Li, Zheng, & Kelly, 2013), activated sludge (Villar-Navarro, Baena-Nogueras, Paniw, Perales, & Lara-Martín, 2018), biofilms (Mohamed El-hadi & Alamri, 2018), membrane bioreactors (Song et al., 2018), and so on (Krzeminski et al., 2019). Apart from these processes, some physicochemical approaches are also being employed for the removal of sludge-like coagulation/flocculation (Teh, Budiman, Shak, & Wu, 2016), sedimentation (Wen, Yi, & Zeng, 2016), reverse osmosis (Lopera, Ruiz, & Alonso, 2019), precipitation (Brandl, Bertrand, Lima, & Langer, 2015), sorption (Lucas et al., 2018), ozonation (Xiao, Xie, & Cao, 2015), advanced oxidation processes (Deng & Zhao, 2015), and so on. These methods have proved to be very helpful and promising for the removal of sludge and contaminants. Apart from these procedures, some biological treatments have also been employed to treat the wastewater produced from the paper and pulp industry. Bioremediation commonly includes treating fungi (Siddiquee et al., 2015) and bacteria (Yuan, Guo, Wei, & Yang, 2016). The mentioned procedures are very helpful for treating wastewater obtained from the paper and pulp industry on a massive scale.

16.5.1 Removal of contaminants through primary treatment

The most commonly employed technique for environmental remediation through removing paper and pulp industry contaminants is the primary treatment of the sludge. The sludge obtained from the paper and pulp industry mainly consists of chips, wood barks, plastics, paper, cloth, and so on. This sludge is treated with multiple devices in two steps (Rathi, Kumar, & Show, 2021). In the first step, preliminary treatment of the sludge is performed, followed by the primary sedimentation. The primary treatment principle is to slow down the flow of wastewater up to the velocity of 1−2 feet/min. This will cause the denser sludge material to settle down at the bottom, while water will float at the upper surface. The commonly present inorganic and organic contaminants of the industrial wastewater can efficiently be removed by the primary treatment (Rivera-Utrilla, Sánchez-Polo, Ferro-García, Prados-Joya, & Ocampo-Pérez, 2013). Boulaadjoul, Zemmouri, Bendjama, and Drouiche (2018) performed the primary treatment of the industrial sludge obtained from the paper and pulp industry. The operation was carried out in the presence of *Moringa oleifera* and alum. The results obtained showed good efficiency in removing contaminants with a turbidity reduction of 96.02% and 97.1% for *Moringa* and alum, respectively. While the COD removal obtained for *Moringa* and alum was 97.28% and 92.67%, respectively.

16.5.2 Removal of contaminants through secondary treatment

At times, primary treatment is not enough to remove contaminants and causes the production of secondary sludge. The secondary sludge can be treated by performing secondary treatment of the wastewater. The secondary treatment relies on utilizing different bioreactors to treat wastewater (Behera, Kim, Oh, & Park, 2011). As mentioned above, the most commonly utilized form of paper and pulp industrial sludge is based on the activated sludge process among different types of bioreactors. The activated sludge process consists of two steps, which are aeration and sedimentation. In the first step, the wastewater is aerated along with the treatment of microorganisms at high concentration. The retention time of the aeration period may vary according to the level of contamination. The aerated water is then sent to the sedimentation section, which separates the sludge onto the sediment basin and sent it back to the aeration zone. The activated sludge-based secondary treatment of wastewater has proved to be a satisfactory method for reducing the COD of the wastewater (Katsou, Malamis, & Loizidou, 2011). Sarkar et al. (2017) analyzed the radiotracer investigation carried out in an industrial sludge process. The purpose of the study was to analyze the mean hydraulic retention time. In the performed study, Iodine-131 was used as a radiotracer. The computer modeling and studies were carried out to check the efficiency of the activated sludge process. Poojamnong, Tungsudjawong, Khongnakorn, and Jutaporn (2020) also studied the efficiency of the membrane bioreactor for the treatment of wastewater obtained from the paper and pulp industry. The effect of the membrane fouling was also analyzed while performing the studies. The results showed a satisfactory response for the removal of industrial effluents with 100% recovery.

16.5.3 Removal of contaminants through coagulation/flocculation/sedimentation

The inefficiency of the primary and secondary treatment of the wastewater has led to the development of physicochemical processes with better efficiency in removing contaminants (Zodi et al., 2011). One of such processes is coagulation or flocculation, which makes use of different coagulants/flocculants for the removal of contaminants. Normally, Aluminum chloride is used as a coagulant for contaminants removal, while modified natural polymer (starch-g-PAM-gPDMC) is used as a flocculent to treat the wastewater obtained after primary or secondary treatment. Similarly, sedimentation and flotation strategies have also been followed for the removal of contaminants. But the efficiency of these processes is lowered due to the high costs and production of secondary pollutants (Akcil, Erust, Ozdemiroglu, Fonti, & Beolchini, 2015). Hugar and Marol (2020) studied electrocoagulation using iron electrodes for the treatment of wastewater obtained from the paper and pulp industry. The operating parameters like voltage, time, and so on, were optimized, and the results showed excellent efficiency. The turbidity and COD removal values obtained for the performed procedure were noted to be 65% and 70%, respectively. Gao, Sun, Kong, Liu, and Fatehi (2020) also studied the treatment of spent liquor from the pulping process by sedimentation. The lignin and hemicelluloses from the spent liquor were removed by treating with poly diallyl dimethylammonium chloride. The results showed an excellent efficiency of the process for the removal of industrial contaminants. The maximum removal efficiency of 57% for lignin and 36% for hemicelluloses was obtained.

16.5.4 Sorption/membrane-based removal of contaminants

A little more advanced and promising step towards removing contaminants was made with the development of membranes for the removal of contaminants. The membrane-based removal of wastewater contaminants has proved to help improve wastewater quality with the minimum intake of sources. The membrane filtration has been quite helpful for the removal of large molecular-sized organic contaminants as well as inorganic substances (Mohmood et al., 2013). The most commonly used membrane filtration techniques for removing industrial contaminants include reverse osmosis (Yangali-Quintanilla et al., 2011) and ultra-filtration techniques (Duan et al., 2017). A more advanced, low cost, simple, and feasible approach for the removal of contaminants has led to the exploitation of sorption for the removal of contaminants. The sorption has greatly been employed to remove dyes, chemicals, organic and inorganic contaminants, and control the COD and BOD. A number of natural and synthetic substances have been used as sorbents depending upon the needs and availability of the resources. The natural substances used include clays (Yadav, Gadi, & Kalra, 2019), chitin (Krupadam, Sridevi, & Sakunthala, 2011), chitosan (Vidal & Moraes, 2019), lignin (Naseer et al., 2019), cellulose (Goetz, Naseri, Nair, Karim, & Mathew, 2018), fly ash (Olabemiwo et al., 2017), and so on, while the synthetic sorbents include activated carbon (Goher et al., 2015), zeolites (Meng, Chen, Lin, Lin, & Sun, 2017), metal oxides (Gusain, Gupta, Joshi, & Khatri, 2019), etc. The sorptive removal of contaminants has emerged as one of the most promising techniques for the contaminant's removal based upon its specificity and simplicity (Santos, Ungureanu, Boaventura, & Botelho, 2015). Chen et al. (2021) utilized the poly (vinylidene fluoride) ultrafiltration membrane for the treatment of wastewater obtained from the paper and pulp industry. The phenomenon of fouling was also analyzed while performing the study. The fouling of the membrane was released using various foulants and the excellent efficiency of the membrane was retrieved. Houa et al. (2020) performed the sorption of contaminants in the wastewater obtained from the paper and pulp industry onto the modified activated carbon. The sorption process was further strengthened by microwave irradiation. The results showed an excellent efficiency of the process with 79.7% COD removal. The obtained results were further analyzed by applying the sorption isotherms.

16.5.5 Advanced oxidation processes and ozonation

The advancement in technology and the higher need for the removal of contaminants from industrial wastewater have led to the development of more advanced techniques. One of the advanced approaches is the ozonation of contaminants by exploiting the oxidizing nature of ozone. The ozone is thus used for oxidizing the electron-rich substances (both organic and inorganic contaminants), most with carbon double bonds or aromatic alcohols (Gomes, Costa, Quinta-Ferreira, & Martins, 2017). The ozone also causes the production of hydroxyl radicals. The ozone treatment converts the contaminants into biodegradable substances, which can further be treated through the biological process for complete vanishing. The paper and pulp contaminants have also significantly been removed by this process. Another approach for contaminants removal is the advanced oxidation processes (Barrera-Díaz, Canizares, Fernández, Natividad, & Rodrigo, 2014). The advanced oxidation processes in the generation of strong oxidizing agents like hydroxyl radicals oxidizes complex contaminants into simpler ones. The commonly followed advanced oxidation processes include Fenton's reagent UV/H_2O_2 treatment, H_2O_2 treatment, photochemical processes, etc. (Brillas, 2020). Sethupathy, Arun, Sivashanmugam, and Kumar (2020) also performed the biosolids' treatment obtained from the paper and pulp industry through the ozonation combined with hydrolytic enzymes. Various factors like ozone dosage, pretreatment time, and so on, were optimized and satisfactory results were obtained with a methane gas production potential of 324 L/kg VS. Can-Güven, Guvenc, Kavan, and Varank (2021) studied the COD and turbidity removal of wastewater obtained from the paper and pulp industry by the Fe^{2+}/heat-activated persulfate oxidation process. Various operating conditions (pH, reaction time, PS/COD, Fe^{2+} concentration for Fe^{2+}-activated PS, and temperature) were optimized for obtaining better results. The results showed 56.1 and 98.9% COD and turbidity removal for the effluents in the wastewater.

16.5.6 Bioremediation of wastewater from the paper and pulp industry

The chemical and physical treatment methods are more costly and technical, which require a high energy intake. Such treatment methods can be replaced by biological methods, which are more environmentally friendly and green approach. The biological methods use plants/microbes or enzymes to remove contaminants from the wastewater obtained from the paper and pulp industry (Paździor, Bilińska, & Ledakowicz, 2019). The most commonly employed sources for the treatment of contaminants are fungi and bacteria. The fungi commonly produce extracellular ligninolytic enzymes (laccase, MnP, and LiP), which cause the degradation of the contaminants. Similarly, bacterial treatment of the contaminants has also proved to be a promising process for the removal of contaminants including various species of bacteria like *Pseudomonas putida*, *Micrococcus luteus*, *Pseudomonas aeruginosa*, *Bacillus subtilis*, *Bacillus cereus*, *Acinetobacter calcoaceticus*, *Ancylobacter*, *Methylobacterium*, and so on, have been utilized for treating the wastewater bodies (Popat, Nidheesh, Singh, & Kumar, 2019). Sen, Raut, Gaur, and Raut (2020) analyzed the biodegradation of lignin present in the wastewater obtained from the paper and pulp industry wastewater. The bacterial consortia were used as the biological agent for the removal of lignin at a higher rate. At optimized operating conditions, lignin biodegradation efficiency of 94.83% was obtained. Costa et al. (2017) performed the biodegradation of lignin from the paper and pulp industry using white-rot fungi *B. adusta* and *P. crysosporium*. The operating conditions were optimized, and lignin degradation of 97% and 74% was obtained by *B. adusta* and *P. crysosporium*, respectively (Table 16.1),

Table 16.1 Paper and pulp industrial contaminants removal by different routes.

Treatment process	Source		Contaminants removal efficiency			References
			COD/BOD	Color	Other suspended compounds	
Tertiary treatment	Wastewater from paper and pulp industry		80%	80%		Zhang, Wang, Gong, and Shen (2017)
Physicochemical treatment	Effluents from paper mill		84%	89%	90%	Mehmood et al. (2019)
Physicochemical and biological processes	Paper mill contaminants		99.81%			Liang et al. (2021)
Electrocoagulation	Pulp and paper industry effluents		82%	99%		Kumar and Sharma (2019)
Fenton oxidation	Organic compounds from Paper and pulp mill	Recycle mill effluent	63%			Brink, Sheridan, and Harding (2017)
		Neutral sulfite semichemical mill effluent	44%			
Phytoremediation	Paper and pulp wastewater		66.1%	55.8%	87.2%	Yusoff, Rozaimah, Hassimi, Hawati, and Habibah (2019)
Electrocoagulation	Paper-recycling wastewater		79.5%	98.5%	83.4%	Izadi, Hosseini, Darzi, Bidhendi, and Shariati (2018)
Enhanced primary treatment	Paper and pulp industry contaminants	*M. oleifera* seeds	97.28%			Boulaadjoul et al. (2018)
		Alum	92.67%			
Biostimulation	Effluents from paper and pulp industry					Chandra, Sharma, Yadav, and Tripathi (2018)

(Continued)

Table 16.1 (Continued)

Treatment process	Source	Contaminants removal efficiency			References
		COD/BOD	Color	Other suspended compounds	
Phytoremediation	Sludge from paper and pulp industry	12301 $\pm$ 402.97			Sharma et al. (2021a, 2021b)
Advanced treatment	Pulp and paper secondary wastewater	84.3%		93%	Chu, Wang, and Liu (2016)
Biological treatment	Paper and pulp industrial wastewater	81%/71%		65%	Mazhar et al. (2019)
Physico-chemical analysis	Paper and pulp industrial wastewater				Yadav and Chandra (2018)
Electrocoagulation	Paper and pulp industrial wastewater	85%			Azadi Aghdam, Kariminia, and Safari (2016)
Advanced oxidation process	Paper and pulp industry contaminants	74.8%	58.3%		Abedinzadeh, Shariat, Monavari, and Pendashteh (2018)
Aerobic granular sludge technology	Lignin and tannin from paper and pulp industry	90%			Vashi, Iorhemen, and Tay (2018)
Electrocoagulation and adsorption	Wastewater from paper and pulp industry			93%	Barhoumi et al. (2019)

Abbreviations: BOD, biological oxygen demand; COD, chemical oxygen demand.

16.6 Conclusion

The industrial sector worldwide is constantly growing to cope with the rising population and their increasing demands. Among many industries working at massive scales, one is the paper and pulp industry. The paper and pulp industry is considered a huge industry due to the high demand for paper. As the paper and pulp industry's production rate is very high, so is the byproduct released in different forms. The production of contaminants from the paper and pulp industry has a negative impact on the surrounding environment. The beauty of nature is being ruined with deforestation and ultimate effluents released from the paper and pulp industry. Many researchers are taking steps towards developing promising techniques for environmental remediation from the hazardous effects caused by the paper and pulp industry. The chapter covers the impact of the paper and pulp industry on the environment and the steps being taken for its remediation.

Acknowledgement

Consejo Nacional de Ciencia y Tecnología (MX) is thankfully acknowledged for partially supporting this work under Sistema Nacional de Investigadores (SNI) program awarded to Hafiz M.N. Iqbal (CVU: 735340).

Conflict of interests

The author(s) declare no conflicting interests.

References

Abedinzadeh, N., Shariat, M., Monavari, S. M., & Pendashteh, A. (2018). Evaluation of color and COD removal by Fenton from biologically (SBR) pre-treated pulp and paper wastewater. *Process Safety and Environmental Protection*, *116*, 82–91.

Akcil, A., Erust, C., Ozdemiroglu, S., Fonti, V., & Beolchini, F. (2015). A review of approaches and techniques used in aquatic contaminated sediments: metal removal and stabilization by chemical and biotechnological processes. *Journal of Cleaner Production*, *86*, 24–36.

Ali, N., Ahmad, S., Khan, A., Khan, S., Bilal, M., Ud Din, S., ... Khan, H. (2020). Selenide-chitosan as high-performance nanophotocatalyst for accelerated degradation of pollutants. *Chemistry—An Asian Journal*, *15*(17), 2660–2673.

Ali, N., Azeem, S., Khan, A., Khan, H., Kamal, T., & Asiri, A. M. (2020). Experimental studies on removal of arsenites from industrial effluents using tridodecylamine supported liquid membrane. *Environmental Science and Pollution Research*, 1–12.

Ali, N., Bilal, M., Khan, A., Ali, F., & Iqbal, H. M. (2020). Design, engineering and analytical perspectives of membrane materials with smart surfaces for efficient oil/water separation. *TrAC Trends in Analytical Chemistry*, 115902.

Ali, N., Bilal, M., Khan, A., Ali, F., Yang, Y., Khan, M., ... Iqbal, H. M. (2020). Dynamics of oil-water interface demulsification using multifunctional magnetic hybrid and assembly materials. *Journal of Molecular Liquids*, 113434.

Ali, N., Bilal, M., Khan, A., Ali, F., Yang, Y., Malik, S., ... Iqbal, H. M. (2021). Deployment of metal-organic frameworks as robust materials for sustainable catalysis and remediation of pollutants in environmental settings. *Chemosphere*, 129605.

Ali, N., Bilal, M., Nazir, M. S., Khan, A., Ali, F., & Iqbal, H. M. (2020). Thermochemical and electrochemical aspects of carbon dioxide methanation: A sustainable approach to generate fuel via waste to energy theme. *Science of The Total Environment*, *712*, 136482.

Ali, N., Ismail, M., Khan, A., Khan, H., Haider, S., & Kamal, T. (2018). Spectrophotometric methods for the determination of urea in real samples using silver nanoparticles by standard addition and 2nd order derivative methods. *Spectrochimica Acta Part A: Molecular and Biomolecular Spectroscopy*, *189*, 110–115.

Ali, N., Kamal, T., Ul-Islam, M., Khan, A., Shah, S. J., & Zada, A. (2018). Chitosan-coated cotton cloth supported copper nanoparticles for toxic dye reduction. *International Journal of Biological Macromolecules*, *111*, 832–838.

Ali, N., Khan, A., Bilal, M., Malik, S., Badshah, S., & Iqbal, H. (2020). Chitosan-based bio-composite modified with thiocarbamate moiety for decontamination of cations from the aqueous media. *Molecules*, *25*(1), 226.

Ali, N., Khan, A., Malik, S., Badshah, S., Bilal, M., & Iqbal, H. M. (2020). Chitosan-based green sorbent material for cations removal from an aqueous environment. *Journal of Environmental Chemical Engineering*, *8*(5), 104064.

Ali, N., Khan, A., Nawaz, S., Bilal, M., Malik, S., Badshah, S., & Iqbal, H. M. (2020). Characterization and deployment of surface-engineered chitosan-triethylenetetramine nanocomposite hybrid nano-adsorbent for divalent cations decontamination. *International Journal of Biological Macromolecules*, *152*, 663–671.

Ali, N., Naz, N., Shah, Z., Khan, A., & Nawaz, R. (2020). Selective transportation of molybdenum from model and ore through poly inclusion membrane. *Bulletin of the Chemical Society of Ethiopia*, *34*(1), 93–104.

Ali, N., Uddin, S., Khan, A., Khan, S., Khan, S., Ali, N., ... Bilal, M. (2020). Regenerable chitosan-bismuth cobalt selenide hybrid microspheres for mitigation of organic pollutants in an aqueous environment. *International Journal of Biological Macromolecules*, *161*, 1305–1317.

Ali, N., Zada, A., Zahid, M., Ismail, A., Rafiq, M., Riaz, A., & Khan, A. (2019). Enhanced photodegradation of methylene blue with alkaline and transition-metal ferrite nanophotocatalysts under direct sun light irradiation. *Journal of the Chinese Chemical Society*, *66*(4), 402–408.

Ávila, C., Reyes, C., Bayona, J. M., & García, J. (2013). Emerging organic contaminant removal depending on primary treatment and operational strategy in horizontal subsurface flow constructed wetlands: influence of redox. *Water Research*, *47*(1), 315–325.

Azadi Aghdam, M., Kariminia, H. R., & Safari, S. (2016). Removal of lignin, COD, and color from pulp and paper wastewater using electrocoagulation. *Desalination and Water Treatment*, *57*(21), 9698–9704.

Azevedo, A. R. G., Marvila, T. M., Fernandes, W. J., Alexandre, J., Xavier, G. C., Zanelato, E. B., ... Mendes, Bc (2019). Assessing the potential of sludge generated by the pulp and paper industry in assembling locking blocks. *Journal of Building Engineering*, *23*, 334–340.

Aziz, A., Ali, N., Khan, A., Bilal, M., Malik, S., Ali, N., & Khan, H. (2020). Chitosan-zinc sulfide nanoparticles, characterization and their photocatalytic degradation efficiency for azo dyes. *International Journal of Biological Macromolecules*, *15*(153), 502–512.

Barhoumi, A., Ncib, S., Chibani, A., Brahmi, K., Bouguerra, W., & Elaloui, E. (2019). High-rate humic acid removal from cellulose and paper industry wastewater by combining electrocoagulation process with adsorption onto granular activated carbon. *Industrial Crops and Products*, *140*, 111715.

Barrera-Díaz, C., Canizares, P., Fernández, F. J., Natividad, R., & Rodrigo, M. A. (2014). Electrochemical advanced oxidation processes: an overview of the current applications to actual industrial effluents. *Journal of the Mexican Chemical Society*, *58*(3), 256–275.

Beckline, M., Yujun, S., Eric, Z., & Kato, M. S. (2016). Paper consumption and environmental impact in an emerging economy. *Journal of Energy, Environment and Chemical Engineering*, *1*(1), 13–18.

Behera, S. K., Kim, H. W., Oh, J. E., & Park, H. S. (2011). Occurrence and removal of antibiotics, hormones and several other pharmaceuticals in wastewater treatment plants of the largest industrial city of Korea. *Science of the Total Environment*, *409*(20), 4351–4360.

Boulaadjoul, S., Zemmouri, H., Bendjama, Z., & Drouiche, N. (2018). A novel use of *Moringa oleifera* seed powder in enhancing the primary treatment of paper mill effluent. *Chemosphere*, *206*, 142–149.

Branco, R. H., Serafim, L. S., & Xavier, A. M. (2019). Second generation bioethanol production: on the use of pulp and paper industry wastes as feedstock. *Fermentation*, *5*(1), 4.

Brandl, F., Bertrand, N., Lima, E. M., & Langer, R. (2015). Nanoparticles with photoinduced precipitation for the extraction of pollutants from water and soil. *Nature Communications*, *6*(1), 1−10.

Brillas, E. (2020). A review on the photoelectro-Fenton process as efficient electrochemical advanced oxidation for wastewater remediation. Treatment with UV light, sunlight, and coupling with conventional and other photo-assisted advanced technologies. *Chemosphere*, *250*, 126198.

Brink, A., Sheridan, C., & Harding, K. (2018). Combined biological and advance oxidation processes for paper and pulp effluent treatment. *South African Journal of Chemical Engineering*, *25*, 116−122.

Brink, A., Sheridan, C. M., & Harding, K. G. (2017). The Fenton oxidation of biologically treated paper and pulp mill effluents: a performance and kinetic study. *Process Safety and Environmental Protection*, *107*, 206−215.

CB, S., & Babu, B.R. (2018). Studies on electrochemical treatment of pulp and paper mill waste water. In *Proceedings of First International Conference on Energy and Environment: Global Challenges*.

Can-Güven, E., Guvenc, S. Y., Kavan, N., & Varank, G. (2021). Paper mill wastewater treatment by Fe2 + and heat-activated persulfate oxidation: Process modeling and optimization. *Environmental Progress & Sustainable Energy*, *40*(2), e13508.

Chakraborty, D., Shelvapulle, S., Reddy, K. R., Kulkarni, R. V., Puttaiahgowda, Y. M., Naveen, S., & Raghu, A. V. (2019). Integration of biological pre-treatment methods for increased energy recovery from paper and pulp biosludge. *Journal of Microbiological Methods*, *160*, 93−100.

Chandra, R., Sharma, P., Yadav, S., & Tripathi, S. (2018). Biodegradation of endocrine-disrupting chemicals and residual organic pollutants of pulp and paper mill effluent by biostimulation. *Frontiers in microbiology*, *9*, 960.

Chen, M., Ding, W., Zhou, M., Zhang, H., Ge, C., Cui, Z., & Xing, W. (2021). Fouling mechanism of PVDF ultrafiltration membrane for secondary effluent treatment from paper mills. *Chemical Engineering Research and Design*, *167*, 37−45.

Chu, H., Wang, Z., & Liu, Y. (2016). Application of modified bentonite granulated electrodes for advanced treatment of pulp and paper mill wastewater in three-dimensional electrode system. *Journal of Environmental Chemical Engineering*, *4*(2), 1810−1817.

Costa, S., Dedola, D. G., Pellizzari, S., Blo, R., Rugiero, I., Pedrini, P., & Tamburini, E. (2017). Lignin biodegradation in pulp-and-paper mill wastewater by selected white rot fungi. *Water*, *9*(12), 935.

de Azevedo, A. R., Alexandre, J., Pessanha, L. S. P., da ST Manhães, R., de Brito, J., & Marvila, M. T. (2019). Characterizing the paper industry sludge for environmentally-safe disposal. *Waste Management*, *95*, 43−52.

Deeba, F., Pruthi, V., & Negi, Y. S. (2018). *Effect of emerging contaminants from paper mill industry into the environment and their control. Environmental Contaminants* (pp. 391−408). Singapore: Springer.

Deng, Y., & Zhao, R. (2015). Advanced oxidation processes (AOPs) in wastewater treatment. *Current Pollution Reports*, *1*(3), 167−176.

Didone, M., Saxena, P., Brilhuis-Meijer, E., Tosello, G., Bissacco, G., Mcaloone, T. C., ... Howard, T. J. (2017). Moulded pulp manufacturing: overview and prospects for the process technology. *Packaging Technology and Science*, *30*(6), 231−249.

Duan, W., Chen, G., Chen, C., Sanghvi, R., Iddya, A., Walker, S., ... Jassby, D. (2017). Electrochemical removal of hexavalent chromium using electrically conducting carbon nanotube/polymer composite ultrafiltration membranes. *Journal of Membrane Science*, *531*, 160−171.

Elakkiya, E., & Niju, S. (2020). Application of microbial fuel cells for treatment of paper and pulp industry wastewater: opportunities and challenges. *Environmental Biotechnology*, *2*, 125−149.

Gao, W., Sun, Y., Kong, F., Liu, Z., & Fatehi, P. (2020). Aggregation and sedimentation performance of lignin and hemicellulose derived flocs in the spent liquor of thermomechanical pulping process. *Waste and Biomass Valorization*, 1−14.

Goetz, L. A., Naseri, N., Nair, S. S., Karim, Z., & Mathew, A. P. (2018). All cellulose electrospun water purification membranes nanotextured using cellulose nanocrystals. *Cellulose*, *25*(5), 3011−3023.

Goher, M. E., Hassan, A. M., Abdel-Moniem, I. A., Fahmy, A. H., Abdo, M. H., & El-sayed, S. M. (2015). Removal of aluminum, iron and manganese ions from industrial wastes using granular activated carbon and Amberlite IR-120H. *The Egyptian Journal of Aquatic Research*, *41*(2), 155–164.

Gomes, J., Costa, R., Quinta-Ferreira, R. M., & Martins, R. C. (2017). Application of ozonation for pharmaceuticals and personal care products removal from water. *Science of the Total Environment*, *586*, 265–283.

Gonzalo, A., Bimbela, F., Sánchez, J. L., Labidi, J., Marín, F., & Arauzo, J. (2017). Evaluation of different agricultural residues as raw materials for pulp and paper production using a semichemical process. *Journal of Cleaner Production*, *156*, 184–193.

Gupta, G. K., & Shukla, P. (2020). Insights into the resources generation from pulp and paper industry wastes: challenges, perspectives and innovations. *Bioresource Technology*, *297*, 122496.

Gusain, R., Gupta, K., Joshi, P., & Khatri, O. P. (2019). Adsorptive removal and photocatalytic degradation of organic pollutants using metal oxides and their composites: A comprehensive review. *Advances in Colloid and Interface Science*, *272*, 102009.

Houa, R., Lia, H., Chena, H., Yuana, R., Wanga, F., Chenb, Z., & Zhoua, B. (2020). Tertiary treatment of biologically treated effluents from pulp and paper industry by microwave modified activated carbon adsorption. *Desalination and Water Treatment*, *182*, 118–126.

Hugar, G. M., & Marol, C. K. (2020). Feasibility of electro coagulation using iron electrodes in treating paper industry wastewater. *Sustainable Water Resources Management*, *6*(4), 1–10.

Hussain, S., Ullah, Z., Gul, S., Khattak, R., Kazmi, N., Rehman, F., … Khan, A. (2016). Adsorption characteristics of magnesium-modified bentonite clay with respect to acid blue 129 in aqueous media. *Polish Journal of Environmental Studies*, *25*(5), 1947–1953.

Izadi, A., Hosseini, M., Darzi, G. N., Bidhendi, G. N., & Shariati, F. P. (2018). Treatment of paper-recycling wastewater by electrocoagulation using aluminum and iron electrodes. *Journal of Environmental Health Science and Engineering*, *16*(2), 257–264.

Katsou, E., Malamis, S., & Loizidou, M. (2011). Performance of a membrane bioreactor used for the treatment of wastewater contaminated with heavy metals. *Bioresource Technology*, *102*(6), 4325–4332.

Khan, A., Ali, N., Bilal, M., Malik, S., Badshah, S., & Iqbal, H. (2019). Engineering functionalized chitosan-based sorbent material: characterization and sorption of toxic elements. *Applied Sciences*, *9*(23), 5138.

Khan, A., Badshah, S., & Airoldi, C. (2011). Biosorption of some toxic metal ions by chitosan modified with glycidylmethacrylate and diethylenetriamine. *Chemical Engineering Journal*, *171*(1), 159–166.

Khan, A., Badshah, S., & Airoldi, C. (2015). Environmentally benign modified biodegradable chitosan for cation removal. *Polymer Bulletin*, *72*(2), 353–370.

Khan, A., Begum, S., Ali, N., Khan, S., Hussain, S., & Sotomayor, M. D. P. T. (2017). Preparation of crosslinked chitosan magnetic membrane for cations sorption from aqueous solution. *Water Science and Technology*, *75*(9), 2034–2046.

Khan, A., Shah, S. J., Mehmood, K., Ali, N., & Khan, H. (2019). Synthesis of potent chitosan beads a suitable alternative for textile dye reduction in sunlight. *Journal of Materials Science: Materials in Electronics*, *30*(1), 406–414.

Khan, A., Wahid, F., Ali, N., Badshah, S., & Airoldi, C. (2015). Single-step modification of chitosan for toxic cations remediation from aqueous solution. *Desalination and Water Treatment*, *56*(4), 1099–1109.

Khan, H., Gul, K., Ara, B., Khan, A., Ali, N., Ali, N., & Bilal, M. (2020). Adsorptive removal of acrylic acid from the aqueous environment using raw and chemically modified alumina: Batch adsorption, kinetic, equilibrium and thermodynamic studies. *Journal of Environmental Chemical Engineering*, 103927.

Khan, H., Khalil, A. K., & Khan, A. (2019). Photocatalytic degradation of alizarin yellow in aqueous medium and real samples using chitosan conjugated tin magnetic nanocomposites. *Journal of Materials Science: Materials in Electronics*, *30*(24), 21332–21342.

Khan, H., Khalil, A. K., Khan, A., Saeed, K., & Ali, N. (2016). Photocatalytic degradation of bromophenol blue in aqueous medium using chitosan conjugated magnetic nanoparticles. *Korean Journal of Chemical Engineering*, *33*(10), 2802–2807.

Khan, S., Khan, A., Ali, N., Ahmad, S., Ahmad, W., Malik, S., … Bilal, M. (2021). Degradation of carcinogenic Congo red dye using ternary metal selenide-chitosan microspheres as robust and reusable catalysts. *Environmental Technology & Innovation*, 101402.

Khan, S. U., Khan, F. U., Khan, I. U., Muhammad, N., Badshah, S., Khan, A., ... Nasrullah, A. (2016). Biosorption of nickel (II) and copper (II) ions from aqueous solution using novel biomass derived from Nannorrhops ritchiana (Mazri Palm). *Desalination and Water Treatment*, *57*(9), 3964–3974.

Kong, L., Hasanbeigi, A., & Price, L. (2016). Assessment of emerging energy-efficiency technologies for the pulp and paper industry: a technical review. *Journal of Cleaner Production*, *122*, 5–28.

Krupadam, R. J., Sridevi, P., & Sakunthala, S. (2011). Removal of endocrine disrupting chemicals from contaminated industrial groundwater using chitin as a biosorbent. *Journal of Chemical Technology & Biotechnology*, *86*(3), 367–374.

Krzeminski, P., Tomei, M. C., Karaolia, P., Langenhoff, A., Almeida, C. M. R., Felis, E., ... Fatta-Kassinos, D. (2019). Performance of secondary wastewater treatment methods for the removal of contaminants of emerging concern implicated in crop uptake and antibiotic resistance spread: A review. *Science of the Total Environment*, *648*, 1052–1081.

Kumar, D., & Sharma, C. (2019). Remediation of pulp and paper industry effluent using electrocoagulation process. *Journal of Water Resource and Protection*, *11*(03), 296.

Kumar, H., & Christopher, L. P. (2017). Recent trends and developments in dissolving pulp production and application. *Cellulose*, *24*(6), 2347–2365.

Li, H., Legere, S., He, Z., Zhang, H., Li, J., Yang, B., ... Ni, Y. (2018). Methods to increase the reactivity of dissolving pulp in the viscose rayon production process: a review. *Cellulose*, *25*(7), 3733–3753.

Li, X., Zheng, W., & Kelly, W. R. (2013). Occurrence and removal of pharmaceutical and hormone contaminants in rural wastewater treatment lagoons. *Science of the Total Environment*, *445*, 22–28.

Liang, J., Mai, W., Wang, J., Li, X., Su, M., Du, J., ... Wei, Y. (2021). Performance and microbial communities of a novel integrated industrial-scale pulp and paper wastewater treatment plant. *Journal of Cleaner Production*, *278*, 123896.

Lindholm-Lehto, P. C., Knuutinen, J. S., Ahkola, H. S., & Herve, S. H. (2015). Refractory organic pollutants and toxicity in pulp and paper mill wastewaters. *Environmental Science and Pollution Research*, *22*(9), 6473–6499.

Lopera, A. E. C., Ruiz, S. G., & Alonso, J. M. Q. (2019). Removal of emerging contaminants from wastewater using reverse osmosis for its subsequent reuse: pilot plant. *Journal of Water Process Engineering*, *29*, 100800.

Lucas, D., Castellet-Rovira, F., Villagrasa, M., Badia-Fabregat, M., Barceló, D., Vicent, T., ... Rodríguez-Mozaz, S. (2018). The role of sorption processes in the removal of pharmaceuticals by fungal treatment of wastewater. *Science of the Total Environment*, *610*, 1147–1153.

Ma, X., Zhai, Y., Zhang, R., Shen, X., Zhang, T., Ji, C., ... Hong, J. (2019). Energy and carbon coupled water footprint analysis for straw pulp paper production. *Journal of Cleaner Production*, *233*, 23–32.

Mazhar, S., Ditta, A., Bulgariu, L., Ahmad, I., Ahmed, M., & Nadiri, A. A. (2019). Sequential treatment of paper and pulp industrial wastewater: Prediction of water quality parameters by Mamdani Fuzzy Logic model and phytotoxicity assessment. *Chemosphere*, *227*, 256–268.

Mehmood, K., Rehman, S. K. U., Wang, J., Farooq, F., Mahmood, Q., Jadoon, A. M., ... Ahmad, I. (2019). Treatment of pulp and paper industrial effluent using physicochemical process for recycling. *Water*, *11*(11), 2393.

Meng, Q., Chen, H., Lin, J., Lin, Z., & Sun, J. (2017). Zeolite A synthesized from alkaline assisted pre-activated halloysite for efficient heavy metal removal in polluted river water and industrial wastewater. *Journal of Environmental Sciences*, *56*, 254–262.

Mohamed El-hadi, A., & Alamri, H. R. (2018). The new generation from biomembrane with green technologies for wastewater treatment. *Polymers*, *10*(10), 1174.

Mohammadi, A., Sandberg, M., Venkatesh, G., Eskandari, S., Dalgaard, T., Joseph, S., & Granström, K. (2019). Environmental performance of end-of-life handling alternatives for paper-and-pulp-mill sludge: Using digestate as a source of energy or for biochar production. *Energy*, *182*, 594–605.

Mohmood, I., Lopes, C. B., Lopes, I., Ahmad, I., Duarte, A. C., & Pereira, E. (2013). Nanoscale materials and their use in water contaminants removal—a review. *Environmental Science and Pollution Research*, *20*(3), 1239–1260.

Naseer, A., Jamshaid, A., Hamid, A., Muhammad, N., Ghauri, M., Iqbal, J., ... Shah, N. S. (2019). Lignin and lignin based materials for the removal of heavy metals from waste water-an overview. *Zeitschrift für Physikalische Chemie*, *233*(3), 315–345.

Nawaz, A., Khan, A., Ali, N., Ali, N., & Bilal, M. (2020). Fabrication and characterization of new ternary ferrites-chitosan nanocomposite for solar-light driven photocatalytic degradation of a model textile dye. *Environmental Technology & Innovation, 20*, 101079.

Neelofar, N., Ali, N., Khan, A., Amir, S., Khan, N. A., & Bilal, M. (2017). Synthesis of Schiff bases derived from 2-hydroxy-1-naphth-aldehyde and their tin (II) complexes for antimicrobial and antioxidant activities. *Bulletin of the Chemical Society of Ethiopia, 31*(3), 445–456.

Novais, R. M., Carvalheiras, J., Senff, L., & Labrincha, J. A. (2018). Upcycling unexplored dregs and biomass fly ash from the paper and pulp industry in the production of eco-friendly geopolymer mortars: A preliminary assessment. *Construction and Building Materials, 184*, 464–472.

Olabemiwo, F. A., Tawabini, B. S., Patel, F., Oyehan, T. A., Khaled, M., & Laoui, T. (2017). Cadmium removal from contaminated water using polyelectrolyte-coated industrial waste fly ash. *Bioinorganic Chemistry and Applications, 2017*, 7298351.

Paździor, K., Bilińska, L., & Ledakowicz, S. (2019). A review of the existing and emerging technologies in the combination of AOPs and biological processes in industrial textile wastewater treatment. *Chemical Engineering Journal, 376*, 120597.

Poojamnong, K., Tungsudjawong, K., Khongnakorn, W., & Jutaporn, P. (2020). Characterization of reversible and irreversible foulants in membrane bioreactor (MBR) for eucalyptus pulp and paper mill wastewater treatment using fluorescence regional integration. *Journal of Environmental Chemical Engineering, 8*(5), 104231.

Popat, A., Nidheesh, P. V., Singh, T. A., & Kumar, M. S. (2019). Mixed industrial wastewater treatment by combined electrochemical advanced oxidation and biological processes. *Chemosphere, 237*, 124419.

Rahim, M., Ullah, I., Khan, A., & Haris, M. R. H. M. (2016). Health risk from heavy metals via consumption of food crops in the vicinity of District Shangla. *Journal of the Chemical Society of Pakistan, 38*(1).

Ramírez-Malule, H., Quiñones-Murillo, D. H., & Manotas-Duque, D. (2020). Emerging contaminants as global environmental hazards. A bibliometric analysis. *Emerging Contaminants, 6*, 179–193.

Rathi, B. S., Kumar, P. S., & Show, P. L. (2021). A review on effective removal of emerging contaminants from aquatic systems: Current trends and scope for further research. *Journal of Hazardous Materials, 409*, 124413.

Rivera-Utrilla, J., Sánchez-Polo, M., Ferro-García, M. Á., Prados-Joya, G., & Ocampo-Pérez, R. (2013). Pharmaceuticals as emerging contaminants and their removal from water. A review. *Chemosphere, 93*(7), 1268–1287.

Saeed, K., Sadiq, M., Khan, I., Ullah, S., Ali, N., & Khan, A. (2018). Synthesis, characterization, and photocatalytic application of Pd/ZrO 2 and Pt/ZrO 2. *Applied Water Science, 8*(2), 60.

Saeli, M., Tobaldi, D. M., Seabra, M. P., & Labrincha, J. A. (2019). Mix design and mechanical performance of geopolymeric binders and mortars using biomass fly ash and alkaline effluent from paper-pulp industry. *Journal of Cleaner Production, 208*, 1188–1197.

Santos, S., Ungureanu, G., Boaventura, R., & Botelho, C. (2015). Selenium contaminated waters: an overview of analytical methods, treatment options and recent advances in sorption methods. *Science of the Total Environment, 521*, 246–260.

Sarkar, M., Sangal, V. K., Sharma, V. K., Samantray, J., Bhunia, H., Bajpai, P. K., ... Pant, H. J. (2017). Radiotracer investigation and modeling of an activated sludge system in a pulp and paper industry. *Applied Radiation and Isotopes, 130*, 270–275.

Sartaj, S., Ali, N., Khan, A., Malik, S., Bilal, M., Khan, M., ... Khan, S. (2020). Performance evaluation of photolytic and electrochemical oxidation processes for enhanced degradation of food dyes laden wastewater. *Water Science and Technology, 81*(5), 971–984.

Sen, S. K., Raut, S., Gaur, M., & Raut, S. (2020). Biodegradation of lignin from pulp and paper mill effluent: Optimization and toxicity evaluation. *Journal of Hazardous, Toxic, and Radioactive Waste, 24*(4), 04020032.

Sethupathy, A., Arun, C., Sivashanmugam, P., & Kumar, R. R. (2020). Enrichment of biomethane production from paper industry biosolid using ozonation combined with hydrolytic enzymes. *Fuel, 279*, 118522.

Shah, S., ud Din, S., Khan, A., & Shah, S. A. (2018). Green synthesis and antioxidant study of silver nanoparticles of root extract of Sageretia thea and its role in oxidation protection technology. *Journal of Polymers and the Environment, 26*(6), 2323–2332.

Sharma, K., Sarma, N. S., & Devi, A. (2020). *Paper mill effluents: Identification of emerging pollutants in Taranga Beel of Assam, India. Climate Impacts on Water Resources in India* (pp. 89−96). Cham: Springer.

Sharma, P., Tripathi, S., & Chandra, R. (2021a). Highly efficient phytoremediation potential of metal and metalloids from the pulp paper industry waste employing *Eclipta alba* (L) and *Alternanthera philoxeroide* (L): Biosorption and pollution reduction. *Bioresource Technology*, *319*, 124147.

Sharma, P., Tripathi, S., & Chandra, R. (2021b). Metagenomic analysis for profiling of microbial communities and tolerance in metal-polluted pulp and paper industry wastewater. *Bioresource Technology*, *324*, 124681.

Siddiquee, S., Rovina, K., Azad, S. A., Naher, L., Suryani, S., & Chaikaew, P. (2015). Heavy metal contaminants removal from wastewater using the potential filamentous fungi biomass: A review. *Journal of Microbial & Biochemical Technology*, 7(6), 384−393.

Simão, L., Hotza, D., Raupp-Pereira, F., Labrincha, J. A., & Montedo, O. R. K. (2018). Wastes from pulp and paper mills-a review of generation and recycling alternatives. *Cerâmica*, *64*(371), 443−453.

Simão, L., Jiusti, J., Lóh, N. J., Hotza, D., Raupp-Pereira, F., Labrincha, J. A., & Montedo, O. R. K. (2017). Waste-containing clinkers: Valorization of alternative mineral sources from pulp and paper mills. *Process Safety and Environmental Protection*, *109*, 106−116.

Söderholm, P., Bergquist, A. K., & Söderholm, K. (2019). Environmental regulation in the pulp and paper industry: impacts and challenges. *Current Forestry Reports*, *5*(4), 185−198.

Sohni, S., Gul, K., Ahmad, F., Ahmad, I., Khan, A., Khan, N., & Bahadar Khan, S. (2018). Highly efficient removal of acid red-17 and bromophenol blue dyes from industrial wastewater using graphene oxide functionalized magnetic chitosan composite. *Polymer Composites*, *39*(9), 3317−3328.

Song, X., Luo, W., Hai, F. I., Price, W. E., Guo, W., Ngo, H. H., & Nghiem, L. D. (2018). Resource recovery from wastewater by anaerobic membrane bioreactors: Opportunities and challenges. *Bioresource Technology*, *270*, 669−677.

Sun, M., Wang, Y., Shi, L., & Klemeš, J. J. (2018). Uncovering energy use, carbon emissions and environmental burdens of pulp and paper industry: A systematic review and *meta*-analysis. *Renewable and Sustainable Energy Reviews*, *92*, 823−833.

Teh, C. Y., Budiman, P. M., Shak, K. P. Y., & Wu, T. Y. (2016). Recent advancement of coagulation−flocculation and its application in wastewater treatment. *Industrial & Engineering Chemistry Research*, *55*(16), 4363−4389.

Toczyłowska-Mamińska, R. (2017). Limits and perspectives of pulp and paper industry wastewater treatment−A review. *Renewable and Sustainable Energy Reviews*, *78*, 764−772.

Usman, M., Ma, Z., Wasif Zafar, M., Haseeb, A., & Ashraf, R. U. (2019). Are air pollution, economic and non-economic factors associated with per capita health expenditures? Evidence from emerging economies. *International Journal of Environmental Research and Public Health*, *16*(11), 1967.

van Ewijk, S., Stegemann, J. A., & Ekins, P. (2021). Limited climate benefits of global recycling of pulp and paper. *Nature Sustainability*, *4*(2), 180−187.

Vashi, H., Iorhemen, O. T., & Tay, J. H. (2018). Degradation of industrial tannin and lignin from pulp mill effluent by aerobic granular sludge technology. *Journal of Water Process Engineering*, *26*, 38−45.

Vidal, R. R. L., & Moraes, J. S. (2019). Removal of organic pollutants from wastewater using chitosan: a literature review. *International Journal of Environmental Science and Technology*, *16*(3), 1741−1754.

Villar-Navarro, E., Baena-Nogueras, R. M., Paniw, M., Perales, J. A., & Lara-Martín, P. A. (2018). Removal of pharmaceuticals in urban wastewater: High rate algae pond (HRAP) based technologies as an alternative to activated sludge based processes. *Water Research*, *139*, 19−29.

Wahid, F., Mohammadzai, I. U., Khan, A., Shah, Z., Hassan, W., & Ali, N. (2017). Removal of toxic metals with activated carbon prepared from Salvadora persica. *Arabian Journal of Chemistry*, *10*, S2205−S2212.

Wen, J., Yi, Y., & Zeng, G. (2016). Effects of modified zeolite on the removal and stabilization of heavy metals in contaminated lake sediment using BCR sequential extraction. *Journal of Environmental Management*, *178*, 63−69.

Xiao, J., Xie, Y., & Cao, H. (2015). Organic pollutants removal in wastewater by heterogeneous photocatalytic ozonation. *Chemosphere*, *121*, 1−17.

Yadav, S., & Chandra, R. (2018). Detection and assessment of the phytotoxicity of residual organic pollutants in sediment contaminated with pulp and paper mill effluent. *Environmental Monitoring and Assessment*, *190*(10), 1−15.

Yadav, V. B., Gadi, R., & Kalra, S. (2019). Clay based nanocomposites for removal of heavy metals from water: a review. *Journal of Environmental Management*, *232*, 803–817.

Yang, S., Yang, B., Duan, C., Fuller, D. A., Wang, X., Chowdhury, S. P., ... Ni, Y. (2019). Applications of enzymatic technologies to the production of high-quality dissolving pulp: a review. *Bioresource Technology*, *281*, 440–448.

Yang, Y., Ali, N., Khan, A., Khan, S., Khan, S., Khan, H., ... Bilal, M. (2021). Chitosan-capped ternary metal selenide nanocatalysts for efficient degradation of Congo red dye in sunlight irradiation. *International Journal of Biological Macromolecules*, *167*, 169–181.

Yangali-Quintanilla, V., Maeng, S. K., Fujioka, T., Kennedy, M., Li, Z., & Amy, G. (2011). Nanofiltration vs. reverse osmosis for the removal of emerging organic contaminants in water reuse. *Desalination and Water Treatment*, *34*(1–3), 50–56.

Yuan, Q. B., Guo, M. T., Wei, W. J., & Yang, J. (2016). Reductions of bacterial antibiotic resistance through five biological treatment processes treated municipal wastewater. *Environmental Science and Pollution Research*, *23*(19), 19495–19503.

Yusoff, M. F. M., Rozaimah, S. A. S., Hassimi, A. H., Hawati, J., & Habibah, A. (2019). Performance of continuous pilot subsurface constructed wetland using Scirpus grossus for removal of COD, colour and suspended solid in recycled pulp and paper effluent. *Environmental Technology & Innovation*, *13*, 346–352.

Zainith, S., Chowdhary, P., & Bharagava, R. N. (2019). Recent advances in physico-chemical and biological techniques for the management of pulp and paper mill waste. *Emerging and Eco-Friendly Approaches for Waste Management*, 271–297.

Zhang, A., Wang, G., Gong, G., & Shen, J. (2017). Immobilization of white rot fungi to carbohydrate-rich corn cob as a basis for tertiary treatment of secondarily treated pulp and paper mill wastewater. *Industrial Crops and Products*, *109*, 538–541.

Zodi, S., Louvet, J. N., Michon, C., Potier, O., Pons, M. N., Lapicque, F., & Leclerc, J. P. (2011). Electrocoagulation as a tertiary treatment for paper mill wastewater: Removal of non-biodegradable organic pollution and arsenic. *Separation and Purification Technology*, *81*(1), 62–68.

CHAPTER 17

Paper and pulp mill wastewater: characterization, microbial-mediated degradation, and challenges

Adarsh Kumar[1], Ajay Kumar Singh[1], Muhammad Bilal[2], Sonal Prasad[3], K.R. Talluri Rameshwari[4] and Ram Chandra[1]

[1]Department of Environmental Microbiology, School for Environmental Science, Babasaheb Bhimrao Ambedkar University (A Central University), Lucknow, India
[2]School of Life Science and Food Engineering, Huaiyin Institute of Technology, Huai'an, P.R. China
[3]Department of Bio-Sciences, Institute of Bio-Sciences and Technology, Shri Ramswaroop Memorial University, Barabanki, India
[4]Division of Microbiology, Department of Water & Health, Faculty of Life Sciences, JSS Academy of Higher Education and Research, Mysuru, India

17.1 Introduction

One of the significant industrial sectors that uses lignocellulogic materials and water is the pulp and paper industry, which releases chlorinated lignosulfonic acids along with chlorinated phenols, resin acids, and hydrocarbon in the effluent (Kumar, Singh, & Chandra, 2020b).During the manufacturing process, some excessive toxic and unmanageable mixture, such as dibenzofuran and dibenzo-p-dioxin, are established accidentally in the emission of pulp and paper industry (Thakur, 1996). The effluents which remain untreated when come out from paper and pulp industry get dispensed into water bodies, damage the standard characteristics of water, making it unfit for consumption (Table 17.1). Also, the life of aquatic organisms gets a possible threat from the toxic contaminated undiluted effluents exhibiting a strong mutagenic effect. The released paper and pulp industry effluent, most of the physical, chemical, and biological methods are used for removing color. But it is observed that physical and chemical techniques are very costly and they are known to eliminate constituents such as color, high molecular-weight chlorinated lignin, suspended solids, toxicants, and chemical oxygen demand (COD) from the effluent in a highly effective manner unlike biological oxygen demand (BOD) and other low molecular-weight compound which cannot be separated effectively (Kumar, Singh, Ahmad, & Chandra, 2020a, b). The biological color removal process minimizes low molecular-weight chlorolignins with BOD which makes it highly attractive when it adds color and COD (Singh, Kumar, & Chandra, 2020a, b). Microorganisms are better known for degrading some chemicals thus removing them from the atmosphere, while there are some other chemicals that

Nanotechnology in Paper and Wood Engineering
DOI: https://doi.org/10.1016/B978-0-323-85835-9.00011-8

Table 17.1 Pollutants released in different steps of papermaking process.

Debarking process	Remove soil, bark, and dirt from the wood, chips and bark are extracted, and the wood is cleaned with water. As a result, this source's effluent includes suspended solids, BOD, clay, grit, fibers, and other contaminants.
Pulping process	Initially discharge effluent from the digester referred to as "black liquor." The cooking chemicals along with lignin and other wood extractives were found in kraft spent cooking "black liquor" containing resins, fatty acids, color, BOD, COD, and VOCs (terpenes, methanol, chloroform alcohols, acetone, etc.) are contained in effluent.
Pulp washing	Pulp washing released dark brown color effluent containing high pH, BOD, COD, and suspended solids.
Pulp bleaching	Dissolved lignin, carbohydrate, color, COD, organic and inorganic chlorinated compounds, that is, chlorophenols, furans, dioxins, VOCs, chloromethane, trichloroethane, acetone, chloroform, methylene chloride, carbon disulfide, among others are included in the bleaching effluent.
Papermaking	Organic and inorganic compounds, dyes, particulate waste, acetone, COD, and other contaminants are found in papermaking effluent.

Notes: *AOX*, adsorbable organic halogen; *BOD*, biological oxygen demand; *COD*, chemical oxygen demand; *VOC*, volatile organic compounds.

does not degrade rapidly and get accumulated in the environment increasing the toxicity (Singh et al., 2020a, b). Biodegradation is a process which signifies some advanced methods while decomposing complicated and biologically damaged pollutants into simpler substances with the application of microorganisms. The theory behind the mechanism of biodegradation is optimizing nutrient ratios (the growth of some selected microorganisms are encouraged to degrade the toxic pollutants) and applying the strains isolated from some of the selected microorganisms with highly effective degradation abilities (Kucerova, 2006). The unavailability of suitable microorganism in biodegradation, deprivation of genetic capability in unfavorable environmental conditions, and development of unmanageable compounds of different structural formulation along with inferior streamlining process for treatment at large scale are some of the factors which does not make the treatment of pulp and paper industry effluent as an effective technique.

However, the physical and chemical methods are usually applied for the treatment of the pulp and paper industry discharge, but are not similar like biological treatment due to a varied difference in cost effectiveness and residual effects. The organic treatments are responsible for the reduction of the biotic load and poisonous effects of kraft industry effluents (Kumar et al., 2020b). Effluents are treated with the microorganism basically by two process, that is, enzymatic action and biosorption mechanism (Kumar et al., 2020a). Various ligninolytic enzymes such as laccase, lignin peroxidase, manganese peroxidase, and versatile peroxidase are involved in degradation and decolorization of paper and pulp industry effluent (Kumar & Chandra, 2020). These enzymes are produced abundantly by the microorganisms, showing their effectiveness in the paper and pulp industry effluent treatment exhibiting, its characteristic of microbial degradation. This chapter has been characteristic of paper and pulp industry effluent and their aquatic and terrestrial toxicity as well as microbial degradation.

17.2 Characteristics of paper and pulp industry effluent

Effluents with large BOD and COD are produced by pulp and paper industry. The color of this effluent is due to the presence of lignin and their derivatives which are generated from raw materials used for papermaking steps and released in the effluents (Kumar and Chandra, 2021). This brown color increases the temperature of water and reduces the photosynthesis due to which dissolved oxygen concentration gets decreased (Ragunathan and Swaminathan). The paper and pulp industry effluent generation and properties depend on the variety of manufacturing process and the level of water which is reused in the papermaking process. Kraft pulping effluent is extremely contaminated and specified by specifications which are distinctive to the wastes such as color with allied organic compound and adsorbable organic halides. The major source of color is the alkaline extraction stage of bleaching plant effluent, which occurs due to the lignin and its derivatives (Singh et al., 2020a, b). Lignin is released in effluent from different steps of the papermaking process, that is, pulping, bleaching, and chemical recovery steps. Lignin is heterogeneous organic polymer composed of oxyphenyl propanoid units. After bleaching, huge amount of chlorine is released and reacts with lignin along with its derivatives, and leads to the formation of excessive poisonous, along with ungovernable, compounds accountable for high BOD and COD. Therefore the paper and pulp industry effluent contains crucial pollutants formed, that is, trichloroguicol, trichlorophenol, dichlorophenol, tetrachloroguicol, pentachlorophenol, and dichlorophenol (Kumar et al., 2020b) (Table 17.2).

17.2.1 Characterization of organic compounds

The majority of organic compounds present in bleached and unbleached paper industry effluents were persistent compounds, do not degrade after secondary treatment process, and were released into the aquatic sources. During the various steps of the papermaking process, including chipping, wood digestion, paper recycling, bleaching, and pulping, polluted effluent was discharged. Paper and pulp industry effluent causes toxicity in flora and fauna (Kumar et al., 2020b) (Table 17.3).

Table 17.2 Physiochemical characteristics of bleached and unbleached paper and pulp industry effluents.

Serial number	Parameters	Bleached paper industry effluent	Unbleached paper industry effluent	Permissible limits (USEPA, 2012)
1.	pH	7.68 ± 0.21	7.95 ± 0.16	5–9
2.	BOD	225.0 ± 2.24	112.0 ± 1.14	40
3.	COD	543.04 ± 1.22	413.5 ± 0.81	120
4.	Total solids	3280.0 ± 1.32	698.0 ± 2.34	300
5.	Dissolved solid	3110.0 ± 2.42	584.0 ± 1.82	500
6.	Suspended solids	164.0 ± 0.42	270.0 ± 0.28	100
7.	Fixed solids	2900.0 ± 2.35	566.0 ± 1.12	–
8.	Volatile solids	380.0 ± 0.31	132.0 ± 0.22	–
9.	Chloride	2350.0 ± 1.14	1740.0 ± 1.10	1500
10.	Phosphate	2.56 ± 0.12	1.1 ± 0.20	–
11.	Sulfate	713.1 ± 1.11	316.71 ± 0.81	250
12.	Nitrate	210.08 ± 2.32	47.17 ± 1.10	50
13.	Color	1275 ± 3.11	4180 ± 2.10	Colorless
14.	Total phenol	13.195 ± 0.52	11.691 ± 0.82	0.50
15.	Lignin	578 ± 0.13	285 ± 0.20	0.05
Heavy metals				
16.	Lead	0.2550 ± 0.12	0.1360 ± 0.42	0.05
17.	Iron	1.8598 ± 0.90	0.4232 ± 1.02	2
18.	Chromium	0.3058 ± 0.01	0.0834 ± 0.11	0.05
19.	Copper	1.3814 ± 0.16	0.8971 ± 0.20	0.5
20.	Cadmium	0.08632 ± 0.13	0.219 ± 0.10	0.01
21.	Zinc	0.0945 ± 1.21	0.0408 ± 0.82	2.00

Notes: All the parameters are in mg/l except pH and color (Pt–Co). *BOD*, Biological oxygen demand; *COD*, chemical oxygen demand.

17.2.2 Environmental impact of paper and pulp industry effluent

17.2.2.1 Phytotoxicity

Paper and pulp industry effluent toxicity assessment through seed germination test (*Phaseolus aureus*) was done. Kumar et al. (2020b) reported paper industry effluent varied concentrations' (10%, 25%, 50%, 75%, and 100%) toxicity on germination and the growth of seedling in context of *P. aureus* seeds. In comparison to controls, when the seedling growth was assessed, lengths of root and shoot of seedling were found to be maximum at 10% (vol./vol.) effluent, but the subsequent hike in effluent concentration

Table 17.3 Some identified organic pollutant in bleached and unbleached paper industry effluents (Kumar et al., 2020b).

	Bleached paper industry effluent		Unbleached paper industry effluent	
Serial number	**Retention time (min)**	**Identified compounds**	**Retention time (min)**	**Identified compounds**
1.	13.29	Dodecane, 1-iodo-	13.28	Nonadecane
2.	14.33	Decane	14.33	Dodecane, 1-iodo-
3.	16.40	Tetracosane	16.52	3,6-Dioxa-2,7-disilaoctane
4.	17.91	Heptacosane	17.90	Tricosane
5.	19.88	Pentan-1,3-dioldiisobutyrate, 2,2,4-trimethyl	20.70	7,7-diphenyl-3,5-dioxo-7-hydroxyheptanenitrile
6.	22.00	Octadecane, 1-iodo-	22.00	Tetradecane
7.	25.68	Heneicosane	25.68	Eicosane
8.	27.82	Hexadecanoic acid, trimethylsilyl ester	27.82	Hexadecanoic acid
9.	29.02	Eicosane	30.69	Octadecanoic acid
10.	30.70	Diethyl 3,4-dihydro-2-naphthyl-phosphonate	31.77	Eicosane
11.	31.78	Hexadecane	37.68	Chloromethyl-3-methyl-5-phenylisoxazole
12.	35.88	1,2,3,4,5-Pentaisopropylbis, cobalticimium	41.18	Perylenetetracarboxydiimide
13.	40.03	Methyl ester compounds	45.00	3,5,7-Tris(trimethylsiloxy)-2-[3,4-di(trimethylsiloxy)phenyl]-4H-1-benzopyran-4-one
14.	43.03	2-(1-Methyl-1H-2-pyrrolyl)quinoline	47.39	(3R)-3-Phenyl-2,3-dihydro-1H-isoindol-1-one
15.	43.79	Cyclotrisiloxane, hexamethyl- (CAS)		
16.	45.02	5,11,17,23-Tetra-t-butyl-25,26,27, 28-tetrahydroxycalix-4-arene		
17.	47.36	a-Fluoro-(p-methyl)chalcone		

displayed a reduction in the seed germination percentage and shoot and root lengths as well. The percentage of phytotoxicity was found to be greater at higher concentration of effluent and lower at lesser effluent concentration. This is because of the appearance of excessive dreadful toxic organic and inorganic pollutants in paper and pulp industry effluent. The effluent which was toxic, mitigated the seedling growth parameters. Excessive toxic organic pollutants along with heavy metals were reported to inhibit several phytohormones (i.e., cytokinins, auxins, and gibberellins) which play a vital role in seeds germination (Chandra, Abhishek, & Sankhwar, 2011). Apart from this, many remarkable changes were observed in physiological parameters revealing more precise and correct information about the detrimental consequences of bleached and unbleached paper pulp industry effluents in the assessment of phytotoxicity (*P. aureus*) (Kumar and Chandra, 2021). The toxic compounds present in the effluent discharged from unbleached and bleached paper industry were exhibited with the speed of germination which was minimized from a lower level of concentration to a higher level of effluents. The toxic nature of unbleached and bleached paper pulp industry effluents was established by seedling mortality rate, which had a higher concentration of effluent. Additionally, daily germination mean, root—shoot ratio, and vigor index of seed ranged in varied concentration from maximum to minimum for lower to higher effluent concentrations. In the study, the germination indices were found to be at 100% at the control and varied from lesser to higher concentrations of the unbleached and bleached paper pulp industry effluents sample. Growths of both shoot and root were observed in the seeds which were treated with control, whereas extremely short root (0.06 and 0.16 cm) and short shoot (0.06 and 0.36 cm) were observed in those seeds which were treated with 100% bleached and unbleached paper mill wastewater (BUPMW) in varied concentrations (Kumar et al., 2020b). The activity of α-amylase was observed in germinating seeds establishing the fact of the toxicity of seeds while treating with unbleached and bleached paper pulp industry effluents on both stages of seed germination and seedling growth (Fig. 17.1).

17.2.2.2 Animal toxicity

Tubifex worm is generally used as the powerful indicator to assess the different level of pollution in aquatic environment, as it is one of the important links in the aquatic food chain life (Kumar et al., 2021b). The unbleached and bleached paper pulp industry effluents were used to assess the level of toxicity in varied concentration with a distinct interval of time. The toxicity was lower at a concentration of more than 25% up to 48-h exposure and greater at a concentration of more than 50% effluents at 48 h of exposure. The effluents which contain various mutagenic and carcinogenic pollutants along with other residual organic pollutants were reported to cause toxicity (Singh et al., 2021). The increased mortality rate of *Tubifex*, its physiological damage, cell bursting, and whole-cell damage were due to the presence of the pollutants and contaminants prevailing in effluents (Singh et al., 2021). These organochlorinated

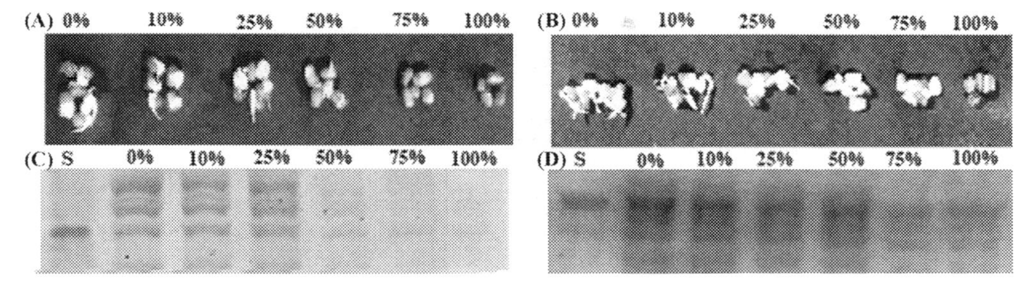

Figure 17.1 Showing the effect of organic pollutants of unbleached and bleached paper pulp industry effluents on *Phaseolus aureus*. (A) Effects of bleached paper industry effluents; (B) effects of unbleached paper industry effluents; (C) effects of bleached paper industry effluents on α-amylase activity; (D) effects of unbleached paper industry effluents on α-amylase activity. Lane S indicates amylase standard and 0% indicates seed treated with tap water (Kumar et al., 2020b).

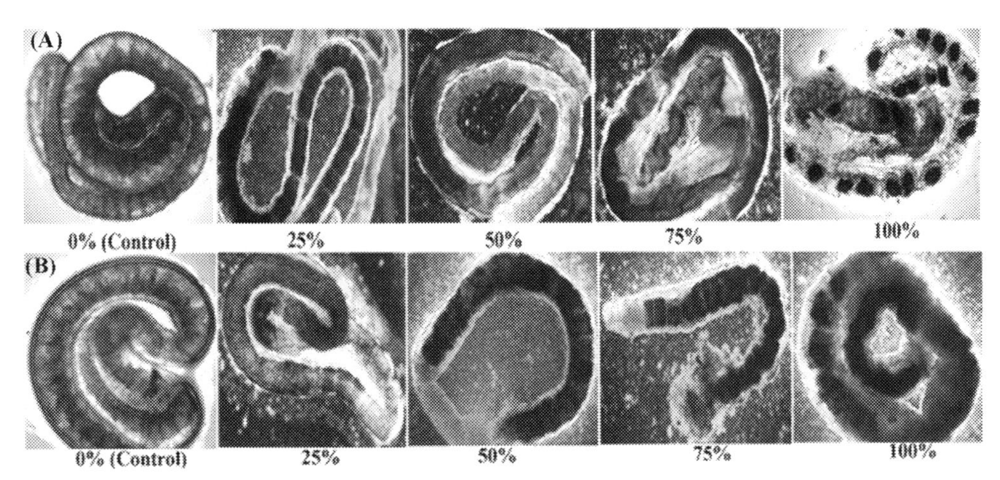

Figure 17.2 Effects of unbleached and bleached paper and pulp industry effluents. (A) Effects of bleached paper and pulp industry effluents and (B) unbleached paper and pulp industry effluents on *Tubifex tubifex* toxicity. Control (0%) worm in healthy conditions and when treated with different concentration of effluents (Kumar et al., 2020b).

pollutants are very detrimental to overall ecosystem affecting human beings along with other living organisms in the environment. In control, during the testing period, *Tubifex* worms remained highly active. They were accumulated at the base of the test container exhibiting the typical *Tubifex* movement. The reason behind this was the presence of excessive dreadful pollutants and toxic heavy metals in unbleached and bleached paper pulp industry effluents causing the toxicity of worm as indicated in the Fig. 17.2. This experimental work indicated that the bleached paper industry effluents were highly toxic in comparison to unbleached paper industry effluents (Kumar et al., 2020b). After 24 h of exposure at the fatal concentrations of effluent, the hemoglobin level was lowered and the rear parts of the body became white along with the decomposition of the body (Table 17.4).

17.3 Microbial-mediated degradation

The pollutant removal from paper and pulp industry effluent is one of the major problems and is a matter of concern on which scientists and researchers are doing lots of studies. Due to the prevalence of lignin derivatives, color is observed as the main problem in pulp and paper industry effluent. Several processes including pulping, bleaching, and recovery sections are responsible for the production of large amount of lignin. Several physio-chemical methods are used for the removal of color from paper and pulp industry effluent (Kumar et al., 2021c). But physical and chemical methods are very costly, removing some chlorinated lignins, toxicity, color, suspended solids, and COD while BOD along with some other chemicals cannot be eliminated efficiently (Kumar et al., 2021c). However, the biological method reduces color and COD along with BOD and other chemicals, thus showing its better characteristics (Kumar et al., 2021b). Bioremediation is a technique which uses biological systems for degradation and transformation of various pollutants into lesser harmful simpler compounds (Kumar et al., 2021a). Therefore bioremediation is applied in treating many industrial effluents like paper and pulp industry effluent, thus exhibiting its beneficial effects (Wu, Xiao, & Yu, 2005; Yang et al., 2008).

17.3.1 Bacterial-mediated degradation of paper and pulp industry effluent

The diverse and flexible paths for disintegrating various aromatic ring structural compounds to generate phenols and lignin macromolecules, which must make it simpler and more easily digested by bacterial metabolic activities. (Koma, Yamanaka, & Mo, 2012). The enzymes produced by bacteria that serve the specific biotechnological interest, exhibit less dependency on mediators of the organic compounds for their working and functioning (Bandounas, Pinkse, Winde, & Ruijssenaars, 2013). A new spectrophotometric assay has been developed for the study of specificity and credible potential of *Rhodococcus jostii* RHA1 and *Pseudomonas putida* mt-2 to degrade and breakdown the lignin component which are generated from the various sources of pulp biomass and wheat straw, miscanthus and pine (Ahmad et al., 2010). Most of the bacteria degrade the lignin part of the pulp and paper industry biomass. Some bacteria were isolated from the different environmental conditions that showed good effects on lignin treatment. The farm soil isolated *Pseudomonas* sp. (PKE117) has properties to destruct the lignin structure as compared to cellulose in *Larix olgensis* (Yang et al., 2008). A lignin degradation bacterial consortium named LDC was screened from the sludge of a reeds pond by a restricted subculture. It could break down 60.9% lignin in reeds at 30°C under conditions of static culture within 15 days. (Wang et al., 2013b).

Table 17.4 Paper and pulp effluent toxicity approaches for measuring effects on diverse life forms.

Toxicity assay	Target organism	Concluding highlights	References
Salmonella/ mammalian microsome assay	*Salmonella typhi*	Increase in His^+ reversion mutations as a function of dose. The mutagenic response of key effluent components was investigated.	Nestmann, Lee, Matula, Douglas, and Mueller (1980)
DCF assay	Human keratinocyte cell line	The fluorescence of DCFDA revealed the cell damage (apoptosis). Cytotoxicity was demonstrated by the development of reactive oxygen species.	Abhishek, Dwivedi, Tandan, and Kumar (2015)
Bioluminescence assay	*Vibrio fischeri Daphnia magna*	End-product was more toxic than total organic carbon even after catalytic wet oxidation assay treatment.	Pintar, Besson, Gallezot, Gibert, and Martin (2004)
Comet assay	*Saccharomyces cerevisiae*	Single- and double-strand breaks, incomplete excision repair sites, and alkali labile sites were all found to have DNA damage.	Singhal and Thakur (2009a, b);
Bioassay of seed germination	*Vigna radiate* and *Vicia faba*	It was discovered that sprouting duration and α-amylase activity were both inhibited.	Haq, Kumar, Kumari, Singh, and Raj (2016)
Genotoxicity assay	*Channa punctatus*	Centromeric holes, chromatid breaks, acentric fragments, pycnosis, and polyploidy were all chromosomal aberrations.	Malik, Kumar, Seth, and Rishi (2009)
Metabolic capacity of liver	*Rainbow trout*	Pasteur's compensatory impact at the enzyme-based electron transport stage, there is a change in aerobic metabolism.	Orrego, Pandelides, Guchardi, and Holdway (2011)
MTT assay, EROD assay, genotoxicity	HuH7 cell line	Overall, the pulp and paper industry sewage sediment adds to a load of toxicants that induce cytotoxicity and genotoxicity by distorting natural cell growth and proliferation.	Das, Budhraja, Mishra, and Thakur (2012)

Notes: DCF, Dichlorofluorescin; *DCFDA*, dichlorofluorescin diacetate; *MMT*, modified thiazolyl blue tetrazolium; *EROD*, ethoxyresorufin-O-deethylase.

The consortium (LDC) contains different bacteria from different genera, that is, *Geovibrio, Desulfomicrobium, Thauera, Acinetobacter, Clostridiales, Azoarcus, Microbacterium, Paenibacillus, Pseudomonas*, and *Cohnella*. In this consortium bacterial species of different genera of *Pseudomonas* and *Microbacterium* controlled the rate of degradation of lignin content and sustained the less cellulose decomposition (Wang et al., 2013b). Apart from this, after the bacterial consortium, hand-sheets of paper made via reed biomass exhibited better physical attributes. These research studies provide a path for the sustainable development of treatment technology by the use of bacterial consortium to degrade and detoxify lignin content present in the paper industry effluent and have a good opportunity for the future research.

17.3.2 Fungal-mediated degradation of paper and pulp industry effluent

According to investigations and research, white-rot fungi is widely distributed and documented as the efficient and good biodegrader of lignin content from paper industry effluent. Mainly the basidiomycetes white-rot fungi including *Pleurotus ostreatus, Chrysosporium, Phanerochaete, Phlebia subserialis, and Ceriporiopsis subvermispora* are reported by various workers for the effective metabolization of the lignin content from various lignocellulosic waste in the environment (Hatakka, 1983; Keller, Hamilton, & Nguyen, 2003; Taniguchi et al., 2005). However, sustainable delignification process is affected because of continuous and prolonged use in treatment, it may lose its carbohydrate fraction due to fungal activity and also have no ability to degrade efficient amount of lignin content (Hatakka, 1983).

Lignin measures the kappa level content and estimated the demand for pulp along with its oxygen level. *Thelephora* sp. and *Coriolus versilor* were registered to reduced kappa number (29% and 63%) because of action on kraft pulp during the biobleaching (Kirkpatrick, Reid, Ziomek, & Paice, 1990; Selvam, Priya, & Arungandhi, 2011). Other studies also noticed for reduced kappa number from kraft pulp when compared with initials points from 12.4 to 10.4 via *Geotrichum candidum* Dec1 after 6 days (Shintani & Shoda, 2013). To help in lignin degradation and mycelium growth in pulp biomass in industry, fungi require additional nutrients. The cosubstrates used in the treatment process may raise treatment cost, limiting its commercial viability (Rahi, Rahi, Pandey, & Rajak, 2009). In comparison to other microbes, the ligninolytic ability of the bacterial enzymes is the new field of research for the sustainable and effective uses in the delignification of paper industry biomass containing lignin polymers (Wang, Yan, & Cui, 2011).

17.3.3 Benefits of microbial ligninolytic potential on pulp treatment

The enhancement in brightness and physical properties of pulp biomass is the most fascinating features in the delignification by bacterial approaches as shown in Table 17.5.

Table 17.5 Bacterial ligninolytic enzymes and their mode of actions.

Bacterial strains	Ligninolytic enzymes	Substrate mediator	Reaction type	References
Streptomyces coelicolor, Streptomyces viridosporus T7 A	Laccase	Phenols, mediators, for example, hydroxybenzotriazole or 2,2′-azino-bis(3-ethylbenzothiazoline-6-sulfonic acid)	$C\alpha$ oxidation, phenols are oxidized to phenoxyl radicals in the presence of mediators	Majumdar et al. (2014)
Stenotrophomonas maltophilia				Yadav and Chandra (2012)
Bacillus subtilis, Bordetella campestris, Caulobacter crescentus, Escherichia coli, Mycobacterium tuberculosum, Pseudomonas syringae, Pseudomonas aeruginosa, Yersinia pestis				Enguita, Martins, Henriques, and Carrondo (2003); Bains, Capalash, and Sharma (2003)
Bacillus sp. CSA105	Lignin peroxidase	Veratryl alcohol, Azure-B	Aromatic ring oxidized to cation radical	Kharayat and Thakur (2012)
Bacillus sp. SHC1, *Ochrobactrum* sp. SCH2, *Leucobacter* sp. SHC3				Rahman, Rahman, Aziz, and Hassan (2013)
B. subtilis LPTK				Renugadevi, Ayyappadas, Preethy, and Savetha (2011)
Serratia liquefaciens	Manganese peroxidase	Mn, organic acids as chelators, thiols, and unsaturated fatty acids	Mn (II) oxidized to Mn (III); chelated Mn (III) oxidizes phenolic compounds to phenoxyl radicals	Haq et al. (2016)
Citrobacter freundii, Citrobacter sp.				Chandra et al. (2011)
Thermobifida fusca	Dye-decolorizing peroxidases	2,4-Dichlorophenol 2,2′-azino-bis (3ethylbenzothiazoline-6-sulfonic acid)	Oxidation of β-aryl ether lignin model compound Mn^{2+} oxidation, $C\alpha-C\beta$ cleavage	Rahmanpour, Rea, Jamshidi, Fülop, and Bugg (2016)
Rhodococcus jostii				Ahmad et al. (2011)
Sphingobacterium sp. T2	Manganese superoxide dismutase	Pyrogallol, organosolv lignin	Inhibition of cytochrome c oxidation, pyrogallol autooxidation, and oxidation of organosolv lignin.	Rashid et al. (2015)

This helps to reduce the use of harmful bleaching chemicals and, as a result, the release of toxic substances into industrial effluents. The bond cleavage linked via carbohydrate and lignin, increases the pulp quality and opening in pulp structure with the use of xylanase and is first mechanism of metabolic activities of bacterial communities in bleaching process of pulp (Kumar et al., 2021a). As a result, more bacterial assault on lignin is facilitated, as is the release of chromophores connected from carbohydrates of part of the pulp (Paice, Bernier, & Jurasek, 1988; Patel, Grabski, & Jeffries, 1993). Using bacterial biocatalysts (xylanase enzymes), a large number of studies on pulp bio-bleaching have recently been conducted. The treatment via 40 IU xylanases produced by *Bacillus* sp. in per gram of pulp has been discovered to reduce kappa number up to 16.2% and brightness (ISO) 25.94% within 8 h. Enzymatic pretreatment of the pulp saved 15% active chlorine charges in single-step and 18.7% in multiple-step chemical-bleaching with brightness level to that of the control (Saleem et al., 2009). Agrawal, Yadav, and Mahajan (2016) found that xylanopectinolytic enzymes released by the *Bacillus pumilus* (AJK) decreases kappa number by 9.4%, chlorine dioxide and chlorine by 23.8% and 25%, respectively. Mostly researchers noted that physical properties of pulp are (tear factor, viscosity, and breaking volume) improved as a result of the above procedure used.

Although xylanase has been used in the treatment of pulp in some paper industries and is specifically mentioned in the scientific studies, there is limited evidence for owing the large-scale application on the paper industries around the world. It has three types namely: (1) time-sucking stages in the development of enzymes from expensive growth substrates; (2) purification of enzymes has more affluent steps; and (3) method inflexibilities in the face of ever-changing industrial conditions (Dhillon, Gupta, Jauhari, & Khanna, 2000).

Since whole-cell biobleaching needs only mild operating conditions, it may be able to resolve the enzymatic treatment limitations. Sharma, Sood, Singh, and Capalash (2015) planned submerged strong fermentation (SSF) for biobleaching of whole-cell of kraft pulp of eucalyptus by *Bacillus halodurans*. The SSF method has the added benefit of lowering the water consumption requirement, allowing for high scale-up flexibility (Kumar et al., 2021a). Under minor operating conditions, preg-rown cell cultures of bacteria were inoculated directly into kraft pulp of eucalyptus for biobleaching (Kumar et al., 2021b). As compared to an uninoculated pulp, bleaching process resulted in 35% reduction in kappa number, 20% reduction in chlorine load, and a 5.8% increase in brightness during submerged fermentation within 48 h.

The method also saw an increase in viscosity of 8.6% with no weight loss in bacterially processed pulp, implying an increase in production of cellulose fraction of pulp. *Planococcus* sp. TRC1 and *Pseudomonas fluorescens* NITDPY reduced the kappa percentage of eucalyptus kraft pulp by 37% and 32%, respectively (Priyadarshinee, Kumar,

Mandal, & Dasguptamandal, 2015). However, impregnation process was followed by a rise in crystallinity of cellulose, but specific size of crystallite of pulp samples remained unchanged, suggesting that there was no negative impact on the pulp's cellulose fiber (Kumar et al., 2021c). These recent studies suggest that novel methods involving ligninolytic bacteria may be used to alleviate the effects of climate change (Kumar and Chnadra, 2021). The use of other chemicals such as bleaching powder, chlorine, etc. would leave the waste in the water and releases the chlorolignins in the effluent of paper and pulp industries.

17.4 Challenges and future expectations

As a result of the wide range of products that can be made from pulp biomass, modern paper and pulp industries have been dubbed "future bio-refineries" (Kamm & Kamm, 2004). Ligninolytic bacteria and fungi will open up new possibilities within this system of resource synthesis and recovery. As a result of bacterial selective ligninolytic ability, notable improvements in pulp biomass strength and properties have yielded promising results. This opens up the possibility of more bacteria being used in the development of viable value-added bioproducts (chemicals, fuels, and energy) from pulp biomass. Yan et al. (2012) used subculture to establish the bacterial consortium BYND-5, which consists of many ligninolytic bacteria. The consortium found that lignin from rice straw degraded the most, compared to cellulose and hemicellulose after a year of enrichment. During the treatment phase, propionic acid, acetic acid, glycerin, butyric acid, biogas, and methane were all detected. In another analysis, *Bacillus* sp. CS-1 was found to extract a significant amount of lignin from rice straw. After that, the biomass was treated with lactic acid bacteria to increase the net sugar yield (Chang, Choi, Takamizawa, & Kikuchi, 2014). By fermenting these sugars biogas and bioethanol along with dihydrogen can be formed (Pawar, Nkemka, Zeidan, Murto, & Niel Ed, 2013). The bioconversions of lignocellulosic biomass were into the several value-added products. The main challenges are (1) delignification kinetics, (2) process configuration, and (3) process improvement such as optimization of culture requirements.

The facultative anaerobic bacterium *P. glucanolyticus* SLM1 was isolated from paper industry sludge and extensively used in degradation of lignin (Mathews, Pawlak, & Grunden, 2014). The content of lignin in softwood and hardwood were also reduced by this bacterium species using as carbon source (Mathews, Grunden, & Pawlak, 2016). Black liquor waste of paper industry is also metabolized through the same ligninolytic bacterium for conversion into value-added products, that is, succinic, malonic acids, ethanol, lactic acids, and propanoic acid (Kumar et al., 2021b). As a result, ligninolytic microorganisms may play a key role in the transition from resource depletion to resource recovery, as well as the production or synthesis of essential bioproducts (Kumar and Chandra, 2020; 2021).

17.5 Conclusion

The aim of this chapter is the characteristics of paper industry effluent and their toxicity as well as microbial mediated treatment. Microbial ligninolytic enzymes can aid in the delignification of pulp biomass as well as the removal of lignin and its related contaminants from effluent, as well as the development of a variety of other useful diversified products. This will provide a wide range of benefits to paper and pulp industry while still meeting environmental goals. To achieve the desired efficiency and competitiveness, further research is needed before full-scale implementation of bacterial-mediated delignification and bioconversion can be realized.

References

Abhishek, A., Dwivedi, A., Tandan, N., & Kumar, U. (2015). Comparative bacterial degradation and detoxification of model and kraft lignin from pulp paper wastewater and its metabolites. *Applied Water Science*, *7*, 757–767. Available from https://doi.org/10.1007/s13201-015-0288-9.

Agrawal, S., Yadav, R. D., & Mahajan, R. (2016). Synergistic effect of xylanopectinolytic enzymes produced by a bacterial isolate in bleaching of plywood industrial waste. *Journal of Cleaner Production*, *118*, 229–233.

Ahmad, M., Roberts, J. N., Hardiman, E. M., Singh, R., Eltis, L. D., & Bugg, T. D. H. (2011). Identification of DypB from *Rhodococcus jostii* RHA1 as a lignin peroxidase. *Biochemistry*, *50*, 5096–5107.

Ahmad, M., Taylor, C. R., Pink, D., Burton, K., Eastwood, D., Bending, G. R., & Bugg, T. D. H. (2010). Development of novel assays for lignin degradation: comparative analysis of bacterial and fungal lignin degraders. *Molecular Biosystems*, *6*, 815–821.

Bains, J., Capalash, N., & Sharma, P. (2003). Laccase from a non-melanogenic, alkalotolerant γ-proteobacterium JB isolated from industrial wastewater drained soil. *Biotechnology Letters*, *25*(14), 1155–1159.

Bandounas, L., Pinkse, M., Winde, J. H. D., & Ruijssenaars, H. J. (2013). Identification of a quinone dehydrogenase from a *Bacillus* sp. involved in the decolourization of the lignin-model dye, azure B. *New Biotechnology*, *30*, 196–204.

Chandra, R., Abhishek, A., & Sankhwar, M. (2011). Bacterial decolorization and detoxification of black liquor from rayon grade pulp manufacturing paper industry and detection of their metabolic products. *Bioresource Technology*, *102*, 6429–6436.

Chang, Y. C., Choi, D. B., Takamizawa, K., & Kikuchi, S. (2014). Isolation of *Bacillus* sp. strains capable of decomposing alkali lignin and their application in combination with lactic acid bacteria for enhancing cellulase performance. *Bioresource Technology*, *152*, 429–436.

Das, M. T., Budhraja, V., Mishra, M., & Thakur, I. S. (2012). Toxicological evaluation of paper mill sewage sediment treated by indigenous dibenzofuran-degrading *Pseudomonas* sp. *Bioresource Technology*, *110*, 71–78.

Dhillon, A., Gupta, J. K., Jauhari, B. M., & Khanna, S. A. (2000). Cellulase-poor, thermostable, alkalitolerant xylanase produced by *Bacillus circulans* AB 16 grown on rice straw and its application in biobleaching of eucalyptus pulp. *Bioresource Technology*, *73*, 273–277.

Enguita, F. J., Martins, L. O., Henriques, A. O., & Carrondo, M. A. (2003). Crystal structure of a bacterial endospore coat component- A laccase with enhanced thermostability properties. *The Journal of Biological Chemistry*, *278*(21), 19416–19425.

Haq, I., Kumar, S., Kumari, V., Singh, S. K., & Raj, A. (2016). Evaluation of bioremediation potentiality of ligninolytic *Serratia liquefaciens* for detoxification of pulp and paper mill effluent. *Journal of Hazardous Materials*, *305*, 190–199.

Hatakka, A. I. (1983). Pretreatment of wheat straw by white-rot fungi for enzymatic saccharification of cellulose. *European Journal of Applied Microbiology and Biotechnology*, *18*, 350−357.
Kamm, B., & Kamm, M. (2004). Principles of biorefineries. *Applied Microbiology and Biotechnology*, *64*(137), 145.
Keller, F. A., Hamilton, J. E., & Nguyen, Q. A. (2003). Microbial pretreatment of biomass potential for reducing severity of thermo-chemical biomass pretreatment. *Applied Biochemistry and Biotechnology*, *105*, 27−41.
Kharayat, Y., & Thakur, I. S. (2012). Isolation of bacterial strain from sediment core of pulp and paper mill industries for production and purification of lignin peroxidase (LiP) enzyme. *Bioremediation Journal*, *16*(2), 125−130.
Kirkpatrick, N., Reid, I. D., Ziomek, E., & Paice, M. G. (1990). Biological bleaching of hardwood kraft pulp using *Trametes (Coriolus) versicolor* immobilized in polyurethane foam. *Applied Microbiology and Biotechnology*, *33*, 105−108.
Koma, D., Yamanaka, H., & Mo, K. (2012). Production of aromatic compounds by metabolically engineered *Escherichia coli* with an expanded shikimate pathway. *Applied and Environmental Microbiology*, *78*(17), 6203.
Kucerova, R. (2006). Application of *Pseudomonas putida* and *Rhodococcus* sp. by biodegradation of PAH(S), PCB(S) and NEL soil samples from the hazardous waste dump in pozdatky(Czech republic). *Rudarsko-Geolosko-Naftni Zbornik*, *1897*, 101.
Kumar, A., & Chandra, R. (2020). Ligninolytic enzymes and its mechanisms for degradation of lignocellulosic waste in environment. *Heliyon*, *6*, e03170.
Kumar, A., & Chandra, R. (2021). Biodegradation and toxicity reduction of pulp paper mill wastewater by isolated laccase producing Bacillus cereus AKRC03. *Cleaner Engineering and Technology*, *4*, 100193. Available from https://doi.org/10.1016/j.clet.2021.100193.
Kumar, A., Saxena, G., Kumar, V., & Chandra, R. (2021b). Environmental contamination, toxicity profile and bioremediation approaches for treatment and detoxification of pulp paper industry effluent. In V Kumar, & M Shah (Eds.), *Bioremediation for Environmental Sustainability Toxicity, Mechanisms of Contaminants Degradation, Detoxification, and Challenges*. Elsevier.
Kumar, A., Singh, A. K., Ahmad, S., & Chandra, R. (2020a). Optimization of laccase production by *Bacillus* sp. strain AKRC01 in presence of agro-waste as effective substrate using response surface methodology. *Journal of Pure and Applied Microbiology*, *14*(1), 351−362.
Kumar, A., Singh, A. K., Bilal, M., & Chandra, R. (2021a). Sustainable production of thermostable laccase from agro-residues waste by Bacillus aquimaris AKRC02. *Catalysis Letters*. Available from https://doi.org/10.1007/s10562-021-03753-y.
Kumar, A., Singh, A. K., & Chandra, R. (2020b). Comparative analysis of residual organic pollutants from bleached and unbleached paper mill wastewater and their toxicity on *Phaseolus aureus* and *Tubifex tubifex*. *Urban Water Journal*, *17*(10), 860−870.
Kumar, A., Singh, A. K., & Chandra, R. (2021c). Recent advances in physicochemical and biological approaches for degradation and detoxification of industrial wastewater. In I. Haq, & A. Kalamdhad (Eds.), *Emerging Treatment Technologies for Waste Management*. Springer.
Majumdar, S., Lukk, T., Bauer, S., Nair, S. K., Cronan, J. E., & Gerlt, J. A. (2014). Roles of small laccases from *Streptomyces* in lignin degradation. *Biochemistry*, *53*, 4047−4058.
Malik, M. K., Kumar, P., Seth, R., & Rishi, S. (2009). Genotoxic effect of paper mill effluent on chromosomes of fish *Channa punctatus*. *Current World Environment*, *4*(2), 353−357.
Mathews, S. L., Grunden, A. M., & Pawlak, J. (2016). Degradation of lignocellulose and lignin by *Paenibacillus glucanolyticus*. *International Biodeterioration & Biodegradation*, *110*, 79−86. Available from https://doi.org/10.1016/j.ibiod.2016.02.012.
Mathews, S. L., Pawlak, J. J., & Grunden, A. M. (2014). Isolation of *Paenibacillus glucanolyticus* from pulp mill sources with potential to deconstruct pulping waste. *Bioresource Technology*, *164*, 100−105.
Nestmann, E. R., Lee, E. G. H., Matula, T. I., Douglas, G. R., & Mueller, J. C. (1980). Mutagenicity of constituents identified in pulp and paper mill effluents using the *Salmonella*/mammalian-microsome assay. *Mutation Research*, *79*(3), 203−212.

Orrego, R., Pandelides, Z., Guchardi, J., & Holdway, D. (2011). Effects of pulp and paper mill effluent extracts on liver anaerobic and aerobic. *Ecotoxicology and Environmental Safety*, *74*, 761–768.

Paice, M. G., Bernier, R., Jr, & Jurasek, L. (1988). Viscosity enhancing bleaching of hardwood kraft pulp with xylanase from a cloned gene. *Biotechnology and Bioengineering*, *32*, 235–239.

Patel, R. N., Grabski, A. C., & Jeffries, T. W. (1993). Chromophore release from kraft pulp by purified *Streptomyces roseiscleroticus* xylanases. *Applied Microbiology and Biotechnology*, *39*, 405–412.

Pawar, S. S., Nkemka, V. N., Zeidan, A. A., Murto, M., & Niel Ed, W. J. V. (2013). *Biohydrogen production from wheat straw hydrolysate using* Caldicellulosiruptor saccharolyticus *followed by biogas production in a two-step uncoupled process*, . *International Journal of Hydrogen Energy* (38, pp. 9121–9130).

Pintar, A., Besson, M., Gallezot, P., Gibert, J., & Martin, D. (2004). Toxicity to *Daphnia magna* and *Vibrio fischeri* of kraft bleach plant effluents treated by catalytic wet-air oxidation. *Water Research*, *38*, 289–300.

Priyadarshinee, R., Kumar, A., Mandal, T., & Dasguptamandal, D. (2015). Improving the perspective of raw eucalyptus kraft pulp for industrial applications through autochthonous bacterial mediated delignification. *Industrial Crops and Products*, *74*, 293–303.

Rahi, D. K., Rahi, S., Pandey, A. K., & Rajak, R. C. (2009). Enzymes from mushrooms and their industrial applications. In M. Rai (Ed.), *Advances in fungal biotechnology I.K* (pp. 136–184). NewDelhi: International Publishing House Pvt Ltd.

Rahman, N. H. A., Rahman, N. A. A., Aziz, S. A., & Hassan, M. A. (2013). Production of ligninolytic enzymes by newly isolated bacteria from palm oil plantation soils. *Bioresources*, *8*(4), 6136–6150.

Rahmanpour, R., Rea, D., Jamshidi, S., Fülop, V., & Bugg, T. D. H. (2016). Structure of *Thermobifida fusca* DyP-type peroxidase and activity towards Kraft lignin and lignin model compounds. *Archives of Biochemistry and Biophysics*, *594*, 54–60.

Rashid, G. M. M., Taylor, C. R., Liu, Y., Zhang, X., Rea, D., Fülöp, V., & Bugg, T. D. H. (2015). Identification of manganese superoxide dismutase from *Sphingobacterium* sp. T2 as a novel bacterial enzyme for lignin oxidation. *ACS Chemical Biology*, *10*, 2286–2294.

Renugadevi, R., Ayyappadas, M. P., Preethy, P. H., & Savetha, S. (2011). Isolation, screening and induction of mutation in strain for extracellular lignin peroxidase producing bacteria from soil and its partial purification. *Journal of Research in Biology*, *4*, 312–318.

Saleem, M., Tabassum, M. R., Yasmin, R., & Imran, M. (2009). Potential of xylanase from thermophilic *Bacillus* sp. XTR-10 in biobleaching of wood kraft pulp. *International Biodeterioration & Biodegradation*, *63*, 1119–1124.

Selvam, K., Priya, M. S., & Arungandhi, K. (2011). Pretreatment of wood chips and pulps with *Thelephora* sp. to reduce chemical consumption in paper industries. *International Journal of ChemTech Research*, *3*, 471–476.

Sharma, P., Sood, C., Singh, G., & Capalash, N. (2015). An eco-friendly process for biobleaching of eucalyptus kraft pulp with xylanase producing *Bacillus halodurans*. *Journal of Cleaner Production*, *87*, 966–970.

Shintani, N., & Shoda, M. (2013). Decolorization of oxygen-delignified bleaching effluent and biobleaching of oxygen-delignified kraft pulp by non-white-rot fungus *Geotrichum candidum* Dec 1. *Journal of Environmental Sciences*, *25*, 164–168.

Singh, A. K., Kumar, A., Bilal, M., & Chandra, R. (2021). Organometallic pollutants of paper mill wastewater and their toxicity assessment on Stinging catfish and Sludge Worm. *Environmental Technology and Innovation*, *24*, 101831. Available from https://doi.org/10.1016/j.eti.2021.101831.

Singh, A. K., Kumar, A., & Chandra, R. (2020a). Detection of refractory organic pollutants from pulp paper mill effluent and their toxicity on *Triticum aestivum*; *Brassica campestris* and *Tubifex-tubifex*. *Journal of Experimental Biology and Agricultural Sciences*, *8*(5), 663–675.

Singh, A. K., Kumar, A., & Chandra, R. (2020b). Residual organic pollutants detected from pulp and paper industry wastewater and their toxicity on Triticum aestivum and Tubifex tubifex worms. *Materials Today: Proceedings*. Available from https://doi.org/10.1016/j.matpr.2020.10.862.

Singhal, A., & Thakur, I. S. (2009a). Decolorization and detoxification of pulp paper effluent by *Cryptococcus* sp. *Biochemical Engineering Journal*, *46*(1), 21–27.

Singhal, A., & Thakur, I. S. (2009b). Decolourization and detoxification of pulp and paper mill effluent by *Emericella nidulans* var. nidulans. *Journal of Hazardous Materials*, *171*, 619–625.

Taniguchi, M., Suzuki, H., Watanabe, D., Sakai, K., Hoshino, K., & Tanaka, T. (2005). Evaluation of pretreatment with *P. ostreatus* for enzyme hydrolysis of rice straw. *Journal of Bioscience and Bioengineering*, *100*, 637–643.

Thakur, I. S. (1996). Use of monoclonal antibodies against dibenzo-p-dioxin degrading *Sphingomonas* sp. strain RW1. *Letters in Applied Microbiology*, *22*, 141–144.

United States Environmental Protection Agency, USEPA. (2012). Endocrine disruptor screening program, universe of chemicals for potential endocrine disruptor screening and testing. *jointly developed by the Office of Chemical Safety & Pollution Prevention, the Office of Water and the Office of Research and Development*, 1–176. <https://www.epa.gov/sites/production/files/2015-07/documents/edsp_chemical_universe_and_general_validations_white_paper_11_12.pdf>.

Wang, W. D., Yan, L., & Cui, Z. J. (2011). characterization of a microbial consortium capable of degrading lignocellulose. *Bioresource Technology*, *120*, 9321–9324.

Wang, Y., Liu, Q., Yan, L., Gao, Y., Wang, Y., & Wang, W. (2013b). A novel lignin degradation bacterial consortium for efficient pulping. *Bioresource Technology*, *139*, 113–119.

Wu, J., Xiao, Y. Z., & Yu, H. Q. (2005). Degradation of lig-nin in pulp mill wastewaters by white-rot fungi on bio-film. *Bioresource Technology*, *96*, 1357–1363.

Yadav, S., & Chandra, R. (2012). Biodegradation of organic compounds of molasses melanoidin (MM) from biomethanated distillery spent wash (BMDS) during the decolorization by a potential bacterial consortium. *Biodegradation*, *23*, 609–620. Available from https://doi.org/10.1007/s10532-012-9537-x.

Yan, L., Gao, Y., Wang, Y., Liu, Q., Sun, Z., Fu, B., ... Wang, W. (2012). Diversity of a mesophilic lignocellulolytic microbial consortium which is useful for enhancement of biogas production. *Bioresource Technology*, *111*, 49–54.

Yang, C., Cao, G., Li, Y., Zhang, X., Ren, H., Wang, X., ... Xu, P. (2008). A constructed alkaline consortium and its dynamics in treating alkaline black liquor with very high pollution load. *PLoS One*, *3*(11), e3777.

CHAPTER 18

Nanocellulose: fascinating and sustainable nanomaterial for papermaking

Ritesh Kumar and Gulshan Kumar
University School of Basic and Applied Sciences, Guru Gobind Singh Indraprastha University, New Delhi, India

18.1 Introduction

In the 21st century, nanotechnology has emerged as a broad multidisciplinary field showing the discovery of novel materials, devices and systems examined, processes development, and fabrication at the nanoscale. Nanotechnology is mainly applied as a novel tool to modify materials to achieve desired properties. Moreover, nanotechnology is important because of less space, less material, faster, and less energy, a diverse array of ultra-small-scale material, and wide-ranging processing method on the molecular scale (El-Samahy, Mohamed, Abdel Rehim, & Mohram, 2017; Kamel, 2007; Mohieldin, Zainudin, Paridah, & Ainun, 2011). Some possible application areas of nanotechnology are electronics, plastics, pharmaceuticals, cosmetics, packaging, etc. (Samyn, Barhoum, Öhlund, & Dufresne, 2018). Packaging materials required higher strength, water and oxygen barrier properties, and antimicrobial property especially for food packaging (Soni, Hassan, Schilling, & Mahmoud, 2016). The pulp and paper industry also required all these property enhancements, so nanotechnology is a better choice to be used in this sector (Klemm, Heublein, Fink, & Bohn, 2005; Samyn et al., 2018). With the increased environmental concerns, nanomaterial from renewable origins has been strongly recommended (Kumar, Rai, & Kumar, 2019). In this sense, cellulose derived from readily available biomass (as produced by all plants and microorganisms including bacteria, plankton, and algae) is the most abundant biomaterial in the biosphere and its worldwide annual production is approximately 75 billion tones and has been used in various domestic and industrial applications. It has superior mechanical and chemical properties compared to any other natural or synthetic polymer (Ferrer, Pal, & Hubbe, 2016; Rana, Mitra, & Banerjee, 1999). For more than 150 years, cellulose and its derivative have been used in various applications such as coating, optical films, pharmaceuticals, pulp and paper, and food and cosmetic additives (shown in Fig. 18.1; Jeevahan & Chandrasekaran, 2019; Sun, Yang, & Wang, 2016;

Nanotechnology in Paper and Wood Engineering
DOI: https://doi.org/10.1016/B978-0-323-85835-9.00001-5

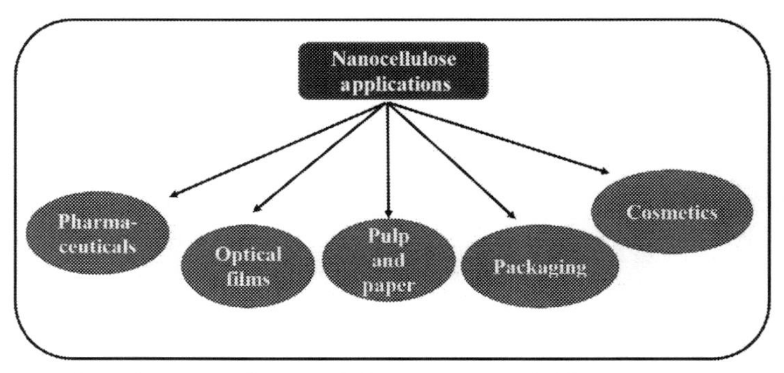

Figure 18.1 Potential applications of nanocellulose in various fields.

Tayeb, Amini, Ghasemi, & Tajvidi, 2018). Nanocellulose derived from cellulose possibly be used for the development of next generation sustainable products and is groomed for commercial-scale use only. Nanocellulose has great potential to be used in the papermaking industry to enhance the quality of paper (Clemons, 2016; Mohieldin et al., 2011). Nanocellulose possesses remarkable properties that include lightweight, a high surface area, unique optical properties, high strength, stiffness, biodegradability, renewability, and sustainability, etc. (Li, Wang, Ma, & Wang, 2018; Toivonen, Kurki-Suonio, & Schacher, 2015). Considering all these remarkable properties and nanocellulose compatibility with the pulp may helpful in improving the barrier and mechanical properties of paper, to use it for various applications like printing media and food packaging. Product developed from nanocellulose has low safety risks, less human or animal health hazard, and a low environmental impact (Eichhorn, Dufresne, & Aranguren, 2010; Moon, Martini, & Nairn, 2011). Paper and board material produced from nanocellulose present higher density and flexibility, have a lower coefficient of thermal expansion, can be optically transparent, low porosity, and excellent barrier properties to oxygen and water(Balea, Fuente, & Monte, 2020; Korhonen & Laine, 2014; Lu, An, & Zhang, 2020). Nowadays, several review article and book chapters have been published about the uses of nanocellulose in papermaking as a strength additive (Samyn et al., 2018), coating agent (Ridgway & Gane, 2012), wet strength aid (Merayo, Balea, & de la Fuente, 2017), for flexographic ink removal (Diab, Curtil, & El-shinnawy, 2015; Sun, Hou, Liu, & Ni, 2015), as linting control agent (Balea, Monte, & de la Fuente, 2017), book restoration (Syverud & Stenius, 2009), or as a vehicle to confer special paper properties, like antimicrobial (Santos, Carbajo, & Gómez, 2017), electric behavior (Eichhorn et al., 2010), and fireproof (Balea et al., 2020). However, cost-related issues associated with production cost and commercialization of cellulose nanofibers are still a challenge. Therefore for the production of nanocellulose on a large scale, further research and development are required to address various cost-related issues (Kim, Shim, & Kim, 2015). However, this chapter aims to gather all relevant information about the technical issues that emerged during the use of nanocellulose in papermaking. It also includes a brief introduction to nanocellulose, bulk addition of nanocellulose, wet-end optimization, modification, functional properties, and market perspective of nanocellulose.

18.2 Chemistry of cellulose

Cellulose is a homopolysaccharide built-up from repeating (β-1,4) linked glucopyranose units and is the most abundant and renewable biopolymer found in nature (Kinloch, Taylor, & Techapaitoon, 2015). In recent years, cellulose has been considered as an alternative biopolymers to the petroleum-based materials. The use of cellulose-based materials has increased at a tremendous rate because of their recyclability, high mechanical strength, and their relatively low price compared to conventional inorganic materials (Ridgway & Gane, 2012). Generally, cellulose consists of both amorphous and crystalline domains. There are three hydroxyl groups in each of the anhydroglucose units at C-2, C-3, and C-6 positions; the alcohol at the C-6 position is a primary alcohol, while the alcohol at positions C-2 and C-3 are secondary alcohols (Fig. 18.2). The hydroxyl group in cellulose structure is available for hydrogen bonding, which affects the physical and chemical properties of cellulose (Abraham, Deepa, & Pothen, 2013; Balea et al., 2017; Cao, Dong, & Li, 2007). Intermolecular and intramolecular hydrogen-bonding effects directly the crystallinity, hydroxyl reactivity, solubility as well as chemically modify the cellulose to produce different derivatives (Souza, Kano, & Bonvent, 2017). The structure of cellulose is extremely complex, it does not exist as an individual molecule. The structure of cellulose consists of hemicellulose, cellulose, lignin, and other extractives such as pectins, mineral matters, waxes, and proteins in varying amounts depending on their origin (Jeevahan & Chandrasekaran, 2019). Lignin is a three-dimensional polymer network that binds cellulose and hemicellulose to form a tight and compact structure. Hemicellulose acts as an interface between hydrophilic cellulose and hydrophobic lignin, and it links with lignin and cellulose in plant fiber cells (Ferrer et al., 2016). Cellulose is the major component of cell walls in plants and represents about 35%–50% of the dry weight of lignocellulosic biomass and hemicellulose and lignin present at about 20%–35% and

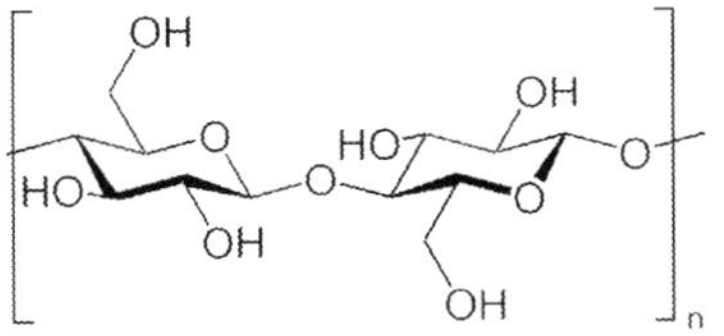

OH OH HO O O HO O O O OH OH n

Figure 18.2 Structure of cellulose (https://en.wikipedia.org/wiki/Cellulose#/media/File:Cellulose_Sessel.svg).

10%–25% of the dry weight of biomass, respectively (Sharma, Thakur, & Bhattacharya, 2019). By introducing different substituents and by changing the hydrogen-bonding network cellulose can be chemically functionalized and cellulose derivatives have been designed and adjusted to obtain certain desired properties. Cellulose can occur in four polymorphic forms: cellulose I, cellulose II, cellulose III, and cellulose IV (Islam, Alam, & Zoccola, 2013). Cellulose I is a naturally occurring polymorphic form of cellulose and composed of two distinct crystalline forms, Iα (triclinic one chain unit cell) and Iβ (monoclinic two chain unit cell). Among all the polymorphs cellulose II is rarely found in nature and a more stable form of cellulose can be produced by mercerization of cellulose I. Cellulose III and IV are also present and they can be prepared by chemical treatment of either cellulose I or II (Abdul Khalil, Davoudpour, & Saurabh, 2016). Cellulose is isolated from the source by large-scale chemical pulping by removing unwanted hemicellulose, lignin, and other extractives such as waxes, pectins, minerals matters, and proteins (Li et al., 2018). Different raw materials, physical and chemical modification, as well as different bottom-up and top-down production methods, have been used to isolate nanocellulose with different properties, such as morphology, aspect ratio, size, degree of crystallinity, and crystal structure (Liu, Yuan, Bhattacharyya, & Easteal, 2010).

18.3 Source of cellulose

Wood is the major source of cellulose, due to its plantation in a major part of the world or availability of wood supply from forests, accounting for 90% of pulp production. It possesses higher cellulose content of about 40% and remained the major source of cellulose (Islam et al., 2013). Nonwood fibers such as flax, jute, and kenaf are seasonally available and have a slightly lower cellulosic content of about 30%, can only contribute 5%–10% of pulp worldwide (Jorfi & Foster, 2015). Apart from plant sources, cellulose can be extracted from nonplant materials such as tunicate, fungi, sea creatures, and bacteria. Generally, pure cellulose can be produced from *Acetobacter xylinum* bacteria with appealing properties (higher purity and crystallinity than plant cellulose) (Kashcheyeva, Gismatulina, & Budaeva, 2019). Despite having good properties, the cost of production is the bigger issue to produce bacterial cellulose on industrial scale. While plant fiber can be the cheaper option to produce cellulose because of its widespread availability and continuous supply of cellulose (Desmet, Takács, Wojnárovits, & Borsa, 2011). From all the plant fibers, cotton has a higher value of cellulosic content (95%–99% cellulose in its original form) (de Morais Teixeira, Corrêa, & Manzoli, 2010). However, it also requires chemical treatment to remove the noncellulosic materials such as organic acids, inorganic salts, and protein. Nanocellulose can also be obtained from the banana rachis, corn cobs, jute, jute plant waste, castor plant waste, palm leaves, and cuttings, coconut husk, and mulberry plant.

Characteristics of cellulose varied according to the change in the degree of polymerization and polydispersity, fiber bundle size, morphology, and so on (Kumar et al., 2019; Wang, Huang, & Lu, 2009).

18.4 Nanocellulose

Cellulose can be isolated from plants, wood, algae, and bacteria. Cellulose microfibrils are the intrinsic ingredient of wood that have 20−100 nm width and 100−200 nm long fiber length. Cellulose consists of both crystalline and amorphous parts. The amorphous part of cellulose is easy to break, while it is impossible to break the crystalline part of cellulose due to the strong hydrogen bonding between hydroxyl groups. Nanocellulose is the crystalline part of cellulose having a diameter <100 nm and length in few micrometers, which can be isolated from natural cellulose fiber by the pretreatment of natural cellulose and breaking of amorphous part of cellulose (De France, Hoare, & Cranston, 2017; Jeevahan & Chandrasekaran, 2019; Lönnberg, Fogelström, & Berglund, 2008). It is light in weight having a density of around 1.6 gm cc^{-1}, a tensile strength of 10 GPa, which is comparable to cast iron, and biodegradable. It also has reactive hydroxyl groups that make it suitable for surface functionalization for use in various applications (Ferrer et al., 2016; Nair, Zhu, Deng, & Ragauskas, 2014). Generally, there are three types of nanocellulose: cellulose nanocrystals, nanofibrillated cellulose, and bacterial cellulose (Kashcheyeva et al., 2019; Shi, Phillips, & Yang, 2013). Schematic representation of isolation of cellulose nanofibers and cellulose nanocrystals from lignocellulosic biomass is shown in Fig. 18.3. Owing to its distinctive properties, such as tunable surface chemistry, crystallinity, high mechanical strength, nontoxicity, barrier properties, and biodegradability, it emerged as a potential renewable green nanomaterial for coatings, fillers in composites, and many more uses which will significantly influence the commercial application (Voisin, Bergström, Liu, & Mathew, 2017).

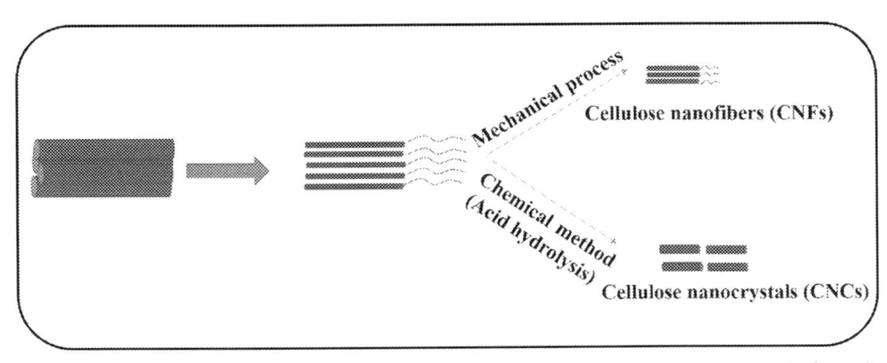

Figure 18.3 Schematic of isolation of cellulose nanofibers and cellulose nanocrystals from lignocellulosic biomass.

18.4.1 Cellulose nanofibers

Cellulose nanofibers consist of aggregates of elementary nanofibrils that possessed alternating amorphous and crystalline domains and exhibit web-like structures. In contrast to cellulose nanocrystals, cellulose nanofibers have more flexibility due to the presence of an amorphous region, which is not present in cellulose nanocrystals (Endes, Camarero-Espinosa, & Mueller, 2016). Cellulose nanofibers should have a width of 20−50 nm and length varies from 500 to 2000 nm which depends mainly on chemical modification and condition used during its preparation (Toivonen et al., 2015). Various sources such as wood pulp, bacterial cellulose, nonwood plants, agriculture, and wood residue can be used, and different mechanical, biological, and chemical methodologies are usually used to isolate cellulose nanofibers. Cryocrushing, high-pressure homogenization, and grinding are a few of the mechanical techniques used to isolate cellulose nanofibers (Serra, González, & Oliver-Ortega, 2017; Vieira, Da Silva, Dos Santos, & Beppu, 2011).

18.4.2 Cellulose nanocrystals

Cellulose nanocrystals exhibited a highly crystalline rod-like shape, typically ranging from 5 to 70 nm in width and 50 to 500 nm in length, and have high rigidity compared to cellulose nanofibers because of the removal of the higher portion of the amorphous region during chemical treatment (Sheltami, Abdullah, & Ahmad, 2012). The preparation of cellulose nanocrystals from biomass begins with the pretreatment (alkaline and bleaching) of biomass followed by strong acid hydrolysis. This process ensures the elimination of the amorphous domain while, the crystalline domain remains intact, giving rise to the production of cellulose nanocrystals (Cao et al., 2007). The cellulose nanocrystals obtained are highly crystalline in the range from 54% to 88%. However, reaction condition, cellulose source, isolation procedure, and pre- and post treatment, the characteristics of the cellulose nanocrystals such as dimension, morphology, yield, crystallinity, surface chemistry, thermal and physicochemical properties can be tailored for specific uses. Cellulose nanocrystals can be used in aerogels, low-calorie food additives, liquid crystals, biosensors, bioimaging, and energy storage (Dammak, Moreau, & Beury, 2013).

18.5 Challenges for nanocellulose in papermaking

Cellulose nanofibers and cellulose nanocrystals are the two main families of nanocellulose. These nanostructures are completely different from each other based on dimensions, structure, morphology, and consequently different properties. Despite having dimensions in the nano range, the use of cellulose nanocrystals in papermaking is found difficult, and hence, only surface coating can be proposed. Therefore cellulose nanofibers become a good choice to be used as an additive or reinforcement to enhance the properties of paper in the paper industry (Das, Ghosh, & Das, 2019; Merayo et al., 2017). Cellulose nanofibers show promising results in biomedical applications due to their antimicrobial and antiinflammatory effects. Surprisingly, very less publications have been found on the use of cellulose nanofibers in papermaking. Studies have shown that like printing media and food packaging the use of cellulose nanofibers in papermaking also improves the barrier and mechanical properties of paper (Feng, Chua, & Zhao, 2017; Hassan, Hassan, & Oksman, 2011). In papermaking, cellulose nanofibers could be mixed with pulp before paper formation. The enhancement in internal cohesion of paper with cellulose nanofibers, tensile strength, burst index is because of large number of hydroxyl groups' presence on the surface of cellulose nanofibers, which increases the hydrogen bonding with fibers. However, this could decrease brightness, increases the drainage time of pulp, and increases the energy consumption for an efficient cellulose nanofibers' dispersion. Also, cellulose nanofibers' small size and high swelling ability show the high impact on barrier properties and porosity, respectively. The presence of cellulose nanofibers on paper decrease porosity, which improves barrier properties; but the high cellulose nanofiber's swelling ability increases the water-vapor transmission rate of paper that affects the barrier properties (Boufi, González, & Delgado-aguilar, 2016; Klemm, Kramer, & Moritz, 2011). The main drawbacks of cellulose nanofibers for industrial use are their high cost due to the high cost of chemical and huge energy requirement, which cannot be tolerated for large-scale production. However, chemical or enzymatic pretreatment before mechanical treatment has been suggested by researchers to reduce energy consumption during processing. The lack of cost-effective, reliable, fast, and accurate measurement tools capable to characterize nanocellulose at large-scale production is also a bigger challenge in the path of nanocellulose to be used at an industrial scale. Nanocellulose properties such as their stability over time and dispersion, are critical in several large-scale unit operations, like pumping, storage, filtering, mixing, as well as for the improvement of quality of the final product. Therefore the gap between production cost, quality, and application must be filled in to develop a valuable product for society.

18.6 Application of cellulose nanofibers into the papermaking

Nowadays, nanocellulose became a potential biomaterial for many applications to develop the quality-based product. So, in papermaking direct addition of cellulose nanofibers is used in the beginning or onto the final paper as coating material. Thus with the use of cellulose nanofibers, a modified paper sheet can be developed with enhanced physicomechanical properties (Fig. 18.4). Table 18.1 shows the application of cellulose nanofibers based on their type and way of addition.

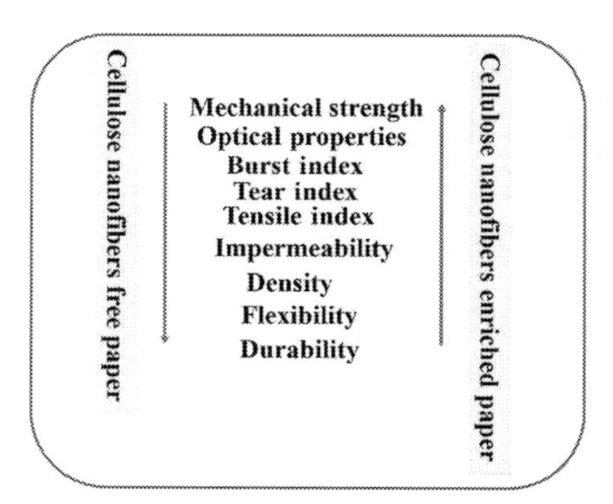

Figure 18.4 Effect of cellulose nanofibers on the properties of the developed paper sheet.

18.6.1 Direct reinforcement of cellulose nanofibers into the pulp suspension

The addition of an adequate amount of cellulose nanofibers and addition points are the key points to improve paper mechanical properties, improvement of the filler retention as well as water vapor transmission rate. Fig. 18.5 shows the schematic representation of direct addition of cellulose nanofibers into base paper. The presence of a hydrogen bond between each cellulose nanofibers increases the strength of the fiber network. Thus flexibility, strength, small size, and high surface area may be responsible to enhance strength of the developed composite material (Hassan et al., 2011; Merayo et al., 2017). Hassan et al. (2011) studied the effect of the addition of cellulose nanofibers on the tensile strength, opacity, burst strength, porosity, and tear resistance of paper sheets. The addition of cellulose nanofibers into bagasse pulp suspension enhanced the dry and wet tensile strength. However, tear resistance and burst strength decrease with an increase in cellulose nanofibers' concentration. Sehaqui, Zhou, and Berglund (2013) have observed the enhancement in density for softwood kraft pulp sheets by the addition of 10% homogenized cellulose nanofibers by 30%–50%. In the same trend, grinder-produced cellulose nanofibers' addition to softwood pulp enhances the density up to 20%. Taipale, Österberg, and Nykänen (2010) have reported a great decrement in water permeability from 1450 to 450 mL min^{-1} with the increasing percentage of cellulose nanofibers from 0% to 30%. The addition of cellulose nanofibers can enhance the breaking length of the paper sheet. Higher the percentage of cellulose nanofibers results in a higher breaking length of the paper sheet. Ahola, Österberg, and Laine (2008) have reported that the addition of 6% cellulose nanofibers to softwood pulp increases the tensile strength up to 100%. Tensile strength

Table 18.1 Application of nanocellulose based on their type and way of addition in papermaking.

Nanocellulose type	Content(wt. %)	Additives added	Pulp for paper sheet	Increase in tensile index (%)	References
Mechanical (CNF)	6	None	Beaten and unbeaten bleached eucalyptus kraft	300	Boufi et al. (2016)
Carboxymethylated CNF	20	None	TMP	15	Brodin and Eriksen (2015)
CNF	4	None	Beaten TMP	21	Brodin and Eriksen (2015)
CNF	10	None	Unbeaten bleached softwood craft	40	Diab et al. (2015)
CNF	20	None	Beaten softwood and beaten baggase	60	Afra, Yousefi, Mahdi, and Nishino (2013)
CNF	6	None	Handsheets (60 g/m^2)	26–30	Hassan et al. (2011)
CNF	0–15	None	Handsheets (60 g/m^2)	20	Madani, Kiiskinen, Olson, and Martinez (2011)
CNF	3–10	None	50–250 g/m^2 Laminated paper	20–30	Mörseburg and Chinga-Carrasco (2009)
CNF	0.1–2 g/m^2	None	Multilayer board	20	Satam, Irvin, and Lang (2018)
TEMPO oxidized (neutral pH)	12	Cationic starch (CS)	Unbeaten unbleached hardwood kraft	169	Boufi et al. (2016)
TEMPO oxidized (neutral pH)	4.5	Cationic starch (CS)	Biobeaten bleached hardwood kraft	68	González, Vilaseca, and Alcalá (2013)
CNF	5	C-PAM	Bagasse pulp	40	Rahman, Petroudy, and Syverud (2013)
TEMPO oxidized (basic pH)	4.5	Cationic starch (CS)	Deinked old newspaper	82	Delgado-Aguilar, Recas, and Puig (2015)
TEMPO oxidized (alkaline pH)	6	Cationic starch (CS)	Stone groundwood hardwood	135	Boufi et al. (2016)
CNF(acetylated)	10	C-PAM	Unbeaten softwood	17	Mashkour, Afra, Resalati, and Mashkour (2015)

Notes: CNF, cellulose nanofiber; *C-PAM*, cationic polyacrylamide; *TEMPO*, 2,2,6,6-tetramethylpiperidine-1-oxyl radical; *TMP*, thermomechanical pulp.

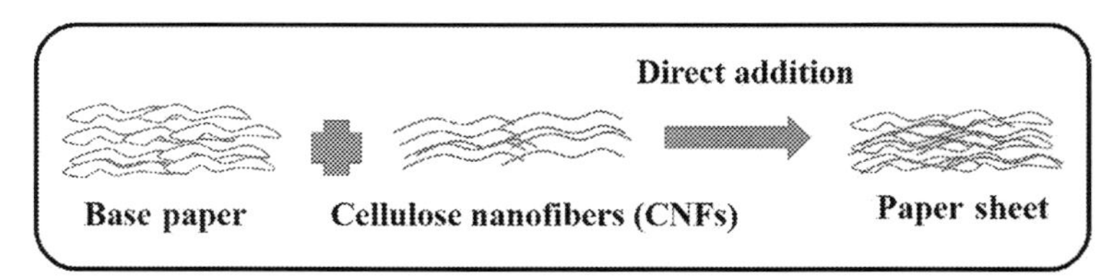

Figure 18.5 Schematic of direct addition of cellulose nanofibers into base paper.

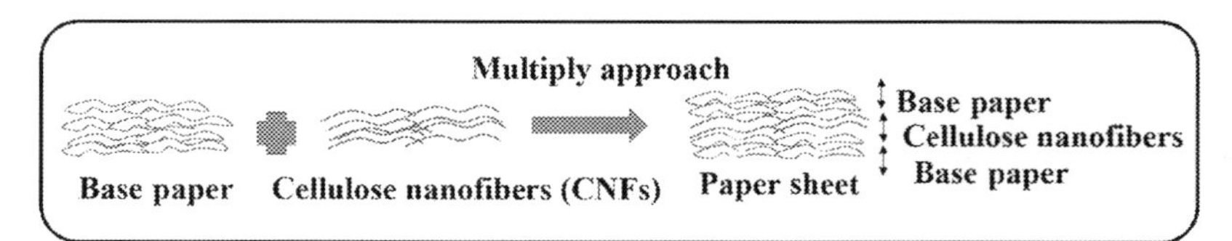

Figure 18.6 Schematic of multiply approach for paper sheet fabrication with the addition of cellulose nanofibers.

of paper depends mainly on shape, length, orientation, bond area, bond strength, and distribution of fiber. The effect of cellulose nanofibers on tensile strength depends on the dosage of cellulose nanofibers, degree of pulp refining, the intensity of fibrillation, strategy, and type of addition of retention agent. The tensile index increment mainly depends on the amount of cellulose nanofibers content in the paper sheet (Iwamoto, Nakagaito, & Yano, 2007; Mörseburg & Chinga-Carrasco, 2009). In the market different quality of cellulose nanofibers are available that affect the desired result.

18.6.2 Multiply strategy

Compared to the direct addition of cellulose nanofibers in the base paper, Mörseburg and Chinga-Carrasco (2009) reported the use of cellulose nanofibers as a single ply in a paper sheet. Fig. 18.6 shows the schematic representation of the multiply approach for the fabrication of paper sheets. The study was based on the interaction between cellulose nanofibers, clay, and thermomechanical pulp in an oriented layered laboratory sheet. Based on roughness, strength, gloss, tensile index, and light scattering, they found that the best sheet construction was obtained when cellulose nanofibers layer placed in the center. Cellulose nanofibers suspension ranging from 0.1 to 2 $g\,m^{-2}$ (as dry) can be deposited between the two plies of the board surface using a spray coating technique. The developed material with enhanced properties can be used as food packaging boards. However, there are several drawbacks of using these strategies to use cellulose nanofibers that include relatively higher cost as compared to wet-end process, poor retention of cellulose nanofibers in the fibrous materials, and showed negative impact on drainage and retention (Kangas, Lahtinen, & Sneck, 2014; Korhonen & Laine, 2014).

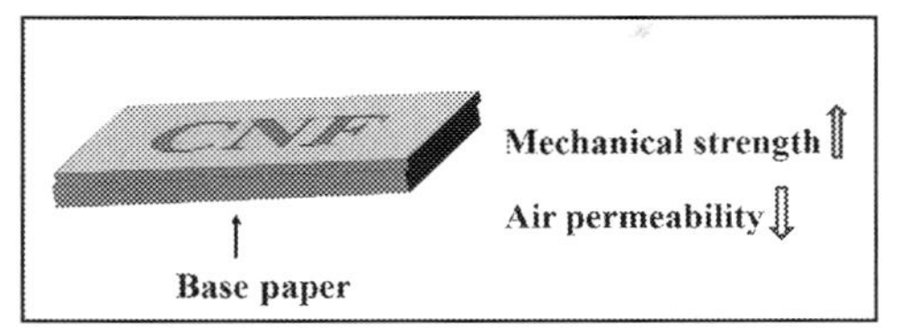

Figure 18.7 Schematic representation of coating of cellulose nanofibers (CNFs) over the base paper.

18.6.3 Pure cellulose nanofibers coating

As a coating material, Syverud and Stenius (2009) in 2009 reported the preparation of cellulose nanofibers coated paper. Fig. 18.7 shows the schematic representation of cellulose nanofibers' coating over the base paper. They found that the coating of cellulose nanofibers on the surface of base paper significantly enhanced the mechanical strength of the paper sheet and reduced its air permeability. The prepared paper sheets can be used as biodegradable and transparent packaging films having high mechanical strength and higher barrier properties. However, directly coating paper and board with cellulose nanofibers show high overall cost, process issues, sensitivity to humidity, and least improvement in properties. For the coating of cellulose nanofibers on the paper surface, various techniques such as spray, roll, bar, and size press coating have been used. Aulin, Salazar-Alvarez, and Lindström (2012) reported the strong impact of coating cellulose nanofibers on the reduction of air permeability. They coated two different papers, that is, greaseproof paper and kraft paper and found that the air permeability significantly decreased for both papers. Bardet and Bras (2014) reported that the cellulose nanofibers' ratio of 6%–8% is required for homogeneous deposition of a layer of cellulose nanofibers, irrespective of the basic weight of the paper. Thus thin cellulose nanofibers layer exerts a positive impact on liquid and gas barrier properties. Syverud and Stenius (2009) have studied that the cellulose nanofibers' coating on the unrefined softwood pulp results in higher tensile index and also improved the barrier properties. Lavoine, Desloges, Dufresne, and Bras (2012) have used the bar coating method for coating cellulose nanofibers on cardboard and found that the compressive strength and stiffness have been increased while barrier properties of cardboard decrease in this study. Ridgway and Gane (2012) have reported that the cellulose nanofibers coated cardboard or paperboard shows higher blending stiffness. Chinga-carrasco and Syverud (2012) have reported a lower oxygen transmission rate for 2,2,6,6-tetramethylpiperidine-1-oxyl radical (TEMPO)-oxidized cellulose nanofibers film at 50% relative humidity. Aulin et al. (2012) have reported that cellulose nanofibers coated paper shows higher oil resistance in comparison to without cellulose nanofibers coated base paper. Cellulose nanofibers blocked the pores of base paper resulted in reduction of oil transmission capacity of the paper.

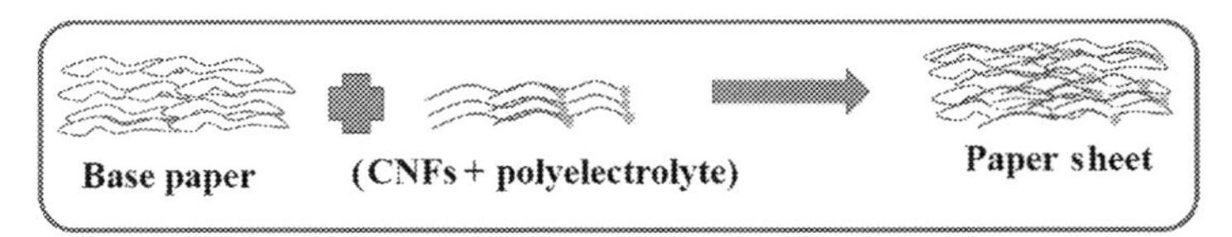

Figure 18.8 Schematic of paper sheet preparation with the addition of cellulose nanofibers (CNFs) with polyelectrolyte.

18.6.4 Wet-end optimization (cellulose nanofiber + polyelectrolyte)

Solely addition of cellulose nanofibers into pulp suspension leads to an increase in total surface area of pulp suspension, thus increasing the drainage time by increasing the hydrogen bonding with the fibers due to the presence of hydroxyl groups on the surface of cellulose nanofibers. Hence to improve the wet-end process, research has mainly focused on the combination of cellulose nanofibers with polyelectrolyte, such as polyacrylamide (PAM) improves the mechanical properties and retention of the fillers without increasing the drainage process (Ferrer et al., 2016; Merayo et al., 2017; Mousavi, Afra, Tajvidi, & Bousfield, 2017). Fig. 18.8 shows the schematic representation of the addition of cellulose nanofibers with polyelectrolyte into the paper to enhance its properties. Taipale et al. (2010) confirmed that the use of 3% dry weight of cellulose nanofibers in papermaking increases the mechanical strength of the paper while decreases the drainage rate of pulp suspension. To compensate for this constraint, along with cellulose nanofibers an optimum dosage of polyelectrolyte [cationic PAM (C-PAM) and polyelectrolytes cationic starch (CH)] and process conditions (pH and salt addition) have been found to enhance the mechanical properties without increasing the drainage process. Manninen, Puumalainen, Talja, and Pettersson (2002) studied the effect of cellulose nanofibers along with cationic starch (CS) in papermaking. They found that with the addition of cellulose nanofibers (5%–20%) decrease in dimension stability was revealed. However, the drying shrinkage and the hygroexpansion could be improved with the addition of polyelectrolyte cationic starch. Finally, they found that the combination of cellulose nanofibers and cationic starch led to paper sheets with improved mechanical properties, dimensional stability, and drying shrinkage. Rahman et al. (2013) reported that by combining cellulose nanofibers and C-PAM, it is possible to achieve a high tensile index without increasing the drainage time. Merayo et al. (2017) also proved that with the combination of cellulose nanofibers, C-PAM, and bentonite, both tensile index and drainage rate can be enhanced.

18.7 Modification of nanocellulose

To improve retention of cellulose nanofibers, a chemical pretreatment of cellulose nanofibers during production or by means of cellulose nanofibers' reaction with convenient reagent is proposed to modify surface charge density by different authors. Since retention systems are always cationic, the approach of cationizing cellulose nanofibers can avoid or minimize the necessity of other synthetic retention agents and flocculants. Using this method, retention and mechanical properties of cellulose nanofibers can be enhanced (Merayo et al., 2017).

Korhonen and Laine (2014) cationized the cellulose nanofibers with glycidaltrialkylammoniumchloride (GTMAC) and observed that the increase of mechanical properties with cationic cellulose nanofibers could be higher than unmodified cellulose nanofibers. Lu et al. (2020) observed that the cellulose nanofibers modified with GTMAC increased the density (0.218 vs 0.265 $g\ m^{-3}$) and basic weight (76.2 vs 78.7 $g\ m^{-2}$) of the paper sheet which was due to the improved fine retention in the paper structure. Cationic cellulose nanofibers also enhance the mechanical strength of the paper sheet without affecting the dewatering rate. Huang, Zhou, and Dong (2017) found that the young modulus of paper made of softwood pulp significantly increased when 10% of fibers were replaced by cationic cellulose nanofibers instead of untreated or TEMPO-oxidized cellulose nanofibers. This might be possible due to the strong bonding between fibers and cationic cellulose nanofibers due to strong electrostatic force. Liu, Liu, and Hui (2019) proved that 0.4% cationic cellulose nanofibers enhanced precipitated calcium carbonate (PCC) and tobacco pulp retention by 31.8% and 81.6%, respectively. Air and bulk permeabilities of paper sheets increased by 41.8% and 6.8% and drainage of tobacco pulp was also improved (Fig. 18.9). Carboxymethylation is also a good process to increase the charge of the cellulose nanofibers by the incorporation of carboxymethyl groups in the hydroxyl groups of cellulose that may affect the interaction with fiber (Merayo et al., 2017). The use of carboxymethylated cellulose nanofibers can result in the enhancement of filler retention and tensile strength by comparing with that for TEMPO-oxidized cellulose nanofibers. Xu, Li, and Cheng (2014) found that the combination of carboxymethyl cellulose nanofibers with C-PAM resulted in the improvement of retention and drainage rate of the paper sheet.

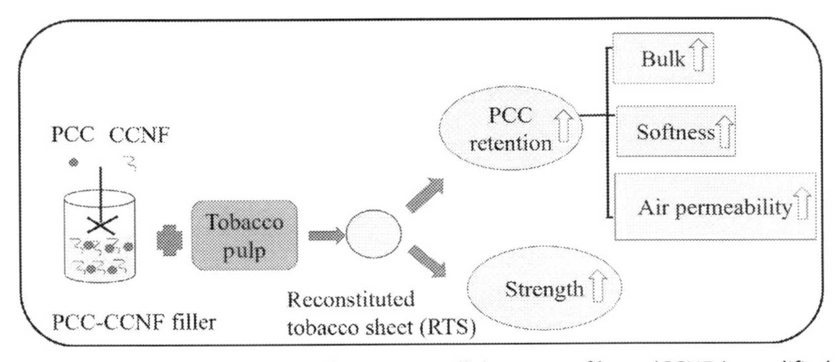

Figure 18.9 Schematic of fabrication of cationic cellulose nanofibers (CCNFs) modified reconstituted tobacco sheet. *PCC*, Precipitated calcium carbonate.

18.8 Functional properties of cellulose nanofibers

Arbatan, Zhang, Fang, and Shen (2012) use the formulation of PCC, alkyl ketene dimer, and cellulose nanofibers to prepare the superhydrophobic paper. It was found that the cellulose nanofibers can act as the essential component to the formation of the PCC coating layer required for the superhydrophobicity of the paper. Martin, Freire, Pinto, and Silvestre (2012) studied the antimicrobial property of silver nanoparticles and cellulose nanofibers hybrids and used it as a filler material in starch coating formulation for paper sheets. Chiappone, Nair, and Gerbaldi (2011) developed thin and flexible lithium electrolyte membranes based on cellulose materials reinforced with cellulose nanofibers with excellent mechanical stability. In this way, value-added products based on cellulose nanofibers will be a great opportunity to develop a new value-added market.

18.9 Market perspectives of nanocellulose

Nanocellulose is widely used in packaging (36% of the demand), filtration (19% of the demand), composites (25%), and paper products. There are various types of nanocellulose with different characteristics and cost is available in the market depending on production method and reaction condition (Heggset et al., 2018; Kyle, Jessop, & Al-sabah, 2018). Shatkin, Wegner, and Bilek (2014) estimated the worldwide use of nanocellulose in paper and board industry (4.56 million metric tons), automotive components (4.16 million metric tons), packaging sector (11.8 million metric tons), cement (4.13 million metric tons), hygiene and absorbent products (3.24 million metric tons), plastic film replacement (3.37 million metric tons), and textile for clothing (2.54 million metric tons). Around 20 years ago production of cellulose nanofibers on an industrial scale was limited to a few companies and research centers mainly from Japan, France, and North America. Nowadays, many companies from worldwide such as Fiberlean technologies (United Kingdom), Borregaard (Norway), GranBio (United States), Nippon paper industries Co. Ltd. (Japan), celluForce (Canada), Kruger (Canada), etc. are the largest producer of cellulose nanofibers (Balea et al., 2020).

18.10 Conclusion

The use of nanocellulose in papermaking has been established as a potential additive to improve the overall performance of paper. After nanocellulose addition, the density and strength of paper are enhanced, whereas opacity and porosity of paper are reduced. The use of nanocellulose as coating material also developed high-quality paper with enhanced strength and barrier properties. Nanocellulose combined with different materials shows a positive impact on the properties of paper. The use of materials such as clay or calcium carbonate along with nanocellulose improved the deposition of nanocellulose coating onto base paper. It improves the barrier and strength properties of the paper. Moreover, the commercial cost of production of nanocellulose is lower or equivalent to petrol-based materials (latex, strengthening agent, etc.), which can be used as an alternative. However, there are few challenges such as minimizing the production cost of nanocellulose, drainage rate is reduced that may increase drying cost, and cellulose nanofibers' coating onto base paper contains higher water. These challenges suggest that a lot of research is required in coming years to explore the all-possible potential of nanocellulose in a wide range of paper-making applications.

References

Abdul Khalil, H. P. S., Davoudpour, Y., Saurabh, C. K., et al. (2016). A review on nanocellulosic fibres as new material for sustainable packaging: Process and applications. *Renewable and Sustainable Energy Reviews*, *64*, 823–836. Available from https://doi.org/10.1016/j.rser.2016.06.072.

Abraham, E., Deepa, B., Pothen, L. A., et al. (2013). A brief review on extraction of nanocellulose and its application. *Carbohydrate Polymers*, *4*, 81–87. Available from https://doi.org/10.1109/ISBEIA.2012.6423003.

Afra, E., Yousefi, H., Mahdi, M., & Nishino, T. (2013). Comparative effect of mechanical beating and nanofibrillation of cellulose on paper properties made from bagasse and softwood pulps. *Carbohydrate Polymers*, *97*, 725–730. Available from https://doi.org/10.1016/j.carbpol.2013.05.032.

Ahola, S., Österberg, M., & Laine, J. (2008). Cellulose nanofibrils - Adsorption with poly(amideamine) epichlorohydrin studied by QCM-D and application as a paper strength additive. *Cellulose*, *15*, 303–314. Available from https://doi.org/10.1007/s10570-007-9167-3.

Arbatan, T., Zhang, L., Fang, X., & Shen, W. (2012). Cellulose nanofibers as binder for fabrication of superhydrophobic paper. *Chemical Engineering Journal*, *210*, 74–79. Available from https://doi.org/10.1016/j.cej.2012.08.074.

Aulin, C., Salazar-Alvarez, G., & Lindström, T. (2012). High strength, flexible and transparent nanofibrillated cellulose–nanoclay biohybrid films with tunable oxygen and water vapor permeability. *Nanoscale*, *4*, 6622. Available from https://doi.org/10.1039/c2nr31726e.

Balea, A., Fuente, E., Monte, M. C., et al. (2020). Industrial application of nanocelluloses in papermaking: A review of challenges, technical solutions, and market perspectives. *Molecules (Basel, Switzerland)*, *25*, 526. Available from https://doi.org/10.3390/molecules25030526.

Balea, A., Monte, M. C., de la Fuente, E., et al. (2017). Application of cellulose nanofibers to remove water-based flexographic inks from wastewaters. *Environmental Science and Pollution Research*, *24*, 5049–5059. Available from https://doi.org/10.1007/s11356-016-8257-x.

Bardet, R., & Bras, J. (2014). Cellulose nanofibers and their use in paper industry. *Materials and energy; Handbook of green materials* (pp. 207–232). River Edge, NJ: World Scientific.

Boufi, S., González, I., Delgado-aguilar, M., et al. (2016). Nanofibrillated cellulose as an additive in papermaking process: A review. *Carbohydrate Polymers*, *154*, 151–166. Available from https://doi.org/10.1016/j.carbpol.2016.07.117.

Brodin, F. W., & Eriksen, Ø. (2015). Preparation of individualised lignocellulose microfibrils based on thermomechanical pulp and their effect on paper properties. *Nordic Pulp and Paper Research Journal*, *30*, 443–451. Available from https://doi.org/10.3183/npprj-2015-30-03-p443-451.

Cao, X., Dong, H., & Li, C. M. (2007). New nanocomposite materials reinforced with flax cellulose nanocrystals in waterborne polyurethane. *Biomacromolecules*, *8*, 899–904. Available from https://doi.org/10.1021/bm0610368.

Chiappone, A., Nair, J. R., Gerbaldi, C., et al. (2011). Microfibrillated cellulose as reinforcement for Li-ion battery polymer electrolytes with excellent mechanical stability. *Journal of Power Sources*, *196*, 10280–10288. Available from https://doi.org/10.1016/j.jpowsour.2011.07.015.

Chinga-carrasco, G., & Syverud, K. (2012). On the structure and oxygen transmission rate of biodegradable cellulose nanobarriers. *Nanoscale Research Letters*, *7*, 2–7. <http://www.nanoscalereslett.com/content/7/1/192>.

Clemons, C. (2016). Nanocellulose in spun continuous fibers: A Review and future outlook. *Journal of Renewable Materialsz*, *4*, 327–339. Available from https://doi.org/10.7569/JRM.2016.634112.

Dammak, A., Moreau, C., Beury, N., et al. (2013). Elaboration of multilayered thin films based on cellulose nanocrystals and cationic xylans: application to xylanase activity detection. *Holzforschung*, *67*, 579–586. Available from https://doi.org/10.1515/hf-2012-0176.

Das, T. K., Ghosh, P., & Das, N. C. (2019). Preparation, development, outcomes, and application versatility of carbon fiber-based polymer composites: a review. *Advanced Composites and Hybrid Materials*, *2*, 214–233. Available from https://doi.org/10.1007/s42114-018-0072-z.

De France, K. J., Hoare, T., & Cranston, E. D. (2017). Review of hydrogels and aerogels containing nanocellulose. *Chemistry of Materials: a Publication of the American Chemical Society*, *29*, 4609–4631. Available from https://doi.org/10.1021/acs.chemmater.7b00531.

de Morais Teixeira, E., Corrêa, A. C., Manzoli, A., et al. (2010). Cellulose nanofibers from white and naturally colored cotton fibers. *Cellulose*, *17*, 595–606. Available from https://doi.org/10.1007/s10570-010-9403-0.

Delgado-Aguilar, M., Recas, E., Puig, J., et al. (2015). Aplicación de celulosa nanofibrilada, en masa y superficie, a la pulpa mecánica de muela de piedra: una sólida alternativa al tratamiento clásico de refinado. *Maderas: Ciencia y Tecnologia*, *17*. Available from https://doi.org/10.4067/S0718-221X2015005000028, 0–0.

Desmet, G., Takács, E., Wojnárovits, L., & Borsa, J. (2011). Cellulose functionalization via high-energy irradiation-initiated grafting of glycidyl methacrylate and cyclodextrin immobilization. *Radiation Physics and Chemistry*, *80*, 1358–1362. Available from https://doi.org/10.1016/j.radphyschem.2011.07.009.

Diab, M., Curtil, D., El-shinnawy, N., et al. (2015). Biobased polymers and cationic microfibrillated cellulose as retention and drainage aids in papermaking: Comparison between softwood and bagasse pulps. *Industrial Crops and Products*, *72*, 1–12. Available from https://doi.org/10.1016/j.indcrop.2015.01.072.

Eichhorn, S. J., Dufresne, A., Aranguren, M., et al. (2010). Review: Current international research into cellulose nanofibres and nanocomposites. *Journal of Material Science*, *45*, 1–33. Available from https://doi.org/10.1007/s10853-009-3874-0.

El-Samahy, M. A., Mohamed, S. A. A., Abdel Rehim, M. H., & Mohram, M. E. (2017). Synthesis of hybrid paper sheets with enhanced air barrier and antimicrobial properties for food packaging. *Carbohydrate Polymers*, *168*, 212–219. Available from https://doi.org/10.1016/j.carbpol.2017.03.041.

Endes, C., Camarero-Espinosa, S., Mueller, S., et al. (2016). A critical review of the current knowledge regarding the biological impact of nanocellulose. *Journal of Nanobiotechnology*, *14*, 78. Available from https://doi.org/10.1186/s12951-016-0230-9.

Feng, T., Chua, H. J., & Zhao, Y. (2017). Reduction-responsive carbon dots for real-time ratiometric monitoring of anticancer prodrug activation in living cells. *ACS Biomaterials Science and Engineering*, *3*, 1535–1541. Available from https://doi.org/10.1021/acsbiomaterials.7b00264.

Ferrer, A., Pal, L., & Hubbe, M. (2016). Nanocellulose in packaging: Advances in barrier layer technologies. *Industrial Crops and Products*, *95*, 574–582. Available from https://doi.org/10.1016/j.indcrop.2016.11.012.

González, I., Vilaseca, F., Alcalá, M., et al. (2013). Effect of the combination of biobeating and NFC on the physico-mechanical properties of paper. *Cellulose*, *20*, 1425–1435. Available from https://doi.org/10.1007/s10570-013-9927-1.

Hassan, E. A., Hassan, M. L., & Oksman, K. (2011). Improving bagasse pulp paper sheet properties with microfibrillated cellulose isolated from xylanase-treated bagasse. *Wood and Fiber Science: Journal of the Society of Wood Science and Technology*, *43*, 76–82.

Heggset, E., Strand, B. L., Sundby, K. W., Simon, S., Chinga-Carrasco, G., & Syverud, K. (2018). Viscoelastic properties of nanocellulose based inks for 3D printing and mechanical properties of CNF/alginate biocomposite gels. *Cellulose*, *26*, 581–595. Available from https://doi.org/10.1007/s10570-018-2142-3.

Huang, J., Zhou, Y., Dong, L., et al. (2017). Enhancement of mechanical and electrical performances of insulating presspaper by introduction of nanocellulose. *Composites Science and Technology*, *138*, 40–48. Available from https://doi.org/10.1016/j.compscitech.2016.11.020.

Islam, M. T., Alam, M. M., & Zoccola, M. (2013). Review on modification of nanocellulose for application in composites. *Internation Journal of Innovative Research in Science, Engineering and Technology*, *2*, 5444–5451.

Iwamoto, S., Nakagaito, A. N., & Yano, H. (2007). Nano-fibrillation of pulp fibers for the processing of transparent nanocomposites. *Applied Physics A*, *89*, 461–466. Available from https://doi.org/10.1007/s00339-007-4175-6.

Jeevahan, J., & Chandrasekaran, M. (2019). Nanoedible films for food packaging: A review. *Journal of Material Science*, *54*, 12290–12318. Available from https://doi.org/10.1007/s10853-019-03742-y.

Jorfi, M., & Foster, E. J. (2015). Recent advances in nanocellulose for biomedical applications. *Journal of Applied Polymer Science*, *132*. Available from https://doi.org/10.1002/app.41719, n/a-n/a.

Kamel, S. (2007). Nanotechnology and its applications in lignocellulosic composites, a mini review. *eXPRESS Polymer Letters*, *1*, 546–575. Available from https://doi.org/10.3144/expresspolymlett.2007.78.

Kangas, H., Lahtinen, P., Sneck, A., et al. (2014). Characterization of fibrillated celluloses. A short review and evaluation of characteristics with a combination of methods. *Nordic Pulp and Paper Research Journal*, *29*, 129–143. Available from https://doi.org/10.3183/NPPRJ-2014-29-01-p129-143.

Kashcheyeva, E. I., Gismatulina, Y. A., & Budaeva, V. V. (2019). Pretreatments of non-woody cellulosic feedstocks for bacterial cellulose synthesis. *Polymers (Basel)*, *11*, 1645. Available from https://doi.org/10.3390/polym11101645.

Kim, J. H., Shim, B. S., Kim, H. S., et al. (2015). Review of nanocellulose for sustainable future materials. *International Journal of Precision Engineering and Manufacturing-Green Technology*, *2*, 197–213. Available from https://doi.org/10.1007/s40684-015-0024-9.

Kinloch, A. J., Taylor, A. C., Techapaitoon, M., et al. (2015). Tough, natural-fibre composites based upon epoxy matrices. *Journal of Material Science*, *50*, 6947–6960. Available from https://doi.org/10.1007/s10853-015-9246-z.

Klemm, D., Heublein, B., Fink, H. P., & Bohn, A. (2005). Cellulose: Fascinating biopolymer and sustainable raw material. *Angewandte Chemie International Edition*, *44*, 3358–3393. Available from https://doi.org/10.1002/anie.200460587.

Klemm, D., Kramer, F., Moritz, S., et al. (2011). Nanocelluloses: A new family of nature-based materials. *Angewandte Chemie International Edition*, *50*, 5438–5466. Available from https://doi.org/10.1002/anie.201001273.

Korhonen, M., & Laine, J. (2014). Flocculation and retention of fillers with nanocelluloses. *Nordic Pulp and Paper Research Journal*, *29*, 119–128. Available from https://doi.org/10.3183/npprj-2014-29-01-p119-128.

Kumar, R., Rai, B., & Kumar, G. (2019). A simple approach for the synthesis of cellulose nanofiber reinforced chitosan/PVP bio nanocomposite film for packaging. *Journal of Polymers and the Environment*, *27*, 2963–2973. Available from https://doi.org/10.1007/s10924-019-01588-8.

Kyle, S., Jessop, Z. M., Al-sabah, A., et al. (2018). Characterization of pulp derived nanocellulose hydrogels using AVAP ® technology. *Carbohydrate Polymers*, *198*, 270–280. Available from https://doi.org/10.1016/j.carbpol.2018.06.091.

Lavoine, N., Desloges, I., Dufresne, A., & Bras, J. (2012). Microfibrillated cellulose – Its barrier properties and applications in cellulosic materials: A review. *Carbohydrate Polymers*, *90*, 735–764. Available from https://doi.org/10.1016/j.carbpol.2012.05.026.

Li, Y.-Y., Wang, B., Ma, M.-G., & Wang, B. (2018). Review of recent development on preparation, properties, and applications of cellulose-based functional materials. *International Journal of Polymer Science*, *2018*, 1–18. Available from https://doi.org/10.1155/2018/8973643.

Liu, D. Y., Yuan, X. W., Bhattacharyya, D., & Easteal, A. J. (2010). Characterisation of solution cast cellulose nanofibre - Reinforced poly(lactic acid). *eXPRESS Polymer Letters*, *4*, 26–31. Available from https://doi.org/10.3144/expresspolymlett.2010.5.

Liu, H., Liu, Z., Hui, L., et al. (2019). Cationic cellulose nanofibers as sustainable flocculant and retention aid for reconstituted tobacco sheet with high performance. *Carbohydrate Polymers*, *210*, 372–378. Available from https://doi.org/10.1016/j.carbpol.2019.01.065.

Lönnberg, H., Fogelström, L., Berglund, L., et al. (2008). Surface grafting of microfibrillated cellulose with poly(ε-caprolactone) - Synthesis and characterization. *European Polymer Journal*, *44*, 2991–2997. Available from https://doi.org/10.1016/j.eurpolymj.2008.06.023.

Lu, Z., An, X., Zhang, H., et al. (2020). Cationic cellulose nano-fibers (CCNF) as versatile flocculants of wood pulp for high wet web performance. *Carbohydrate Polymers, 229*, 115434. Available from https://doi.org/10.1016/j.carbpol.2019.115434.

Madani, A., Kiiskinen, H., Olson, J. A., & Martinez, D. M. (2011). Fractionation of microfibrillated cellulose and its effects on tensile index and elongation of paper. *Nordic Pulp and Paper Research Journal, 26*, 306–311.

Manninen, J., Puumalainen, T., Talja, R., & Pettersson, H. (2002). Energy aspects in paper mills utilising future technology. *Applied Thermal Engineering, 22*, 929–937. Available from https://doi.org/10.1016/S1359-4311(02)00010-8.

Martins, N. C. T., Freire, S. R., Pinto, R. J. B., & Silvestre, A. J. D. (2012). Electrostatic assembly of Ag nanoparticles onto nanofibrillated cellulose for antibacterial paper products. *Cellulose, 19*, 1425–1436. Available from https://doi.org/10.1007/s10570-012-9713-5.

Mashkour, M., Afra, E., Resalati, H., & Mashkour, M. (2015). Moderate surface acetylation of nanofibrillated cellulose for the improvement of paper strength and barrier properties. *RSC Advances, 5*, 60179–60187. Available from https://doi.org/10.1039/C5RA08161K.

Merayo, N., Balea, A., de la Fuente, E., et al. (2017). Interactions between cellulose nanofibers and retention systems in flocculation of recycled fibers. *Cellulose, 24*, 677–692. Available from https://doi.org/10.1007/s10570-016-1138-0.

Mohieldin, S. D., Zainudin, E. S., Paridah, M. T., & Ainun, Z. M. (2011). Nanotechnology in pulp and paper industries: A review. *Key Engineering Materials, 471–472*, 251–256. Available from https://doi.org/10.4028/http://www.scientific.net/KEM.471-472.251.

Moon, R. J., Martini, A., Nairn, J., et al. (2011). Cellulose nanomaterials review: Structure, properties and nanocomposites. *Chemical Society Reviews, 40*, 3941. Available from https://doi.org/10.1039/c0cs00108b.

Mörseburg, K., & Chinga-Carrasco, G. (2009). Assessing the combined benefits of clay and nanofibrillated cellulose in layered TMP-based sheets. *Cellulose, 16*, 795–806. Available from https://doi.org/10.1007/s10570-009-9290-4.

Mousavi, S. M. M., Afra, E., Tajvidi, M., & Bousfield, D. W. (2017). Cellulose nanofiber / carboxymethyl cellulose blends as an efficient coating to improve the structure and barrier properties of paperboard. *Cellulose, 24*. Available from https://doi.org/10.1007/s10570-017-1299-5.

Nair, S. S., Zhu, J., Deng, Y., & Ragauskas, A. J. (2014). High performance green barriers based on nanocellulose. *Sustainable Chemical Processes, 2*, 23. Available from https://doi.org/10.1186/s40508-014-0023-0.

Rahman, S., Petroudy, D., Syverud, K., et al. (2013). Effects of bagasse microfibrillated cellulose and cationic polyacrylamide on key properties of bagasse paper. *Carbohydrate Polymers, 99*, 311–318. Available from https://doi.org/10.1016/j.carbpol.2013.07.073.

Rana, aK., Mitra, B. C., & Banerjee, aN. (1999). Short jute fiber-reinforced polypropylene composites: Dynamic mechanical study. *Journal of Applied Polymer Science, 71*, 531–539, https://doi.org/10.1002/(SICI)1097–4628(19990124)71:4 < 531::AID-APP2 > 3.0.CO;2-I.

Ridgway, C. J., & Gane, P. A. C. (2012). Constructing NFC-pigment composite surface treatment for enhanced paper stiffness and surface properties. *Cellulose, 19*, 547–560. Available from https://doi.org/10.1007/s10570-011-9634-8.

Samyn, P., Barhoum, A., Öhlund, T., & Dufresne, A. (2018). Review: Nanoparticles and nanostructured materials in papermaking. *Journal of Material Science, 53*, 146–184. Available from https://doi.org/10.1007/s10853-017-1525-4.

Santos, S. M., Carbajo, J. M., Gómez, N., et al. (2017). Paper reinforcing by in situ growth of bacterial cellulose. *Journal of Material Science, 52*, 5882–5893. Available from https://doi.org/10.1007/s10853-017-0824-0.

Satam, C. C., Irvin, C. W., Lang, A. W., et al. (2018). Spray-coated multilayer cellulose nanocrystal—Chitin nanofiber films for barrier applications. *ACS Sustainable Chemistry & Engineering, 6*, 10637–10644. Available from https://doi.org/10.1021/acssuschemeng.8b01536.

Sehaqui, H., Zhou, Q., & Berglund, L. A. (2013). Nanofibrillated cellulose for enhancement of strength in high-density paper structures. *Nordic Pulp and Paper Research Journal, 28*, 182–189. Available from https://doi.org/10.3183/npprj-2013-28-02-p182-189.

Serra, A., González, I., Oliver-Ortega, H., et al. (2017). Reducing the amount of catalyst in TEMPO-oxidized cellulose nanofibers: Effect on properties and cost. *Polymers (Basel)*, *9*, 557. Available from https://doi.org/10.3390/polym9110557.

Sharma, A., Thakur, M., Bhattacharya, M., et al. (2019). Commercial application of cellulose nano-composites—A review. *Biotechnology Reports*, *21*, e00316. Available from https://doi.org/10.1016/j.btre.2019.e00316.

Shatkin, J. O. A., Wegner, T. H., & Bilek, E. (2014). Market projections of cellulose nanomaterial-enabled products- Part 1: Applications. *Tappi Journal*, *13*, 9–16.

Sheltami, R. M., Abdullah, I., Ahmad, I., et al. (2012). Extraction of cellulose nanocrystals from mengkuang leaves (*Pandanus tectorius*). *Carbohydrate Polymers*, *88*, 772–779. Available from https://doi.org/10.1016/j.carbpol.2012.01.062.

Shi, Z., Phillips, G. O., & Yang, G. (2013). Nanocellulose electroconductive composites. *Nanoscale*, *5*, 3194–3201. Available from https://doi.org/10.1039/c3nr00408b.

Soni, B., Hassan, E. B., Schilling, M. W., & Mahmoud, B. (2016). Transparent bionanocomposite films based on chitosan and TEMPO-oxidized cellulose nanofibers with enhanced mechanical and barrier properties. *Carbohydrate Polymers*, *151*, 779–789. Available from https://doi.org/10.1016/j.carbpol.2016.06.022.

Souza, A. G. D., Kano, F. S., & Bonvent, J. J. (2017). Cellulose nanostructures obtained from waste paper industry: A comparison of acid and mechanical isolation methods. *Materials Research*, *20*, 209–214. Available from https://doi.org/10.1590/1980-5373-MR-2016-0863.

Sun, B., Hou, Q., Liu, Z., & Ni, Y. (2015). Sodium periodate oxidation of cellulose nanocrystal and its application as a paper wet strength additive. *Cellulose*, *22*, 1135–1146. Available from https://doi.org/10.1007/s10570-015-0575-5.

Sun, Y., Yang, Q., & Wang, H. (2016). Synthesis and characterization of nanodiamond reinforced chitosan for bone tissue engineering. *Journal of Functional Biomaterials*, 7, 27. Available from https://doi.org/10.3390/jfb7030027.

Syverud, K., & Stenius, P. (2009). Strength and barrier properties of MFC films. *Cellulose*, *16*, 75–85. Available from https://doi.org/10.1007/s10570-008-9244-2.

Taipale, T., Österberg, M., Nykänen, A., et al. (2010). Effect of microfibrillated cellulose and fines on the drainage of kraft pulp suspension and paper strength. *Cellulose*, *17*, 1005–1020. Available from https://doi.org/10.1007/s10570-010-9431-9.

Tayeb, A., Amini, E., Ghasemi, S., & Tajvidi, M. (2018). Cellulose nanomaterials—Binding properties and applications: A review. *Molecules (Basel, Switzerland)*, *23*, 2684. Available from https://doi.org/10.3390/molecules23102684.

Toivonen, M. S., Kurki-Suonio, S., Schacher, F. H., et al. (2015). Water-resistant, transparent hybrid Nanopaper by physical cross-linking with chitosan. *Biomacromolecules*, *16*, 1062–1071. Available from https://doi.org/10.1021/acs.biomac.5b00145.

Vieira, M. G. A., Da Silva, M. A., Dos Santos, L. O., & Beppu, M. M. (2011). Natural-based plasticizers and biopolymer films: A review. *European Polymer Journal*, *47*, 254–263. Available from https://doi.org/10.1016/j.eurpolymj.2010.12.011.

Voisin, H., Bergström, L., Liu, P., & Mathew, A. (2017). Nanocellulose-based materials for water purification. *Nanomaterials*, 7, 57. Available from https://doi.org/10.3390/nano7030057.

Wang, H., Huang, L., & Lu, Y. (2009). Preparation and characterization of micro- and nano-fibrils from jute. *Fibers and Polymers*, *10*, 442–445. Available from https://doi.org/10.1007/s12221-009-0442-9.

Xu, Q. H., Li, W. G., Cheng, Z. L., et al. (2014). TEMPO/NaBr/NaClO-mediated surface oxidation of nanocrystalline cellulose and its microparticulate retention system with cationic polyacrylamide. *BioResources*, *9*, 994–1006. Available from https://doi.org/10.15376/biores.9.1.994-1006.

CHAPTER 19

Utilization of nanocellulose fibers, nanocrystalline cellulose and bacterial cellulose in biomedical and pharmaceutical applications

Nurul Huda Abd Kadir[1], Masita Mohammad[2], Mahboob Alam[3], Mohammad Torkashvand[4], Thayvee Geetha Bharathi Silvaragi[5] and Sarminiyy Lenga Gururuloo[6]

[1]Faculty of Science and Marine Environment, Universiti Malaysia Terengganu, Kuala Nerus, Malaysia
[2]Solar Energy Research Institute (SERI), National University of Malaysia, Bangi Selangor, Malaysia
[3]Division of Chemistry and Biotechnology, Dongguk University, Gyeongju, Republic of Korea
[4]Fouman Faculty of Engineering, College of Engineering University of Tehran, Tehran, Iran
[5]Integrative Medicine Cluster, Advanced Medical and Dental Institute, Universiti Sains Malaysia, Kepala Batas, Malaysia
[6]Infectomics Cluster, Advanced Medical and Dental Institute Universiti Sains Malaysia, Kepala Batas, Malaysia

19.1 Introduction

Nanocrystalline cellulose (NCC) is an organic compound consist of β(1-4)-linked D-glucose units, it is an important structural component part of plants that are naturally occurring biodegradable and renewable polymers in the world. As a cellulose derivative, it is a nondigestible constituent of insoluble dietary fiber, so it will not cause any harm to the human body and has thus become a popular bulking agent in a variety of consumables, including food and pharmaceutical products. Rising demand for pharmaceutical, food and beverage as well as personal care industries has made NCC to be used widely in the market in the region. According to Transparency Market Research, the global microcrystalline cellulose (MCC) market is expected to reach a valuation of US$936.3 million by 2020 (Roman, 2015). The market has followed a steady growth trajectory in recent years and is expected to achieve close to 50% growth within the last six years of the ongoing decade, even with the fast development on the market use of NCC.

NCC can be obtained from various kinds of materials such as wood pulp, algal cellulose, and MCC. They are commonly prepared from acid hydrolysis of virgin cellulose as its amorphous region can be easily hydrolyzed (Roman, 2015). A key issue for a sustainable agriculture is the rational use of residues and byproducts from the agricultural processes. Thus the development of new strategies for the use of cellulose can minimize their environmental impact and at the same time, adds value to this residue. NCCs are highly

Nanotechnology in Paper and Wood Engineering
DOI: https://doi.org/10.1016/B978-0-323-85835-9.00025-8

crystalline cellulose nanostructures giving remarkable properties, such as low density, surface reactivity and surface area, high biocompatibility, and biodegradability (Trache et al., 2020). NCCs are rod-shaped and they are typically reported to have diameters of 5–20 nm and length distribution of 100–600 nm (Rasheed, Jawaid, Parveez, Zuriyati, & Khan, 2020).

NCC is originated from lignocellulosics through acidic extraction method, thus it possess the structure of crystallites, which is a kind of nanodimensional cellulose (Habibi, Lucia, & Rojas, 2010). Due to its negative charge and large surface area, NCCs act as an excipient in drug delivery. This is proven that it allows the binding of large quantities of drugs to its surface and thus it owns the high potential in payloads and dosage control (Shaikh, Birdi, Qutubuddin, Lakatosh, & Baskaran, 2007) compared to MCC. Although MCC has also been applied in the drug-delivery systems due to its inert and biocompatible properties, it does not participate in the molecular level of drug release control directly through the binding interactions of drugs. This is due to the small amount of hydroxyl residues, which leads to the limited amount of negative charges on its large surface, causing the adsorption of major drugs not easy (Guo et al., 2021). In the aspect of toxicology, Domingues, Gomes, and Reis (2014) stated that NCC is considered low toxic in the majority of both animal and plant cells due to its natural biopolymers source, that is, cellulose (Zoppe et al., 2014). Moreover, Domingues et al. (2014) also stated that NCC becomes cytotoxic when appearing in high concentrations, which is low compared to the toxic concentration in carbon nanotubes and crocidolite asbestos. This enables NCC to have an edge over other materials in various applications.

On the other hand, cellulose has been one of the well-known excipients exploited due to its medicinal properties especially in producing tablets. In spite of the prolonged history of cellulose in producing pills, scientists are still researching on various types of cellulose such as nanofibrils, nanocrystals, microcrystals, and microfibrils extracted from different species (Phanthong et al., 2018). Hence, NCC was recognized to be used in the drug delivery systems (Lin & Dufresne, 2014). This is due to the functional properties on the outer layer of NCC that acts as a binder to the loaded drug. One of the procedures of producing the drug tablet is to grind both the drug and the excipients together before direct compression should be done. Next, nanocrystalline has been utilized in the tissue engineering field too. Tissue engineering is done by culturing to repair the wounded and damaged tissues. Those cells that undergo cell culture protocols need a surface for them to attach naturally. Hence, nontoxic NCC and its high tensile strength aided the attachment of cells for skin preparation (Luo et al., 2019).

19.2 Chemical and physical properties of nanocellulose

Nanocellulose (NC) is a natural, sustainable material derived from different sources such as trees, bacteria, or synthesized in vitro (Pylkkänen et al., 2020). From vegetables, NC is generally extracted through a hydrolysis process that preferentially cleaves the amorphous regions of the cellulose fibers, leaving high aspect-ratio crystalline rigid rod-shaped particles (De France, Hoare, & Cranston, 2017). In these few years, the development of environmental-friendly materials had risen due to the expansion of environmental issues. Cellulose is made up of β-1,4-linked anhydro-D-glucose units, which is the richest natural polymer on the earth (Kasiri & Fathi, 2018). Due to its biocompatibility, biodegradability, nontoxicity, and renewability excellent properties, cellulose has been known as a potential alternative material to be used in the production industry (Trache et al., 2020). According to Siqueira, Abdillahi, Bras, and Dufresne (2010) and Yildirim and Shaler (2017), cellulose nanomaterials (CNs) can be obtained from the sources such as animals, plants, or mineral plants. There are two general classes of CNs which are cellulose nanocrystals (NCCs) and cellulose nanofibrils (NFCs). NCC also can be named as cellulose nanowhiskers (CNW) is one of the examples from cellulose nanocrystals, while nanofibrillated cellulose (NFC) is classified under NFCs. NCC is utilized as reinforcement in hydrophilic polymers due to the existence of a vast quantity of hydroxyl group (-OH) on its surface which forms strong hydrogen bonding (Klemm et al., 2011). Moreover, there are few journals stipulated that the way of extractions and the source of cellulose will affect the performance of NCC as a nano-reinforcing agent indirectly due to the differences in characteristics and morphologies of the nanoparticles (NPs) (Kalia et al., 2011). However, there is lack of study regarding the different properties of NCC that can be contributed in various fields of application.

19.3 Mechanical and reinforcement properties of nanocellulose in pharmaceutical applications

Cellulose fibrils are composed of two parts, which are the amorphous and crystalline parts (Hanif, Ahmed, Shin, Kim, & Um, 2014). NCC can be prepared from many sources of biomass under certain parameters of acid hydrolysis by removing the amorphous regions forming a rigid rod-shaped structure (Lin & Dufresne, 2014). According to Hanif et al. (2014), the shape and size of nanocrystal were different depending on the preparation method that was applied to the fibrils. For example, fibrils that were hydrolyzed by sulfuric acid will form a rod-shaped nanocrystalline particle (Hanif et al., 2014). This material can be a good resource as it has properties such as low density, high aspect ratio, high surface area, and modifiable surface properties (Brinchi, Cotana, Fortunati, & Kenny, 2013).

NCC has similar physical and chemical properties as that of MCC since they came from the same resources (Trache et al., 2020). NCC is actively being investigated as a material for use in the packing, aerospace, automotive, and cosmetic industries due to its large surface area, stiffness, strength, and phase behavior (Jackson et al., 2011). Nowadays, NC has gained much attention in biomedical fields because of its physical properties, special surface chemistry, and excellent biological properties for biocompatibility, biodegradability, and low toxicity (Lin & Dufresne, 2014).

19.4 Biological properties of nanocellulose (that make it suitable in pharmaceutical applications)

NCC is nanosized material that is also called in other names like nanowhisker, nanocrystals, nanofibrils, or cellulose nanocrystals. There are many methods used to produce NCC and the most frequently used is acid hydrolysis. NCC is the product obtained from the disaggregation of MCC during the acid hydrolysis (Zhao et al., 2020). MCC structure contains both crystalline and amorphous regions. Amorphous region in the MCC structure has weak bonds. Acids, such as sulfuric acid, used will eradicate the amorphous sections from the crystalline part. The final structure of NCC will be formed as elongated cylindrical shape as shown in Fig. 19.1 (Bhat, Dasan, Khan, Soleimani, & Usmani, 2017). The length dimension of NCC comprises from 100 to 600 nm (Halib et al., 2017), whereas the diameter is around 2–20 nm (Bhat et al., 2017). Apparently, the length and diameter of NCC demonstrate that it is smaller in size than MCC. The smaller size of NCC possesses a bigger surface area (Halib et al., 2017) where these excipients can bind with more compounds surrounding it. The structure, size, and crystallinity of NCC vary with the method used in preparing the sample. Moreover, the weight of NCC was obviously less in density, which is approximately 1.566 g cm^{-3} due to its nanosize. NCCs are very strong particles as they have tremendous resistance from breaking under tension and a good heat conductor (Bhat et al., 2017).

NCC is also known for its biocompatible properties (Azeredo, Rosa, & Mattoso, 2017). NCCs are capable of being a stabilizing agent because generally, cellulose is from amphiphilic family (Buffiere et al., 2017). Thus they can play their role as stabilizer in both hydrophobic and hydrophilic conditions. Tablets that are coated with NCC can hold optimum dosage of drug needed by certain disease because of the porosity of NCC that compact the drug together within it (Song, Leng Chew, Yaw Choong, & Tan, 2016). The tablet that is taken orally will be digested in the gastrointestinal tract, where the tablets become into smaller sizes. At this moment NCC acts as disintegrant by absorbing more water to aid the dissolution of the drug. The drug will be then delivered directly to cells at a precise time by NCC due to its permeability properties toward membranes (Choudhury, Sahoo, & Gohil, 2020). This remarkable property of NCC is due to its stiffness, biodegradability, and disintegrant which potentially become biomaterials for medical applications.

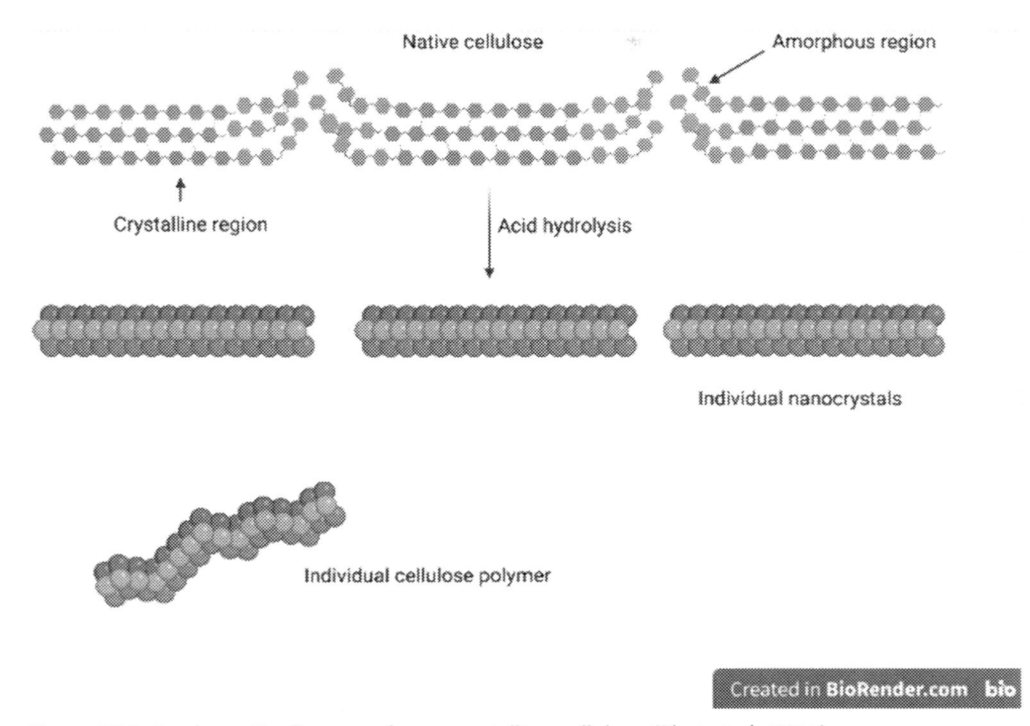

Figure 19.1 A schematic diagram of nanocrystalline cellulose (Bhat et al., 2017).

19.5 Biocompatibility and cytotoxicity of nanocellulose

Cellulose has been considered as effective, safe, and sustainable biopolymers. However, some modifications were made on the cellulose might give some effect on the cellulose in terms of toxicity even though they came from the same polysaccharide (Kollar et al., 2011). Previous study by Nie, Wang, Wang, Wei, and Zheng (2011) reported that nanocrystalline prepared from commercial MCC by high pressure torsion technique showed less cytotoxicity. Another study conducted by Kollar et al. (2011), where the cytotoxicity of modified types of MCC and NCC were tested on THP-1 cells, showed less toxicity. According to Kollar et al. (2011), there was no significant changes on the cell growth, and cells' viability was observed despite different concentration of modified crystalline cellulose were used.

NCCs are plant-derived, elongated NPs that have gained attention for their comparatively low cost, high strength and stiffness, and tendency to form colloidal liquid crystalline phases (Dong et al., 2016). The results of this study provided a first indication that the tested NCCs produced from grape pomace are not toxic, and thus are appropriate for the use in the food or pharmaceutical applications (Coelho et al., 2018). The cells cultured in the presence of nanofibers' extract showed a morphological pattern similar to the negative control cells, forming a confluent monolayer of spread cells. Also, the morphology and basic cell functions appear not to be affected by contact with the extract medium, since the morphology displays features of fibroblast cells. These morphological results corroborate the cellular activity (MTT), evidenced through the cell viability test, confirming that the nanofibers produced did not exhibit any cytotoxic behavior.

19.6 Nanocellulose-based pharmaceutical applications

Historically, cellulose has been one of the well-known excipients exploited due to its medicinal properties especially in producing tablets. In spite of the prolonged history of cellulose in producing pills, scientists are still researching on various types of cellulose such as nanofibrils, nanocrystals, microcrystals, and microfibrils extracted from different species (Bhat et al., 2017). Hence, NCC is commonly used in drug delivery system (Bhat et al., 2017; Lin & Dufresne, 2014).

19.6.1 Drug delivery

Over the past few years, there has been large-scale research on developing feasible drug delivery systems. In conjunction with the advancement in drug delivery system, the potential pharmaceutical application of NC due to its unique properties such as renewability, biodegradability, and nontoxicity is also widely being investigated in drug delivery for the improvement of treatment efficacy. NC and its derivatives are generally used in drug delivery either as an excipient of particular function or as a carrier in which a drug can be loaded. It has been reported in numerous recent studies that the application of all different categories of NC in various drug delivery systems is highly capable of overcoming the limitations possessed by each and every drug delivery routes such as oral delivery, and ocular delivery.

19.6.2 Rapid drug delivery

Unlike other NC-based drug delivery systems on which numerous recent studies have been carried out, limited pieces of evidences are available in regard to the application of NC in the rapid drug delivery systems. The pharmaceutical applications of NFC and NCC as one of the tablet ingredients and NP preparation material, have shown to result in a rapid drug delivery system. In a study conducted by Kolakovic et al. (2011), NFC was used as one of the tableting excipients to study its potency as a substitute for two grades of commercial MCC, namely Avicel PH101 and Avicel PH102 for the production of paracetamol loaded tablets using direct compression method. The results revealed that the tablets composed of NFC disintegrated slightly faster resulting in rapid drug release in comparison to that made of Avicel PH102. Further, it has also been reported that the MCC tablets were incapable of releasing the drugs completely mainly due to their adsorptive characteristics and large particle surface area. In another study by Akhlaghi, Berry, and Tam (2013), NCC was grafted with chitosan oligosaccharide via the oxidation of primary alcohol groups present on NCC into carboxylic groups, which then reacts with amino groups on chitosan oligosaccharide. In this study, grafted NCC/chitosan oligosaccharide NPs were used to encapsulate procaine hydrochloride. In accordance with the findings from the previous study, these NPs were also reported to release the drug relatively faster, within an hour at pH 8 substantiating that NCC has a great potential to be used as a nanomaterial in drug delivery for topical applications. Table 19.1 summarizes the latest reports on NC-based rapid drug delivery systems.

19.6.3 Controlled and sustained drug delivery

In recent times, a wide range of NC-based drug delivery systems has shown controlled or sustained drug release which is considered as one of the main benefits of the pharmaceutical application of NC in drug delivery. These systems include formulation of composite (Bayoumi, Sarg, Fahmy, Mohamed, & El-Zawawy, 2020; Hong et al., 2019), membrane (Saïdi, Vilela, Oliveira, Silvestre, & Freire, 2017), film (Poonguzhali, Basha, & Kumari, 2018), hydrogel (Ilkar Erdagi, Asabuwa Ngwabebhoh, & Yildiz, 2020; Karzar Jeddi & Mahkam, 2019; Kopač, Krajnc, & Ručigaj, 2021; Patel, Dutta, Hexiu, Ganguly, & Lim, 2020), aerogel (Liang, Zhu, Wang, He, & Wang, 2020), cryogel (Li, Wang, et al., 2019), microsphere (Kadry, 2019), and NPs (Abo-Elseoud et al., 2018). As an example, in 2017, Poonguzhali et al. (2018) investigated the effect of NC and alginate nanocomposite film on in vitro delivery of ampicillin. The study exhibited an efficient drug delivery system with prolonged drug release as well as improved tensile strength of the film substantiating its pharmaceutical role in improving the drug delivery system. On the other hand, a great deal of research has also been carried out investigating the impact of crosslinking agents such as N,N'-methylenebis (acrylamide) (MBA) (Saïdi et al., 2017), glutaraldehyde (Kadry, 2019), genipin (Ilkar Erdagi et al., 2020) for controlled drug delivery preparation. Bacterial cellulose (BC)-based drug delivery system in the form of nanocomposite membrane has been reported. Saïdi et al. (2017) investigated the in vitro release of diclofenac sodium salt (DCF) from poly(N-methacryloyl glycine)/nanocellulose (PMGly/BC) nanocomposites. The study aimed to design pH-sensitive nanocomposite materials of PMGly and BC in order to ensure a sustained delivery of diclofenac. For this purpose, several nanocomposite formulations were prepared via in situ free radical polymerization of MGly with MBA as crosslinker within the swollen BC network under green reaction conditions. The results demonstrated that the release of drug from PMGly/BC nanocomposites was pH sensitive in which the DCF release rate was slower at pH 2.1 compared to that of at pH 7.4. The authors also observed no cytotoxic effects on

Table 19.1 Nanocellulose-based drug delivery systems and toxicological assessments for rapid drug delivery.

Drug delivery system	Material component		Model drug	Drug uses	Drug delivery system results	Toxicological experiment	Toxicology results	References
	Nanocellulose	Comaterial						
Tablet preparation	NFC		Paracetamol	Painkiller	Rapid drug release	—	—	Kolakovic et al. (2011)
NP/NCC nanocomposite	NCC	Chitosan	Procaine hydrochloride	Anesthetic drug	Rapid drug release	—	—	Akhlaghi et al. (2013)

NCC, Nanocrystalline cellulose; *NFC*, nanofibrillated cellulose; *NP*, nanoparticle.

HaCat cell lines proving that BC is nontoxic and has a great potential as a nanocomposite material for both transdermal and oral drug delivery systems based on the pH responsive and sustained delivery behavior exhibited in the study. Also, oxidized NCCs (OXNCCs) were investigated for controlled drug delivery. Hydrogen bonding established between NCC and OXNCC, and the drug namely repaglinide (RPG) affecting the swelling were reported to be the factor behind the controlled drug delivery (Abo-Elseoud et al., 2018). Li, Liu, et al. (2019) developed NFC/quercetin nanoformulation for potential application in controlled drug delivery. There are several other NC-based advanced drug delivery systems that successfully showed controlled drug delivery including oral, ocular, topical, intratumoral, and transdermal drug delivery system (TDDS), as shown in Table 19.2.

19.6.4 Oral delivery

Löbmann, Wohlert, Müllertz, Wågberg, and Svagan (2017) utilized NFC drug dispersion for the oral delivery of poorly soluble indomethacin. Their study aimed to study the effects of changes in the hierarchical structure of the delivery matrix on the drug release profile and the versatility of NFC-based drug delivery systems from sustained drug release to a rapid drug release. For this purpose, nanopapers and nanofoams utilizing NFC/indomethacin were tested. In comparison to nanopapers which showed rapid drug release profile (10−20 min), nanofoam exhibited a relatively controlled drug release (≈ 24 h) due to the longer amount of time needed for the drugs to diffuse through the NFC-based cell walls as well as due to the foams' structure which appears approximately 60 times thicker than the corresponding films used in the same study. In another interesting study, pH sensitive poly(acrylamido glycolic acid) nanocomposite (PAGA-NC) hydrogel composed of NCCs were found to be potential oral drug carriers for controlled drug release (Rao, Kumar, & Han, 2017). They observed a sustained and pH responsive drug release for PAGA-NC. PAGA-NC hydrogels were also shown to possess nontoxic nature based on the MTT assay performed on NIH-3T3 fibroblast cell line. Abo-Elseoud et al. (2018) synthesized chitosan NPs and NCC NPs for controlled drug delivery of repaglinide (RPG); the drug encapsulation efficiency (DEE) was 98%. In this study, NCCs were also modified through oxidation to introduce carboxylic group on their surface (OXNCC) to further investigate its potentiality in a controlled drug delivery system. The results revealed that the drug release was pH dependent, and burst release was demonstrated at first 2 h followed by a sustained drug release for chitosan nanoparticle/ nanocrystalline cellulose (CHNP/ NCC) NPs. In contrast to CHNP/NCC/RPG, further decrease in RPG release was reported for CHNP/OXNCC/RPG due to the presence of OXNCC in CHNP which may possibly be attributed to the greater number of hydrogen bonding formed between the drug's functional group and that of OXNCC. These CHNP/NCC/

Table 19.2 Nanocellulose-based drug delivery systems and toxicological assessments for controlled and sustained drug delivery.

Drug delivery system	Material component		Model drug	Drug uses	Drug delivery system results	Toxicological experiment	Toxicology results	References
	Nanocellulose	Comaterial						
Alg/NC nanocomposite films	NC (originated from *Hibiscus sabdariffa*)	Alginate (Alg)	Ampicillin	Antibiotic against bacteria caused infections	Good mechanical properties Prolonged release pattern	—	—	Poonguzhali et al. (2018)
PMGly/BC nanocomposite membranes	BC	Poly N-methacryloylglycine (PMGly)	Diclofenac sodium salt	Reduce inflammation and pain	pH sensitive nature Controlled drug release	MTT assay toward HaCat cell line	No toxicity	Saïdi et al. (2017)
NP/NCC nanocomposite	NCC (originated from palm fruit stalks), oxidized NCC	Chitosan nanoparticles, pentasodium tripolyphosphate (TPP)	Repaglinide	Antihyperglycemic	High drug encapsulation efficiency Sustained drug delivery	—	—	Abo-Elseoud et al. (2018)
NFC/drug nanoformulation	NFC (originated from (*Populus ussuriensis*)		Quercetin	Antioxidant	High drug entrapment efficiency Sustained drug release	—	—	Li, Liu, et al. (2019)
CMCNF CIP-MMT composite	Carboxymethylated cellulose nano fibril (CMCNF)	Montmorillonite (MMT)	Ciprofloxacin (CIP)	Antibiotics	Good mechanical properties Sustained drug release	—	—	Hong et al. (2019)
Hydrogel	Magnetic nano carboxymethyl cellulose (MNCMC)	Alginate/chitosan hydrogel beads	Dexamethasone	Treatment of acute and chronic ocular diseases	Improved drug loading capacity Controlled and sustained drug release pH responsive-based drug release	—	—	Karzar Jeddi and Mahkam (2019)

Microsphere	Gelatin/sodium carboxymethyl cellulose (NaCMC) and gelatin/sodium carboxymethyl nanocellulose (NaCMNC) (originated from rice straw)	Gelatin/glutaraldehyde	Tramadol	Analgesic	Controlled and sustained drug release	—	—	Kadry (2019)
Nanocellulose/gelatin composite cryogel	NFC (originated from softwood pulp)	Gelatin/dialdehyde starch	5-Fluorouracil	Anticancer	High drug loading capacity Sustained drug release	—	—	Li, Wang, et al. (2019)
Nanocellulose–drug composite	NCC/MC (originated from rice straw)		Gentamicin sulfate	Inflammation antibiotics drug	Sustained drug release was observed for NCC Rapid drug delivery was observed for MC	—	—	Bayoumi et al. (2020)
Hydrogel	Carboxylated modified nanocellulose	Gelatin/genipin/diosgenin	Neomycin		Good mechanical integrity Improved drug loading capacity Controlled and sustained drug release pH responsive drug delivery	MTT assay toward human dermal fibroblast	No toxicity	Ilkar Erdagi et al. (2020)
Hydrogel	Tempo-modified nanocellulose	Sodium alginate			Controlled and sustained drug release	—	—	Kopač et al. (2021)

(Continued)

Table 19.2 (Continued)

Drug delivery system	Material component		Model drug	Drug uses	Drug delivery system results	Toxicological experiment	Toxicology results	References
	Nanocellulose	Comaterial						
NFC-PEI-NIPAM aerogel	NFC (originated from bagasse pulp)	Polyethyleneimine—N-isopropylacrylamide (PEI-NIPAM)	Doxorubicin	Antibiotic, used to treat skin infections	Cumulative drug release rate increased with temperature and decreased with increasing pH Controlled drug release pH- and temperature-based drug delivery	In vitro cytotoxicity Skin sensitization test Systemic toxicity test	Excellent biocompatibility (nontoxic)	Liang et al. (2020)
Hydrogel	NCC (originated from rice husk)	Chitosan			Increased swelling potential Improved mechanical strength Thermo responsive property Sustained drug release	WST-1 assay toward bone marrow stem cells	Enhanced cell viability (nontoxic)	Patel, Dutta, Ganguly, and Lim (2021)

BC, Bacterial cellulose; *NC*, nanocellulose; *NCC*, nanocrystalline cellulose; *NFC*, nanofibrillated cellulose; *NP*, nanoparticle.

RPG and CHNP/OXNCC/RPG are desirable for antidiabetic controlled drug delivery systems. In another study, Badshah et al. (2018) studied the effects of surface-modified BC matrices developed via different drying techniques, namely oven technique and freeze-drying technique in drug delivery. In this study, two model drugs of different aqueous solubility (poorly water-soluble famotidine and highly water soluble tizanidine) were tested. This study shows the versatility of surface modified BC-based drug delivery with different drying techniques from exhibiting immediate drug release through oven drying to demonstrating sustained drug release through freeze drying technique. The rapid drug release of famotidine attributed to the adsorption of drug content on the oven dried matrices as well as the lower path length for drug release. At the same time, the rapid drug release of tizanidine could be due to the hydrophilic nature of the drug itself. Whereas, the sustaining effect of drug as a result of freeze drying technique attributed to the reduction in BC microfibrils interstitial fluids voids due to the increased mutual contact among nanofibers. Alginate-NCC hybrid NPs (ALG-NCC NPs) for in vitro release of rifampicin (RIF) was studied by Thomas, Latha, and Thomas (2018). They synthesized ALG-NCC NPs by a green method. Several parameters such as RIF: ALG ratio, NCC:ALG ratio, and surfactant concentration were taken into account for the preparation of best formulation for the formation of NPs, where the formulation with 1% surfactant, a 1:6 ratio of NCC:ALG, and a 1:4 ratio of RIF:ALG was used in the study. The potentiality of ALG-NCC NPs as a drug carrier for the in vitro release of RIF resulted the NPs to be swelled and caused poor delivery at pH 1 and pH 2. Interestingly, the release rate of ALG-NCC NPs was found to be efficient at pH 6.8 and pH 7.4. On the other hand, the negative charge and large surface area of NCC was reported to be facilitating the binding to the drug molecule and resulting in an improved loading efficiency. Recently, in a different strategy, Xiao et al. (2020) developed berberine hydrochloride (BBH)/β-cyclodextrin (β-CD) inclusion complex (IC) loaded BC as a potential carrier for antiinfection oral administration. The results revealed that pH dependent drug release as well as sustained drug release were observed for (BBH)/(β-CD) hydrogel loaded BC suggesting BC/IC hydrogels could be utilized as an antioral administration medicine. Table 19.3 summarizes recent reports on NC-based oral drug delivery systems.

19.6.5 Ocular delivery

Ophthalmic delivery is still considered relatively challenging due to the unique structure of the eye (human cornea made up of epithelium, substantia propria, and endothelium). In addition, several protective and complex barriers such as tear dilution, protein binding, etc. are available to inhibit the entry of foreign substances including drug molecules into the eye which limits the bioavailability of drugs $<5\%$ upon application, resulting in frequent drug administration (Zhu, Wang, & Li, 2018). In this regard, several recent studies have been carried out with various approaches having prolonged ocular residence time as

Table 19.3 Nanocellulose-based drug delivery systems and toxicological assessments for oral drug delivery.

Drug delivery system	Material component		Model drug	Drug uses	Drug delivery system results	Toxicological experiment	Toxicology results	References
	Nanocellulose	Comaterial						
NFC/drug dispersion (films and foams)	NFC	Ethanol, water	Indomethacin	Used to treat osteoarthritis	Nanopapers—rapid drug release Nanofoam—sustained drug release	—	—	Löbmann et al. (2017)
Hydrogel	NCC	Poly(acrylamido glycolic acid) based nanocomposite, MBA	Diclofenac sodium	Reduces inflammation and pain	pH responsive drug delivery Percentage of DCF decreased with increasing NCCs	MTT assay toward NIH-3T3 fibroblast cell line	No toxicity	Rao et al. (2017)
NCC/NP nanocomposite	NCC (originated from palm fruit)	Chitosan nanoparticles, TPP	Repaglinide	Antihyperglycemic	High drug encapsulation efficiency Sustained drug delivery	—	—	Abo-Elseoud et al. (2018)
BC/drug matrices	Surface modified BC (Bacterial strain of *Gluconacetobacter xylinus*)		Famotidine/tizanidine	Famotidine—to treat heartburn and other symptoms Tizanidine—to treat muscle spasms	Oven drying technique—immediate drug release Freeze drying technique—sustaining effect of drugs	—	—	Badshah et al. (2018)
Hybrid nanoparticles	NCC (originated from banana fiber)	Alginate/honey/calcium chloride	Rifampicin	Antibiotic used against *Mycobacterium tuberculosis*	pH-dependent drug release behavior Sustained drug release	MTT assay toward L929 fibrobalst cell line	No toxicity	Thomas et al. (2018)
Hydrogels	BC (originated from	β-Cyclodextrin	Berberine hydrochloride	Antibacterial, antiinflammatory, antitumor, and antiinsulin resistance	pH dependent drug release behavior Sustained drug release	—	—	Xiao et al. (2020)

BC, Bacterial cellulose; *NC*, nanocellulose; *NCC*, nanocrystalline cellulose; *NFC*, nanofibrillated cellulose; *NP*, nanoparticle.

an objective of their study. In an interesting study, in situ gel NPs loaded NCC-poly (vinyl) alcohol (PVA) hydrogel lenses were developed for controlled ophthalmic drug delivery (Åhlén, Tummala, & Mihranyan, 2018). The results revealed that the NPs and NCC gelation is capable of preventing the leaching by interlocking the particles to the NCC, demonstrating NCC as a promising candidate in controlled ophthalmic drug delivery. Similarly, another group of researchers investigated the effects of NCC on the *in situ* gelation behavior of triblock poloxamer copolymer (PM) and in vitro release of pilocarpine hydrochloride (PL) from the nanocomposites (Orasugh, Sarkar, et al., 2019). For this purpose, a series of nanocomposite formulations with 16.6% (wt./vol.) PM were developed, including those without NCC (called M1) and others containing 0.8%, 1.0%, and 1.2% (wt./vol.) NCC (called M2, M3, and M4). The results revealed that the addition of NCC decreased the gelation temperature, increased the gel strength, and prolonged the release of PL over a period of 20 h. Also, the cumulative percentages of PL release from M1, M2, M3, and M4 were 87.26%, 52.89%, 40.43%, and 34.57% respectively, 420 min posttest, thus demonstrating the potentiality of NCC is modulating drug release kinetics. The nanocomposites were also shown to possess nontoxic nature, which is desirable in ophthalmic drug delivery. In a subsequent study by Orasugh, Dutta, et al. (2019) used NFC and nanocollagen (CG) based NFC grafted collagen (CGC) nanocomposites consist of PM for ophthalmic drug delivery. This study showed consistent results with the previous in which controlled release of ketorolac tromethamine and increased mechanical strength of hydrogel. The sustained drug release reported in the study attributed to the interlocking of the drug into the matrix, the electrostatic interaction between the drug molecules and CGC particles. Table 19.4 summarizes recent reports on NC-based ocular drug delivery systems.

19.6.6 Intratumoral delivery

A novel carrier consisting tris(2-aminoethyl) amine functionalized NCC coating as the shell and magnetic NPs as the core for methotrexate (MTX) delivery was developed in a study conducted by Rahimi et al. (2017). The carrier showed high loading capacity as a result of interior cellulose network structure and ionic interaction between MTX and nanocarrier itself. Additionally, pH responsive drug release behavior was reported in which 29% and 79% MTX was released from the nanocarrier at pH 7.4 and pH 5.4 respectively, affirming the potential of this system as a good drug delivery vehicle for the tumor's environment, that is, acidic. In another study, Cacicedo et al. (2018) used BC hydrogel loaded with lipid NPs (NLCs) for localized cancer treatment using doxorubicin (Dox) as a model drug. The study was aimed to allow high drug concentrations at the tumor site as well as to minimize the adverse effects resulting from the administration. In this approach, NLCs loaded with cationic Dox (NLCs-H) or neutral Dox (NLCs-N) were administered in vivo into an orthotopic breast cancer mouse model. The results

Table 19.4 Nanocellulose-based drug delivery systems and toxicological assessments for ocular drug delivery.

Drug delivery system	Material component		Model drug	Drug uses	Drug delivery system results	Toxicological experiment	Toxicology results	References
	Nanocellulose	Comaterial						
Hydrogel-based contact lens	NCC	Polyvinyl alcohol			Slow nanoparticles loaded hydrogel disintegration rate	—	—	Åhlén et al. (2018)
In situ nanocomposite thermo responsive gels	NCC (originated from jute fibers)	Triblock poloxamer	Pilocarpine hydrochloride	Reduce pressure inside the eye	Improved gel mechanical strength Sustained drug release	Lactate dehydrogenase and hemolysis assay	No toxicity	Orasugh, Sarkar, et al. (2019)
Poloxamer407NFC-g nanocollagen	NFC (originated from jute fiber)	Poloxamer 407, collagen	Ketorolac methamine	Nonsteroidal antiinflammatory drug	Increased hydrogel mechanical strength Controlled drug release Thermo responsive behavior	Lactate dehydrogenase and hemolysis assay	No toxicity	Orasugh, Dutta, et al. (2019)

NCC, Nanocrystalline cellulose; *NFC*, nanofibrillated cellulose.

revealed that NLCs-H possesses lower encapsulation efficiency (48%) and rapid release of drug in comparison to NLCs-N with relatively higher encapsulation (97%) and controlled drug release fashion. In vivo study of the system showed biocompatibility with no side effects and increased antitumor efficacy over free Dox. Anirudhan, Chithra Sekhar, and Thomas (2019) investigated the effects of layer-by-layer assembled carboxylic acid functionalized NCC and aminated nanodextran (AND) polyelectrolyte complexes (PECs) on the surface of modified graphene oxide (MGO) in developing a pH and light responsive drug delivery carrier. Curcumin (CUR) is used as a model chemotherapeutic drug in this study due to its wide range of therapeutic roles as antitumor, antioxidant, antibacterial, etc. MGO-AND/NCC nanocomposite revealed an increase in swelling with the increase in time, pH responsive drug release behavior, and controlled drug release over a period of 48 h. These results showed that MGO-AND/NCC complex has a great potential for intratumoral drug delivery system. In addition, a study conducted by Ning, You, Zhang, Li, and Wang (2020) used pH responsive hydrogel-based on NCCs for the delivery of paclitaxel. In consistent with the previous findings, the results of this study showed that surface modified NC gel also possesses pH responsive behavior and sustained drug release manner at the slight acidic environment. Further, the hydrogel enhanced the antitumor efficacy by improving the distribution of paclitaxel in cancer cells intracellularly, demonstrating this system would be another good drug delivery vehicle for cancer treatment. In another recent report by Moghaddam et al. (2020), to overcome the limitations of combination therapy including MTX and CUR such as lack of bioavailability, nanocarrier composed of NCC NPs along with amino acid L-lysine was developed. The pH sensitive nature, nontoxic nature, sustained drug release manner, and enhanced antitumor efficiency indicated in the study, demonstrate the nanocarrier composed of NCC as a promising candidate in the field of multi-drug delivery for combination cancer therapy. Table 19.5 summarizes recent reports on NC-based intratumoral drug delivery systems.

19.6.7 Topical delivery

Alkhatib et al. (2017) developed a bacterial NC (BNC)-based drug delivery system which incorporated poloxamer 338 and 407 of various concentrations for long-term dermal wound treatment. The study aimed to provide controlled extended octenidine release from the gel with good biocompatibility as well as improved mechanical strength. BNC has been previously used as a drug delivery system, targeting rapid drug release for octenidine without incorporation of any other additional comaterials. Thus in this study, poloxamer was used to improve the aforementioned properties for the gel prepared at the same time to ensure the sustained drug release pattern. The controlled release studies showed sustained drug release for around 8 days when 10% poloxamer 338 and 5% poloxamer were added compared to the free octenidine,

Table 19.5 Nanocellulose-based drug delivery systems and toxicological assessments for intratumoral drug delivery.

Drug delivery system	Material component		Model drug	Drug uses	Drug delivery system results	Toxicological experiment	Toxicology results	References
	Nanocellulose	Comaterial						
Magnetic NPs	NCC	Tris(2-aminoethyl)amine, Fe_3O_4 NPs	Methotrexate	Anticancer	Improved drug loading Good binding ability Direct target to cancer cells pH responsive drug release Controlled and sustained drug release			Rahimi et al. (2017)
BC hydrogel loaded lipid nanoparticles	BC	Grodamol MM, Gro Damol GTCC-LQ, Pluronic F68	Doxorubicin	Anticancer	High drug loading No adverse effects Controlled drug release Good antitumor efficacy	MTT assay	No toxicity	Cacicedo et al. (2018)
MGO-AND/NCC Nanocomposite	Carboxylated nanocellulose	Aminated nanodextran (AND)/modified graphene oxide (MGO)	Curcumin	Anticancer	pH-responsive drug release behavior Sustained drug release	MTT assay	No toxicity	Anirudhan et al. (2019)
Surface modified nanocellulose hydrogel loaded with paclitaxel	Cellulose nanocrystals		Paclitaxel	Anticancer	pH-responsive drug release behavior Sustained drug release	–	–	Ning et al. (2020)
	Nanocrystalline cellulose	Amino acid L-lysine	Methotrexate Curcumin (combination therapy)	Anticancer	pH-responsive drug release behavior Sustained drug release	MTT assay	No toxicity	Moghaddam et al. (2020)

BC, Bacterial cellulose; *NC*, nanocellulose; *NCC*, nanocrystalline cellulose; *NP*, nanoparticle.

attributing to the rise in number of micelles and drop in aqueous channels upon the addition of poloxamer. Also, the incorporation of poloxamer in BNC network was found to result in good water binding ability, improved mechanical strength, besides possessing nontoxic nature which makes it a good candidate for topical drug delivery. In another study, Hasan et al. (2020) prepared a composite film composed of chitosan (CH), polyvinylpyrrolidone (PVP) with incorporated cellulose nano whiskers (CNWs) for wound treatment applications. Despite being used as wound dressing materials for years, polymers like CH and PVP are still known for their limitations in terms of poor controlled drug release. CNWs were utilized in this study to overcome the limitation and to improve the mechanical strength and swelling behavior of the film. The results revealed sustained release of curcumin (CUR) with low cumulative percentage of drug release of 46.12% over a period of 56 h post the addition of 5% CNW in the formulation. This result suggests that CH-PVP-CNWs is a good delivery system, and the poor drug release pattern of chitosan and PVP can be adjusted by incorporating with NCWs. Bajpai, Ahuja, Chand, and Bajpai (2017) developed cellulose nanocrystals loaded chitosan films with CUR and silver NPs incorporated for wound healing treatment. The skin irritation test revealed that the film did not show any irritation besides demonstrating 98% wound reduction, which is significantly higher than that of not treated with any drug and CH/NCC(CUR_y). Table 19.6 summarizes recent reports on NC-based topical drug delivery systems.

19.6.8 Transdermal delivery

The TDDS is often considered a replacement for oral and parenteral administration where the drugs are delivered through the skin into the systemic circulation. Hence, the drug avoids the gastrointestinal tract as well as the liver metabolism, which could result in side effects otherwise. There are several recent studies that discussed the application of NC in TDDS. In a recent research, PEC films reinforced with gold NPs and NC (GNP-NC) nanofillers were prepared for the transdermal drug delivery of diltiazem hydrochloride. The PEC membrane was developed using cationic guar gum and PVA. The results revealed that the incorporation of GNP-NC nanofillers into the PEC has several advantages, such as improved tensile strength of the film, reduced water permeability, improved thermal stability, and enhanced DEE. The DEE was found to increase with increasing nanofiller concentrations as the matrix-filler interaction is enhanced through hydrogen bond linkages. In addition, the cytotoxicity test and skin irritation test revealed cell viability above 80% and no toxic signs, confirming that the NC is a promising candidate in TDDS. In another study by Medhi et al. (2017), biodegradable microneedles constructed using fish scale biopolymer-NCs were utilized for delivering lidocaine in vitro through transdermal drug delivery. The purpose of the study was to allow controlled skin permeation to deliver the drug as well

Table 19.6 Nanocellulose-based drug delivery systems and toxicological assessments for topical drug delivery.

Drug delivery system	Material component		Model drug	Drug uses	Drug delivery system results	Toxicological experiment	Toxicology results	References
	Nanocellulose	Comaterial						
Bacterial cellulose (BC)/ octenidine/poloxamer hybrid system	BC	Poloxamer 338 and 407	Octenidine	Antiseptic	Improved mechanical strength Sustained drug release	Shell-less hen's egg model	No toxicity	Alkhatib et al. (2017)
Chitosan-polyvinylpyrrolidone-CNW film	CNW	Chitosan Polyvinylpyrrolidone	Curcumin	Wound healing agent	Improved thermal and mechanical strength of the film Improved swelling behavior of film Sustained drug release Good biocompatibility Good antibacterial effect	MTT assay	No toxicity	Hasan et al. (2017)
CS/NCC (Ag Np_x/ CUR_y) film	Cellulose nanocrystals (NCC)	Chitosan (CS), silver nanoparticles (Ag NPs)	Curcumin	Anticancer, antibacterial, antioxidant, trauma healing, antiinflammatory agent	Good film stability Improved wound reduction percentage	Primary skin irritation studies	No irritation	Bajpai et al. (2017)
Amoxicillin-grafted bacterial cellulose sponges	Regenerated bacterial cellulose		Amoxicillin	Antibiotic	Good porosity Good swelling behaviors Enhanced wound healing effect (in vivo evaluation)	MTT assay toward HEK293 cell line	No toxicity	Ye et al. (2018)
BC/CMC biocomposite	Bacterial cellulose	Carboxymethylcellulose	Methotrexate	Antipsoriasis	Sustained drug release	—	—	de Lima Fontes et al. (2018)

Silver nanoparticles with curcumin cyclodextrins loaded into bacterial cellulose based hydrogel	Bacterial cellulose	Silver nanoparticles, cyclodextrins	Curcumin	Wound healing agent	Significant antimicrobial activity Possesses hemolytic properties	MTT assay	No toxicity	Gupta et al. (2020)
Bacteria nanocellulose loaded with *Boswellia serrata* extract	Bacterial nanocellulose		*B. serrata* extract	To treat skin diseases	Improved water holding capacity Improved mechanical stability of the matrices	Shell-less hen's egg test	No toxicity	Karl et al. (2020)
Nanocellulose film coated with honey and polyvinylpyrrolidone (PVP)	Cellulose nanocrystals	Polyvinylpyrrolidone	Honey	Wound dressing agent	Sustained drug (honey) release Significant antimicrobial efficacy	—	—	Taher et al. (2020)
Cetyltrimethylammonium Bromide-Nanocrystalline cellulose- based microemulsions	Nanocrystalline cellulose	Cetyltrimethylammonium bromide	Curcumin	Wound healing, antiinflammatory, anticarcinogenic	Enhanced skin uptake of curcumin Higher permeation rate of curcumin	MTT assay toward L929 cells	No toxicity	Zainuddin, Ahmad, Zulfakar, Kargarzadeh, and Ramli (2021)

as to lessen the local anesthesia needed. The microneedles prepared in this study possessed uniform needle lengths that enabled them to pierce through the stratum corneum at a depth below than that could cause pain. Also, despite not having significant swelling capability, the microneedles are able to release lidocaine due to the adhesive property of gelation that absorbs moisture and leads to the dissociation of polymer matrix for drug release. Similarly, Sarkar et al. (2017) used NFC and chitosan-based transdermal film for sustained release of ketorolac tromethamine. The developed film showed promising results such as controlled drug release fashion and increased mechanical strength of the film with the increasing NFC content. These results suggest that NFCs have a great potential to be used in TDDS. Saïdi et al. (2017) developed pH-sensitive systems using poly(N-methacryloyl glycine)/BNC composites for controlled release of diclofenac. This study showed good thermal, mechanical, and viscoelastic properties, nontoxic nature, pH sensitive nature, and good water-uptake capacity of the film, which are desired for efficient TDDS. Table 19.7 summarizes recent reports on NC-based TDDSs.

19.7 Advanced nanomaterials for tissue engineering, wound healing, repair and regeneration

NC materials have proven their ability and functionality in a number of different forms of medical applications including film (Liu, Sui, & Bhattacharyya, 2014), spheres, core-shell (Han et al., 2019), hydrogels, aerogel (Chen et al., 2021), and with potentially broad applications (Dufresne, 2019). Meanwhile, mixed materials of cellulose and crystalline nanotubes (CNTs) can have various functions and take many forms, such as composites, aerogels, papers (Lengowski, Bonfatti Júnior, Nishidate Kumode, Carneiro, & Satyanarayana, 2019), films, or fibers depending on their combinations, as shown in Fig. 19.2. These mixed materials not only exhibit the excellent characteristics of cellulose, NC, and CNTs, but are also renewable, biodegradable, recyclable, and energy saving.

The antimicrobial NC for biomedical applications can also be fabricated by the addition of silver NPs to BNC (Barud et al., 2011). BNC could be made extremely suitable for biomedical applications by overcoming its disadvantages like lack of antimicrobial nature, solubility issue during processing as well as insufficient pore size and porosity. Within the tissue engineering context, the possibility of tailoring porous three-dimensional (3D) architectures in a straightforward fashion while maintaining structural integrity, is what makes NC particularly relevant for such a purpose.

As a result of the property of biocompatibility and the right mechanical properties similar to natural tissue, NC-based materials can be used as special tissue bioscaffold. BC (that considered as a promising and cost-effective natural nanomaterial for biomedical uses) (Carvalho, Guedes, Sousa, Freire, & Santos, 2019; Sharma & Bhardwaj,

Table 19.7 Nanocellulose-based drug delivery systems and toxicological assessments for transdermal drug delivery.

Drug delivery system	Material component		Model drug	Drug uses	Drug delivery system results	Toxicological experiment	Toxicology results	References
	Nanocellulose	Comaterial						
Preparation of GNP-NC reinforced PEC film containing DH	Nanocellulose (NC)	Gold nanoparticle (GNP), cationic guar gum	Diltiazem hydrochloride (DH)	To treat hypertension	Improved tensile strength of the film Reduced water vapor permeability Increased thermal stability Enhanced drug encapsulation efficiency	MTT assay Skin irritation test	No toxicity No signs of irritation	Anirudhan, Nair, and Chithra Sekhar (2017)
Fish-scale nanocellulose biopolymer composite microneedles	Bacterial cellulose nanofibers	Silicone elastomer	Lidocaine	Used for numbing tissue	Increased drug permeation rate Negligible swellability Good tissue insertion	—	—	Medhi et al. (2017)
pH-sensitive Poly(N-methacryloyl glycine)/nanocellulose composites	Bacterial cellulose (BC)	Poly(N-methacryloyl glycine)			Good thermal, mechanical, and viscoelastic properties Noncytotoxic and pH sensitive High water uptake capacity	MTT assay	No toxicity	Saïdi et al. (2017)
NFC/chitosan transdermal film	Nanofibrillated cellulose (NFC) (originated from jute fibers)	Chitosan	Ketorolectromethamine	Nonsteroidal analgesic	Sustained drug release Increased mechanical strength of transdermal film Swelling behavior based drug release mechanism	—	—	Sarkar et al. (2017)

(*Continued*)

Table 19.7 (Continued)

Drug delivery system	Material component		Model drug	Drug uses	Drug delivery system results	Toxicological experiment	Toxicology results	References
	Nanocellulose	Comaterial						
Bacterially derived cellulose-composite double-layer membrane	BC	Hydroxypropyl-β-cyclodextrin, piperine	Curcumin	Antioxidant, antiinflammatory agent	Good skin permeation High amount of drug loading	—	—	(Jantarat, Sirathanarun, Boonmee, Meechoosin, & Wangpittaya, 2018)
NCC reinforced MC nanocomposites	Cellulose nanocrystals (NCC) (originated from jute card waste)	Methylcellulose (MC)	Ketorolac Tromethamine	Antiinflammatory, analgesic	Good mechanical barrier Good thermal stability Good water absorption Good ultraviolet resistance Sustained drug release	Lactate dehydrogenase assay, hemolysis assay, and skin irritation test	No toxicity	Orasugh, Saha, Rana, et al. (2018)
Cellulose nanofibrils/hydroxypropyl methylcellulose nanocomposite	NFC	Hydroxypropyl methylcellulose	Ketorolac tromethamine	Antiinflammatory, analgesic	Controlled drug release	—	—	Orasugh, Saha, Sarkar, et al. (2018)
BC membranes	BC		Crocin	Antioxidant	Stable and prolonged drug permeation Good uptake of the drug into BC membranes	—	—	Abba et al. (2019)
Anisotropic nanocellulose gel membranes (transdermal drug delivery patches)	NFC		Piroxicam	Nonsteroidal antiinflammatory drug	Great surface area, small average pore, and tunable surface charge properties Good skin adsorption Controlled drug release	—	—	Plappert, Liebner, Konnerth, and Nedelec (2019)

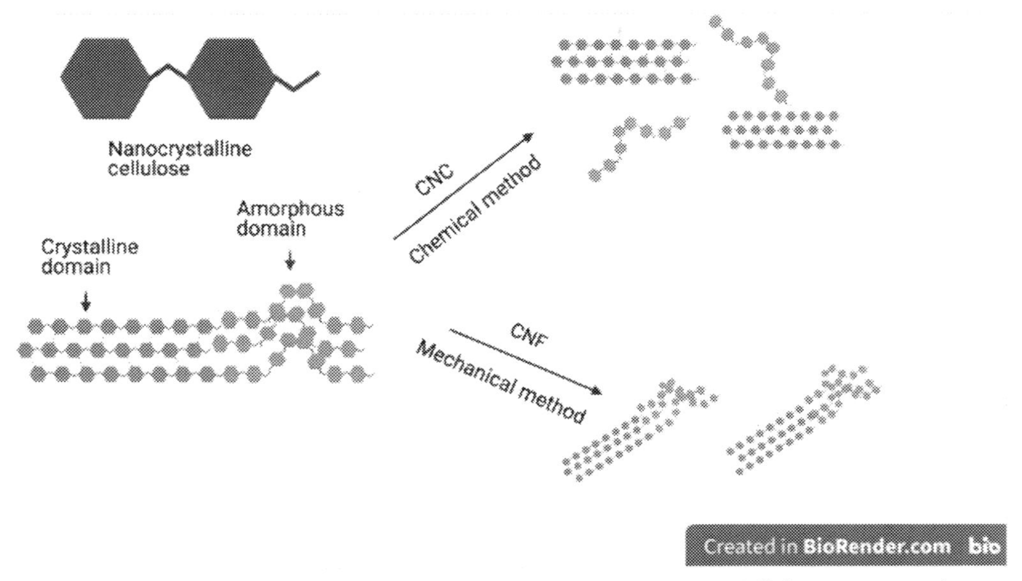

Figure 19.2 Ilustration of paper making (Lengowski et al., 2019). *CNC*, Cellulose nanocrystals; *CNF*, cellulose nanofibers.

2019), secreted by *Gluconacetobacter xylinus*, was explored as a novel scaffold material due to its unusual biocompatibility, light transmittance, and material properties.

Bacterial or microbial cellulose is high aspect ratio cellulose, apparently produced/secreted by bacteria in the form of ribbon-shaped fibril of 20–100 nm diameter and around 100 lm length. Distinctly, as compared to plant-based cellulose, it has a very long chain of cellulose (high degree of polymerization) and a high crystallinity up to 90%; and it is almost pure form of cellulose free from functional groups (carbonyl, carboxyl, sulfate, etc.) and other materials (lignin, hemicelluloses, pectin, etc.), as illustrated in Fig. 19.3 (Picheth et al., 2017).

The corneal stromal cell grown on BC scaffold shows a regular isodirectional arrangement, while the biocompatible BC tubes showed excellent mechanical properties capable of being utilized as vascular grafts for large animals. In addition, BC tubes can have different dimensions. The mechanical properties of BC biomaterials can be regulated by the effective cellulose contents. By the study, it was proved that BC is a promising material to achieve the mechanical properties of ear cartilage replacement, and can be produced in patient-specific ear shapes (Nimeskern et al., 2013).

NC also could be used as a substitute/medical biomaterial because of its excellent biocompatibility, mechanical properties, nontoxicity to cells, and so on. By several in vitro and in vivo studies, it has been demonstrated that NC-based materials are non- or minimally cytotoxic; biocompatibility testing for specific applications is still largely missing (Xu et al., 2019). Furthermore, material biocompatibility is highly

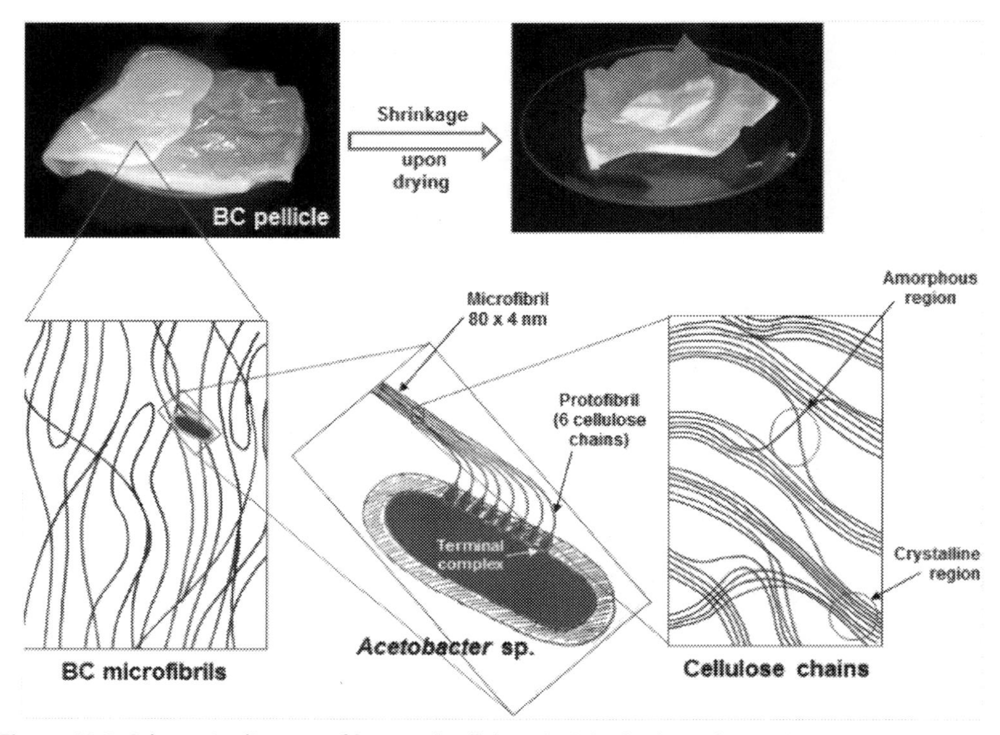

Figure 19.3 Schematic diagram of bacterial cellulose (BC) (Picheth et al., 2017).

dependent on the route of internalization, with inhalation representing one of the major concerns due to the high aspect ratio of NCs (Table 19.8).

NC such as NCC, NFC, and BNC prepared by freeze-drying, solvent casting, 3D printing, and electrospinning have been widely used to encounter problems regarding functional loss of tissues and organs. A new-era approach of tissue engineering serves consumers with repair, improve, and replace damaged tissues, as well as organ replacement (as stated in Table 19.1). The technology provides repairing and regeneration of skin, cartilage, bones, vascular tissues, and wound healing by mimicking the functionality of the biomaterials with the natural extracellular matrix of native tissues involved in biological activities of cell proliferation, growth, and differentiation. Furthermore, NCC, NFC, and BNC can form scaffolds that exhibited different properties including pore size, hydrophilicity, bioactivity, and mechanical properties. The biomaterials also exhibit biocompatibility that enables them to attach to cells, tissues, and organs.

19.7.1 Diagnostic devices

Currently, biomaterials catch the eye of researchers in the biomedical field due to their biocompatibility features that exhibit a promising approach for better diagnostic

Table 19.8 Nanocrystalline prepared from different methods for various functions and advantages.

Types of nanocellulouse	Polymers	Method of preparation	Function	Advantages	References
Skin tissue engineering and wound Healing					
NCC	Poly(lactic-co-glycolic) acid and NCC	Electrospinning	Cytocompatibility and facilitate fibroblast adhesion, proliferation for skin tissue engineering	Biocompatibility, biodegradable, stable functionality, and lower toxicity	Mo et al. (2015)
NFC	Brown algae cellulose nanofibers with quarternized B-chitin and organic rectorite	Freeze drying	Promote collagen synthesis, neovascularization, and accelerating wound healing	Effective wound dressings due to its high porosity and liquid absorption Antibacterial property	Gao et al. (2019)
NCC	NCC with polyvinyl alcohol	Freeze drying	Biocompatibility with human fibroblast skin, facilitate cell spreading, adhesion, and proliferation for biomedical applications	Biocompatibility, high porosity, provides strength and stiffness that enhances mechanical property	Lam, Chollakup, Smitthipong, Nimchua, and Sukyai (2017)
NFC	NFC cross linking with 1,4-butanediol diglycidyl ether	3D printing	Promotes fibroblasts proliferation, support structures of cellular process of cells during wound healing, regeneration, and tissue repair	Provides mechanical strength and increase rigidity of the fibroblast cells	Xu et al. (2018)
NFC	NFC with gelatin methacrylate	3D printing	Promotes survival, attachment, adhesion, and proliferation of fibroblast cells that are involved in wound healing for soft tissue regeneration	Cytocompatible High fidelity and stability Provides mechanical strength and nontoxic	Xu et al. (2019)

(*Continued*)

Table 19.8 (Continued)

Types of nanocellulouse	Polymers	Method of preparation	Function	Advantages	References
NFC	NFC with chitosan	Solvent casting	Mechanical property of the material in wet condition match with human skin and potentially use as artificial skin and wound dressings	Improve mechanical property, increase Young's modulus, similar texture to the elastic tissue in skin.	Wu, Farnood, O'Kelly, and Chen (2014)
NCC	NCC with collagen	Solvent casting	Good effect on cell morphology, viability, proliferation, and compatible in skin tissue engineering applications	Good biocompatibility Provides mechanical property	Li et al. (2014)
NCC	NCC with mixture of chitosan, gelatin, and calcium peroxide	Solvent casting	Enhance proliferation of human fibroblast cells and potentially as antibacterial wound dressing	Improve mechanical property, reduce water vapor transmission rate Antibacterial Nontoxic to the cells	Akhavan-Kharazian and Izadi-Vasafi (2019)
NFC	NFC with lysozymes nanofibers	Solvent casting	Biocompatible L929 fibroblasts cells. Promotes cell migration, proliferation and possess antioxidant property that could be used as wound healing patches	Good mechanical performance, increase Young's modulus, antimicrobial	Silva et al. (2020)
BNC (bacterial nanocellulouse)	BNC with tannic acid, chitosan, and gelatin	Solvent casting	Promotes epithelialization, wound closure, and contraction of Winstar rats wounds	Improve mechanical property	Taheri, Jahanmardi, Koosha, and Abdi (2020)

NCC	NCC with poly (caprolactone) and gelatin nanofibers	Electrospinning	Promotes cell growth, cell differentiation, and proliferation of NIH/3T3 cells. Enhance wound closure and healing of Balb/c mice wounds	Good biocompatibility Increase Yong's modulus and tensile strength Nontoxic	Hivechi, Bahrami, Siegel, Milan, and Amoupour (2020)
BNC	BNC mixed dexpanthenol, coated with chitosan and alginate	Electrospinning	Promotes cell migration with full occlusion of wound scratch, and increase cell viability of HaCaT cells. Inhibit *Staphylococcus aureus* growth. Potentially use as wound healing applications	High mechanical performance Increase Young's modulus Good moisture uptake Antibacterial Nontoxic	Fonseca et al. (2020)
BNC	BNC with lactoferrin and collagen	Electrophoresis	Promotes epithelial cell migration, adhesion, and proliferation of fibroblast cells. Gives moisture. Inhibit bacterial growth. Potentially used as wound dressing.	Improve water holding capacity and water vapor transmission rate. Anti-bacterial. Enhance therapeutic effect in rat model of wound healing.	Yuan, Chen, and Hong (2021)
BNC	BNC with acrylic acid hydrogel	Electron beam irradiation		Nonbiodegradable material that promotes newly formed skin microstructure	Loh et al. (2020)

(*Continued*)

Table 19.8 (Continued)

Types of nanocellulouse	Polymers	Method of preparation	Function	Advantages	References
Vascular tissue engineering					
BNC	Air dried BNC layer	Solvent casting and air drying	Cytocompatibility, hemocompatibility, and facilitate patency of blood circulation in white rabbits especially for blood vessel applications.	Biocompatibility, higher mechanical strength, suture retention, and increase Young's modulus	Bao, Hong, Li, Hu, and Chen (2021)
BNC	Surface formed in air contact (SAC) BNC	Growth in a hollow glass cylinder	SAC BNC grafts promote low loss of platelets, leukocytes, and low activation of coagulation system as vascular grafts	Hemocompatibility, cytocompatibility, and improves biocompatibility	Wacker et al. (2020)
BNC	BNC-3D biomaterials	3D bioprinting	Imitate structure of organs' microducts for tissue regenerative applications	Biocompatibility	Osorio et al. (2020)
Neural tissue engineering					
NCC	Cellulose nanofibrils (NFC)	3D bioprinting	Promote neural cells differentiation	Biocompatibility and could be the model of neural network	Bordoni et al. (2020)
BNC	BC—graphene scaffold	3D bioprinting	Supports neural stem cell growth, adhesion, maintains the cells stemness, and enhances proliferation rates of the cells	Biocompatibility, supports neural stem cell culture and neural network formation and enhances network activities of cortisol neurons	Guo et al. (2021)

NCC	Cellulose nanofibrils	3D printing	Neural cells able to attach, proliferate and differentiate on the cellulosic 3D printed structure	Acts as ink substitute to mimic the actual neural tissue	Kuzmenko, Karabulut, Pernevik, Enoksson, and Gatenholm (2018)
NCC	Chitosan–cellulose nanofiber hydrogel	Solvents casting	Provides suitable environment for cell differentiation, proliferation, enhances self-healing, and promotes central nerve system functionality.	Biocompatibility and biodegradability. Provide better performance in neural tissue regeneration	Cheng, Huang, Wei, and Hui Hsu (2019)
NCC	Cellulose nanofibrous scaffold	Electrospinning	Support growth, differentiation of SH-SY5Y cells and improve neural network formation	Could be as a biomaterial for neural tissue regeneration	Kuzmenko et al. (2016)
Skeletal muscle and bone tissue engineering					
NCC	Alginate/gelatin/cellulose nanocrystals hydrogels	3D printing	Promotes cell proliferation, adhesion, nutrients exchange, and matrix mineralization for bone tissue engineering	Biodegradable and provides mechanical strength	Dutta, Hexiu, Patel, Ganguly, and Lim (2021)
NCC	Poly(lactic acid)/cellulose nanocrystals scaffold	Electrospinning	Improve osteogenesis and enhance bone regeneration	Enhance thermal and mechanical strength	Patel et al. (2020)
NCC	Gelatin–nanocellulose hydrogels	Freeze drying	Promotes osteogenesis differentiation for bone tissue engineering	Cytocompatible and no adverse effect	Carlström et al. (2020)

(*Continued*)

Table 19.8 (Continued)

Types of nanocellulouse	Polymers	Method of preparation	Function	Advantages	References
NCC	Regenerated cellulose nanofibers in chitosan hydrogel scaffold	Solvent castings	Enhance MC3T3-E1 cell adhesion, proliferation, and adhesion for bone engineering	Improve mechanical properties, water absorption, and swelling capacity	Carlström et al. (2020)
NCC	Cellulose nanocrystals–poly (ε-caprolactone) nanocomposites	3D printing	Able to induce biomineralization and enhance mechanical strength	Improve mechanical properties and induce calcium phosphate formation. Increase Young's modulus and tensile strength. Nontoxic to MC3T3 preosteoblasts cells.	Hong, Cooke, Whittington, and Roman (2021)
Ocular tissue treatment					
BNC	Bacterial nanocrystalline cellulose	Solvents casting	BNC meets basic preclinical requirement as bandage material to treat ocular surface disorder.	Biocompatibility. It can be heat sterilized and stable at room temperature	Anton-Sales et al. (2020)
BNC	Bacterial nanocrystalline cellulose hydrogel	Solvents casting	Allows attachment of hESC-LSC cells on the BNC membrane and potentially becomes cell carrier.	Mechanical stability and thermal stability.	Anton-Sales, Koivusalo, Skottman, Laromaine, and Roig (2021)
NCC	Nanocellulose-reinforced poly (vinyl alcohol) hydrogels	Solvents casting	Similar mechanical properties collagenous soft ocular tissues	Provides mechanical strength	Tummala, Joffre, Rojas, Persson, and Mihranyan (2017)

NCC	Nanocellulose-reinforced poly (vinyl alcohol) hydrogels	3D printing	Potential as biomaterial for cornea regeneration	Biocompatibility toward corneal epithelial cells	Tummala, Lopes, Mihranyan, and Ferraz (2019)
Soft tissue—ligament, meniscus, and cartilage replacements					
NCC	Nanocellulose/alginate bioink	3D printing	Increase in cell growth and cell survival within cartilaginous tissue. Potential as a future treatment to repair damaged cartilage in joints	Biocompatibility	Nguyen et al. (2017)
NCC	Alginate sulfate–nanocellulose bioink	3D printing	High cell viability and growth	High shapre fidelity allows 3D printing of the biomaterial structure	Müller, Öztürk, Arlov, Gatenholm, and Zenobi-Wong (2017)
BNC	Bacterial nanocellulose bioink	3D printing	Enhance structural integrity and chondrocyte proliferation. Suitable solution for cartilage repair.	Exhibits biocompatibility. Promotes mechanical stability and structural integrity	Apelgren et al. (2019)

3D, three dimensional; *NCC*, nanocrystalline cellulose; *NFC*, nanofibrillated cellulose.

devices discovery (Turky, Moussa, Hasanin, El-Sayed, & Kamel, 2021). Effective biomedical diagnostic device discoveries are really demanding due to the nosocomial infections that usually originate from the hospital environment itself. Nosocomial infections on the biomedical devices have caused the device to become a pathogen transmission agent which led to various types of diseases (Shehabeldine & Hasanin, 2019; Turky et al., 2021). Throughout the diagnosis and monitoring in clinical operations, there are many types of equipment that need direct contact with patients to treat them. This indirectly increases the possibility of contaminant transmission from the device to patients (Turky et al., 2021). Hence, biosensors associated with biomaterials can be a strong diagnostic device for many other diseases despite infectious diseases (Kamel & Khattab, 2020; Li et al., 2020; Sharma & Bhardwaj, 2019). Biosensors are fast reaction affordable methods that can transform the biochemical information into an electrical signal (Kamel & Khattab, 2020), and largely used bioactive elements are enzymes, antibodies, and receptors (Fontenot et al., 2017; Sharma & Bhardwaj, 2019). Optical, electrical, and mechanical are the three different types of biosensors that are associated with many detection methods. Types of detection methods include colorimetric, fluorescent, bioluminescent, electrochemical, conductometric, electronic, gravimetric, and pyroelectric methods (Kamel & Khattab, 2020). Biomaterials such as NC and BC with and without metallic NP biomodification have been widely used in the biomedical field to cure and prevent infectious and noninfectious diseases (Kamel & Khattab, 2020; Loynachan et al., 2018; Sharma & Bhardwaj, 2019). NC usage in point-of-care (POC) immunoassays, application of graphene-based nanomaterials, and bacterial NC in diagnostic and monitoring will be mainly reviewed here. Most of the diagnostic devices are immunoassays because they are widely used in hospitals. Immunoassays are antigen—antibody based detection that determines the concentrations of the analytes (Li et al., 2020). Generally, immunoassays generate detection signals primarily based on labels used. Optical, electrochemical, and electrochemiluminescence immunoassays are categorized by referring to the labels used as well as the different methods used in signal detection (Li et al., 2020).

19.7.1.1 Cellulose nanofibers substrate in paper-based point-of-care immunoassays with metallic nanoparticles conjugated antibodies

The paper used in the assays is made up of different types of materials such as filter paper, nitrocellulose membrane, and glass fibers. All the listed paper materials had setbacks of having high porosity and rough surface (Hoeng, Denneulin, & Bras, 2016; Li et al., 2020), which affected the efficiency of assays. Hence, nanopaper made up of cellulose nanofibers from plants was invented and it was a promising substrate material for the paper-based electrodes due to its properties of being easily foldable. This substrate material was further doped with silver nanowires and carbon tubes to produce hybrids to enhance the efficiency of paper-based POC immunoassays (Hsieh, Koga,

Suganuma, & Nogi, 2017; Li et al., 2020; Sanjay, Dou, Sun, & Li, 2016). There are a few detection methods under optical biosensing such as colorimetric, fluorescence, chemiluminescence, surface plasmon enhancement, chemical, and electrochemiluminescent methods. Lateral flow immunoassay is one of the paper-based POC families which falls under colorimetric detections and is reported to use metallic NPs especially gold NP-based immunoassays for cardiovascular diseases (CVDs), malaria, and dengue fever diagnosis (Li et al., 2020). In the CVDs tests, biomarkers are used to identify acute myocardial infarction, heart failures (Gong et al., 2017), and other abnormal heart conditions. This reaction is called as a biomolecule—NP interaction because the analyte will bind to the NP conjugated antibody (Yahaya et al., 2017), as shown in Fig. 19.4. Calorimetric immunoassay has been also broadly used in the detection of the human immunodeficiency virus (HIV) (Loynachan et al., 2018), sperm activity (Nosrati, Gong, Gabriel, & Pedraza, 2016), adulterated whiskey (Cardoso et al., 2017), and glucose (Gao et al., 2018). Moreover, to reduce the elevated mortality rate due to poor diagnosis captivates thorough research on paper-based POC that could control infectious diseases such as malaria, HIV, hepatitis C virus, hepatitis B virus, dengue, pneumonia, and many more viruses (Blanco-Covián et al., 2013; Channon et al., 2018; Kurdekar et al., 2016; Li et al., 2020).

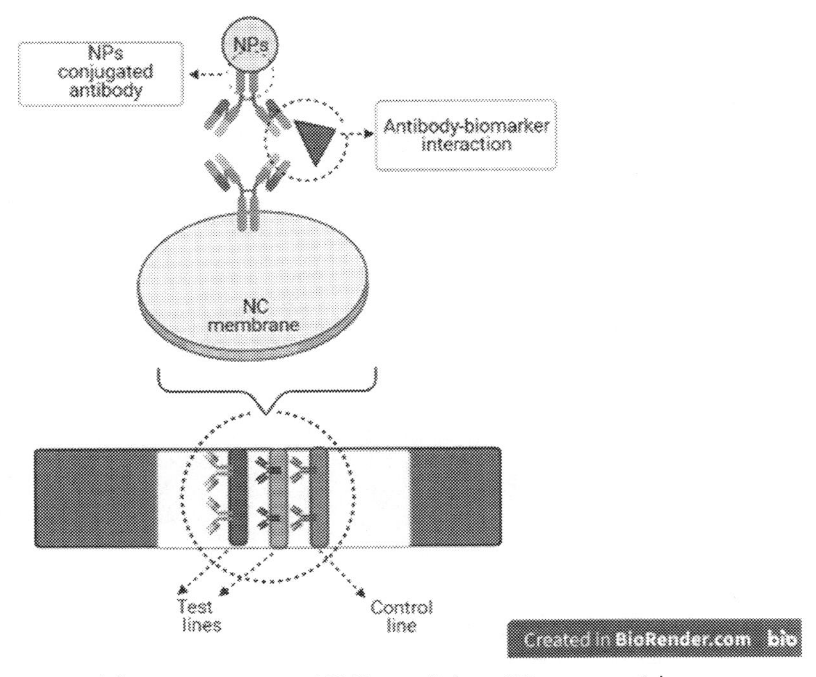

Figure 19.4 Lateral flow immunoassay. *NC*, Nanocellulose; *NPs*, nanoparticles.

19.7.1.2 Bacterial nanocellulose in biosensing

Application of BNC in diagnostic was reported in many studies (Baldikova et al., 2017; Sharma & Bhardwaj, 2019), especially the biomodified BNC, such as BNC conjugated with Ag (silver nanocomposite). These Ag-BNC analyzed many biomolecules such as amino acids, L-glutamine, and l-phenylalanine (Kamel & Khattab, 2020; Sharma & Bhardwaj, 2019) exhibiting to be a promising biosensor. Next, glucose oxidase biomolecule detection was tested with the external lamella BNC (Eisele, Ammon, Kindervater, Gröbe, & Göpel, 1994). Nanofibers BNC with conjugate Au nanomaterial was also able to detect glucose content in the blood (Sharma & Bhardwaj, 2019).

19.7.1.3 Graphene based nanomaterials in biosensing

There are various types of graphene derivatives that are commonly used in biosensor devices, for instance, pristine graphene made up of single-crystalline grain, polycrystalline graphene, graphene quantum dots known as nanocrystal, and reduced graphene oxide (rGO) (Morales-Narváez, Baptista-Pires, Zamora-Gálvez, & Merkoçi, 2017; Peña-Bahamonde, Nguyen, Fanourakis, & Rodrigues, 2018). Out of all nano-based biosensors, graphenes that are fabricated with nanomaterials are the ones reported to be a promising approach in diagnostic as they actively responded by enhancing signals (Peña-Bahamonde et al., 2018). Graphene in nanomaterial-based exhibits an outstanding biocompatibility property by binding to biological molecules such as antibodies, as in Fig. 19.5, DNA, cells, enzymes, and proteins. In parallel with the excellent binding properties, graphene-based nanomaterial reported being multidetector for biomolecules and cells (Wang, Li, Wang, Li, & Lin, 2011). Nanomaterials conjugated with graphene act as transducers by exchanging the chemical interaction of receptor and organic or inorganic targeted molecules into a measurable signal (Peña-Bahamonde et al., 2018). Many antibody biosensing kits have been recently invented with various graphene-based nanomaterials to produce multipurpose immunoassays which

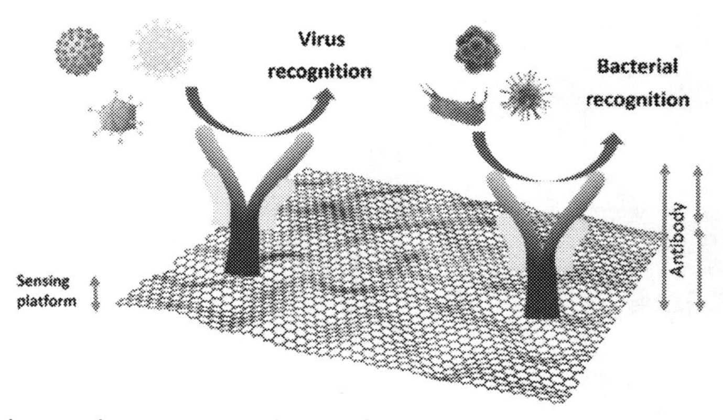

Figure 19.5 Pathogen detection using biomodified graphene with antibody (Peña-Bahamonde et al., 2018).

can promptly identify diseases. For instance, the GO cellulose nanopaper that targets the *Escherichia coli* using photoluminescence (Cheeveewattanagul et al., 2017) and electrical detecting method (Akbari, Buntat, Afroozeh, Zeinalinezhad, & Nikoukar, 2015; Thakur et al., 2018), GO-Ag nanocomposite that targets *Salmonella typhimurium* using cyclic voltammetry detecting method (Sign & Sumana, 2016), and graphene that targets Zika virus using electrical detection method (Afsahi et al., 2018). Moreover, for GO, it has been used in different detection methods to detect various pathogens such as dengue virus, via electrochemical impedance spectroscopy detecting method (Navakul et al., 2017), whereas Rotavirus has been tested with photoluminescence detecting method (Sign & Sumana, 2016). Despite targeting pathogens, graphene-based nanomaterials had also contributed to the diagnosis of cancer with electrochemical and electrochemical impedance spectroscopy (Peña-Bahamonde et al., 2018; Wang et al., 2011; Zhou et al., 2017).

19.7.2 Immobilization and recognition of enzyme/protein

Proteins are biomacromolecules that play critical roles in a variety of life processes, including regulation of the metabolic process, exchange of cellular information, cell-cycle control, molecular transport, and environmental protection. Proteins, for example, are extremely valuable biomarkers of disease in biomedicine (Amiri-dashatan, Koushki, Abbaszadeh, & Rostami-nejad, 2018). In the field of biotechnology, for instance, the role of enzymes as biocatalysts is the subject of numerous studies and which have the potential to increase the rate of almost all chemical reactions in a cell without being permanently altered or consumed by the reactions, as well as without affecting the equilibrium between the reactants and products (Sheldon & Woodley, 2018). Reaction rates are multiplied by over a million, so reactions that would take years to complete in the absence of catalysis can now be completed in fractions of a second when the appropriate enzyme is present. In the absence of enzymes, the progress of most biochemical reactions will be quite slow, making them no longer capable of sustaining complex life. However, the use of enzyme is often correlated with other flaws such as vulnerability to process conditions, poor stability, or a proclivity to be hampered by high concentrations of reaction components. Most enzymes are very fragile, and their industrial application is often hindered by a lack of long-term operational stability as well as a technically difficult recovery and enzyme reuse procedure. Furthermore, the interactions between the sequence of proteins and the residues in the protein core are not necessarily completely optimized and only meet the minimum requirements for proper functioning. These shortcomings provide plenty of room for improvement of the efficiency of enzymes. Immobilization is one of the methods that have been implemented to reduce costs in order to improve the use of enzymes in biotechnology processes. The term "enzyme/protein immobilization" is used to describe the process of immobilizing enzymes/proteins (Mohamad, Marzuki,

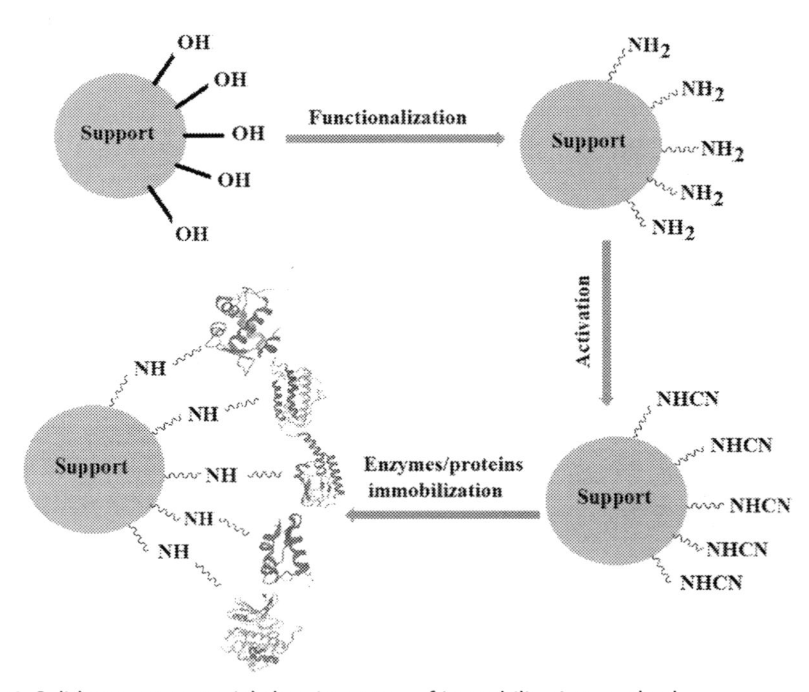

Figure 19.6 Solid support material showing steps of immobilization method.

Buang, Huyop, & Wahab, 2015). In general terms, immobilization refers to the physical containment or localization of enzyme molecules or proteins in a given region of space while their catalytic activities are retained (Mahmoud & Helmy, 2009). The following methods can be used to do this. Use of water-insoluble inorganic (silica-based and other oxide-based materials) and organic (polymers and biopolymers) material supports for physical adsorption of enzymes/proteins (Fig. 19.6).

1. Immobilization by entrapping (Molina-Fernandez & Luis, 2021) means the capture of enzymes within a polymeric array or polymer microcapsules that allow substrates and products to cross but retain the enzyme (Fig. 19.7).
2. Covalent binding of the enzyme to water-insoluble support (Kuan, Lai, & Lee, 2021) (Fig. 19.8).

Enzymes or proteins are required to immobilize for a variety of reasons, including increased stability, cost, and ease of down streaming. Enzymes are immobilized in their most stable functional forms in an inert carrier, and their resistance to pH and temperature variations is reduced because the sites for sensing these variations are locked in and can be utilized for batch for reaction and easily separated. These properties can be placed as a reason for increased stability, cost-effectiveness, and separation from the support system. Table 19.9 summarizes the classic materials and enzyme types that can be immobilized using these supports, as well as information about immobilization

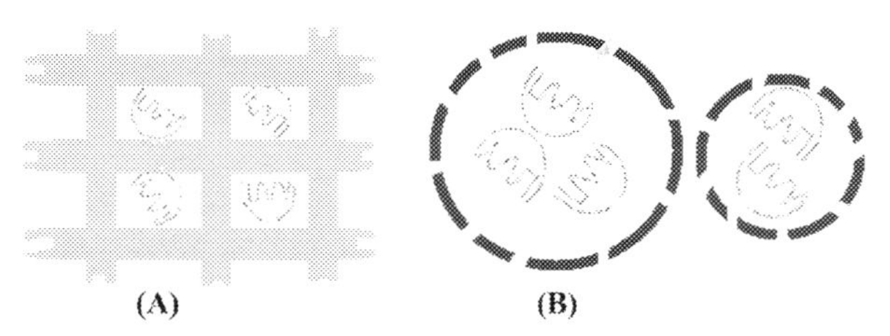

Figure 19.7 Enzyme entrapped in (A) matrix and (B) droplets.

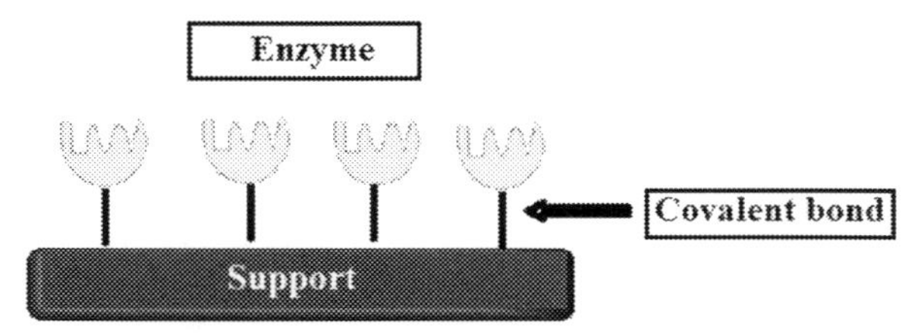

Figure 19.8 Enzyme is covalently bound to a water-insoluble support.

type, crosslinking agents, and binding group (Zdarta, Meyer, Jesionowski, & Pinelo, 2018, references cited therein).

19.7.2.1 Methods of enzyme/protein immobilization

There are many ways to classify different types of immobilization processes. The classification is determined by the type of interaction that causes immobilization. Several methods for enzyme immobilization have been tested over the last few decades, as well as modern methods for remediation of the former. Immobilization processes are classified as shown in Figs. 19.9 and 19.10.

The binding reaction is carried out under moderate conditions in all enzyme immobilization techniques so that the protein is not denatured. The amino, carboxyl, sulfhydryl, hydroxyl, or phenolic groups of the amino acids in the proteins' residual side chains are used to couple with the reactive groups of the polymer in covalent binding. Here are a few selected examples of different types of immobilization.

19.7.2.2 Physical immobilization methods

The procedures used in the physical methods are designed to localize the enzyme without relying on the formation of covalent bonds. Immobilization in this category is based on physical forces (for example, electrostatic interactions), trapping enzymes within microcompartments, or confining the catalyst in prefabricated membrane devices for containing enzymes. In principle, physical methods are reversible since the enzyme should be able to be completely recovered in its active state (Liu, Ma, & Shi, 2020). However, in reality, either recovery is incomplete or any inactivation occurs.

Table 19.9 Summary and examples of materials, both inorganic and organic materials used to immobilize enzymes.

Support material	Binding groups	Crosslinking agent	Immobilization type	Immobilized enzyme
Sol-gel silica	–OH	—	Adsorption	Lipase from *Aspergillus niger*
Silica gel	–OH –C=O	Glutaraldehyde	Covalent binding	Commercial lipase
Montmorillonite	–OH	3-Aminopropyl triethoxysilane	Covalent binding	Glucoamylase from *A. niger*
ZrO_2	–OH	—	Adsorption	α-Amylase from *Bacillus subtilis*
Hydroxyapatite	–OH	—	Adsorption	Glucose oxidase from *A. niger*
Bentonite	–OH –NH2	Tetramethyl ammonium hydroxide	Covalent binding	Glucose oxidase from *A. niger*
Activated carbon	–OH –C=O	—	Adsorption	Cellulose from *A. niger*
Activated charcoal	–OH –C=O	—	Adsorption	Papain
Activated charcoal	–OH –C=O –COOH	—	Adsorption	Amyloglucosidase
Polyaniline	–N–H –C=O	Glutaraldehyde	Covalent binding	α-Amylase
Agarose	–OH	—	Entrapment	α-Amylase
Chitosan	–OH –NH2	—	Entrapment	Lipase from *Candida rugosa*
Sponges	–COOH	—	Adsorption	Lipase from *A. niger*
Polystyrene	C=O epoxy groups	Poly(glycidyl methacrylate)	Covalent binding	Lipase
Poly(vinyl) alcohol	–OH –C=O	Glutaraldehyde	Covalent binding	Laccase from *Trametes versicolor*
Luffa cylindrica	–OH –C=O	—	Adsorption	Lipase from *A. niger*
Cellulose nanocrystals	–OH	—	Adsorption	Lipase from *C. rugosa*

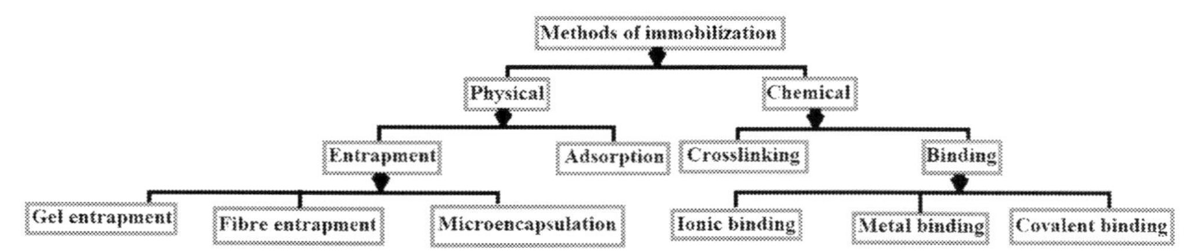

Figure 19.9 The methods for immobilization.

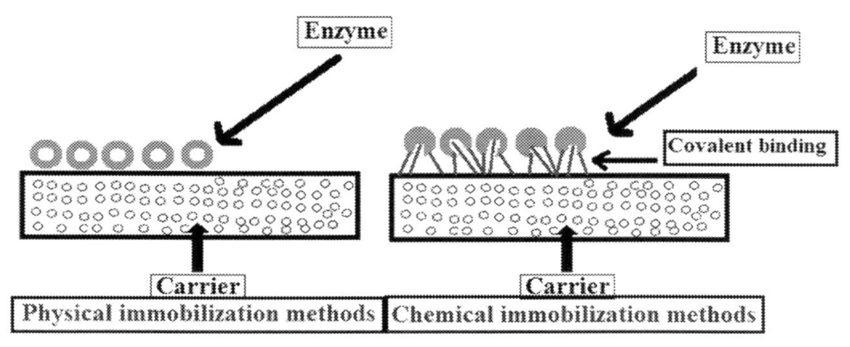

Figure 19.10 Graphical presentation of physical and chemical immobilization methods.

However, since the binding forces between the enzyme protein and the carrier are weak, the adsorbed enzyme can leak from the carrier during use. Another drawback is that substrate diffusion issues can impair the enzyme's efficiency. The enzyme concentration available to the unit surface of the carrier during the stabilization process is a major factor influencing the amount of enzyme adsorbed on a solid support. The pH, experimental variables such as the nature of the solvent, ionic strength, protein concentration, and adsorption, time, and temperature play a role in the adsorption of an enzyme on a water-insoluble substance. This has been accomplished by the use of a number of systems. Both inorganic and organic materials have been used as carriers. Examples of inorganic supports include alumina, clay, silica gel, titania, calcium phosphate gel, porous glass, zeolite, zirconium oxide, and others. Adsorption techniques have also been used to immobilize enzymes using organic supports such as activated carbon, sugar, chitosan, cellulose derivatives, gluten, and so on. Adsorption is one of the physical immobilization methods and is the most fundamental and oldest method for immobilizing an enzyme on a water-insoluble carrier. Since its discovery, the adsorption method has been used to treat a wide range of enzymes and whole cells. It has been known for much longer than the adsorption of single enzymes that whole cells can be made to adhere to suitable solid bodies; for example, bacteria attached to wood shavings were already used in the production of vinegar (Hartmeier, 1988).

Entrapment

Many compounds involved in entrapment immobilization that contain double bonds can polymerize, but monomers with only one double bond per molecule can only form linear polymers that are useless for enzyme entrapment. Entrapment immobilization has been accomplished using either sphere encapsulation vesicles or a matrix entrapment lattice made of organic polymers or silica sol-gel. Both immobilization preparations have several advantages and also disadvantages in terms of leakage and mass transfer resistance to the substrate (Trelles & Rivero, 2013).

Gel entrapment

In this method, protein molecules are physically entrapped within a highly crosslinked polymer network formed in the presence of protein. Since the enzyme molecules are stuck within the matrix, any leakage from the matrix can be avoided. However, such systems are plagued by diffusion issues. Diffusion of reactants or products is difficult if the protein is deep within the solid matrix. The enzyme activity can also be inhibited by the retained reaction products and most widely used crosslinked polymers for protein entrapment is polyacrylamide. In addition, naturally occurring polymers including agar, alginate, gelatin, and so on can be dissolved in water and combined with the enzyme solution to make a gel for gel entrapment system. On the other hand, the enzyme—polymer mixture is gelled by mixing it with a solvent that is immiscible with water in order to enhance the enzyme-polymer binding. The literature describes a wide range of gel entrapment systems (Kato, Lee, & Nagata, 2020; Liang, Li, & Yang, 2000).

Fiber entrapment

Previously, fiber entrapment immobilization process was used to meet the structural requirements of various reactors, where, an enzyme solution is emulsified in an organic solvent of cellulose acetate and sprayed into fibers, which are woven into cloth or formed into other shapes. As the fiber is so fine, the specific area is very large and the trapping capacity is high, making this method more advantageous than gel trapping. The fibers are also resistant to weak acids and alkalis as well as certain organic solvents. They can be resistant to microbial attack depending on the polymer used. Due to its low cost and strong biological and chemical resistance, cellulose acetate is one of the most widely used polymers for fiber entrapment (Zhu, Cheng, et al., 2020).

19.7.2.3 Microencapsulation

This technique is similar to the entrapment method. In this method, microcapsule entrapment encloses enzymes inside a semipermeable polymer membrane microcapsule or molecular cavity (Bodade & Bodade, 2020). Microcapsules with diameters ranging

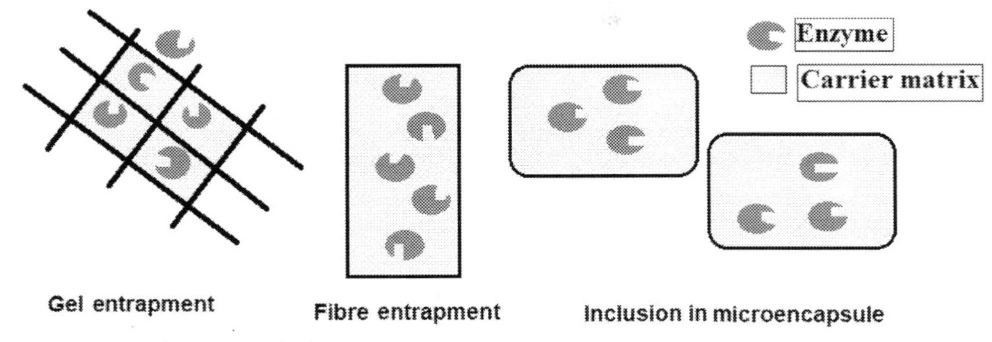

Figure 19.11 Different methods of enzyme entrapment.

from a few micrometers to several hundred micrometers are either permanent or nonpermanent semipermeable membranes. Nonpermanent membranes are made with the appropriate surfactants, additives, and hydrocarbons, whereas permanent membranes are made with an interfacial polycondensation or phase separation process (Fig. 19.11).

This method can provide a very large specific surface area. Alternatively, semipermeable membranes can be made to form a membrane reactor. The micropores of the membrane allow small molecular substrates and products to move through, but enzymes and other large molecular compounds cannot. It is the ideal method for immobilizing agents for the purpose of delivering and releasing them to remote and targeted locations. As a result, it is mostly used in medical and pharmaceutical applications. The following methods are used to achieve microcapsules and are illustrated with figures.

1. Phase separation (Fig. 19.12A);
2. Interfacial polymerization (Fig. 19.12B);
3. Liquid drying (Fig. 19.12C); and
4. Liquid surfactant membrane method (Fig. 19.12D) (Fig. 19.12 was drawn as mentioned in the literature) (Reji Kumar, 1997).

Chemical binding

Chemical immobilization is accomplished by forming chemical bonds between the biorecognition element's functional group (and any side chains not needed for catalytic activity) and the transducer's surface. Chemical reagents (such as glutaraldehyde or carbodiimide) or a preactivated membrane added to the transducer surface form chemical bonds on the active transducer surface. Covalent bonding and crosslinking are the most common chemical immobilization methods; however, both methods are often combined to provide stronger binding and activity (Wahab, Elias, Abdullah, & Ghoshal, 2020).

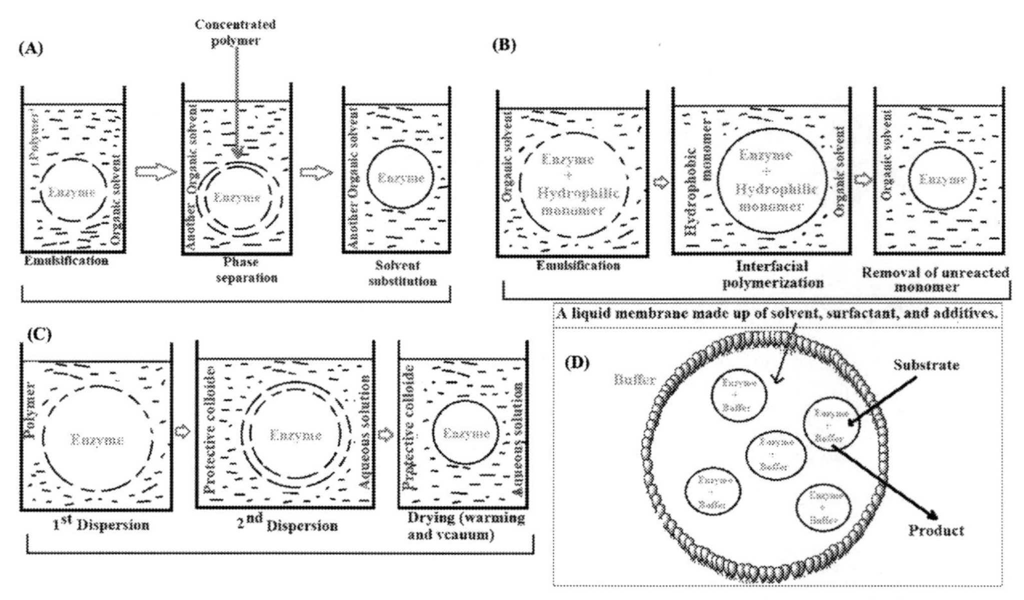

Figure 19.12 Microencapsulation by various techniques (A) phase separation, (B) interfacial polymerization, (C) liquid drying, and (D) liquid surfactant membrane method.

Crosslinking

It is an irreversible way of immobilizing enzymes (Fig. 19.13A). It differs from other methods in that it does not necessitate the use of support for immobilization. This method of immobilization is focused on the formation of chemical bonds, similar to the covalent binding method, but it does not use water-insoluble carriers. The formation of intermolecular cross-linkages between enzyme molecules is accomplished using bi- or multifunctional reagents to immobilize enzymes. Immobilization by crosslinking has been performed using crosslinking reagents such as glutaraldehyde, isocyanate derivatives, bis(diazo)benzidine, N,N'-polymethylene-bis(iodoacetamide), and N,N-ethyiene bis-maleimide. Crosslinking reagents may often alter the conformation of active centers of enzymes, resulting in a substantial loss of enzyme activity (Zdarta et al., 2018).

Ionic binding

Ionic binding immobilization is a simple reversible process that relies on the ionic binding of enzyme molecules to solid supports with ion exchange residues. In some cases, both ionic binding and physical adsorption can occur at the same time (Zhu, Chen, Shao, Jia, & Zhang, 2020). Ionic bonding and physical adsorption vary mostly in the strength of the enzyme-support bonds, which are much stronger for ionic bonding but weaker than covalent bonding. Ionic bonding is performed using

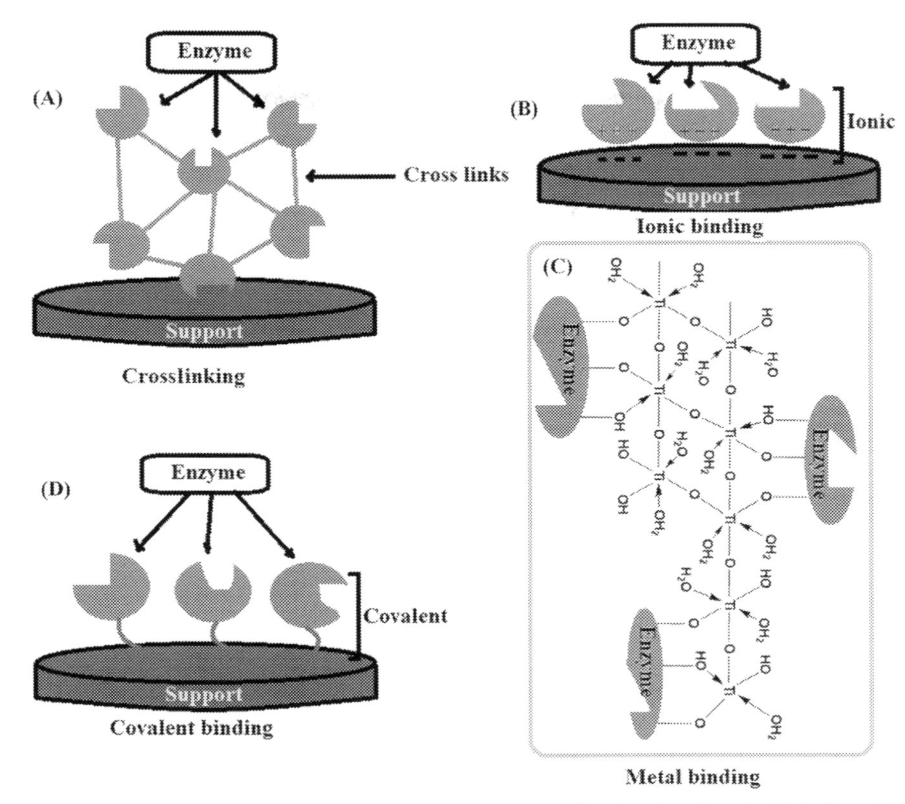

Figure 19.13 Various techniques to achieve enzyme immobilization by chemical binding; (A) cross-linking, (B) ionic binding, (C) metal binding, and (D) covalent binding.

polysaccharides and synthetic polymers with ion exchange residues (Fig. 19.14B). Although it is simple and only involves simple binding reversal inputs, the drawback of this process is that it causes enzyme desorption when the pH, electrolyte concentration, or temperature changes. Mitz was the first to announce the immobilization of an enzyme using this process in 1956. Catalase was immobilized on DEAE-cellulose in this study. Since then, a wide range of ionic binding carriers have been used, and the specifics are addressed in *Wisema's Handbook of Enzyme Biotechnology*.

Metal binding

Another common form of reversible enzyme immobilization is chelation, or metal binding, which is often used as a chromatographic process (Mokhtar et al., 2020). This approach is less used in industries because it is relatively costly and involves safety concerns. In this method, it was thought that activation of support materials with transition metal compounds, typically titanium (IV) chloride, produced active transition metal chelates capable of binding enzymatic molecules by replacing the metal chloride

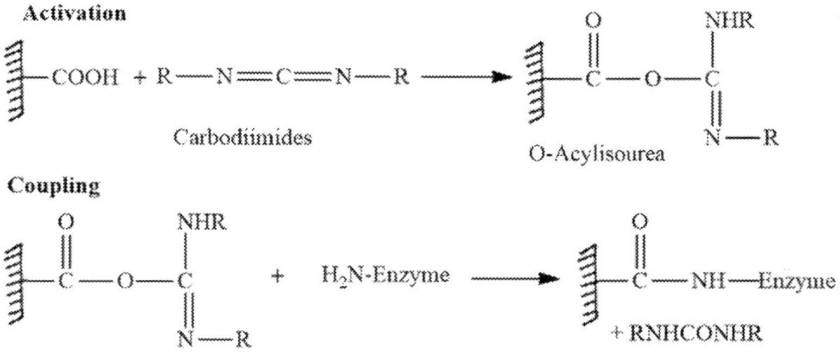

Figure 19.14 Immobilization of the enzyme with the activated support followed by coupling.

bonds with groups present in the enzyme molecule. *Wiseman's Handbook of Enzyme Biotechnology* discussed enzyme immobilization by metal binding (Fig. 19.13C).

Covalent binding

There is at least one covalent or partially covalent bond between the amino acid residue of the protein and a water-insoluble polymer in this immobilization process. In this method, covalent bonds are extremely stable and ensure that the enzyme is firmly attached to the support (Guisan, López-Gallego, Bolivar, Rocha-Martín, & Fernandez-Lorente, 2020; Nguyen & Kim, 2017).

Since 1973, it has been used in various industries including the production of 6-aminopenicillanic acid from penicillin G, which uses penicillin acylase covalently linked to Sephadex G-200. When compared to covalent binding immobilization, physical immobilization techniques have the disadvantage of slowly leaking protein from the bulk of the material. This disadvantage is overcome by the covalent attachment of the enzyme. However, this can cause changes in the protein's conformation as well as its active centers. As a result, there is a significant decrease in biological activity or a change in substrate specificity. In the presence of a competitive inhibitor or substrate, enzymes may also be covalently bound to the carrier to avoid coupling at active sites. Binding is usually done with one of the reactions mentioned:

1. Peptide bond formation,
2. Alkylation and arylation
3. Diazo linkage,
4. Isourea linkage, and
5. Other reactions.

The majority of polymeric supports contain unactivated hydroxyl, amine, amide, and carboxyl groups that can be activated prior to use. In order to prevent denaturation of the enzyme—protein, the coupling reaction must be performed under carefully regulated conditions. The schematic representation of one of the selected and applied methods is shown in Fig. 19.14 as an outline (Reji Kumar, 1997).

In this technique (Kuijpers et al., 1999), activation of the carboxyl group of the polymer must be accomplished using various methods documented in the literature followed by activation for completion of the immobilizing enzyme as shown in Fig. 19.14, showing reaction between carbodiimides and carboxyl group. In brief, there are numerous methods for immobilizing enzymes, but some of the immobilizing materials are expensive, so finding appropriate supports at a lower cost is critical. Table 19.10 lists the benefits and drawbacks of enzyme immobilization (Zucca & Sanjust, 2014, references cited therein).

Table 19.10 The advantages and disadvantages of the most common methods of enzyme immobilization.

Immobilization method	Advantages	Disadvantages
Adsorption	There is no chemical modification of the enzyme. It is simple and inexpensive to carry out.	Leakage of enzymes The reaction has a low specificity (i.e., adsorption and ion-exchange could overlap)
Encapsulation/ entrapment	There is no chemical modification of the enzyme. Under the conditions of polymerization/transition of the support, the enzyme should retain catalytic activity.	Leakage of enzymes Mass transfer issues
Ionic binding	There is no chemical modification of the enzyme. Simple to carry out	Leakage of enzymes Low specificity of the reaction (i.e., adsorption and ion-exchange could overlap)
Enzyme crosslinking	There is no need for support. Stabilization of the enzyme Minimize catalyst leakage	Significant chemical modification of the enzyme is a possibility. Complicated experimental processes Mass transfer issues
Covalent binding	Binding strength Minimize catalyst leakage Enzyme stabilization	Possibility ineffective modifications of the enzyme Enzymatic activity may be reduced. Chemical modifications of the support are needed In most cases, the attachment is irreversible, preventing the support from being reused.

19.7.3 Antimicrobial nanomaterials

NPs have emerged as a novel option for combating bacterial multidrug resistance, which is a global problem caused by antibiotic misuse (Hemeg, 2017). The microbicidal design of NPs results from direct contact with the bacterial cell wall (Liao, Li, & Tjong, 2019), without the need to penetrate the cell, so their use as antimicrobial agents could overcome mechanisms of bacterial resistance. Antimicrobial activity was described as the ability of chemical substances to destroy or inhibit the growth of bacteria (bactericidal or bacteriostatic) and fungi (Mahira et al., 2019). The majority of antimicrobial agents are chemically synthesized or derived naturally from different natural resources such as aminoglycosides and antibiotics like sulfonamides. Antimicrobial resistance to different metal NPs has been less than that of antibiotics, suggesting that NPs could be used as antimicrobial theranostics in medicine (Singh, Garg, Pandit, Mokkapati, & Mijakovic, 2018). NPs are 3D materials with a basic unit in the nanometer scale (1−100 nm) or at least one dimension in this range (Khan, Saeed, & Khan, 2019). Antimicrobial activity of nanomaterials against fungi, mycobacteria, Gram-positive and Gram-negative bacteria, has been demonstrated (Wang, Hu, & Shao, 2017). The antibacterial activity of the NPs differs depending on the type of NP. Despite the fact that the antimicrobial activity of the NPs is still unknown, it has been suggested that multiple mechanisms may contribute to the antimicrobial mechanisms. The surface characteristics and inner structure of NPs may have inherent antibacterial properties due to their membrane damaging abrasiveness, as seen in GO NPs (Díez-Pascual, 2020). Another mechanism that has been proposed is the enhanced release of antibacterial metal ions from the surface of NPs. NPs with a lower surface-to-volume ratio have more antimicrobial activity, resulting in more contact between nanomaterials and the environment (Sánchez-López et al., 2020). The most important variables influencing antibacterial activity are chemistry, particle size, particle shape, and zeta potential, which can be verified using different spectroscopic instruments. Metals have been used as antimicrobial agents in many countries for decades. The Egyptians used copper salt as a constringent in 1500 BC. The Greeks, Egyptians, Persians, Romans, and Indians have used silver and copper to clean water and preserve food. Metal NPs with a smaller size and a greater surface area have been shown to have better antimicrobial properties. Several experiments have been performed to synthesize and analyze metal NPs with improved antimicrobial properties since the advent of nanotechnology. For example, ZnO has been shown to be highly effective against pathogenic microbes and viruses. However, due to their small size, ZnO NPs are much more effective against pathogens than zinc oxide. The morphology, average size (nm), antimicrobial activity in terms of minimum inhibitory concentration and minimum bactericidal/ fungicidal concentration values of some NPs synthesized using environmentally friendly techniques and their antimicrobial activities were shown in Table 19.11, as mentioned in the literature (Goswami, Sarkar, & Ghosh, 2013; Sánchez-López et al., 2020).

Table 19.11 Biosynthesis of alternative metal-based nanoparticles with potential antibacterial activity at minimum inhibitory concentration (MIC) and minimum bactericidal/ fungicidal concentration (MBC/MFC) values.

Species	Sources	NPs	Morphology	Size (nm)	Antimicrobial activity	MIC and MBC/ MFC values
Alternanthera bettzickiana	Plant extract	AuNPs	Spherical	80–120	*Salmonella typhi; Pseudomonas aeruginosa; Enterobacter aerogenes; Staphylococcus aureus; Bacillus subtilis; Micrococcus luteus*	MIC values (expressed in μL of AuNPs): 10 μL *B. subtilis* 20 μL *S. aureus* 30 μL *M. luteus* 40 μL *E. aerogenes*, *S. typhi*, and *P. aeruginosa*
Stoechospermum marginatum	Algae	AuNPs	Spherical to irregular	18.7–93.7	*P. aeruginosa; Vibrio cholerae; Vibrio parahaemoluticus; Salmonella paratyphi; Proteus vulgaris; S. typhi; Klebsiella pneumoniae; Klebsiella. oxytoca; Enterococcus faecalis*(+);	AuNPs more effective against *E. faecalis* > *K. pneumoniae*. Noneffective against *Escherichia coli*
Shewanella loihica PV-4	Bacteria	CuNPs	Spherical	10–16	*E. coli*	100 μg/mL Cu-NPs inhibits 86% of the bacteria
Cystoseira trinodis	Algae	CuONPs	Spherical	6–7.8	*E. coli; S. typhi; E. faecalis; S. aureus; B. subtilis; Streptococcus* \\ *faecalis*	*E. coli* and *S. aureus* MIC: 2.5 μg/mL E. faecalis MIC: 5 μg/mL *Salmonella typhimurium* MIC: 10 μg/mL
Glycosmis pentaphylla	Plant extract	ZnONPs		32–36	*Shigella dysenteriae; S. paratyphi; S. aureus; Bacillus cereus*	At 100 μg/mL maximum inhibition is observed
Suaeda aegyptiaca	Plant extract	ZnONPs		~60	*P. aeruginosa;* *E. coli;* *S. aureus; B. subtilis*	*P. aeruginosa* MIC and MBC: 0.19–0.78 mg/mL *E. coli* MIC: 1.56–12.50 mg/mL MBC: 6.25–12.50 mg/mL *S. aureus* MIC and MBC: 0.39–1.56 mg/mL *B. subtilis* MIC: 0.19–0.39 mg/mL MBC: 0.78–12.50 mg/mL
Penicillium citrinum 1	Fungi	AgNPS	Spherical	20–40	*Schizosaccharomyces pombe* *E. coli; B. subtilis*	8 μg/mL, 16 μg/mL 4 μg/mL, 8 μg/mL; 8 μg/mL, 32 μg/ml

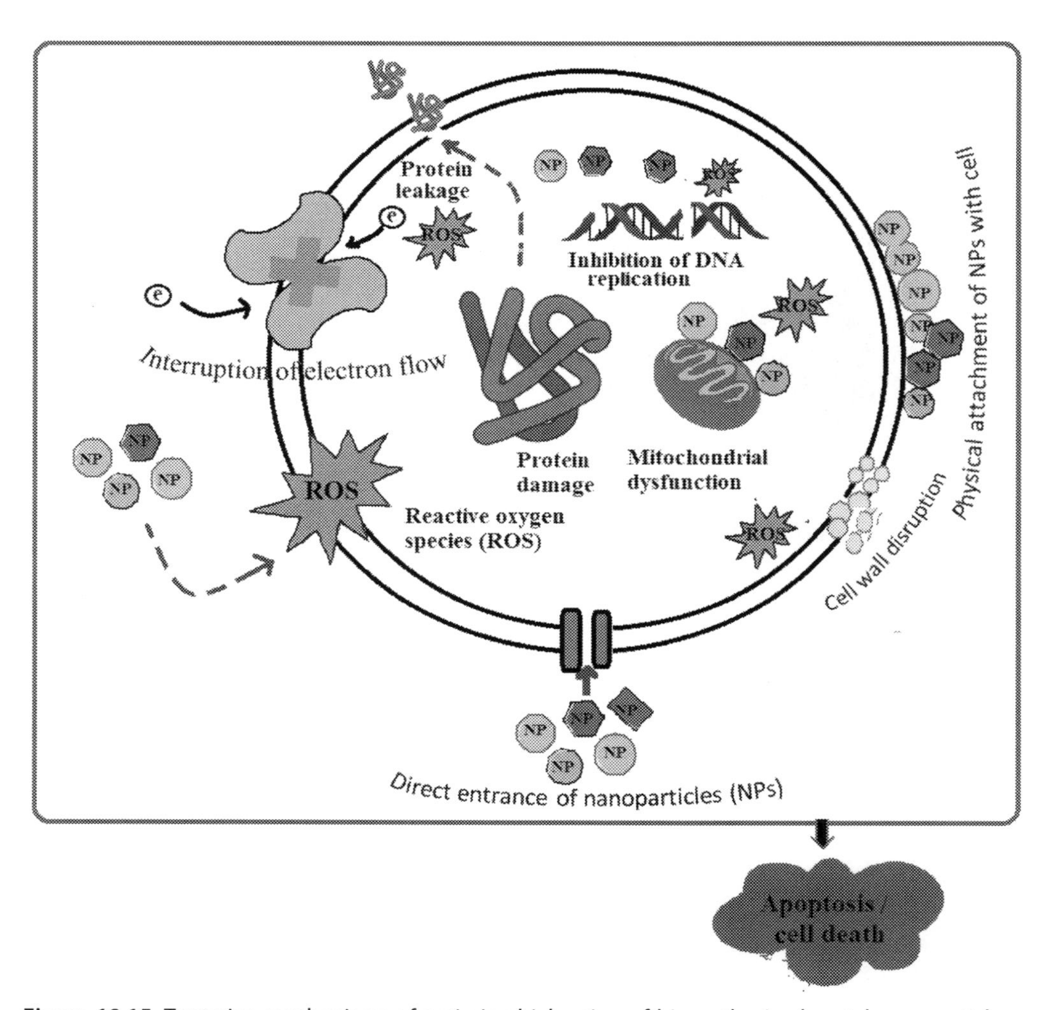

Figure 19.15 Tentative mechanisms of antimicrobial action of biosynthesized metal nanoparticles.

NP antimicrobial mechanism in which NPs disrupt these soft bases in DNA, causing targeted cell death. In pathogens, the interaction of NPs with sulfur and phosphorous in DNA can cause problems with DNA replication. NPs have also been found to affect bacterial cell signaling by phosphorylating protein substrates and changing the phosphotyrosine profile of bacterial peptides. The tentative antibacterial mechanism of action of biosynthesized metal NPs is depicted in Fig. 19.15, based on literature (Nisar, Ali, Rahman, Ali, & Shinwari, 2019).

19.8 Conclusions and remarks/prospects

NCFs, NCC, BC, and other NPs have been proven widely used in medical and pharmaceutical applications due to their biocompatibility, mechanical strength, functionality as well as nontoxic properties.

In this chapter, we have classified NCFs, NCC, BC based on their method of preparation, polymers build, functionality, and advantages. Biocompatibility, Young's modulus, and mechanical strength are important factors that the biomaterials are potential to be utilized in tissue engineering for repairing tissue damage and tissue replacement. Besides, biomaterials are also considered as one of the ingredients in tablet formulation due to their nontoxic properties and involvement in the drug delivery systems. Despite of NCC and NFC efficiently exhibit drug delivery by prolonging drug release, NCC and NFC considered to possess a pH-dependent manner for drug delivery. Biomaterials structural modification should be carried out to allow NCC and NFC to become excellent drug delivery agents.

Enzyme immobilization is a commonly used method in a variety of industries, including food, pharmaceuticals, bioremediation, detergents, and textiles. As a result of the technological and economic advantages of this approach, a large number of enzymes have been immobilized and used in various large-scale processes. This immobilization method will reduce the cost of the enzyme and also provide the enzyme's operational stability. NPs are a feasible alternative to antibiotics and seem to have a lot of promise in terms of solving the issue of multiple drug resistance of bacteria. Antibacterial mechanisms can support the development of effective antibacterial NPs as well as prevent cytotoxicity using NPs.

References

Abba, M., Ibrahim, Z., Chong, C. S., Zawawi, N. A., Kadir, M. R. A., Yusof, A. H. M., & Razak, S. I. A. (2019). Transdermal delivery of crocin using bacterial nanocellulose membrane. *Fibers and Polymers*, *20*, 2025–2031. Available from https://doi.org/10.1007/s12221-019-9076-8.

Abo-Elseoud, W. S., Hassan, M. L., Sabaa, M. W., Basha, M., Hassan, E. A., & Fadel, S. M. (2018). Chitosan nanoparticles/cellulose nanocrystals nanocomposites as a carrier system for the controlled release of repaglinide. *International Journal of Biological Macromolecules*, *111*, 604–613. Available from https://doi.org/10.1016/j.ijbiomac.2018.01.044.

Afsahi, S., Lerner, M. B., Goldstein, J. M., Lee, J., Tang, X., Bagarozzi, D. A., ... Goldsmith, B. R. (2018). Novel graphene-based biosensor for early detection of Zika virus infection. *Biosensors & Bioelectronics*, *100*, 85–88. Available from https://doi.org/10.1016/j.bios.2017.08.051.

Åhlén, M., Tummala, G. K., & Mihranyan, A. (2018). Nanoparticle-loaded hydrogels as a pathway for enzyme-triggered drug release in ophthalmic applications. *International Journal of Pharmaceutics*, *536*, 73–81. Available from https://doi.org/10.1016/j.ijpharm.2017.11.053.

Akbari, E., Buntat, Z., Afroozeh, A., Zeinalinezhad, A., & Nikoukar, A. (2015). *Escherichia coli* bacteria detection by using graphene-based biosensor. *IET Nanobiotechnology*, *9*, 273–279. Available from https://doi.org/10.1049/iet-nbt.2015.0010.

Akhavan-Kharazian, N., & Izadi-Vasafi, H. (2019). Preparation and characterization of chitosan/gelatin/nanocrystalline cellulose/calcium peroxide films for potential wound dressing applications.

International Journal of Biological Macromolecules, *133*, 881–891. Available from https://doi.org/10.1016/j.ijbiomac.2019.04.159.

Akhlaghi, S. P., Berry, R. C., & Tam, K. C. (2013). Surface modification of cellulose nanocrystal with chitosan oligosaccharide for drug delivery applications. *Cellulose*, *20*, 1747–1764. Available from https://doi.org/10.1007/s10570-013-9954-y.

Alkhatib, Y., Dewaldt, M., Moritz, S., Nitzsche, R., Kralisch, D., & Fischer, D. (2017). Controlled extended octenidine release from a bacterial nanocellulose/Poloxamer hybrid system. *European Journal of Pharmaceutics and Biopharmaceutics: Official Journal of Arbeitsgemeinschaft fur Pharmazeutische Verfahrenstechnik e.V*, *112*, 164–176. Available from https://doi.org/10.1016/j.ejpb.2016.11.025.

Amiri-dashatan, N., Koushki, M., Abbaszadeh, H., & Rostami-nejad, M. (2018). Proteomics applications in health: Biomarker and drug discovery and food industry. *Iranian Journal of Pharmaceutical Research*, *17*, 1523–1536.

Anirudhan, T. S., Nair, S. S., & Chithra Sekhar, V. (2017). Deposition of gold-cellulose hybrid nanofiller on a polyelectrolyte membrane constructed using guar gum and poly(vinyl alcohol) for transdermal drug delivery. *Journal of Membrane Science*, *539*, 344–357. Available from https://doi.org/10.1016/j.memsci.2017.05.054.

Anirudhan, T. S., Chithra Sekhar, V. S. F., & Thomas, J. P. (2019). Effect of dual stimuli responsive dextran/nanocellulose polyelectrolyte complexes for chemophotothermal synergistic cancer therapy. *International Journal of Biological Macromolecules*, *135*, 776–789. Available from https://doi.org/10.1016/j.ijbiomac.2019.05.218.

Anton-Sales, I., D'Antin, J. C., Fernández-Engroba, J., Charoenrook, V., Laromaine, A., Roig, A., & Michael, R. (2020). Bacterial nanocellulose as a corneal bandage material: A comparison with amniotic membrane. *Biomaterials Science*, *8*, 2921–2930. Available from https://doi.org/10.1039/d0bm00083c.

Anton-Sales, I., Koivusalo, L., Skottman, H., Laromaine, A., & Roig, A. (2021). Limbal stem cells on bacterial nanocellulose carriers for ocular surface regeneration. *Small (Weinheim an der Bergstrasse, Germany)*, *17*, 1–11. Available from https://doi.org/10.1002/smll.202003937.

Apelgren, P., Karabulut, E., Amoroso, M., Mantas, A., Martínez Ávila, H., Kölby, L., ... Gatenholm, P. (2019). In vivo human cartilage formation in three-dimensional bioprinted constructs with a novel bacterial nanocellulose bioink. *ACS Biomaterials Science and Engineering*, *5*, 2482–2490. Available from https://doi.org/10.1021/acsbiomaterials.9b00157.

Azeredo, H. M. C., Rosa, M. F., & Mattoso, L. H. C. (2017). Nanocellulose in bio-based food packaging applications. *Industrial Crops and Products*, *97*, 664–671. Available from https://doi.org/10.1016/j.indcrop.2016.03.013.

Badshah, M., Ullah, H., Khan, A. R., Khan, S., Park, J. K., & Khan, T. (2018). Surface modification and evaluation of bacterial cellulose for drug delivery. *International Journal of Biological Macromolecules*, *113*, 526–533. Available from https://doi.org/10.1016/j.ijbiomac.2018.02.135.

Bajpai, S. K., Ahuja, S., Chand, N., & Bajpai, M. (2017). Nano cellulose dispersed chitosan film with Ag NPs/Curcumin: An in vivo study on Albino rats for wound dressing. *International Journal of Biological Macromolecules*, *104*, 1012–1019. Available from https://doi.org/10.1016/j.ijbiomac.2017.06.096.

Baldikova, E., Pospiskova, K., Ladakis, D., Kookos, I. K., Koutinas, A. A., Safarikova, M., & Safarik, I. (2017). Magnetically modified bacterial cellulose: A promising carrier for immobilization of affinity ligands, enzymes, and cells. *Materials Science and Engineering C*, *71*, 214–221. Available from https://doi.org/10.1016/j.msec.2016.10.009.

Bao, L., Hong, F. F., Li, G., Hu, G., & Chen, L. (2021). Implantation of air-dried bacterial nanocellulose conduits in a small-caliber vascular prosthesis rabbit model. *Materials Science and Engineering C*, *122*, 111922. Available from https://doi.org/10.1016/j.msec.2021.111922.

Barud, H. S., Regiani, T., Marques, R. F. C., Lustri, W. R., Messaddeq, Y., & Ribeiro, S. J. L. (2011). Antimicrobial bacterial cellulose-silver nanoparticles composite membranes. *Journal of Nanomaterials*, *2011*. Available from https://doi.org/10.1155/2011/721631.

Bayoumi, A., Sarg, M. T., Fahmy, T. Y. A., Mohamed, N. F., & El-Zawawy, W. K. (2020). The behavior of natural biomass materials as drug carriers in releasing loaded gentamicin sulphate. *Arabian Journal of Chemistry*, *13*, 8920–8934. Available from https://doi.org/10.1016/j.arabjc.2020.10.018.

Bhat, A. H., Dasan, Y. K., Khan, I., Soleimani, H., & Usmani, A. (2017). *Application of nanocrystalline cellulose: Processing and biomedical applications. Cellulose-reinforced nanofibre composites (pp. 215–240)*. Elsevier Ltd. Available from 10.1016/B978-0-08-100957-4.00009-7.

Blanco-Covián, L., Montes-García, V., Girard, A., Teresa Fernández-Abedul, M., Pérez-Juste, J., Pastoriza-Santos, I., ... Carmen Blanco-López, M. (2013). Au@Ag SERRS tags coupled to a lateral flow immunoassay for the sensitive detection of Pneumolysin. *Nanoscale*, *9*, 1–3. Available from https://doi.org/10.1039/x0xx00000x.

Bodade, R. G., & Bodade, A. G. (2020). *Microencapsulation of bioactive compounds and enzymes for therapeutic applications. Biopolymer-based formulations (pp.381-404)*. Elsevier Inc. Available from 10.1016/B978-0-12-816897-4.00017-5.

Bordoni, M., Karabulut, E., Kuzmenko, V., Fantini, V., Pansarasa, O., Cereda, C., & Gatenholm, P. (2020). 3D Printed conductive nanocellulose scaffolds for the differentiation of human neuroblastoma cells. *Cells*, *9*. Available from https://doi.org/10.3390/cells9030682.

Brinchi, L., Cotana, F., Fortunati, E., & Kenny, J. M. (2013). Production of nanocrystalline cellulose from lignocellulosic biomass: Technology and applications. *Carbohydrate Polymers*, *94*, 154–169. Available from https://doi.org/10.1016/j.carbpol.2013.01.033.

Buffiere, J., Balogh-Michels, Z., Borrega, M., Geiger, T., Zimmermann, T., & Sixta, H. (2017). The chemical-free production of nanocelluloses from microcrystalline cellulose and their use as Pickering emulsion stabilizer. *Carbohydrate Polymers*, *178*, 48–56. Available from https://doi.org/10.1016/j.carbpol.2017.09.028.

Cacicedo, M. L., Islan, G. A., León, I. E., Álvarez, V. A., Chourpa, I., Allard-Vannier, E., ... Castro, G. R. (2018). Bacterial cellulose hydrogel loaded with lipid nanoparticles for localized cancer treatment. *Colloids Surfaces B Biointerfaces.*, *170*, 596–608. Available from https://doi.org/10.1016/j.colsurfb.2018.06.056.

Cardoso, T. M. G., Channon, R. B., Adkins, J. A., Talhavini, M., Coltro, W. K. T., & Henry, C. S. (2017). A paper-based colorimetric spot test for the identification of adulterated whiskeys. *Chemical Communications*, *53*, 7957–7960. Available from https://doi.org/10.1039/c7cc02271a.

Carlström, I. E., Rashad, A., Campodoni, E., Sandri, M., Syverud, K., Bolstad, A. I., & Mustafa, K. (2020). Cross-linked gelatin-nanocellulose scaffolds for bone tissue engineering. *Materials Letters*, *264*, 1–5. Available from https://doi.org/10.1016/j.matlet.2020.127326.

Carvalho, T., Guedes, G., Sousa, F. L., Freire, C. S. R., & Santos, H. A. (2019). Latest advances on bacterial cellulose-based materials for wound healing, delivery systems, and tissue engineering. *Biotechnology Journal*, *14*, 1–19. Available from https://doi.org/10.1002/biot.201900059.

Channon, R. B., Nguyen, M. P., Scorzelli, A. G., Henry, E. M., Volckens, J., Dandy, D. S., & Henry, C. S. (2018). Rapid flow in multilayer microfluidic paper-based analytical devices. *Lab on a Chip*, *18*, 793–802. Available from https://doi.org/10.1039/c7lc01300k.

Cheeveewattanagul, N., Morales-Narváez, E., Hassan, A. R. H. A., Bergua, J. F., Surareungchai, W., Somasundrum, M., & Merkoçi, A. (2017). Straightforward immunosensing platform based on graphene oxide-decorated nanopaper: A highly sensitive and fast biosensing approach. *Advanced Functional Materials*, *27*, 1–8. Available from https://doi.org/10.1002/adfm.201702741.

Chen, Y., Zhang, L., Yang, Y., Pang, B., Xu, W., Duan, G., ... Zhang, K. (2021). Recent progress on nanocellulose aerogels: Preparation, modification, composite fabrication, applications. *Advanced Materials*, *33*. Available from https://doi.org/10.1002/adma.202005569.

Cheng, K. C., Huang, C. F., Wei, Y., & Hui Hsu, S. (2019). Novel chitosan–cellulose nanofiber self-healing hydrogels to correlate self-healing properties of hydrogels with neural regeneration effects. *NPG Asia Materials*, *11*. Available from https://doi.org/10.1038/s41427-019-0124-z.

Choudhury, R. R., Sahoo, S. K., & Gohil, J. M. (2020). Potential of bioinspired cellulose nanomaterials and nanocomposite membranes thereof for water treatment and fuel cell applications. *Cellulose*, *27*, 6719–6746. Available from https://doi.org/10.1007/s10570-020-03253-z.

Coelho, C. C. S., Michelin, M., Cerqueira, M. A., Gonçalves, C., Tonon, R. V., Pastrana, L. M., ... Teixeira, J. A. (2018). Cellulose nanocrystals from grape pomace: Production, properties and cytotoxicity assessment. *Carbohydrate Polymers*, *192*, 327–336. Available from https://doi.org/10.1016/j.carbpol.2018.03.023.

De France, K. J., Hoare, T., & Cranston, E. D. (2017). Review of hydrogels and aerogels containing nanocellulose. *Chemistry of Materials: A Publication of the American Chemical Society*, *29*, 4609−4631. Available from https://doi.org/10.1021/acs.chemmater.7b00531.

Díez-Pascual, A. M. (2020). Antibacterial action of nanoparticle loaded nanocomposites based on graphene and its derivatives: A mini-review. *International Journal of Molecular Sciences*, *21*. Available from https://doi.org/10.3390/ijms21103563.

Domingues, R. M. A., Gomes, M. E., & Reis, R. L. (2014). The potential of cellulose nanocrystals in tissue engineering strategies. *Biomacromolecules*, *15*, 2327−2346. Available from https://doi.org/10.1021/bm500524s.

Dong, H., Xie, Y., Zeng, G., Tang, L., Liang, J., He, Q., ... Wu, Y. (2016). The dual effects of carboxymethyl cellulose on the colloidal stability and toxicity of nanoscale zero-valent iron. *Chemosphere*, *144*, 1682−1689. Available from https://doi.org/10.1016/j.chemosphere.2015.10.066.

Dufresne, A. (2019). Nanocellulose processing properties and potential applications. *Current Forestry Reports*. Available from https://doi.org/10.1007/s40725-019-00088-1.

Dutta, S. D., Hexiu, J., Patel, D. K., Ganguly, K., & Lim, K. T. (2021). 3D-Printed bioactive and biodegradable hydrogel scaffolds of alginate/gelatin/cellulose nanocrystals for tissue engineering. *International Journal of Biological Macromolecules*, *167*, 644−658. Available from https://doi.org/10.1016/j.ijbiomac.2020.12.011.

Eisele, S., Ammon, H. P. T., Kindervater, R., Gröbe, A., & Göpel, W. (1994). Optimized biosensor for whole blood measurements using a new cellulose based membrane. *Biosensors & Bioelectronics*, *9*, 119−124. Available from https://doi.org/10.1016/0956-5663(94)80102-9.

Fonseca, D. F. S., Carvalho, J. P. F., Bastos, V., Oliveira, H., Moreirinha, C., Almeida, A., ... Freire, C. S. R. (2020). Antibacterial multi-layered nanocellulose-based patches loaded with dexpanthenol for wound healing applications. *Nanomaterials*, *10*, 1−16. Available from https://doi.org/10.3390/nano10122469.

Fontenot, K. R., Edwards, J. V., Haldane, D., Pircher, N., Liebner, F., Condon, B. D., ... Yager, D. (2017). Designing cellulosic and nanocellulosic sensors for interface with a protease sequestrant wound-dressing prototype: Implications of material selection for dressing and protease sensor design. *Journal of Biomaterials Applications*, *32*, 622−637. Available from https://doi.org/10.1177/0885328217735049.

Gao, H., Zhong, Z., Xia, H., Hu, Q., Ye, Q., Wang, Y., ... Zhang, L. (2019). Construction of cellulose nanofibers/quaternized chitin/organic rectorite composites and their application as wound dressing materials. *Biomaterials Science*, *7*, 2571−2581. Available from https://doi.org/10.1039/c9bm00288j.

Gao, W., Hu, Y., Xu, L., Liu, M., Wu, H., & He, B. (2018). Dual pH and glucose sensitive gel gated mesoporous silica nanoparticles for drug delivery. *Chinese Chemical Letters*, *29*, 1795−1798. Available from https://doi.org/10.1016/j.cclet.2018.05.022.

Gong, Y., Hu, J., Choi, J. R., You, M., Zheng, Y., Xu, B., ... Xu, F. (2017). Improved LFIAs for highly sensitive detection of BNP at point-of-care. *International Journal of Nanomedicine*, *12*, 4455−4466. Available from https://doi.org/10.2147/IJN.S135735.

Goswami, A. M., Sarkar, T. S., & Ghosh, S. (2013). An ecofriendly synthesis of silver nano-bioconjugates by penicillium citrinum (MTCC9999) and its antimicrobial effect. *AMB Express*, *3*, 1−9. Available from https://doi.org/10.1186/2191-0855-3-16.

Guisan, J. M., López-Gallego, F., Bolivar, J. M., Rocha-Martín, J., & Fernandez-Lorente, G. (2020). The science of enzyme immobilization. *Methods in Molecular Biology*, *2100*, 1−26. Available from https://doi.org/10.1007/978-1-0716-0215-7_1.

Guo, R., Li, J., Chen, C., Xiao, M., Liao, M., Hu, Y., ... Tang, M. (2021). Biomimetic 3D bacterial cellulose-graphene foam hybrid scaffold regulates neural stem cell proliferation and differentiation. *Colloids Surfaces B Biointerfaces*, *200*, 111590. Available from https://doi.org/10.1016/j.colsurfb.2021.111590.

Gupta, A., Briffa, S. M., Swingler, S., Gibson, H., Kannappan, V., Adamus, G., ... Radecka, I. (2020). Synthesis of silver nanoparticles using curcumin-cyclodextrins loaded into bacterial cellulose-based hydrogels for wound dressing applications. *Biomacromolecules*, *21*, 1802−1811. Available from https://doi.org/10.1021/acs.biomac.9b01724.

Habibi, Y., Lucia, L. A., & Rojas, O. J. (2010). Cellulose nanocrystals: Chemistry, self-assembly, and applications. *Chemical Reviews*, *110*, 3479–3500. Available from https://doi.org/10.1021/cr900339w.

Halib, N., Perrone, F., Cemazar, M., Dapas, B., Farra, R., Abrami, M., ... Grassi, M. (2017). Potential applications of nanocellulose-containing materials in the biomedical field. *Materials (Basel)*, *10*, 1–31. Available from https://doi.org/10.3390/ma10080977.

Han, J., Wang, S., Zhu, S., Huang, C., Yue, Y., Mei, C., ... Xia, C. (2019). Electrospun core-shell nanofibrous membranes with nanocellulose-stabilized carbon nanotubes for use as high-performance flexible supercapacitor electrodes with enhanced water resistance, thermal stability, and mechanical toughness. *ACS Applied Materials & Interfaces*, *11*, 44624–44635. Available from https://doi.org/10.1021/acsami.9b16458.

Hanif, Z., Ahmed, F. R., Shin, S. W., Kim, Y. K., & Um, S. H. (2014). Size- and dose-dependent toxicity of cellulose nanocrystals (CNC) on human fibroblasts and colon adenocarcinoma. *Colloids Surfaces B Biointerfaces*, *119*, 162–165. Available from https://doi.org/10.1016/j.colsurfb.2014.04.018.

Hartmeier, W. (1988). *Methods of immobilization. Immobilized biocatalysts* (pp. 22–50.). Berlin, Heidelberg: Springer. Available from 10.1007/978-3-642-73364-2_2.

Hasan, A., Waibhaw, G., Tiwari, S., Dharmalimgam, K., Shukla, I., & Pandey, L. M. (2017). Fabrication and characterization of chitosan, polyvinylpyrrolidone, and cellulose nanowhiskers nanocomposite films for wound healing drug delivery application. *Journal of Biomedical Materials Research*, *105*, 2391–2404. Available from https://doi.org/10.1002/jbm.a.36097.

Hasan, N., Rahman, L., Kim, S. H., Cao, J., Arjuna, A., Lallo, S., ... Yoo, J. W. (2020). Recent advances of nanocellulose in drug delivery systems. *Journal of Pharmaceutical Investigation*, *50*, 553–572. Available from https://doi.org/10.1007/s40005-020-00499-4.

Hemeg, H. A. (2017). Nanomaterials for alternative antibacterial therapy. *International Journal of Nanomedicine*, *12*, 8211–8225. Available from https://doi.org/10.2147/IJN.S132163.

Hivechi, A., Bahrami, S. H., Siegel, R. A., Milan, P. B., & Amoupour, M. (2020). In vitro and in vivo studies of biaxially electrospun poly(caprolactone)/gelatin nanofibers, reinforced with cellulose nanocrystals, for wound healing applications. *Cellulose*, *27*, 5179–5196. Available from https://doi.org/10.1007/s10570-020-03106-9.

Hoeng, F., Denneulin, A., & Bras, J. (2016). Use of nanocellulose in printed electronics: A review. *Nanoscale*, *8*, 13131–13154. Available from https://doi.org/10.1039/c6nr03054h.

Hong, H. J., Kim, J., Kim, D. Y., Kang, I., Kang, H. K., & Ryu, B. G. (2019). Synthesis of carboxymethylated nanocellulose fabricated ciprofloxacine – Montmorillonite composite for sustained delivery of antibiotics. *International Journal of Pharmaceutics*, *567*, 118502. Available from https://doi.org/10.1016/j.ijpharm.2019.118502.

Hong, J. K., Cooke, S. L., Whittington, A. R., & Roman, M. (2021). Bioactive cellulose nanocrystal-poly(ε-caprolactone) nanocomposites for bone tissue engineering applications. *Frontiers in Bioengineering and Biotechnology*, *9*. Available from https://doi.org/10.3389/fbioe.2021.605924.

Hsieh, M. C., Koga, H., Suganuma, K., & Nogi, M. (2017). Hazy transparent cellulose nanopaper. *Scientific Reports*, 7, 1–7. Available from https://doi.org/10.1038/srep41590.

Ilkar Erdagi, S., Asabuwa Ngwabebhoh, F., & Yildiz, U. (2020). Genipin crosslinked gelatin-diosgenin-nanocellulose hydrogels for potential wound dressing and healing applications. *International Journal of Biological Macromolecules*, *149*, 651–663. Available from https://doi.org/10.1016/j.ijbiomac.2020.01.279.

Jackson, J. K., Letchford, K., Wasserman, B. Z., Ye, L., Hamad, W. Y., & Burt, H. M. (2011). The use of nanocrystalline cellulose for the binding and controlled release of drugs. *International Journal of Nanomedicine*, *6*, 321–330. Available from https://doi.org/10.2147/ijn.s16749.

Jantarat, C., Sirathanarun, P., Boonmee, S., Meechoosin, W., Wangpittaya, H., et al. (2018). Effect of piperine on skin permeation of curcumin from a bacterially derived cellulose-composite double-layer membrane for transdermal curcumin delivery. *Scientia Pharmaceutica*, *86*(3). Available from https://doi.org/10.3390/scipharm86030039.

Kadry, G. (2019). Comparison between gelatin/carboxymethyl cellulose and gelatin/carboxymethyl nanocellulose in tramadol drug loaded capsule. *Heliyon*, *5*, e02404. Available from https://doi.org/10.1016/j.heliyon.2019.e02404.

Kalia, S., Dufresne, A., Cherian, B. M., Kaith, B. S., Avérous, L., Njuguna, J., & Nassiopoulos, E. (2011). Cellulose-based bio- and nanocomposites: A review. *International Journal of Polymer Science, 2011*. Available from https://doi.org/10.1155/2011/837875.

Kamel, S., & Khattab, T. (2020). Recent advances in cellulose-based biosensors for. *Biosensors, 10*, 1–26.

Karl, B., Alkhatib, Y., Beekmann, U., Bellmann, T., Blume, G., Steiniger, F., ... Fischer, D. (2020). Development and characterization of bacterial nanocellulose loaded with *Boswellia serrata* extract containing nanoemulsions as natural dressing for skin diseases. *International Journal of Pharmaceutics, 587*, 119635. Available from https://doi.org/10.1016/j.ijpharm.2020.119635.

Karzar Jeddi, M., & Mahkam, M. (2019). Magnetic nano carboxymethyl cellulose-alginate/chitosan hydrogel beads as biodegradable devices for controlled drug delivery. *International Journal of Biological Macromolecules, 135*, 829–838. Available from https://doi.org/10.1016/j.ijbiomac.2019.05.210.

Kasiri, N., & Fathi, M. (2018). Production of cellulose nanocrystals from pistachio shells and their application for stabilizing Pickering emulsions. *International Journal of Biological Macromolecules, 106*, 1023–1031. Available from https://doi.org/10.1016/j.ijbiomac.2017.080.112.

Kato, K., Lee, S., & Nagata, F. (2020). Efficient enzyme encapsulation inside sol-gel silica sheets prepared by poly-L-lysine as a catalyst. *Journal of Asian Ceramic Societies, 8*, 396–406. Available from https://doi.org/10.1080/21870764.2020.1747167.

Khan, I., Saeed, K., & Khan, I. (2019). Nanoparticles: Properties, applications and toxicities. *Arabian Journal of Chemistry, 12*, 908–931. Available from https://doi.org/10.1016/j.arabjc.2017.05.011.

Klemm, D., Kramer, F., Moritz, S., Lindström, T., Ankerfors, M., Gray, D., & Dorris, A. (2011). Nanocelluloses: A new family of nature-based materials. *Angewandte Chemie International Edition, 50*, 5438–5466. Available from https://doi.org/10.1002/anie.201001273.

Kolakovic, R., Peltonen, L., Laaksonen, T., Putkisto, K., Laukkanen, A., & Hirvonen, J. (2011). Spray-dried cellulose nanofibers as novel tablet excipient. *AAPS PharmSciTech, 12*, 1366–1373. Available from https://doi.org/10.1208/s12249-011-9705-z.

Kollar, P., Závalová, V., Hošek, J., Havelka, P., Sopuch, T., Karpíšek, M., ... Suchý, P. (2011). Cytotoxicity and effects on inflammatory response of modified types of cellulose in macrophage-like THP-1 cells. *International Immunopharmacology, 11*, 997–1001. Available from https://doi.org/10.1016/j.intimp.2011.02.016.

Kopač, T., Krajnc, M., & Ručigaj, A. (2021). A mathematical model for pH-responsive ionically crosslinked TEMPO nanocellulose hydrogel design in drug delivery systems. *International Journal of Biological Macromolecules, 168*, 695–707. Available from https://doi.org/10.1016/j.ijbiomac.2020.11.126.

Kuan, W. C., Lai, J. W., & Lee, W. C. (2021). Covalent binding of glutathione on magnetic nanoparticles: Application for immobilizing small fragment ubiquitin-like-specific protease 1. *Enzyme and Microbial Technology, 143*, 109697. Available from https://doi.org/10.1016/j.enzmictec.2020.109697.

Kuijpers, A. J., Engbers, G. H. M., Feijen, J., De Smedt, S. C., Meyvis, T. K. L., Demeester, J., ... Dankert, J. (1999). Characterization of the network structure of carbodiimide cross-linked gelatin gels. *Macromolecules, 32*, 3325–3333. Available from https://doi.org/10.1021/ma981929v.

Kurdekar, A., Chunduri, L. A. A., Bulagonda, E. P., Haleyurgirisetty, M. K., Kamisetti, V., & Hewlett, I. K. (2016). Comparative performance evaluation of carbon dot-based paper immunoassay on Whatman filter paper and nitrocellulose paper in the detection of HIV infection. *Microfluidics and Nanofluidics, 20*. Available from https://doi.org/10.1007/s10404-016-1763-9.

Kuzmenko, V., Kalogeropoulos, T., Thunberg, J., Johannesson, S., Hägg, D., Enoksson, P., & Gatenholm, P. (2016). Enhanced growth of neural networks on conductive cellulose-derived nanofibrous scaffolds. *Materials Science and Engineering C, 58*, 14–23. Available from https://doi.org/10.1016/j.msec.2015.08.012.

Kuzmenko, V., Karabulut, E., Pernevik, E., Enoksson, P., & Gatenholm, P. (2018). Tailor-made conductive inks from cellulose nanofibrils for 3D printing of neural guidelines. *Carbohydrate Polymers, 189*, 22–30. Available from https://doi.org/10.1016/j.carbpol.2018.01.097.

Lam, N. T., Chollakup, R., Smitthipong, W., Nimchua, T., & Sukyai, P. (2017). Utilizing cellulose from sugarcane bagasse mixed with poly(vinyl alcohol) for tissue engineering scaffold fabrication. *Industrial Crops and Products, 100*, 183–197. Available from https://doi.org/10.1016/j.indcrop.2017.02.031.

Lengowski, E. C., Bonfatti Júnior, E. A., Nishidate Kumode, M. M., Carneiro, M. E., & Satyanarayana, K. G. (2019). Nanocellulose in the paper making. In S. Thomas, R. Kumar Mishra, & A. Asiri (Eds.), *Sustainable polymer composites and nanocomposites*. Cham: Springer. Available from 10.1007/978-3-030-05399-4_36.

Li, F., You, M., Li, S., Hu, J., Liu, C., Gong, Y., ... Xu, F. (2020). Paper-based point-of-care immunoassays: Recent advances and emerging trends. *Biotechnology Advances*, *39*, 107442. Available from https://doi.org/10.1016/j.biotechadv.2019.107442.

Li, J., Wang, Y., Zhang, L., Xu, Z., Dai, H., & Wu, W. (2019). Nanocellulose/gelatin composite cryogels for controlled drug release. *ACS Sustainable Chemistry & Engineering*, *7*, 6381−6389. Available from https://doi.org/10.1021/acssuschemeng.9b00161.

Li, W., Guo, R., Lan, Y., Zhang, Y., Xue, W., & Zhang, Y. (2014). Preparation and properties of cellulose nanocrystals reinforced collagen composite films. *Journal of Biomedical Materials Research Part A*, *102*, 1131−1139. Available from https://doi.org/10.1002/jbm.a.34792.

Li, X., Liu, Y., Yu, Y., Chen, W., Liu, Y., & Yu, H. (2019). Nanoformulations of quercetin and cellulose nanofibers as healthcare supplements with sustained antioxidant activity. *Carbohydrate Polymers*, *207*, 160−168. Available from https://doi.org/10.1016/j.carbpol.2018.11.084.

Liang, J. F., Li, Y. T., & Yang, V. C. (2000). Biomedical application of immobilized enzymes. *Journal of Pharmaceutical Sciences*, *89*, 979−990. Available from https://doi.org/10.1002/1520-6017(200008)89:8 < 979::AID-JPS2 > 3.0.CO;2-H.

Liang, Y., Zhu, H., Wang, L., He, H., & Wang, S. (2020). Biocompatible smart cellulose nanofibres for sustained drug release via pH and temperature dual-responsive mechanism. *Carbohydrate Polymers*, *249*, 116876. Available from https://doi.org/10.1016/j.carbpol.2020.116876.

Liao, C., Li, Y., & Tjong, S. C. (2019). Bactericidal and cytotoxic properties of silver nanoparticles. *International Journal of Molecular Sciences*, *20*. Available from https://doi.org/10.3390/ijms20020449.

de Lima Fontes, M., Meneguin, A. B., Tercjak, A., Gutierrez, J., Cury, B. S. F., dos Santos, A. M., ... Barud, H. S. (2018). Effect of in situ modification of bacterial cellulose with carboxymethylcellulose on its nano/microstructure and methotrexate release properties. *Carbohydrate Polymers*, *179*, 126−134. Available from https://doi.org/10.1016/j.carbpol.2017.09.061.

Lin, N., & Dufresne, A. (2014). Nanocellulose in biomedicine: Current status and future prospect. *European Polymer Journal*, *59*, 302−325. Available from https://doi.org/10.1016/j.eurpolymj.2014.07.025.

Liu, D. Y., Sui, G. X., & Bhattacharyya, D. (2014). Synthesis and characterisation of nanocellulose-based polyaniline conducting films. *Composites Science and Technology*, *99*, 31−36. Available from https://doi.org/10.1016/j.compscitech.2014.05.001.

Liu, J., Ma, R. T., & Shi, Y. P. (2020). "Recent advances on support materials for lipase immobilization and applicability as biocatalysts in inhibitors screening methods"—A review. *Analytica Chimica Acta*, *1101*, 9−22. Available from https://doi.org/10.1016/j.ac.2019.11.073.

Löbmann, K., Wohlert, J., Müllertz, A., Wågberg, L., & Svagan, A. J. (2017). Cellulose nanopaper and nanofoam for patient-tailored drug delivery. *Advanced Materials Interfaces*, *4*. Available from https://doi.org/10.1002/admi.201600655.

Loh, E. Y. X., Fauzi, M. B., Ng, M. H., Ng, P. Y., Ng, S. F., & Amin, M. C. I. Mohd (2020). Insight into delivery of dermal fibroblast by non-biodegradable bacterial nanocellulose composite hydrogel on wound healing. *International Journal of Biological Macromolecules*, *159*, 497−509. Available from https://doi.org/10.1016/j.ijbiomac.2020.05.011.

Loynachan, C. N., Thomas, M. R., Gray, E. R., Richards, D. A., Kim, J., Miller, B. S., ... Stevens, M. M. (2018). Platinum nanocatalyst amplification: Redefining the gold standard for lateral flow immunoassays with ultrabroad dynamic range. *ACS Nano*, *12*, 279−288. Available from https://doi.org/10.1021/acsnano.7b06229.

Luo, H., Cha, R., Li, J., Hao, W., Zhang, Y., & Zhou, F. (2019). Advances in tissue engineering of nanocellulose-based scaffolds: A review. *Carbohydrate Polymers*, *224*, 115144. Available from https://doi.org/10.1016/j.carbpol.2019.115144.

Mahira, S., et al. (2019). *Antimicrobial materials—An overview. Antimicrobial materials for biomedical applications*. Royal Society of Chemistry. Available from 10.1039/9781788012638-00001.

Mahmoud, D. A. R., & Helmy, W. A. (2009). Potential application of immobilization technology in enzyme and biomass production (review article). *Journal of Applied Sciences Research*, *5*, 2466–2476.

Medhi, P., Olatunji, O., Nayak, A., Uppuluri, C. T., Olsson, R. T., Nalluri, B. N., & Das, D. B. (2017). Lidocaine-loaded fish scale-nanocellulose biopolymer composite microneedles. *AAPS PharmSciTech*, *18*, 1488–1494. Available from https://doi.org/10.1208/s12249-017-0758-5.

Mo, Y., Guo, R., Liu, J., Lan, Y., Zhang, Y., Xue, W., & Zhang, Y. (2015). Preparation and properties of PLGA nanofiber membranes reinforced with cellulose nanocrystals. *Colloids Surfaces B Biointerfaces.*, *132*, 177–184. Available from https://doi.org/10.1016/j.colsurfb.2015.05.029.

Moghaddam, S. V., Abedi, F., Alizadeh, E., Baradaran, B., Annabi, N., Akbarzadeh, A., & Davaran, S. (2020). Lysine-embedded cellulose-based nanosystem for efficient dual-delivery of chemotherapeutics in combination cancer therapy. *Carbohydrate Polymers*, *250*, 116861. Available from https://doi.org/10.1016/j.carbpol.2020.116861.

Mohamad, N. R., Marzuki, N. H. C., Buang, N. A., Huyop, F., & Wahab, R. A. (2015). An overview of technologies for immobilization of enzymes and surface analysis techniques for immobilized enzymes. *Biotechnology & Biotechnological Equipment*, *29*, 205–220. Available from https://doi.org/10.1080/13102818.2015.1008192.

Mokhtar, N. F., N., R., R., Z., Rahman, A., D., N., & Noor, M. (2020). The immobilization of lipases on porous support by adsorption and hydrophobic interaction method. *Catalysts*, *10*(7), 1–17.

Molina-Fernandez, C., & Luis, P. (2021). Immobilization of carbonic anhydrase CO2 capture and its industrial implementation: A review. *Journal of CO2 Utilization*, *47*, 101475.

Morales-Narváez, E., Baptista-Pires, L., Zamora-Gálvez, A., & Merkoçi, A. (2017). Graphene-based biosensors: Going simple. *Advanced Materials*, *29*. Available from https://doi.org/10.1002/adma.201604905.

Müller, M., Öztürk, E., Arlov, Ø., Gatenholm, P., & Zenobi-Wong, M. (2017). Alginate sulfate–nanocellulose bioinks for cartilage bioprinting applications. *Annals of Biomedical Engineering*, *45*, 210–223. Available from https://doi.org/10.1007/s10439-016-1704-5.

Navakul, K., Warakulwit, C., Thai Yenchitsomanus, P., Panya, A., Lieberzeit, P. A., & Sangma, C. (2017). A novel method for dengue virus detection and antibody screening using a graphene-polymer based electrochemical biosensor. *Nanomedicine: Nanotechnology, Biology, and Medicine*, *13*, 549–557. Available from https://doi.org/10.1016/j.nano.2016.08.009.

Nguyen, D., Hgg, D. A., Forsman, A., Ekholm, J., Nimkingratana, P., Brantsing, C., ... Simonsson, S. (2017). Cartilage tissue engineering by the 3D bioprinting of iPS cells in a nanocellulose/alginate bioink. *Scientific Reports*, *7*, 1–10. Available from https://doi.org/10.1038/s41598-017-00690-y.

Nguyen, H. H., & Kim, M. (2017). An overview of techniques in enzyme immobilization. *Applied Science and Convergence Technology*, *26*, 157–163. Available from https://doi.org/10.5757/asct.2017.26.6.157.

Nie, F. L., Wang, S. G., Wang, Y. B., Wei, S. C., & Zheng, Y. F. (2011). Comparative study on corrosion resistance and in vitro biocompatibility of bulk nanocrystalline and microcrystalline biomedical 304 stainless steel. *Dental Materials: Official Publication of the Academy of Dental Materials*, *27*, 677–683. Available from https://doi.org/10.1016/j.dental.2011.03.009.

Nimeskern, L., Martínez Ávila, H., Sundberg, J., Gatenholm, P., Müller, R., & Stok, K. S. (2013). Mechanical evaluation of bacterial nanocellulose as an implant material for ear cartilage replacement. *Journal of the Mechanical Behavior of Biomedical Materials*, *22*, 12–21. Available from https://doi.org/10.1016/j.jmbbm.2013.03.005.

Ning, L., You, C., Zhang, Y., Li, X., & Wang, F. (2020). Synthesis and biological evaluation of surface-modified nanocellulose hydrogel loaded with paclitaxel. *Life Sciences*, *241*, 117137. Available from https://doi.org/10.1016/j.lfs.2019.117137.

Nisar, P., Ali, N., Rahman, L., Ali, M., & Shinwari, Z. K. (2019). Antimicrobial activities of biologically synthesized metal nanoparticles: An insight into the mechanism of action. *Journal of Biological Inorganic Chemistry: JBIC: A Publication of the Society of Biological Inorganic Chemistry*, *24*, 929–941. Available from https://doi.org/10.1007/s00775-019-01717-7.

Nosrati, R., Gong, M. M., Gabriel, M. C. S., & Pedraza, C. E. (2016). Paper-based quantification of male fertility potential. *Clinical Chemistry*, *465*, 458–465. Available from https://doi.org/10.1373/clinchem.2015.250282.

Orasugh, J. T., Saha, N. R., Sarkar, G., Rana, D., Mishra, R., Mondal, D., ... Chattopadhyay, D. (2018). Synthesis of methylcellulose/cellulose nano-crystals nanocomposites: Material properties and study of sustained release of ketorolac tromethamine. *Carbohydrate Polymers*, *188*, 168–180. Available from https://doi.org/10.1016/j.carbpol.2018.01.108.

Orasugh, J. T., Saha, N. R., Rana, D., Sarkar, G., Mollick, M. M. R., Chattoapadhyay, A., ... Chattopadhyay, D. (2018). Jute cellulose nano-fibrils/hydroxypropylmethylcellulose nanocomposite: A novel material with potential for application in packaging and transdermal drug delivery system. *Industrial Crops and Products*, *112*, 633–643. Available from https://doi.org/10.1016/j.indcrop.2017.12.069.

Orasugh, J. T., Sarkar, G., Saha, N. R., Das, B., Bhattacharyya, A., Das, S., ... Chattopadhyay, D. (2019). Effect of cellulose nanocrystals on the performance of drug loaded in situ gelling thermo-responsive ophthalmic formulations. *International Journal of Biological Macromolecules*, *124*, 235–245. Available from https://doi.org/10.1016/j.ijbiomac.2018.11.217.

Orasugh, J. T., Dutta, S., Das, D., Pal, C., Zaman, A., Das, S., ... Chattopadhyay, D. (2019). Sustained release of ketorolac tromethamine from poloxamer 407/cellulose nanofibrils graft nanocollagen based ophthalmic formulations. *International Journal of Biological Macromolecules*, *140*, 441–453. Available from https://doi.org/10.1016/j.ijbiomac.2019.08.143.

Osorio, M., Martinez, E., Kooten, T. V., Gañán, P., Naranjo, T., Ortiz, I., & Castro, C. (2020). Biomimetics of microducts in three-dimensional bacterial nanocellulose biomaterials for soft tissue regenerative medicine. *Cellulose*, *27*, 5923–5937. Available from https://doi.org/10.1007/s10570-020-03175-w.

Patel, D. K., Dutta, S. D., Hexiu, J., Ganguly, K., & Lim, K. T. (2020). Bioactive electrospun nanocomposite scaffolds of poly(lactic acid)/cellulose nanocrystals for bone tissue engineering. *International Journal of Biological Macromolecules*, *162*, 1429–1441. Available from https://doi.org/10.1016/j.ijbiomac.2020.07.246.

Patel, D. K., Dutta, S. D., Ganguly, K., & Lim, K. T. (2021). Multifunctional bioactive chitosan/cellulose nanocrystal scaffolds eradicate bacterial growth and sustain drug delivery. *International Journal of Biological Macromolecules*, *170*, 178–188. Available from https://doi.org/10.1016/j.ijbiomac.2020.12.145.

Peña-Bahamonde, J., Nguyen, H. N., Fanourakis, S. K., & Rodrigues, D. F. (2018). Recent advances in graphene-based biosensor technology with applications in life sciences. *Journal of Nanobiotechnology*, *16*, 1–17. Available from https://doi.org/10.1186/s12951-018-0400-z.

Phanthong, P., Reubroycharoen, P., Hao, X., Xu, G., Abudula, A., & Guan, G. (2018). Nanocellulose: Extraction and application. *Carbon Resources Conversion*, *1*, 32–43. Available from https://doi.org/10.1016/j.crcon.2018.05.004.

Picheth, G. F., Pirich, C. L., Sierakowski, M. R., Woehl, M. A., Sakakibara, C. N., de Souza, C. F., ... de Freitas, R. A. (2017). Bacterial cellulose in biomedical applications: A review. *International Journal of Biological Macromolecules*, *104*, 97–106. Available from https://doi.org/10.1016/j.ijbiomac.2017.05.171.

Plappert, S. F., Liebner, F. W., Konnerth, J., & Nedelec, J. M. (2019). Anisotropic nanocellulose gel–membranes for drug delivery: Tailoring structure and interface by sequential periodate–chlorite oxidation. *Carbohydrate Polymers*, *226*, 115306. Available from https://doi.org/10.1016/j.carbpol.2019.115306.

Poonguzhali, R., Basha, S. K., & Kumari, V. S. (2018). Synthesis of alginate/nanocellulose bionanocomposite for in vitro delivery of ampicillin. *International Journal of Drug Delivery*, *9*, 107. Available from https://doi.org/10.5138/09750215.2168.

Pylkkänen, R., Mohammadi, P., Arola, S., De Ruijter, J. C., Sunagawa, N., Igarashi, K., & Penttilä, M. (2020). In vitro synthesis and self-assembly of cellulose II nanofibrils catalyzed by the reverse reaction of *Clostridium thermocellum* cellodextrin phosphorylase. *Biomacromolecules*, *21*, 4355–4364. Available from https://doi.org/10.1021/acs.biomac.0c01162.

Rahimi, M., Shojaei, S., Safa, K. D., Ghasemi, Z., Salehi, R., Yousefi, B., & Shafiei-Irannejad, V. (2017). Biocompatible magnetic tris(2-aminoethyl)amine functionalized nanocrystalline cellulose as a novel nanocarrier for anticancer drug delivery of methotrexate. *New Journal of Chemistry*, *41*, 2160–2168. Available from https://doi.org/10.1039/C6NJ03332F.

Rao, K. M., Kumar, A., & Han, S. S. (2017). Poly(acrylamidoglycolic acid) nanocomposite hydrogels reinforced with cellulose nanocrystals for pH-sensitive controlled release of diclofenac sodium. *Polymer Testing*, *64*, 175–182. Available from https://doi.org/10.1016/j.polymertesting.2017.10.006.

Rasheed, M., Jawaid, M., Parveez, B., Zuriyati, A., & Khan, A. (2020). Morphological, chemical and thermal analysis of cellulose nanocrystals extracted from bamboo fibre. *International Journal of Biological Macromolecules*, *160*, 183–191. Available from https://doi.org/10.1016/j.ijbiomac.2020.05.170.

S. Reji Kumar *Physicochemical studies of immobilized enzyme systems* (Doctoral thesis). Department of Chemistry, Maharaja Sayajirao University of Baroda, Gujarat, India. 1997. http://hdl.handle.net/10603/60005.

Roman, M. (2015). Toxicity of cellulose nanocrystals: A review. *Industrial Biotechnology*, *11*, 25–33. Available from https://doi.org/10.1089/ind.2014.0024.

Saïdi, L., Vilela, C., Oliveira, H., Silvestre, A. J. D., & Freire, C. S. R. (2017). Poly(N-methacryloyl glycine)/nanocellulose composites as pH-sensitive systems for controlled release of diclofenac. *Carbohydrate Polymers*, *169*, 357–365. Available from https://doi.org/10.1016/j.carbpol.2017.04.030.

Sánchez-López, E., Gomes, D., Esteruelas, G., Bonilla, L., Lopez-Machado, A. L., Galindo, R., ... Souto, E. B. (2020). Metal-based nanoparticles as antimicrobial agents: An overview. *Nanomaterials*, *10*, 1–39. Available from https://doi.org/10.3390/nano10020292.

Sanjay, S. T., Dou, M., Sun, J., & Li, X. (2016). A paper/polymer hybrid microfluidic microplate for rapid quantitative detection of multiple disease biomarkers. *Scientific Reports.*, *6*, 1–10. Available from https://doi.org/10.1038/srep30474.

Sarkar, G., Orasugh, J. T., Saha, N. R., Roy, I., Bhattacharyya, A., Chattopadhyay, A. K., ... Chattopadhyay, D. (2017). Cellulose nanofibrils/chitosan based transdermal drug delivery vehicle for controlled release of ketorolac tromethamine. *New Journal of Chemistry.*, *41*, 15312–15319. Available from https://doi.org/10.1039/c7nj02539d.

Shaikh, S., Birdi, A., Qutubuddin, S., Lakatosh, E., & Baskaran, H. (2007). Controlled release in transdermal pressure sensitive adhesives using organosilicate nanocomposites. *Annals of Biomedical Engineering*, *35*, 2130–2137. Available from https://doi.org/10.1007/s10439-007-9369-8.

Sharma, C., & Bhardwaj, N. K. (2019). Bacterial nanocellulose: Present status, biomedical applications and future perspectives. *Materials Science and Engineering C*, *104*, 109963. Available from https://doi.org/10.1016/j.msec.2019.109963.

Shehabeldine, A., & Hasanin, M. (2019). Green synthesis of hydrolyzed starch–chitosan nano-composite as drug delivery system to gram negative bacteria. *Environmental Nanotechnology Monitoring & Management*, *12*, 100252. Available from https://doi.org/10.1016/j.enmm.2019.100252.

Sheldon, R. A., & Woodley, J. M. (2018). Role of biocatalysis in sustainable chemistry. *Chemical Reviews*, *118*, 801–838. Available from https://doi.org/10.1021/acs.chemrev.7b00203.

Sign, C., & Sumana, G. (2016). Antibody conjugated graphene nanocomposites for pathogen detection. *Journal of Physics Conference Series*, *704*. Available from https://doi.org/10.1088/1742-6596/704/1/012014.

Silva, N. H. C. S., Garrido-Pascual, P., Moreirinha, C., Almeida, A., Palomares, T., Alonso-Varona, A., ... Freire, C. S. R. (2020). Multifunctional nanofibrous patches composed of nanocellulose and lysozyme nanofibers for cutaneous wound healing. *International Journal of Biological Macromolecules*, *165*, 1198–1210. Available from https://doi.org/10.1016/j.ijbiomac.2020.09.249.

Singh, P., Garg, A., Pandit, S., Mokkapati, V. R. S. S., & Mijakovic, I. (2018). Antimicrobial effects of biogenic nanoparticles. *Nanomaterials*, *8*, 1–19. Available from https://doi.org/10.3390/nano8121009.

Siqueira, G., Abdillahi, H., Bras, J., & Dufresne, A. (2010). High reinforcing capability cellulose nanocrystals extracted from *Syngonanthus nitens* (Capim Dourado). *Cellulose*, *17*, 289–298. Available from https://doi.org/10.1007/s10570-009-9384-z.

Song, Y. K., Leng Chew, I. M., Yaw Choong, T. S., & Tan, K. W. (2016). Nanocrystalline cellulose, an environmental friendly nanoparticle for pharmaceutical application - A quick study. *MATEC Web of Conferences*, *60*, 1–5. Available from https://doi.org/10.1051/matecconf/20166001006.

Taher, M. A., Zahan, K. A., Rajaie, M. A., Ring, L. C., Rashid, S. A., Mohd Nor Hamin, N. S., ... Yenn, T. W. (2020). Nanocellulose as drug delivery system for honey as antimicrobial wound dressing. *Materials Today: Proceedings*, *31*, 14–17. Available from https://doi.org/10.1016/j.matpr.2020.01.076.

Taheri, P., Jahanmardi, R., Koosha, M., & Abdi, S. (2020). Physical, mechanical and wound healing properties of chitosan/gelatin blend films containing tannic acid and/or bacterial nanocellulose. *International Journal of Biological Macromolecules*, *154*, 421–432. Available from https://doi.org/10.1016/j.ijbiomac.2020.03.114.

Thakur, B., Zhou, G., Chang, J., Pu, H., Jin, B., Sui, X., ... Chen, J. (2018). Rapid detection of single *E. coli* bacteria using a graphene-based field-effect transistor device. *Biosensors & Bioelectronics, 110*, 16–22. Available from https://doi.org/10.1016/j.bios.2018.03.014.

Thomas, D., Latha, M. S., & Thomas, K. K. (2018). Synthesis and in vitro evaluation of alginate-cellulose nanocrystal hybrid nanoparticles for the controlled oral delivery of rifampicin. *Journal of Drug Delivery Science and Technology, 46*, 392–399. Available from https://doi.org/10.1016/j.jddst.2018.06.004.

Trache, D., Tarchoun, A. F., Derradji, M., Hamidon, T. S., Masruchin, N., Brosse, N., & Hussin, M. H. (2020). Nanocellulose: From fundamentals to advanced applications. *Frontiers in Chemistry, 8*, 392. Available from https://doi.org/10.3389/fchem.2020.00392.

Trelles, J. A., & Rivero, C. W. (2013). Whole cell entrapment techniques. In J. Guisan (Ed.), *Immobilization of enzymes and cells (Vol. 1051)* (pp. 365–374). . Available from 10.1007/978-1-62703-550-7.

Tummala, G. K., Joffre, T., Rojas, R., Persson, C., & Mihranyan, A. (2017). Strain-induced stiffening of nanocellulosereinforced poly(vinyl alcohol) hydrogels mimicking collagenous soft tissues. *Soft Matter, 13*, 3936–3945. Available from https://doi.org/10.1039/c7sm00677b.

Tummala, G. K., Lopes, V. R., Mihranyan, A., & Ferraz, N. (2019). Biocompatibility of nanocellulose-reinforced PVA hydrogel with human corneal epithelial cells for ophthalmic applications. *Journal of Functional Biomaterials., 10*. Available from https://doi.org/10.3390/jfb10030035.

Turky, G., Moussa, M. A., Hasanin, M., El-Sayed, N. S., & Kamel, S. (2021). Carboxymethyl cellulose-based hydrogel: Dielectric study, antimicrobial activity and biocompatibility. *Arabian Journal for Science and Engineering., 46*, 17–30. Available from https://doi.org/10.1007/s13369-020-04655-8.

Wacker, M., Kießwetter, V., Slottosch, I., Awad, G., Paunel-Görgülü, A., Varghese, S., ... Scherner, M. (2020). In vitro hemo- And cytocompatibility of bacterial nanocelluose small diameter vascular grafts: Impact of fabrication and surface characteristics. *PLoS One, 15*, 1–19. Available from https://doi.org/10.1371/journal.pone.0235168.

Wahab, R. A., Elias, N., Abdullah, F., & Ghoshal, S. K. (2020). On the taught new tricks of enzymes immobilization: An all-inclusive overview. *Reactive & Functional Polymers, 152*, 104613. Available from https://doi.org/10.1016/j.reactfunctpolym.2020.104613.

Wang, L., Hu, C., & Shao, L. (2017). The-antimicrobial-activity-of-nanoparticles—present-situation and prospects for the future. *International Journal of Nanomedicine, 12*, 1227–1249. Available from https://www.ncbi.nlm.nih.gov/pmc/articles/PMC5317269/pdf/ijn-12-1227.pdf.

Wang, Y., Li, Z., Wang, J., Li, J., & Lin, Y. (2011). Graphene and graphene oxide: Biofunctionalization and applications in biotechnology. *Trends in Biotechnology, 29*, 205–212. Available from https://doi.org/10.1016/j.tibtech.2011.01.008.

Wu, T., Farnood, R., O'Kelly, K., & Chen, B. (2014). Mechanical behavior of transparent nanofibrillar cellulose-chitosan nanocomposite films in dry and wet conditions. *Journal of the Mechanical Behavior of Biomedical Materials, 32*, 279–286. Available from https://doi.org/10.1016/j.jmbbm.2014.01.014.

Xiao, L., Poudel, A. J., Huang, L., Wang, Y., Abdalla, A. M. E., & Yang, G. (2020). Nanocellulose hyperfine network achieves sustained release of berberine hydrochloride solubilized with β-cyclodextrin for potential anti-infection oral administration. *International Journal of Biological Macromolecules, 153*, 633–640. Available from https://doi.org/10.1016/j.ijbiomac.2020.03.030.

Xu, C., Zhang Molino, B., Wang, X., Cheng, F., Xu, W., Molino, P., ... Wallace, G. (2018). 3D Printing of nanocellulose hydrogel scaffolds with tunable mechanical strength towards wound healing application. *Journal of Materials Chemistry B., 6*, 7066–7075. Available from https://doi.org/10.1039/c8tb01757c.

Xu, W., Molino, B. Z., Cheng, F., Molino, P. J., Yue, Z., Su, D., ... Wallace, G. G. (2019). On low-concentration inks formulated by nanocellulose assisted with gelatin methacrylate (GelMA) for 3D printing toward wound healing application. *ACS Applied Materials & Interfaces., 11*, 8838–8848. Available from https://doi.org/10.1021/acsami.8b21268.

Yahaya, M. L., Zakaria, D., Noordin, R., Razak, K. A., Lukman Yahaya, M., Zakaria, D., ... Razak, K. A. (2017). *Multiplexing of nanoparticles-based lateral flow immunochromatographic strip: A review. Advanced materials and their applications: Micro to nano scale* (pp. 112–139)). One Central Press (OCP).

Ye, S., Jiang, L., Wu, J., Su, C., Huang, C., Liu, X., & Shao, W. (2018). Flexible amoxicillin-grafted bacterial cellulose sponges for wound dressing: In Vitro and in Vivo Evaluation. *ACS Applied Materials & Interfaces.*, *10*, 5862–5870. Available from https://doi.org/10.1021/acsami.7b16680.

Yildirim, N., & Shaler, S. (2017). A study on thermal and nanomechanical performance of cellulose nanomaterials (CNs). *Materials (Basel)*, *10*. Available from https://doi.org/10.3390/ma10070718.

Yuan, H., Chen, L., & Hong, F. F. (2021). Homogeneous and efficient production of a bacterial nanocellulose-lactoferrin-collagen composite under an electric field as a matrix to promote wound healing. *Biomaterials Science*, *9*, 930–941. Available from https://doi.org/10.1039/d0bm01553a.

Zainuddin, N., Ahmad, I., Zulfakar, M. H., Kargarzadeh, H., & Ramli, S. (2021). Cetyltrimethylammonium bromide-nanocrystalline cellulose (CTAB-NCC) based microemulsions for enhancement of topical delivery of curcumin. *Carbohydrate Polymers*, *254*, 117401. Available from https://doi.org/10.1016/j.carbpol.2020.117401.

Zdarta, J., Meyer, A. S., Jesionowski, T., & Pinelo, M. (2018). A general overview of support materials for enzyme immobilization: Characteristics, properties, practical utility. *Catalysts*, *8*. Available from https://doi.org/10.3390/catal8020092.

Zhao, Y., Wan, S., Li, L., Li, S., Shi, Y., & Pan, L. (2020). Nanocellulose and nanohydrogel for energy, environmental, and biomedical applications. *Sustainable nanocellulose and nanohydrogels from natural sources*, 33–64. Available from https://doi.org/10.1016/b978-0-12-816789-2.00002-x.

Zhou, L., Mao, H., Wu, C., Tang, L., Wu, Z., Sun, H., ... Zhao, J. (2017). Label-free graphene biosensor targeting cancer molecules based on non-covalent modification. *Biosensors & Bioelectronics*, *87*, 701–707. Available from https://doi.org/10.1016/j.bios.2016.09.025.

Zhu, G., Cheng, L., Qi, R., Zhang, M., Zhao, J., Zhu, L., & Dong, M. (2020). A metal-organic zeolitic framework with immobilized urease for use in a tapered optical fiber urea biosensor. *Microchimica Acta.*, *187*. Available from https://doi.org/10.1007/s00604-019-4026-0.

Zhu, M., Wang, J., & Li, N. (2018). A novel thermo-sensitive hydrogel-based on poly(N-isopropylacrylamide)/hyaluronic acid of ketoconazole for ophthalmic delivery. *Artificial Cells*, *46*, 1282–1287. Available from https://doi.org/10.1080/21691401.2017.1368024.

Zhu, Y., Chen, Q., Shao, L., Jia, Y., & Zhang, X. (2020). Microfluidic immobilized enzyme reactors for continuous biocatalysis. *Reaction Chemistry & Engineering.*, *5*, 9–32. Available from https://doi.org/10.1039/c9re00217k.

Zoppe, J. O., Ruottinen, V., Ruotsalainen, J., Rönkkö, S., Johansson, L. S., Hinkkanen, A., ... Seppälä, J. (2014). Synthesis of cellulose nanocrystals carrying tyrosine sulfate mimetic ligands and inhibition of alphavirus infection. *Biomacromolecules*, *15*, 1534–1542. Available from https://doi.org/10.1021/bm500229d.

Zucca, P., & Sanjust, E. (2014). Inorganic materials as supports for covalent enzyme immobilization: Methods and mechanisms. *Molecules (Basel, Switzerland)*, *19*, 14139–14194. Available from https://doi.org/10.3390/molecules190914139.

CHAPTER 20

Nano-driven processes toward the treatment of paper and pulp industrial effluent: from the view of resource recovery and circular economy

G. Madhubala[1], S. Abiramasundari[1], Nelson Pynadathu Rumjit[2], V.C. Padmanaban[1] and Chin Wei Lai[2]
[1]Centre for Research, Department of Biotechnology, Kamaraj College of Engineering and Technology, Madurai, India
[2]Nanotechnology & Catalysis Research Centre (NANOCAT), Institute for Advanced Studies (IAS), University of Malaya (UM), Kuala Lumpur, Malaysia

20.1 Introduction

Paper and pulp industries are the class of industries that convert wood or recycled fiber into paper and pulp. In general, the paper mills manufacture paper and other converted paper products from wood pulp. In contrast, the pulp mills are used to separate fibers present in the wood to produce a clean source of pulp. The paper and pulp industry is one of the rapidly growing industries (Mandeep, Gupta, Liu, & Shukla, 2019). Among the significant marketing industries worldwide, paper and pulp mill is considered to be one of the water-intensive industries. It is reported that the water used in industrial sectors is about 34 billion m^3 per day and is expected to increase by four folds by the end of 2050 (Saadia & Ashfaq, 2010). The paper and pulp industry is a vast industry as the world's demand for paper keeps rising 5%–6% every year. In 2012, the paper consumption globally was recorded 58 kg per capital and in India as 9.3 kg per capital (Mandeep et al., 2019). The literature survey states that the annual global paper production was roughly about 490 million tonne in 2020 and is expected to increase to 596 million tonne by 2031 (Gómez-Pastora et al., 2017). The Indian paper and pulp mill stands out to be the 15th largest paper and pulp industry in the world. The Indian government interprets the paper and pulp industry as one of the 35 most prioritized industries in the country. In India currently, 759 paper mills manufacture about 10.9 Mt per annum (Singh, Kaur, Khatri, & Arya, 2019). In the present 20th century, paper production has almost doubled every 20 years and is estimated at 400 million tonnes annually. It can also be forecasted that in 2030, paper production would globally hit a mark of 495 million metric tonnes, and the global shipping

Nanotechnology in Paper and Wood Engineering
DOI: https://doi.org/10.1016/B978-0-323-85835-9.00026-X

Table 20.1 Technological advances in paper and pulp industries.

Period	Technological advance
Preindustrial	Sunlight + potash (or urine)
1700s	The multistage process with lye, sunlight, and lactic acid
1756	Sulfuric acid shortens the bleaching process
1774	Scheele discovers chlorine gas
1785	Berthollet suggests chlorine as a bleaching agent
1799	Introduction of bleaching powder
1815	Davy synthesizes chlorine dioxide
1920s	Chlorine gas reintroduced as a bleaching agent
1946	The first commercial chlorine dioxide plant opens
1970	The first commercial oxygen plant started
1972	The first plant with ozone bleaching started
1990s	Chlorine-free bleaching processes introduced

volume is expected to cross a mark beyond 100 billion parcels. Moreover, these paper and pulp mills also are a major employment platform for more than 2 lakh people. The estimated capacities of these mills are also known to increase to 19.7 million tonnes in 2030 from the earlier estimates of 14 million tonnes in 2020 (Kesalkar, Khedikar, & Sudame, 2012).

The paper mills in India can be classified into three categories based on raw material: wood/forest-based mills, agro-based residue mills, and wastepaper mills. The production from wood/forest-based mills is about 70%, and the remaining 30% from agro-residue based mills and waste paper mill (Bhatnagar, 2015; Kesalkar et al., 2012). Though paper and pulp mill is considered one of the important industrial sectors of the global economy, this mill remains one of the most polluting industries that are also water intensive.

Even though paper and pulp mill is regarded as one among the rapidly growing industries, it is known to discharge about 20−40 m^3 of wastewater per tonne of pulp and is considered the most energy and water sweeping industrial sector in the world. Table 20.1 outlines the various technological advances in the paper and pulp industries.

20.2 Characteristics of paper and pulp industry effluents

Based on the production of effluents, the paper and pulp industries are categorized at sixth position in the list of high-risk polluting industries worldwide and as one among the 20 most prominent polluting industries according to the Ministry of Environment and Forest, Government of India. The expulsion of the untreated and partially treated effluents from the industries to the water bodies is the major cause responsible for environmental pollution (Mandeep et al., 2019). The quantity and characteristics of discharged wastewater mainly depend upon the type of production process adopted and the scope toward water reuse. In general, the paper and pulp mill is characterized by a very high biochemical oxygen demand (BOD), strong color, high suspended solid, and a high chemical oxygen demand (COD)/BOD ratio.

The main processes in paper and pulp industries are divided into five major groups, including various pulping methods such as mechanical, chemical, chemomechanical, thermomechanical pulping, and papermaking process discharges an elevated amount of wastewater with distinct characteristics (Mussey, 1955).

20.2.1 Raw material preparation (Barker bearing cooling water)

The raw material for the paper and pulp industry may be hardwood, softwood, or nonfiber (straw, bagasse, bamboo, kenaf, and so on) in nature (Kumar, Saha, & Sharma, 2015). This plant fiber is converted into small pieces known as chips by mechanical or hydraulically methods. During this process, unwanted materials such as bark, dirt, grids, fiber, and suspended solids are removed. High suspended solids in the effluent result in high COD, BOD, and turbidity (Abn & Lim, 2009). Tanin, resin acid, fatty acid, phytosterol, and retene are a few of the important components present in that effluent. Mostly, characteristic of the effluent depends on the type of raw material they used. Hardwood (oaks maples, birches) extractives consist of lipophilic components, such as sterol, wax, glyceride, sterol ester, long-chain aliphatic acid, and alcohol (Cabrera, 2017). Softwoods like cedar, pines, firs, spruces, and hemlock have high resin acid than hardwood. Total solid present in the effluent can be categorized as settle or nonsettle solids. Each of these categories has both organic and inorganic components (Singh, Srivastava, Jagadish, & Upadhyay, 2019).

20.2.2 Pulping (black liquor)

In pulping, chips are converted into pulp by using different mechanical, semichemical, and chemical (kraft, sulfite) pulping. Among these, chemical pulping produces a large volume of polluted wastewater, but it is the most efficient method for pulping. Paper produced by the chemical pulping method has more mechanical strength because the cellulose content is not damaged during the process. In this process majority of lignin and hemicelluloses are removed (Singh, Srivastava, et al., 2019). The effluent generated from the pulping process is called black liquor, dark brown because of lignin and its derivative (Kumar et al., 2015). The effluent also contains adsorbable organic halides (AOX), color, organic polymeric components such as lignin, carbohydrates and their derivatives, BOD, COD, resins, fatty acid, volatile organic compounds (VOC) such as terpenes, alcohol, phenol, acetone, chloroform, etc. (Devi, Yadav, Shihua, Singh, & Belagali, 2011; Kumar et al., 2015).

20.2.3 Washing (wash water)

Normally three to five washing stages were present in the pulp and paper industry. The washing process may take place as single-stage washing, multistage washing, and multistage countercurrent washing. Among these, multistage countercurrent washing is the ideal one. The effluent from washing contains a high concentration of pH, BOD, COD, and solids suspended because of the processing and temperature conditions (Afonso & De Pinho, 1997).

20.2.4 Bleaching (bleach plant wash water)

Bleaching agents such as chlorine, chlorine dioxide, hydrogen peroxide, and ozone aim to increase the quality and whiteness of paper by removing lignin present in pulp (Singh & Ambika, 2018). Chlorine reacts with lignin and its derivatives during this process, producing extremely toxic, xenobiotic, and recalcitrant components like trichlorophenol, trichloroguicol, tetrachloroguicol, dichlorophenol, dichoroguicol, and pentachlorophenol, which are responsible for high COD and BOD content in the effluent (Kumar et al., 2015). Washing removes the bleaching agent from the pulp. The effluent contains dissolved lignin, carbohydrate, color, COD, AOX, organic compounds (dioxin, furan, and chlorophenols), inorganic chlorine compounds such as chlorate $\left(ClO_3^-\right)$, and VOCs (methylene chloride, carbon disulfide, chloroform, chloromethane, acetone, and trichloroethanes) (Afonso & De Pinho, 1997). Around 85% of the total effluent of the paper and pulp industry comes from the bleaching process (Cabrera, 2017). The major important step is chlorine oxidation, resulting in wastewater containing AOX and chlorinated organic compounds (Ince, Cetecioglu, & Ince, 2011).

20.2.5 Paper manufacturing (white water)

The effluent from the formation of the cellulose sheet is called white water. Even though the process water appears in brown due to dye, black liquor is still called white water because of fiber fine, filler, and air bubble that scatter light present in it. White water from paper producing machine was reused many times, mainly for dilution of incoming fiber. White water also contains all chemicals, originated during the drying of the sheet, which includes organic compounds, fiber, polysaccharides, calcium, sodium hydroxide, sodium sulfide, and inorganic dyes COD, acetone, acetic acid, propionic acid, and so on (Afonso & De Pinho, 1997).

The proportion of the wastewater depends on raw material, type of paper machine, manner the pulp was washed, adopted pulping and bleaching method. The composition of the white water should maintain stability for a long time to produce a more uniform product and run the paper machine system more efficiently.

20.3 Key challenges in pulp and paper industry

Water is one of the primary elements of the papermaking industry. Being one of the most predominant industrial water consumers, paper production is impossible without water requirement. During the emergence of paper making industries, paper was produced with high water consumption rates. Though various treatment technologies have evolved to treat effluents and wastewater recycling, water consumption remains an invincible challenge. About 20–40 m^3 of wastewater per tonne of pulp is discharged from the paper and pulp mills. About 40% is utilized in pulp production, 35% in the paper machine, 15% in raw material preparation, and 5% for stock preparation and utilities. The paper and pulp industry is also regarded as one of the most renowned polluting industries worldwide. As a result, a notable amount of wastewater, solid waste, and gaseous waste are discharged. The wastewater generated from the paper and pulp mills is known to cause severe damage to the aquatic life forms and give rise to diverse health imputations. The wastewater consists of suspended solids such as bark particles, dirt, and pigments, dissolved inorganic compounds like sodium hydrogen, sodium silicate, and some bleaching agents (Hooda, Bhardwaj, & Singh, 2018). The wastewater discharged also comprises various color, dyes, and toxic elements.

The wastewater is generated significantly in major quantities from two major papermaking processes: pulping and bleaching. The untreated effluents from these processes are usually expelled into the water bodies, thereby damaging the water quality in various water systems. Since the process of bleaching involves the use of chlorine and chlorine-based chemicals, which when combined with various hydrocarbons, result in the formation of organochlorine compounds, termed as AOX, these components, when mixed with other bleaching components, have been held accountable to cause several health disorders such as skin irritation, respiratory diseases and inflammation, DNA damage, and reproductive damages in terrestrial as well as ariel life forms (Khan et al., 2019; Mandeep et al., 2019).

The wastewater from the paper and pulp mills contains high BOD and COD rates. The wastewater also sustains other compounds such as phosphate, nitrogen, chlorine, and sulfur, ultimately resulting in eutrophication and poisoning of the aquatic life forms when discharged into water bodies. The direct dumping of effluents is dangerous to aquatic organisms and poses a great challenge for bioremediation and hence gets accumulated in the food chain. The dense metal-polluted wastewater used for agricultural practices leads to phytotoxicity in plants, causing various health defects when consumed. When consumed by the primary producers and the succeeding consumers, they are not easily expelled to the external environment and instead accumulate at each successive trophic level, resulting in biological magnification. The wastewater from the pulp and paper industry has been reported to cause phytotoxicity in *Allium cepa* (Yadav & Chandra, 2018).

Table 20.2 Conventional treatment strategies used in paper and industrial pulp effluent.

Treatment type	Limitations
Distillation	High requirement of energy and water. Pollutant boiling point >100°C challenging to remove
Biological treatment	Microorganisms difficult to control and byproduct damages cells, not cost-effective, and time taking
Ultraviolet treatment	Expensive method and inactivated due to turbidity, ineffective due to heavy metals, and ineffective in inorganic contaminants' removal
Ultrafiltration	Do not remove dissolved inorganics, high energy requirement, and difficult in cleaning
Chemical transformation	Excess reagents needed, low-quality mixture, inactive adverse conditions, and selective method
Coagulation and flocculation	Low efficiency and pH dependent

Various researchers have also reported the presence of many androgenic and carcinogenic compounds in the effluent. In India, about 100 million kg of harmful and dangerous effluents are released from these industries yearly; hence, the Ministry of Environment and Forest, Government of India, has categorized the paper and pulp industry as one among the most 20 polluting industries (Mandeep et al., 2019). The paper and pulp mills produce about 225–320 m^3 tonnes of waste than the total amount of paper produced. The effluents majorly consist of organic and inorganic compounds such as sulfur oxides, nitrogen oxides, chlorine and chlorine dioxide, chloroform, dioxins and furans, hydrogen chloride (as part of particulate matter), methanol, phenols, total reduced sulfur compounds, VOCs, AOX, pulping liquors, and bleaching effluents. In addition to this, the wastewater effluent also contains trace metals such as mercury. When the concentration of these metals increases in the effluents, they are known to cause harmful and hazardous effects on human health (Jitendra, Srivastava, Pachauri, & Srivastava, 2014). The present conventional methods used for the treatment of the effluents are listed in Table 20.2.

20.4 Nano-driven processes for the remediation of paper and pulp industry effluent

The efficiency of these conventional methods is much less for the complete removal of the wastewater effluents and possess certain limitations. Thus there is a need to shift toward a more efficient method through the application of nanotechnology. Nanotechnology can be defined as studying designs and characterization of devices and systems by configuring the shape and size at a nanoscale. Nanotechnology can increase the levels of energy consumption, helps to clean the environment, and provides the solution to major health problems. Moreover, the nanotech claims that the nanotechnology product will be a nanoscale product that is cheaper, reliable, and more functional. Nanotechnology plays an important role in various sectors such as health and sanitation, drug delivery, use of nanosensors on crops and nanoparticles in fertilizers, and growth of crops in hostile conditions. Nanoscale filters, nanocomposites and nanoparticles could clean the environment and supports in renewable and sustainable energy to help the environment. The study of nano-based treatment technologies has been a wide area of research from the past. The various nanomaterials based treatment methods have been tabulated in Table 20.3 (Yaqoob, Parveen, Umar, & Ibrahim, 2020).

20.4.1 Photocatalysis based treatment of paper and pulp mill effluents

The paper and pulp mill effluents contain a suspension of various complex organic and inorganic compounds, which when untreated and are discharged into water systems, might have a negative impact on a wider scale (Kumar, Kumar, Bhardwaj, & Choudhary, 2011). The effluents are released mostly from two stages of the paper and pulp making process, namely the pulping and bleaching processes. Paper bleaching plays an important role in contributing to the color, organic content levels, and the toxicity of the effluents expelled out from the industry. The use of chlorine and chlorine-based compounds during the bleaching process might results in the formation of toxic organic compounds such as phenol, dioxins, resins, and fatty acids.

Hence, biological and other conventional treatment technologies are not sufficient for the complete removal of the undesirable compounds present in the effluent. Thus the advanced oxidation process is used to treat the effluents released from paper and pulp industries. There are many advanced oxidation types, namely photocatalysis, photo-oxidation, ozonation, and fenton type reactions, to eliminate the chromophoric and nonchromophoric elements present in the paper and pulp mill effluent (Ince et al., 2011). TiO_2 photocatalysis is an important substitute as it results in the complete mineralization of all organic and inorganic elements. TiO_2 photocatalysis has been considered as the best method with high efficacy to reduce the BOD and COD levels present in the effluents (Ince et al., 2011). In this process, the TiO_2 nanoparticles are used wherein the target analyte is organic pollutants and the positive aspects of this method include less toxicity, water insolubility, and photostability. When a semiconducting material such as TiO_2 absorbs the light received from the source, which is either equal or greater than the TiO_2 band gap (3.2 eV or more), it carries to the electron−hole pairs ($e^-−h^+$). Suppose the process of separation of charges is continued. In that case, the electron−hole pair moves on to the catalyst's surface, thereby contributing along with the sorbed species in redox reactions. In particular, h_{+vb} (hall in the

Table 20.3 Various nanomaterials-based treatment methods.

Nanoparticles (NPs)	Target analyte	Treatment mechanism	Limitation	Positive aspect
TiO_2	Organic pollutants	Photocatalysis	Higher operational cost, tough to recover, sludge production	Less toxic, water insolubility, photostability
Fe	Heavy metals, anions, organic pollutants	Reduction, adsorption	Tough to recover, sludge production, difficult sludge disposal, health risk	In situ water remediation, less cost, harmless to handle
Bimetallic NPs	Dichlorination	Reduction, adsorption	Tough to recover, sludge production	Higher reactivity
Nanofiltration and nanomembranes	Organic and inorganic substances	Nanofiltration	High cost, membrane fouling	Low pressure
Magnetite NPs	Heavy metals, organic compounds	Adsorption	Outside magnetic field needed for separation	Easy separation, no sludge production
Metalsorbing vesicles	Heavy metals	Adsorption	Difficult to maintain stability	Re-use option, higher selective uptake profile, better metal affinity
Micelles	Organic pollutants	Adsorption	Costly	In situ treatment, an excellent affinity for hydrophobic
Dendrimers	Heavy metals, organic pollutants	Encapsulation	Costly	Easy separation, renewable, high binding, no sludge production
Nanotube	Heavy metals, anions, organic pollutants	Adsorption	High cost, lower adsorption process, tough to recovery, sludge production, dangerous health risk	Dealing with pollution from water, good mechanical properties, exclusive electrical properties, good chemical stability
Nanoclay	Heavy metals, anions, organic pollutants	Adsorption	Sludge production	Lower cost, exclusive structures, long stability, recycle, higher sorption capacity, informal recovery, better surface and pore volume

valence band)_ reacts with the surface-bound water molecules to generate hydroxyl ions. At the same time, e_{-cb} (electron in the conduction band) picked up by the oxygen to produce superoxide radicals. The mineralization and degradation pathway is designated in the following Eq. 20.1–20.3:

$$TiO_2 + h\nu \rightarrow e_{-cb} + h_{+vb} \tag{20.1}$$

$$O_2 + e_{-cb} \rightarrow O_2^- \tag{20.2}$$

$$H_2O + h_{+vb} \rightarrow OH + H+ \tag{20.3}$$

Once the pollutants are degraded in the presence of light, the degraded products are analyzed using high performance liquid chromatography coupled with mass spectrometry or gas chromatography coupled with mass spectrometry methodologies to confirm the toxicity of the reduced products. In comparison, the rate of mineralization is less when compared to degradation. The probable reason behind the low rates of mineralization is stable intermediates during photocatalytic activity. Therefore it strictly requires long-term irradiation for the complete removal of total organic carbon (TOC). The TOC is defined as the total quantity of bounded carbons in any organic compound and measured by the TOC analyzer.

Lignin is the second-largest macromolecule, a polymer of phenolic compounds, and undergoes a similar photocatalytic degradation as phenol. The applicable wavelength range for lignin degradation is between 300 and 400 nm (Awungacha Lekelefac, Busse, Herrenbauer, & Czermak, 2015; Tanaka, Calanag, & Hisanaga, 1999). The higher energy of ultraviolet (UV) radiations and the short wavelength range result in two different mechanism pathways: the *electron*–hole reaction and OH radical oxidation. Mostly, preferably used nanophotocatalysts for lignin degradation is TiO_2 in the presence of light. The efficiency of applying only UV radiation in the degradation pathway is minimum, whereby the usage of nanophotocatalysts such as TiO_2 shows a significant decrease in dissolved organic carbon (DOC) and American Dye Manufacture Institute value (ADMI) (Ma, Chang, Chiang, Sung, & Chao, 2008). In the lignin degradation process, TiO_2 (anatase) is used as the nanophotocatalyst. First, the required quantity of TiO_2 is suspended in an aqueous solution of lignin in a reaction vessel containing a mercury lamp of high pressure in the center of the reaction vessel. The entire system is then jacketed by a pyrex glass tube through which the cooling water runs. Wavelength less than 310 nm is cut-off by the water jacket. Finally, the parameters are reported through various methods such as absorption spectrum, TOC analyzer, and gel permeation chromatogram, and further analysis of the efficiency of degradation is observed through nuclear magnetic resonance spectroscopy and Fourier transform infrared spectroscopy techniques (MacHado et al., 2000). The three major generated species in the photocatalytic mechanism include singlet oxygen and superoxide and hydroxyl radicals, wherein hydroxyl radical plays an important role. The following equations represent the degradation pathway where L represents the lignin substrate Eq. 20.4–20.10.

$$TiO_2 + h\nu \rightarrow TiO_2\left(e^-/h^+\right) \quad (20.4)$$

$$TiO_2\left(h^+\right) + H_2O \rightarrow TiO_2 + HO^{\cdot} + H^+ \quad (20.5)$$

$$TiO_2\left(h^+\right) + L \rightarrow TiO_2 + L^{+\cdot} \quad (20.6)$$

$$TiO_2\left(h^+\right) + OH^- \rightarrow TiO_2 + HO^{\cdot} \quad (20.7)$$

$$TiO_2(e^-) + O_2 \rightarrow TiO_2 + O_2^{-\cdot} \quad (20.8)$$

$$OH^- + O_2 \rightarrow HO^{\cdot} + O_2^{-\cdot} \quad (20.9)$$

$$HO^{\cdot} + O_2^{-\cdot} \rightarrow OH^- + O_2 \quad (20.10)$$

Photocatalysis remains one of the most successful methods for removing cellulose and cellulose-based effluents discharged from the paper and pulp mills. Advanced oxidation with ozone or UV or a combination of both and semiconductor photo-assisted catalysis applied to cellulosic effluents have been reported earlier (Cristina Yeber, Rodríguez, Freer, Durán, & Mansilla, 2000). For cellulose degradation, mostly TiO_2 catalysts are preferred, and these catalysts are immobilized on Raschig rings and are further activated by light (Cristina Yeber et al., 2000). Photocatalysis is used to reduce COD levels and decrease the level of toxicity present in the wastewater effluents. A similar procedure to that of lignin degradation is followed where ZnO proves to be an effective catalyst for the decrease in the levels of COD (Hasegawa, Daniel, Takashima, Batista, & Da Silva, 2014). Most of the VOCs are oxidizable; hence, photocatalytic degradation appears to be a favorable method of degradation when compared to other treatment technologies (Lin et al., 2013). The VOCs are mainly composed of some halogenated hydrocarbons and benzene, toluene, ethylbenzene, and o-xylene (Li, Li, Ao, Lee, & Hou, 2005). The most used photocatalyst for the degradation of VOC is $Ln^{3+}-TiO_2$ catalysts. The degradation of organic acids such as formic acid has been reported using Fe-doped and calcined TiO_2. In this process, the Fe^{2+} ions remain in the solution after degradation, move back to the TiO_2 surface, and are oxidized by the newly photogenerated holes. Hence in this way, the catalyst is recycled and is used for the next degradation process to be continued. In general, for the degradation of organic acids, the photocatalytic characters of the TiO_2 photocatalyst can be enhanced by doping it with chromium and iron to make the photocatalyst more active (Araña et al., 2001). Other organic contaminants are photo degraded using the emission characteristics of fluorescence light of ZnO photocatalysts. First, the nanocrystalline ZnO photocatalysts are prepared, and the organic contaminants are detected directly after the quenching of fluorescence detected from the ZnO semiconductor films. Moreover, this ZnO can also act as a nonspecific sensor that could also detect other inorganic contaminants present in the wastewater. The major organic contaminants detected from the wastewater include aromatic chloro compounds and solvents and aliphatic chloro compound (Hariharan, 2006). Some recent studies have confirmed that ZnO is a better photocatalyst than TiO_2 for the better degradation of some organic contaminants and dyes. For the effective degradation of furans, ZnO and mixed-phase TiO_2 containing either rutile or anatase are used as a photocatalyst.

On comparing the various photocatalysts, ZnO proves to be more efficient toward the complete elimination of furans, especially carbofurans and their organic intermediates formed during the degradation pathway (Fenoll, Hellín, Flores, Martínez, & Navarro, 2013). Recent studies have reported the photocatalytic degradation of carbofurans in aqueous solutions wherein the TiO_2 acts as a photocatalyst. However, as mentioned earlier, the efficiency of degradation due to TiO_2 remains much less than that of the degradation due to ZnO. The formed intermediates during the pathways were also studied to get a clear idea about the mineralization pathway.

20.4.2 Nanomembrane based treatment of paper and pulp mill effluents

In recent years membrane-based technology like microfiltration, nanofiltration, ultrafiltration, and reverse osmosis are the advanced approaches used in many industries for the treatment of wastewater effluent because of its high efficiency in the removal of pathogens, monovalent or divalent ions, and other suspended solid substance with consumption of less energy (Pizzichini, Russo, & Di Meo, 2005) and low cost (Abdelbasir & Shalan, 2019; Qingping, 2018; Timmer, 2001). Around the 1950s, membrane technology was used to treat water, and they play a major role in reducing the pollutant and giving better-purified water (Qingping, 2018).

In recent years, nanomembrane-based technology proves to be a promising, sustainable technology applied to treat wastewater and drinking water. The membrane acts as a physical barrier for the removal of contaminants (Butt, 2020). In many industries, nanomaterial fabricated membrane filtration has become an integral part, and a most effective strategy in reducing the pollutants present in the wastewater discharged from these industries (Anjum, Miandad, Waqas, Gehany, & Barakat, 2016; El Saliby,

Shon, Kandasamy, & Vigneswaran, 1999). Nanofiltration method is one of the membrane filtration techniques that has gained wide attention due to various advantages such as catalytic activity, low maintenance cost, high permeability, as an effective disinfectant, the requirement of low space, fouling resistance, simple design, easy handling, cost effectiveness, and high withholding ability of multivalent ion and organic molecules (Anjum et al., 2016; Madhura & Singh, 2018). Nanofiltration is a pressure-driven process that separates or removes ions, molecules, particles ranging from 0.5 to 0.1 nm by charge-based repulsion mechanism and whose properties lie between ultrafiltration and reverse osmosis (Gawaad, Sharma, & Sambi, 2011).

During the 1970s and 1980s, nanofiltration membranes were used as an intermediate material in ultrafiltration and reverse osmosis (El Saliby et al., 1999) because of their diffusion mechanism and electrostatic charge effects. Nanofiltration as a pretreatment to reverse osmosis prevents reverse osmosis from membrane fouling by organic and inorganic compounds (Negaresh & Leslie, 2012).

The size exclusion property of nanomembrane depends upon the amount of molecular weight cut-off of the membrane. As a result of this property, if a compound with molecular weight higher than the molecular cut-off of the membrane, then the compound does not pass through the membrane and gets separated. Size exclusion is responsible for the elimination of uncharged solute. Next to size exclusion, electrostatic interaction plays an important role in nanofiltration in eliminating effluents. Electrostatic interaction takes place between the surface of the membrane and charged compounds. In general, nanomembranes are negatively charged, and for this reason, negatively charged components are eliminated more than neutral compounds. The rise in pH results in increased elimination rates of the effluents. Membrane fouling complicates the process of nanofiltration and can be reduced by microfiltration as a pretreatment (Negaresh & Leslie, 2012). Nanomembranes do not allow any solid particles to pass through them due to their pore size of 0.006 μm (Yaqoob et al., 2020). Nanomembranes-based treatment is highly effective for the removal of heavy metals, dyes, and other components at low operating pressure (Madhura & Singh, 2018). Charge-based repulsion mechanism is responsible for the removal of metals from wastewater (Butt, 2020). It is also a preferred method to reduce the odor, hardness, and color of the effluent.

In comparison with ultrafiltration membranes, nanofiltration membranes have a high rejection capacity of organic compounds with low operating pressure (Madhura & Singh, 2018). These membranes can be handled at high concentration and a wide range of pH, from acidic to basic (Gawaad et al., 2011). Membrane fabricated with carbon nanofiber at high pressure shows excellent selective filtration and removal efficiency. Assembling beta cyclodextrins (Abdelbasir & Shalan, 2019) in carbon nanofiber membranes can remove phenolphthalein and fuchsin acid (Anjum et al., 2016). Carbon nanotube (CNT) can be categorized into two types, namely single-walled CNT (SWCNT) with diameters ranging from 1.0 to 1.4 nm and another multi-walled CNT (MWCNT) with diameters ranging from 10 to 50 nm (Anjum et al., 2016).

They have low mass density, high strength, high flexibility and antimicrobial property. CNTs can remove various organic and inorganic pollutants, including volatile organic compounds (Mishra, 2014).

CNT has antimicrobial property, low membrane fouling and biofilm formation (Yaqoob et al., 2020), and can reduce membrane biofouling. Polymeric membrane surface with silver nanoparticles can deactivate the virus, inhibit bacteria adherence biofouling (Abdelbasir & Shalan, 2019). Modification of nanomembranes using different types of nanoparticles with polymeric or inorganic membranes increases the membrane filtration (Gehrke, Geiser, & Somborn-Schulz, 2015). Examples of such commonly used nanoparticles include hydrophilic metal oxide nanoparticle, antimicrobial nanoparticles, and photocatalytic nanomaterials. Metal oxide nanoparticles in the membrane show an increase in its permeability, antifouling character, and hydrophilicity of the membrane surface. In addition to it, when these nanoparticles are embedded with polymeric membranes, it results in the development of mechanically and thermally stable polymeric membranes (Butt, 2020).

Furthermore, the use of polymeric nanofibre membrane results in the absorption of toxic heavy metal such as As, Cr, and Pb (Abdelbasir & Shalan, 2019). Most nanofiltration membranes are composite membranes that comprise mixed matrix membranes (Gehrke et al., 2015) and surface-functionalized membranes. These two membranes are the most novel and reliable materials in the separation process. The composite membrane increases the hydrophilic and tensile mechanical properties, which increase the rate of removal of the selected pollutant (Anjum et al., 2016). Membranes with TiO_2 nanoparticle have the efficiency in degrading organic components and can deactivate certain types of microorganism, as they have excellent surface characteristics (Madhura & Singh, 2018). Silver nanoparticles with small size and large surface area have been incorporated into ultrafiltration to give excellent antimicrobial property and mechanical strength to the membrane (Butt, 2020). In the development of multifunctional membrane, nanohybrid membrane plays a major role (Anjum et al., 2016). Within the nanofiltration membrane, a centrifugation pump is used for the circulation of wastewater. Most nanofiltration nanocomposite materials are nanocomposite in which carbon-based material is incorporated into polymeric matrices (Gawaad et al., 2011; Yaqoob et al., 2020). The electrospinning method is used to produce an inorganic nanofiller, which can be embedded in the matrix material of membranes (Gehrke et al., 2015). Often nanoparticles are used as a filler in the nanocomposite.

Nanocomposites are thermally stable, biodegradable, resistant to the solvent, effective barrier to organic and inorganic contaminants from wastewater (Singh & Ambika, 2018). Nanofillers incorporate with metal oxide nanoparticles produce mechanically and thermally stable membrane, and permeate flux of the nanomembrane increases.

Nanofillers have a large specific surface area, and their diameter is in the range of nanometer (Butt, 2020). Silver ZnO nanoparticles as nanofillter incorporated in ultrafiltration enhance their properties like water permeability, antimicrobial activity, and so on. Semimetallic graphene fabricated nanocomposites have few special characteristics like a smooth surface plane and edges, large surface area, two-dimensional single-layer structure, and efficiently removes ion Mg^{2+}, Ca^{2+}, Na^{+}, and Li^{+} (Anjum et al., 2016; Chi, Wang, Liu, & Yang, 2018). Reduction of total dissolved solids, VOCs (organic chemicals), heavy metals, nitrates and sulfates, color, tannins, and turbidity is possible by nanofiltration with lower operating costs, energy costs, and discharge.

In the paper and pulp industry, membrane technology has been used since the late 1960s (Negaresh & Leslie, 2012). Nanofiltration plays a major role in removing pollutants discharged by the paper and pulp industries. In the paper and pulp industry, the nanofiltration-based technique reduces wastewater effluent and facilitates the reuse of treated water. Highly hazardous and low molecular weight components from the bleaching effluent can be removed by nanofiltration (Lastra et al., 2004). The high pH from bleaching effluent induces a negative charge on the nanomembrane's surface and eliminates organochlorinated compounds present in the wastewater effluent (Afonso & De Pinho, 1997). Nanomembrane filtration is a feasible method to remove organic components from bleaching effluent and gives demineralized water (Lastra et al., 2004). By combining two membrane filtration methods, namely, ultrafiltration followed by nanofiltration in series, lignin content has been reduced effectively due to increased flux. Above 95% of lignin, retention occurs when there is a rise in nanomembrane density, but the flux was insufficient (Arkell, Olsson, & Wallberg, 2014). In recent years there is a modification in membrane functionality and specificity for efficient treatment. Studies carried out by Dafinov, Font, and Garcia-Valls (2005) gave the result as lignin and cellulose content present in black liquor can be removed by using nanofiltration. According to Jönsson, Nordin, and Wallberg (2008), the nanofiltration process can be used to treat effluent from the pulping process. The authors indicated that during nanofiltration, the retentate contained a high amount of lignin. Lignin is the major organic component present in the effluent obtained during pulping (black liquor) (Jönsson et al., 2008). According to Daniel Humpert and Rufus Ziezig, Tomani, and Theliander (2014), membrane-based filtration is the most economical, versatile, and environmentally sustainable method used to treat effluent from the paper and pulp industry with a high recovery of lignin (Lastra et al., 2004). Different components from the black liquor can be removed by membrane-based technology without using chemicals (Arkell et al., 2014).

Currently, the levels of different harmful organic compounds present in the wastewater effluent have increased, and they cause damage to the ecosystem by creating environmental pollution.

20.4.3 Nanosorption-based treatment of paper and pulp mill effluents

In recent years, nanotechnology has become a promising, sustainable technology that provides tremendous opportunities to treat wastewater and drinking water (Gehrke et al., 2015; Khan et al., 2019). Nanomaterials are excellent adsorbing agent because of their large specific surface, high porosity, small size, short intraparticle diffusion distance, and surface multifunctionality, which means they can react and bind chemically to different adjacent atom and molecules (Bhatnagar & Sillanpää, 2015; El-sayed, 2020). In addition to the good adsorbent, they are reusable, withstand for long term and considerable metal binding capacities (Bhatnagar & Sillanpää, 2015; Sharma, Srivastava, Singh, Kaul, & Weng, 2009). In recent years demand for nanoparticles rapidly increases due to their application in various fields. Nanoadsorbent is a nanoscale organic or inorganic material that has the efficiency in removing contaminants by attachment of contaminants on its surface (Bhatnagar & Sillanpää, 2015; Gehrke et al., 2015). Nanoadsorbent is the best one that has the potential to overcome the conventional adsorption method due to its high specificity and adsorption capacity (Sharma et al., 2009). In recent years much research has been carried out to find the absorption property of different nanomaterials. The nanoparticle has a high ratio of surface to volume. Small size nanoparticle possesses a larger surface area, increases durability and adsorption capacity of the nanoparticle (nanoadsorbent). Nanoparticle fabricated nanoadsorbent is created to produce some special functions, including water repellent, antiodor, antimicrobial, UV-protection, flexible, and durable (Onoghwarite & Ikechukwu, 2018). Nano absorbent has the potential to adsorb metallic contaminants like Cr (VI), Cu (II), Co (II), Cd (II), As (V), and As (III), and that adsorbent should have the properties of absorbing metals with different valency (higher to lower or zero valency), low toxicity and cost (Gawaad et al., 2011; Khan et al., 2019).

20.4.3.1 Carbon-based nanoadsorbent

As a result of the selectivity of organic solutes, adsorption, thermal, antioxidants, and electrical properties, carbon-based materials such as CNTs, carbon nanoparticles, and carbon nanosheets attract attention in recent years. The CNTs are structurally modified cylindrical carbon material with a highly reactive surface and numerous adsorption point. In CNT, the metal pollutant is getting adsorb by chemical bond and electrostatic force (Butt, 2020). CNT's notable advantages are super adsorption capacity, large surface area, small size, fast kinetics, and ease to renew by modifying operating conditions (El-sayed, 2020). CNT with metal or metal oxide nanoparticles shows an increase in adsorption capacity and mechanical strength. It requires a supporting medium or matrix. It has hydrophobic surface properties and removes Cu^{2+}, Cd^{2+}, Zn^{2+}, Pb^{2+}, dichlorobenzene, and ethylbenzene (Lu et al., 2016).

20.4.3.2 Metallic and nonmetallic nanoparticles

The silver nanoparticle has broad-spectrum solid antimicrobial property, which means the capacity to eliminate different microbial contaminants with low toxicity. It is also used as a disinfectant and antibiofouling agent for wastewater. Titanium nanoparticles have unique characteristics like chemical stability, efficiency to remove different contaminants, low toxicity level, and durability. Also, it is less expensive when compared with other metals (El-sayed, 2020). Magnetite nanorods are especially used for the removal of Fe^{2+}, Pb^{2+}, Cd^{2+}, and Cu^{2+} (Khan et al., 2019). Besides, magnetic nanoparticles also remove various pathogens, sediments and chemical effluents (Patil, 2015). They have a large surface area with superparamagnetic properties. Magnetic nanosorbents are prepared by ligands coating with magnetic nanoparticles to remove different organic pollutants from wastewater (Yaqoob et al., 2020). Removal of this magnetic material was uncomplicated. The magnetic material can be directed and recovered by applying an external magnetic field (Chen, Chu, He, Hu, & Yang, 2011).

Metallic oxide nanoparticles include iron oxide, copper oxide, nickel oxide, ZnO, titanium oxide, and manganese oxide, which are used to treat wastewater (Butt, 2020). Oxidized metal has high adsorption capacity, small intraparticle diffusion distance, and faster kinetics using various functionalized groups like carbonyl. This metal oxide nanoadsorbent can remove As, Cr, Cu, Pb, Cd, Zn, and Ni in their ionic form.

MgO has a high adsorption affinity to heavy metals. Therefore it will remove different heavy metals from effluent water (Anjum et al., 2016). In many industries, porous nanostructure ZnO nanoparticles play a major role in the purification of wastewater as a nanosorbent. Adsorption of heavy metals like Cu^{2+}, is possible by ZnO nanoparticles because of the larger surface area (Yang, Zhang, & Hu, 2013) with special modified structures like nanoassemblies, nanoplates, microspheres with nanosheets and hierarchical ZnO nanorods. Magnetite (Fe_3O_4), Maghamite (γ-Fe_2O_3), and Hematite (α-Fe_2O_3) are the different forms of iron oxide nanoparticles with peculiar characteristics like low toxicity, surface modifiability, chemical inertness, biocompatibility, chemical inertness, large surface area, high adsorption capability, and superparamagnetism (Neyaz, Siddiqui, & Nair, 2014). Manganese oxide is used as the adsorbent to remove organic pollutant, synthesis when Mn element reacts with $KMnO_4$.

Magnetic manganese oxide nanostructures prepared by oleic acid coated with Fe_3O_4 nanoparticles in the presence of $KMnO_4$ solution at low temperature. It can adsorb organic components in effluent water. In addition to that, complete recovery of this material is possible using an external magnetic field (Chen et al., 2011). Without damaging or decreasing the surface area, nanometals and nanometal oxides can be compressed or converted into powder or porous pellets for more suitable effluent treatment (El-sayed, 2020).

Dendrite polymers include random hyperbranched polymers, dendrigraft polymers, dendrons, and dendrimers, these are polymer-based nanosorbent. Besides, they have the potential to replace conventional adsorbents. Large surface area, mechanical rigidity, and adaptable surface property are the few advantages of dendrite polymeric nano adsorbents. Further, they reduce the amount of organic and inorganic pollutants by the adsorption method of effluent treatment. External branches adsorb heavy metals, and internal hydrophobic fractions mainly adsorb organic compounds (El-sayed, 2020). Dendrimers' size ranges from 2 to 20 nm with an inner core consisting of multifunctional dense shell and surface with branching sites and terminal groups (Tyagi, Singh, Vats, & Kumar, 2012).

Aluminosilicate mineral zeolite has many electrostatic holes on the surface, occupied by cation and water (Wang & Peng, 2010). Compare with natural, traditional microscale zeolites, nano zeolites within the range of 10−500 nm has a large surface area and huge adsorption sites (El-sayed, 2020). In the past decade, zeolite an adsorbent, a promising technique for purifying water and wastewater. Generally, zeolite can remove cation only, but wastewater contains anion and organic addition to cation. For the removal of anion, two methods were adopted. One uses cation surfactant, which modified the zeolite's surface to integrate with anion and organic compounds. Another is surface precipitated with heavy metals adsorb anions effectively (Wang & Peng, 2010). In zeolite, several silver, gold, and copper nanoparticles are implanted to give better adsorption technology (Yaqoob et al., 2020). Zeolite has antimicrobial property and an excellent affinity for heavy metals ions like Cr (III), Ni (II), Zn (II), Cu (II) and Cd (II) (Tyagi et al., 2012). Since the 1980s, silver embedded in porous zeolites nanomaterial structure has been used.

20.5 Future perspectives

Nanotechnology, a progressing research area, is more promising toward one of its many applications—water treatment strategies. Though several methodologies for the specific treatment of the pulp and paper mill effluents have been discussed above, yet still are not in practical application. Recent advancements have been made in the sector of nanophotocatalysis, which led to a remarkable achievement in the development and design of reactors to modify the nanophotocatalysts according to the needs. The two major challenges in nanophotocatalysis still yet to be solved are the higher energy consumption and the mass transfer limitations. However, the electron−hole recombination problem has been solved due to nanocomposites' innovative use and nanophotocatalytic reactor structures. Microfluidic reactors are the class of reactors used to study the synthesis and reaction phases (Lin, Wang, Wang, & Tseng, 2009). Still, it remains difficult for the process of nanophotocatalysis to be applied on a wide scale in treating wastewater effluents, and further studies are yet to be made.

In the contemporary era, nano and micromotors have gained attention as a mechanism in which the energy is converted from various sources into a machine-driven force. These motors can be self-propelled by using fuel or without the use of fuel through various means such as electric and magnetic fields. The advantages of these motors are enhanced power and speed, self-mix capacity, and specific control movement. Nano and micromotors are considered as an excellent option because the nano-based materials are more effective and have budding properties, making them more efficient in converting toxic pollutants to nontoxic ones (Yaqoob et al., 2020). These nanomotors provide various advantages compared to the traditional remediation methodologies in terms of cost and time efficiency, simultaneously abiding by the in-situ and ex-situ remediation rules (Safdar, Simmchen, & Jänis, 2017). The constant movement of the nanoscale particles can be used to transfer reactive nanomaterials for the elimination of undesirable components present in the wastewater, for the discharge of the remediation to long distances, etc. The present-day technologies face a major challenge of scaling-up, wherein using nanomotors, and these processes are even more simplified. The energetic mixing overcomes the diffusion boundaries in these motors due to the self-propelling mechanisms. These self-propelling mechanisms stimulate the efficiency of the treatment process by merging them with nanoparticles. These motors can do into the detentions by applying a magnetic field, one of the major cons of conventional treatment technologies. As these motors offer various advantages, still there exist some cons.

The efficiency in biological applications is less though some photocatalytic and biocatalytic nanomotors are presently industrialized (Yaqoob et al., 2020). Yet, these motors require some advancement in future to treat the paper and pulp mill effluents. Also, the lifetime of functional nanomotors depends upon the residual components. Nano and micromotors are still considered techniques that require some advancement, but this field of research remains an upcoming and emerging research area. The novel materials used must be combined with these motors, such as graphene, to remove wastewater effluents present in the paper and pulp mills. Thus nanomotors' development would improvise the standard of degradation methods, thereby providing a better technique for treating paper and pulp mill effluents. Most of the research is carried out only on a laboratory scale. The use of nanomembrane at the industrial scale was still challenging one, huge ideal, and research work needed.

Novel nanomembrane has to be created with reasonable cost, less toxicity, improved mechanical stability, high resistance to membrane fouling, high specificity, sustainable, and multifunctional membrane.

Further investigation and research have to be done to improve the lifetime and rejection efficiency of the membrane. Further experimental development should be essential for commercialization. The scientific community must improve the proper adsorption mechanism and recovery, degradation, elimination of nanoparticles with less toxicity and side effects at the industrial-level effluent treatment. Enough knowledge and research are required to find the behavior of different nanoparticles in varying pH, temperature and pressure in addition to the natural environment. Further investigation and test at the pilot level are required to integrate different nanomembrane and nanoadsorbent in paper and pulp industry effluent treatment.

20.6 Conclusion

Nanotechnology thus has provided a scope for the control and degradation of wastewater effluents. Nanotechnology-based treatment methods offer various advantages over other methods like sedimentation, flocculation, coagulation, and activated carbon used to remove water pollutants. Various nano-based methodologies such as nanofiltration, nanoadsorption, and nanophotocatalysis have proved to be quite successful for the effective treatment of wastewater effluents, especially in paper and pulp mills. Moreover, the above-listed technologies offer various advantages: efficiency, cost effectiveness, less energy requirement, and ecofriendly. It has also been reported that the combination of two or more methods would be useful for the effective removal of effluents. Metal oxide-based nanomaterials are the very effective catalyst used in various oxidation and reduction reactions.

Acknowledgments

The authors would like to thank the University of Malaya for funding this research work. This work was financially supported by impact-oriented Interdisciplinary Research Grant (No. IIRG018A-2019), and Global Collaborative Programme-SATU Joint Research Scheme (No. ST012-2019).

References

Abdelbasir, S. M., & Shalan, A. E. (2019). An overview of nanomaterials for industrial wastewater treatment. *Korean Journal of Chemical Engineering*, *36*, 1209–1225. Available from https://doi.org/10.1007/s11814-019-0306-y.

Abn, J.-W., & Lim, M.-H. (2009). Characteristics of wastewater from the pulp.paper industry and its biological treatment technologies. *Journal of Korean Institute of Resources Recycling*, *18*, 16–29.

Afonso, M. D., & De Pinho, M. N. (1997). Nanofiltration of bleaching pulp and paper effluents in tubular polymeric membranes. *Seperation Science and Technology*, *32*, 2641–2658. Available from https://doi.org/10.1080/01496399708006961.

Anjum, M., Miandad, R., Waqas, M., Gehany, F., & Barakat, M. A. (2016). Remediation of wastewater using various nanomaterials. *Arabian Journal of Chemistry*, *12*. Available from https://doi.org/10.1016/j.arabjc.2016.10.004.

Araña, J., González Díaz, O., Miranda Saracho, M., Doa Rodríguez, J. M., Herrera Melián, J. A., & Pérez Pea, J. (2001). Photocatalytic degradation of formic acid using Fe/TiO2 catalysts: The role of Fe3+/Fe2+ ions in the degradation mechanism. *Applied Catalysis B Environmental*, *32*, 49–61. Available from https://doi.org/10.1016/S0926-3373(00)00289-7.

Arkell, A., Olsson, J., & Wallberg, O. (2014). Process performance in lignin separation from softwood black liquor by membrane filtration. *Chemical Engineering Research and Design*, *92*, 1792–1800. Available from https://doi.org/10.1016/j.cherd.2013.12.018.

Awungacha Lekelefac, C., Busse, N., Herrenbauer, M., & Czermak, P. (2015). Photocatalytic based degradation processes of lignin derivatives. *International Journal of Photoenergy*, *2015*. Available from https://doi.org/10.1155/2015/137634.

Bhatnagar, A. (2015). Assessment of physico-chemical characteristics of paper industry effluents. *Rasayan Journal of Chemistry*, *8*, 143–145.

Bhatnagar, A., & Sillanpää, M. (2015). *Application of nanoadsorbents in water treatment. Nanomaterials for environmental protection* (pp. 237–247). Wiley. Available from https://doi.org/10.1002/9781118845530.ch15.

Butt, B. Z. (2020). Nanotechnology and waste water treatment. In S. Javad (Ed.), *Nanoagronomy* (pp. 153–177). Cham: Springer. Available from doi:10.1007/978-3-030-41275-3_9.

Cabrera, M. N. (2017). *Pulp mill wastewater: Characteristics and treatment. Biological wastewater treatment and resource recovery*. Croatia: IntechOpen. Available from https://doi.org/10.5772/67537.

Chen, H., Chu, P. K., He, J., Hu, T., & Yang, M. (2011). Porous magnetic manganese oxide nanostructures: Synthesis and their application in water treatment. *Journal of Colloid and Interface Science*, *359*, 68–74. Available from https://doi.org/10.1016/j.jcis.2011.03.089.

Chi, Z., Wang, Z., Liu, Y., & Yang, G. (2018). Preparation of organosolv lignin-stabilized nano zero-valent iron and its application as granular electrode in the tertiary treatment of pulp and paper wastewater. *Chemical Engineering Journal*, *331*, 317–325. Available from https://doi.org/10.1016/j.cej.2017.08.121.

Cristina Yeber, M., Rodríguez, J., Freer, J., Durán, N., & Mansilla, H. D. (2000). Photocatalytic degradation of cellulose bleaching effluent by supported TiO_2 and ZnO. *Chemosphere*, *41*, 1193–1197. Available from https://doi.org/10.1016/S0045-6535(99)00551-2.

Dafinov, A., Font, J., & Garcia-Valls, R. (2005). Processing of black liquors by UF/NF ceramic membranes. *Desalination*, *173*, 83–90. Available from https://doi.org/10.1016/j.desal.2004.07.044.

Devi, N. L., Yadav, I. C., Shihua, Q. I., Singh, S., & Belagali, S. L. (2011). Physicochemical characteristics of paper industry effluents - A case study of South India Paper Mill (SIPM). *Environmental Monitoring and Assessment*, *177*, 23–33. Available from https://doi.org/10.1007/s10661-010-1614-1.

El Saliby, I., Shon, H., Kandasamy, J., & Vigneswaran, S. (1999). Nanotechnology for wastewater treatment: In brief. *Water and Wastewater Treatment Technologies*. Available from https://www.eolss.net/Sample-Chapters/C05/E6-144-23.pdf.

El-sayed, M. E. A. (2020). Nanoadsorbents for water and wastewater remediation. *The Science of the Total Environment*, *739*, 139903. Available from https://doi.org/10.1016/j.scitotenv.2020.139903.

Fenoll, J., Hellín, P., Flores, P., Martínez, C. M., & Navarro, S. (2013). Degradation intermediates and reaction pathway of carbofuran in leaching water using TiO_2 and ZnO as photocatalyst under natural sunlight. *Journal of Photochemistry And Photobiology A Chemistry*, *251*, 33–40. Available from https://doi.org/10.1016/j.jphotochem.2012.10.012.

Gawaad, R. S., Sharma, S. K., & Sambi, S. S. (2011). Sodium sulphate recovery from industrial wastewater using nano-membranes: A review. *International Review of Chemical Engineering*, *3*, 392–398.

Gehrke, I., Geiser, A., & Somborn-Schulz, A. (2015). Innovations in nanotechnology for water treatment. *Nanotechnology, Science, and Applications*, *8*, 1–17. Available from https://doi.org/10.2147/NSA.S43773.

Gómez-Pastora, J., Dominguez, S., Bringas, E., Rivero, M. J., Ortiz, I., & Dionysiou, D. D. (2017). Review and perspectives on the use of magnetic nanophotocatalysts (MNPCs) in water treatment. *Chemical Engineering Journal*, *310*, 407–427. Available from https://doi.org/10.1016/j.cej.2016.04.140.

Hariharan, C. (2006). Photocatalytic degradation of organic contaminants in water by ZnO nanoparticles: Revised. *Applied Catalysis A General*, *304*, 55–61. Available from https://doi.org/10.1016/j.apcata.2006.02.020.

Hasegawa, M. C., Daniel, J. F. D. S., Takashima, K., Batista, G. A., & Da Silva, S. M. C. P. (2014). COD removal and toxicity decrease from tannery wastewater by zinc oxide-assisted photocatalysis: A case study. *Environmental Technology (United Kingdom)*, *35*, 1589–1595. Available from https://doi.org/10.1080/09593330.2013.874499.

Hooda, R., Bhardwaj, N. K., & Singh, P. (2018). *Brevibacillus parabrevis* MTCC 12105: A potential bacterium for pulp and paper effluent degradation. *World Journal of Microbiology and Biotechnology*, *34*. Available from https://doi.org/10.1007/s11274-018-2414-y.

Ince, B. K., Cetecioglu, Z., & Ince, O. (2011). *Pollution prevention in the pulp and paper Industries. Environmental management in practice*. Intech. Available from https://doi.org/10.5772/23709.

Jitendra, G., Srivastava, A., Pachauri, S., & Srivastava, P. C. (2014). Effluents from paper and pulp industries and thier impact on soil properties and chemical composition of plants in Uttarakhand, India. *Journal of Environment and Waste Management*, *1*, 26–30.

Jönsson, A. S., Nordin, A. K., & Wallberg, O. (2008). Concentration and purification of lignin in hardwood kraft pulping liquor by ultrafiltration and nanofiltration. *Chemical Engineering Research and Design*, *86*, 1271–1280. Available from https://doi.org/10.1016/j.cherd.2008.06.003.

Kesalkar, V. P., Khedikar, I. P., & Sudame, A. M. (2012). Physico-chemical characteristics of wastewater from Paper Industry. *International Journal of Engineering Research and Applications*, *2*, 137–143.

Khan, N. A., Khan, S. U., Ahmed, S., Farooqi, I. H., Dhingra, A., Hussain, A., & Changani, F. (2019). Applications of nanotechnology in water and wastewater treatment: A review. *Asian Journal of Water, Environment and Pollution*, *16*, 81–86. Available from https://doi.org/10.3233/AJW190051.

Kumar, P., Kumar, S., Bhardwaj, N. K., & Choudhary, A. K. (2011). Advanced oxidation of pulp and paper industry effluent. *Engineering*, *15*, 9.

Kumar, S., Saha, T., & Sharma, S. (2015). Treatment of pulp and paper mill effluents using novel biodegradable polymeric flocculants based on anionic polysaccharides: A new way to treat the waste water. *Journal of Water, Environment and Pollution*, *2*, 1415–1428.

Lastra, A., Gómez, D., Romero, J., Francisco, J. L., Luque, S., & Álvarez, J. R. (2004). Removal of metal complexes by nanofiltration in a TCF pulp mill: Technical and economic feasibility. *Journal of Membrane Science*, *242*, 97–105. Available from https://doi.org/10.1016/j.memsci.2004.05.012.

Li, F. B., Li, X. Z., Ao, C. H., Lee, S. C., & Hou, M. F. (2005). Enhanced photocatalytic degradation of VOCs using Ln3 + -TiO 2 catalysts for indoor air purification. *Chemosphere*, *59*, 787–800. Available from https://doi.org/10.1016/j.chemosphere.2004.11.019.

Lin, L., Chai, Y., Zhao, B., Wei, W., He, D., He, B., & Tang, Q. (2013). Photocatalytic oxidation for degradation of VOCs. *Open Journal of Inorganic Chemistry*, *03*, 14–25. Available from https://doi.org/10.4236/ojic.2013.31003.

Lin, W.-Y., Wang, Y., Wang, S., & Tseng, H.-R. (2009). Integrated microfluidic reactors. *Nano Today*, *4*(6), 470–481. Available from https://doi.org/10.1016/j.nantod.2009.10.007.

Lu, H., Wang, J., Stoller, M., Wang, T., Bao, Y., & Hao, H. (2016). An overview of nanomaterials for water and wastewater treatment. *Advances in Material Science and Engineering*, *2016*, 1–10. Available from https://doi.org/10.1155/2016/4964828.

Ma, Y. S., Chang, C. N., Chiang, Y. P., Sung, H. F., & Chao, A. C. (2008). Photocatalytic degradation of lignin using Pt/TiO_2 as the catalyst. *Chemosphere*, *71*, 998–1004. Available from https://doi.org/10.1016/j.chemosphere.2007.10.061.

MacHado, A. E. H., Furuyama, A. M., Falone, S. Z., Ruggiero, R., Perez, D. D. S., & Castellan, A. (2000). Photocatalytic degradation of lignin and lignin models, using titanium dioxide: The role of the hydroxyl radical. *Chemosphere*, *40*, 115–124. Available from https://doi.org/10.1016/S0045-6535(99)00269-6.

Madhura, L., & Singh, S. (2018). *A review on the advancements of nanomembranes for water treatment*, . *Nanotechnology in environmental science* (Vol. 1–2, pp. 391–412). Wiley. Available from https://doi.org/10.1002/9783527808854.ch12.

Mandeep., Gupta, G. K., Liu, H., & Shukla, P. (2019). Pulp and paper industry–based pollutants, their health hazards and environmental risks. *Current Opinion in Environmental Science & Health*, *12*, 48–56. Available from https://doi.org/10.1016/j.coesh.2019.09.010.

Mishra, A. K. (2014). *Application of nanotechnology in water research* (pp. 1–522). Wiley. Available from https://doi.org/10.1002/9781118939314.

Mussey, O. D. (1955). *Water Requirements of Pulp and paper industry*. U.S. Gov. Print. Off. Available from doi:10.3133/wsp1330A.

E. Negaresh, G. Leslie, Reclamation of newsprint mill effluent using membrane technology (Ph.D. thesis). Sydney, Australia: The University of New South Wales. (2012) 256. <http://unsworks.unsw.edu.au/fapi/datastream/unsworks:10475/SOURCE02?view = true>

Neyaz, N., Siddiqui, W. A., & Nair, K. K. (2014). Application of surface functionalized iron oxide nanomaterials as a nanosorbents in extraction of toxic heavy metals from ground water: A review. *International Journal of Environmental Sciences*, *4*, 472–483. Available from https://doi.org/10.6088/ijes.2014040400004.

Onoghwarite, O. E., & Ikechukwu, O. P. (2018). Emerging trends in nanoabsorbents absorption applications. *International Journal of Advances in Scientific Research and Engineering*, *4*, 201–206. Available from https://doi.org/10.31695/ijasre.2018.32963.

Pizzichini, M., Russo, C., & Di Meo, C. D. (2005). Purification of pulp and paper wastewater, with membrane technology, for water reuse in a closed loop. *Desalination*, *178*, 351–359. Available from https://doi.org/10.1016/j.desal.2004.11.045.

Qingping, W. (2018). Application of nanotechnology in wastewater treatment. *Nanoscience & Nanotechnology*, *1*. Available from https://doi.org/10.18063/nn.v1i1.495.

Saadia, A., & Ashfaq, A. (2010). Environmental management in pulp and paper industry. *Journal of Industrial Pollution Control*, *26*, 71–77.

Safdar, M., Simmchen, J., & Jänis, J. (2017). Light-driven micro- and nanomotors for environmental remediation. *Environmental Science Nano.*, *4*, 1602–1616. Available from https://doi.org/10.1039/c7en00367f.

Sharma, Y. C., Srivastava, V., Singh, V. K., Kaul, S. N., & Weng, C. H. (2009). Nano-adsorbents for the removal of metallic pollutants from water and wastewater. *Environmental Technology*, *30*, 583–609. Available from https://doi.org/10.1080/09593330902838080.

Singh, G., Kaur, S., Khatri, M., & Arya, S. K. (2019). Biobleaching for pulp and paper industry in India: Emerging enzyme technology. *Biocatalyst and Agricultural Biotechnology*, *17*, 558–565. Available from https://doi.org/10.1016/j.bcab.2019.01.019.

Singh, P., Srivastava, N., Jagadish, R., & Upadhyay, A. (2019). Effect of toxic pollutants from pulp & paper mill on water and soil quality and its remediation. *International Journal of Lakes and Rivers.*, *12*, 1–20.

Singh, P. P., & Ambika. (2018). *Environmental remediation by nanoadsorbents-based polymer nanocomposite. New polymer nanocomposites for environmental remediation* (pp. 223–241)). Elsevier. Available from https://doi.org/10.1016/B978-0-12-811033-1.00010-X.

Tambe Patil, B. B. (2015). Wastewater treatment using nanoparticles. *Journal of Advanced Chemical Engineering*, *5*. Available from https://doi.org/10.4172/2090-4568.1000131.

Tanaka, K., Calanag, R. C. R., & Hisanaga, T. (1999). Photocatalyzed degradation of lignin on TiO_2. *Journal of Molecular Catalysis A Chemistry*, *138*, 287–294. Available from https://doi.org/10.1016/S1381-1169(98)00161-7.

J.M.K. Timmer, Properties of nanofiltration membranes: Model development and industrial application, *Technische Universiteit Eindhoven* (2001) 154. Available from https://doi.org/10.6100/IR545659.

P.K. Tyagi, R. Singh, S. Vats, D. Kumar, Nanomaterials use in wastewater treatment, In: International conference on nanotechnology and chemical engineering (ICNCS'2012), Bangkok, Thailand. (2012) 5.

Wang, S., & Peng, Y. (2010). Natural zeolites as effective adsorbents in water and wastewater treatment. *Chemical Engineering Journal*, *156*, 11–24. Available from https://doi.org/10.1016/j.cej.2009.10.029.

Yadav, S., & Chandra, R. (2018). Detection and assessment of the phytotoxicity of residual organic pollutants in sediment contaminated with pulp and paper mill effluent. *Environmental Monitoring and Assessment*, *190*. Available from https://doi.org/10.1007/s10661-018-6947-1.

Yang, Y., Zhang, C., & Hu, Z. (2013). Impact of metallic and metal oxide nanoparticles on wastewater treatment and anaerobic digestion. *Environmental Science: Processing and Impacts*, *15*, 39–48. Available from https://doi.org/10.1039/c2em30655g.

Yaqoob, A. A., Parveen, T., Umar, K., & Ibrahim, M. N. M. (2020). Role of nanomaterials in the treatment of wastewater: A review. *Water (Switzerland)*, *12*. Available from https://doi.org/10.3390/w12020495.

Ziezig, R., Tomani, P., & Theliander, H. (2014). Production of a pure lignin product part 2: Separation of lignin from membrane filtration permeates of black liquor. *Cellulose Chemistry and Technology*, *48*, 805–811.

CHAPTER 21

Future perspective of pulp and paper industry

Muhammad Bilal[1], Tuan Anh Nguyen[2] and Hafiz M.N. Iqbal[3]
[1]School of Life Science and Food Engineering, Huaiyin Institute of Technology, Huai'an, P.R. China
[2]Institute for Tropical Technology, Vietnam Academy of Science and Technology, Hanoi, Vietnam
[3]Tecnologico de Monterrey, School of Engineering and Sciences, Monterrey, Mexico

21.1 Introduction

In recent years, the manufacturing and demand for paper and paper products are rising dramatically around the globe. This leads to high energy consumption and emission of CO_2 in the paper and pulp industry for increased production. A large amount of waste generation as a result of industrial activities for producing paper is a significant problem. The pulp and paper industry waste is categorized in rejects, like bleaching, pulping, washing, primary, and secondary sludge. All these types of wastes are likely to pollute soil, air, and water matrices (Kumar, Singh, & Chandra, 2020; Singh, Kumar, & Chandra, 2020). Air contamination is considered a significant threat due to the emission of many reduced sulfur-containing components, including hydrogen sulfide, methyl mercaptan, dimethyl sulfide, particulate matter, and oxides of nitrogen and sulfur by various processes (Kumar et al., 2011). In addition, the solid waste infiltrates the food chain resulting in a large number of disorders such as cancer, respiratory disorder, irritation to eyes and skin, headache, nausea and heart complications. Therefore gainful utilization of this waste to produce high-value products has been appreciated as a fascinating approach for sustainable energy and a greener, cleaner, and contamination-free environment (Gupta & Shukla, 2020).

In the wake of increasing demand for renewable energy resources, pulp and paper industries have committed to boosting their energy adequacy. The production of green energy, sorbents development, and clinker preparation from pulp industry waste is an innovative and revolutionary concept for environmental sustainability, economic development, and reducing its adverse impacts on human health and the ecosystem. Recently, some technologies have demonstrated the technically feasible production of biogas from waste sludge from the pulp mill (Kamali, Gameiro, Costa, & Capela, 2016). In contrast to electrochemical and thermochemical methods, the production of biohydrogen gas by biological means has emerged as a promising and environmentally friendly way. Additionally, methane production and energy recovery by the anaerobic dark fermentation process has been commercially utilized for the treatment of waste

Nanotechnology in Paper and Wood Engineering
DOI: https://doi.org/10.1016/B978-0-323-85835-9.00017-9

materials and industrial sludge (Hay, Wu, Juan, & Jahim, 2015). The conversion of waste effluents into valuable bioproducts (i.e., biofuels) using biorefinery approaches like pyrolysis, fermentation, incineration, gasification, and anaerobic digestion is of paramount importance (Jagadevan et al., 2018).

21.2 Economic feasibility and environmental regulation

The economic feasibility of resource generation from waste through valorization techniques is an issue of high apprehension. Some techniques analyzing the cost analysis have corroborated the viable biotransformation of waste materials into valuable resources. Effluents such as pulp sludge, waste sludge, and black liquor can be utilized as feedstocks to produce ammonia via the gasification process. One study in Canada has reported the manufacturing cost of ammonia ($\sim$\$743–\$748 t^{-1}) via this process that appeared cost-effective with respect to the current market price. The market value of ammonia was reported to be \$976 t^{-1} in 2013 (Akbari, Oyedun, & Kumar, 2018). Production of formic acid, levulinic acid, biochar, and furfural from waste has been found cost competitive and can be sold at high market price. Likewise, the conversion of activated carbon into 750 metric tons of biochar was recorded enough to meet the electrical and thermal energy that is required for biorefinery industries. Generally, the market price of specialty chemicals and products from pulp mill waste effluents lies between \$1.5 and 2 kg^{-1} and between \$2 and 3 kg^{-1}, which is lower than the current market value (Gunukula, Klein, Pendse, DeSisto, & Wheeler, 2018). The market price of lignin-derived jet fuel is ranged from \$6.35 to \$1.76 gal^{-1} and depends on the raw feedstock, capacity, and production rate (Shen et al., 2019). All the reports mentioned above represent the feasibility of cost analysis of resources for some industrial sectors. Using these studies, additional studies can be planned for making other technologies economical for sustainable and renewable production.

Environmental regulations have a notable influence on the contemporary practices and upcoming opportunities of the paper and pulp industry because they rely on export and compete for the international markets. Though stringent regulatory protocols are indispensable for mitigating the adverse environmental consequences of current maneuvers, but long authorizing processes and strict standards may also lead to future business uncertainties and even impede incentives for pursuing sustainable technological transition (Korhonen, Pätäri, Toppinen, & Tuppura, 2015; Söderholm, Bergquist, & Söderholm, 2019). The exclusive use of forest-based raw materials for producing pulp and paper products could also produce green chemicals (e.g., organic acids), low-carbon fuels (e.g., biodiesel), and numerous substances useful for construction purposes (e.g., ligninconcrete mixes) (Hansen & Coenen, 2017). It can be inferred that rigorous environmental regulations are likely to create many opportunities, as well as pose threats to achieve future environmental perfections.

21.3 Challenges, perspectives, and innovations

As discussed above, the controlled or uncontrolled generation of various pollution forms in all key spheres of the environment, for example, biosphere, hydrosphere, anthroposphere, or technosphere, and others are key challenges. Undoubtedly, the pulp and paper industry sector contributes massively to fulfill the ever-growing demand for paper and related products alone around the globe. Regardless of such needful practice, this also leads to consuming a massive amount of water and energy, which generates a big pile of waste in the form of chemical-loaded wastewater effluents and CO_2 emissions (Szabó, Soria, Forsström, Keränen, & Hytönen, 2009). The unrestricted CO_2 release and waste disposal are the main challenges in the pulp and paper industry. Owing to the steps involved in the papermaking, that is, (1) pulping, (2) bleaching, and (3) papermaking or finishing, wastewater in the form of heavily loaded with chemicals is generated, which is discharged into different water matrices and cause serious environmental and health hazards (Gupta, Liu, & Shukla, 2019; Sonkar, Kumar, Dutt, & Kumar, 2019). The bleaching involved in the pulp and paper industry is considered the most polluting section, which generates most of the chemical-loaded wastewater effluents (Pokhrel & Viraraghavan, 2004; Sonkar, Kumar, & Dutt, 2021) (an important limitation with a significant environmental pollution issue).

Waste to value-based strategic theme offers unique perspective and potentialities to limit or lessen the waste generation. For instance, as compared to the traditional process, the integration of the biorefinery concept was a unique solution not only to reduce the waste but also to help to produce valuable energy (Fig. 21.1). This, in turn, also facilitates the energy requirements of the industry. Further to this, zero-waste or zero-emission strategies are supporting the sustainable development goals (SDGs). The energy generated via zero-waste or zero-emission strategy couple with integrated biorefinery will make the industries, including the pulp and paper industry, highly efficient and sustainable by reducing the load on nonrenewable and/or petrochemical sources for energy generation. Moreover, the treatment and utilization of wastewater effluents, which significantly pose serious environmental and health hazards if discharge untreated, in integrated biorefinery will lower/limit the contaminants from the pulp and paper industry waste effluents and thus minimize the environmental pollution with particular reference to the water matrices pollution.

From the perspective of the innovation, pulp and paper industry has entered modern revolutionization and is thus considered as a large-scale sector. Such an industry where innovations are escalating with reference to environmentally friendly processes, along with product-based or business-oriented model implementation. Considering this significant rise and technological hike in pulp and paper industry, future investigations should essence the green chemistry adaptations by fulfilling the SDGs. Particular focus needs to be given on specific SDGs, for example, Goal #6 (clean water and sanitation), Goal #7

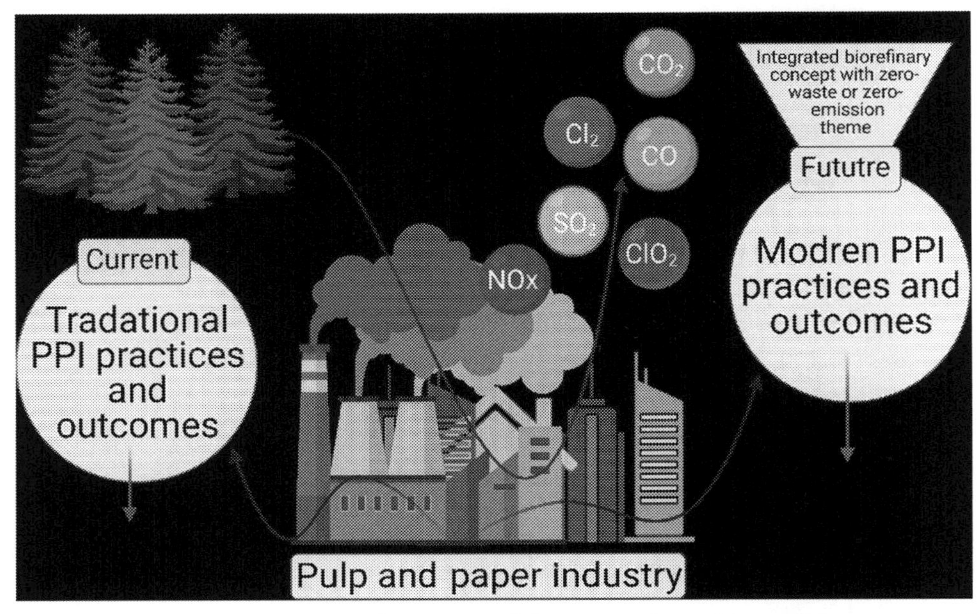

Figure 21.1 Comparative overview of current and future practices with possible outcomes. *PPI*, pulp and paper industry.

(affordable and clean energy), Goal #9 (industry innovation and infrastructure), Goal #12 (responsible consumption and production), and Goal # 13 (climate action).

21.4 Concluding note

In conclusion, the insights and prospects given above related to the pulp and paper industry reflects socio-economic viability and environmental regulation measures. More specifically, this short overview study provides a way forward to apprehension to regulate the strategic measures/regulations going beyond the traditional environmental influence valuation and cost-effective assessment strategies. The future trends should foster the industrial processes by considering the European Union directive regulations and international trade in recycled paper. This will further enrich the shelf-life and/or life-cycle characteristics while taking into account the environmental influences/impacts of pulp and paper industry products which are either made of pristine materials or recycled paper.

Acknowledgment

Consejo Nacional de Ciencia y Tecnología (CONACYT) is thankfully acknowledged for partially supporting this work under Sistema Nacional de Investigadores (SNI) program awarded to Hafiz M. N. Iqbal (CVU: 735340).

Conflict of interest

The author(s) declare no conflicting interests.

References

Akbari, M., Oyedun, A. O., & Kumar, A. (2018). Ammonia production from black liquor gasification and co-gasification with pulp and waste sludges: A techno-economic assessment. *Energy*, *151*, 133–143.

Gunukula, S., Klein, S. J., Pendse, H. P., DeSisto, W. J., & Wheeler, M. C. (2018). Techno-economic analysis of thermal deoxygenation based biorefineries for the coproduction of fuels and chemicals. *Applied Energy*, *214*, 16–23.

Gupta, G. K., & Shukla, P. (2020). Insights into the resources generation from pulp and paper industry wastes: Challenges, perspectives and innovations. *Bioresource Technology*, *297*, 122496.

Gupta, G. K., Liu, H., & Shukla, P. (2019). Pulp and paper industry–based pollutants, their health hazards and environmental risks. *Current Opinion in Environmental Science & Health*, *12*, 48–56.

Hansen, T., & Coenen, L. (2017). Unpacking resource mobilisation by incumbents for biorefineries: The role of micro-level factors for technological innovation system weaknesses. *Technology Analysis & Strategic Management*, *29*(5), 500–513.

Hay, J. X. W., Wu, T. Y., Juan, J. C., & Jahim, J. M. (2015). Improved biohydrogen production and treatment of pulp and paper mill effluent through ultrasonication pretreatment of wastewater. *Energy Conversion and Management*, *106*, 576–583.

Jagadevan, S., Banerjee, A., Banerjee, C., Guria, C., Tiwari, R., Baweja, M., & Shukla, P. (2018). Recent developments in synthetic biology and metabolic engineering in microalgae towards biofuel production. *Biotechnology for Biofuels*, *11*(1), 1–21.

Kamali, M., Gameiro, T., Costa, M. E. V., & Capela, I. (2016). Anaerobic digestion of pulp and paper mill wastes–An overview of the developments and improvement opportunities. *Chemical Engineering Journal*, *298*, 162–182.

Korhonen, J., Pätäri, S., Toppinen, A., & Tuppura, A. (2015). The role of environmental regulation in the future competitiveness of the pulp and paper industry: The case of the sulfur emissions directive in Northern Europe. *Journal of Cleaner Production*, *108*, 864–872.

Kumar, A., Singh, A. K., & Chandra, R. (2020). Comparative analysis of residual organic pollutants from bleached and unbleached paper mill wastewater and their toxicity on *Phaseolus aureus* and *Tubifex tubifex*. *Urban Water Journal*, *17*(10), 860–870.

Kumar, P. S., Abhinaya, R. V., Lashmi, K. G., Arthi, V., Pavithra, R., Sathyaselvabala, V., ... Sivanesan, S. (2011). Adsorption of methylene blue dye from aqueous solution by agricultural waste: Equilibrium, thermodynamics, kinetics, mechanism and process design. *Colloid Journal*, *73*(5), 651–661.

Pokhrel, D., & Viraraghavan, T. (2004). Treatment of pulp and paper mill wastewater—A review. *Science of the Total Environment*, *333*(1–3), 37–58.

Shen, R., Tao, L., & Yang, B. (2019). Techno-economic analysis of jet-fuel production from biorefinery waste lignin. *Biofuels, Bioproducts and Biorefining*, *13*(3), 486–501.

Singh, A. K., Kumar, A., & Chandra, R. (2020). Residual organic pollutants detected from pulp and paper industry wastewater and their toxicity on *Triticum aestivum* and *Tubifex-tubifex* worms. *Materials Today: Proceedings*. Available from https://doi.org/10.1016/j.matpr.2020.10.862.

Söderholm, P., Bergquist, A. K., & Söderholm, K. (2019). Environmental regulation in the pulp and paper industry: Impacts and challenges. *Current Forestry Reports*, *5*(4), 185–198.

Sonkar, M., Kumar, M., Dutt, D., & Kumar, V. (2019). Treatment of pulp and paper mill effluent by a novel bacterium *Bacillus* sp. IITRDVM-5 through a sequential batch process. *Biocatalysis and Agricultural Biotechnology*, *20*, 101232.

Sonkar, M., Kumar, V., & Dutt, D. (2021). A novel sequence batch treatment of wastewater using *Bacillus* sp. IITRDVM-5 mixing with paper mill and sewage sludge powders. *Environmental Technology & Innovation*, *21*, 101288.

Szabó, L., Soria, A., Forsström, J., Keränen, J. T., & Hytönen, E. (2009). A world model of the pulp and paper industry: Demand, energy consumption and emission scenarios to 2030. *Environmental Science & Policy*, *12*(3), 257–269.

Index

Note: Page numbers followed by "*f*" and "*t*" referred to figures and tables, respectively.

M

N

Q

R

S

T

Printed in the United States
by Baker & Taylor Publisher Services